- Joint probability of two mutually exclusive events:
$$P(A \text{ and } B) = 0$$

- Addition rule for mutually nonexclusive events:
$$P(A \text{ or } B) = P(A) + P(B) - P(A \text{ and } B)$$

- Addition rule for mutually exclusive events:
$$P(A \text{ or } B) = P(A) + P(B)$$

- n factorial: $n! = n(n-1)(n-2)\ldots 3 \cdot 2 \cdot 1$

- Number of combinations of n items selected x at a time:
$${}_nC_x = \frac{n!}{x!(n-x)!}$$

- Number of permutations of n items selected x at a time:
$${}_nP_x = \frac{n!}{(n-x)!}$$

Chapter 5 • Discrete Random Variables and Their Probability Distributions

- Mean of a discrete random variable x: $\mu = \Sigma x P(x)$

- Standard deviation of a discrete random variable x:
$$\sigma = \sqrt{\Sigma x^2 P(x) - \mu^2}$$

- Binomial probability formula: $P(x) = {}_nC_x\, p^x\, q^{n-x}$

- Mean and standard deviation of the binomial distribution:
$$\mu = np \quad \text{and} \quad \sigma = \sqrt{npq}$$

- Hypergeometric probability formula:
$$P(x) = \frac{{}_rC_x\ {}_{N-r}C_{n-x}}{{}_NC_n}$$

- Poisson probability formula: $P(x) = \dfrac{\lambda^x e^{-\lambda}}{x!}$

- Mean, variance, and standard deviation of the Poisson probability distribution:
$$\mu = \lambda, \quad \sigma^2 = \lambda, \quad \text{and} \quad \sigma = \sqrt{\lambda}$$

Chapter 6 • Continuous Random Variables and the Normal Distribution

- z value for an x value: $z = \dfrac{x - \mu}{\sigma}$

- Value of x when μ, σ, and z are known: $x = \mu + z\sigma$

Chapter 7 • Sampling Distributions

- Mean of $\bar{x}$: $\mu_{\bar{x}} = \mu$

- Standard deviation of $\bar{x}$ when $n/N \leq .05$: $\sigma_{\bar{x}} = \sigma/\sqrt{n}$

- z value for $\bar{x}$: $z = \dfrac{\bar{x} - \mu}{\sigma_{\bar{x}}}$

- Population proportion: $p = X/N$
- Sample proportion: $\hat{p} = x/n$
- Mean of $\hat{p}$: $\mu_{\hat{p}} = p$
- Standard deviation of $\hat{p}$ when $n/N \leq .05$: $\sigma_{\hat{p}} = \sqrt{pq/n}$

- z value for $\hat{p}$: $z = \dfrac{\hat{p} - p}{\sigma_{\hat{p}}}$

Chapter 8 • Estimation of the Mean and Proportion

- Point estimate of $\mu = \bar{x}$

- Confidence interval for μ using the normal distribution when σ is known:
$$\bar{x} \pm z\sigma_{\bar{x}} \quad \text{where} \quad \sigma_{\bar{x}} = \sigma/\sqrt{n}$$

- Confidence interval for μ using the t distribution when σ is not known:
$$\bar{x} \pm ts_{\bar{x}} \quad \text{where} \quad s_{\bar{x}} = s/\sqrt{n}$$

- Margin of error of the estimate for μ:
$$E = z\sigma_{\bar{x}} \quad \text{or} \quad ts_{\bar{x}}$$

- Determining sample size for estimating μ:
$$n = z^2\sigma^2/E^2$$

- Confidence interval for p for a large sample:
$$\hat{p} \pm zs_{\hat{p}} \quad \text{where} \quad s_{\hat{p}} = \sqrt{\hat{p}\hat{q}/n}$$

- Margin of error of the estimate for p:
$$E = zs_{\hat{p}} \quad \text{where} \quad s_{\hat{p}} = \sqrt{\hat{p}\hat{q}/n}$$

- Determining sample size for estimating p:
$$n = z^2pq/E^2$$

Chapter 9 • Hypothesis Tests about the Mean and Proportion

- Test statistic z for a test of hypothesis about μ using the normal distribution when σ is known:
$$z = \frac{\bar{x} - \mu}{\sigma_{\bar{x}}} \quad \text{where} \quad \sigma_{\bar{x}} = \frac{\sigma}{\sqrt{n}}$$

- Test statistic for a test of hypothesis about μ using the t distribution when σ is not known:
$$t = \frac{\bar{x} - \mu}{s_{\bar{x}}} \quad \text{where} \quad s_{\bar{x}} = \frac{s}{\sqrt{n}}$$

- Test statistic for a test of hypothesis about p for a large sample:
$$z = \frac{\hat{p} - p}{\sigma_{\hat{p}}} \quad \text{where} \quad \sigma_{\hat{p}} = \sqrt{\frac{pq}{n}}$$

Chapter 10 • Estimation and Hypothesis Testing: Two Populations

- Mean of the sampling distribution of $\bar{x}_1 - \bar{x}_2$:

$$\mu_{\bar{x}_1 - \bar{x}_2} = \mu_1 - \mu_2$$

- Confidence interval for $\mu_1 - \mu_2$ for two independent samples using the normal distribution when σ_1 and σ_2 are known:

$$(\bar{x}_1 - \bar{x}_2) \pm z\sigma_{\bar{x}_1 - \bar{x}_2} \quad \text{where} \quad \sigma_{\bar{x}_1 - \bar{x}_2} = \sqrt{\frac{\sigma_1^2}{n_1} + \frac{\sigma_2^2}{n_2}}$$

- Test statistic for a test of hypothesis about $\mu_1 - \mu_2$ for two independent samples using the normal distribution when σ_1 and σ_2 are known:

$$z = \frac{(\bar{x}_1 - \bar{x}_2) - (\mu_1 - \mu_2)}{\sigma_{\bar{x}_1 - \bar{x}_2}}$$

- For two independent samples taken from two populations with equal but unknown standard deviations:
  Pooled standard deviation:

$$s_p = \sqrt{\frac{(n_1 - 1)s_1^2 + (n_2 - 1)s_2^2}{n_1 + n_2 - 2}}$$

  Estimate of the standard deviation of $\bar{x}_1 - \bar{x}_2$:

$$s_{\bar{x}_1 - \bar{x}_2} = s_p\sqrt{\frac{1}{n_1} + \frac{1}{n_2}}$$

  Confidence interval for $\mu_1 - \mu_2$ using the t distribution:

$$(\bar{x}_1 - \bar{x}_2) \pm ts_{\bar{x}_1 - \bar{x}_2}$$

  Test statistic using the t distribution:

$$t = \frac{(\bar{x}_1 - \bar{x}_2) - (\mu_1 - \mu_2)}{s_{\bar{x}_1 - \bar{x}_2}}$$

- For two independent samples selected from two populations with unequal and unknown standard deviations:

  Degrees of freedom: $$df = \frac{\left(\dfrac{s_1^2}{n_1} + \dfrac{s_2^2}{n_2}\right)^2}{\dfrac{\left(\dfrac{s_1^2}{n_1}\right)^2}{n_1 - 1} + \dfrac{\left(\dfrac{s_2^2}{n_2}\right)^2}{n_2 - 1}}$$

  Estimate of the standard deviation of $\bar{x}_1 - \bar{x}_2$:

$$s_{\bar{x}_1 - \bar{x}_2} = \sqrt{\frac{s_1^2}{n_1} + \frac{s_2^2}{n_2}}$$

  Confidence interval for $\mu_1 - \mu_2$ using the t distribution:

$$(\bar{x}_1 - \bar{x}_2) \pm ts_{\bar{x}_1 - \bar{x}_2}$$

  Test statistic using the t distribution:

$$t = \frac{(\bar{x}_1 - \bar{x}_2) - (\mu_1 - \mu_2)}{s_{\bar{x}_1 - \bar{x}_2}}$$

- For two paired or matched samples:
  Sample mean for paired differences: $\bar{d} = \sum d/n$
  Sample standard deviation for paired differences:

$$s_d = \sqrt{\frac{\sum d^2 - \dfrac{(\sum d)^2}{n}}{n - 1}}$$

  Mean and standard deviation of the sampling distribution of $\bar{d}$

$$\mu_{\bar{d}} = \mu_d \quad \text{and} \quad s_{\bar{d}} = s_d/\sqrt{n}$$

  Confidence interval for μ_d using the t distribution:

$$\bar{d} \pm ts_{\bar{d}} \quad \text{where} \quad s_{\bar{d}} = s_d/\sqrt{n}$$

  Test statistic for a test of hypothesis about μ_d using the t distribution:

$$t = \frac{\bar{d} - \mu_d}{s_{\bar{d}}}$$

- For two large and independent samples, confidence interval for $p_1 - p_2$:

$$(\hat{p}_1 - \hat{p}_2) \pm zs_{\hat{p}_1 - \hat{p}_2} \quad \text{where} \quad s_{\hat{p}_1 - \hat{p}_2} = \sqrt{\frac{\hat{p}_1\hat{q}_1}{n_1} + \frac{\hat{p}_2\hat{q}_2}{n_2}}$$

- For two large and independent samples, for a test of hypothesis about $p_1 - p_2$ with H_0: $p_1 - p_2 = 0$:
  Pooled sample proportion:

$$\bar{p} = \frac{x_1 + x_2}{n_1 + n_2} \quad \text{or} \quad \frac{n_1\hat{p}_1 + n_2\hat{p}_2}{n_1 + n_2}$$

  Estimate of the standard deviation of $\hat{p}_1 - \hat{p}_2$:

$$s_{\hat{p}_1 - \hat{p}_2} = \sqrt{\bar{p}\bar{q}\left(\frac{1}{n_1} + \frac{1}{n_2}\right)}$$

  Test statistic: $$z = \frac{(\hat{p}_1 - \hat{p}_2) - (p_1 - p_2)}{s_{\hat{p}_1 - \hat{p}_2}}$$

Chapter 11 • Chi-Square Tests

- Expected frequency for a category for a goodness-of-fit test:

$$E = np$$

- Degrees of freedom for a goodness-of-fit test:

$$df = k - 1 \quad \text{where } k \text{ is the number of categories}$$

- Expected frequency for a cell for an independence or homogeneity test:

$$E = \frac{(\text{Row total})(\text{Column total})}{\text{Sample size}}$$

- Degrees of freedom for a test of independence or homogeneity:

$$df = (R - 1)(C - 1)$$

where R and C are the total number of rows and columns, respectively, in the contingency table

WileyPLUS

Now with: ORION, An Adaptive Experience

WileyPLUS is a research-based online environment for effective teaching and learning.

WileyPLUS builds students' confidence because it takes the guesswork out of studying by providing students with a clear roadmap:

- what to do
- how to do it
- if they did it right

It offers interactive resources along with a complete digital textbook that help students learn more. With *WileyPLUS*, students take more initiative so you'll have greater impact on their achievement in the classroom and beyond.

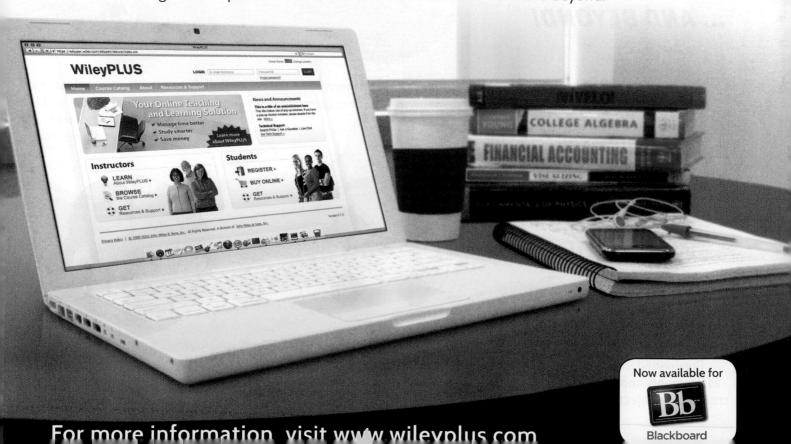

For more information, visit www.wileyplus.com

WileyPLUS

ALL THE HELP, RESOURCES, AND PERSONAL SUPPORT YOU AND YOUR STUDENTS NEED!

www.wileyplus.com/resources

1st DAY OF CLASS ... AND BEYOND!

2-Minute Tutorials and all of the resources you and your students need to get started

WileyPLUS

Student Partner Program

Student support from an experienced student user

Wiley Faculty Network

Collaborate with your colleagues, find a mentor, attend virtual and live events, and view resources
www.WhereFacultyConnect.com

WileyPLUS

Quick Start

Pre-loaded, ready-to-use assignments and presentations created by subject matter experts

Technical Support 24/7
FAQs, online chat, and phone support
www.wileyplus.com/support

© Courtney Keating/iStockphoto

Your *WileyPLUS* Account Manager, providing personal training and support

MANN'S
INTRODUCTORY STATISTICS
GLOBAL EDITION

PREM S. MANN

EASTERN CONNECTICUT STATE UNIVERSITY

WILEY

To the memory of my parents

ISBN: 978-1-119-24894-1

Printed in Markono Print Media Pte Ltd

10 9 8 7 6 5 4 3

PREFACE

Introductory Statistics is written for a one- or two-semester first course in applied statistics. This book is intended for students who do not have a strong background in mathematics. The only prerequisite for this text is knowledge of elementary algebra.

Today, college students from almost all fields of study are required to take at least one course in statistics. Consequently, the study of statistical methods has taken on a prominent role in the education of students from a variety of backgrounds and academic pursuits. The goal of **Introductory Statistics** has been to make the subject of statistics interesting and accessible to a wide and varied audience. Three major elements of this text support this goal:

1. Realistic content of its examples and exercises, drawing from a comprehensive range of applications from all facets of life
2. Clarity and brevity of presentation
3. Soundness of pedagogical approach

These elements are developed through the interplay of a variety of significant text features. The feedback received from the users of the original editions of **Introductory Statistics** has been very supportive and encouraging. Positive experiences reported by instructors and students have served as evidence that this text offers an interesting and accessible approach to statistics— the author's goal from the very first edition. The author has pursued the same goal through the refinements and updates in this edition, so that **Introductory Statistics** can continue to provide a successful experience in statistics to a growing number of students and instructors.

New to This Edition

The following are some of the changes made in the ninth edition:

■ New to this edition, are the videos that are accessible via the *WileyPLUS* course associated with this text. These videos provide step-by-step solutions to selected examples in the book.

■ A large number of the examples and exercises are new or revised, providing contemporary and varied ways for students to practice statistical concepts.

■ Coverage of sample surveys, sampling techniques, and design of experiments has been moved from Appendix A to Chapter 1.

■ In Chapter 3, the discussions of weighted mean, trimmed mean, and coefficient of variation have been moved from the exercises to the main part of the chapter.

■ The majority of the case studies are new or revised, drawing on current uses of statistics in areas of student interest.

■ New data are integrated throughout, reinforcing the vibrancy of statistics and the relevance of statistics to student lives right now.

■ The *Technology Instructions* sections have been updated to support the use of the latest versions of TI-84 Color/TI-84, Minitab, and Excel.

■ Many of the *Technology Assignments* at the end of each chapter are either new or have been updated.

- The data sets posted on the book companion Web site and *WileyPLUS* have been updated.
- Most of the *Uses and Misuses* sections at the end of each chapter have been updated or replaced.
- Many of the *Mini-Projects*, which are now located on the book companion Web site, are either new or have been updated.
- Many of the *Decide for Yourself* sections, also located on the book companion Web site, are either new or have been updated.

Hallmark Features of This Text

Clear and Concise Exposition The explanation of statistical methods and concepts is clear and concise. Moreover, the style is user-friendly and easy to understand. In chapter introductions and in transitions from section to section, new ideas are related to those discussed earlier.

Thorough Examples The text contains a wealth of examples. The examples are usually presented in a format showing a problem and its solution. They are well sequenced and thorough, displaying all facets of concepts. Furthermore, the examples capture students' interest because they cover a wide variety of relevant topics. They are based on situations that practicing statisticians encounter every day. Finally, a large number of examples are based on real data taken from sources such as books, government and private data sources and reports, magazines, newspapers, and professional journals.

Step-by-Step Solutions A clear, concise solution follows each problem presented in an example. When the solution to an example involves many steps, it is presented in a step-by-step format. For instance, examples related to tests of hypothesis contain five steps that are consistently used to solve such examples in all chapters. Thus, procedures are presented in the concrete settings of applications rather than as isolated abstractions. Frequently, solutions contain highlighted remarks that recall and reinforce ideas critical to the solution of the problem. Such remarks add to the clarity of presentation.

Titles for Examples Each example based on an application of concepts now contains a title that describes to what area, field, or concept the example relates.

Margin Notes for Examples A margin note appears beside each example that briefly describes what is being done in that example. Students can use these margin notes to assist them as they read through sections and to quickly locate appropriate model problems as they work through exercises.

Frequent Use of Diagrams Concepts can often be made more understandable by describing them visually with the help of diagrams. This text uses diagrams frequently to help students understand concepts and solve problems. For example, tree diagrams are used a few times in Chapters 4 and 5 to assist in explaining probability concepts and in computing probabilities. Similarly, solutions to all examples about tests of hypothesis contain diagrams showing rejection regions, nonrejection regions, and critical values.

Highlighting Definitions of important terms, formulas, and key concepts are enclosed in colored boxes so that students can easily locate them.

Cautions Certain items need special attention. These may deal with potential trouble spots that commonly cause errors, or they may deal with ideas that students often overlook. Special emphasis is placed on such items through the headings *Remember*, *An Observation*, or *Warning*. An icon is used to identify such items.

Real World Case Studies These case studies, which appear in most of the chapters, provide additional illustrations of the applications of statistics in research and statistical analysis. Most of these case studies are based on articles or data published in journals, magazines, newspapers, or Web sites. Almost all case studies are based on real data.

Variety of Exercises The text contains a variety of exercises, including technology assignments. Moreover, a large number of these exercises contain several parts. Exercise sets appearing at the end of each section (or sometimes at the end of two or three sections) include problems on the topics of that section. These exercises are divided into two parts: **Concepts and Procedures** that emphasize key ideas and techniques and **Applications** that use these ideas and techniques in concrete settings. Supplementary exercises appear at the end of each chapter and contain exercises on all sections and topics discussed in that chapter. A large number of these exercises are based on real data taken from varied data sources such as books, government and private data sources and reports, magazines, newspapers, and professional journals. Not only do the exercises given in the text provide practice for students, but the real data contained in the exercises provide interesting information and insight into economic, political, social, psychological, and other aspects of life. The exercise sets also contain many problems that demand critical thinking skills. The answers to selected odd-numbered exercises appear in the *Answers* section at the back of the book. **Optional exercises** are indicated by an asterisk (*).

Advanced Exercises All chapters have a set of exercises that are of greater difficulty. Such exercises appear under the heading *Advanced Exercises* after the *Supplementary Exercises*.

Uses and Misuses This feature toward the end of each chapter (before the Glossary) points out common misconceptions and pitfalls students will encounter in their study of statistics and in everyday life. Subjects highlighted include such diverse topics as *do not feed the animals*.

Decide for Yourself This feature is accessible online at **www.wiley.com/college/mann**. Each Decide for Yourself discusses a real-world problem and raises questions that readers can think about and answer.

Glossary Each chapter has a glossary that lists the key terms introduced in that chapter, along with a brief explanation of each term.

Self-Review Tests Each chapter contains a *Self-Review Test*, which appears immediately after the *Supplementary* and *Advanced Exercises*. These problems can help students test their grasp of the concepts and skills presented in respective chapters and monitor their understanding of statistical methods. The problems marked by an asterisk (*) in the *Self-Review Tests* are **optional**. The answers to almost all problems of the *Self-Review Tests* appear in the *Answer* section.

Technology Usage At the end of each chapter is a section covering uses of three major technologies of statistics and probability: the TI-84 Color/TI-84, Minitab, and Excel. For each technology, students are guided through performing statistical analyses in a step-by-step fashion, showing them how to enter, revise, format, and save data in a spreadsheet, workbook, or named and unnamed lists, depending on the technology used. Illustrations and screen shots demonstrate the use of these technologies. Additional detailed technology instruction is provided in the technology manuals that are online at www.wiley.com/college/mann.

Technology Assignments Each chapter contains a few technology assignments that appear at the end of the chapter. These assignments can be completed using any of the statistical software.

Mini-projects Associated with each chapter of the text are Mini-projects posted online at **www.wiley.com/college/mann**. These Mini-projects are either very comprehensive exercises or they ask students to perform their own surveys and experiments. They provide practical applications of statistical concepts to real life.

Data Sets A large number of data sets appear on the book companion Web site at **www.wiley.com/college/mann**. These large data sets are collected from various sources, and they contain information on several variables. Many exercises and assignments in the text are based on these data sets. These large data sets can also be used for instructor-driven analyses using a wide variety of statistical software packages as well as the TI-84. **These data sets are available on the Web site of the text in numerous formats, including Minitab and Excel.**

Videos New for this Edition, videos for each text section illustrate concepts related to the topic covered in that section to more deeply engage the students. These videos are accessible via *WileyPLUS*.

GAISE Report Recommendations Adopted

In 2003, the American Statistical Association (ASA) funded the Guidelines for Assessment and Instruction in Statistics Education (GAISE) Project to develop ASA-endorsed guidelines for assessment and instruction in statistics for the introductory college statistics course. The report, which can be found at www.amstat.org/education/gaise, resulted in the following series of recommendations for the first course in statistics and data analysis.

1. Emphasize statistical literacy and develop statistical thinking.
2. Use real data.
3. Stress conceptual understanding rather than mere knowledge of procedures.
4. Foster active learning in the classroom.
5. Use technology for developing concepts and analyzing data.
6. Use assessments to improve and evaluate student learning.

Here are a few examples of how this Introductory Statistics text can assist in helping you, the instructor, in meeting the GAISE recommendations.

1. Many of the exercises require interpretation, not just answers in terms of numbers. Graphical and numeric summaries are combined in some exercises in order to emphasize looking at the whole picture, as opposed to using just one graph or one summary statistic.
2. The *Uses and Misuses* and online *Decide for Yourself* features help to develop statistical thinking and conceptual understanding.
3. All of the data sets listed in Appendix A are available on the book's Web site. They have been formatted for a variety of statistical software packages. This eliminates the need to enter data into the software. A variety of software instruction manuals also allow the instructor to spend more time on concepts and less time teaching how to use technology.
4. The *online Mini-projects* help students to generate their own data by performing an experiment and/or taking random samples from the large data sets mentioned in Appendix A.

We highly recommend that all statistics instructors take the time to read the GAISE report. There is a wealth of information in this report that can be used by everyone.

Web Site

www.wiley.com/college/mann

After you go to the page exhibited by the above URL, click on *Visit the Companion Sites*. Then click on the site that applies to you out of the two choices. This Web site provides additional resources for instructors and students. The following items are available for instructors on this Web site:

- Key Formulas
- Printed Test Bank
- Mini-Projects
- Decide for Yourself
- Power Point Lecture Slides
- Instructor's Solutions Manual
- Data Sets (see Appendix A for a complete list of these data sets)
- Chapter 14: Multiple Regression
- Chapter 15: Nonparametric Methods
- Technology Resource Manuals:
  - TI Graphing Calculator Manual
  - Minitab Manual
  - Excel Manual

These manuals provide step-by-step instructions, screen captures, and examples for using technology in the introductory statistics course. Also provided are exercise lists and indications of which exercises from the text best lend themselves to the use of the package presented.

Using *WileyPLUS*

SUCCESS: *WileyPLUS* helps to ensure that each study session has a positive outcome by putting students in control. Through instant feedback and study objective reports, students know **if they did it right** and where to focus next, so they achieve the strongest results.

Our efficacy research shows that with *WileyPLUS,* students improve their outcomes by as much as one letter grade. *WileyPLUS* helps students take more initiative, so you will have greater impact on their achievement in the classroom and beyond.

What Do Students Receive with *WileyPLUS*?

• The complete digital textbook, saving students up to 60% off the cost of a printed text.
• Question assistance, including links to relevant sections in the online digital textbook.
• Immediate feedback and proof of progress, 24/7.
• Integrated, multimedia resources—including videos—that provide multiple study paths and encourage more active learning.

What Do Instructors Receive with *WileyPLUS*?

• Reliable resources that reinforce course goals inside and outside of the classroom.
• The ability to easily identify those students who are falling behind.
• Media-rich course materials and assessment content, including Instructor's Solutions Manual, PowerPoint slides, Learning Objectives, Printed Test Bank, and much more.
www.wileyplus.com. Learn More.

Acknowledgments

I thank the following reviewers of the ninth edition of this book, whose comments and suggestions were invaluable in improving the text.

D. P. Adhikari
Marywood University
Wendy Ahrendsen
South Dakota State University
Nan Hutchins Bailey
The University of Texas at Tyler
Les Barnhouse
Athabasca University
Hossein Behforooz
Utica College
Joleen Beltrami
University of the Incarnate Word
Bill Burgin
Gaston College
David Bush
Villanova University
Ferry Butar Butar
Sam Houston State University
Chris Chappa
University of Texas at Tyler

Jerry Chen
*Suffolk County Community College,
  Ammerman Campus*
A. Choudhury
Illinois State University
Robert C. Forsythe
Frostburg State University
Daesung Ha
Marshall University
Rhonda Hatcher
Texas Christian University
Joanna Jeneralczuk
University of Massachusetts, Amherst
Annette Kanko
University of the Incarnate Word
Mohammad Kazemi
*University of North Carolina,
  Charlotte*
Janine Keown-Gerrard
Athabasca University

Hoon Kim
California State Polytechnic
University—Pomona
Matthew Knowlen
Horry Georgetown Technical College
Maurice LeBlanc
Kennesaw State University
Min-Lin Lo
California State University, San Bernardino
Natalya Malakhova
Johnson County Community College
Nola McDaniel
McNeese State University
Stéphane Mechoulan
Dalhousie University
Robert L. Nichols
Florida Gulf Coast University

Jonathan Oaks
Macomb Community
College—South Campus
Marty Rhoades
West Texas A&M University
Mary Beth Rollick
Kent State University—Kent
Seema Sehgal
Athabasca University
Linda Simonsen
University of Washington, Bothell
Criselda Toto
Chapman University
Viola Vajdova
Benedictine University
Daniel Weiner
Boston University

I express my thanks to the following for their contributions to earlier editions of this book that made it better in many ways: Chris Lacke (Rowan University), Gerald Geissert (formerly of Eastern Connecticut State University), Daniel S. Miller (Central Connecticut State University), and David Santana-Ortiz (Rand Organization).

I extend my special thanks to Doug Tyson, James Bush, and Chad Cross, who contributed to this edition in many significant ways. I take this opportunity to thank Mark McKibben for working on the solutions manuals and preparing the answer section and Julie M. Clark for checking the solutions for accuracy.

It is of utmost importance that a textbook be accompanied by complete and accurate supplements. I take pride in mentioning that the supplements prepared for this text possess these qualities and much more. I thank the authors of all these supplements.

It is my pleasure to thank all the professionals at John Wiley & Sons with whom I enjoyed working during this revision. Among them are Laurie Rosatone (Vice President and Director), Joanna Dingle (Acquisitions Editor), Wendy Lai (Senior Designer), Billy Ray (Senior Photo Editor), Valerie Zaborski (Senior Content Manager), Ken Santor (Senior Production Editor), Ellen Keohane (Project Editor), Jennifer Brady (Sponsoring Editor), Carolyn DeDeo (Market Solutions Assistant), David Dietz (Senior Product Designer), and John LaVacca (Marketing Manager). I also want to thank Jackie Henry (Full Service Manager) for managing the production process and Lisa Torri (Art Development Editor) for her work on the case study art.

Any suggestions from readers for future revisions would be greatly appreciated. Such suggestions can be sent to the author at **premmann@yahoo.com** or **mann@easternct.edu**. I recently retired from Eastern Connecticut State University and, hence, will prefer that readers email me at **premmann@yahoo.com**.

Prem S. Mann
December 2015

CONTENTS

CHAPTER 1

Oleksiy Mark/Shutterstock

Introduction

Are you, as an American, thriving in your life? Or are you struggling? Or, even worse, are you suffering? A poll of 176,903 American adults, aged 18 and older, was conducted January 2 to December 30, 2014, as part of the Gallup-Healthways Well-Being Index survey. The poll found that while 54.1% of these Americans said that they were thriving, 42.1% indicated that they were struggling, and 3.8% mentioned that they were suffering. (See Case Study 1–2.)

The study of statistics has become more popular than ever over the past four decades. The increasing availability of computers and statistical software packages has enlarged the role of statistics as a tool for empirical research. As a result, statistics is used for research in almost all professions, from medicine to sports. Today, college students in almost all disciplines are required to take at least one statistics course. Almost all newspapers and magazines these days contain graphs and stories on statistical studies. After you finish reading this book, it should be much easier to understand these graphs and stories.

Every field of study has its own terminology. Statistics is no exception. This introductory chapter explains the basic terms and concepts of statistics. These terms and concepts will bridge our understanding of the concepts and techniques presented in subsequent chapters.

1.1 | Statistics and Types of Statistics

In this section we will learn about statistics and types of statistics.

1.1.1 What Is Statistics?

The word **statistics** has two meanings. In the more common usage, *statistics* refers to numerical facts. The numbers that represent the income of a family, the age of a student, the percentage of passes completed by the quarterback of a football team, and the starting salary of a typical college graduate are examples of statistics in this sense of the word. A 1988 article in *U.S. News & World Report* mentioned that "Statistics are an American obsession."[1] During the 1988 baseball World Series between the Los Angeles Dodgers and the Oakland A's, the then NBC commentator Joe Garagiola reported to the viewers numerical facts about the players' performances. In response, fellow commentator Vin Scully said, "I love it when you talk statistics." In these examples, the word *statistics* refers to numbers.

The following examples present some statistics:

1. During March 2014, a total of 664,000,000 hours were spent by Americans watching March Madness live on TV and/or streaming (*Fortune* Magazine, March 15, 2015).
2. Approximately 30% of Google's employees were female in July 2014 (*USA TODAY*, July 24, 2014).
3. According to an estimate, an average family of four living in the United States needs $130,357 a year to live the American dream (*USA TODAY*, July 7, 2014).
4. Chicago's O'Hare Airport was the busiest airport in 2014, with a total of 881,933 flight arrivals and departures.
5. In 2013, author James Patterson earned $90 million from the sale of his books (*Forbes*, September 29, 2014).
6. About 22.8% of U.S. adults do not have a religious affiliation (*Time*, May 25, 2015).
7. Yahoo CEO Marissa Mayer was the highest paid female CEO in America in 2014, with a total compensation of $42.1 million.

Statistics is the science of collecting, analyzing, presenting, and interpreting data, as well as of making decisions based on such analyses.

The second meaning of *statistics* refers to the field or discipline of study. In this sense of the word, *statistics* is defined as follows.

Every day we make decisions that may be personal, business related, or of some other kind. Usually these decisions are made under conditions of uncertainty. Many times, the situations or problems we face in the real world have no precise or definite solution. Statistical methods help us make scientific and intelligent decisions in such situations. Decisions made by using statistical methods are called *educated guesses*. Decisions made without using statistical (or scientific) methods are *pure guesses* and, hence, may prove to be unreliable. For example, opening a large store in an area with or without assessing the need for it may affect its success.

Like almost all fields of study, statistics has two aspects: theoretical and applied. *Theoretical* or *mathematical statistics* deals with the development, derivation, and proof of statistical theorems, formulas, rules, and laws. *Applied statistics* involves the applications of those theorems, formulas, rules, and laws to solve real-world problems. This text is concerned with applied statistics and not with theoretical statistics. By the time you finish studying this book, you will have learned how to think statistically and how to make educated guesses.

1.1.2 Types of Statistics

Broadly speaking, applied statistics can be divided into two areas: **descriptive statistics** and **inferential statistics**.

---

[1]"The Numbers Racket: How Polls and Statistics Lie," *U.S. News & World Report*, July 11, 1988, pp. 44–47.

2014 Lobbying Spending by Selected Companies

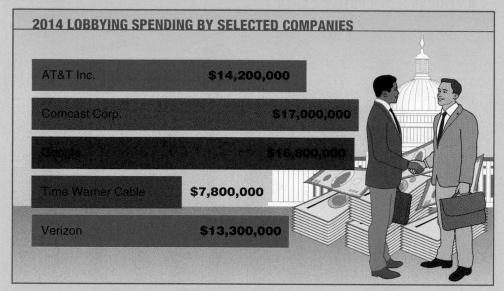

2014 LOBBYING SPENDING BY SELECTED COMPANIES

AT&T Inc.	$14,200,000
Comcast Corp.	$17,000,000
Google	$16,800,000
Time Warner Cable	$7,800,000
Verizon	$13,300,000

Data source: Fortune Magazine, June 1, 2015

The accompanying chart shows the lobbying spending by five selected companies during 2014. Many companies spend millions of dollars to win favors in Washington. According to *Fortune* Magazine of June 1, 2015, "Comcast has remained one of the biggest corporate lobbyists in the country." In 2014, Comcast spent $17 million, Google spent $16.8 million, AT&T spent $14.2 million, Verizon spent $13.3 million, and Time Warner Cable spent $7.8 million on lobbying. These numbers simply describe the total amounts spent by these companies on lobbying. We are not drawing any inferences, decisions, or predictions from these data. Hence, this data set and its presentation is an example of descriptive statistics.

Descriptive Statistics

Suppose we have information on the test scores of students enrolled in a statistics class. In statistical terminology, the whole set of numbers that represents the scores of students is called a **data set**, the name of each student is called an **element**, and the score of each student is called an **observation**. (These terms are defined in more detail in Section 1.2.)

Many data sets in their original forms are usually very large, especially those collected by federal and state agencies. Consequently, such data sets are not very helpful in drawing conclusions or making decisions. It is easier to draw conclusions from summary tables and diagrams than from the original version of a data set. So, we summarize data by constructing tables, drawing graphs, or calculating summary measures such as averages. The portion of statistics that helps us do this type of statistical analysis is called **descriptive statistics**.

Chapters 2 and 3 discuss descriptive statistical methods. In Chapter 2, we learn how to construct tables and how to graph data. In Chapter 3, we learn how to calculate numerical summary measures, such as averages.

Case Study 1–1 presents an example of descriptive statistics.

> **Descriptive statistics** consists of methods for organizing, displaying, and describing data by using tables, graphs, and summary measures.

Inferential Statistics

In statistics, the collection of all elements of interest is called a **population**. The selection of a portion of the elements from this population is called a **sample**. (Population and sample are discussed in more detail in Section 1.5.)

A major portion of statistics deals with making decisions, inferences, predictions, and forecasts about populations based on results obtained from samples. For example, we may make some decisions about the political views of all college and university students based on the political views of 1000 students selected from a few colleges and universities. As another example, we may want to find the starting salary of a typical college graduate. To do so, we may select 2000

Americans' Life Outlook, 2014

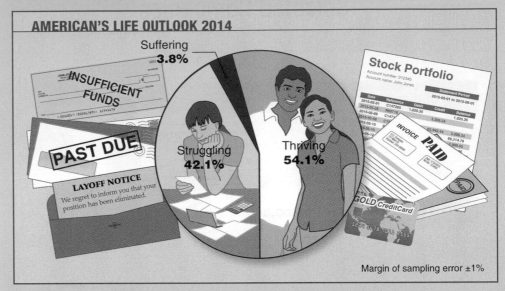

AMERICAN'S LIFE OUTLOOK 2014

Suffering **3.8%**

Struggling **42.1%**

Thriving **54.1%**

INSUFFICIENT FUNDS

PAST DUE

LAYOFF NOTICE
We regret to inform you that your position has been eliminated.

Stock Portfolio

INVOICE PAID

GOLD CreditCard

Margin of sampling error ±1%

Data source: Gallup-Healthways Well-Being Index

A poll of 176,903 American adults, aged 18 and older, was conducted January 2 to December 30, 2014, as part of the Gallup-Healthways Well-Being Index survey. Gallup and Healthways have been "tracking Americans' life evaluations daily" since 2008. According to this poll, in 2014, Americans' outlook on life was the best in seven years, as 54.1% "rated their lives highly enough to be considered thriving," 42.1% said they were struggling, and 3.8% mentioned that they were suffering. As mentioned in the chart, the margin of sampling error was ±1%. In Chapter 8, we will discuss the concept of margin of error, which can be combined with these percentages when making inferences. As we notice, the results described in the chart are obtained from a poll of 176,903 adults. We will learn in later chapters how to apply these results to the entire population of adults. Such decision making about the population based on sample results is called inferential statistics.

recent college graduates, find their starting salaries, and make a decision based on this information. The area of statistics that deals with such decision-making procedures is referred to as **inferential statistics**. This branch of statistics is also called *inductive reasoning* or *inductive statistics*.

Case Study 1–2 presents an example of inferential statistics. It shows the results of a survey in which American adults were asked about their opinions about their lives.

Chapters 8 through 15 and parts of Chapter 7 deal with inferential statistics.

Probability, which gives a measurement of the likelihood that a certain outcome will occur, acts as a link between descriptive and inferential statistics. Probability is used to make statements about the occurrence or nonoccurrence of an event under uncertain conditions. Probability and probability distributions are discussed in Chapters 4 through 6 and parts of Chapter 7.

> **Inferential statistics** consists of methods that use sample results to help make decisions or predictions about a population.

EXERCISES

CONCEPTS AND PROCEDURES

1.1 What are the different meanings of the word statistics?

1.2 What are the two major types of statistics?

APPLICATIONS

1.3 Is the following an example of descriptive statistics or inferential statistics? Explain.

    a. In a survey by *Fortune* Magazine and SurveyMonkey, participants were asked what was the most important factor when purchasing groceries (*Fortune*, June 1, 2015). The following

table lists the summary of the responses of these participants. Assume that the maximum margin of error is ±1.5%.

Factor	Percent of Respondents
Price	42.4
Nutrition	36.0
Absence of additives	16.4
Number of calories	3.8
Carbon footprint	1.5

1.2 | Basic Terms

It is very important to understand the meaning of some basic terms that will be used frequently in this text. This section explains the meaning of an element (or member), a variable, an observation, and a data set. An element and a data set were briefly defined in Section 1.1. This section defines these terms formally and illustrates them with the help of an example.

Table 1.1 gives information, based on *Forbes* magazine, on the total wealth of the world's eight richest persons as of March 2015. Each person listed in this table is called an **element** or a **member** of this group. Table 1.1 contains information on eight elements. Note that elements are also called **observational units**.

> An **element** or **member** of a sample or population is a specific subject or object (for example, a person, firm, item, state, or country) about which the information is collected.

Table 1.1   Total Wealth of the World's Eight Richest Persons

Name	Total Wealth (billions of dollars)
Bill Gates	79.2
Carlos Slim Helu	77.1
Warren Buffett	72.7
Amancio Ortega	64.5
Larry Ellison	54.3
Charles Koch	42.9
David Koch	42.9
Christy Walton	41.7

Source: *Forbes*, March 23, 2015.

(Variable ← Total Wealth; An element or member → Warren Buffett; An observation or measurement → 72.7)

The total wealth in our example is called a variable. The total wealth is a characteristic of these persons on which information is collected.

A few other examples of variables are household incomes, the number of houses built in a city per month during the past year, the makes of cars owned by people, the gross profits of companies, and the number of insurance policies sold by a salesperson per day during the past month.

> A **variable** is a characteristic under study that assumes different values for different elements. In contrast to a variable, the value of a *constant* is fixed.

In general, a variable assumes different values for different elements, as illustrated by the total wealth for the eight persons in Table 1.1. For some elements in a data set, however, the values of the variable may be the same. For example, if we collect information on incomes of households, these households are expected to have different incomes, although some of them may have the same income.

A variable is often denoted by x, y, or z. For instance, in Table 1.1, the total wealth for persons may be denoted by any one of these letters. Starting with Section 1.7, we will begin to use these letters to denote variables.

Each of the values representing the total wealths of the eight persons in Table 1.1 is called an **observation** or **measurement**.

> The value of a variable for an element is called an **observation** or **measurement**.

From Table 1.1, the total wealth of Warren Buffett was $72.7 billion. The value $72.7 billion is an observation or a measurement. Table 1.1 contains eight observations, one for each of the eight persons.

The information given in Table 1.1 on the total wealth of the eight richest persons is called the data or a **data set**.

Other examples of data sets are a list of the prices of 25 recently sold homes, test scores of 15 students, opinions of 100 voters, and ages of all employees of a company.

> A **data set** is a collection of observations on one or more variables.

EXERCISES

CONCEPTS AND PROCEDURES

1.4 Define the following terms in statistics: an element, a variable, an observation and a data set.

APPLICATIONS

1.5 The following table gives the number of dog bites reported to the police last year in six cities.

City	Number of Bites
Center City	52
Elm Grove	32
Franklin	43
Bay City	44
Oakdale	12
Sand Point	2

Briefly explain the meaning of a member, a variable, a measurement, and a data set with reference to this table.

1.6 The following table lists the number of billionaires in eight countries as of February 2011, as reported in *The New York Times* of July 27, 2011.

Country	Number of Billionaires
United States	413
China	115
Russia	101
India	55
Germany	52
Britain	32
Brazil	30
Japan	26

Source: Forbes, International Monetary Fund.

Briefly explain the meaning of a member, a variable, a measurement, and a data set with reference to this table.

1.7 For problems 1.5 and 1.6 also answer the following:

a. What is the variable for this data set?
b. How many observations are in this data set?
c. How many elements does this data set contain?

1.3 | Types of Variables

In Section 1.2, we learned that a variable is a characteristic under investigation that assumes different values for different elements. Family income, height of a person, gross sales of a company, price of a college textbook, make of the car owned by a family, number of accidents, and status (freshman, sophomore, junior, or senior) of a student enrolled at a university are examples of variables.

A variable may be classified as quantitative or qualitative. These two types of variables are explained next.

1.3.1   Quantitative Variables

Some variables (such as the price of a home) can be measured numerically, whereas others (such as hair color) cannot. The price of a home is an example of a **quantitative variable** while hair color is an example of a **qualitative variable**.

Income, height, gross sales, price of a home, number of cars owned, and number of accidents are examples of quantitative variables because each of them can be expressed numerically. For instance, the income of a family may be $81,520.75 per year, the gross sales for a company may be $567 million for the past year, and so forth. Such quantitative variables may be classified as either *discrete variables* or *continuous variables*.

> A variable that can be measured numerically is called a **quantitative variable**. The data collected on a quantitative variable are called **quantitative data**.

Discrete Variables

The values that a certain quantitative variable can assume may be countable or noncountable. For example, we can count the number of cars owned by a family, but we cannot count the height of a family member, as it is measured on a continuous scale. A variable that assumes countable values is called a **discrete variable**. Note that there are no possible intermediate values between consecutive values of a discrete variable.

For example, the number of cars sold on any given day at a car dealership is a discrete variable because the number of cars sold must be 0, 1, 2, 3, . . . and we can count it. The number of cars sold cannot be between 0 and 1, or between 1 and 2. Other examples of discrete variables are

> A variable whose values are countable is called a **discrete variable**. In other words, a discrete variable can assume only certain values with no intermediate values.

the number of people visiting a bank on any day, the number of cars in a parking lot, the number of cattle owned by a farmer, and the number of students in a class.

Continuous Variables

Some variables assume values that cannot be counted, and they can assume any numerical value between two numbers. Such variables are called **continuous variables**.

The time taken to complete an examination is an example of a continuous variable because it can assume any value, let us say, between 30 and 60 minutes. The time taken may be 42.6 minutes, 42.67 minutes, or 42.674 minutes. (Theoretically, we can measure time as precisely as we want.) Similarly, the height of a person can be measured to the tenth of an inch or to the hundredth of an inch. Neither time nor height can be counted in a discrete fashion. Other examples of continuous variables are the weights of people, the amount of soda in a 12-ounce can (note that a can does not contain exactly 12 ounces of soda), and the yield of potatoes (in pounds) per acre. Note that any variable that involves money and can assume a large number of values is typically treated as a continuous variable.

> A variable that can assume any numerical value over a certain interval or intervals is called a **continuous variable**.

1.3.2 Qualitative or Categorical Variables

Variables that cannot be measured numerically but can be divided into different categories are called **qualitative** or **categorical variables**.

For example, the status of an undergraduate college student is a qualitative variable because a student can fall into any one of four categories: freshman, sophomore, junior, or senior. Other examples of qualitative variables are the gender of a person, the make of a computer, the opinions of people, and the make of a car.

Figure 1.1 summarizes the different types of variables.

> A variable that cannot assume a numerical value but can be classified into two or more nonnumeric categories is called a **qualitative** or **categorical variable.** The data collected on such a variable are called **qualitative data**.

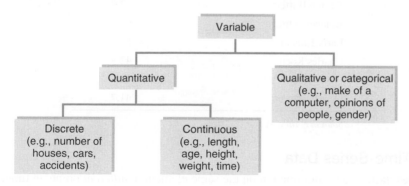

Figure 1.1 Types of variables.

EXERCISES

CONCEPTS AND PROCEDURES

1.8 Differentiate between the following:

    a. Qualitative and quantitative variable
    b. Continuous and discrete variable
    c. Qualitative and quantitative data

APPLICATIONS

1.9 Indicate which of the following variables are quantitative and which are qualitative.

    a. Number of children under 18 years of age in a family
    b. Colors of cars

    c. Age of people
    d. Time to commute from home to work
    e. Number of errors in a person's credit report

1.10 Indicate which of the following variables are quantitative and which are qualitative. Classify the quantitative variables as discrete or continuous.

    a. Women's favorite TV programs
    b. Salaries of football players
    c. Number of pets owned by families
    d. Favorite breed of dog for each of 20 persons

1.11 A survey of families living in a certain city was conducted to collect information on the following variables: age of the oldest person in the family, number of family members, number of males in the family, number of females in the family, whether or not they own a house, income of the family, whether or not the family took vacations during the past one year, whether or not they are happy with their financial situation, and the amount of their monthly mortgage or rent.

Identify the quantitative variables and separate them as discrete and continuous variables.

1.4 | Cross-Section Versus Time-Series Data

Based on the time over which they are collected, data can be classified as either cross-section or time-series data.

1.4.1 Cross-Section Data

> Data collected on different elements at the same point in time or for the same period of time are called **cross-section data**.

Cross-section data contain information on different elements of a population or sample for the *same* period of time. The information on incomes of 100 families for 2015 is an example of cross-section data. All examples of data already presented in this chapter have been cross-section data.

Table 1.1, reproduced here as Table 1.2, shows the total wealth of each of the eight richest persons in the world. Because this table presents data on the total wealth of eight persons for the same period, it is an example of cross-section data.

Table 1.2   Total Wealth of World's Eight Richest Persons

Name	Total Wealth (billions of dollars)
Bill Gates	79.2
Carlos Slim Helu	77.1
Warren Buffett	72.7
Amancio Ortega	64.5
Larry Ellison	54.3
Charles Koch	42.9
David Koch	42.9
Christy Walton	41.7

Source: *Forbes*, March 23, 2015.

1.4.2 Time-Series Data

> Data collected on the same element for the same variable at different points in time or for different periods of time are called **time-series data**.

Time-series data contain information on the same element at *different* points in time. Information on U.S. exports for the years 2001 to 2015 is an example of time-series data.

The data given in Table 1.3 are an example of time-series data. This table lists the average tuition and fee charges in 2014 dollars for in-state students at four-year public institutions in the United States for selected years.

Table 1.3   Average Tuition and Fees in 2014 Dollars at Four-Year Public Institutions

Years	Tuition and Fee (Dollars)
1974–75	2469
1984–85	2810
1994–95	4343
2004–05	6448
2014–15	9139

Source: The College Board.

EXERCISES

CONCEPTS AND PROCEDURES

1.12 What is the difference between time series data and cross-section data? Give an example of each of these two types of data.

APPLICATIONS

1.13 Classify the following as cross-section or time-series data.

a. Average prices of houses in 100 cities
b. Salaries of 50 employees
c. Number of cars sold each year by General Motors from 1980 to 2012
d. Number of employees employed by a company each year from 1985 to 2012
e. Number of armed robberies in each year in Baltimore from 1998 to 2009
f. Number of gyms in 50 cities on December 31, 2016

1.5 | Population Versus Sample

We will encounter the terms *population* and *sample* on almost every page of this text. Consequently, understanding the meaning of each of these two terms and the difference between them is crucial.

Suppose a statistician is interested in knowing the following:

1. The percentage of all voters in a city who will vote for a particular candidate in an election
2. Last year's gross sales of all companies in New York City
3. The prices of all homes in California

In these examples, the statistician is interested in *all* voters in a city, *all* companies in New York City, and *all* homes in California. Each of these groups is called the **population** for the respective example. In statistics, a population does not necessarily mean a collection of people. It can, in fact, be a collection of people or of any kind of item such as houses, books, television sets, or cars. The population of interest is usually called the **target population**.

Most of the time, decisions are made based on portions of populations. For example, the election polls conducted in the United States to estimate the percentages of voters who favor various candidates in any presidential election are based on only a few hundred or a few thousand voters selected from across the country. In this case, the population consists of all registered voters in the United States. The sample is made up of a few hundred or few thousand voters who are included in an opinion poll. Thus, the collection of a number of elements selected from a population is called a **sample**. Figure 1.2 illustrates the selection of a sample from a population.

> A **population** consists of all elements—individuals, items, or objects—whose characteristics are being studied. The population that is being studied is also called the **target population**.

> A portion of the population selected for study is referred to as a **sample**.

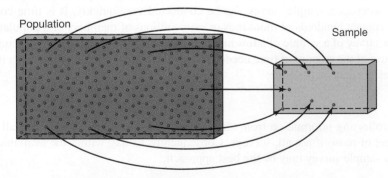

Figure 1.2   Population and sample.

The collection of information from the elements of a sample is called a **sample survey**. When we collect information on all elements of the target population, it is called a **census**. Often the target population is very large. Hence, in practice, a census is rarely taken because it is expensive and time-consuming. In many cases, it is even impossible to identify each element

> A survey that includes every member of the population is called a **census**. A survey that includes only a portion of the population is called a **sample survey**.

of the target population. Usually, to conduct a survey, we select a sample and collect the required information from the elements included in that sample. We then make decisions based on this sample information. As an example, if we collect information on the 2015 incomes of all families in Connecticut, it will be referred to as a census. On the other hand, if we collect information on the 2015 incomes of 50 families selected from Connecticut, it will be called a sample survey.

The purpose of conducting a sample survey is to make decisions about the corresponding population. It is important that the results obtained from a sample survey closely match the results that we would obtain by conducting a census. Otherwise, any decision based on a sample survey will not apply to the corresponding population. As an example, to find the average income of families living in New York City by conducting a sample survey, the sample must contain families who belong to different income groups in almost the same proportion as they exist in the population. Such a sample is called a **representative sample**. Inferences derived from a representative sample will be more reliable.

> A sample that represents the characteristics of the population as closely as possible is called a **representative sample**.

A sample may be selected with or without replacement. In sampling **with replacement**, each time we select an element from the population, we put it back in the population before we select the next element. Thus, in sampling with replacement, the population contains the same number of items each time a selection is made. As a result, we may select the same item more than once in such a sample. Consider a box that contains 25 marbles of different colors. Suppose we draw a marble, record its color, and put it back in the box before drawing the next marble. Every time we draw a marble from this box, the box contains 25 marbles. This is an example of sampling with replacement. The experiment of rolling a die many times is another example of sampling with replacement because every roll has the same six possible outcomes.

Sampling **without replacement** occurs when the selected element is not replaced in the population. In this case, each time we select an item, the size of the population is reduced by one element. Thus, we cannot select the same item more than once in this type of sampling. Most of the time, samples taken in statistics are without replacement. Consider an opinion poll based on a certain number of voters selected from the population of all eligible voters. In this case, the same voter is not selected more than once. Therefore, this is an example of sampling without replacement.

1.5.1   Why Sample?

Most of the time surveys are conducted by using samples and not a census of the population. Three of the main reasons for conducting a sample survey instead of a census are listed next.

Time

In most cases, the size of the population is quite large. Consequently, conducting a census takes a long time, whereas a sample survey can be conducted very quickly. It is time-consuming to interview or contact hundreds of thousands or even millions of members of a population. On the other hand, a survey of a sample of a few hundred elements may be completed in much less time. In fact, because of the amount of time needed to conduct a census, by the time the census is completed, the results may be obsolete.

Cost

The cost of collecting information from all members of a population may easily fall outside the limited budget of most, if not all, surveys. Consequently, to stay within the available resources, conducting a sample survey may be the best approach.

Impossibility of Conducting a Census

Sometimes it is impossible to conduct a census. First, it may not be possible to identify and access each member of the population. For example, if a researcher wants to conduct a survey about homeless people, it is not possible to locate each member of the population and include him or her in the survey. Second, sometimes conducting a survey means destroying the items included in the survey. For example, to estimate the mean life of lightbulbs would necessitate burning out all the bulbs included in the survey. The same is true about finding the average life of batteries. In such cases, only a portion of the population can be selected for the survey.

1.5.2 Random and Nonrandom Samples

Depending on how a sample is drawn, it may be a **random sample** or a **nonrandom sample**.

Suppose we have a list of 100 students and we want to select 10 of them. If we write the names of all 100 students on pieces of paper, put them in a hat, mix them, and then draw 10 names, the result will be a random sample of 10 students. However, if we arrange the names of these 100 students alphabetically and pick the first 10 names, it will be a nonrandom sample because the students who are not among the first 10 have no chance of being selected in the sample.

A random sample is usually a representative sample. Note that for a random sample, each member of the population may or may not have the same chance of being included in the sample.

Two types of nonrandom samples are a *convenience sample* and a *judgment sample*. In a **convenience sample**, the most accessible members of the population are selected to obtain the results quickly. For example, an opinion poll may be conducted in a few hours by collecting information from certain shoppers at a single shopping mall. In a **judgment sample**, the members are selected from the population based on the judgment and prior knowledge of an expert. Although such a sample may happen to be a representative sample, the chances of it being so are small. If the population is large, it is not an easy task to select a representative sample based on judgment.

The so-called **pseudo polls** are examples of nonrepresentative samples. For instance, a survey conducted by a magazine that includes only its own readers does not usually involve a representative sample. Similarly, a poll conducted by a television station giving two separate telephone numbers for *yes* and *no* votes is not based on a representative sample. In these two examples, respondents will be only those people who read that magazine or watch that television station, who do not mind paying the postage or telephone charges, or who feel compelled to respond.

Another kind of sample is the **quota sample**. To select such a sample, we divide the target population into different subpopulations based on certain characteristics. Then we select a subsample from each subpopulation in such a way that each subpopulation is represented in the sample in exactly the same proportion as in the target population. As an example of a quota sample, suppose we want to select a sample of 1000 persons from a city whose population has 48% men and 52% women. To select a quota sample, we choose 480 men from the male population and 520 women from the female population. The sample selected in this way will contain exactly 48% men and 52% women. Another way to select a quota sample is to select from the population one person at a time until we have exactly 480 men and 520 women.

Until the 1948 presidential election in the United States, quota sampling was the most commonly used sampling procedure to conduct opinion polls. The voters included in the samples were selected in such a way that they represented the population proportions of voters based on age, sex, education, income, race, and so on. However, this procedure was abandoned after the 1948 presidential election, in which the underdog, Harry Truman, defeated Thomas E. Dewey, who was heavily favored based on the opinion polls. First, the quota samples failed to be representative because the interviewers were allowed to fill their quotas by choosing voters based on their own judgments. This caused the selection of more upper-income and highly educated people, who happened to be Republicans. Thus, the quota samples were unrepresentative of the population because Republicans were overrepresented in these samples. Second, the results of the opinion polls based on quota sampling happened to be false because a large number of factors differentiate voters, but the pollsters considered only a few of those factors. A quota sample based on a few factors will skew the results. A random sample (one that is not based on quotas) has a much better chance of being representative of the population of all voters than a quota sample based on a few factors.

> A **random sample** is a sample drawn in such a way that each member of the population has some chance of being selected in the sample. In a **nonrandom sample**, some members of the population may not have any chance of being selected in the sample.

1.5.3   Sampling and Nonsampling Errors

The results obtained from a sample survey may contain two types of errors: sampling and non-sampling errors. The sampling error is also called the chance error, and nonsampling errors are also called the systematic errors.

Sampling or Chance Error

> The **sampling error** is the difference between the result obtained from a sample survey and the result that would have been obtained if the whole population had been included in the survey.

Usually, all samples selected from the same population will give different results because they contain different elements of the population. Moreover, the results obtained from any one sample will not be exactly the same as the ones obtained from a census. The difference between a sample result and the result we would have obtained by conducting a census is called the **sampling error**, assuming that the sample is random and no nonsampling error has been made.

The sampling error occurs because of chance, and it cannot be avoided. A sampling error can occur only in a sample survey. It does not occur in a census. Sampling error is discussed in detail in Chapter 7, and an example of it is given there.

Nonsampling Errors or Biases

> The errors that occur in the collection, recording, and tabulation of data are called **nonsampling errors or biases**.

Nonsampling errors or biases can occur both in a sample survey and in a census. Such errors occur because of human mistakes and not chance. Nonsampling errors are also called **systematic errors** or **biases**.

Nonsampling errors can be minimized if questions are prepared carefully and data are handled cautiously. Many types of systematic errors or biases can occur in a survey, including selection error, nonresponse error, response error, and voluntary response error. Figure 1.3 shows the types of errors.

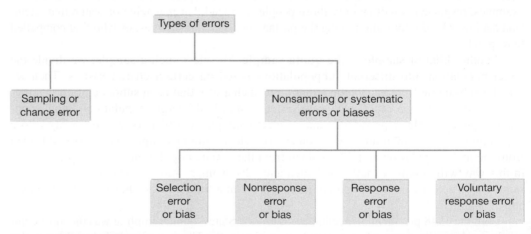

Figure 1.3   Types of errors.

(i) Selection Error or Bias

When we need to select a sample, we use a list of elements from which we draw a sample, and this list usually does not include many members of the target population. Most of the time it is not feasible to include every member of the target population in this list. This list of members of the population that is used to select a sample is called the **sampling frame**. For example, if we use a telephone directory to select a sample, the list of names that appears in this directory makes the sampling frame. In this case we will miss the people who are not listed in the telephone directory. The people we miss, for example, will be poor people (including homeless people) who do not have telephones and people who do not want to be listed in the directory. Thus, the sampling frame that is used to select a sample may not be representative of the population. This may cause the sample results to be different from the population results. The error that occurs because the sampling frame is not representative of the population is called the **selection error or bias**.

If a sample is nonrandom (and, hence, nonrepresentative), the sample results may be quite different from the census results.

(ii) Nonresponse Error or Bias

Even if our sampling frame and, consequently, the sample are representative of the population, **nonresponse error** may occur because many of the people included in the sample may not respond to the survey.

This type of error occurs especially when a survey is conducted by mail. A lot of people do not return the questionnaires. It has been observed that families with low and high incomes do not respond to surveys by mail. Consequently, such surveys overrepresent middle-income families. This kind of error occurs in other types of surveys, too. For instance, in a face-to-face survey where the interviewer interviews people in their homes, many people may not be home when the interviewer visits their homes. The people who are home at the time the interviewer visits and the ones who are not home at that time may differ in many respects, causing a bias in the survey results. This kind of error may also occur in a telephone survey. Many people may not be home when the interviewer calls. This may distort the results. To avoid the nonresponse error, every effort should be made to contact all people included in the sample.

(iii) Response Error or Bias

The **response error or bias** occurs when the answer given by a person included in the survey is not correct. This may happen for many reasons. One reason is that the respondent may not have understood the question. Thus, the wording of the question may have caused the respondent to answer incorrectly. It has been observed that when the same question is worded differently, many people do not respond the same way. Usually such an error on the part of respondents is not intentional.

Sometimes the respondents do not want to give correct information when answering a question. For example, many respondents will not disclose their true incomes on questionnaires or in interviews. When information on income is provided, it is almost always biased in the upward direction.

Sometimes the race of the interviewer may affect the answers of respondents. This is especially true if the questions asked are about race relations. The answers given by respondents may differ depending on the race of the interviewer.

(iv) Voluntary Response Error or Bias

Another source of systematic error is a survey based on a voluntary response sample.

The polls conducted based on samples of readers of magazines and newspapers suffer from **voluntary response error** or **bias**. Usually only those readers who have very strong opinions about the issues involved respond to such surveys. Surveys in which the respondents are required to call some telephone numbers also suffer from this type of error. Here, to participate, many times a respondent has to pay for the call, and many people do not want to bear this cost. Consequently, the sample is usually neither random nor representative of the target population because participation is voluntary.

1.5.4   Random Sampling Techniques

There are many ways to select a random sample. Four of these techniques are discussed next.

Simple Random Sampling

From a given population, we can select a large number of samples of the same size. If each of these samples has the same probability of being selected, then it is called **simple random sampling**.

The list of members of the target population that is used to select a sample is called the sampling frame. The error that occurs because the sampling frame is not representative of the population is called the **selection error or bias**.

The error that occurs because many of the people included in the sample do not respond to a survey is called the **nonresponse error or bias**.

The **response error or bias** occurs when people included in the survey do not provide correct answers.

Voluntary response error or bias occurs when a survey is not conducted on a randomly selected sample but on a questionnaire published in a magazine or newspaper and people are invited to respond to that questionnaire.

In this sampling technique, each sample of the same size has the same probability of being selected. Such a sample is called a **simple random sample**.

One way to select a simple random sample is by a lottery or drawing. For example, if we need to select 5 students from a class of 50, we write each of the 50 names on a separate piece of paper. Then, we place all 50 names in a hat and mix them thoroughly. Next, we draw 1 name randomly from the hat. We repeat this experiment four more times. The 5 drawn names make up a simple random sample.

The second procedure to select a simple random sample is to use a table of random numbers, which has become an outdated procedure. In this age of technology, it is much easier to use a statistical package, such as Minitab, to select a simple random sample.

Systematic Random Sampling

> In **systematic random sampling**, we first randomly select one member from the first k units of the list of elements arranged based on a given characteristic where k is the number obtained by dividing the population size by the intended sample size. Then every kth member, starting with the first selected member, is included in the sample.

The simple random sampling procedure becomes very tedious if the size of the population is large. For example, if we need to select 150 households from a list of 45,000, it is very time-consuming either to write the 45,000 names on pieces of paper and then select 150 households or to use a table of random numbers. In such cases, it is more convenient to use **systematic random sampling**.

The procedure to select a systematic random sample is as follows. In the example just mentioned, we would arrange all 45,000 households alphabetically (or based on some other characteristic). Since the sample size should equal 150, the ratio of population to sample size is $45,000/150 = 300$. Using this ratio, we randomly select one household from the first 300 households in the arranged list using either method. Suppose by using either of the methods, we select the 210th household. We then select the 210th household from every 300 households in the list. In other words, our sample includes the households with numbers 210, 510, 810, 1110, 1410, 1710, and so on.

Stratified Random Sampling

> In a **stratified random sample**, we first divide the population into subpopulations, which are called strata. Then, one sample is selected from each of these strata. The collection of all samples from all strata gives the stratified random sample.

Suppose we need to select a sample from the population of a city, and we want households with different income levels to be proportionately represented in the sample. In this case, instead of selecting a simple random sample or a systematic random sample, we may prefer to apply a different technique. First, we divide the whole population into different groups based on income levels. For example, we may form three groups of low-, medium-, and high-income households. We will now have three *subpopulations*, which are usually called **strata**. We then select one sample from each subpopulation or stratum. The collection of all three samples selected from the three strata gives the required sample, called the **stratified random sample**. Usually, the sizes of the samples selected from different strata are proportionate to the sizes of the subpopulations in these strata. Note that the elements of each stratum are identical with regard to the possession of a characteristic.

Thus, whenever we observe that a population differs widely in the possession of a characteristic, we may prefer to divide it into different strata and then select one sample from each stratum. We can divide the population on the basis of any characteristic, such as income, expenditure, gender, education, race, employment, or family size.

Cluster Sampling

> In **cluster sampling**, the whole population is first divided into (geographical) groups called clusters. Each cluster is representative of the population. Then a random sample of clusters is selected. Finally, a random sample of elements from each of the selected clusters is selected.

Sometimes the target population is scattered over a wide geographical area. Consequently, if a simple random sample is selected, it may be costly to contact each member of the sample. In such a case, we divide the population into different geographical groups or clusters and, as a first step, select a random sample of certain clusters from all clusters. We then take a random sample of certain elements from each selected cluster. For example, suppose we are to conduct a survey of households in the state of New York. First, we divide the whole state of New York into, say, 40 regions, which are called **clusters** or **primary units**. We make sure that all clusters are similar and, hence, representative of the population. We then select at random, say, 5 clusters from 40. Next, we randomly select certain households from each of these 5 clusters and conduct a survey of these selected households. This is called **cluster sampling**. Note that all clusters must be representative of the population.

EXERCISES

CONCEPTS AND PROCEDURES

1.14 Define the following terms in statistics: population, sample, representative sample, sampling with replacement, and sampling without replacement.

1.15 Differentiate between sampling with replacement and without replacement. Give an example of each of these two types of sampling.

1.16 What are the merits of conducting a sample survey compared to conducting a census?

1.17 Differentiate between the following:

    a. Random and nonrandom samples
    b. Convenience, judgement and quota samples

1.18 What are the four sampling techniques? Explain each in detail.

1.19 Name the sampling technique in which all samples of the same size selected from a population have the same chance of being selected.

APPLICATIONS

1.20 Explain whether each of the following constitutes data collected from a population or a sample.

    a. Opinions on a certain issue obtained from all adults living in a city.
    b. The price of a gallon of regular unleaded gasoline on a given day at each of 35 gas stations in the Miami, Florida, metropolitan area.
    c. Credit card debts of 100 families selected from San Antonio, Texas.
    d. The percentage of all U.S. registered voters in each state who voted in the 2012 Presidential election.
    e. The number of left-handed students in each of 50 classes selected from a given university.

1.21 A professor is teaching a large class that has 247 students. He wants to select a sample of 15 students to do a study on the habits of his students. For each of the following sampling methods, explain what kind of method is used (e.g., is it a random sample, nonrandom sample, sample with replacement, sample without replacement, convenience sample, judgment sample, quota sample).

    a. There are 15 sociology majors in his class. He selects these 15 students to include in the sample.
    b. He goes through the roster and finds that he knows 30 of the 247 students. Using his knowledge about these students, he selects 15 students from these 30 to include in the sample.
    c. He enters names of all 247 students in a spreadsheet on his computer. He uses a statistical software (such as Minitab) to select 15 students.

1.22 A statistics professor wanted to find the average GPA (grade point average) of all students at her university. The professor obtains a list of all students enrolled at the university from the registrar's office and then selects 250 students at random from this list using a statistical software package such as Minitab.

    a. Is this sample a random or a nonrandom sample? Explain.
    b. What kind of sample is it? In other words, is it a simple random sample, a systematic sample, a stratified sample, a cluster sample, a convenience sample, a judgment sample, or a quota sample? Explain.
    c. Do you think any systematic error will be made in this case? Explain.

1.23 A statistics professor wanted to select 25 students from his class of 400 students to collect detailed information on the profiles of his students. The professor enters the names of all students enrolled in his class on a computer. He then selects a sample of 25 students at random using a statistical software package such as Minitab.

    a. Is this sample a random or a nonrandom sample? Explain.
    b. What kind of sample is it? In other words, is it a simple random sample, a systematic sample, a stratified sample, a cluster sample, a convenience sample, a judgment sample, or a quota sample? Explain.
    c. Do you think any systematic error will be made in this case? Explain.

1.24 A company has 1200 employees, of whom 60% are men and 40% are women. The research department at the company wanted to conduct a quick survey by selecting a sample of 50 employees and asking them about their opinions on an issue. They divided the population of employees into two groups, men and women, and then selected 29 men and 21 women from these respective groups. The interviewers were free to choose any 29 men and 21 women they wanted. What kind of sample is it? Explain.

1.25 A magazine published a questionnaire for its readers to fill out and mail to the magazine's office. In the questionnaire, cell phone owners were asked how much they would have to be paid to do without their cell phones for one month. The magazine received responses from 4568 cell phone owners.

    a. What type of sample is this? Explain.
    b. To what kind(s) of systematic error, if any, would this survey be subject?

1.26 A researcher wanted to conduct a survey of major companies to find out what benefits are offered to their employees. She mailed questionnaires to 1800 companies and received questionnaires back from 396 companies. What kind of systematic error does this survey suffer from? Explain.

1.27 In a survey, based on a random sample, conducted by an opinion poll agency the interviewers called the parents included in the sample and asked them the following questions:

    i. Do you believe in punishing children?
    ii. Have you ever punished your children?
    iii. If the answer to the second question is yes, how often?

What kind of systematic error, if any, does this survey suffer from? Explain.

1.28 A survey based on a random sample taken from one borough of Los Angeles City showed that 58% of the people living there would prefer to live somewhere other than Los Angeles City if they had the opportunity to do so. Based on this result, can the researcher say that 58% of people living in Los Angeles City would prefer to live somewhere else if they had the opportunity to do so? Explain.

1.6 | Design of Experiments

To use statistical methods to make decisions, we need access to data. Consider the following examples about decision making:

1. A government agency wants to find the average income of households in the United States.
2. A company wants to find the percentage of defective items produced by a machine.
3. A researcher wants to know if there is an association between eating unhealthy food and cholesterol level.
4. A pharmaceutical company has developed a new medicine for a disease and it wants to check if this medicine cures the disease.

All of these cases relate to decision making. We cannot reach a conclusion in these examples unless we have access to data. Data can be obtained from observational studies, experiments, or surveys. This section is devoted mainly to controlled experiments. However, it also explains observational studies and how they differ from surveys.

Suppose two diets, Diet 1 and Diet 2, are being promoted by two different companies, and each of these companies claims that its diet is successful in reducing weight. A research nutritionist wants to compare these diets with regard to their effectiveness for losing weight. Following are the two alternatives for the researcher to conduct this research.

1. The researcher contacts the persons who are using these diets and collects information on their weight loss. The researcher may contact as many persons as she has the time and financial resources for. Based on this information, the researcher makes a decision about the comparative effectiveness of these diets.
2. The researcher selects a sample of persons who want to lose weight, divides them randomly into two groups, and assigns each group to one of the two diets. Then she compares these two groups with regard to the effectiveness of these diets.

The first alternative is an example of an **observational study**, and the second is an example of a **controlled experiment**. In an experiment, we exercise control over some factors when we design the study, but in an observational study we do not impose any kind of control.

In an observational study the investigator does not impose a *treatment* on subjects or elements included in the study. For instance, in the first alternative mentioned above, the researcher simply collects information from the persons who are currently using these diets. In this case, the persons were not assigned to the two diets at random; instead, they chose the diets voluntarily. In this situation the researcher's conclusion about the comparative effectiveness of the two diets may not be valid because the effects of the diets will be **confounded** with many other factors or variables. When the effects of one factor cannot be separated from the effects of some other factors, the effects are said to be confounded. The persons who chose Diet 1 may be completely different with regard to age, gender, and eating and exercise habits from the persons who chose Diet 2. Thus, the weight loss may not be due entirely to the diet but to other factors or variables as well. Persons in one group may aggressively manage both diet and exercise, for example, whereas persons in the second group may depend entirely on diet. Thus, the effects of these other variables will get mixed up (confounded) with the effect of the diets.

Under the second alternative, the researcher selects a group of people, say 100, and randomly assigns them to two diets. One way to make random assignments is to write the name of each of these persons on a piece of paper, put them in a hat, mix them, and then randomly draw 50 names from this hat. These 50 persons will be assigned to one of the two diets, say Diet 1. The remaining 50 persons will be assigned to the second diet, Diet 2. This procedure is called **randomization**. Note that random assignments can also be made by using other methods such as a table of random numbers or technology.

When people are assigned to one or the other of two diets at random, the other differences among people in the two groups almost disappear. In this case these groups will not differ very much with regard to such factors as age, gender, and eating and exercise habits. The two groups will be very similar to each other. By using the random process to assign people to one or the

> A condition (or a set of conditions) that is imposed on a group of elements by the experimenter is called a **treatment**.

> The procedure in which elements are assigned to different groups at random is called **randomization**.

other of two diets, we have *controlled* the other factors that can affect the weights of people. Consequently, this is an example of a **designed experiment**.

As mentioned earlier, a condition (or a set of conditions) that is imposed on a group of elements by the experimenter is called a treatment. In the example on diets, each of the two diet types is called a treatment. The experimenter randomly assigns the elements to these two treatments. Again, in such cases the study is called a designed experiment.

The group of people who receive a treatment is called the **treatment group**, and the group of people who do not receive a treatment is called the **control group**. In our example on diets, both groups are treatment groups because each group is assigned to one of the two types of diet. That example does not contain a control group.

> When the experimenter controls the (random) assignment of elements to different treatment groups, the study is said to be a **designed experiment**. In contrast, in an **observational study** the assignment of elements to different treatments is voluntary, and the experimenter simply observes the results of the study.

EXAMPLE 1–1	**Testing a Medicine**

An example of an observational study.

Suppose a pharmaceutical company has developed a new medicine to cure a disease. To see whether or not this medicine is effective in curing this disease, it will have to be tested on a group of humans. Suppose there are 100 persons who have this disease; 50 of them voluntarily decide to take this medicine, and the remaining 50 decide not to take it. The researcher then compares the cure rates for the two groups of patients. Is this an example of a designed experiment or an observational study?

> The group of elements that receives a treatment is called the **treatment group**, and the group of elements that does not receive a treatment is called the **control group**.

Solution   This is an example of an observational study because 50 patients voluntarily joined the treatment group; they were not randomly selected. In this case, the results of the study may not be valid because the effects of the medicine will be confounded with other variables. All of the patients who decided to take the medicine may not be similar to the ones who decided not to take it. It is possible that the persons who decided to take the medicine are in the advanced stages of the disease. Consequently, they do not have much to lose by being in the treatment group. The patients in the two groups may also differ with regard to other factors such as age, gender, and so on.

EXAMPLE 1–2	**Testing a Medicine**

An example of a designed experiment.

Reconsider Example 1–1. Now, suppose that out of the 100 people who have this disease, 50 are selected at random. These 50 people make up one group, and the remaining 50 belong to the second group. One of these groups is the treatment group, and the second is the control group. The researcher then compares the cure rates for the two groups of patients. Is this an example of a designed experiment or an observational study?

Solution   In this case, the two groups are expected to be very similar to each other. Note that we do not expect the two groups to be exactly identical. However, when randomization is used, the two groups are expected to be very similar. After these two groups have been formed, one group will be given the actual medicine. This group is called the treatment group. The other group will be administered a placebo (a dummy medicine that looks exactly like the actual medicine). This group is called the control group. This is an example of a designed experiment because the patients are assigned to one of two groups—the treatment or the control group—randomly.

Usually in an experiment like the one in Example 1–2, patients do not know which group they belong to. Most of the time the experimenters do not know which group a patient belongs to. This is done to avoid any bias or distortion in the results of the experiment. When neither patients nor experimenters know who is taking the real medicine and who is taking the placebo, it is called a **double-blind experiment**. For the results of the study to be unbiased and valid, an experiment must be a double-blind designed experiment. Note that if either experimenters or patients or both have access to information regarding which patients belong to treatment or control groups, it will no longer be a double-blind experiment.

The use of a placebo in medical experiments is very important. A placebo is just a dummy pill that looks exactly like the real medicine. Often, patients respond to any kind of medicine. Many studies have shown that even when the patients were given sugar pills (and did not know it), many of them indicated a decrease in pain. Patients respond to a placebo because they have confidence in their physicians and medicines. This is called the **placebo effect**.

Note that there can be more than two groups of elements in an experiment. For example, an investigator may need to compare three diets for people with regard to weight loss. Here, in a designed experiment, the people will be randomly assigned to one of the three diets, which are the three treatments.

In some instances we have to base our research on observational studies because it is not feasible to conduct a designed experiment. For example, suppose a researcher wants to compare the starting salaries of business and psychology majors. The researcher will have to depend on an observational study. She will select two samples, one of recent business majors and another of recent psychology majors. Based on the starting salaries of these two groups, the researcher will make a decision. Note that, here, the effects of the majors on the starting salaries of the two groups of graduates will be confounded with other variables. One of these other factors is that the business and psychology majors may be different in regard to intelligence level, which may affect their salaries. However, the researcher cannot conduct a designed experiment in this case. She cannot select a group of persons randomly and ask them to major in business and select another group and ask them to major in psychology. Instead, persons voluntarily choose their majors.

In a survey we do not exercise any control over the factors when we collect information. This characteristic of a survey makes it very close to an observational study. However, a survey may be based on a probability sample, which differentiates it from an observational study.

If an observational study or a survey indicates that two variables are related, it does not mean that there is a cause-and-effect relationship between them. For example, if an economist takes a sample of families, collects data on the incomes and rents paid by these families, and establishes an association between these two variables, it does not necessarily mean that families with higher incomes pay higher rents. Here the effects of many variables on rents are confounded. A family may pay a higher rent not because of higher income but because of various other factors, such as family size, preferences, or place of residence. We cannot make a statement about the cause-and-effect relationship between incomes and rents paid by families unless we control for these other variables. The association between incomes and rents paid by families may fit any of the following scenarios.

1. These two variables have a cause-and-effect relationship. Families that have higher incomes do pay higher rents. A change in incomes of families causes a change in rents paid.

2. The incomes and rents paid by families do not have a cause-and-effect relationship. Both of these variables have a cause-and-effect relationship with a third variable. Whenever that third variable changes, these two variables change.

3. The effect of income on rent is confounded with other variables, and this indicates that income affects rent paid by families.

If our purpose in a study is to establish a cause-and-effect relationship between two variables, we must control for the effects of other variables. In other words, we must conduct a designed study.

EXERCISES

CONCEPTS AND PROCEDURES

1.29 What is the difference between a survey and an experiment?

1.30 What is the difference between an observational study and an experiment?

APPLICATIONS

1.31 A pharmaceutical company developed a new medicine for compulsive behavior. To test this medicine on humans, the company advertised for volunteers who were suffering from this disease and wanted to participate in the study. As a result, 1820 persons responded. Using their own judgment, the group of physicians who were conducting this study assigned 910 of these patients to the treatment group and the remaining 910 to the control group. The patients in the treatment group were administered the actual medicine, and the patients in the control group were given a placebo. Six months later the conditions of the patients in the two groups were examined and compared. Based on this comparison, the physicians concluded that this medicine improves the condition of patients suffering from compulsive behavior.

   a. Comment on this study and its conclusion.
   b. Is this an observational study or a designed experiment? Explain.
   c. Is this a double-blind study? Explain.

1.32 A pharmaceutical company has developed a new medicine for compulsive behavior. To test this medicine on humans, the physicians conducting this study obtained a list of all patients suffering from compulsive behavior who were being treated by doctors in all hospitals in the country. Further assume that this list is representative of the population of all such patients. The physicians then randomly selected 1820 patients from this list. Of these 1820, randomly selected 910 patients were assigned to the treatment group, and the remaining 910 patients were assigned to the control group. The patients did not know which group they belonged to, but the doctors had access to such information. Six months later the conditions of the patients in the two groups were examined and compared. Based on this comparison, the physicians concluded that this medicine improves the condition of patients suffering from compulsive behavior.

   a. Comment on this study and its conclusion.
   b. Is this an observational study or a designed experiment? Explain.
   c. Is this a double-blind study? Explain.

1.33 A federal government think tank wanted to investigate whether a job training program helps the families who are on welfare to get off the welfare program. The researchers at this agency selected 5000 volunteer families who were on welfare and offered the adults in those families free job training. The researchers selected another group of 5000 volunteer families who were on welfare and did not offer them such job training. After 3 years the two groups were compared in regard to the percentage of families who got off welfare. Is this an observational study or a designed experiment? Explain.

1.34 A pharmaceutical company has developed a new medicine for compulsive behavior. To test this medicine on humans, the physicians conducting this study obtained a list of all patients suffering from compulsive behavior who were being treated by doctors in all hospitals in the country. Assume that this list is representative of the population of all such patients. The physicians then randomly selected 1820 patients from this list. Of these 1820 patients, randomly selected 910 patients were assigned to the treatment group, and the remaining 910 patients were assigned to the control group. Neither patients nor doctors knew what group the patients belonged to. Six months later the conditions of the patients in the two groups were examined and compared. Based on this comparison, the physicians concluded that this medicine improves the condition of patients suffering from compulsive behavior.

   a. Is this an observational study or a designed experiment? Explain.
   b. Is this a double-blind study? Explain.

1.35 A federal government think tank wanted to investigate whether a job training program helps the families who are on welfare to get off the welfare program. The researchers at this agency selected 5000 volunteer families who were on welfare and offered the adults in those families free job training. The researchers selected another group of 5000 volunteer families who were on welfare and did not offer them such job training. After three years the two groups were compared in regard to the percentage of families who got off welfare. Based on that study, the researchers concluded that the job training program causes (helps) families who are on welfare to get off the welfare program. Do you agree with this conclusion? Explain.

1.36 A federal government think tank wanted to investigate whether a job training program helps the families who are on welfare to get off the welfare program. The researchers at this agency selected 10,000 families at random from the list of all families that were on welfare. Of these 10,000 families, the researchers randomly selected 5000 families and offered the adults in those families free job training. The remaining 5000 families were not offered such job training. After three years the two groups were compared in regard to the percentage of families who got off welfare. Based on that study, the researchers concluded that the job training program causes (helps) families who are on welfare to get off the welfare program. Do you agree with this conclusion? Explain.

1.7 | Summation Notation

Sometimes mathematical notation helps express a mathematical relationship concisely. This section describes the **summation notation** that is used to denote the sum of values.

   Suppose a sample consists of five books, and the prices of these five books are \$175, \$80, \$165, \$97, and \$88, respectively. The variable, *price of a book*, can be denoted by x. The prices of the five books can be written as follows:

$$\text{Price of the first book} = x_1 = \$175$$

$$\uparrow$$

Subscript of x denotes the number of the book

Similarly,

$$\text{Price of the second book} = x_2 = \$80$$
$$\text{Price of the third book} = x_3 = \$165$$
$$\text{Price of the fourth book} = x_4 = \$97$$
$$\text{Price of the fifth book} = x_5 = \$88$$

In this notation, x represents the price, and the subscript denotes a particular book.

Now, suppose we want to add the prices of all five books. We obtain

$$x_1 + x_2 + x_3 + x_4 + x_5 = 175 + 80 + 165 + 97 + 88 = \$605$$

The uppercase Greek letter Σ (pronounced *sigma*) is used to denote the sum of all values. Using Σ notation, we can write the foregoing sum as follows:

$$\Sigma x = x_1 + x_2 + x_3 + x_4 + x_5 = \$605$$

The notation Σx in this expression represents the sum of all values of x and is read as "sigma x" or "sum of all values of x."

EXAMPLE 1–3   Annual Salaries of Workers

Using summation notation: one variable.

Annual salaries (in thousands of dollars) of four workers are 75, 90, 125, and 61, respectively. Find

(a) Σx (b) $(\Sigma x)^2$ (c) Σx^2

Solution Let $x_1, x_2, x_3,$ and x_4 be the annual salaries (in thousands of dollars) of the first, second, third, and fourth worker, respectively. Then,

$$x_1 = 75, \quad x_2 = 90, \quad x_3 = 125, \quad \text{and} \quad x_4 = 61$$

(a) $\Sigma x = x_1 + x_2 + x_3 + x_4 = 75 + 90 + 125 + 61 = 351 = \textbf{\$351,000}$

(b) Note that $(\Sigma x)^2$ is the square of the sum of all x values. Thus,

$$(\Sigma x)^2 = (351)^2 = \textbf{123,201}$$

(c) The expression Σx^2 is the sum of the squares of x values. To calculate Σx^2, we first square each of the x values and then sum these squared values. Thus,

$$\Sigma x^2 = (75)^2 + (90)^2 + (125)^2 + (61)^2$$
$$= 5625 + 8100 + 15{,}625 + 3721 = \textbf{33,071}$$

EXAMPLE 1–4

Using summation notation: two variables.

The following table lists four pairs of m and f values:

m	12	15	20	30
f	5	9	10	16

Compute the following:

(a) Σm (b) Σf^2 (c) Σmf (d) $\Sigma m^2 f$

Solution   We can write

$$m_1 = 12 \quad m_2 = 15 \quad m_3 = 20 \quad m_4 = 30$$
$$f_1 = 5 \quad f_2 = 9 \quad f_3 = 10 \quad f_4 = 16$$

(a) $\Sigma m = 12 + 15 + 20 + 30 = \textbf{77}$

(b) $\Sigma f^2 = (5)^2 + (9)^2 + (10)^2 + (16)^2 = 25 + 81 + 100 + 256 = \textbf{462}$

(c) To compute Σmf, we multiply the corresponding values of m and f and then add the products as follows:

$$\Sigma mf = m_1 f_1 + m_2 f_2 + m_3 f_3 + m_4 f_4$$
$$= 12(5) + 15(9) + 20(10) + 30(16) = \mathbf{875}$$

(d) To calculate $\Sigma m^2 f$, we square each m value, then multiply the corresponding m^2 and f values, and add the products. Thus,

$$\Sigma m^2 f = (m_1)^2 f_1 + (m_2)^2 f_2 + (m_3)^2 f_3 + (m_4)^2 f_4$$
$$= (12)^2 (5) + (15)^2 (9) + (20)^2 (10) + (30)^2 (16) = \mathbf{21{,}145}$$

The calculations done in parts (a) through (d) to find the values of Σm, Σf^2, Σmf, and $\Sigma m^2 f$ can be performed in tabular form, as shown in Table 1.4.

Table 1.4

m	f	f^2	mf	$m^2 f$
12	5	$5 \times 5 = 25$	$12 \times 5 = 60$	$12 \times 12 \times 5 = 720$
15	9	$9 \times 9 = 81$	$15 \times 9 = 135$	$15 \times 15 \times 9 = 2025$
20	10	$10 \times 10 = 100$	$20 \times 10 = 200$	$20 \times 20 \times 10 = 4000$
30	16	$16 \times 16 = 256$	$30 \times 16 = 480$	$30 \times 30 \times 16 = 14{,}400$
$\Sigma m = 77$	$\Sigma f = 40$	$\Sigma f^2 = 462$	$\Sigma mf = 875$	$\Sigma m^2 f = 21{,}145$

The columns of Table 1.4 can be explained as follows.

1. The first column lists the values of m. The sum of these values gives $\Sigma m = 77$.
2. The second column lists the values of f. The sum of this column gives $\Sigma f = 40$.
3. The third column lists the squares of the f values. For example, the first value, 25, is the square of 5. The sum of the values in this column gives $\Sigma f^2 = 462$.
4. The fourth column records products of the corresponding m and f values. For example, the first value, 60, in this column is obtained by multiplying 12 by 5. The sum of the values in this column gives $\Sigma mf = 875$.
5. Next, the m values are squared and multiplied by the corresponding f values. The resulting products, denoted by $m^2 f$, are recorded in the fifth column. For example, the first value, 720, is obtained by squaring 12 and multiplying this result by 5. The sum of the values in this column gives $\Sigma m^2 f = 21{,}145$.

EXERCISES

CONCEPTS AND PROCEDURES

1.37 The following table lists six pairs of x and y values.

x	7	11	8	4	14	28
y	5	15	7	10	9	19

Compute the value of each of the following:

   a. Σy **b.** Σx^2 **c.** Σxy **d.** $\Sigma x^2 y$ **e.** Σy^2

APPLICATIONS

1.38 Seven randomly selected customers at a local grocery store spent the following amounts on groceries in a single visit: \$137, \$35, \$92, \$105, \$175, \$21, and \$57, respectively. Let y denote the amount spent on groceries in a single visit. Find:

   a. Σy **b.** $(\Sigma y)^2$ **c.** Σy^2

1.39 The number of pizzas delivered to a college campus on six randomly selected nights is 48, 103, 95, 188, 286, and 136, respectively. Let x denote the number of pizzas delivered to this college campus on any given night. Find:

   a. Σx **b.** $(\Sigma x)^2$ **c.** Σx^2

USES AND MISUSES...

JUST A SAMPLE PLEASE

Have you ever read an article or seen an advertisement on television that claims to have found some important result based on a scientific study? Or maybe you have seen some results based on a poll conducted by a research firm. Have you ever wondered how samples are selected for such studies, or just how valid the results really are?

One of the most important considerations in conducting a study is to ensure that a *representative sample* has been taken in such a way that the sample represents the underlying population of interest. So, if you see a claim that a recent study found that a new drug was a *magic bullet* for curing cancer, how might we investigate how valid this claim is?

It is very important to consider how the population was sampled in order to understand both how useful the result is and if the result is really *generalizable*. In research, a result that is generalizable means that it applies to the general population of interest, not just to the sample that was taken or to only a small segment of the population.

One of the most common sampling strategies is the *simple random sample* in which each subject in a population has some chance of being chosen for the study. Though there are numerous ways to take a valid statistical sample, simple random samples ensure that there is equal representation of members of the population of interest, and therefore we see this type of sampling strategy employed quite often. But, how many samples taken for studies are truly random? It is not uncommon in survey research, for example, to send people out to a mall or to a busy street corner with clipboard and survey in hand to ask questions of those willing to take the survey. It is also not uncommon for polling firms to call from a random set of telephone numbers—in fact, many of you may have received a call to participate in a poll. In either of these scenarios, it is likely that those willing to take the time to answer a survey may not represent everyone in the population of interest. Additionally, if a telephone poll is used, it is impossible to sample those without a telephone or those that do not answer the call, and hence this segment of the population may not be available to be included in the sample.

One of the major consequences of having a poor sample is the risk of creating *bias* in the results. Bias causes the results to potentially deviate from what they would be under ideal sampling circumstances. In the examples just presented, we have an example of *volunteer bias* in which the results are skewed to represent only those willing to participate. There could also be *observer bias* in which the person conducting a poll only approaches certain people—perhaps those that seem friendly or more likely to participate—in which case a large segment of the potential population is ignored. Results from such studies would certainly be suspect.

The take-home message is that, when reading the results of studies or polls, it is very important to stop and consider what population was actually sampled, how representative that sample really is, and if the results may be biased in such a way that the results are not generalizable to the larger population. One of the great things about taking a statistics course is that you will learn many tools that will allow you to look at the world of data more critically so that you can make informed decisions about the studies that you read or hear about in the media. So the next time you hear, "9 out of 10 people say…," consider if they asked only 10 people or if they asked a million people. Learn to be good consumers of data!

GLOSSARY

Census A survey conducted by including every element of the population.

Cluster A subgroup (usually geographical) of the population that is representative of the population.

Cluster sampling A sampling technique in which the population is divided into clusters and a sample is chosen from one or a few clusters.

Continuous variable A (quantitative) variable that can assume any numerical value over a certain interval or intervals.

Control group The group on which no condition is imposed.

Convenience sample A sample that includes the most accessible members of the population.

Cross-section data Data collected on different elements at the same point in time or for the same period of time.

Data or data set Collection of observations or measurements on a variable.

Descriptive statistics Collection of methods for organizing, displaying, and describing data using tables, graphs, and summary measures.

Designed experiment A study in which the experimenter controls the assignment of elements to different treatment groups.

Discrete variable A (quantitative) variable whose values are countable.

Double-blind experiment An experiment in which neither the doctors (or researchers) nor the patients (or members) know to which group a patient (or member) belongs.

Element or member A specific subject or object included in a sample or population.

Experiment A method of collecting data by controlling some or all factors.

Inferential statistics Collection of methods that help make decisions about a population based on sample results.

Judgment sample A sample that includes the elements of the population selected based on the judgment and prior knowledge of an expert.

Nonresponse error The error that occurs because many of the people included in the sample do not respond.

Nonsampling or systematic errors The errors that occur in the collection, recording, and tabulation of data.

Observation or measurement The value of a variable for an element.

Observational study A study in which the assignment of elements to different treatments is voluntary, and the researcher simply observes the results of the study.

Population or target population The collection of all elements whose characteristics are being studied.

Qualitative or categorical data   Data generated by a qualitative variable.

Qualitative or categorical variable   A variable that cannot assume numerical values but is classified into two or more categories.

Quantitative data   Data generated by a quantitative variable.

Quantitative variable   A variable that can be measured numerically.

Quota sample   A sample selected in such a way that each group or subpopulation is represented in the sample in exactly the same proportion as in the target population.

Random sample   A sample drawn in such a way that each element of the population has some chance of being included in the sample.

Randomization   The procedure in which elements are assigned to different (treatment and control) groups at random.

Representative sample   A sample that contains the characteristics of the population as closely as possible.

Response error   The error that occurs because people included in the survey do not provide correct answers.

Sample   A portion of the population of interest.

Sample survey   A survey that includes elements of a sample.

Sampling frame   The list of elements of the target population that is used to select a sample.

Sampling or chance error   The difference between the result obtained from a sample survey and the result that would be obtained from the census.

Selection error   The error that occurs because the sampling frame is not representative of the population.

Simple random sampling   If all samples of the same size selected from a population have the same chance of being selected, it is called simple random sampling. Such a sample is called a simple random sample.

Statistics   Science of collecting, analyzing, presenting, and interpreting data and making decisions.

Stratified random sampling   A sampling technique in which the population is divided into different strata and a sample is chosen from each stratum.

Stratum   A subgroup of the population whose members are identical with regard to the possession of a characteristic.

Survey   Collecting data from the elements of a population or sample.

Systematic random sampling   A sampling method used to choose a sample by selecting every kth unit from the list.

Target population   The collection of all subjects of interest.

Time-series data   Data that give the values of the same variable for the same element at different points in time or for different periods of time.

Treatment   A condition (or a set of conditions) that is imposed on a group of elements by the experimenter. This group is called the treatment group.

Variable   A characteristic under study or investigation that assumes different values for different elements.

Voluntary response error   The error that occurs because a survey is not conducted on a randomly selected sample, but people are invited to respond voluntarily to the survey.

SUPPLEMENTARY EXERCISES

1.40 The following table lists the number of reports filed with the U.S. Department of Transportation about mishandled baggage during the first nine months of 2010, as reported in the *USA TODAY* of July 14, 2011.

Airline	Mishandled Baggage Reports
AirTran	30,801
Alaska	36,525
American	205,247
Delta	247,660
JetBlue	41,174
Hawaiian	11,987

Explain the meaning of a member, a variable, a measurement, and a data set with reference to this table.

1.41 Is the following an example of descriptive statistics or inferential statistics? Explain.

The following table gives the earnings of the world's top seven female professional athletes for the year 2014 (ceoworld.biz).

Female Professional Athlete	2014 Earnings (millions of dollars)
Maria Sharapova	24.4
Li Na	23.6
Serena Williams	22.0
Kim Yuna	16.3
Danica Patrick	15.0
Victoria Azarenka	11.1
Caroline Wozniacki	10.8

1.42 Indicate whether each of the following examples refers to a population or to a sample.
   a. Salaries of CEOs of all companies in New York City
   b. Four hundred houses selected from a city
   c. Gross sales for 2009 of three pizza chains
   d. Annual incomes of all 28 employees of a restaurant

1.43 State which of the following is an example of sampling with replacement and which is an example of sampling without replacement.
   a. Selecting ten cities to market a new shampoo
   b. Selecting a high school teacher to drive students to a lecture in September, then selecting a teacher from the same group to chaperone a dance in November

1.44 The number of restaurants in each of five small towns is 4, 14, 8, 10, and 7, respectively. Let y denote the number of restaurants in a small town. Find:
   a. Σy **b.** $(\Sigma y)^2$ **c.** Σy^2

1.45 The following table lists five pairs of m and f values.

m	3	16	11	9	20
f	7	32	17	12	34

Compute the value of each of the following:

a. $\sum m$ **b.** $\sum f^2$ **c.** $\sum mf$ **d.** $\sum m^2 f$ **e.** $\sum m^2$

1.46 A statistics professor wanted to find out the average GPA (grade point average) for all students at her university. She used all students enrolled in her statistics class as a sample and collected information on their GPAs to find the average GPA.

a. Is this sample a random or a nonrandom sample? Explain.

b. What kind of sample is it? In other words, is it a simple random sample, a systematic sample, a stratified sample, a cluster sample, a convenience sample, a judgment sample, or a quota sample? Explain.

c. What kind of systematic error, if any, will be made with this kind of sample? Explain.

1.47 A researcher advertised for volunteers to study the relationship between the amount of meat consumed and cholesterol level. In response to this advertisement, 3476 persons volunteered. The researcher collected information on the meat consumption and cholesterol level of each of these persons. Based on these data, the researcher concluded that there is a very strong positive association between these two variables.

a. Is this an observational study or a designed experiment? Explain.

b. Based on this study, can the researcher conclude that consumption of meat increases cholesterol level? Explain why or why not.

1.48 In March 2005, *The New England Journal of Medicine* published the results of a 10-year clinical trial of low-dose aspirin therapy for the cardiovascular health of women (*Time*, March 21, 2005). The study was based on 40,000 healthy women, most of whom were in their 40s and 50s when the trial began. Half of these women were administered 100 mg of aspirin every other day, and the others were given a placebo. The study looked at the incidences of heart attacks in the two groups of women. Overall the study did not find a statistically significant difference in heart attacks between the two groups of women. However, the study noted that among women who were at least 65 years old when the study began, there was a lower incidence of heart attack for those

who took aspirin than for those who took a placebo. Suppose that some medical researchers want to study this phenomenon more closely. They recruit 2000 healthy women aged 65 years and older, and randomly divide them into two groups. One group takes 100 mg of aspirin every other day, and the other group takes a placebo. The women do not know to which group they belong, but the doctors who are conducting the study have access to this information.

a. Is this an observational study or a designed experiment? Explain.

b. Is this a double-blind study? Explain.

1.49 In March 2005, *The New England Journal of Medicine* published the results of a 10-year clinical trial of low-dose aspirin therapy for the cardiovascular health of women (*Time*, March 21, 2005). The study was based on 40,000 healthy women, most of whom were in their 40s and 50s when the trial began. Half of these women were administered 100 mg of aspirin every other day, and the others were given a placebo. The study noted that among women who were at least 65 years old when the study began, there was a lower incidence of heart attack for those who took aspirin than for those who took placebo. Some medical researchers want to study this phenomenon more closely. They recruit 2000 healthy women aged 65 years and older, and randomly divide them into two groups. One group takes 100 mg of aspirin every other day, and the other group takes a placebo. Neither patients nor doctors know which group patients belong to.

a. Is this an observational study or a designed experiment? Explain.

b. Is this study a double-blind study? Explain.

1.50 A federal government think tank wanted to investigate whether a job training program helps the families who are on welfare to get off the welfare program. The researchers at this agency selected 10,000 families at random from the list of all families that were on welfare. Of these 10,000 families, the agency randomly selected 5000 families and offered them free job training. The remaining 5000 families were not offered such job training. After 3 years the two groups were compared in regard to the percentage of families who got off welfare. Is this an observational study or a designed experiment? Explain.

ADVANCED EXERCISES

1.51 A college mailed a questionnaire to all 5432 of its alumni who graduated in the last 5 years. One of the questions was about the current annual incomes of these alumni. Only 1620 of these alumni returned the completed questionnaires, and 1240 of them answered that question. The current average annual income of these 1240 respondents was $61,200.

a. Do you think $61,200 is likely to be an unbiased estimate of the current average annual income of all 5432 alumni? If so, explain why.

b. If you think that $61,200 is probably a biased estimate of the current average annual income of all 5432 alumni, what sources of systematic errors do you think are present here?

c. Do you expect the estimate of $61,200 to be above or below the current average annual income of all 5432 alumni? Explain.

1.52 A group of veterinarians wants to test a new canine vaccine for Lyme disease. (Lyme disease is transmitted by the bite of an infected deer tick.) One hundred dogs are randomly selected to receive the vaccine (with their owners' permission) from an area that has a high incidence of Lyme disease. These dogs are examined by veterinarians for symptoms of Lyme disease once a month for a period of 12 months. During this 12-month period, 10 of these 100 dogs are diagnosed with Lyme disease. During the same 12-month period, 18% of the unvaccinated dogs in the area are found to have contracted Lyme disease.

a. Does this experiment have a control group?

b. Is this a double-blind experiment?

c. Identify any potential sources of bias in this experiment.

d. Explain how this experiment could have been designed to reduce or eliminate the bias pointed out in part ç,

SELF-REVIEW TEST

1. A population in statistics means a collection of all

a. men and women

b. subjects or objects of interest

c. people living in a country

2. A sample in statistics means a portion of the

a. people selected from the population of a country

b. people selected from the population of an area

c. population of interest

3. Indicate which of the following is an example of a sample with replacement and which is a sample without replacement.

a. Five friends go to a livery stable and select five horses to ride (each friend must choose a different horse).

b. A box contains five marbles of different colors. A marble is drawn from this box, its color is recorded, and it is put back into the box before the next marble is drawn. This experiment is repeated 12 times.

4. Explain the following types of samples: random sample, nonrandom sample, representative sample, convenience sample, judgement sample, quota sample, sample with replacement, and sample without replacement.

5. Explain the following concepts: sampling or chance error, nonsampling or systematic error, selection bias, nonresponse bias, response bias, and voluntary response bias.

6. Explain the following sampling techniques: simple random sampling, systematic random sampling, stratified random sampling, and cluster sampling.

7. Explain the following concepts: an observational study, a designed experiment, randomization, treatment group, control group, double-blind experiment, and placebo effect.

8. A psychologist needs 10 pigs for a study of the intelligence of pigs. She goes to a pig farm where there are 40 young pigs in a large pen. Assume that these pigs are representative of the population of all pigs. She selects the first 10 pigs she can catch and uses them for her study.
  a. Do these 10 pigs make a random sample?
  b. Are these 10 pigs likely to be representative of the entire population? Why or why not?
  c. If these 10 pigs do not form a random sample, what type of sample is it?
  d. Can you suggest a better procedure for selecting a sample of 10 from the 40 pigs in the pen?

9. A newspaper wants to conduct a poll to estimate the percentage of its readers who favor a gambling casino in their city. People register their opinions by placing a phone call that costs them $1.
  a. Is this method likely to produce a random sample?
  b. Which, if any, of the types of biases listed in this appendix are likely to be present and why?

10. The Centre for Nutrition and Food Research at Queen Margaret University College in Edinburgh studied the relationship between sugar consumption and weight gain (*Fitness*, May 2002). All the people who participated in the study were divided into two groups, and both of these groups were put on low-calorie, low-fat diets. The diet of the people in the first group was low in sugar, but the people in the second group received as much as 10% of their calories from sucrose. Both groups stayed on their respective diets for 8 weeks. During these 8 weeks, participants in both groups lost 1/2 to 3/4 pound per week.
  a. Was this a designed experiment or an observational study?
  b. Was there a control group in this study?
  c. Was this a double-blind experiment?

11. A researcher wanted to study if regular exercise can control diabetes. He took a sample of 1200 adults who were under treatment for diabetes. He asked them if they exercise regularly or not and whether their diabetes was under control or not. Based on this information he reached his conclusion that regular exercise can keep diabetes under control. Is this an observational study or a designed experiment?

12. A researcher wanted to know if using regular fertilizer or organic fertilizer will affect the yield of potatoes. He divided a large parcel of land into 10 small parcels and then randomly selected five parcels to grow potatoes using regular fertilizer and the other five parcels using organic fertilizer. Is this an observational study or a randomized experiment?

13. The following table lists the age (rounded to the nearest year) and price (rounded to the nearest thousand dollars) for each of the seven cars of the same model.

Age (x) (years)	Price (y) (thousand dollars)
3	28.4
7	17.2
5	21.6
9	13.9
12	6.3
8	16.8
10	9.4

Let age be denoted by x and price be denoted by y. Calculate
  a. Σx **b.** Σy **c.** Σx^2 **d.** Σxy **e.** $\Sigma x^2 y$ **f.** $(\Sigma y)^2$

MINI-PROJECTS

Note: The Mini-Projects are located on the text's Web site, www.wiley.com/college/mann.

DECIDE FOR YOURSELF

Note: The Decide for Yourself feature is located on the text's Web site, www.wiley.com/college/mann.

TECHNOLOGY INSTRUCTIONS   CHAPTER 1

Note: Complete TI-84, Minitab, and Excel manuals are available for download at the textbook's Web site, www.wiley.com/college/mann.

TI-84 Color/TI-84

The TI-84 Color Technology Instructions feature of this text is written for the TI-84 Plus C color graphing calculator running the 4.0 operating system. Some screens, menus, and functions will be slightly different in older operating systems. The TI-84 and TI-84 Plus can perform all of the same functions but will not have the "Color" option referenced in some of the menus.

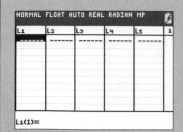

Screen 1.1

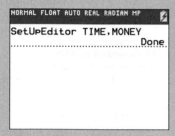

Screen 1.2

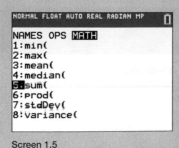

Entering and Editing Data

1. On the TI-84, variables are referred to as lists.

2. When entering data that does not need to be saved for later use, you can use the default lists, which are named **L1**, **L2**, **L3**, **L4**, **L5**, and **L6**.

3. To set up the editor to view these lists, select **STAT > EDIT > SetUpEditor**. Press **ENTER**.

4. To view these lists in the editor, select **STAT > EDIT > Edit**. (See **Screen 1.1**.)

5. To enter data in a list, type each data value and then press **ENTER** after each data value. (See **Screen 1.2**.)

6. To edit data values in a list, use the arrow keys to move to the cell you wish to edit, type the new value, and then press **ENTER**.

Creating a Named List

1. When entering data that will be saved for later use, you can create named lists. List names may contain a maximum of five characters, including letters or numbers.

2. Select **STAT > EDIT > SetUpEditor** and then type the names of your lists, separated by commas. Press **ENTER**. (See **Screen 1.3**.)

   Note: To type list names, select **ALPHA** and the letter (in green on your keypad) or select **2ⁿᵈ > ALPHA** to engage the **A-LOCK** (alpha-lock). To disengage the **A-LOCK**, press **ALHPA**.

3. To view these lists in the editor, select **STAT > EDIT > Edit**. (See **Screen 1.4**.)

4. To return the editor to the default lists, select **STAT > EDIT > SetUpEditor**. Press **ENTER**.

Operations with Lists

1. To calculate the sum of the values in a list select **2ⁿᵈ > STAT > MATH > sum**, enter the name of the list (e.g., **2ⁿᵈ > 1** for **L1**), and then press the **)** key. Press **ENTER**. (See **Screens 1.5** and **1.6**.)

2. To calculate the square of the sum of the values in a list, denoted by $(\sum x)^2$, use the same instructions as in step 1, but press the x^2 key before you press **ENTER**. (See **Screen 1.6**.)

3. To calculate the square of each value in a list, select **2ⁿᵈ > STAT**, select the name of your list, press the x^2 key, and then press **ENTER**. (See **Screen 1.6**.)

4. To calculate the sum of the squares of the values in a list, denoted by $\sum x^2$, apply the same instructions as in step 1 to the list created in step 3. **Screen 1.6** shows the appearance of these two processes. (See **Screen 1.6**.)

Screen 1.3

Screen 1.4

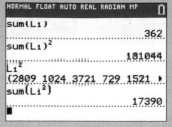

Screen 1.5

Screen 1.6

Selecting Random Samples

1. The following data represent the ages, in years, at which 10 patients were first diagnosed with diabetes. Enter these data in **L1** in the order they appear below. We will select a random sample of 4 data values from these 10.

    18, 7, 46, 32, 62, 65, 51, 40, 28, 72

2. Select 2nd > **MODE** to reach the home screen.

3. Select **MATH** > **PROB** > **randIntNoRep**.

4. Use the following settings in the **randIntNoRep** menu:

    • Type 1 at the **lower** prompt.

    • Type 10 at the **upper** prompt.

    • Type 10 at the **n** prompt.

    Note: Always enter the total number of data values for the **upper** and **n** prompt.

5. Highlight **Paste** and press **ENTER**.

6. Press **STO >**.

7. Select 2nd > **STAT** > **NAMES** > **L2**.

8. Press **ENTER**. (See **Screen 1.7**.)

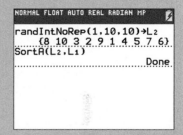

Screen 1.7

    Note: The numbers 1 through 10 may be in a different order on your screen than they appear in **Screen 1.7** because they were randomly sampled.

9. Select 2nd > **STAT** > **OPS** > **SortA**.

10. Select 2nd > **2** to display **L2**.

11. Press **,** (the comma key).

12. Select 2nd > **1** to display **L1**.

13. Press **)** (the right parenthesis key).

14. Press **ENTER**. (See **Screen 1.7**.)

15. Select **STAT** > **EDIT** > **Edit** to view your lists. (See **Screen 1.8**.)

16. The top four numbers in **L1** are the random sample of 4 data values.

    Note: The numbers in **L1** may be in a different order on your screen than they appear in **Screen 1.8** because the data were randomly sorted by the numbers in **L2**.

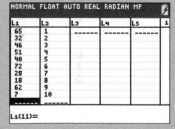

Screen 1.8

Minitab

The Minitab Technology Instructions feature of this text is written for Minitab version 17. Some screens, menus, and functions will be slightly different in older versions of Minitab.

Entering and Saving Data

1. In Minitab you will see the screen divided into two parts:

    • The top half is called the **Session** window, which will contain numeric output.

    • The bottom half is called the **Worksheet**, which looks similar to a spreadsheet, where you will enter your data. Minitab allows you to have multiple worksheets within a project.

Screen 1.9

	C1-D	C2	C3-T
	Date	Sales	Employee
1	January 10	36	J. Molesky
2	May 28	41	C. Andreasen
3	June 10	35	C. Henriksen
4	October 28	31	M. Costello

Worksheet 1 ***

2. Use the mouse or the arrow keys to select where you want to start entering your data in the worksheet. Each column in the worksheet corresponds to a variable, so you can enter only one kind of data into a given column. Data can be numeric, text, or date/time. If a column contains text data, Minitab will add "-T" to the column heading and if a column contains a date/time, Minitab will add "-D" to the column heading. (See **Screen 1.9**.)

3. The blank row between the column labels and the first row is for variable names. In these blank cells, you can type the names of variables.

4. The direction in which data is entered can be changed by clicking the direction arrow at the top left of the worksheet (See **Screen 1.9**.)

5. Click on a cell and begin typing. Press **ENTER** when you are finished an entry.

6. If you need to revise an entry, go to that cell with the mouse or the arrow keys and begin typing. Press **ENTER** to put the revised entry into the cell.

7. When you are finished, select **File > Save Project As** to save your work as a file on your computer. Minitab will automatically assign the file extension *.mpj* to your work after the filename. Or you can also save just the worksheet by selecting **File > Save Current Worksheet**. Minitab will add *.mtb* extensions to your saved file.

8. As an exercise, enter the data shown below into Minitab. Name the columns *Month*, *Cost*, and *Increase*. Save the result as the file *test.mpj*.

Month	Cost	Increase
January	52	0.08
February	48	0.06
March	49	0.07

9. To open a saved file, select **File > Open** and select the file name.

10. If you are already in Minitab and you want to start a new worksheet, select **File > New** and choose Worksheet. Whenever you save a project, Minitab will automatically save all of the worksheets in the project.

Creating New Columns from Existing Columns

Note: This example calculates the squares of values in a column but other computations can be performed, such as computing $\sum x^2$ or $\sum xy$.

1. To calculate a column containing the squares of the values in the column *Sales* as shown in Screen 1.9, select **Calc > Calculator**.

2. In the **Store result in variable** box, type the name of the column to contain the new values (such as C4).

3. Click inside the **Expression** box, click **C2 Sales** in the column to the left of the **Expression** box, and then click **Select**. You can also type **C2** in the **Expression** box.

4. Click on the exponentiation (^) button on the keypad in this dialog box. Type **2** after the two ^ in the **Expression** box. Click **OK**. (See **Screen 1.10**.)

Screen 1.10

5. The numbers 1296, 1681, 1225, and 961 should appear in column C4 of the worksheet.

Calculating the Sum of the Values in a Column

1. To calculate the sum of the values in the *Sales* column from Screen 1.9, select **Calc > Column Statistics**, which will produce a dialog box.

2. From the **Statistic** list, select **Sum**.

3. Click in the **Input Variable** box. The list of variables will appear in the left portion of the dialog box.

Screen 1.11

Screen 1.12

4. Click on the column *Sales*, then click **Select**. You can also type **C2** in the **Input variable** box. (See **Screen 1.11**.)

5. Click **OK**. The result will appear in the **Session** window.

Selecting Random Samples

1. The following data represent the ages, in years, at which 10 patients were first diagnosed with diabetes. Enter these data in column **C1** in the order they appear below. We will select a random sample of 4 data values from these 10.

18,   7,   46,   32,   62,   65,   51,   40,   28,   72

2. Select **Calc > Random Data > Sample from Columns**.

3. Use the following settings in the dialog box that appears on screen: (See **Screen 1.12**.)

   • Type 4 in the **Number of rows to sample** box.

   • Type C1 in the **From columns** box.

   • Type C2 in the **Store samples in** box.

4. Click **OK**.

5. Four data values from C1 are randomly selected without replacement and stored in C2. (See **Screen 1.13**.)

   Note: The numbers in **C2** may be different on your screen than they appear in **Screen 1.13** because the data were randomly sampled.

▦ Worksheet 1 ***		
↓	**C1**	**C2**
1	18	40
2	7	51
3	46	18
4	32	65
5	62	
6	65	
7	51	
8	40	
9	28	
10	72	
11		

Screen 1.13

Excel

The Excel Technology Instructions feature of this text is written for Excel 2013. Some screens, menus, and functions will be slightly different in older versions of Excel. To enable some of the advanced Excel functions, you must enable the Data Analysis Add-In for Excel.

Entering and Saving Data

1. In Excel, use the mouse or the arrow keys to select where you want to start entering your data. Data can be numeric or text. The rectangles are called cells, and the cells are collectively known as a spreadsheet.

2. You can format your data by selecting the cells that you want to format, then clicking **HOME** and choosing the formatting option from the **NUMBER** group.

3. If you need to revise an entry, go to that cell with the mouse or the arrow keys. You can retype the entry or you can edit it. To edit it, double-click on the cell and use the arrow keys and the backspace key to help you revise the entry, then press **ENTER** to place the revised entry into the cell.

4. When you are done, select **Save As** from the **FILE** tab to save your work as a file on your computer. Excel will automatically assign the file extension *.xlsx* after the filename.

5. As an exercise, enter the following data into Excel, format the second column as currency, and then format the third column as percent. Save the result as the file *test.xlsx*. (See **Screen 1.14**.)

Month	Cost	Increase
January	52	0.08
February	48	0.06
March	49	0.07

	A	B	C
1	Month	Cost	Increase
2	January	$52.00	8%
3	February	$48.00	6%
4	March	$49.00	7%

Screen 1.14

6. To open a saved file, select **Open** from the **FILE** tab and select the filename.

Creating New Columns from Existing Columns

Note: This example calculates the squares of values in a column but other computations can be performed, such as computing $\sum x^2$ or $\sum xy$.

1. Open the file *test.xlsx*. Click on cell **D2**.

2. Type **=B2^2** and then press **ENTER**. The value $2704.00 should appear, which is the square of $52.00, the value in cell **B2**. (See **Screen 1.15**.)

SUM	▾	:	✕	✓	f_x	=B2^2

	A	B	C	D
1	Month	Cost	Increase	
2	January	$52.00	8%	=B2^2
3	February	$48.00	6%	
4	March	$49.00	7%	

Screen 1.15

3. Highlight cell **D2** and then select **Copy** from the **Clipboard** group on the **HOME** tab. Highlight cells **D3** and **D4** and then select **Paste** from the **Clipboard** group on the **HOME** tab.

4. The numbers $2304.00 and $2401.00 should appear in cells **D3** and **D4**, respectively. (See **Screen 1.16**.)

Calculating the Sum of the Values in a Column

1. Open the file *test.xlsx*. Click on cell **B5**.

2. Select **FORMULAS > Function Library > Σ**. Excel should automatically enter the Sum function along with the range of cells involved in the sum. (*Note*: If the range of cells is incorrect, you can type any changes.) (See **Screen 1.16**.)

3. Press **ENTER**. The number $149.00 should appear in cell **B5**.

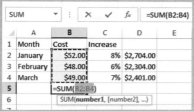

SUM	▾	:	✕	✓	f_x	=SUM(B2:B4)

	A	B	C	D	E
1	Month	Cost	Increase		
2	January	$52.00	8%	$2,704.00	
3	February	$48.00	6%	$2,304.00	
4	March	$49.00	7%	$2,401.00	
5		=SUM(B2:B4)			
6		SUM(number1, [number2], ...)			

Screen 1.16

Selecting Random Samples

1. The following data represent the age, in years, at which 10 patients were first diagnosed with diabetes. Enter these data into cells A1 through A10 in the order they appear below. We will select a random sample of 4 data values from these 10.

$$18, \quad 7, \quad 46, \quad 32, \quad 62, \quad 65, \quad 51, \quad 40, \quad 28, \quad 72$$

2. Select **DATA > Data Analysis**.

3. From the dialog box that appears on screen, select **Sampling**.

4. Use the following settings in the dialog box that appears on screen (See **Screen 1.17**):

- Type A1:A10 in the **Input Range** box.

- Select Random in the **Sampling Method** box.

- Type 4 in the **Number of Samples** box.

- Select New Worksheet Ply in the **Output options** box.

5. Click **OK**.

6. Four data values from A1 to A10 are randomly selected without replacement and stored in a new worksheet.

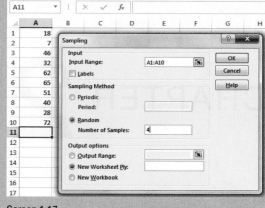

Screen 1.17

Enabling the Data Analysis Add-In

1. Click the **FILE > Options**.

2. From the resulting dialog box, click **Add-ins**.

3. In the **Manage** box, select **Excel Add-ins** and then click **Go**.

4. In the **Add-Ins** dialog box, select the **Analysis ToolPak** check box, and then click **OK**.

5. If **Analysis ToolPak** is not listed in the Add-Ins available box, click **Browse** to locate it.

6. If you are prompted that the **Analysis ToolPak** is not currently installed on your computer, click **Yes** to install it.

TECHNOLOGY ASSIGNMENTS

TA1.1 The following table gives the names, hours worked, and salary for the past week for five workers.

Name	Hours Worked	Salary
John	42	$ 725
Shannon	33	1583
Kathy	28	1255
David	47	1090
Steve	40	820

a. Enter these data into the spreadsheet. Save the data file as WORKER. Exit the session or program. Then restart the program or software and retrieve the file WORKER.

b. Print a hard copy of the spreadsheet containing data you entered.

TA1.2 Refer to data on total revenues for 2010 of six companies given in table below.

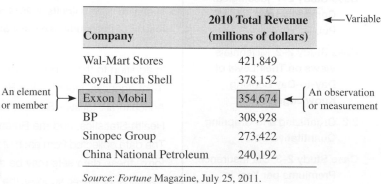

Company	2010 Total Revenue (millions of dollars)
Wal-Mart Stores	421,849
Royal Dutch Shell	378,152
Exxon Mobil	354,674
BP	308,928
Sinopec Group	273,422
China National Petroleum	240,192

Source: Fortune Magazine, July 25, 2011.

Enter these data into the spreadsheet and save this file as REVENUES.

CHAPTER 2

Alistair Berg/Getty Images, Inc.

Organizing and Graphing Data

What is your political ideology? Do you classify yourself a consistently liberal person or a consistently conservative person, a mostly liberal person or a mostly conservative person, or are you someone who belongs to a group called "mixed"? Pew Research Center conducted a national survey of 10,013 adults in 2014 to find the political views of adults in the United States. To see the results of this study, see Case Study 2–1.

In addition to thousands of private organizations and individuals, a large number of U.S. government agencies (such as the Bureau of the Census, the Bureau of Labor Statistics, the National Agricultural Statistics Service, the National Center for Education Statistics, the National Center for Health Statistics, and the Bureau of Justice Statistics) conduct hundreds of surveys every year. The data collected from each of these surveys fill hundreds of thousands of pages. In their original form, these data sets may be so large that they do not make sense to most of us. Descriptive statistics, however, supplies the techniques that help to condense large data sets by using tables, graphs, and summary measures. We see such tables, graphs, and summary measures in newspapers and magazines every day. At a glance, these tabular and graphical displays present information on every aspect of life. Consequently, descriptive statistics is of immense importance because it provides efficient and effective methods for summarizing and analyzing information.

This chapter explains how to organize and display data using tables and graphs. We will learn how to prepare frequency distribution tables for qualitative and quantitative data; how to construct bar graphs, pie charts, histograms, and polygons for such data; and how to prepare stem-and-leaf displays.

2.1 | Organizing and Graphing Qualitative Data

This section discusses how to organize and display qualitative (or categorical) data. Data sets are organized into tables and displayed using graphs. First we discuss the concept of raw data.

2.1.1 Raw Data

When data are collected, the information obtained from each member of a population or sample is recorded in the sequence in which it becomes available. This sequence of data recording is random and unranked. Such data, before they are grouped or ranked, are called **raw data**.

Suppose we collect information on the ages (in years) of 50 students selected from a university. The data values, in the order they are collected, are recorded in Table 2.1. For instance, the first student's age is 21, the second student's age is 19 (second number in the first row), and so forth. The data in Table 2.1 are quantitative raw data.

> Data recorded in the sequence in which they are collected and before they are processed or ranked are called **raw data**.

Table 2.1   Ages of 50 Students

21	19	24	25	29	34	26	27	37	33
18	20	19	22	19	19	25	22	25	23
25	19	31	19	23	18	23	19	23	26
22	28	21	20	22	22	21	20	19	21
25	23	18	37	27	23	21	25	21	24

Suppose we ask the same 50 students about their student status. The responses of the students are recorded in Table 2.2. In this table, F, SO, J, and SE are the abbreviations for freshman, sophomore, junior, and senior, respectively. This is an example of qualitative (or categorical) raw data.

Table 2.2   Status of 50 Students

J	F	SO	SE	J	J	SE	J	J	J
F	F	J	F	F	F	SE	SO	SE	J
J	F	SE	SO	SO	F	J	F	SE	SE
SO	SE	J	SO	SO	J	J	SO	F	SO
SE	SE	F	SE	J	SO	F	J	SO	SO

The data presented in Tables 2.1 and 2.2 are also called **ungrouped data**. An ungrouped data set contains information on each member of a sample or population individually. If we rank the data of Table 2.1 from lowest to the highest age, they will still be ungrouped data but not raw data.

2.1.2 Frequency Distributions

The Gallup polling agency recently surveyed randomly selected 1015 adults aged 18 and over from all 50 U.S. states and the District of Columbia. These adults were asked, "Please tell me how concerned you are right now about each of the following financial matters, based on your current financial situation—are you very worried, moderately worried, not too worried, or not worried at all." Among a series of financial situations, one such situation was not having enough money to pay their normal monthly bills. Table 2.3 lists the responses of these adults. The Gallup report contained the percent of adults belonging to each category, which we have converted to numbers in the table. In this table, the variable is how much are adults worried about not having enough money to pay normal monthly bills. The categories representing this variable are listed in the first column of the table. Note that these categories are mutually exclusive. In other words, each of the 1015 adults belongs to one and only one of these categories. The number of adults who belong to a certain category is called the frequency of that category. A **frequency distribution**

> A **frequency distribution** of a qualitative variable lists all categories and the number of elements that belong to each of the categories.

exhibits how the frequencies are distributed over various categories. Table 2.3 is called a *frequency distribution table* or simply a *frequency table.*

Table 2.3 Worries About Not Having Enough Money to Pay Normal Monthly Bills

Response	Number of Adults
Very worried	162
Moderately worried	203
Not too worried	305
Not worried at all	325
Others	20
	Sum = 1015

Variable → Response; Category → Not too worried; Frequency column → Number of Adults; Frequency → 305

Source: Gallup Poll.

Example 2–1 illustrates how a frequency distribution table is constructed for a qualitative variable.

EXAMPLE 2–1 What Variety of Donuts Is Your Favorite?

Constructing a frequency distribution table for qualitative data.

A sample of 30 persons who often consume donuts were asked what variety of donuts is their favorite. The responses from these 30 persons are as follows:

glazed	filled	other	plain	glazed	other
frosted	filled	filled	glazed	other	frosted
glazed	plain	other	glazed	glazed	filled
frosted	plain	other	other	frosted	filled
filled	other	frosted	glazed	glazed	filled

Construct a frequency distribution table for these data.

© Jack Puccio/iStockphoto

Solution Note that the variable in this example is *favorite variety of donut.* This variable has five categories (varieties of donuts): glazed, filled, frosted, plain, and other. To prepare a frequency distribution, we record these five categories in the first column of Table 2.4. Then we read each response (each person's favorite variety of donut) from the given information and mark a *tally*, denoted by the symbol |, in the second column of Table 2.4 next to the corresponding category. For example, the first response is *glazed.* We show this in the frequency table by marking a tally in the second column next to the category *glazed.* Note that the tallies are marked in blocks of five for counting convenience. Finally, we record the total of the tallies for each category in the third column of the table. This column is called the *column of frequencies* and is usually denoted by *f*. The sum of the entries in the frequency column gives the sample size or total frequency. In Table 2.4, this total is 30, which is the sample size.

Table 2.4 Frequency Distribution of Favorite Donut Variety

Donut Variety	Tally	Frequency (f)
Glazed	~~||||~~ |||	8
Filled	~~||||~~ ||	7
Frosted	~~||||~~	5
Plain	|||	3
Other	~~||||~~ ||	7
		Sum = 30

2.1.3   Relative Frequency and Percentage Distributions

The **relative frequency** of a category is obtained by dividing the frequency of that category by the sum of all frequencies. Thus, the relative frequency shows what fractional part or proportion of the total frequency belongs to the corresponding category. A *relative frequency distribution* lists the relative frequencies for all categories.

Calculating Relative Frequency of a Category

$$\text{Relative frequency of a category} = \frac{\text{Frequency of that category}}{\text{Sum of all frequencies}}$$

The **percentage** for a category is obtained by multiplying the relative frequency of that category by 100. A *percentage distribution* lists the percentages for all categories.

Calculating Percentage

$$\text{Percentage} = (\text{Relative frequency}) \cdot 100\%$$

EXAMPLE 2–2   What Variety of Donuts Is Your Favorite?

Constructing relative frequency and percentage distributions.

Determine the relative frequency and percentage distributions for the data in Table 2.4.

Solution   The relative frequencies and percentages from Table 2.4 are calculated and listed in Table 2.5. Based on this table, we can state that 26.7% of the people in the sample said that glazed donut is their favorite. By adding the percentages for the first two categories, we can state that 50% of the persons included in the sample said that glazed or filled donut is their favorite. The other numbers in Table 2.5 can be interpreted in similar ways.

Table 2.5   Relative Frequency and Percentage Distributions of Favorite Donut Variety

Donut Variety	Relative Frequency	Percentage
Glazed	8/30 = .267	.267(100) = 26.7
Filled	7/30 = .233	.233(100) = 23.3
Frosted	5/30 = .167	.167(100) = 16.7
Plain	3/30 = .100	.100(100) = 10.0
Other	7/30 = .233	.233(100) = 23.3
	Sum = 1.000	Sum = 100%

Notice that the sum of the relative frequencies is always 1.00 (or approximately 1.00 if the relative frequencies are rounded), and the sum of the percentages is always 100 (or approximately 100 if the percentages are rounded).

2.1.4   Graphical Presentation of Qualitative Data

All of us have heard the adage "a picture is worth a thousand words." A graphic display can reveal at a glance the main characteristics of a data set. The *bar graph* and the *pie chart* are two types of graphs that are commonly used to display qualitative data.

Ideological Composition of the U.S. Public, 2014

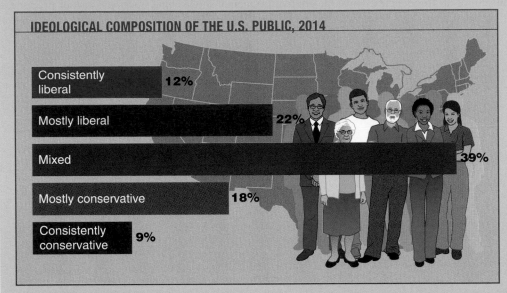

IDEOLOGICAL COMPOSITION OF THE U.S. PUBLIC, 2014

Consistently liberal — 12%

Mostly liberal — 22%

Mixed — 39%

Mostly conservative — 18%

Consistently conservative — 9%

Data source: Pew Research Center

Pew Research Center conducted a national survey of 10,013 adults January 23 to March 16, 2014, to find the political views of adults in the United States. As the above bar chart shows, 12% of the adults polled said that they were consistently liberal, 22% indicated that they were mostly liberal, and so on. In this survey, Pew Research Center also found that, overall, the percentage of Americans who indicated that they were consistently conservative or consistently liberal has increased from 10% to 21% during the past two decades. Note that in this chart, the bars are drawn horizontally.

Source: Pew Research Center, June, 2014 Report: *Political Polarization in the American Public.*

Bar Graphs

A graph made of bars whose heights represent the frequencies of respective categories is called a **bar graph**.

To construct a **bar graph** (also called a *bar chart*), we mark the various categories on the horizontal axis as in Figure 2.1. Note that all categories are represented by intervals of the same width. We mark the frequencies on the vertical axis. Then we draw one bar for each category such that the height of the bar represents the frequency of the corresponding category. We leave a small gap between adjacent bars. Figure 2.1 gives the bar graph for the frequency distribution of Table 2.4.

The bar graphs for relative frequency and percentage distributions can be drawn simply by marking the relative frequencies or percentages, instead of the frequencies, on the vertical axis.

Sometimes a bar graph is constructed by marking the categories on the vertical axis and the frequencies on the horizontal axis. Case Study 2–1 presents such an example.

Figure 2.1 Bar graph for the frequency distribution of Table 2.4.

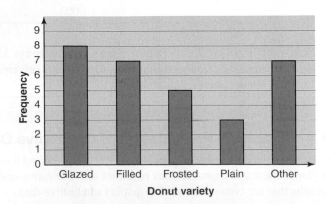

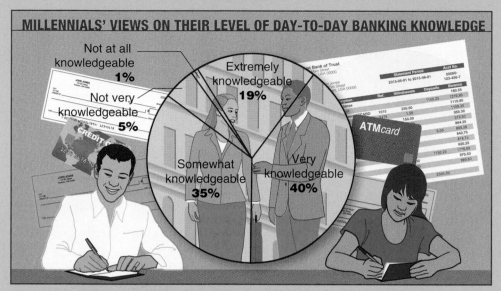

Pareto Chart

To obtain a Pareto chart, we arrange (in a descending order) the bars in a bar graph based on their heights (frequencies, relative frequencies, or percentages). Thus, the bar with the largest height appears first (on the left side) in a bar graph and the one with the smallest height appears at the end (on the right side) of the bar graph.

Figure 2.2 shows the Pareto chart for the frequency distribution of Table 2.4. It is the same bar chart that appears in Figure 2.1 but with bars arranged based on their heights.

> A **Pareto chart** is a bar graph with bars arranged by their heights in descending order. To make a Pareto chart, arrange the bars according to their heights such that the bar with the largest height appears first on the left side, and then subsequent bars are arranged in descending order with the bar with the smallest height appearing last on the right side.

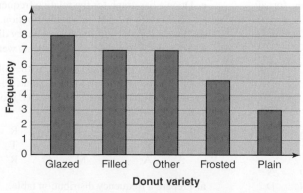

Figure 2.2 Pareto chart for the frequency distribution of Table 2.4.

37

Pie Charts

> A circle divided into portions that represent the relative frequencies or percentages of a population or a sample belonging to different categories is called a **pie chart**.

A **pie chart** is more commonly used to display percentages, although it can be used to display frequencies or relative frequencies. The whole pie (or circle) represents the total sample or population. Then we divide the pie into different portions that represent the different categories.

Figure 2.3 shows the pie chart for the percentage distribution of Table 2.5.

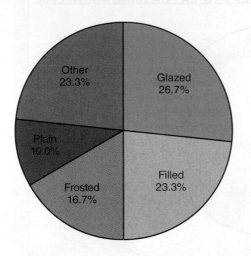

Figure 2.3   Pie chart for the percentage distribution of Table 2.5.

EXERCISES

CONCEPTS AND PROCEDURES

2.1   What are the benefits of grouping data in the form of a frequency table?

2.2   Explain how to find the relative frequencies and percentages of categories from frequencies of categories with the help of an example.

APPLICATIONS

2.3   The following data give the results of a sample survey. The letters P, Q, and R represent the three categories.

R	Q	Q	P	P	P	Q	P	R	P
P	P	P	P	Q	P	P	Q	Q	P
Q	P	P	Q	R	Q	P	P	P	P
P	P	Q	Q	P	P	Q	Q	R	P

    a. Prepare a frequency distribution table.
    b. Calculate the relative frequencies and percentages for all categories.
    c. What percentage of the elements in this sample belong to category P?
    d. What percentage of the elements in this sample belong to category Q or R?
    e. Draw a pie chart for the percentage distribution.
    f. Make a Pareto chart for the percentage distribution.

2.4   The following data give the results of a sample survey. The letters Y, N, and D represent the three categories.

N	N	N	Y	Y	Y	N	Y	D	N
Y	Y	Y	Y	N	Y	Y	N	N	D
D	Y	Y	D	D	N	N	N	Y	N
Y	Y	N	N	Y	Y	N	N	D	Y

    a. Prepare a frequency distribution table.
    b. Calculate the relative frequencies and percentages for all categories.
    c. What percentage of the elements in this sample belong to category Y?
    d. What percentage of the elements in this sample belong to category N or D?
    e. Draw a pie chart for the percentage distribution.

2.5   The following data give the political party of each of the first 30 U.S. presidents. In the data, D stands for Democrat, DR for Democratic Republican, F for Federalist, R for Republican, and W for Whig.

F	F	DR	DR	DR	DR	D	D	W	W
D	W	W	D	D	R	D	R	R	R
R	D	R	D	R	R	R	D	R	R

    a. Prepare a frequency distribution table for these data.
    b. Calculate the relative frequency and percentage distributions.
    c. Draw a bar graph for the relative frequency distribution and a pie chart for the percentage distribution.
    d. Make a Pareto chart for the frequency distribution.
    e. What percentage of these presidents were Whigs?

2.6   Thirty adults were asked which of the following conveniences they would find most difficult to do without: television (T), refrigerator (R), air conditioning (A), computer (C), or mobile phone (M). Their responses are listed below.

R	A	R	A	A	T	R	M	C	A
A	R	R	T	C	C	T	R	A	A
R	A	A	T	M	C	R	A	C	R

    a. Prepare a frequency distribution table.
    b. Calculate the relative frequencies and percentages for all categories.

c. What percentage of these adults named refrigerator or air conditioning as the convenience that they would find most difficult to do without?

d. Draw a bar graph for the relative frequency distribution.

2.7   In a May 4, 2011 Quinnipiac University poll, a random sample of New York City residents were asked, "How serious is the problem of police officers fixing tickets: very serious, somewhat serious, not too serious, or not at all serious?" (Note: In 2010 to 2011, New York City investigated the widespread problem of traffic ticket fixing by police officers. Many police officers were charged with this crime after the investigation.) The following table summarizes residents' responses.

Response	Percentage of Responses
Very serious	38
Somewhat serious	26
Not too serious	17
Not at all serious	8

Source: www.quinnipiac.edu.

Note that these percentages add up to 89%. The remaining respondents stated that they did not know or had no opinion. Assume that 11% belong to the category *did not know*. Draw a pie chart for this percentage distribution.

2.2 | Organizing and Graphing Quantitative Data

In the previous section we learned how to group and display qualitative data. This section explains how to group and display quantitative data.

2.2.1   Frequency Distributions

Table 2.6 gives the weekly earnings of 100 employees of a large company. The first column lists the *classes*, which represent the (quantitative) variable *weekly earnings*. For quantitative data, an interval that includes all the values that fall on or within two numbers—the lower and upper limits—is called a **class**. Note that the classes always represent a variable. As we can observe, the classes are nonoverlapping; that is, each value for earnings belongs to one and only one class. The second column in the table lists the number of employees who have earnings within each class. For example, 4 employees of this company earn $801 to $1000 per week. The numbers listed in the second column are called the **frequencies,** which give the number of data values that belong to different classes. The frequencies are denoted by *f*.

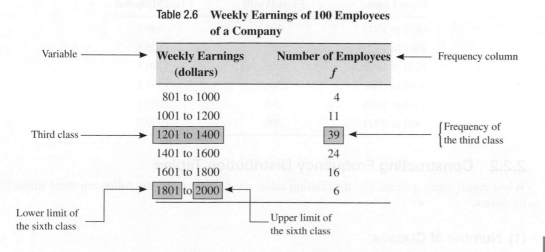

Table 2.6   **Weekly Earnings of 100 Employees of a Company**

Weekly Earnings (dollars)	Number of Employees *f*
801 to 1000	4
1001 to 1200	11
1201 to 1400	39
1401 to 1600	24
1601 to 1800	16
1801 to 2000	6

For quantitative data, the frequency of a class represents the number of values in the data set that fall in that class. Table 2.6 contains six classes. Each class has a *lower limit* and an *upper limit*. The values 801, 1001, 1201, 1401, 1601, and 1801 give the lower limits, and the values 1000, 1200, 1400, 1600, 1800, and 2000 are the upper limits of the six classes, respectively. The data presented in Table 2.6 are an illustration of a **frequency distribution table** for quantitative data. Whereas the data that list individual values are called ungrouped data, the data presented in a frequency distribution table are called **grouped data**.

> A **frequency distribution** for quantitative data lists all the classes and the number of values that belong to each class. Data presented in the form of a frequency distribution are called **grouped data**.

The difference between the lower limits of two consecutive classes gives the **class width**. The class width is also called the **class size**.

Finding Class Width

To find the width of a class, subtract its lower limit from the lower limit of the next class. Thus:

Width of a class = Lower limit of the next class − Lower limit of the current class

---

Thus, in Table 2.6,

$$\text{Width of the first class} = 1001 - 801 = 200$$

The class widths for the frequency distribution of Table 2.6 are listed in the second column of Table 2.7. Each class in Table 2.7 (and Table 2.6) has the same width of 200.

The **class midpoint** or **mark** is obtained by dividing the sum of the two limits of a class by 2.

Calculating Class Midpoint or Mark

$$\text{Class midpoint or mark} = \frac{\text{Lower limit} + \text{Upper limit}}{2}$$

Thus, the midpoint of the first class in Table 2.6 or Table 2.7 is calculated as follows:

$$\text{Midpoint of the first class} = \frac{801 + 1000}{2} = 900.5$$

The class midpoints for the frequency distribution of Table 2.6 are listed in the third column of Table 2.7.

Table 2.7   Class Widths and Class Midpoints for Table 2.6

Class Limits	Class Width	Class Midpoint
801 to 1000	200	900.5
1001 to 1200	200	1100.5
1201 to 1400	200	1300.5
1401 to 1600	200	1500.5
1601 to 1800	200	1700.5
1801 to 2000	200	1900.5

2.2.2   Constructing Frequency Distribution Tables

When constructing a frequency distribution table, we need to make the following three major decisions.

(1) Number of Classes

Usually the number of classes for a frequency distribution table varies from 5 to 20, depending mainly on the number of observations in the data set.[1] It is preferable to have more classes as the size of a data set increases. The decision about the number of classes is arbitrarily made by the data organizer.

---

[1]One rule to help decide on the number of classes is Sturge's formula:

$$c = 1 + 3.3 \log n$$

where c is the number of classes and n is the number of observations in the data set. The value of log n can be obtained by using a calculator.

(2) Class Width

Although it is not uncommon to have classes of different sizes, most of the time it is preferable to have the same width for all classes. To determine the class width when all classes are the same size, first find the difference between the largest and the smallest values in the data. Then, the approximate width of a class is obtained by dividing this difference by the number of desired classes.

Calculation of Class Width

$$\text{Approximate class width} = \frac{\text{Largest value} - \text{Smallest value}}{\text{Number of classes}}$$

Usually this approximate class width is rounded to a convenient number, which is then used as the class width. Note that rounding this number may slightly change the number of classes initially intended.

(3) Lower Limit of the First Class or the Starting Point

Any convenient number that is equal to or less than the smallest value in the data set can be used as the lower limit of the first class.

Example 2–3 illustrates the procedure for constructing a frequency distribution table for quantitative data.

EXAMPLE 2–3 **Values of Baseball Teams, 2015**

Constructing a frequency distribution table for quantitative data.

The following table gives the value (in million dollars) of each of the 30 baseball teams as estimated by *Forbes* magazine (*source: Forbes* Magazine, April 13, 2015). Construct a frequency distribution table.

Values of Baseball Teams, 2015

Team	Value (millions of dollars)	Team	Value (millions of dollars)
Arizona Diamondbacks	840	Milwaukee Brewers	875
Atlanta Braves	1150	Minnesota Twins	895
Baltimore Orioles	1000	New York Mets	1350
Boston Red Sox	2100	New York Yankees	3200
Chicago Cubs	1800	Oakland Athletics	725
Chicago White Sox	975	Philadelphia Phillies	1250
Cincinnati Reds	885	Pittsburgh Pirates	900
Cleveland Indians	825	San Diego Padres	890
Colorado Rockies	855	San Francisco Giants	2000
Detroit Tigers	1125	Seattle Mariners	1100
Houston Astros	800	St. Louis Cardinals	1400
Kansas City Royals	700	Tampa Bay Rays	605
Los Angeles Angels of Anaheim	1300	Texas Rangers	1220
Los Angeles Dodgers	2400	Toronto Blue Jays	870
Miami Marlins	650	Washington Nationals	1280

Solution   In these data, the minimum value is 605, and the maximum value is 3200. Suppose we decide to group these data using six classes of equal width. Then,

$$\text{Approximate width of each class} = \frac{3200 - 605}{6} = 432.5$$

Now we round this approximate width to a convenient number, say 450. The lower limit of the first class can be taken as 605 or any number less than 605. Suppose we take 601 as the lower limit of the first class. Then our classes will be

601–1050,   1051–1500,   1501–1950,   1951–2400,   2401–2850,   and   2851–3300

We record these five classes in the first column of Table 2.8.

Table 2.8   Frequency Distribution of the Values of Baseball Teams, 2015

Value of a Team (in million $)	Tally	Number of Teams (f)
601–1050	||||| ||||| |||||		16
1051–1500	||||| ||||	9
1501–1950			1
1951–2400	|||	3
2401–2850		0
2851–3300			1
		$\Sigma f = 30$

Now we read each value from the given data and mark a tally in the second column of Table 2.8 next to the corresponding class. The first value in our original data set is 840, which belongs to the 601–1050 class. To record it, we mark a tally in the second column next to the 601–1050 class. We continue this process until all the data values have been read and entered in the tally column. Note that tallies are marked in blocks of five for counting convenience. After the tally column is completed, we count the tally marks for each class and write those numbers in the third column. This gives the column of frequencies. These frequencies represent the number of baseball teams with values in the corresponding classes. For example, 16 of the teams have values in the interval $601–$1050 million.

Using the Σ notation (see Section 1.7 of Chapter 1), we can denote the sum of frequencies of all classes by Σf. Hence,

$$\Sigma f = 16 + 9 + 1 + 3 + 0 + 1 = 30$$

The number of observations in a sample is usually denoted by n. Thus, for the sample data, Σf is equal to n. The number of observations in a population is denoted by N. Consequently, Σf is equal to N for population data. Because the data set on the values of baseball teams in Table 2.8 is for all 30 teams, it represents a population. Therefore, in Table 2.8 we can denote the sum of frequencies by N instead of Σf.

Note that when we present the data in the form of a frequency distribution table, as in Table 2.8, we lose the information on individual observations. We cannot know the exact value of any team from Table 2.8. All we know is that 16 teams have values in the interval $601–$1050 million, and so forth.

2.2.3   Relative Frequency and Percentage Distributions

Using Table 2.8, we can compute the relative frequency and percentage distributions in the same way as we did for qualitative data in Section 2.1.3. The relative frequencies and percentages for a quantitative data set are obtained as follows. Note that relative frequency is the same as proportion.

Calculating Relative Frequency and Percentage

$$\text{Relative frequency of a class} = \frac{\text{Frequency of that class}}{\text{Sum of all frequencies}} = \frac{f}{\Sigma f}$$

$$\text{Percentage} = (\text{Relative frequency}) \cdot 100\%$$

Example 2–4 illustrates how to construct relative frequency and percentage distributions.

EXAMPLE 2–4   Values of Baseball Teams, 2015

Constructing relative frequency and percentage distributions.

Calculate the relative frequencies and percentages for Table 2.8.

Solution   The relative frequencies and percentages for the data in Table 2.8 are calculated and listed in the second and third columns, respectively, of Table 2.9.

Table 2.9   Relative Frequency and Percentage Distributions of the Values of Baseball Teams

Value of a Team (in million $)	Relative Frequency	Percentage
601–1050	16/30 = .533	53.3
1051–1500	9/30 = .300	30.0
1501–1950	1/30 = .033	3.3
1951–2400	3/30 = .100	10.0
2401–2850	0/30 = .000	0.0
2851–3300	1/30 = .033	3.3
	Sum = .999	Sum = 99.9%

Using Table 2.9, we can make statements about the percentage of teams with values within a certain interval. For example, from Table 2.9, we can state that about 53.3% of the baseball teams had estimated values in the interval $601 million to $1050 million in April 2015. By adding the percentages for the first two classes, we can state that about 83.3% of the baseball teams had estimated values in the interval $601 million to $1500 million in April 2015. Similarly, by adding the percentages for the last three classes, we can state that about 13.3% of the baseball teams had estimated values in the interval $1951 million to $3300 million in April 2015.

2.2.4   Graphing Grouped Data

Grouped (quantitative) data can be displayed in a *histogram* or a *polygon*. This section describes how to construct such graphs. We can also draw a pie chart to display the percentage distribution for a quantitative data set. The procedure to construct a pie chart is similar to the one for qualitative data explained in Section 2.1.4; it will not be repeated in this section.

Histograms

A **histogram** can be drawn for a frequency distribution, a relative frequency distribution, or a percentage distribution. To draw a histogram, we first mark classes on the horizontal axis

CASE STUDY 2–3

Car Insurance Premiums Per Year in 50 States

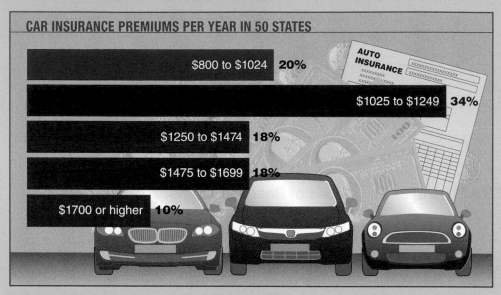

CAR INSURANCE PREMIUMS PER YEAR IN 50 STATES

$800 to $1024	**20%**
$1025 to $1249	**34%**
$1250 to $1474	**18%**
$1475 to $1699	**18%**
$1700 or higher	**10%**

Data source: www.insure.com

The above histogram shows the percentage distribution of annual car insurance premiums in 50 states. The data used to make this distribution and histogram are based on estimates made by insure.com. They collected data from six large insurance companies in 10 ZIP codes for each state. The rates were obtained "for the same full-coverage policy for the same driver—a 40-year-old man with a clean driving record and good credit." The rates used in this histogram are the averages "for the 20 best-selling vehicles in the U.S." As the histogram shows, in 20% of the states the car insurance rates were in the interval $800 to $1024, and so on. Note that the last class ($1700 or higher) has no upper limit. Such a class is called an *open-ended* class.

Source: www.insure.com, April 13, 2015.

> A **histogram** is a graph in which classes are marked on the horizontal axis and the frequencies, relative frequencies, or percentages are marked on the vertical axis. The frequencies, relative frequencies, or percentages are represented by the heights of the bars. In a histogram, the bars are drawn adjacent to each other.

and frequencies (or relative frequencies or percentages) on the vertical axis. Next, we draw a bar for each class so that its height represents the frequency of that class. The bars in a histogram are drawn adjacent to each other with no gap between them. A histogram is called a **frequency histogram**, a **relative frequency histogram**, or a **percentage histogram** depending on whether frequencies, relative frequencies, or percentages are marked on the vertical axis.

Figures 2.4 and 2.5 show the frequency and the percentage histograms, respectively, for the data of Tables 2.8 and 2.9 of Sections 2.2.2 and 2.2.3. The two histograms look alike

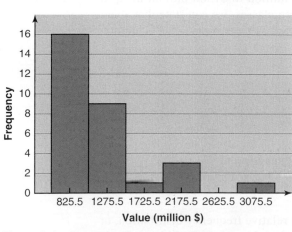

Figure 2.4 Frequency histogram for Table 2.8.

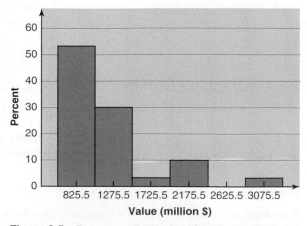

Figure 2.5 Percentage distribution histogram for Table 2.9.

Hours Worked in a Typical Week by Full-Time U.S. Workers

HOURS WORKED IN A TYPICAL WEEK BY FULL-TIME U.S. WORKERS

Less than 40 hours

60 or more **18%**   8%

50 to 59 **21%**

40 to 49 **53%**

Data source: www.gallup.com

The above pie chart shows the percentage distribution of hours worked in a typical week by full-time workers in the United States. The data are based on a recent Gallup poll of 1271 workers. As the numbers in the pie chart show, 8% of these workers said they work for less than 40 hours a week, 53% work for 40 to 49 hours a week, and so on. As you can observe, two of the classes are open-ended classes in this chart.

Source: www.gallup.com, August 29, 2014.

because they represent the same data. A relative frequency histogram can be drawn for the relative frequency distribution of Table 2.9 by marking the relative frequencies on the vertical axis.

In Figures 2.4 and 2.5, we used class midpoints to mark classes on the horizontal axis. However, we can show the intervals on the horizontal axis by using the class limits instead of the class midpoints.

Polygons

A **polygon** is another device that can be used to present quantitative data in graphic form. To draw a **frequency polygon**, we first mark a dot above the midpoint of each class at a height equal to the frequency of that class. This is the same as marking the midpoint at the top of each bar in a histogram. Next we include two more classes, one at each end, and mark their midpoints. Note that these two classes have zero frequencies. In the last step, we join the adjacent dots with straight lines. The resulting line graph is called a frequency polygon or simply a polygon.

A polygon with relative frequencies marked on the vertical axis is called a *relative frequency polygon*. Similarly, a polygon with percentages marked on the vertical axis is called a *percentage polygon*.

Figure 2.6 shows the frequency polygon for the frequency distribution of Table 2.8.

> A graph formed by joining the midpoints of the tops of successive bars in a histogram with straight lines is called a **polygon**.

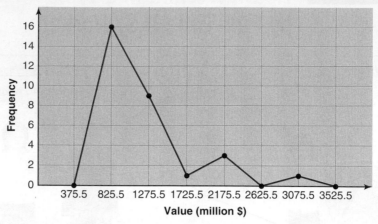

Figure 2.6 Frequency polygon for Table 2.8.

For a very large data set, as the number of classes is increased (and the width of classes is decreased), the frequency polygon eventually becomes a smooth curve. Such a curve is called a *frequency distribution curve* or simply a *frequency curve*. Figure 2.7 shows the frequency curve for a large data set with a large number of classes.

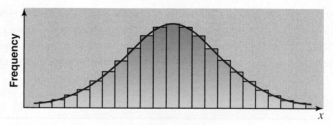

Figure 2.7 Frequency distribution curve.

2.2.5 More on Classes and Frequency Distributions

This section presents two alternative methods for writing classes to construct a frequency distribution for quantitative data.

Less-Than Method for Writing Classes

The classes in the frequency distribution given in Table 2.8 for the data on values of baseball teams were written as 601–900, 901–1200, and so on. Alternatively, we can write the classes in a frequency distribution table using the *less-than* method. The technique for writing classes shown in Table 2.8 is used for data sets that do not contain fractional values. The *less-than* method is more appropriate when a data set contains fractional values. Example 2–5 illustrates the *less-than* method.

EXAMPLE 2–5	**Federal and State Tax on Gasoline as of April 1, 2015**

Constructing a frequency distribution using the less-than method.

Based on the information collected by American Petroleum Institute, Table 2.10 lists the total of federal and state taxes (in cents per gallon) on gasoline for each of the 50 states as of April 1, 2015 (www.api.org).

Table 2.10 Total Federal and State Tax on Gasoline as of April 1, 2015

State	Gasoline Tax	State	Gasoline Tax
Alabama	39.3	Montana	46.2
Alaska	29.7	Nebraska	44.9
Arizona	37.4	Nevada	51.6
Arkansas	40.2	New Hampshire	42.2
California	66.0	New Jersey	32.9
Colorado	40.4	New Mexico	37.3
Connecticut	59.3	New York	62.9
Delaware	41.4	North Carolina	54.7
Florida	54.8	North Dakota	41.4
Georgia	44.9	Ohio	46.4
Hawaii	62.1	Oklahoma	35.4
Idaho	43.4	Oregon	49.5
Illinois	52.5	Pennsylvania	70.0
Indiana	51.3	Rhode Island	51.4
Iowa	50.4	South Carolina	35.2
Kansas	42.4	South Dakota	48.4
Kentucky	44.4	Tennessee	39.8
Louisiana	38.4	Texas	38.4
Maine	48.4	Utah	42.9
Maryland	48.7	Vermont	48.9
Massachusetts	44.9	Virginia	40.8
Michigan	51.5	Washington	55.9
Minnesota	47.0	West Virginia	53.0
Mississippi	37.2	Wisconsin	51.3
Missouri	35.7	Wyoming	42.4

Construct a frequency distribution table. Calculate the relative frequencies and percentages for all classes.

Solution The minimum value in the data set of Table 2.10 is 29.7, and the maximum value is 70. Suppose we decide to group these data using five classes of equal width. Then,

$$\text{Approximate class width} = \frac{70 - 29.7}{5} = 8.06$$

We round this number to a more convenient number—say 9—and take 9 as the width of each class. We can take the lower limit of the first class equal to 29.7 or any number lower than 29.7. If we start the first class at 27, the classes will be written as 27 to less than 36, 36 to less than 45, and so on. The five classes, which cover all the data values of Table 2.10, are recorded in the first column of Table 2.11. The second column in Table 2.11 lists the frequencies of these classes. A value in the data set that is 27 or larger but less than 36 belongs to the first class, a value that is 36 or larger but less than 45 falls into the second class, and so on. The relative frequencies and percentages for classes are recorded in the third and fourth columns, respectively, of Table 2.11. Note that this table does not contain a column of tallies.

Table 2.11   **Frequency, Relative Frequency, and Percentage Distributions of the Total Federal and State Tax on Gasoline**

Federal and State Tax (in cents)	Frequency	Relative Frequency	Percentage
27 to less than 36	5	.10	10
36 to less than 45	21	.42	42
45 to less than 54	16	.32	32
54 to less than 63	6	.12	12
63 to less than 72	2	.04	4
	Sum = 50	Sum = 1.00	Sum = 100

Note that in Table 2.11, the first column lists the class intervals using the **boundaries**, and not the limits. When we use the *less than method* to write classes, we call the two end-points of a class the **lower and upper boundaries**. For example, in the first class, which is *27 to less than 36*, 27 is the lower boundary and 36 is the upper boundary. The difference between the two boundaries gives the width of the class. Thus, the width of the first class is 36 − 27 = 9. All classes in Table 2.11 have the same width, which is 9.

CASE STUDY 2–5 How Many Cups of Coffee Do You Drink a Day?

Data source: Gallup poll of U.S. adults aged 18 and older conducted July 9–12, 2012

In a Gallup poll conducted by telephone interviews on July 9–12, 2012, U.S. adults of age 18 years and older were asked, "How many cups of coffee, if any, do you drink on an average day?" According to the results of the poll, shown in the accompanying pie chart, 36% of these adults said that they drink no coffee (represented by zero cups in the chart), 26% said that they drink one cup of coffee per day, and so on. The last class is open-ended class that indicates that 10% of these adults drink four or more cups of coffee a day. This class has no upper limit. Since the values of the variable (cups of coffee) are discrete and the variable assumes only a few possible values, the first four classes are single-valued classes.

Source: http://www.gallup.com/poll/156116/Nearly-Half-Americans-Drink-Soda-Daily.aspx.

A histogram and a polygon for the data of Table 2.11 can be drawn the same way as for the data of Tables 2.8 and 2.9.

Single-Valued Classes

If the observations in a data set assume only a few distinct (integer) values, it may be appropriate to prepare a frequency distribution table using *single-valued classes*—that is, classes that are made of single values and not of intervals. This technique is especially useful in cases of discrete data with only a few possible values. Example 2–6 exhibits such a situation.

| EXAMPLE 2–6 | Number of Vehicles Owned by Households |

Constructing a frequency distribution using single-valued classes.

The administration in a large city wanted to know the distribution of the number of vehicles owned by households in that city. A sample of 40 randomly selected households from this city produced the following data on the number of vehicles owned.

© Jorge Salcedo/iStockphoto

5	1	1	2	0	1	1	2	1	1
1	3	3	0	2	5	1	2	3	4
2	1	2	2	1	2	2	1	1	1
4	2	1	1	2	1	1	4	1	3

Construct a frequency distribution table for these data using single-valued classes.

Solution   The observations in this data set assume only six distinct values: 0, 1, 2, 3, 4, and 5. Each of these six values is used as a class in the frequency distribution in Table 2.12, and these six classes are listed in the first column of that table. To obtain the frequencies of these classes, the observations in the data that belong to each class are counted, and the results are recorded in the second column of Table 2.12. Thus, in these data, 2 households own no vehicle, 18 own one vehicle each, 11 own two vehicles each, and so on.

The data of Table 2.12 can also be displayed in a bar graph, as shown in Figure 2.8. To construct a bar graph, we mark the classes, as intervals, on the horizontal axis with a little gap between consecutive intervals. The bars represent the frequencies of respective classes.

The frequencies of Table 2.12 can be converted to relative frequencies and percentages the same way as in Table 2.9. Then, a bar graph can be constructed to display the relative frequency or percentage distribution by marking the relative frequencies or percentages, respectively, on the vertical axis.

Table 2.12   **Frequency Distribution of the Number of Vehicles Owned**

Vehicles Owned	Number of Households (f)
0	2
1	18
2	11
3	4
4	3
5	2
	$\Sigma f = 40$

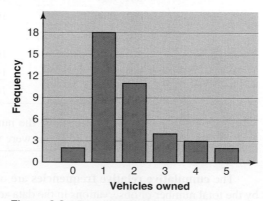

Figure 2.8   Bar graph for Table 2.12.

2.2.6   Cumulative Frequency Distributions

Consider again Example 2–3 of Section 2.2.2 about the values of baseball teams. Suppose we want to know how many baseball teams had values of $1500 million or less in 2015. Such a question can be answered by using a **cumulative frequency distribution**. Each class in a cumulative

> A **cumulative frequency distribution** gives the total number of values that fall below the upper boundary of each class.

frequency distribution table gives the total number of values that fall below a certain value. A cumulative frequency distribution is constructed for quantitative data only.

In a cumulative frequency distribution table, each class has the same lower limit but a different upper limit. Example 2–7 illustrates the procedure for preparing a cumulative frequency distribution.

EXAMPLE 2–7   Values of Baseball Teams, 2015

Constructing a cumulative frequency distribution table.

Using the frequency distribution of Table 2.8, reproduced here, prepare a cumulative frequency distribution for the values of the baseball teams.

Value of a Team (in million $)	Number of Teams (*f*)
601–1050	16
1051–1500	9
1501–1950	1
1951–2400	3
2401–2850	0
2851–3300	1

Solution   Table 2.13 gives the cumulative frequency distribution for the values of the baseball teams. As we can observe, 601 (which is the lower limit of the first class in Table 2.8) is taken as the lower limit of each class in Table 2.13. The upper limits of all classes in Table 2.13 are the same as those in Table 2.8. To obtain the cumulative frequency of a class, we add the frequency of that class in Table 2.8 to the frequencies of all preceding classes. The cumulative frequencies are recorded in the second column of Table 2.13.

Table 2.13   Cumulative Frequency Distribution of Values of Baseball Teams, 2015

Class Limits	Cumulative Frequency
601–1050	16
601–1500	16 + 9 = 25
601–1950	16 + 9 + 1 = 26
601–2400	16 + 9 + 1 + 3 = 29
601–2850	16 + 9 + 1 + 3 + 0 = 29
601–3300	16 + 9 + 1 + 3 + 0 + 1 = 30

From Table 2.13, we can determine the number of observations that fall below the upper limit of each class. For example, 26 teams were valued between $601 and $1950 million.

The **cumulative relative frequencies** are obtained by dividing the cumulative frequencies by the total number of observations in the data set. The **cumulative percentages** are obtained by multiplying the cumulative relative frequencies by 100.

Calculating Cumulative Relative Frequency and Cumulative Percentage

$$\text{Cumulative relative frequency} = \frac{\text{Cumulative frequency of a class}}{\text{Total observations in the data set}}$$

$$\text{Cumulative percentage} = (\text{Cumulative relative frequency}) \cdot 100\%$$

Table 2.14 contains both the cumulative relative frequencies and the cumulative percentages for Table 2.13. We can observe, for example, that 90% of the teams were valued between $601 and $1800 million.

Table 2.14   **Cumulative Relative Frequency and Cumulative Percentage Distributions for Values of Baseball Teams, 2015**

Class Limits	Cumulative Relative Frequency	Cumulative Percentage
601–1050	16/30 = .5333	53.33
601–1500	25/30 = .8333	83.33
601–1950	26/30 = .8667	86.67
601–2400	29/30 = .9667	96.67
601–2850	29/30 = .9667	96.67
601–3300	30/30 = 1.000	100.00

2.2.7 Shapes of Histograms

A histogram can assume any one of a large number of shapes. The most common of these shapes are

1. Symmetric
2. Skewed
3. Uniform or rectangular

A **symmetric histogram** is identical on both sides of its central point. The histograms shown in Figure 2.9 are symmetric around the dashed lines that represent their central points.

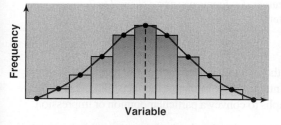

Figure 2.9   Symmetric histograms.

A **skewed histogram** is nonsymmetric. For a skewed histogram, the tail on one side is longer than the tail on the other side. A **skewed-to-the-right histogram** has a longer tail on the right side (see Figure 2.10a). A **skewed-to-the-left histogram** has a longer tail on the left side (see Figure 2.10b).

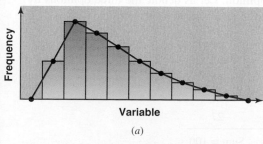

(a)

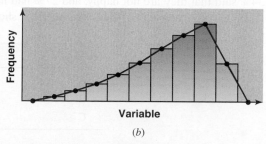

(b)

Figure 2.10   (a) A histogram skewed to the right. (b) A histogram skewed to the left.

A **uniform** or **rectangular histogram** has the same frequency for each class. Figure 2.11 is an illustration of such a case.

Figure 2.11 A histogram with uniform distribution.

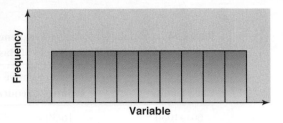

Figures 2.12*a* and 2.12*b* display symmetric frequency curves. Figures 2.12*c* and 2.12*d* show frequency curves skewed to the right and to the left, respectively.

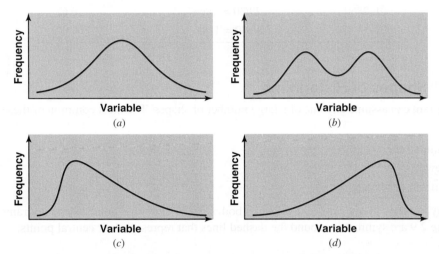

(a) (b) (c) (d)

Figure 2.12 (*a*), (*b*) Symmetric frequency curves. (*c*) Frequency curve skewed to the right. (*d*) Frequency curve skewed to the left.

2.2.8 Truncating Axes

Describing data using graphs gives us insights into the main characteristics of the data. But graphs, unfortunately, can also be used, intentionally or unintentionally, to distort the facts and deceive the reader. The following are two ways to manipulate graphs to convey a particular opinion or impression.

1. *Changing the scale* either on one or on both axes—that is, shortening or stretching one or both of the axes.
2. *Truncating the frequency axis*—that is, starting the frequency axis at a number greater than zero.

Suppose 400 randomly selected adults were asked whether or not they are happy with their jobs. Of them, 156 said that they are happy, 136 said that they are not happy, and 108 had no opinion. Converting these numbers to percentages, 39% of these adults said that they are happy, 34% said that they are not happy, and 27% had no opinion. Let us denote the three opinions by A, B, and C, respectively. The following table shows the results of this survey.

Opinion	Percentage
A	39
B	34
C	27
	Sum = 100

Now let us make two bar graphs—one showing the complete vertical axis and the second using a truncated vertical axis. Figure 2.13 shows a bar graph with the complete vertical axis. By looking at this bar graph, we can observe that the opinions represented by three categories in fact differ by small percentages. But now look at Figure 2.14 in which the vertical axis has been truncated to start at 25%. By looking at this bar chart, if we do not pay attention to the vertical axis, we may erroneously conclude that the opinions represented by three categories vary by large percentages.

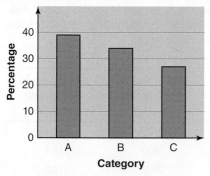

Figure 2.13   Bar graph without truncation of the vertical axis.

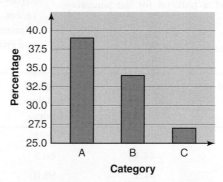

Figure 2.14   Bar graph with truncation of the vertical axis.

When interpreting a graph, we should be very cautious. We should observe carefully whether the frequency axis has been truncated or whether any axis has been unnecessarily shortened or stretched.

EXERCISES

CONCEPTS AND PROCEDURES

2.8   What are the three main decisions that have to be made to group a data set in the form of a frequency distribution table?

2.9   Explain how to find the relative frequencies and percentages of classes from frequencies of classes with the help of an example.

2.10   Explain the following methods for grouping quantitative data into classes and give one example for each.

   a. writing classes using limits
   b. the less than method
   c. grouping data using single-valued classes

APPLICATIONS

2.11   The following data give the number of text messages sent on 40 randomly selected days during 2015 by a high school student:

32	33	33	34	35	36	37	37	37	37
38	39	40	41	41	42	42	42	43	44
44	45	45	45	47	47	47	47	47	48
48	49	50	50	51	52	53	54	59	61

   a. Construct a frequency distribution table. Take 32 as the lower limit of the first class and 6 as the class width.
   b. Calculate the relative frequency and percentage for each class.
   c. Construct a histogram for the frequency distribution of part a.
   d. On what percentage of these 40 days did this student send 44 or more text messages?

   e. Prepare the cumulative frequency, cumulative relative frequency, and cumulative percentage distributions.

2.12   A data set on money spent on lottery tickets during the past year by 200 households has a lowest value of $1 and a highest value of $983. Suppose we want to group these data into six classes of equal widths.

   a. Assuming that we take the lower limit of the first class as $1 and the width of each class equal to $200, write the class limits for all six classes.
   b. What are the class boundaries and class midpoints?

2.13   The following data give the repair costs (in dollars) for 30 cars randomly selected from a list of cars that were involved in collisions

2300	750	2500	410	555	1576
2460	1795	2108	897	989	1866
2105	335	1344	1159	1236	1395
6108	4995	5891	2309	3950	3950
6655	4900	1320	2901	1925	6896

   a. Construct a frequency distribution table. Take $1 as the lower limit of the first class and $1400 as the width of each class.
   b. Compute the relative frequencies and percentages for all classes.
   c. Draw a histogram and a polygon for the relative frequency distribution.
   d. What are the class boundaries and the width of the fourth class?

2.14   Nixon Corporation manufactures computer monitors. The following data are the numbers of computer monitors produced at the company for a sample of 30 days.

30	34	27	23	33	33	26	25	24	28
21	26	31	22	27	33	27	23	28	21
31	35	34	22	26	25	23	35	31	27

a. Construct a frequency distribution table using the classes 21–23, 24–26, 27–29, 30–32, and 33–35.

b. Calculate the relative frequencies and percentages for all classes.

c. Construct a histogram and a polygon for the percentage distribution.

d. For what percentage of the days is the number of computer monitors produced in the interval 27–29?

Exercises 2.15 through 2.19 are based on the following data

The following table gives the age-adjusted cancer incidence rates (new cases) per 100,000 people for three of the most common types of cancer contracted by both females and males: colon and rectum cancer, lung and bronchus cancer, and non-Hodgkin lymphoma. The rates given are for the District of Columbia and 26 states east of the Mississippi River for the years 2003 to 2007, which are the most recent data available from the American Cancer Society. Age-adjusted rates take into account the percentage of people in different age groups within each state's population.

State	Colon and Rectum (Males)	Colon and Rectum (Females)	Lung and Bronchus (Males)	Lung and Bronchus (Females)	Non-Hodgkin Lymphoma (Males)	Non-Hodgkin Lymphoma (Females)
Alabama	60.8	41.6	106.2	53.4	20.5	13.8
Connecticut	59.4	44.4	80.5	60.3	26.0	18.1
Delaware	61.4	44.0	98.0	70.7	23.9	16.6
D.C.	58.1	47.9	79.4	46.3	22.9	13.4
Florida	53.1	40.4	86.7	59.4	21.5	15.2
Georgia	56.9	41.2	98.8	53.9	21.1	14.3
Illinois	65.6	47.3	91.2	59.4	24.2	16.2
Indiana	61.3	45.2	102.4	63.9	22.9	17.0
Kentucky	67.6	48.9	131.3	78.2	23.5	17.1
Maine	61.6	47.2	99.1	66.6	24.6	18.8
Maryland	54.4	41.3	81.5	57.9	20.9	14.4
Massachusetts	60.5	43.9	82.2	63.1	24.5	16.9
Michigan	57.1	43.4	91.9	62.5	25.7	18.7
Mississippi	63.5	45.9	113.3	55.0	20.4	14.0
New Hampshire	56.0	43.1	82.5	62.4	23.5	18.1
New Jersey	62.6	46.0	78.3	56.3	25.6	17.7
New York	58.4	44.3	78.2	54.3	25.0	17.5
North Carolina	56.0	40.9	101.0	57.6	21.9	15.4
Ohio	60.0	44.5	96.1	59.7	23.1	16.4
Pennsylvania	63.9	47.4	90.0	57.1	25.0	17.5
Rhode Island	61.8	45.7	92.6	61.9	24.9	17.4
South Carolina	58.5	42.8	100.2	53.7	20.8	14.4
Tennessee	57.9	40.8	93.6	54.5	22.8	16.3
Vermont	49.4	42.9	84.5	61.1	23.8	18.3
Virginia	54.2	41.0	88.5	53.8	20.8	13.9
West Virginia	68.0	48.7	116.3	71.3	24.0	17.3
Wisconsin	54.6	42.2	76.8	53.8	25.5	18.7

Source: American Cancer Society, www.cancer.org/downloads/STT/2008CAFFfinalsecured.pdf.

2.15 a. Prepare a frequency distribution table for colon and rectum cancer rates for women using six classes of equal width.

b. Construct the relative frequency and percentage distribution columns.

2.16 a. Prepare a frequency distribution table for colon and rectum cancer rates for men using six classes of equal width.

b. Construct the relative frequency and percentage distribution columns.

2.17 a. Prepare a frequency distribution table for lung and bronchus cancer rates for women.

b. Construct the relative frequency and percentage distribution columns.

c. Draw a histogram and polygon for the relative frequency distribution.

2.18 a. Prepare a frequency distribution table for lung and bronchus cancer rates for men.
b. Construct the relative frequency and percentage distribution columns.
c. Draw a histogram and polygon for the relative frequency distribution.

2.19 a. Prepare a frequency distribution table for non-Hodgkin lymphoma rates for women.
b. Construct the relative frequency and percentage distribution columns.
c. Draw a histogram and polygon for the relative frequency distribution.

2.20 The following table lists the number of strikeouts per game (K/game) for each of the 30 Major League baseball teams during the 2010 regular season. Note that Florida Marlins are now Miami Marlins.

Team	K/game	Team	K/game
Arizona Diamondbacks	9.44	Milwaukee Brewers	7.51
Atlanta Braves	7.04	Minnesota Twins	5.97
Baltimore Orioles	6.52	New York Mets	6.76
Boston Red Sox	7.04	New York Yankees	7.01
Chicago Cubs	7.63	Oakland Athletics	6.55
Chicago White Sox	5.69	Philadelphia Phillies	6.57
Cincinnati Reds	7.52	Pittsburgh Pirates	7.45
Cleveland Indians	7.31	San Diego Padres	7.30
Colorado Rockies	7.86	San Francisco Giants	6.78
Detroit Tigers	7.08	Seattle Mariners	7.31
Florida Marlins	8.49	St. Louis Cardinals	6.34
Houston Astros	6.33	Tampa Bay Rays	7.98
Kansas City Royals	5.59	Texas Rangers	6.09
Los Angeles Angels	6.60	Toronto Blue Jays	7.19
Los Angeles Dodgers	7.31	Washington Nationals	7.53

Source: MLB.com.

a. Construct a frequency distribution table. Take 5.50 as the lower boundary of the first class and .8 as the width of each class.
b. Prepare the relative frequency and percentage distribution columns for the frequency distribution table of part a.

2.21 The following data give the number of turnovers (fumbles and interceptions) by a college football team for each game in the past two seasons.

3	1	1	4	0	4	2	3	0	3	2	3
0	2	3	1	4	1	3	2	4	0	0	2

a. Prepare a frequency distribution table for these data using single-valued classes.
b. Calculate the relative frequencies and percentages for all classes.
c. In how many games did the team commit two or more turnovers?
d. Draw a bar graph for the frequency distribution of part a.

2.22 The following table gives the frequency distribution for the numbers of parking tickets received on the campus of a university during the past week by 200 students.

Number of Tickets	Number of Students
0	69
1	40
2	37
3	26
4	28

Draw two bar graphs for these data, the first without truncating the frequency axis and the second by truncating the frequency axis. In the second case, mark the frequencies on the vertical axis starting with 25. Briefly comment on the two bar graphs.

2.3 | Stem-and-Leaf Displays

In a **stem-and-leaf display** of quantitative data, each value is divided into two portions—a stem and a leaf. The leaves for each stem are shown separately in a display.

Another technique that is used to present quantitative data in condensed form is the **stem-and-leaf display**. An advantage of a stem-and-leaf display over a frequency distribution is that by preparing a stem-and-leaf display we do not lose information on individual observations. A stem-and-leaf display is constructed only for quantitative data.

Example 2–8 describes the procedure for constructing a stem-and-leaf display.

EXAMPLE 2–8   Scores of Students on a Statistics Test

Constructing a stem-and-leaf display for two-digit numbers.

The following are the scores of 30 college students on a statistics test.

75	52	80	96	65	79	71	87	93	95
69	72	81	61	76	86	79	68	50	92
83	84	77	64	71	87	72	92	57	98

Construct a stem-and-leaf display.

Solution To construct a stem-and-leaf display for these scores, we split each score into two parts. The first part contains the first digit of a score, which is called the *stem*. The second part contains the second digit of a score, which is called the *leaf*. Thus, for the score of the first student, which is 75, 7 is the stem and 5 is the leaf. For the score of the second student, which is 52, the stem is 5 and the leaf is 2. We observe from the data that the stems for all scores are 5, 6, 7, 8, and 9 because all these scores lie in the range 50 to 98. To create a stem-and-leaf display, we draw a vertical line and write the stems on the left side of it, arranged in increasing order, as shown in Figure 2.15.

Figure 2.15 Stem-and-leaf display.

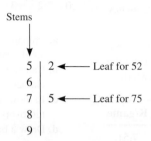

After we have listed the stems, we read the leaves for all scores and record them next to the corresponding stems on the right side of the vertical line. For example, for the first score we write the leaf 5 next to stem 7; for the second score we write the leaf 2 next to stem 5. The recording of these two scores in a stem-and-leaf display is shown in Figure 2.15.

Now, we read all scores and write the leaves on the right side of the vertical line in the rows of corresponding stems. The complete stem-and-leaf display for scores is shown in Figure 2.16.

```
5 | 2 0 7
6 | 5 9 1 8 4
7 | 5 9 1 2 6 9 7 1 2
8 | 0 7 1 6 3 4 7
9 | 6 3 5 2 2 8
```

Figure 2.16 Stem-and-leaf display of test scores.

By looking at the stem-and-leaf display of Figure 2.16, we can observe how the data values are distributed. For example, the stem 7 has the highest frequency, followed by stems 8, 9, 6, and 5.

The leaves for each stem of the stem-and-leaf display of Figure 2.16 are *ranked* (in increasing order) and presented in Figure 2.17.

```
5 | 0 2 7
6 | 1 4 5 8 9
7 | 1 1 2 2 5 6 7 9 9
8 | 0 1 3 4 6 7 7
9 | 2 2 3 5 6 8
```

Figure 2.17 Ranked stem-and-leaf display of test scores.

As already mentioned, one advantage of a stem-and-leaf display is that we do not lose information on individual observations. We can rewrite the individual scores of the 30 college students from the stem-and-leaf display of Figure 2.16 or Figure 2.17. By contrast, the information on individual observations is lost when data are grouped into a frequency table.

EXAMPLE 2–9   Monthly Rents Paid by Households

Constructing a stem-and-leaf display for three- and four-digit numbers.

The following data give the monthly rents paid by a sample of 30 households selected from a small town.

880	1081	721	1075	1023	775	1235	750	965	960
1210	985	1231	932	850	825	1000	915	1191	1035
1151	630	1175	952	1100	1140	750	1140	1370	1280

Construct a stem-and-leaf display for these data.

Solution   Each of the values in the data set contains either three or four digits. We will take the first digit for three-digit numbers and the first two digits for four-digit numbers as stems. Then we will use the last two digits of each number as a leaf. Thus for the first value, which is 880, the stem is 8 and the leaf is 80. The stems for the entire data set are 6, 7, 8, 9, 10, 11, 12, and 13. They are recorded on the left side of the vertical line in Figure 2.18. The leaves for the numbers are recorded on the right side.

```
 6 | 30
 7 | 21 75 50 50
 8 | 80 50 25
 9 | 65 60 85 32 15 52
10 | 81 75 23 00 35
11 | 91 51 75 00 40 40
12 | 35 10 31 80
13 | 70
```

Figure 2.18   Stem-and-leaf display of rents.

Sometimes a data set may contain too many stems, with each stem containing only a few leaves. In such cases, we may want to condense the stem-and-leaf display by *grouping the stems*. Example 2–10 describes this procedure.

EXAMPLE 2–10   Number of Hours Spent Working on Computers by Students

Preparing a grouped stem-and-leaf display.

The following stem-and-leaf display is prepared for the number of hours that 25 students spent working on computers during the past month.

```
0 | 6
1 | 1 7 9
2 | 2 6
3 | 2 4 7 8
4 | 1 5 6 9 9
5 | 3 6 8
6 | 2 4 4 5 7
7 |
8 | 5 6
```

Prepare a new stem-and-leaf display by grouping the stems.

Mark Harmel/Stone/Getty Images

Solution   To condense the given stem-and-leaf display, we can combine the first three rows, the middle three rows, and the last three rows, thus getting the stems 0–2, 3–5, and 6–8. The leaves for each stem of a group are separated by an asterisk (*), as shown in Figure 2.19. Thus, the leaf 6 in the first row corresponds to stem 0; the leaves 1, 7, and 9 correspond to stem 1; and leaves 2 and 6 belong to stem 2.

```
0–2 | 6 * 1 7 9 * 2 6
3–5 | 2 4 7 8 * 1 5 6 9 9 * 3 6 8
6–8 | 2 4 4 5 7 * * 5 6
```

Figure 2.19 Grouped stem-and-leaf display.

If a stem does not contain a leaf, this is indicated in the grouped stem-and-leaf display by two consecutive asterisks. For example, in the stem-and-leaf display of Figure 2.19, there is no leaf for 7; that is, there is no number in the 70s. Hence, in Figure 2.19, we have two asterisks after the leaves for 6 and before the leaves for 8.

Some data sets produce stem-and-leaf displays that have a small number of stems relative to the number of observations in the data set and have too many leaves for each stem. In such cases, it is very difficult to determine if the distribution is symmetric or skewed, as well as other characteristics of the distribution that will be introduced in later chapters. In such a situation, we can create a stem-and-leaf display with *split stems*. To do this, each stem is split into two or five parts. Whenever the stems are split into two parts, any observation having a leaf with a value of 0, 1, 2, 3, or 4 is placed in the first split stem, while the leaves 5, 6, 7, 8, and 9 are placed in the second split stem. Sometimes we can split a stem into five parts if there are too many leaves for one stem. Whenever a stem is split into five parts, leaves with values of 0 and 1 are placed next to the first part of the split stem, leaves with values of 2 and 3 are placed next to the second part of the split stem, and so on. The stem-and-leaf display of Example 2–11 shows this procedure.

EXAMPLE 2–11

Stem-and-leaf display with split stems.

Consider the following stem-and-leaf display, which has only two stems. Using the split stem procedure, rewrite this stem-and-leaf display.

```
3 | 1 1 2 3 3 3 4 4 7 8 9 9 9
4 | 0 0 0 1 1 1 1 1 2 2 2 2 3 3 6 6 7
```

Solution To prepare a split stem-and-leaf display, let us split the two stems, 3 and 4, into two parts each as shown in Figure 2.20. The first part of each stem contains leaves from 0 to 4, and the second part of each stem contains leaves from 5 to 9.

```
3 | 1 1 2 3 3 3 4 4
3 | 7 8 9 9 9
4 | 0 0 0 1 1 1 1 1 2 2 2 2 3 3
4 | 6 6 7
```

Figure 2.20 Split stem-and-leaf display.

In the stem-and-leaf display of Figure 2.20, the first part of stem 4 has a substantial number of leaves. So, if we decide to split stems into five parts, the new stem-and-leaf display will look as shown in Figure 2.21.

```
3 | 1 1
3 | 2 3 3 3
3 | 4 4
3 | 7
3 | 8 9 9 9
4 | 0 0 0 1 1 1 1 1
4 | 2 2 2 2 3 3
4 |
4 | 6 6 7
```

Figure 2.21 Split stem-and-leaf display.

There are two important properties to note in the split stem-and-leaf display of Figure 2.21. The third part of split stem 4 does not have any leaves. This implies that there are no observations in the data set having a value of 44 or 45. Since there are observations with values larger than 45, we need to leave an empty part of split stem 4 that corresponds to 44 and 45. Also, there are no observations with values of 48 or 49. However, since there are no values larger than 47 in the data, we do not have to write an empty split stem 4 after the largest value.

EXERCISES

CONCEPTS AND PROCEDURES

2.23 Illustrate how to construct a stem-and-leaf plot for a given data set using an example.

2.24 When is it more appropriate to use a stem-and-leaf display for grouping a data set rather than a frequency distribution? Give one example.

2.25 Write the data set that is represented by the following stem-and-leaf display.

2–3	18 45 56 * 29 67 83 97
4–5	04 27 33 71 * 23 37 51 63 81 92
6–8	22 36 47 55 78 89 * * 10 41

APPLICATIONS

2.26 The following data give the time (in minutes) that each of 20 students waited in line at their bookstore to pay for their textbooks in the beginning of Spring 2009 semester. (*Note:* To prepare a stem-and-leaf display, each number in this data set can be written as a two-digit number. For example, 8 can be written as 08, for which the stem is 0 and the leaf is 8.)

15	8	23	21	4	17	31	22	31	6
5	6	14	17	16	25	27	3	31	8

Construct a stem-and-leaf display for these data. Arrange the leaves for each stem in increasing order.

2.27 The following data give the money (in dollars) spent on text-books by 35 students during the 2009–10 academic year.

565	728	470	620	345	368	610	765	550
845	530	705	490	258	320	505	457	787
617	721	635	438	575	702	538	720	460
540	890	560	570	706	430	268	638	

a. Prepare a stem-and-leaf display for these data using the last two digits as leaves.
b. Condense the stem-and-leaf display by grouping the stems as 2–4, 5–6, and 7–8.

2.28 Refer to the data below on the birth rates for all 56 counties of Montana for the year 2008. The numbers are rounded to the nearest unit.

10	22	16	12	8	3	15	8
14	9	11	9	15	10	14	15
18	23	5	6	20	8	10	15
14	10	10	10	9	9	14	12
11	11	2	10	11	5	7	9
10	14	20	19	11	7	12	12
11	9	8	7	10	10	6	15

a. Prepare a stem-and-leaf display for the data. Arrange the leaves for each stem in increasing order.
b. Prepare a split stem-and-leaf display for the data. Split each stem into two parts. The first part should contains the leaves 0, 1, 2, 3, and 4, and the second part should contains the leaves 5, 6, 7, 8, and 9.
c. Which display (the one in part a or the one in part b) provides a better representation of the features of the distribution? Explain why you believe this.

2.29 These data give the times (in minutes) taken to commute from home to work for 20 workers.

10	50	65	33	48	5	11	23	39	26
26	32	17	7	15	19	29	43	21	22

Construct a stem-and-leaf display for these data. Arrange the leaves for each stem in increasing order.

2.30 The following data give the times served (in months) by 35 prison inmates who were released recently.

37	6	20	5	25	30	24	10	12	20
24	8	26	15	13	22	72	80	96	33
84	86	70	40	92	36	28	90	36	32
72	45	38	18	9					

a. Prepare a stem-and-leaf display for these data.
b. Condense the stem-and-leaf display by grouping the stems as 0–2, 3–5, and 6–9.

2.4 | Dotplots

One of the simplest methods for graphing and understanding quantitative data is to create a dotplot. As with most graphs, statistical software should be used to make a dotplot for large data sets. However, Example 2–12 demonstrates how to create a dotplot by hand.

Values that are very small or very large relative to the majority of the values in a data set are called **outliers** or **extreme values**.

Dotplots can help us detect **outliers** (also called **extreme values**) in a data set. Outliers are the values that are extremely large or extremely small with respect to the rest of the data values.

| EXAMPLE 2–12 | Ages of Students in a Night Class |

Creating a dotplot.

A statistics class that meets once a week at night from 7:00 PM to 9:45 PM has 33 students. The following data give the ages (in years) of these students. Create a dotplot for these data.

34	21	49	37	23	22	33	23	21	20	19
33	23	38	32	31	22	20	24	27	33	19
23	21	31	31	22	20	34	21	33	27	21

Solution To make a dotplot, we perform the following steps.

Step 1. The minimum and maximum values in this data set are 19 and 49 years, respectively. First, we draw a horizontal line (let us call this the *numbers line*) with numbers that cover the given data as shown in Figure 2.22. Note that the numbers line in Figure 2.22 shows the values from 19 to 49.

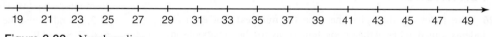

Figure 2.22 Numbers line.

Step 2. Next we place a dot above the value on the numbers line that represents each of the ages listed above. For example, the age of the first student is 34 years. So, we place a dot above 34 on the numbers line as shown in Figure 2.23. If there are two or more observations with the same value, we stack dots above each other to represent those values. For example, as shown in the data on ages, two students are 19 years old. We stack two dots (one for each student) above 19 on the numbers line, as shown in Figure 2.23. After all the dots are placed, Figure 2.23 gives the complete dotplot.

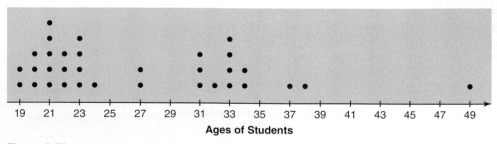

Ages of Students

Figure 2.23 Dotplot for ages of students in a statistics class.

As we examine the dotplot of Figure 2.23, we notice that there are two clusters (groups) of data. Eighteen of the 33 students (which is almost 55%) are 19 to 24 years old, and 10 of the 33 students (which is about 30%) are 31 to 34 years old. There is one student who is 49 years old and is an outlier.

EXERCISES

CONCEPTS AND PROCEDURES

2.31 Illustrate how to construct a dotplot for a given data set using an example.

2.32 What are the benefits of grouping data in the form of a dotplot?

2.33 Prepare a dotplot for the data set below.

1	2	0	5	1	1	3	2	0	5
2	1	2	1	2	0	1	3	1	2

APPLICATIONS

2.34 The following data give the number of text messages sent on 40 randomly selected days during 2015 by a high school student.

32	33	33	34	35	36	37	37	37	37
38	39	40	41	41	42	42	42	43	44
44	45	45	45	47	47	47	47	47	48
48	49	50	50	51	52	53	54	59	61

Create a dotplot for these data.

2.35 The following data give the number of orders received for a sample of 30 hours at the Timesaver Mail Order Company.

34	44	31	52	41	47	38	35	32	39
28	24	46	41	49	53	57	33	27	37
30	27	45	38	34	46	36	30	47	50

Create a dotplot for these data.

2.36 The following table gives the number of times each of the listed players of the 2008 Philadelphia Phillies baseball team was hit by a pitch (HBP). The list includes all players with at least 150 at-bats.

Player	HBP	Player	HBP
C. Utley	27	R. Howard	3
C. Coste	10	P. Burrell	1
S. Victorino	7	G. Dobbs	1
J. Rollins	5	G. Jenkins	1
C. Ruiz	4	P. Feliz	0
J. Werth	4	T. Iguchi	0
E. Bruntlett	3		

Source: Major League Baseball, 2008.

Create a dotplot for these data. Mention any clusters and/or outliers you observe.

2.37 In basketball, a *double-double* occurs when a player accumulates double-digit numbers in any two of five statistical categories—points, rebounds, assists, steals, and blocked shots—in one game. The following table gives the number of times each player of the Miami Heat basketball team scored a double-double (DD) during the 2010–2011 regular season.

Player	DD	Player	DD
Carlos Arroyo	0	Juwan Howard	0
Chris Bosh	28	LeBron James	31
Dwyane Wade	10	Mario Chalmers	0
Eddie House	0	Mike Bibby	0
Erick Dampier	1	Mike Miller	3
James Jones	0	Udonis Haslem	4
Joel Anthony	0	Zydrunas Ilgauskas	3

Source: espn.com.

Create a dotplot for these data. Mention any clusters (groups) and/or outliers you observe.

USES AND MISUSES...

GRAPHICALLY SPEAKING...

Imagine a high school in Middle America, if you will. We shall call it Central High School. In Central High there are a fair number of very talented athletes who play soccer, baseball, football, and basketball. At the annual fundraiser for the school, the cheer squad sells advance tickets to the home games so that the community can be entertained by the local school athletes.

As it happens, the cheer squad was able to sell a total of 175 tickets at this year's fundraising event. The ticket sales for the various sports were as follows:

Sport	Number of Tickets Sold
Soccer	23
Baseball	30
Football	63
Basketball	59

In order to show the results of the ticket sales to the Central High community, some of the students got together and decided to make an attractive graph to display the ticket sales in the local paper. As dutiful students, they carefully investigated types of charts that might work, and ultimately decided to arrange the graph by sport on the horizontal-axis and the number of tickets sold on the vertical-axis. To make the graph a bit more vivid, they embellished it using a *pictogram* (that is, a graph made with pictures). They used the following graph to publish in the local paper.

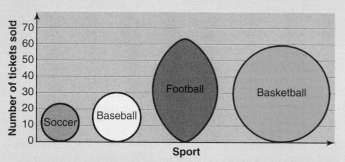

The graph was very well received by the readers of the local paper, and was very attractive to the eye. Further, the football and basketball teams and their supporters were happy to see just how much more popular their respective sports were compared to soccer and baseball. However, there is a major problem with the graph. This graph had the intent of showing the number of tickets sold. This means that we should focus on the *height* of the balls for each sport, which, when read horizontally across to the vertical-axis, do in fact match the number of tickets reported to have been sold in the table above. But the eye is easily deceived. What is immediately clear is that we tend to focus on the relative sizes of the balls, not the heights that they represent. In that context, it seems that a huge number of basketball tickets were sold compared to the other sports because the eye focuses immediately on the circumference or the area of the circle represented by the basketball, which appears much larger than the other balls. In fact, comparing soccer to basketball, the area of the basketball in the graph is 6.6 times larger than that of the soccer ball, whereas the tickets sold for basketball were about 2.6 times more than the soccer tickets sold.

A more conservative approach would have been to use a standard *bar chart*. The following bar graph was constructed using the same information as in the pictogram above.

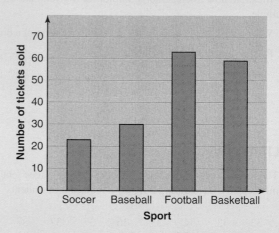

Obviously, this graph does not have the same eye appeal as the one reported in the newspaper; however, it is much clearer to an observer. It is obvious from this graph that the number of soccer tickets and the number of baseball tickets were quite close, as were the number of football and the number of basketball tickets. It is also clear that the number of football tickets is about three times larger than the number of soccer tickets and that the number of football tickets is about two times larger than the number of baseball tickets. These relationships were much more difficult to discern in the pictogram.

In 1954, Darrell Huff wrote a fascinating and entertaining book titled *How to Lie with Statistics* (Huff, Darrell, *How to Lie with Statistics*, 1954, New York: W. W. Norton). That book contains many illustrations of how graphs can be used to deceive—either intentionally or unintentionally—those doing a visual interpretation of data. As Mr. Huff states in his book, "The crooks already know these tricks; honest men must learn them in self-defense." In that regard, it is sometimes useful to learn how *not* to do something in order to understand the importance of doing it correctly.

GLOSSARY

Bar graph A graph made of bars whose heights represent the frequencies of respective categories.

Class An interval that includes all the values in a (quantitative) data set that fall within two numbers, the lower and upper limits of the class.

Class boundary The lower and upper numbers of a class interval in less-than method.

Class frequency The number of values in a data set that belong to a certain class.

Class midpoint or mark The class midpoint or mark is obtained by dividing the sum of the lower and upper limits (or boundaries) of a class by 2.

Class width or size The difference between the two boundaries of a class or the difference between the lower limits of two consecutive classes.

Cumulative frequency The frequency of a class that includes all values in a data set that fall below the upper boundary or limit of that class.

Cumulative frequency distribution A table that lists the total number of values that fall below the upper boundary or limit of each class.

Cumulative percentage The cumulative relative frequency multiplied by 100.

Cumulative relative frequency The cumulative frequency of a class divided by the total number of observations.

Frequency distribution A table that lists all the categories or classes and the number of values that belong to each of these categories or classes.

Grouped data A data set presented in the form of a frequency distribution.

Histogram   A graph in which classes are marked on the horizontal axis and frequencies, relative frequencies, or percentages are marked on the vertical axis. The frequencies, relative frequencies, or percentages of various classes are represented by the heights of bars that are drawn adjacent to each other.

Outliers or Extreme values   Values that are very small or very large relative to the majority of the values in a data set.

Pareto chart   A bar graph in which bars are arranged in decreasing order of heights.

Percentage   The percentage for a class or category is obtained by multiplying the relative frequency of that class or category by 100.

Pie chart   A circle divided into portions that represent the relative frequencies or percentages of different categories or classes.

Polygon   A graph formed by joining the midpoints of the tops of successive bars in a histogram by straight lines.

Raw data   Data recorded in the sequence in which they are collected and before they are processed.

Relative frequency   The frequency of a class or category divided by the sum of all frequencies.

Skewed-to-the-left histogram   A histogram with a longer tail on the left side.

Skewed-to-the-right histogram   A histogram with a longer tail on the right side.

Stem-and-leaf display   A display of data in which each value is divided into two portions—a stem and a leaf.

Symmetric histogram   A histogram that is identical on both sides of its central point.

Ungrouped data   Data containing information on each member of a sample or population individually.

Uniform or rectangular histogram   A histogram with the same frequency for all classes.

SUPPLEMENTARY EXERCISES

2.38 A whatjapanthinks.com survey asked residents of Japan to name their favorite pizza topping. The possible responses included the following choices: pig-based meats, for example, bacon or ham (PI); seafood, for example, tuna, crab, or cod roe (S); vegetables and fruits (V); poultry (PO); beef (B); and cheese (C). The following data represent the responses of a random sample of 36 people.

V	PI	B	PI	V	PO	S	PI	V	S	V	S
PI	S	V	V	V	PI	S	S	V	PI	C	V
V	V	C	V	S	PO	V	PI	S	PI	PO	PI

    a. Prepare a frequency distribution table.
    b. Calculate the relative frequencies and percentages for all categories.
    c. What percentage of the respondents mentioned *vegetables and fruits*, *poultry*, or *cheese*?
    d. Make a Pareto chart for the relative frequency distribution.

2.39 In an April 18, 2010 Pew Research Center report entitled *Distrust, Discontent, Anger and Partisan Rancor—The People and Their Government*, 2505 U.S. adults were asked, "Which is the bigger problem with the Federal government?" Of the respondents, 38% said that Federal government has the wrong priorities (W), 50% said that it runs programs inefficiently (I), and 12% had no opinion or did not know (N). Recently 44 randomly selected people were asked the same question, and their responses were as follows:

I	I	W	I	W	W	W	I	I	W	W
N	I	I	W	N	N	W	I	W	W	I
I	W	N	I	I	W	W	I	N	W	W
W	W	W	I	W	I	W	W	I	N	W

    a. Prepare a frequency distribution for these data.
    b. Calculate the relative frequencies and percentages for all classes.
    c. Draw a bar graph for the frequency distribution and a pie chart for the percentage distribution.
    d. What percentage of these respondents mentioned "Federal government has the wrong priorities" as the bigger problem?

2.40 A local gas station collected data from the day's receipts, recording the gallons of gasoline each customer purchased. The following table lists the frequency distribution of the gallons of gas purchased by all customers on this one day at this gas station.

Gallons of Gas	Number of Customers
0 to less than 4	31
4 to less than 8	78
8 to less than 12	49
12 to less than 16	81
16 to less than 20	117
20 to less than 24	13

    a. How many customers were served on this day at this gas station?
    b. Find the class midpoints. Do all of the classes have the same width? If so, what is this width? If not, what are the different class widths?
    c. Prepare the relative frequency and percentage distribution columns.
    d. What percentage of the customers purchased 12 gallons or more?
    e. Explain why you cannot determine exactly how many customers purchased 10 gallons or less.
    f. Prepare the cumulative frequency, cumulative relative frequency, and cumulative percentage distributions using the given table.

2.41 The following data give the amounts spent on video rentals (in dollars) during 2016 by 36 households randomly selected from those who rented videos in 2016.

595	24	6	100	100	40	622	405	90
55	155	760	405	90	205	70	180	88
808	100	240	127	83	310	350	160	22
111	70	15	22	101	83	121	18	436

    a. Construct a frequency distribution table. Take $1 as the lower limit of the first class and $200 as the width of each class.
    b. Calculate the relative frequencies and percentages for all classes.
    c. What percentage of the households in this sample spent more than $400 on video rentals in 2016?

2.42 The following data give the amounts (in dollars) spent on refreshments by 30 spectators randomly selected from those who patronized the concession stands at a recent Major League Baseball game.

4.95	27.99	8.00	5.80	4.50	2.99	4.85	6.00
9.00	15.75	9.50	3.05	5.65	21.00	16.60	18.00
21.77	12.35	7.75	10.45	3.85	28.45	8.35	17.70
19.50	11.65	11.45	3.00	6.55	16.50		

a. Construct a frequency distribution table using the *less-than* method to write classes. Take $0 as the lower boundary of the first class and $6 as the width of each class.
b. Calculate the relative frequencies, and percentages for all classes.
c. Draw a histogram for the frequency distribution.
d. Prepare the cumulative frequency, cumulative relative frequency, and cumulative percentage distributions.

2.43 The following data give the times served (in months) by 35 prison inmates who were released recently.

37	4	20	5	25	28	24	10	12	20
24	8	26	15	13	22	72	80	96	33
84	86	70	40	92	36	24	90	36	30
72	45	38	10	9					

Prepare a stem-and-leaf display for these data.

2.44 Following are the total yards gained rushing during the 2009 season by 16 running backs of 16 college football teams.

781	921	1133	1024	848	775	800	834
1009	1275	857	838	1145	967	995	1072

Prepare a stem-and-leaf display. Arrange the leaves for each stem in increasing order.

2.45 The National Highway Traffic Safety Administration collects data on fatal accidents that occur on roads in the United States. The following data represent the number of vehicle fatalities for 39 counties in South Carolina for 2012 (www-fars.nhtsa.dot.gov/States).

4	48	9	9	31	22	26	17
20	12	6	5	14	9	16	27
3	33	9	20	68	13	51	13
48	23	12	13	10	15	8	1
2	4	17	16	6	52	50	

Make a dotplot for these data.

2.46 The following data give the times (in minutes) taken by 50 students to complete a statistics examination that was given a maximum time of 75 minutes to finish.

41	28	45	60	53	69	70	50	63	68
37	44	42	38	74	53	66	65	52	64
26	45	66	35	43	44	39	55	64	54
38	52	58	72	67	65	43	65	68	27
64	49	71	75	45	69	56	73	53	72

Create a dotplot for these data.

ADVANCED EXERCISES

2.47 The following frequency distribution table gives the age distribution of drivers who were at fault in auto accidents that occurred during a 1-week period in a city.

Age (years)	f
18 to less than 20	7
20 to less than 25	12
25 to less than 30	18
30 to less than 40	14
40 to less than 50	15
50 to less than 60	16
60 and over	35

a. Draw a relative frequency histogram for this table.
b. In what way(s) is this histogram misleading?
c. How can you change the frequency distribution so that the resulting histogram gives a clearer picture?

2.48 Refer to the data presented in Exercise 2.47. Note that there were 50% more accidents in the *25 to less than 30* age group than in the *20 to less than 25* age group. Does this suggest that the older group of drivers in this city is more accident-prone than the younger group? What other explanation might account for the difference in accident rates?

2.49 Stem-and-leaf displays can be used to compare distributions for two groups using a back-to-back stem-and-leaf display. In such a display, one group is shown on the left side of the stems, and the other group is shown on the right side. When the leaves are ordered, the leaves increase as one moves away from the stems. The following stem-and-leaf display shows the money earned per tournament entered for the top 30 money winners in the 2008–09 Professional Bowlers Association men's tour and for the top 21 money winners in the 2008–09 Professional Bowlers Association women's tour.

Women's		Men's
8	0	
8871	1	
65544330	2	334456899
840	3	03344678
52	4	011237888
21	5	9
	6	9
5	7	
	8	7
	9	5

The leaf unit for this display is 100. In other words, the data used represent the earnings in hundreds of dollars. For example, for the women's tour, the first number is 08, which is actually 800. The second number is 11, which actually is 1100.

a. Do the top money winners, as a group, on one tour (men's or women's) tend to make more money per tournament played than on the other tour? Explain how you can come to this conclusion using the stem-and-leaf display.

b. What would be a typical earnings level amount per tournament played for each of the two tours?

c. Do the data appear to have similar spreads for the two tours? Explain how you can come to this conclusion using the stem-and-leaf display.

d. Does either of the tours appears to have any outliers? If so, what are the earnings levels for these players?

2.50 The pie chart in Figure 2.24 shows the percentage distribution of ages (i.e., the percentages of all prostate cancer patients falling in various age groups) for men who were recently diagnosed with prostate cancer.

a. Are more or fewer than 50% of these patients in their 50s? How can you tell?

b. Are more or fewer than 75% of these patients in their 50s and 60s? How can you tell?

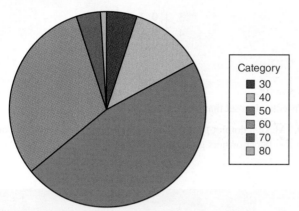

Figure 2.24 Pie chart of age groups.

c. A reporter looks at this pie chart and says, "Look at all these 50-year-old men who are getting prostate cancer. This is a major concern for a man once he turns 50." Explain why the reporter cannot necessarily conclude from this pie chart that there are a lot of 50-year-old men with prostate cancer. Can you think of any other way to present these cancer cases (both graph and variable) to determine if the reporter's claim is valid?

2.51 Back-to-back stem-and-leaf displays can be used to compare the distribution of a variable for two different groups. Consider the following data, which give the alcohol content by volume (%) for different beers produced by the Flying Dog Brewery and the Sierra Nevada Brewery.

Flying Dog Brewery:

| 4.7 | 4.7 | 4.8 | 5.1 | 5.5 | 5.5 | 5.6 | 6.0 | 7.1 |
| 7.4 | 7.8 | 8.3 | 8.3 | 9.2 | 9.9 | 10.2 | 11.5 |

Sierra Nevada Brewery:

| 4.4 | 5.0 | 5.0 | 5.6 | 5.6 | 5.8 | 5.9 | 5.9 | 6.7 |
| 6.8 | 6.9 | 7.0 | 9.6 |

a. Create a back-to-back stem-and-leaf display of these data. Place the Flying Dog Brewery data to the left of the stems.

b. What would you consider to be a typical alcohol content of the beers made by each of the two breweries?

c. Does one brewery tend to have higher alcohol content in its beers than the other brewery? If so, which one? Explain how you reach this conclusion by using the stem-and-leaf display.

d. Do the alcohol content distributions for the two breweries appear to have the same levels of variability? Explain how you reach this conclusion by using the stem-and-leaf display.

SELF-REVIEW TEST

1. Briefly explain the difference between ungrouped and grouped data and give one example of each type.

2. The following table gives the frequency distribution of times (to the nearest hour) that 90 fans spent waiting in line to buy tickets to a rock concert.

Waiting Time (hours)	Frequency
0 to 6	5
7 to 13	27
14 to 20	30
21 to 27	20
28 to 34	8

Circle the correct answer in each of the following statements, which are based on this table.

a. The number of classes in the table is 5, 30, 90.

b. The class width is 6, 7, 34.

c. The midpoint of the third class is 16.5, 17, 17.5.

d. The lower boundary of the second class is 6.5, 7, 7.5.

e. The upper limit of the second class is 12.5, 13, 13.5.

f. The sample size is 5, 90, 11.

g. The relative frequency of the second class is .22, .41, .30.

3. Briefly explain and illustrate with the help of graphs a symmetric histogram, a histogram skewed to the right, and a histogram skewed to the left.

4. Thirty-six randomly selected senior citizens were asked if their net worth is more than or less than $200,000. Their responses are given below, where M stands for more, L represents less, N means do not know or do not want to tell.

M	M	L	L	L	N	M	N	L	L	M	M
L	L	N	L	L	M	M	L	N	L	L	L
M	M	L	M	L	L	M	M	L	N	L	N

a. Make a frequency distribution for these 36 responses.

b. Using the frequency distribution of part a, prepare the relative frequency and percentage distributions.

c. What percentage of these senior citizens claim to have net worth less than $200,000?

d. Make a bar graph for the frequency distribution.

e. Make a Pareto chart for the frequency distribution.

f. Draw a pie chart for the percentage distribution.

5. Forty-eight randomly selected car owners were asked about their typical monthly expense on gas. The following data show the responses of these 48 car owners.

$210	160	430	255	176	135	221	359	380	405	391	477
333	209	267	121	357	87	167	95	347	487	302	545
351	256	492	277	245	367	159	187	253	287	456	64
76	166	304	444	193	479	188	148	53	327	234	110

a. Construct a frequency distribution table. Use the classes 50–149, 150–249, 250–349, 350–449, and 450–549.

b. Calculate the relative frequency and percentage for each class.

c. Construct a histogram for the percentage distribution made in part b.

d. What percentage of the car owners in this sample spend $350 or more on gas per month?

e. Prepare the cumulative frequency, cumulative relative frequency, and cumulative percentage distributions using the table of part a.

6. Thirty customers from all customers who shopped at a large grocery store during a given week were randomly selected and their shopping expenses was noted. The following data show the expenses (in dollars) of these 30 customers.

89.20	145.23	191.54	45.36	67.98	123.67
187.57	56.43	102.45	158.76	134.07	67.49
212.60	165.75	46.50	111.25	64.54	23.67
135.09	193.46	87.65	133.76	156.28	88.64
65.90	120.50	55.45	91.54	153.20	44.39

a. Prepare a frequency distribution table for these data. Use the classes as 20 to less than 60, 60 to less than 100, 100 to less than 140, 140 to less than 180, and 180 to less than 220.

b. What is the width of each class in part a?

c. Using the frequency distribution of part a, prepare the relative frequency and percentage distributions.

d. What percentage of these customers spent less than $140 at this grocery store?

e. Make a histogram for the frequency distribution.

f. Prepare the cumulative frequency, cumulative relative frequency, and cumulative percentage distributions using the table of part a.

7. Construct a stem-and-leaf display for the following data, which give the times (in minutes) that 24 customers spent waiting to speak to a customer service representative when they called about problems with their Internet service provider.

12	15	7	29	32	16	10	14	17	8	19	21
4	14	22	25	18	6	22	16	13	16	12	20

8. Consider this stem-and-leaf display:

3	0 3 7
4	2 4 6 7 9
5	1 3 3 6
6	0 7 7
7	1 9

Write the data set that was used to construct this display.

9. Make a dotplot for the data on typical monthly expenses on gas for the 48 car owners given in Problem 5.

MINI-PROJECTS

Note: The Mini-Projects are located on the text's Web site, www.wiley.com/college/mann.

DECIDE FOR YOURSELF

Note: The Decide for Yourself feature is located on the text's Web site, www.wiley.com/college/mann.

TECHNOLOGY INSTRUCTIONS   CHAPTER 2

Note: Complete TI-84, Minitab, and Excel manuals are available for download at the textbook's Web site, www.wiley.com/college/mann.

TI-84 Color/TI-84

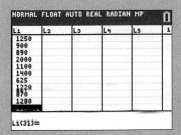

Screen 2.1

The TI-84 Color Technology Instructions feature of this text is written for the TI-84 Plus C color graphing calculator running the 4.0 operating system. Some screens, menus, and functions will be slightly different in older operating systems. The TI-84 and TI-84 Plus can perform all of the same functions but will not have the "Color" option referenced in some of the menus.

TI-84 calculators do not have the options/capabilities to make the bar graph, pie chart, stem-and-leaf display, and dotplot.

Creating a Frequency Histogram for Example 2–3 of the Text

1. Enter the data from Example 2–3 of the text into **L1**. (See **Screen 2.1**.)

2. Select **2nd > Y =** (the **STAT PLOT** menu).

3. If more than one plot is turned on, select **PlotsOff** and then press **ENTER** to turn off the plots.

4. From the **STAT PLOT** menu, select **Plot 1**. Use the following settings (see **Screen 2.2**):

   - Select **On** to turn the plot on.

   - At the **Type** prompt, select the third icon (histogram).

   - At the **Xlist** prompt, enter L_1 by pressing **2nd > 1**.

   - At the **Freq** prompt, enter 1.

   - Select **BLUE** at the **Color** prompt.

   (Skip this step if you do not have the TI-84 Color calculator.)

Screen 2.2

5. Select **ZOOM > ZoomStat** to display the histogram. This function will automatically choose window settings and class boundaries for the histogram. (See **Screen 2.3**.)

6. Press **TRACE** and use the left and right arrow keys to see the class boundaries and frequencies for each class. (See **Screen 2.4**.)

7. To manually change the window settings and the class boundaries, press **WINDOW** and use the following settings:

   - Type 600 at the **Xmin** prompt. **Xmin** is the extreme left value of the window and the lower boundary value of the first class.

   - Type 2700 at the **Xmax** prompt. **Xmax** is the extreme right value of the window.

   - Type 300 at the **Xscl** prompt. **Xscl** is the class width.

   - Type -3 at the **Ymin** prompt. **Ymin** is the extreme bottom value of the window.

   - Type 15 at the **Ymax** prompt. **Ymax** is the extreme top value of the window.

   - Type 3 at the **Yscl** prompt. **Yscl** is the distance between tick marks on the y-axis.

   - Press **GRAPH** to see the histogram with the new settings.

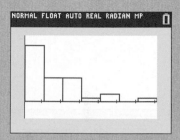

Screen 2.3

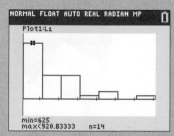

Screen 2.4

Minitab

The Minitab Technology Instructions feature of this text is written for Minitab version 17. Some screens, menus, and functions will be slightly different in older versions of Minitab.

Creating a Bar Graph for Example 2–1 of the Text

Bar Graph for Ungrouped Data

1. Enter the 30 data values of Example 2–1 of the text into column C1.

2. Select **Graph > Bar Chart**.

3. Use the following settings in the resulting dialog box:

   - Select **Counts of unique values**, select **Simple**, and click **OK**.

4. Use the following settings in the new dialog box:

   - Type C1 in the **Categorical variables** box, and click **OK**.

5. The bar graph will appear in a new window. (See **Screen 2.5**.)

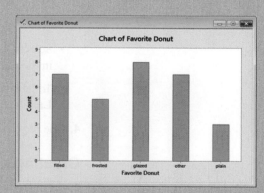

Screen 2.5

Bar Graph for Grouped Data

1. Enter the grouped data from Example 2–1 as shown in Table 2.4 of the text. Put the *Donut Variety* into column C1 and the *Frequency* into column C2.

2. Select **Graph > Bar Chart**.

3. Use the following settings in the resulting dialog box:

   - Select **Values from a table**, select **Simple**, and click **OK**.

4. Use the following settings in the new dialog box:

   - Type C2 in the **Graph variables** box, C1 in the **Categorical variable** box, and click **OK**.

5. The bar graph will appear in a new window. (See **Screen 2.5**.)

To make a Pareto chart, after step 4 above, click on **Chart Options** in the same dialog box. Select **Decreasing Y** in the next dialog box. Click **OK** in both dialog boxes. The Pareto chart will appear in a new window.

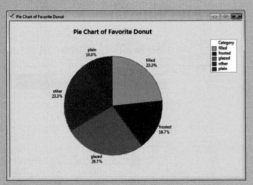

Screen 2.6

Creating a Pie Chart for Example 2–1 of the Text

Pie Chart for Ungrouped Data

1. Enter the 30 data values of Example 2–1 of the text into column C1.

2. Select **Graph > Pie Chart**.

3. Use the following settings in the resulting dialog box:

   - Select **Chart counts of unique values**, and type C1 in the **Categorical variables** box.

   - To label slices of the pie chart with percentages or category names, select **Labels > Slice Labels**. Choose the options you prefer and click **OK**.

4. The pie chart will appear in a new window. (See **Screen 2.6**.)

Pie Chart for Grouped Data

1. Enter the grouped data from Example 2–1 as shown in Table 2.4 of the text. Put the *Donut Variety* into column C1 and the *Frequency* into column C2.

2. Select **Graph > Pie Chart**.

3. Use the following settings in the resulting dialog box:

   - Select **Chart values from a table**, type C1 in the **Categorical variable** box, and type C2 in the **Summary variables** box.

   - To label slices of the pie chart with percentages or category names, select **Labels > Slice Labels**. Choose the options you prefer, and click **OK**.

4. The pie chart will appear in a new window. (See **Screen 2.6**.)

Creating a Frequency Histogram for Example 2–3 of the Text

1. Enter the 30 data values of Example 2–3 of the text into column C1.

2. Select **Graph > Histogram**, select **Simple**, and click **OK**.

3. Use the following settings in the resulting dialog box:

   - Type C1 in the **Graph variables** box.

   - If you wish to add labels, titles, or subtitles to the histogram, select **Labels** and choose the options you prefer, and click **OK**.

4. The histogram will appear in a new window. (See **Screen 2.7**.)

5. To change the class boundaries double-click on any of the bars in the histogram and the **Edit Bars** dialog box will appear. Click on the **Binning** tab. Now choose one of the following two options:

- To set the class midpoints, select **Midpoint** under **Interval Type**, select **Midpoint/Cutpoint positions** under the **Interval Definition** box, and then type the class midpoints (separate each value with a space). Click **OK**.

- To set the class boundaries, select **Cutpoint** under **Interval Type**, select **Midpoint/Cutpoint positions**, and then type the left boundary for each interval and the right endpoint of the last interval (separate each value with a space). Click **OK**.

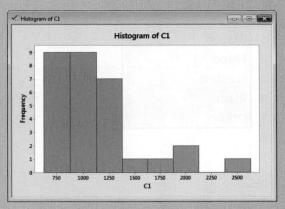

Screen 2.7

Creating a Stem-and-Leaf Display for Example 2–8 of the Text

1. Enter the 30 data values of Example 2–8 of the text into column C1.

2. Select **Graph > Stem-and-Leaf**.

3. Type C1 in the **Graph variables** box and click **OK**.

4. The stem-and-leaf display will appear in the Session window. (See **Screen 2.8**.)

5. If there are too many stems, you can specify an **Increment** for each branch of the stem-and leaf display in the box next to Increment of the above dialog box. For example, the stem-and-leaf display shown in Screen 2.8 has an increment of size 5. In other words, the stem-and-leaf display in Screen 2.8 is a split stem-and-leaf display with each stem split in two.

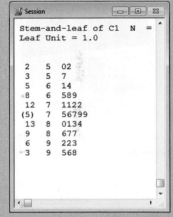

Screen 2.8

Creating a Dotplot for Example 2–12 of the Text

1. Enter the 33 data values from Example 2–12 of the text into column C1.

2. Name the column *Age*.

3. Select **Graph > Dotplot**.

4. Select **Simple** from the **One Y** box and click **OK**.

5. Type C1 in the **Graph variables** box and click **OK**.

6. The dotplot will appear in a new window. (See **Screen 2.9**.)

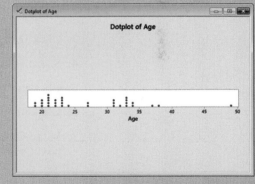

Screen 2.9

Excel

The Excel Technology Instructions feature of this text is written for Excel 2013. Some screens, menus, and functions will be slightly different in older versions of Excel. To enable some of the advanced Excel functions, you must enable the Data Analysis Add-In for Excel.

Excel does not have the capability to make a stem-and-leaf display and dotplot.

Creating a Bar Graph for Example 2–1 of the Text

1. Enter the grouped data from Example 2–1 as shown in Table 2.4 of the text.

	A	B	C
1	Glazed	8	
2	Filled	7	
3	Frosted	5	
4	Plain	3	
5	Other	7	
6			
7			

Screen 2.10

Glazed Filled Frosted Plain Other

Screen 2.11

2. Highlight the cells that contain the donut varieties and the frequencies. (See **Screen 2.10**.)

3. Click **INSERT** and then click the **Insert Column Chart** icon in the **Charts** group.

4. Select the first icon in the **2–D Column** group in the dropdown box.

5. A new graph will appear with the bar graph.

Creating a Pie Chart for Example 2–1 of the Text

1. Enter the grouped data from Example 2–1 as shown in Table 2.4 of the text.

2. Highlight the cells that contain the donut varieties and the frequencies. (See **Screen 2.10**.)

3. Click **INSERT** and then click the **Insert Pie Chart** icon in the **Charts** group.

4. Select the first icon in the **2–D Pie** group in the dropdown box.

5. A new graph will appear with the pie chart. (See **Screen 2.11**.)

Note: You may click in the text box and change "Chart Title" to an appropriate title or you may select the text box and delete it.

Creating a Frequency Histogram for Example 2–3 of the Text

1. Enter the data from Example 2–3 of the text into cells A1 to A30.

2. Determine the class boundaries and type the right boundary for each class in cells B1 to B8. In this example, we will use the right boundary values of 750, 1000, 1250, 1500, 1750, 2000, 2250, and 2500, which are different from the ones used in Example 2–3.

3. Click **DATA** and then click **Data Analysis** from the **Analysis** group.

4. Select **Histogram** from the **Data Analysis** dialog box and click **OK**.

5. Click in the **Input Range** box and then highlight cells A1 to A30.

6. Click in the **Bin Range** box and then highlight cells B1 to B8.

7. In the **Output options** box, select **New Worksheet Ply** and check the **Chart Output** check box. Click **OK**.

8. A new worksheet will appear with the frequency distribution and the corresponding histogram. (See **Screen 2.12**.)

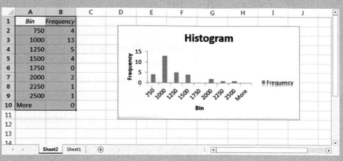

Screen 2.12

TECHNOLOGY ASSIGNMENTS

TA2.1 A July 2011 ESPN SportsNation poll asked, "Which is the best Fourth of July weekend sports tradition?" (http://espn.go.com/espn/fp/flashPollResultsState?sportIndex=frontpage&pollId=116290). The choices were Major League baseball game (B), Nathan's Famous International Hot Dog Eating Contest (H), Breakfast at Wimbledon (W), or NASCAR race at Daytona (N). The following data represent the responses of a random sample of 45 persons who were asked the same question.

H	H	B	W	N	B	H	N	W
N	H	B	W	H	N	N	H	H
B	B	W	H	H	B	W	H	B
H	B	B	H	B	H	B	N	H
B	B	H	H	H	B	H	H	N

Construct a bar graph and a pie chart for the above frequency distribution.

TA2.2 Thirty adults were asked which of the following conveniences they would find most difficult to do without: television (T), refrigerator (R), air conditioning (A), public transportation (P), or microwave (M). Their responses are listed below.

R	A	R	A	A	T	R	M	P	A
A	R	R	T	P	P	T	R	A	A
R	A	A	T	M	P	R	A	P	R

Construct a bar graph and a pie chart for the above frequency distribution.

TA2.3 Nixon Corporation manufactures computer monitors. The following data are the numbers of computer monitors produced at the company for a sample of 30 days.

30	34	27	23	33	33	26	25	24	28
21	26	31	22	27	33	27	23	28	21
31	35	34	22	26	25	23	35	31	27

Create a pie chart for the above data.

TA2.4 According to a survey by the U.S. Public Interest Research Group, about 79% of credit reports contain errors. Suppose in a random sample of 26 credit reports, the number of errors found are as listed below.

1	1	2	3	0	1	0	5	3	1	0	2	1
4	1	2	2	5	3	1	0	0	1	2	2	0

Create a pie chart for the above data.

TA2.5 Refer to Data Set XII on coffeemaker ratings that accompanies this text (see Appendix A). Create a stem-and-leaf display for the overall scores of coffeemakers listed in column 3.

TA2.6 These data give the times (in minutes) taken to commute from home to work for 20 workers.

24	50	65	33	48	5	11	8	39	34
26	30	17	10	15	18	29	43	21	22

Prepare a stem-and-leaf display for the above data.

TA2.7 The following data give the number of times each of the 20 randomly selected female students from the same state university ate at fast-food restaurants during the same 7-day period.

0	0	4	2	4	10	2	5	0	5
6	1	1	4	6	2	4	5	6	0

Make a dotplot for the above data.

TA2.8 In basketball, a *double-double* occurs when a player accumulates double-digit numbers in any two of five statistical categories—points, rebounds, assists, steals, and blocked shots—in one game. The following table gives the number of times each player of the Miami Heat basketball team scored a double-double (DD) during the 2010–2011 regular season.

Player	DD	Player	DD
Carlos Arroyo	0	Juwan Howard	0
Chris Bosh	28	LeBron James	31
Dwyane Wade	10	Mario Chalmers	0
Eddie House	0	Mike Bibby	0
Erick Dampier	1	Mike Miller	3
James Jones	0	Udonis Haslem	4
Joel Anthony	0	Zydrunas Ilgauskas	3

Source: espn.com.

Make a dotplot for the above data.

TA2.9 The following data give the numbers of computer keyboards assembled at the Twentieth Century Electronics Company for a sample of 25 days.

45	49	48	41	54	46	44	42	48	53	51	53	51
48	46	43	52	50	54	47	44	45	50	49	50	

Construct a histogram on the numbers of computer keyboards assembled. Use the classes given in that exercise. Use the midpoints to mark the horizontal axis in the histogram.

TA2.10 Refer to Data Set IV on the Manchester Road Race that accompanies this text (see Appendix A).

a. Create a bar graph for the gender of the runners listed in column 6.

b. Create a histogram for the net time to complete the race listed in column 3.

jokerpro/Shutterstock

CHAPTER 3

Numerical Descriptive Measures

Do you know there can be a big difference in the starting salaries of college graduates with different majors? Whereas engineering majors had an average starting salary of $62,600 in 2013, business majors received an average starting salary of $55,100, math and science majors received $43,000, and humanities and social science majors had an average starting salary of $38,000 in 2013. See Case Study 3–1.

In Chapter 2 we discussed how to summarize data using different methods and to display data using graphs. Graphs are one important component of statistics; however, it is also important to numerically describe the main characteristics of a data set. The numerical summary measures, such as the ones that identify the center and spread of a distribution, identify many important features of a distribution. For example, the techniques learned in Chapter 2 can help us graph data on family incomes. However, if we want to know the income of a "typical" family (given by the center of the distribution), the spread of the distribution of incomes, or the relative position of a family with a particular income, the numerical summary measures can provide more detailed information (see Figure 3.1). The measures that we discuss in this chapter include measures of (1) center, (2) dispersion (or spread), and (3) position.

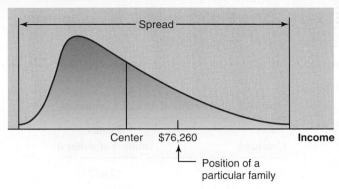

Figure 3.1

3.1 | Measures of Center for Ungrouped Data

We often represent a data set by numerical summary measures, usually called the *typical values*. A **measure of center** gives the center of a histogram or a frequency distribution curve. This section discusses five different measures of center: the mean, the median, the trimmed mean, the weighted mean, and the mode. However, another measure of center, the geometric mean, is explained in an exercise following this section. We will learn how to calculate each of these measures for ungrouped data. Recall from Chapter 2 that the data that give information on each member of the population or sample individually are called *ungrouped data*, whereas *grouped data* are presented in the form of a frequency distribution table.

3.1.1   Mean

The **mean**, also called the *arithmetic mean*, is the most frequently used measure of center. This book will use the words *mean* and *average* synonymously. For ungrouped data, the mean is obtained by dividing the sum of all values by the number of values in the data set:

$$\text{Mean} = \frac{\text{Sum of all values}}{\text{Number of values}}$$

The mean calculated for sample data is denoted by $\bar{x}$ (read as "x bar"), and the mean calculated for population data is denoted by μ (Greek letter *mu*). We know from the discussion in Chapter 2 that the number of values in a data set is denoted by n for a sample and by N for a population. In Chapter 1, we learned that a variable is denoted by x, and the sum of all values of x is denoted by Σx. Using these notations, we can write the following formulas for the mean.

Calculating Mean for Ungrouped Data   The *mean for ungrouped data* is obtained by dividing the sum of all values by the number of values in the data set. Thus,

$$\text{Mean for population data:} \quad \mu = \frac{\Sigma x}{N}$$

$$\text{Mean for sample data:} \quad \bar{x} = \frac{\Sigma x}{n}$$

where Σx is the sum of all values, N is the population size, n is the sample size, μ is the population mean, and $\bar{x}$ is the sample mean.

EXAMPLE 3-1	2014 Profits of 10 U.S. Companies

Calculating the sample mean for ungrouped data.

Table 3.1 lists the total profits (in million dollars) of 10 U.S. companies for the year 2014 (www. fortune.com).

Table 3.1   2014 Profits of 10 U.S. Companies

Company	Profits (million of dollars)
Apple	37,037
AT&T	18,249
Bank of America	11,431
Exxon Mobil	32,580
General Motors	5346
General Electric	13,057
Hewlett-Packard	5113
Home Depot	5385
IBM	16,483
Wal-Mart	16,022

Find the mean of the 2014 profits for these 10 companies.

Solution   The variable in this example is 2014 profits of a company. Let us denote this variable by x. The 10 values of x are given in the above table. By adding these 10 values, we obtain the sum of x values, that is:

$$\Sigma x = 37{,}037 + 18{,}249 + 11{,}431 + 32{,}580 + 5346 + 13{,}057 + 5113 + 5385 + 16{,}483 + 16{,}022$$
$$= 160{,}703$$

Note that the given data include only 10 companies. Hence, it represents a sample with $n = 10$. Substituting the values of Σx and n in the sample formula, we obtain the mean of 2014 profits of 10 companies as follows:

$$\bar{x} = \frac{\Sigma x}{n} = \frac{160{,}703}{10} = 16{,}070.3 = \textbf{\$16,070.3 million}$$

Thus, these 10 companies earned an average of $16,070.3 million profits in 2014.

EXAMPLE 3-2	Ages of Employees of a Company

Calculating the population mean for ungrouped data.

The following are the ages (in years) of all eight employees of a small company:

53   32   61   27   39   44   49   57

Find the mean age of these employees.

Solution   Because the given data set includes *all* eight employees of the company, it represents the population. Hence, $N = 8$. We have

$$\Sigma x = 53 + 32 + 61 + 27 + 39 + 44 + 49 + 57 = 362$$

The population mean is

$$\mu = \frac{\Sigma x}{N} = \frac{362}{8} = \textbf{45.25 years}$$

Thus, the mean age of all eight employees of this company is 45.25 years, or 45 years and 3 months.

Reconsider Example 3–2. If we take a sample of three employees from this company and calculate the mean age of those three employees, this mean will be denoted by $\bar{x}$. Suppose the three values included in the sample are 32, 39, and 57. Then, the mean age for this sample is

$$\bar{x} = \frac{32 + 39 + 57}{3} = 42.67 \text{ years}$$

If we take a second sample of three employees of this company, the value of $\bar{x}$ will (most likely) be different. Suppose the second sample includes the values 53, 27, and 44. Then, the mean age for this sample is

$$\bar{x} = \frac{53 + 27 + 44}{3} = 41.33 \text{ years}$$

Consequently, we can state that the value of the population mean μ is constant. However, the value of the sample mean $\bar{x}$ varies from sample to sample. The value of $\bar{x}$ for a particular sample depends on what values of the population are included in that sample.

Sometimes a data set may contain a few very small or a few very large values. As mentioned in Chapter 2, such values are called **outliers** or **extreme values**.

A major shortcoming of the mean as a measure of center is that it is very sensitive to outliers. Example 3–3 illustrates this point.

EXAMPLE 3–3	**Prices of Eight Homes**

Illustrating the effect of an outlier on the mean.

Following are the list prices of eight homes randomly selected from all homes for sale in a city.

$245,670	176,200	360,280	272,440
450,394	310,160	393,610	3,874,480

Note that the price of the last house is $3,874,480, which is an outlier. Show how the inclusion of this outlier affects the value of the mean.

Solution   If we do not include the price of the most expensive house (the outlier), the mean of the prices of the other seven homes is:

$$\text{Mean without the outlier} = \frac{245{,}670 + 176{,}200 + 360{,}280 + 272{,}440 + 450{,}394 + 310{,}160 + 393{,}610}{7}$$

$$= \frac{2{,}208{,}754}{7} = \$315{,}536.29$$

Now, to see the impact of the outlier on the value of the mean, we include the price of the most expensive home and find the mean price of eight homes. This mean is:

Mean with the outlier

$$= \frac{245{,}670 + 176{,}200 + 360{,}280 + 272{,}440 + 450{,}394 + 310{,}160 + 393{,}610 + 3{,}874{,}480}{8}$$

$$= \frac{6{,}083{,}234}{8} = \$760{,}404.25$$

Thus, when we include the price of the most expensive home, the mean more than doubles, as it increases from $315,536.29 to $760,404.25.

The preceding example should encourage us to be cautious. We should remember that the mean is not always the best measure of center because it is heavily influenced by outliers. Sometimes other measures of center give a more accurate impression of a data set. For example, when a data set has outliers, instead of using the mean, we can use either the trimmed mean or the median as a measure of center.

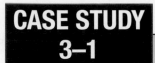

CASE STUDY 3–1 — 2013 Average Starting Salaries for Selected Majors

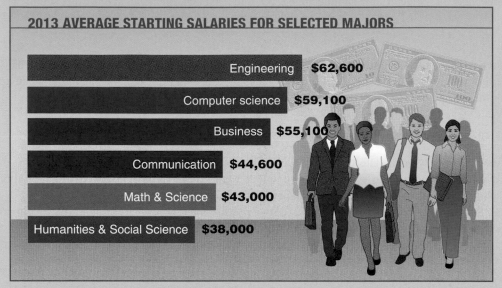

2013 AVERAGE STARTING SALARIES FOR SELECTED MAJORS

Engineering	$62,600
Computer science	$59,100
Business	$55,100
Communication	$44,600
Math & Science	$43,000
Humanities & Social Science	$38,000

Data source: www.forbes.com

The above chart, based on data from Forbes.com, shows the 2013 average starting salaries for a few selected majors. As we can notice, among these six majors, engineering commanded the highest starting salary of $62,600 in 2013, and humanities and social science majors had the lowest starting salary of $38,000 in 2013. Computer science major was the second highest with a starting salary of $59,100. As we can observe, there is a large variation in the 2013 starting salaries for these six majors.

Source: Forbes.com.

3.1.2 Median

Another important measure of center is the **median**. It is defined as follows.

> The **median** is the value that divides a data set that has been ranked in increasing order in two equal halves. If the data set has an odd number of values, the median is given by the value of the middle term in the ranked data set. If the data set has an even number of values, the median is given by the average of the two middle values in the ranked data set.

As is obvious from the definition of the median, it divides a ranked data set into two equal parts. The calculation of the median consists of the following two steps:

1. Rank the given data set in increasing order.
2. Find the value that divides the ranked data set in two equal parts. This value gives the median.[1]

Note that if the number of observations in a data set is *odd*, then the median is given by the value of the middle term in the ranked data. However, if the number of observations is *even*, then the median is given by the average of the values of the two middle terms.

EXAMPLE 3–4 | Compensations of Female CEOs

Calculating the median for ungrouped data: odd number of data values.

Table 3.2 lists the 2014 compensations of female CEOs of 11 American companies (*USA TODAY*, May 1, 2015). (The compensation of Carol Meyrowitz of TJX is for the fiscal year ending in January 2015.)

[1]Note that the middle value in a data set ranked in *decreasing* order will also give the value of the median.

Table 3.2   Compensations of 11 Female CEOs

Company & CEO	2014 Compensation (millions of dollars)
General Dynamics, Phebe Novakovic	19.3
GM, Mary Barra	16.2
Hewlett-Packard, Meg Whitman	19.6
IBM, Virginia Rometty	19.3
Lockheed Martin, Marillyn Hewson	33.7
Mondelez, Irene Rosenfeld	21.0
PepsiCo, Indra Nooyi	22.5
Sempra, Debra Reed	16.9
TJX, Carol Meyrowitz	28.7
Yahoo, Marissa Mayer	42.1
Xerox, Ursula Burns	22.2

Find the median for these data.

Solution   To calculate the median of this data set, we perform the following two steps.

Step 1:   The first step is to rank the given data. We rank the given data in increasing order as follows:

16.2   16.9   19.3   19.3   19.6   21.0   22.2   22.5   28.7   33.7   42.1

Step 2:   The second step is to find the value that divides this ranked data set in two equal parts. Here there are 11 data values. The sixth value divides these 11 values in two equal parts. Hence, the sixth value gives the median as shown below.

16.2   16.9   19.3   19.3   19.6   **21.0**   22.2   22.5   28.7   33.7   42.1

↑
Median

Thus, the median of 2014 compensations for these 11 female CEOs is $21.0 million. Note that in this example, there are 11 data values, which is an odd number. Hence, there is one value in the middle that is given by the sixth term, and its value is the median. Using the value of the median, we can say that half of these CEOs made less than $21.0 million and the other half made more than $21.0 million in 2014.

EXAMPLE 3–5   Cell Phone Minutes Used

Calculating the median for ungrouped data: even number of data values.

The following data give the cell phone minutes used last month by 12 randomly selected persons.

230   2053   160   397   510   380   263   3864   184   201   326   721

Find the median for these data.

Solution   To calculate the median, we perform the following two steps.

Step 1:   In the first step, we rank the given data in increasing order as follows:

160   184   201   230   263   326   380   397   510   721   2053   3864

Step 2:   In the second step, we find the value that divides the ranked data set in two equal parts. This value will be the median. The value that divides 12 data values in two equal parts falls

Education Level and 2014 Median Weekly Earnings

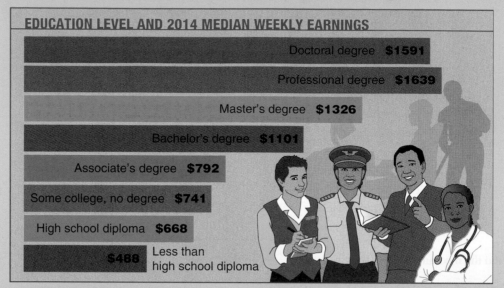

EDUCATION LEVEL AND 2014 MEDIAN WEEKLY EARNINGS

Doctoral degree **$1591**

Professional degree **$1639**

Master's degree **$1326**

Bachelor's degree **$1101**

Associate's degree **$792**

Some college, no degree **$741**

High school diploma **$668**

$488 Less than high school diploma

Data source: U.S. Bureau of Labor Statistics

The above chart shows the 2014 median weekly earnings by education level for persons aged 25 and over who held full-time jobs. These salaries are based on the Current Population Survey conducted by the U.S. Bureau of Labor Statistics. Although this survey is called the Current Population Survey, actually it is based on a sample. Usually the samples taken by the U.S. Bureau of Labor Statistics for these surveys are very large. As shown in the chart, the highest median weekly earning (of $1639) in 2014 was for workers with professional degrees, and the lowest (of $488) was for workers with less than high school diplomas. According to this survey, the median earning for all workers (aged 25 and over with full-time jobs) included in the survey was $839 in 2014, which is not shown in the graph.

Source: www.bls.gov.

between the sixth and the seventh values. Thus, the median will be given by the average of the sixth and the seventh values as follows.

| 160 | 184 | 201 | 230 | 263 | 326 | 380 | 397 | 510 | 721 | 2053 | 3864 |

Median = 353

$$\text{Median} = \text{Average of the two middle values} = \frac{326 + 380}{2} = \textbf{353 minutes}$$

Thus, the median cell phone minutes used last month by these 12 persons was 353. We can state that half of these 12 persons used less than 353 cell phone minutes and the other half used more than 353 cell phone minutes last month. Note that this data set has two outliers, 2053 and 3864 minutes, but these outliers do not affect the value of the median.

The median gives the center of a histogram, with half of the data values to the left of the median and half to the right of the median. The advantage of using the median as a measure of center is that it is not influenced by outliers. Consequently, the median is preferred over the mean as a measure of center for data sets that contain outliers.

Sometime we use the terms resistant and nonresistant with respect to outliers for summary measures. A summary measure that is less affected by outliers is called a **resistant** summary measure, and the one that is affected more by outliers is called a **nonresistant** summary measure. From the foregoing discussion about the mean and median, it is obvious that the mean is a nonresistant summary measure, as it is influenced by outliers. In contrast, the median is a resistant summary measure because it is not influenced by outliers.

3.1.3   Mode

Mode is a French word that means *fashion*—an item that is most popular or common. In statistics, the mode represents the most common value in a data set.

EXAMPLE 3–6   Speeds of Cars

Calculating the mode for ungrouped data.

The following data give the speeds (in miles per hour) of eight cars that were stopped on I-95 for speeding violations.

$$77 \qquad 82 \qquad 74 \qquad 81 \qquad 79 \qquad 84 \qquad 74 \qquad 78$$

Find the mode.

Solution   In this data set, 74 occurs twice, and each of the remaining values occurs only once. Because 74 occurs with the highest frequency, it is the mode. Therefore,

$$\text{Mode} = \textbf{74 miles per hour}$$

A major shortcoming of the mode is that a data set may have none or may have more than one mode, whereas it will have only one mean and only one median. For instance, a data set with each value occurring only once has no mode. A data set with only one value occurring with the highest frequency has only one mode. The data set in this case is called **unimodal** as in Example 3–6 above. A data set with two values that occur with the same (highest) frequency has two modes. The distribution, in this case, is said to be **bimodal** as in Example 3–8 below. If more than two values in a data set occur with the same (highest) frequency, then the data set contains more than two modes and it is said to be **multimodal** as in Example 3–9 below.

EXAMPLE 3–7   Incomes of Families

Data set with no mode.

Last year's incomes of five randomly selected families were $76,150, $95,750, $124,985, $87,490, and $53,740. Find the mode.

Solution   Because each value in this data set occurs only once, this data set contains **no mode**.

EXAMPLE 3–8   Commuting Times of Employees

Data set with two modes.

A small company has 12 employees. Their commuting times (rounded to the nearest minute) from home to work are 23, 36, 14, 23, 47, 32, 8, 14, 26, 31, 18, and 28, respectively. Find the mode for these data.

Solution   In the given data on the commuting times of these 12 employees, each of the values 14 and 23 occurs twice, and each of the remaining values occurs only once. Therefore, this data set has two modes: 14 and 23 minutes.

EXAMPLE 3–9   Ages of Students

Data set with three modes.

The ages of 10 randomly selected students from a class are 21, 19, 27, 22, 29, 19, 25, 21, 22, and 30 years, respectively. Find the mode.

Solution   This data set has three modes: **19, 21**, and **22**. Each of these three values occurs with a (highest) frequency of 2.

One advantage of the mode is that it can be calculated for both kinds of data—quantitative and qualitative—whereas the mean and median can be calculated for only quantitative data.

EXAMPLE 3–10   Status of Students

Finding the mode for qualitative data.

The status of five students who are members of the student senate at a college are senior, sophomore, senior, junior, and senior, respectively. Find the mode.

Solution   Because **senior** occurs more frequently than the other categories, it is the mode for this data set. We cannot calculate the mean and median for this data set.

3.1.4   Trimmed Mean

> After we drop $k\%$ of the values from each end of a ranked data set, the mean of the remaining values is called the $k\%$ **trimmed mean**.

Earlier in this chapter, we learned that the mean as a measure of center is impacted by outliers. When a data set contains outliers, we can use either median or trimmed mean as a measure of the center of a data set.

Thus, to calculate the trimmed mean for a data set, first we rank the given data in increasing order. Then we drop $k\%$ of the values from each end of the ranked data where k is any positive number, such as 5%, 10%, and so on. The mean of the remaining values is called the $k\%$ trimmed mean. Remember, although we drop a total of $2 \times k\%$ of the values, $k\%$ from each end, it is called the $k\%$ trimmed mean. The following example illustrates the calculation of the trimmed mean.

EXAMPLE 3–11   Money Spent on Books by Students

Calculating the trimmed mean.

The following data give the money spent (in dollars) on books during 2015 by 10 students selected from a small college.

| 890 | 1354 | 1861 | 1644 | 87 | 5403 | 1429 | 1993 | 938 | 2176 |

Calculate the 10% trimmed mean.

Solution   To calculate the trimmed mean, first we rank the given data as below.

| 87 | 890 | 938 | 1354 | 1429 | 1644 | 1861 | 1993 | 2176 | 5403 |

To calculate the 10% trimmed mean, we drop 10% of the data values from each end of the ranked data.

$$10\% \text{ of } 10 \text{ values} = 10(.10) = 1$$

Hence, we drop one value from each end of the ranked data. After we drop the two values, one from each end, we are left with the following eight values:

| 890 | 938 | 1354 | 1429 | 1644 | 1861 | 1993 | 2176 |

The mean of these eight values will be called the 10% trimmed mean. Since there are 8 values, $n = 8$. Adding these eight values, we obtain Σx as follows:

$$\Sigma x = 890 + 938 + 1354 + 1429 + 1644 + 1861 + 1993 + 2176 = 12{,}285$$

The 10% trimmed mean will be obtained by dividing 12,285 by 8 as follows:

$$10\% \text{ Trimmed Mean} = \frac{12{,}285}{8} = 1535.625 = \mathbf{\$1535.63}$$

Thus, by dropping 10% of the values from each end of the ranked data for this example, we can state that students spent an average of $1535.63 on books in 2015.

Since in this data set \$87 and \$5403 can be considered outliers, it makes sense to drop these two values and calculate the trimmed mean for the remaining values rather than calculating the mean of all 10 values.

3.1.5 Weighted Mean

In many cases, when we want to find the center of a data set, different values in the data set may have different frequencies or different weights. We will have to consider the weights of different values to find the correct mean of such a data set. For example, suppose Maura bought gas for her car four times during June 2015. She bought 10 gallons at a price of \$2.60 a gallon, 13 gallons at a price of \$2.80 a gallon, 8 gallons at a price of \$2.70 a gallon, and 15 gallons at a price of \$2.75 a gallon. In such a case, the mean of the four prices as calculated in Section 3.1.1 will not give the actual mean price paid by Maura in June 2015. We cannot add the four prices and divide by four to find the average price. That can be done only if she bought the same amount of gas each time. But because the amount of gas bought each time is different, we will have to calculate the weighted mean in this case. The amounts of gas bought will be considered the weight here.

> When different values of a data set occur with different frequencies, that is, each value of a data set is assigned different weight, then we calculate the **weighted mean** to find the center of the given data set.

To calculate the weighted mean for a data set, we denote the variable by x and the weights by w. We add all the weights and denote this sum by $\sum w$. Then we multiply each value of x by the corresponding value of w. The sum of the resulting products gives $\sum xw$. Dividing $\sum xw$ by $\sum w$ gives the weighted mean.

Weighted Mean   The weighted mean is calculated as:

$$\text{Weighted Mean} = \frac{\sum xw}{\sum w}$$

where x and w denote the variable and the weights, respectively.

The following example illustrates the calculation of the weighted mean.

EXAMPLE 3–12   Prices and Amounts of Gas Purchased

Calculating the weighted mean.

Maura bought gas for her car four times during June 2015. She bought 10 gallons at a price of \$2.60 a gallon, 13 gallons at a price of \$2.80 a gallon, 8 gallons at a price of \$2.70 a gallon, and 15 gallons at a price of \$2.75 a gallon. What is the average price that Maura paid for gas during June 2015?

Solution   Here the variable is the price of gas per gallon, and we will denote it by x. The weights are the number of gallons bought each time, and we will denote these weights by w. We list the values of x and w in Table 3.3, and find $\sum w$. Then we multiply each value of x by the corresponding value of w and obtain $\sum xw$ by adding the resulting values. Finally, we divide $\sum xw$ by $\sum w$ to find the weighted mean.

Table 3.3   Prices and Amounts of Gas Purchased

Price (in dollars) x	Gallons of Gas w	xw
2.60	10	26.00
2.80	13	36.40
2.70	8	21.60
2.75	15	41.25
	$\sum w = 46$	$\sum xw = 125.25$

$$\text{Weighted Mean} = \frac{\sum xw}{\sum w} = \frac{125.25}{46} = \mathbf{\$2.72}$$

Thus, Maura paid an average of $2.72 a gallon for the gas she bought in June 2015.

To summarize, we cannot say for sure which of the various measures of center is a better measure overall. Each of them may be better under different situations. Probably the mean is the most-used measure of center, followed by the median. The mean has the advantage that its calculation includes each value of the data set. The median and trimmed mean are better measures when a data set includes outliers. The mode is simple to locate, but it is not of much use in practical applications.

3.1.6   Relationships Among the Mean, Median, and Mode

As discussed in Chapter 2, two of the many shapes that a histogram or a frequency distribution curve can assume are symmetric and skewed. This section describes the relationships among the mean, median, and mode for three such histograms and frequency distribution curves. Knowing the values of the mean, median, and mode can give us some idea about the shape of a frequency distribution curve.

1.   For a **symmetric histogram** and frequency distribution curve with one peak (see Figure 3.2), the values of the mean, median, and mode are identical, and they lie at the center of the distribution.

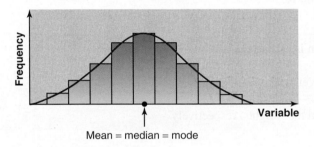

Figure 3.2   Mean, median, and mode for a symmetric histogram and frequency distribution curve.

2.   For a histogram and a frequency distribution curve **skewed to the right** (see Figure 3.3), the value of the mean is the largest, that of the mode is the smallest, and the value of the median lies between these two. (Notice that the mode always occurs at the peak point.) The value of the mean is the largest in this case because it is sensitive to outliers that occur in the right tail. These outliers pull the mean to the right.

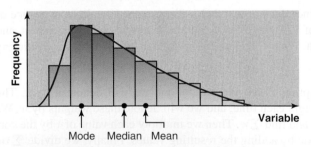

Figure 3.3   Mean, median, and mode for a histogram and frequency distribution curve skewed to the right.

3.   If a histogram and a frequency distribution curve are **skewed to the left** (see Figure 3.4), the value of the mean is the smallest and that of the mode is the largest, with the value of the median lying between these two. In this case, the outliers in the left tail pull the mean to the left.

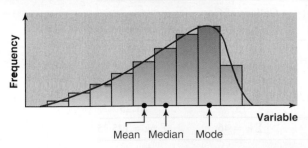

Figure 3.4   Mean, median, and mode for a histogram and frequency distribution curve skewed to the left.

EXERCISES

CONCEPTS AND PROCEDURES

3.1 Explain how the value of the median is determined for a data set that contains an even number of observations and for a data set that contains an odd number of observations.

3.2 Briefly explain the meaning of an outlier. Is the mean or the median a better measure of center for a data set that contains outliers? Illustrate with the help of an example.

3.3 How can outliers affect the value of the mean? Illustrate with the help of an example.

3.4 Which of the five measures of center (the mean, the median, the trimmed mean, the weighted mean, and the mode) can be calculated for quantitative data only, and which can be calculated for both quantitative and qualitative data? Illustrate with examples.

3.5 Which of the five measures of center (the mean, the median, the trimmed mean, the weighted mean, and the mode) can assume more than one value for a data set? Give an example of a data set for which this summary measure assumes more than one value.

3.6 Is a quantitative data set with no mean, no median, or no mode possible? Give an example of a data set for which this summary measure does not exist.

3.7 Explain the relationships among the mean, median, and mode for symmetric and skewed histograms. Illustrate these relationships with graphs.

3.8 Prices of mobile phones have a distribution that is skewed to the right with outliers in the right tail. Which of the measures of center is the best to summarize this data set? Explain.

3.9 The following data set belongs to a population:

$$4 \quad -7 \quad 1 \quad 0 \quad -9 \quad 16 \quad 9 \quad 8$$

Calculate the mean, median, and mode.

APPLICATIONS

3.10 The following table gives the standard deductions and personal exemptions for persons filing with "single" status on their 2009 state income taxes in a random sample of 9 states.

State	Standard Deduction (in dollars)	Personal Exemption (in dollars)
Delaware	3250	110
Hawaii	2000	1040
Kentucky	2100	20
Minnesota	5450	3500
North Dakota	5450	3500
Oregon	1865	169
Rhode Island	5450	3500
Vermont	5450	3500
Virginia	3000	930

Source: TaxFoundation.org.

   a. Find the mean and median for the data on standard deductions for these states.

   b. Calculate the mean and median for the data on personal exemptions for these states.

3.11 The following data give the amounts (in dollars) of electric bills for March 2016 for 12 randomly selected households selected from a small town.

211 265 176 314 243 192 297 357 238 281 342 259

Calculate the mean and median for these data. Do these data have a mode? Explain.

3.12 The following data give the numbers of car thefts that occurred in a city during the past 14 days.

8   3   7   11   4   3   5   4   2   6   4   15   8   3

   a. Calculate the mean, median, and mode for these data.

   b. Calculate the 10% trimmed mean for these data.

3.13 The following data represent the numbers of tornadoes that touched down during 1950 to 1994 in the 12 states that had the most tornadoes during this period. The data for these states are given in the following order: CO, FL, IA, IL, KS, LA, MO, MS, NE, OK, SD, TX.

1113	2009	1374	1137	2110	1086
1166	1039	1673	2300	1139	5490

   a. Calculate the mean and median for these data.

   b. Identify the outlier in this data set. Drop the outlier and recalculate the mean and median. Which of these two summary measures changes by a larger amount when you drop the outlier?

   c. Which is the better summary measure for these data, the mean or the median? Explain.

3.14 Due to antiquated equipment and frequent windstorms, the town of Oak City often suffers power outages. The following data give the numbers of power outages for each of the past 12 months.

4   5   7   3   2   0   2   3   2   1   2   4

   a. Compute the mean, median, and mode for these data.

   b. Calculate the 20% trimmed mean for these data.

*3.15 The following data give the prices (in thousands of dollars) of 20 houses sold recently in a city.

184	324	365	309	245	387	369	438	195	180
323	578	510	679	307	271	318	795	259	590

   a. Calculate the mean, median, and mode for these data.

   b. Calculate the 20% trimmed mean for these data.

3.16 The following data give the annual salaries (in thousand dollars) of 20 randomly selected health care workers.

50	71	57	39	45	64	38	53	35	62
74	40	67	44	77	61	58	55	64	59

   a. Calculate the mean, median, and mode for these data.

   b. Calculate the 15% trimmed mean for these data.

3.17 The following data give the 2015 bonuses (in thousands of dollars) of 10 randomly selected Wall Street managers.

127   82   45   99   153   3261   77   108   68   278

   a. Calculate the mean and median for these data.

   b. Do these data have a mode? Explain why or why not.

   c. Calculate the 10% trimmed mean for these data.

   d. This data set has one outlier. Which summary measures are better for these data?

3.18 The following data represent the systolic blood pressure reading (that is, the *top* number in the standard blood pressure reading) in mmHg for each of 20 randomly selected middle-aged males who were taking blood pressure medication.

139	151	138	153	134	136	141	126	109	144
111	150	107	132	144	116	159	121	127	113

a. Calculate the mean, median, and mode for these data.
b. Calculate the 15% trimmed mean for these data.

3.19 In a survey of 700 parents of young children, 380 said that they will not want their children to play football because it is a very dangerous sport, 230 said that they will let their children play football, and 90 had no opinion. Considering the opinions of these parents, what is the mode?

3.20 A statistics professor based her final grades on quizzes, in-class group work, homework, a midterm exam, and a final exam. However, not all of the assignments contributed equally to the final grade. John received the scores (out of 100 for each assignment) listed in the table below. The instructor weighted each item as shown in the table.

Assignment	John's Score (Total Points)	Percentage of Final Grade Assigned by the Instructor
Quizzes	78	30
In-class group work	55	5
Homework	84	10
Midterm exam	76	15
Final exam	78	40

Calculate John's final grade score (out of 100) in this course. (*Hint*: Note that it is often easier to convert the weights to decimal form for weighting, e.g., 30% = .30; in this way, the sum of weights will equal 1, making the formula easier to apply.)

3.21 A clothing store bought 3000 shirts last year from five different companies. The following table lists the number of shirts bought from each company and the price paid for each shirt.

Company	Number of Shirts Bought	Price per Shirt (dollars)
Best Shirts	200	30
Top Wear	900	45
Smith Shirts	400	40
New Design	1200	35
Radical Wear	3000	50

Calculate the weighted mean that represents the average price paid for these 3000 shirts.

*3.22 One property of the mean is that if we know the means and sample sizes of two (or more) data sets, we can calculate the **combined mean** of both (or all) data sets. The combined mean for two data sets is calculated by using the formula

$$\text{Combined mean} = \bar{x} = \frac{n_1\bar{x}_1 + n_2\bar{x}_2}{n_1 + n_2}$$

where n_1 and n_2 are the sample sizes of the two data sets and $\bar{x}_1$ and $\bar{x}_2$ are the means of the two data sets, respectively. Suppose a sample of 10 statistics books gave a mean price of $140 and a sample of 8 mathematics books gave a mean price of $160. Find the combined mean. (*Hint:* For this example: $n_1 = 10$, $n_2 = 8$, $\bar{x}_1 = \$140$, $\bar{x}_2 = \$160$.)

*3.23 For any data, the sum of all values is equal to the product of the sample size and mean; that is, $\Sigma x = n\bar{x}$. Suppose the average amount of money spent on shopping by 12 persons during a given week is $115.50. Find the total amount of money spent on shopping by these 12 persons.

*3.24 The mean 2016 income for five families was $99,575. What was the total 2016 income of these five families?

*3.25 The mean age of six persons is 52 years. The ages of five of these six persons are 56, 38, 43, 51, and 36 years, respectively. Find the age of the sixth person.

*3.26 Seven airline passengers in economy class on the same flight paid an average of $354 per ticket. Because the tickets were purchased at different times and from different sources, the prices varied. The first five passengers paid $425, $215, $338, $600, and $490. The sixth and seventh tickets were purchased by a couple who paid identical fares. What price did each of them pay?

*3.27 In some applications, certain values in a data set may be considered more important than others. For example, to determine students' grades in a course, an instructor may assign a weight to the final exam that is twice as much as that to each of the other exams. In such cases, it is more appropriate to use the **weighted mean**. In general, for a sequence of n data values $x_1, x_2,..., x_n$ that are assigned weights $w_1, w_2,..., w_n$, respectively, the **weighted mean** is found by the formula

$$\text{Weighted mean} = \frac{\Sigma xw}{\Sigma w}$$

where Σxw is obtained by multiplying each data value by its weight and then adding the products. Suppose an instructor gives two exams and a final, assigning the final exam a weight twice that of each of the other exams. Find the weighted mean for a student who scores 73 and 67 on the first two exams and 85 on the final. (*Hint:* Here, $x_1 = 73$, $x_2 = 67$, $x_3 = 85$, $w_1 = w_2 = 1$, and $w_3 = 2$.)

3.2 | Measures of Dispersion for Ungrouped Data

The measures of center, such as the mean, median, and mode, do not reveal the whole picture of the distribution of a data set. Two data sets with the same mean may have completely different spreads. The variation among the values of observations for one data set may be much larger or smaller than for the other data set. (Note that the words **dispersion**, **spread**, and **variation** have similar meanings.) Consider the following two data sets on the ages (in years) of all workers at each of two small companies.

Company 1:	47	38	35	40	36	45	39
Company 2:	70	33	18	52	27		

The mean age of workers in both these companies is the same, 40 years. If we do not know the ages of individual workers at these two companies and are told only that the mean age of the workers at both companies is the same, we may deduce that the workers at these two companies have a similar age distribution. As we can observe, however, the variation in the workers' ages for each of these two companies is very different. As illustrated in the diagram, the ages of the workers at the second company have a much larger variation than the ages of the workers at the first company.

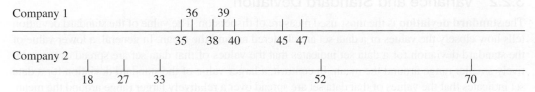

Thus, a summary measure such as the mean, median, or mode by itself is usually not a sufficient measure to reveal the shape of the distribution of a data set. We also need a measure that can provide some information about the variation among data values. The measures that help us learn about the spread of a data set are called the **measures of dispersion**. The measures of center and dispersion taken together give a better picture of a data set than the measures of center alone. This section discusses four measures of dispersion: range, variance, standard deviation, and coefficient of variation.

3.2.1 Range

The **range** is the simplest measure of dispersion to calculate. It is obtained by taking the difference between the largest and the smallest values in a data set.

Finding the Range for Ungrouped Data

$$\text{Range} = \text{Largest value} - \text{Smallest value}$$

EXAMPLE 3–13   Total Areas of Four States

Calculating the range for ungrouped data.

Table 3.4 gives the total areas in square miles of the four western South-Central states of the United States.

Table 3.4

State	Total Area (square miles)
Arkansas	53,182
Louisiana	49,651
Oklahoma	69,903
Texas	267,277

Find the range for this data set.

Solution   The largest total area for a state in this data set is 267,277 square miles, and the smallest area is 49,651 square miles. Therefore,

$$\text{Range} = \text{Largest value} - \text{Smallest value}$$
$$= 267,277 - 49,651 = \textbf{217,626 square miles}$$

Thus, the total areas of these four states are spread over a range of 217,626 square miles.

The range, like the mean, has the disadvantage of being influenced by outliers. In Example 3–13, if the state of Texas with a total area of 267,277 square miles is dropped, the range decreases from 217,626 square miles to $69,903 - 49,651 = 20,252$ square miles. Consequently, the range is not a good measure of dispersion to use for a data set that contains outliers. This indicates that the range is a **nonresistant measure** of dispersion.

Another disadvantage of using the range as a measure of dispersion is that its calculation is based on two values only: the largest and the smallest. All other values in a data set are ignored when calculating the range. Thus, the range is not a very satisfactory measure of dispersion.

3.2.2 Variance and Standard Deviation

The **standard deviation** is the most-used measure of dispersion. The value of the standard deviation tells how closely the values of a data set are clustered around the mean. In general, a lower value of the standard deviation for a data set indicates that the values of that data set are spread over a relatively smaller range around the mean. In contrast, a larger value of the standard deviation for a data set indicates that the values of that data set are spread over a relatively larger range around the mean.

The *standard deviation is obtained by taking the positive square root of the* **variance**. The variance calculated for population data is denoted by σ^2 (read as *sigma squared*),[2] and the variance calculated for sample data is denoted by s^2. Consequently, the standard deviation calculated for population data is denoted by σ, and the standard deviation calculated for sample data is denoted by s. Following are what we will call the *basic formulas* that are used to calculate the variance and standard deviation.[3]

$$\sigma^2 = \frac{\sum(x - \mu)^2}{N} \quad \text{and} \quad s^2 = \frac{\sum(x - \bar{x})^2}{n - 1}$$

$$\sigma = \sqrt{\frac{\sum(x - \mu)^2}{N}} \quad \text{and} \quad s = \sqrt{\frac{\sum(x - \bar{x})^2}{n - 1}}$$

where σ^2 is the population variance, s^2 is the sample variance, σ is the population standard deviation, and s is the sample standard deviation.

The quantity $x - \mu$ or $x - \bar{x}$ in the above formulas is called the *deviation* of the x value from the mean. The sum of the deviations of the x values from the mean is always zero; that is, $\sum(x - \mu) = 0$ and $\sum(x - \bar{x}) = 0$.

For example, suppose the midterm scores of a sample of four students are 82, 95, 67, and 92, respectively. Then, the mean score for these four students is

$$\bar{x} = \frac{82 + 95 + 67 + 92}{4} = 84$$

The deviations of the four scores from the mean are calculated in Table 3.5. As we can observe from the table, the sum of the deviations of the x values from the mean is zero; that is, $\sum(x - \bar{x}) = 0$. For this reason we square the deviations to calculate the variance and standard deviation.

Table 3.5

x	$x - \bar{x}$
82	$82 - 84 = -2$
95	$95 - 84 = +11$
67	$67 - 84 = -17$
92	$92 - 84 = +8$
	$\sum(x - \bar{x}) = 0$

From the computational point of view, it is easier and more efficient to use *short-cut formulas* to calculate the variance and standard deviation. By using the short-cut formulas, we reduce the computation time and round-off errors. Use of the basic formulas for ungrouped data is illustrated in Section A3.1.1 of Appendix 3.1 of this chapter. The short-cut formulas for calculating the variance and standard deviation are given on the next page.

---

[2]Note that Σ is uppercase sigma and σ is lowercase sigma of the Greek alphabet.
[3]From the formula for σ^2, it can be stated that the population variance is the mean of the squared deviations of x values from the mean. However, this is not true for the variance calculated for a sample data set.

Short-Cut Formulas for the Variance and Standard Deviation for Ungrouped Data

$$\sigma^2 = \frac{\Sigma x^2 - \left(\dfrac{(\Sigma x)^2}{N}\right)}{N} \quad \text{and} \quad s^2 = \frac{\Sigma x^2 - \left(\dfrac{(\Sigma x)^2}{n}\right)}{n-1}$$

where σ^2 is the population variance and s^2 is the sample variance.

The standard deviation is obtained by taking the positive square root of the variance.

$$\text{Population standard deviation:} \quad \sigma = \sqrt{\frac{\Sigma x^2 - \left(\dfrac{(\Sigma x)^2}{N}\right)}{N}}$$

$$\text{Sample standard deviation:} \quad s = \sqrt{\frac{\Sigma x^2 - \left(\dfrac{(\Sigma x)^2}{n}\right)}{n-1}}$$

Note that the denominator in the formula for the population variance is N, but that in the formula for the sample variance it is $n - 1$.[4]

Thus, to calculate the variance and standard deviation for a data set, we perform the following three steps: In the first step, we add all the values of x in the given data set and denote this sum by Σx. In the second step, we square each value of x and add these squared values, which is denoted by Σx^2. In the third step, we substitute the values of n, Σx, and Σx^2 in the corresponding formula for variance or standard deviation and simplify.

Examples 3–14 and 3–15 show the calculation of the variance and standard deviation.

EXAMPLE 3–14   Compensations of Female CEOs

Calculating the sample variance and standard deviation for ungrouped data.

Refer to the 2014 compensations of 11 female CEOs of American companies given in Example 3–4. The table from that example is reproduced below.

Company & CEO	2014 Compensation (millions of dollars)
General Dynamics, Phebe Novakovic	19.3
GM, Mary Barra	16.2
Hewlett-Packard, Meg Whitman	19.6
IBM, Virginia Rometty	19.3
Lockheed Martin, Marillyn Hewson	33.7
Mondelez, Irene Rosenfeld	21.0
PepsiCo, Indra Nooyi	22.5
Sempra, Debra Reed	16.9
TJX, Carol Meyrowitz	28.7
Yahoo, Marissa Mayer	42.1
Xerox, Ursula Burns	22.2

Find the variance and standard deviation for these data.

---

[4]The reason that the denominator in the sample formula is $n - 1$ and not n follows: The sample variance underestimates the population variance when the denominator in the sample formula for variance is n. However, the sample variance does not underestimate the population variance if the denominator in the sample formula for variance is $n - 1$. In Chapter 8 we will learn that $n - 1$ is called the degrees of freedom.

Solution　Let x denote the 2014 compensations (in millions of dollars) of female CEOs of American companies. The calculation of $\sum x$ and $\sum x^2$ is shown in Table 3.6.

Table 3.6　Calculation of $\sum x$ and $\sum x^2$

x	x^2
19.3	372.49
16.2	262.44
19.6	384.16
19.3	372.49
33.7	1135.69
21.0	441.00
22.5	506.25
16.9	285.61
28.7	823.69
42.1	1772.41
22.2	492.84
$\sum x = 261.5$	$\sum x^2 = 6849.07$

Calculation of the variance involves the following three steps.

Step 1.　*Calculate $\sum x$.*

The sum of the values in the first column of Table 3.6 gives the value of $\sum x$, which is 261.5.

Step 2.　*Find $\sum x^2$.*

The value of $\sum x^2$ is obtained by squaring each value of x and then adding the squared values. The results of this step are shown in the second column of Table 3.6. Notice that $\sum x^2 = 6849.07$.

Step 3.　*Determine the variance.*

Substitute the values of n, $\sum x$, and $\sum x^2$ in the variance formula and simplify. Because the given data are on the 2014 compensations of 11 female CEOs of American companies, we use the formula for the sample variance using $n = 11$.

$$s^2 = \frac{\sum x^2 - \left(\frac{(\sum x)^2}{n}\right)}{n-1} = \frac{6849.07 - \left(\frac{(261.5)^2}{11}\right)}{11-1} = \frac{6849.07 - 6216.5682}{10} = \mathbf{63.2502}$$

Now to obtain the standard deviation, we take the (positive) square root of the variance. Thus,

$$s = \sqrt{63.2502} = 7.952999 = \textbf{\$7.95 million}$$

Thus, the standard deviation of the 2014 compensations of these 11 female CEOs of American companies is $7.95 million.

Two Observations ➤

1. **The values of the variance and the standard deviation are never negative**. That is, the numerator in the formula for the variance should never produce a negative value. Usually the values of the variance and standard deviation are positive, but if a data set has no variation, then the variance and standard deviation are both zero. For example, if four persons in a group are the same age—say, 35 years—then the four values in the data set are

   35　　35　　35　　35

   If we calculate the variance and standard deviation for these data, their values will be zero. This is because there is no variation in the values of this data set.

2. **The measurement units of the variance are always the square of the measurement units of the original data**. This is so because the original values are squared to calculate the variance. In Example 3–14, the measurement units of the original data are millions of dollars. However, the measurement units of the variance are squared millions of dollars, which, of course, does not make any sense. Thus, the variance of the 2014 compensations of 11 female CEOs of American companies is 63.2502 squared million dollars. But the standard deviation of these compensations is $7.95 million. The measurement units of the standard deviation are the same as the measurement units of the original data because the standard deviation is obtained by taking the square root of the variance.

Earlier in this chapter, we explained the difference between resistant and nonresistant summary measures. The variance and standard deviation are **nonresistant** summary measures as their values are sensitive to the outliers. The existence of outliers in a data set will increase the values of the variance and standard deviation.

EXAMPLE 3–15	Earnings of Employees

Calculating the population variance and standard deviation for ungrouped data.

The following data give the 2015 earnings (in thousands of dollars) before taxes for all six employees of a small company.

$$88.50 \qquad 108.40 \qquad 65.50 \qquad 52.50 \qquad 79.80 \qquad 54.60$$

Calculate the variance and standard deviation for these data.

Solution Let x denote the 2015 earnings before taxes of an employee of this company. The values of Σx and Σx^2 are calculated in Table 3.7.

Table 3.7

x	x^2
88.50	7832.25
108.40	11,750.56
65.50	4290.25
52.50	2756.25
79.80	6368.04
54.60	2981.16
$\Sigma x = 449.30$	$\Sigma x^2 = 35,978.51$

Because the data in this example are on earnings of *all* employees of this company, we use the population formula to compute the variance using $N = 6$. Thus, the variance is

$$\sigma^2 = \frac{\Sigma x^2 - \left(\dfrac{(\Sigma x)^2}{N}\right)}{N} = \frac{35,978.51 - \left(\dfrac{(449.30)^2}{6}\right)}{6} = \textbf{388.90}$$

The standard deviation is obtained by taking the (positive) square root of the variance:

$$\sigma = \sqrt{\frac{\Sigma x^2 - \left(\dfrac{(\Sigma x)^2}{N}\right)}{N}} = \sqrt{388.90} = \textbf{19.721 thousand} = \textbf{\$19,721}$$

Thus, the standard deviation of the 2015 earnings of all six employees of this company is $19,721.

Note that Σx^2 is not the same as $(\Sigma x)^2$. The value of Σx^2 is obtained by squaring the x values and then adding them. The value of $(\Sigma x)^2$ is obtained by squaring the value of Σx.

◀ *Warning*

The uses of the standard deviation are discussed in Section 3.4. Later chapters explain how the mean and the standard deviation taken together can help in making inferences about the population.

3.2.3 Coefficient of Variation

One disadvantage of the standard deviation as a measure of dispersion is that it is a measure of absolute variability and not of relative variability. Sometimes we may need to compare the variability for two different data sets that have different units of measurement. In such cases, a measure of relative variability is preferable. One such measure is the **coefficient of variation**.

Coefficient of Variation

The coefficient of variation, denoted by CV, expresses standard deviation as a percentage of the mean and is computed as follows.

For population data:

$$CV = \frac{\sigma}{\mu} \times 100\%$$

For sample data:

$$CV = \frac{s}{\bar{x}} \times 100\%$$

Note that the coefficient of variation does not have any units of measurement, as it is always expressed as a percent.

Thus, to calculate the coefficient of variation, we perform the following steps: First we calculate the mean and standard deviation for the given data set. Then we divide the standard deviation by the mean and multiply the answer by 100%.

The following example shows the calculation of the coefficient of variation.

EXAMPLE 3–16 Salaries and Education

Calculating the coefficient of variation.

The yearly salaries of all employees working for a large company have a mean of $72,350 and a standard deviation of $12,820. The years of schooling (education) for the same employees have a mean of 15 years and a standard deviation of 2 years. Is the relative variation in the salaries higher or lower than that in years of schooling for these employees? Answer the question by calculating the coefficient of variation for each variable.

Solution Because the two variables (salary and years of schooling) have different units of measurement (dollars and years, respectively), we cannot directly compare the two standard deviations. Hence, we calculate the coefficient of variation for each of these data sets.

$$CV \text{ for salaries} = \frac{\sigma}{\mu} \times 100\% = \frac{12,820}{72,350} \times 100\% = \mathbf{17.72\%}$$

$$CV \text{ for years of schooling} = \frac{\sigma}{\mu} \times 100\% = \frac{2}{15} \times 100\% = \mathbf{13.33\%}$$

Thus, the standard deviation for salaries is **17.72%** of its mean and that for years of schooling is **13.33%** of its mean. Since the coefficient of variation for salaries has a higher value than the coefficient of variation for years of schooling, the salaries have a higher relative variation than the years of schooling.

Note that the coefficient of variation for salaries in the above example is 17.72%. This means that if we assume that the mean of salaries for these employees is 100, then the standard deviation of salaries is 17.72. Similarly, if the mean of years of schooling for these employees is 100, then the standard deviation of years of schooling is 13.33.

3.2.4 Population Parameters and Sample Statistics

A numerical measure such as the mean, median, mode, range, variance, or standard deviation calculated for a population data set is called a *population parameter*, or simply a **parameter**. A summary measure calculated for a sample data set is called a *sample statistic*, or simply a **statistic**. Thus, μ and σ are population parameters, and $\bar{x}$ and s are sample statistics. As an illustration, $\bar{x} = \$16{,}070.3$ million in Example 3–1 is a sample statistic, and $\mu = 45.25$ years in Example 3–2 is a population parameter. Similarly, $s = \$7.95$ million in Example 3–14 is a sample statistic, whereas $\sigma = \$19{,}721$ in Example 3–15 is a population parameter.

EXERCISES

CONCEPTS AND PROCEDURES

3.28 The range, as a measure of spread, has the disadvantage of being influenced by outliers. Illustrate this with an example.

3.29 Is it possible to have a data set with a negative value for standard deviation? Explain.

3.30 Can the standard deviation of a data set be zero? Explain with an example.

3.31 Differentiate between a population parameter and a sample statistic. Give one example of each.

3.32 The following data set belongs to a sample:

$$14 \quad 16 \quad -10 \quad 8 \quad 8 \quad -18$$

Calculate the range, variance, and standard deviation.

APPLICATIONS

3.33 The following data give the prices of seven textbooks randomly selected from a university bookstore.

$$\$112 \quad \$170 \quad \$93 \quad \$113 \quad \$56 \quad \$161 \quad \$123$$

a. Find the mean for these data. Calculate the deviations of the data values from the mean. Is the sum of these deviations zero?
b. Calculate the range, variance, and standard deviation.

3.34 The following data give the number of shoplifters apprehended during each of the past 10 weeks at a large department store.

$$7 \quad 10 \quad 8 \quad 3 \quad 15 \quad 12 \quad 6 \quad 11 \quad 10 \quad 8$$

a. Find the mean for these data. Calculate the deviations of the data values from the mean. Is the sum of these deviations zero?
b. Calculate the range, variance, and standard deviation.

3.35 The data below shows the numbers of tornadoes that touched down in 12 states that had the most tornadoes during the period 1950 to 1994. The data for these states are given in the following order: CO, FL, IA, IL, KS, LA, MO, MS, NE, OK, SD, TX.

$$1113 \quad 2009 \quad 1374 \quad 1137 \quad 2110 \quad 1086$$
$$1166 \quad 1039 \quad 1673 \quad 2300 \quad 1139 \quad 5490$$

a. Compute the range, variance, and standard deviation for these data.
b. Calculate the coefficient of variation.

3.36 The following data represent the total points scored in each of the NFL championship games played from 2000 through 2009 in that order.

$$39 \quad 41 \quad 37 \quad 69 \quad 61 \quad 45 \quad 31 \quad 46 \quad 31 \quad 50$$

a. Calculate the range, variance, and standard deviation for these data.
b. Calculate the coefficient of variation.

3.37 The following data give the numbers of hours spent partying by 10 randomly selected college students during the past week.

$$7 \quad 14 \quad 5 \quad 0 \quad 9 \quad 7 \quad 10 \quad 4 \quad 0 \quad 8$$

a. Calculate the range, variance, and standard deviation for these data.
b. Calculate the coefficient of variation.

3.38 Following are the temperatures (in degrees Fahrenheit) observed during eight wintry days in a midwestern city:

$$23 \quad 14 \quad 6 \quad -7 \quad -2 \quad 11 \quad 16 \quad 19$$

a. Calculate the range, variance, and standard deviation for these data.
b. Calculate the coefficient of variation.

3.39 The following data give the number of hot dogs consumed by 10 participants in a hot-dog-eating contest.

$$21 \quad 17 \quad 32 \quad 5 \quad 25 \quad 15 \quad 17 \quad 21 \quad 9 \quad 24$$

a. Calculate the range, variance, and standard deviation for these data.
b. Calculate the coefficient of variation.

3.40 Attacks by stinging insects, such as bees or wasps, may become medical emergencies if either the victim is allergic to venom or multiple stings are involved. The following data give the number of patients treated each week for such stings in a large regional hospital during 13 weeks last summer.

$$1 \quad 5 \quad 2 \quad 3 \quad 0 \quad 4 \quad 1 \quad 7 \quad 0 \quad 1 \quad 2 \quad 0 \quad 1$$

Calculate the range, variance, and standard deviation for these data.

3.41 The following data represent the 2011 guaranteed salaries (in thousands of dollars) of the head coaches of the final eight teams in the 2011 NCAA Men's Basketball Championship. The data represent the 2011 salaries of basketball coaches of the following universities, entered in that order: Arizona, Butler, Connecticut, Florida, Kansas, Kentucky, North Carolina, and Virginia Commonwealth. (*Source:* www.usatoday.com).

| 1950 | 434 | 2300 | 3575 | 3376 | 3800 | 1655 | 418 |

a. Calculate the range, variance, and standard deviation for these data.

b. Calculate the coefficient of variation.

3.42 The following data give the daily wage rates of eight employees of a company.

| $94 | 94 | 94 | 94 | 94 | 94 | 94 | 94 |

Calculate the standard deviation. Is its value zero? If yes, why?

3.43 The yearly salaries of all employees who work for a company have a mean of $61,850 and a standard deviation of $6740.

The years of experience for the same employees have a mean of 15 years and a standard deviation of 2 years. Is the relative variation in the salaries larger or smaller than that in years of experience for these employees?

*3.44 The SAT scores of 100 students have a mean of 1020 and a standard deviation of 115. The GPAs of the same 100 students have a mean of 3.21 and a standard deviation of .26. Is the relative variation in SAT scores larger or smaller than that in GPAs?

*3.45 Consider the following two data sets.

| Data Set I: | 4 | 8 | 15 | 9 | 11 |
| Data Set II: | 12 | 24 | 45 | 27 | 33 |

Note that each value of the second data set is obtained by multiplying the corresponding value of the first data set by 3. Calculate the standard deviation for each of these two data sets using the formula for population data. Comment on the relationship between the two standard deviations.

3.3 | Mean, Variance, and Standard Deviation for Grouped Data

In Sections 3.1.1 and 3.2.2, we learned how to calculate the mean, variance, and standard deviation for ungrouped data. In this section, we will learn how to calculate the mean, variance, and standard deviation for grouped data.

3.3.1 Mean for Grouped Data

We learned in Section 3.1.1 that the mean is obtained by dividing the sum of all values by the number of values in a data set. However, if the data are given in the form of a frequency table, we no longer know the values of individual observations. Consequently, in such cases, we cannot obtain the sum of individual values. We find an approximation for the sum of these values using the procedure explained in the next paragraph and example. The formulas used to calculate the mean for grouped data follow.

Calculating Mean for Grouped Data

$$\text{Mean for population data:} \quad \mu = \frac{\sum mf}{N}$$

$$\text{Mean for sample data:} \quad \bar{x} = \frac{\sum mf}{n}$$

where m is the midpoint and f is the frequency of a class.

To calculate the mean for grouped data, first find the midpoint of each class and then multiply the midpoints by the frequencies of the corresponding classes. The sum of these products, denoted by $\sum mf$, gives an approximation for the sum of all values. To find the value of the mean, divide this sum by the total number of observations in the data.

| **EXAMPLE 3–17** | **Daily Commuting Times for Employees** |

Calculating the population mean for grouped data.

Table 3.8 gives the frequency distribution of the daily one-way commuting times (in minutes) from home to work for *all* 25 employees of a company.

Table 3.8

Daily Commuting Time (minutes)	Number of Employees (f)
0 to less than 10	4
10 to less than 20	9
20 to less than 30	6
30 to less than 40	4
40 to less than 50	2

Calculate the mean of the daily commuting times.

Solution   Note that because the data set includes *all* 25 employees of the company, it represents the population. Table 3.9 shows the calculation of Σmf. Note that in Table 3.9, m denotes the midpoints of the classes.

Table 3.9

Daily Commuting Time (minutes)	f	m	mf
0 to less than 10	4	5	20
10 to less than 20	9	15	135
20 to less than 30	6	25	150
30 to less than 40	4	35	140
40 to less than 50	2	45	90
	$N = 25$		$\Sigma mf = 535$

To calculate the mean, we first find the midpoint of each class. The class midpoints are recorded in the third column of Table 3.9. The products of the midpoints and the corresponding frequencies are listed in the fourth column. The sum of the fourth column values, denoted by Σmf, gives the approximate total daily commuting time (in minutes) for all 25 employees. The mean is obtained by dividing this sum by the total frequency. Therefore,

$$\mu = \frac{\Sigma mf}{N} = \frac{535}{25} = \textbf{21.40 minutes}$$

Thus, the employees of this company spend an average of 21.40 minutes a day commuting from home to work.

What do the numbers 20, 135, 150, 140, and 90 in the column labeled mf in Table 3.9 represent? We know from this table that 4 employees spend 0 to less than 10 minutes commuting per day. If we assume that the time spent commuting by these 4 employees is evenly spread in the interval 0 to less than 10, then the midpoint of this class (which is 5) gives the mean time spent commuting by these 4 employees. Hence, $4 \times 5 = 20$ is the approximate total time (in minutes) spent commuting per day by these 4 employees. Similarly, 9 employees spend 10 to less than 20 minutes commuting per day, and the total time spent commuting by these 9 employees is approximately 135 minutes a day. The other numbers in this column can be interpreted in the same way. Note that these numbers give the approximate commuting times for these employees based on the assumption of an even spread within classes. The total commuting time for all 25 employees is approximately 535 minutes. Consequently, 21.40 minutes is an approximate and not the exact value of the mean. We can find the exact value of the mean only if we know the exact commuting time for each of the 25 employees of the company.

EXAMPLE 3–18   Number of Orders Received

Calculating the sample mean for grouped data.

Table 3.10 gives the frequency distribution of the number of orders received each day during the past 50 days at the office of a mail-order company.

Table 3.10

Number of Orders	Number of Days (f)
10–12	4
13–15	12
16–18	20
19–21	14

Calculate the mean.

Solution   Because the data set includes only 50 days, it represents a sample. The value of Σmf is calculated in Table 3.11.

Table 3.11

Number of Orders	f	m	mf
10–12	4	11	44
13–15	12	14	168
16–18	20	17	340
19–21	14	20	280
	$n = 50$		$\Sigma mf = 832$

The value of the sample mean is

$$\bar{x} = \frac{\Sigma mf}{n} = \frac{832}{50} = \textbf{16.64 orders}$$

Thus, this mail-order company received an average of 16.64 orders per day during these 50 days.

3.3.2   Variance and Standard Deviation for Grouped Data

Following are what we will call the *basic formulas* that are used to calculate the population and sample variances for grouped data:

$$\sigma^2 = \frac{\Sigma f(m - \mu)^2}{N} \quad \text{and} \quad s^2 = \frac{\Sigma f(m - \bar{x})^2}{n - 1}$$

where σ^2 is the population variance, s^2 is the sample variance, and m is the midpoint of a class.

In either case, the standard deviation is obtained by taking the positive square root of the variance.

Again, the *short-cut formulas* are more efficient for calculating the variance and standard deviation. Section A3.1.2 of Appendix 3.1 at the end of this chapter shows how to use the basic formulas to calculate the variance and standard deviation for grouped data.

Short-Cut Formulas for the Variance and Standard Deviation for Grouped Data

$$\sigma^2 = \frac{\Sigma m^2 f - \left(\dfrac{(\Sigma mf)^2}{N} \right)}{N} \quad \text{and} \quad s^2 = \frac{\Sigma m^2 f - \left(\dfrac{(\Sigma mf)^2}{n} \right)}{n - 1}$$

where σ^2 is the population variance, s^2 is the sample variance, and m is the midpoint of a class.

The standard deviation is obtained by taking the positive square root of the variance.

$$\text{Population standard deviation:}\quad \sigma = \sqrt{\dfrac{\sum m^2 f - \left(\dfrac{(\sum mf)^2}{N}\right)}{N}}$$

$$\text{Sample standard deviation:}\quad s = \sqrt{\dfrac{\sum m^2 f - \left(\dfrac{(\sum mf)^2}{n}\right)}{n-1}}$$

Examples 3–19 and 3–20 illustrate the use of these formulas to calculate the variance and standard deviation.

EXAMPLE 3–19   Daily Commuting Times of Employees

Calculating the population variance and standard deviation for grouped data.

The following data, reproduced from Table 3.8 of Example 3–17, give the frequency distribution of the daily one-way commuting times (in minutes) from home to work for all 25 employees of a company.

Daily Commuting Time (minutes)	Number of Employees (f)
0 to less than 10	4
10 to less than 20	9
20 to less than 30	6
30 to less than 40	4
40 to less than 50	2

Calculate the variance and standard deviation.

Solution   All four steps needed to calculate the variance and standard deviation for grouped data are shown after Table 3.12.

Table 3.12

Daily Commuting Time (minutes)	f	m	mf	$m^2 f$
0 to less than 10	4	5	20	100
10 to less than 20	9	15	135	2025
20 to less than 30	6	25	150	3750
30 to less than 40	4	35	140	4900
40 to less than 50	2	45	90	4050
	$N = 25$		$\sum mf = 535$	$\sum m^2 f = 14{,}825$

Step 1.   *Calculate the value of $\sum mf$.*

To calculate the value of $\sum mf$, first find the midpoint m of each class (see the third column in Table 3.12) and then multiply the corresponding class midpoints and class frequencies (see the fourth column). The value of $\sum mf$ is obtained by adding these products. Thus,

$$\sum mf = 535$$

Step 2. *Find the value of $\sum m^2 f$.*

To find the value of $\sum m^2 f$, square each m value and multiply this squared value of m by the corresponding frequency (see the fifth column in Table 3.12). The sum of these products (that is, the sum of the fifth column) gives $\sum m^2 f$. Hence,

$$\sum m^2 f = 14,825$$

Step 3. *Calculate the variance.*

Because the data set includes all 25 employees of the company, it represents the population. Therefore, we use the formula for the population variance:

$$\sigma^2 = \frac{\sum m^2 f - \left(\dfrac{(\sum mf)^2}{N}\right)}{N} = \frac{14,825 - \left(\dfrac{(535)^2}{25}\right)}{25} = \frac{3376}{25} = \mathbf{135.04}$$

Step 4. *Calculate the standard deviation.*

To obtain the standard deviation, take the (positive) square root of the variance.

$$\sigma = \sqrt{\sigma^2} = \sqrt{135.04} = \mathbf{11.62\ minutes}$$

Thus, the standard deviation of the daily commuting times for these employees is 11.62 minutes.

Note that the values of the variance and standard deviation calculated in Example 3–19 for grouped data are approximations. The exact values of the variance and standard deviation can be obtained only by using the ungrouped data on the daily commuting times of these 25 employees.

EXAMPLE 3–20	**Number of Orders Received**

Calculating the sample variance and standard deviation for grouped data.

The following data, reproduced from Table 3.10 of Example 3–18, give the frequency distribution of the number of orders received each day during the past 50 days at the office of a mail-order company.

Number of Orders	f
10–12	4
13–15	12
16–18	20
19–21	14

Calculate the variance and standard deviation.

Solution All the information required for the calculation of the variance and standard deviation appears in Table 3.13.

Table 3.13

Number of Orders	f	m	mf	$m^2 f$
10–12	4	11	44	484
13–15	12	14	168	2352
16–18	20	17	340	5780
19–21	14	20	280	5600
	$n = 50$		$\sum mf = 832$	$\sum m^2 f = 14{,}216$

Because the data set includes only 50 days, it represents a sample. Hence, we use the sample formulas to calculate the variance and standard deviation. By substituting the values into the formula for the sample variance, we obtain

$$s^2 = \frac{\sum m^2 f - \left(\dfrac{(\sum mf)^2}{n}\right)}{n - 1} = \frac{14{,}216 - \left(\dfrac{(832)^2}{50}\right)}{50 - 1} = 7.5820$$

Hence, the standard deviation is

$$s = \sqrt{s^2} = \sqrt{7.5820} = \textbf{2.75 orders}$$

Thus, the standard deviation of the number of orders received at the office of this mail-order company during the past 50 days is 2.75.

EXERCISES

CONCEPTS AND PROCEDURES

3.46 Are the values of the mean and standard deviation that are calculated using grouped data exact or approximate values of the mean and standard deviation, respectively? Explain.

3.47 Using the population formulas, calculate the mean, variance, and standard deviation for the following grouped data.

x	2–4	5–7	8–10	11–13	14–16
f	3	10	14	8	5

3.48 Using the sample formulas, find the mean, variance, and standard deviation for the grouped data displayed in the following table.

x	f
0 to less than 4	17
4 to less than 8	23
8 to less than 12	15
12 to less than 16	11
16 to less than 20	8
20 to less than 24	5

APPLICATIONS

3.49 The following table gives the distribution of the amounts of rainfall (in inches) for July 2015 for 50 cities.

Rainfall	Number of Cities
0 to less than 2	6
2 to less than 4	10
4 to less than 6	20
6 to less than 8	7
8 to less than 10	4
10 to less than 12	3

Find the mean, variance, and standard deviation. Are the values of these summary measures population parameters or sample statistics?

3.50 The following table gives the grouped data on the weights of all 100 babies born at a hospital in 2009.

Weight (pounds)	Number of Babies
3 to less than 5	3
5 to less than 7	35
7 to less than 9	45
9 to less than 11	15
11 to less than 13	2

Find the mean, variance, and standard deviation. Give a brief interpretation of the values in the column labeled mf in your table of calculations. What does $\sum mf$ represent?

3.51 The following table gives information on the amounts (in dollars) of electric bills for July 2016 for a sample of 50 families.

Amount of Electric Bill (dollars)	Number of Families
0 to less than 60	3
60 to less than 120	17
120 to less than 180	13
180 to less than 240	11
240 to less than 300	6

Find the mean, variance, and standard deviation. Give a brief interpretation of the values in the column labeled mf in your table of calculations. What does $\sum mf$ represent?

3.52 The following data give the amounts (in dollars) spent last month on food by 400 randomly selected families.

Food Expenditure (in dollars)	Number of Families
200 to less than 500	38
500 to less than 800	105
800 to less than 1100	130
1100 to less than 1400	60
1400 to less than 1700	42
1700 to less than 2000	18
2000 to less than 2300	7

Calculate the mean, variance, and standard deviation.

3.53   For 50 airplanes that arrived late at an airport during a week, the time by which they were late was observed. In the following table,

x denotes the time (in minutes) by which an airplane was late, and f denotes the number of airplanes.

x	f
0 to less than 20	14
20 to less than 40	18
40 to less than 60	9
60 to less than 80	5
80 to less than 100	4

Find the mean, variance, and standard deviation.

3.4 | Use of Standard Deviation

By using the mean and standard deviation, we can find the proportion or percentage of the total observations that fall within a given interval about the mean. This section briefly discusses Chebyshev's theorem and the empirical rule, both of which demonstrate this use of the standard deviation.

3.4.1   Chebyshev's Theorem

Chebyshev's theorem gives a lower bound for the area under a curve between two points that are on opposite sides of the mean and at the same distance from the mean.

> **Chebyshev's Theorem**   For any number k greater than 1, at least $(1 - 1/k^2)$ of the data values lie within k standard deviations of the mean.

Note that k gives the distance between the mean and a point in terms of the number of standard deviations. To find the area under a curve between two points using Chebyshev's theorem, we perform the following steps: First we find the distance between the mean and each of the two given points. Note that both these distances must be the same to apply Chebyshev's theorem. Then we divide the distance calculated above by the standard deviation. This gives the value of k. Finally we substitute the value of k in the formula $1 - \frac{1}{k^2}$ and simplify. Multiply this answer by 100 to obtain the percentage. This gives the minimum area (in percent) under the distribution curve between the two points.

Figure 3.5 illustrates Chebyshev's theorem.

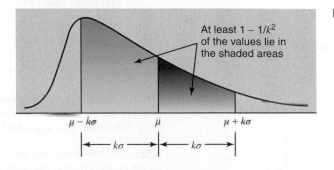

At least $1 - 1/k^2$ of the values lie in the shaded areas

Figure 3.5   Chebyshev's theorem.

Table 3.14 lists the areas under a distribution curve for different values of k using Chebyshev's theorem.

Table 3.14   Areas Under the Distribution Curve Using Chebyshev's Theorem

k	Interval	$1 - \dfrac{1}{k^2}$	Minimum Area Within k Standard Deviations
1.5	$\mu \pm 1.5\sigma$	$1 - \dfrac{1}{1.5^2} = 1 - .44 = .56$	56%
2.0	$\mu \pm 2\sigma$	$1 - \dfrac{1}{2^2} = 1 - .25 = .75$	75%
2.5	$\mu \pm 2.5\sigma$	$1 - \dfrac{1}{2.5^2} = 1 - .16 = .84$	84%
3.0	$\mu \pm 3\sigma$	$1 - \dfrac{1}{3.0^2} = 1 - .11 = .89$	89%

For example, using Table 3.14, if $k = 2$, then according to Chebyshev's theorem, at least .75, or 75%, of the values of a data set lie within two standard deviations of the mean. This is shown in Figure 3.6.

Figure 3.6   Percentage of values within two standard deviations of the mean for Chebyshev's theorem.

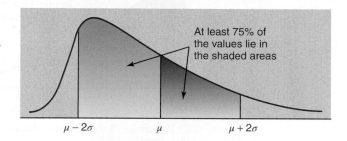

At least 75% of the values lie in the shaded areas

$\mu - 2\sigma$ μ $\mu + 2\sigma$

If $k = 3$, then according to Chebyshev's theorem, at least .89, or 89%, of the values fall within three standard deviations of the mean. This is shown in Figure 3.7.

Figure 3.7   Percentage of values within three standard deviations of the mean for Chebyshev's theorem.

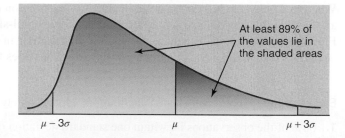

At least 89% of the values lie in the shaded areas

$\mu - 3\sigma$ μ $\mu + 3\sigma$

Although in Figures 3.5 through 3.7 we have used the population notation for the mean and standard deviation, the theorem applies to both sample and population data. Note that Chebyshev's theorem is applicable to a distribution of any shape. However, Chebyshev's theorem can be used only for $k > 1$. This is so because when $k = 1$, the value of $1 - 1/k^2$ is zero, and when $k < 1$, the value of $1 - 1/k^2$ is negative.

EXAMPLE 3–21   Blood Pressure of Women

Applying Chebyshev's theorem.

The average systolic blood pressure for 4000 women who were screened for high blood pressure was found to be 187 mm Hg with a standard deviation of 22. Using Chebyshev's theorem, find the minimum percentage of women in this group who have a systolic blood pressure between 143 and 231 mm Hg.

Solution   Let μ and σ be the mean and the standard deviation, respectively, of the systolic blood pressures of these women. Then, from the given information,

$$\mu = 187 \quad \text{and} \quad \sigma = 22$$

To find the percentage of women whose systolic blood pressures are between 143 and 231 mm Hg, the first step is to determine k. As shown below, each of the two points, 143 and 231, is 44 units away from the mean.

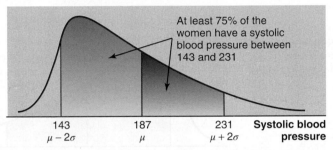

$$
\begin{array}{ccc}
\lvert \longleftarrow 44 \longrightarrow \rvert \longleftarrow 44 \longrightarrow \rvert \\
143 \qquad \mu = 187 \qquad 231
\end{array}
$$

The value of k is obtained by dividing the distance between the mean and each point by the standard deviation. Thus,

$$k = 44/22 = 2$$

From Table 3.14, for $k = 2$, the area under the curve is at least 75%. Hence, according to Chebyshev's theorem, at least 75% of the women have a systolic blood pressure between 143 and 231 mm Hg. This percentage is shown in Figure 3.8.

At least 75% of the women have a systolic blood pressure between 143 and 231

| 143 | 187 | 231 | **Systolic blood** |
| $\mu - 2\sigma$ | μ | $\mu + 2\sigma$ | **pressure** |

Figure 3.8   Percentage of women with systolic blood pressure between 143 and 231.

3.4.2 Empirical Rule

Whereas Chebyshev's theorem is applicable to a distribution of any shape, the **empirical rule** applies only to a specific shape of distribution called a **bell-shaped distribution**, as shown in Figure 3.9. More will be said about such a distribution in Chapter 6, where it is called a *normal curve*. In this section, only the following three rules for the curve are given.

Empirical Rule   For a bell-shaped distribution, approximately

1. 68% of the observations lie within one standard deviation of the mean.
2. 95% of the observations lie within two standard deviations of the mean.
3. 99.7% of the observations lie within three standard deviations of the mean.

Table 3.15 lists the areas under a bell-shaped distribution for the above mentioned three intervals.

Table 3.15   Approximate Areas under a Bell-Shaped Distribution Using the Empirical Rule

Interval	Approximate Area
$\mu \pm 1\sigma$	68%
$\mu \pm 2\sigma$	95%
$\mu \pm 3\sigma$	99.7%

Does Spread Mean the Same As Variability and Dispersion?

In any discipline, there is terminology that one needs to learn in order to become fluent. Accounting majors need to learn the difference between credits and debits, chemists need to know how an ion differs from an atom, and physical therapists need to know the difference between abduction and adduction. Statistics is no different. Failing to learn the difference between the mean and the median makes much of the remainder of this book very difficult to understand.

Another issue with terminology is the use of words other than the terminology to describe a specific concept or scenario. Sometimes the words one chooses to use can be vague or ambiguous, resulting in confusion. One debate in the statistics community involves the use of the word "spread" in place of the words "dispersion" and "variability." In a 2012 article, "Lexical ambiguity: making a case against *spread*," authors Jennifer Kaplan, Neal Rogness, and Diane Fisher point out that the Oxford English Dictionary has more than 25 definitions for the word spread, many of which students know coming into a statistics class. As a result of knowing some of the meanings of spread, students who use the word spread in place of variability or dispersion "do not demonstrate strong statistical meanings of the word spread at the end of a one-semester statistics course."

In order to examine the extent of this issue, the authors of the article designed a study in which they selected 160 undergraduate students taking an introductory statistics course from 14 different professors at three different universities and in the first week of the semester asked them to write sentences and definitions for spread using its primary meaning. Then, at the end of the semester, the same students were asked to write sentences and definitions for spread using its primary meaning in statistics. The authors found that responses of only one-third of the students related spread to the concept of variability, which has to do with how the data vary around the center of a distribution. A slightly larger percentage of students gave responses that "defined spread as 'to cover evenly or in a thin layer,'" while approximately one in eight responded with a definition that was synonymous with the notion of range. Seven other definitions were given by at least three students in the study.

Although more of the definitions and sentences provided at the end of the course had something to do with statistics, the authors did not see an increase in the percentage of definitions that associated spread with the concept of variability. Hence, they suggested that the ambiguity of the term spread is sufficient enough to stop using it in place of the terms variability and dispersion.

Source: Kaplan, J. J., Rogness, N. T., and Fisher, D. G., "Lexical ambiguity: making a case against *spread*," *Teaching Statistics*, 2011, 34, (2), pp. 56–60. © 2011 Teaching Statistics Trust.

Figure 3.9 illustrates the empirical rule. Again, the empirical rule applies to both population data and sample data.

Figure 3.9   Illustration of the empirical rule.

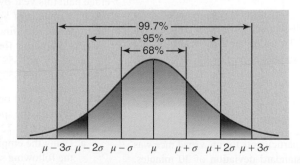

$$\mu - 3\sigma \quad \mu - 2\sigma \quad \mu - \sigma \quad \mu \quad \mu + \sigma \quad \mu + 2\sigma \quad \mu + 3\sigma$$

EXAMPLE 3–22   Age Distribution of Persons

Applying the empirical rule.

The age distribution of a sample of 5000 persons is bell-shaped with a mean of 40 years and a standard deviation of 12 years. Determine the approximate percentage of people who are 16 to 64 years old.

Solution   We use the empirical rule to find the required percentage because the distribution of ages follows a bell-shaped curve. From the given information, for this distribution,

$$\bar{x} = 40 \text{ years} \quad \text{and} \quad s = 12 \text{ years}$$

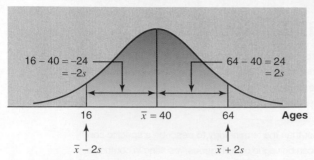

Figure 3.10   Percentage of people who are 16 to 64 years old.

Each of the two points, 16 and 64, is 24 units away from the mean. Dividing 24 by 12, we convert the distance between each of the two points and the mean in terms of the number of standard deviations. Thus, the distance between 16 and 40 and that between 40 and 64 is each equal to $2s$. Consequently, as shown in Figure 3.10, the area from 16 to 64 is the area from $\bar{x} - 2s$ to $\bar{x} + 2s$.

Because, from Table 3.15, the area within two standard deviations of the mean is approximately 95% for a bell-shaped curve, approximately **95%** of the people in the sample are 16 to 64 years old.

EXERCISES

CONCEPTS AND PROCEDURES

3.54   Explain the significance of the Chebyshev's theorem and its application in statistics.

3.55   Briefly explain the empirical rule. To what kind of distribution is it applied?

3.56   A sample of 2000 observations has a mean of 74 and a standard deviation of 12. Using Chebyshev's theorem, find the minimum percentage of the observations that fall in the intervals $\bar{x} \pm 2s$, $\bar{x} \pm 2.5s$, and $\bar{x} \pm 3s$. Note that $\bar{x} \pm 2s$ represents the interval $\bar{x} - 2s$ to $\bar{x} + 2s$, and so on.

3.57   A large population has a mean of 230 and a standard deviation of 41. Using Chebyshev's theorem, find the minimum percentage of the observations that fall in the intervals $\mu \pm 2\sigma$, $\mu \pm 2.5\sigma$, and $\mu \pm 3\sigma$.

3.58   A large population has a bell-shaped distribution with a mean of 310 and a standard deviation of 37. Using the empirical rule, find the approximate percentage of the observations that fall in the intervals $\mu \pm 1\sigma$, $\mu \pm 2\sigma$, and $\mu \pm 3\sigma$.

APPLICATIONS

3.59   The mean time taken by all participants to run a road race was found to be 250 minutes with a standard deviation of 30 minutes. Using Chebyshev's theorem, find the percentage of runners who ran this road race in
    a. 190 to 310 minutes   **b.** 160 to 340 minutes
    c. 175 to 325 minutes

3.60   The 2016 gross sales of all companies in a large city have a mean of $2.3 million and a standard deviation of $.5 million. Using Chebyshev's theorem, find at least what percentage of companies in this city had 2016 gross sales of
    a. $1.3 to $3.3 million   **b.** $1.05 to $3.55 million
    c. $.8 to $3.8 million

3.61   According to the National Center for Education Statistics (www.nces.ed.gov), the amounts of all loans, including Federal Parent PLUS loans, granted to students during the 2007–2008 academic year had a distribution with a mean of $8109.65. Suppose that the standard deviation of this distribution is $2412.
    a. Using Chebyshev's theorem, find at least what percentage of students had 2007–2008 such loans between
        i. $2079.65 and $14,139.65   **ii.** $3285.65 and $12,933.65
    *b. Using Chebyshev's theorem, find the interval that contains the amounts of 2007–2008 such loans for at least 89% of all students.

3.62   According to the Kaiser Family Foundation, U.S. workers who had employer-provided health insurance paid an average premium of $4129 for family coverage during 2011 (*USA TODAY*, October 10, 2011). Suppose that the premiums for such family coverage paid this year by all such workers have a bell-shaped distribution with a mean of $4129 and a standard deviation of $600. Using the empirical rule, find the approximate percentage of such workers who pay premiums for such family coverage between
    a. $2329 and $5929   **b.** $3529 and $4729
    c. $2929 and $5329

3.63   Suppose that on a certain section of I-95 with a posted speed limit of 65 mph, the speeds of all vehicles have a bell-shaped distribution with a mean of 72 mph and a standard deviation of 3 mph.
    a. Using the empirical rule, find the percentage of vehicles with the following speeds on this section of I-95.
        i. 63 to 81 mph   **ii.** 69 to 75 mph
    *b. Using the empirical rule, find the interval that contains the speeds of 95% of vehicles traveling on this section of I-95.

3.64   The one-way commuting times from home to work for all employees working at a large company have a mean of 34 minutes and a standard deviation of 8 minutes.
    a. Using Chebyshev's theorem, find the minimum percentage of employees at this company who have one-way commuting times in the following intervals.
        i. 14 to 54 minutes   **ii.** 18 to 50 minutes

*b. Using Chebyshev's theorem, find the interval that contains one-way commuting times of at least 89% of the employees at this company.

3.65 The prices of all college textbooks follow a bell-shaped distribution with a mean of $180 and a standard deviation of $30.

a. Using the empirical rule, find the (approximate) percentage of all college textbooks with their prices between
  i. $150 and $210   **ii.** $120 and $240

*b. Using the empirical rule, find the interval that contains the prices of (approximate) 99.7% of college textbooks.

3.5 | Measures of Position

A **measure of position** determines the position of a single value in relation to other values in a sample or a population data set. There are many measures of position; however, only quartiles, percentiles, and percentile rank are discussed in this section.

3.5.1   Quartiles and Interquartile Range

Quartiles are the summary measures that divide a ranked data set into four equal parts. Three measures will divide any data set into four equal parts. These three measures are the **first quartile** (denoted by Q_1), the **second quartile** (denoted by Q_2), and the **third quartile** (denoted by Q_3). The data should be ranked in increasing order before the quartiles are determined. The quartiles are defined as follows. Note that Q_1 and Q_3 are also called the lower and the upper quartiles, respectively.

Figure 3.11 describes the positions of the three quartiles.

> **Quartiles** are three values that divide a ranked data set into four equal parts. The second quartile is the same as the median of a data set. The first quartile is the median of the observations that are less than the median, and the third quartile is the median of the observations that are greater than the median.

Figure 3.11   Quartiles.

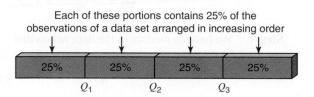

Each of these portions contains 25% of the observations of a data set arranged in increasing order

| 25% | 25% | 25% | 25% |

Q_1 Q_2 Q_3

Approximately 25% of the values in a ranked data set are less than Q_1 and about 75% are greater than Q_1. The second quartile, Q_2, divides a ranked data set into two equal parts; hence, the second quartile and the median are the same. Approximately 75% of the data values are less than Q_3 and about 25% are greater than Q_3.

The difference between the third quartile and the first quartile for a data set is called the **interquartile range (IQR)**, which is a measure of dispersion.

Calculating Interquartile Range   The difference between the third and the first quartiles gives the **interquartile range**; that is,

$$IQR = \text{Interquartile range} = Q_3 - Q_1$$

Examples 3–23 and 3–24 show the calculation of the quartiles and the interquartile range.

EXAMPLE 3–23 | Commuting Times for College Students

Finding quartiles and the interquartile range.

A sample of 12 commuter students was selected from a college. The following data give the typical one-way commuting times (in minutes) from home to college for these 12 students.

| 29 | 14 | 39 | 17 | 7 | 47 | 63 | 37 | 42 | 18 | 24 | 55 |

(a) Find the values of the three quartiles.

(b) Where does the commuting time of 47 fall in relation to the three quartiles?

(c) Find the interquartile range.

Solution

(a) We perform the following steps to find the three quartiles.

Step 1.   First we rank the given data in increasing order as follows:

$$7 \quad 14 \quad 17 \quad 18 \quad 24 \quad 29 \quad 37 \quad 39 \quad 42 \quad 47 \quad 55 \quad 63$$

Step 2.   We find the second quartile, which is also the median. In a total of 12 data values, the median is between sixth and seventh terms. Thus, the median and, hence, the second quartile is given by the average of the sixth and seventh values in the ranked data set, that is the average of 29 and 37. Thus, the second quartile is:

$$Q_2 = \frac{29 + 37}{2} = 33$$

Note that $Q_2 = 33$ is also the value of the median.

Step 3.   We find the median of the data values that are smaller than Q_2, and this gives the value of the first quartile. The values that are smaller than Q_2 are:

$$7 \quad 14 \quad 17 \quad 18 \quad 24 \quad 29$$

The value that divides these six data values in two equal parts is given by the average of the two middle values, 17 and 18. Thus, the first quartile is:

$$Q_1 = \frac{17 + 18}{2} = 17.5$$

Step 4.   We find the median of the data values that are larger than Q_2, and this gives the value of the third quartile. The values that are larger than Q_2 are:

$$37 \quad 39 \quad 42 \quad 47 \quad 55 \quad 63$$

The value that divides these six data values in two equal parts is given by the average of the two middle values, 42 and 47. Thus, the third quartile is:

$$Q_3 = \frac{42 + 47}{2} = 44.5$$

Now we can summarize the calculation of the three quartiles in the following figure.

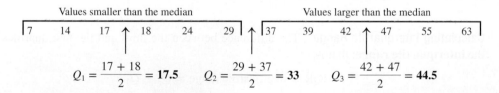

The value of $Q_1 = 17.5$ minutes indicates that 25% of these 12 students in this sample commute for less than 17.5 minutes and 75% of them commute for more than 17.5 minutes. Similarly, $Q_2 = 33$ indicates that half of these 12 students commute for less than 33 minutes and the other half of them commute for more than 33 minutes. The value of $Q_3 = 44.5$ minutes indicates that 75% of these 12 students in this sample commute for less than 44.5 minutes and 25% of them commute for more than 44.5 minutes.

(b) By looking at the position of 47 minutes, we can state that this value lies in the **top 25%** of the commuting times.

(c) The interquartile range is given by the difference between the values of the third and the first quartiles. Thus,

$$\text{IQR} = \text{Interquartile range} = Q_3 - Q_1 = 44.5 - 17.5 = \textbf{27 minutes}$$

EXAMPLE 3–24   Ages of Employees

Finding quartiles and the interquartile range.

The following are the ages (in years) of nine employees of an insurance company:

47   28   39   51   33   37   59   24   33

(a) Find the values of the three quartiles. Where does the age of 28 years fall in relation to the ages of these employees?

(b) Find the interquartile range.

Solution

(a) First we rank the given data in increasing order. Then we calculate the three quartiles as follows:

Finding quartiles for an odd number of data values.

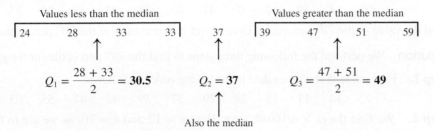

Thus the values of the three quartiles are

$$Q_1 = \textbf{30.5 years}, \quad Q_2 = \textbf{37 years}, \quad \text{and} \quad Q_3 = \textbf{49 years}$$

The age of 28 falls in the **lowest 25%** of the ages.

(b) The interquartile range is

Finding the interquartile range.

$$\text{IQR} = \text{Interquartile range} = Q_3 - Q_1 = 49 - 30.5 = \textbf{18.5 years}$$

3.5.2   Percentiles and Percentile Rank

Percentiles are the summary measures that divide a ranked data set into 100 equal parts. Each (ranked) data set has 99 percentiles that divide it into 100 equal parts. The data should be ranked in increasing order to compute percentiles. The kth percentile is denoted by P_k, where k is an integer in the range 1 to 99. For instance, the 25th percentile is denoted by P_{25}. Figure 3.12 shows the positions of the 99 percentiles.

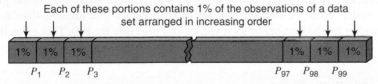

Figure 3.12   Percentiles.

Thus, the kth percentile, P_k, can be defined as a value in a data set such that about $k\%$ of the measurements are smaller than the value of P_k and about $(100 - k)\%$ of the measurements are greater than the value of P_k.

The approximate value of the kth percentile is determined as explained next.

Calculating Percentiles   The (approximate) value of the **kth percentile**, denoted by P_k, is

$$P_k = \text{Value of the } \left(\frac{k \times n}{100}\right) \text{th term in a ranked data set}$$

where k denotes the number of the percentile and n represents the sample size.

If the value of $\dfrac{k \times n}{100}$ is fractional, always round it up to the next higher whole number.

Thus, to calculate the kth percentile, first we rank the given data set in increasing order. Then, to find the kth percentile, we find the $(k \times n/100)$th term where n is the total number of data values in the given data set. If the value of $(k \times n/100)$ is fractional, we round it up to the next higher whole number. The value of this $(k \times n/100)$th term in the ranked data set gives the kth percentile, P_k.

Example 3–25 describes the procedure to calculate the percentiles.

EXAMPLE 3–25   Commuting Times for College Students

Finding the percentile for a data set.

Refer to the data on one-way commuting times (in minutes) from home to college of 12 students given in Example 3–23, which is reproduced below.

29   14   39   17   7   47   63   37   42   18   24   55

Find the value of the 70th percentile. Give a brief interpretation of the 70th percentile.

Solution   We perform the following three steps to find the 70th percentile for the given data.

Step 1.   First we rank the given data in increasing order as follows:

7   14   17   18   24   29   37   39   42   47   55   63

Step 2.   We find the $(k \times n/100)$th term. Here $n = 12$ and $k = 70$, as we are to find the 70th percentile.

$$\frac{k \times n}{100} = \frac{70 \times 12}{100} = 8.4 = 9\text{th term}$$

Thus, the 70th percentile, P_{70}, is given by the value of the 9th term in the ranked data set. Note that we rounded 8.4 up to 9, which is always the case when calculating a percentile.

Step 3.   We find the value of the 9th term in the ranked data. This gives the value of the 70th percentile, P_{70}.

$$P_{70} = \text{Value of the 9th term} = \textbf{42 minutes}$$

Thus, we can state that approximately 70% of these 12 students commute for less than or equal to 42 minutes.

We can calculate the **percentile rank** for a particular value x_i of a data set by using the formula given below. The percentile rank of x_i gives the percentage of values in the data set that are less than x_i. To find the percentile rank for a data value x_i, we first rank the given data set in increasing order. Then we find the number of data values that are less than x_i. Finally, we divide this number by the number of total values and multiply by 100. This gives the percentile rank for x_i.

Finding Percentile Rank of a Value

$$\text{Percentile rank of } x_i = \frac{\text{Number of values less than } x_i}{\text{Total number of values in the data set}} \times 100\%$$

Example 3–26 shows how the percentile rank is calculated for a data value.

EXAMPLE 3–26   Commuting Times for College Students

Finding the percentile rank for a data value.

Refer to the data on one-way commuting times (in minutes) from home to college of 12 students given in Example 3–23, which is reproduced below.

29   14   39   17   7   47   63   37   42   18   24   55

Find the percentile rank of 42 minutes. Give a brief interpretation of this percentile rank.

Solution   We perform the following three steps to find the percentile rank of 42.

Step 1.   First we rank the given data in increasing order as follows:

$$7 \quad 14 \quad 17 \quad 18 \quad 24 \quad 29 \quad 37 \quad 39 \quad 42 \quad 47 \quad 55 \quad 63$$

Step 2.   Find how many data values are less than 42.

In the above ranked data, there are eight data values that are less than 42.

Step 3.   Find the percentile rank of 42 as follows given that 8 of the 12 values in the given data set are smaller than 42:

$$\text{Percentile rank of } 42 = \frac{8}{12} \times 100\% = \mathbf{66.67\%}$$

Rounding this answer to the nearest integral value, we can state that about 67% of the students in this sample commute for less than 42 minutes.

Most statistical software packages use slightly different methods to calculate quartiles and percentiles. Those methods, while more precise, are beyond the scope of this text.

EXERCISES

CONCEPTS AND PROCEDURES

3.66   How do you find the three quartiles of a data set? Illustrate by calculating the three quartiles for two examples, the first with an odd number of observations and the second with an even number of observations.

3.67   How do you find the interquartile range of a data set? Give one example.

3.68   How do you find the percentile of a data set?

3.69   What is meant by the percentile rank for an observation of a data set?

APPLICATIONS

3.70   The following data give the weights (in pounds) lost by 15 members of a health club at the end of 2 months after joining the club.

$$\begin{array}{cccccccc} 4 & 10 & 8 & 7 & 24 & 12 & 5 & 13 \\ 11 & 10 & 20 & 9 & 8 & 9 & 18 \end{array}$$

  a. Compute the values of the three quartiles and the interquartile range.
  b. Calculate the (approximate) value of the 82nd percentile.
  c. Find the percentile rank of 10.

3.71   The following data give the speeds of 14 cars (in mph) measured by radar, traveling on I-84.

$$\begin{array}{ccccccc} 73 & 75 & 69 & 68 & 78 & 74 & 74 \\ 76 & 65 & 79 & 69 & 77 & 71 & 72 \end{array}$$

  a. Find the values of the three quartiles and the interquartile range.
  b. Calculate the (approximate) value of the 35th percentile.
  c. Compute the percentile rank of 71.

3.72   Nixon Corporation manufactures computer monitors. The following data give the numbers of computer monitors produced at the company for a sample of 30 days.

$$\begin{array}{cccccccccc} 24 & 32 & 27 & 23 & 33 & 33 & 29 & 25 & 23 & 36 \\ 26 & 26 & 31 & 20 & 27 & 33 & 27 & 23 & 28 & 29 \\ 31 & 35 & 34 & 22 & 37 & 28 & 23 & 35 & 31 & 43 \end{array}$$

  a. Calculate the values of the three quartiles and the interquartile range. Where does the value of 31 lie in relation to these quartiles?
  b. Find the (approximate) value of the 65th percentile. Give a brief interpretation of this percentile.
  c. For what percentage of the days was the number of computer monitors produced 32 or higher? Answer by finding the percentile rank of 32.

3.73   According to www.money-zine.com, the average FICO score in the United States was around 692 in December 2011. Suppose the following data represent the credit scores of 22 randomly selected loan applicants.

$$\begin{array}{cccccccccccc} 494 & 728 & 468 & 533 & 747 & 639 & 430 & 690 & 604 & 422 & 356 \\ 805 & 749 & 600 & 797 & 702 & 628 & 625 & 617 & 647 & 772 & 572 \end{array}$$

  a. Calculate the values of the three quartiles and the interquartile range. Where does the value 617 fall in relation to these quartiles?
  b. Find the approximate value of the 30th percentile. Give a brief interpretation of this percentile.
  c. Calculate the percentile rank of 533. Give a brief interpretation of this percentile rank.

3.74   The following data give the numbers of text messages sent by a high school student on 40 randomly selected days during 2012:

$$\begin{array}{cccccccccc} 32 & 33 & 33 & 34 & 35 & 36 & 37 & 37 & 37 & 32 \\ 38 & 39 & 40 & 41 & 41 & 42 & 42 & 42 & 43 & 44 \\ 44 & 45 & 45 & 45 & 47 & 58 & 47 & 47 & 41 & 48 \\ 48 & 60 & 50 & 50 & 51 & 36 & 53 & 54 & 59 & 61 \end{array}$$

  a. Calculate the values of the three quartiles and the interquartile range. Where does the value 49 fall in relation to these quartiles?
  b. Determine the approximate value of the 91st percentile. Give a brief interpretation of this percentile.
  c. For what percentage of the days was the number of text messages sent 40 or higher? Answer by finding the percentile rank of 40.

3.6 | Box-and-Whisker Plot

A plot that shows the center, spread, and skewness of a data set. It is constructed by drawing a box and two whiskers that use the median, the first quartile, the third quartile, and the smallest and the largest values in the data set between the lower and the upper inner fences.

A **box-and-whisker plot** gives a graphic presentation of data using five measures: the median, the first quartile, the third quartile, and the smallest and the largest values in the data set between the lower and the upper inner fences. (The inner fences are explained in Example 3–27.) A box-and-whisker plot can help us visualize the center, the spread, and the skewness of a data set. It also helps detect outliers. We can compare different distributions by making box-and-whisker plots for each of them.

Note that a box-and-whisker plot is made using the following five-number summary values:

$$\text{Minimum value} \quad Q_1 \quad \text{Median} \quad Q_3 \quad \text{Maximum value}$$

The minimum and the maximum values used here must be within the lower and the upper inner fences that will be explained below. Making a box-and-whisker plot involves five steps that are explained in the example below.

EXAMPLE 3–27   Incomes of Households

Constructing a box-and-whisker plot.

The following data are the incomes (in thousands of dollars) for a sample of 12 households.

$$75 \quad 69 \quad 84 \quad 112 \quad 74 \quad 104 \quad 81 \quad 90 \quad 94 \quad 144 \quad 79 \quad 98$$

Construct a box-and-whisker plot for these data.

Solution   The following five steps are performed to construct a box-and-whisker plot.

Step 1.   First, rank the data in increasing order and calculate the values of the median, the first quartile, the third quartile, and the interquartile range. The ranked data are

$$69 \quad 74 \quad 75 \quad 79 \quad 81 \quad 84 \quad 90 \quad 94 \quad 98 \quad 104 \quad 112 \quad 144$$

For these data,

$$\text{Median} = (84 + 90)/2 = 87$$
$$Q_1 = (75 + 79)/2 = 77$$
$$Q_3 = (98 + 104)/2 = 101$$
$$\text{IQR} = Q_3 - Q_1 = 101 - 77 = 24$$

Step 2.   Find the points that are $1.5 \times \text{IQR}$ below Q_1 and $1.5 \times \text{IQR}$ above Q_3. These two points are called the **lower** and the **upper inner fences**, respectively.

$$1.5 \times \text{IQR} = 1.5 \times 24 = 36$$
$$\text{Lower inner fence} = Q_1 - 36 = 77 - 36 = 41$$
$$\text{Upper inner fence} = Q_3 + 36 = 101 + 36 = 137$$

Step 3.   Determine the smallest and the largest values in the given data set within the two inner fences. These two values for our example are as follows:

$$\text{Smallest value within the two inner fences} = 69$$
$$\text{Largest value within the two inner fences} = 112$$

Step 4.   Draw a horizontal line and mark the income levels on it such that all the values in the given data set are covered. Above the horizontal line, draw a box with its left side at the position of the first quartile and the right side at the position of the third quartile. Inside the box, draw a vertical line at the position of the median. The result of this step is shown in Figure 3.13.

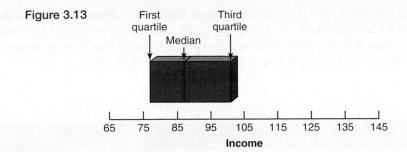

Figure 3.13

Step 5. By drawing two lines, join the points of the smallest and the largest values within the two inner fences to the box. These values are 69 and 112 in this example as listed in Step 3. The two lines that join the box to these two values are called **whiskers**. A value that falls outside the two inner fences is shown by marking an asterisk and is called an outlier. This completes the box-and-whisker plot, as shown in Figure 3.14.

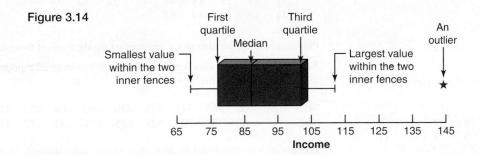

Figure 3.14

In Figure 3.14, about 50% of the data values fall within the box, about 25% of the values fall on the left side of the box, and about 25% fall on the right side of the box. Also, 50% of the values fall on the left side of the median and 50% lie on the right side of the median. The data of this example are skewed to the right because the lower 50% of the values are spread over a smaller range than the upper 50% of the values.

The observations that fall outside the two inner fences are called outliers. These outliers can be classified into two kinds of outliers—mild and extreme outliers. To do so, we define two outer fences—a **lower outer fence** at $3.0 \times$ IQR below the first quartile and an **upper outer fence** at $3.0 \times$ IQR above the third quartile. If an observation is outside either of the two inner fences but within the two outer fences, it is called a **mild outlier**. An observation that is outside either of the two outer fences is called an **extreme outlier**. For Example 3–27, the two outer fences are calculated as follows.

$$3 \times \text{IQR} = 3 \times 24 = 72$$
$$\text{Lower outer fence} = Q_1 - 72 = 77 - 72 = 5$$
$$\text{Upper outer fence} = Q_3 + 72 = 101 + 72 = 173$$

Because 144 is outside the upper inner fence but inside the upper outer fence, it is a mild outlier.

Using the box-and-whisker plot, we can conclude whether the distribution of our data is symmetric, skewed to the right, or skewed to the left. If the line representing the median is in the middle of the box and the two whiskers are of about the same length, then the data have a symmetric distribution. If the line representing the median is not in the middle of the box and/or the two whiskers are not of the same length, then the distribution of data values is skewed. The distribution is skewed to the right if the median is to the left of the center of the box with the right side whisker equal to or longer than the whisker on the left side, or if the median is in the center of the box but the whisker on the right side is longer than the one on the left side. The distribution is skewed to the left if the median is to the right of the center of the box with the left side whisker equal to or longer than the whisker on the right side, or if the median is in the center of the box but the whisker on the left side is longer than the one on the right side.

EXERCISES

CONCEPTS AND PROCEDURES

3.75 Explain what summary measures are used to construct a box-and-whisker plot.

3.76 Prepare a box-and-whisker plot for the following data:

| 11 | 8 | 26 | 31 | 62 | 19 | 7 | 3 | 14 | 75 |
| 33 | 30 | 42 | 15 | 18 | 23 | 29 | 13 | 16 | 6 |

Does this data set contain any outliers?

APPLICATIONS

3.77 The following data give the numbers of new cars sold at a dealership during a 20-day period.

| 8 | 5 | 12 | 3 | 9 | 10 | 6 | 12 | 8 | 8 |
| 4 | 16 | 10 | 11 | 7 | 7 | 3 | 5 | 9 | 11 |

Make a box-and-whisker plot. Comment on the skewness of these data.

3.78 Nixon Corporation manufactures computer monitors. The following are the numbers of computer monitors produced at the company for a sample of 30 days:

24	32	27	23	33	33	29	25	23	28
21	26	31	20	27	33	27	23	28	29
31	35	34	22	26	28	23	35	31	27

Prepare a box-and-whisker plot. Comment on the skewness of these data.

3.79 The following data give the numbers of computer keyboards assembled at the Twentieth Century Electronics Company for a sample of 25 days.

45	52	48	41	56	46	44	42	48	53
51	53	51	48	46	43	52	50	54	47
44	47	50	49	52					

Prepare a box-and-whisker plot. Comment on the skewness of these data.

3.80 The following data represent the credit scores of 22 randomly selected loan applicants.

| 494 | 728 | 468 | 533 | 747 | 639 | 430 | 690 | 604 | 422 | 356 |
| 805 | 749 | 600 | 797 | 702 | 628 | 625 | 617 | 647 | 772 | 572 |

Prepare a box-and-whisker plot. Are these data skewed in any direction?

3.81 The following data give the number of students suspended for bringing weapons to schools in the Tri-City School District for each of the past 10 weeks.

| 15 | 9 | 8 | 7 | 6 | 9 | 14 | 3 | 6 | 9 |

Make a box-and-whisker plot. Comment on the skewness of these data.

USES AND MISUSES...

WHAT DOES IT REALLY "MEAN"?

People are fascinated by data and, as a result, there are statistical records on just about everything you can imagine. Weather is no exception. Annual precipitation data over many decades are available for most major cities. For example, Washington, DC has reliable precipitation data dating back to 1871. (Data source: Capital Climate, capitalclimate.blogspot.com.)

One of the most frequently used summary statistics is the *mean* or *arithmetic average*. This statistic has many interesting uses and is one of the most commonly reported summary statistics for quantitative data. The mean allows us to summarize large sets of data into a single number that is readily understandable.

For example, examine the line graph of precipitation data for Washington, DC for the years 1871–2009 presented in Figure 3.15. Note that precipitation is measured in inches. The data points on the graph represent the mean precipitation for each decade. At first inspection, we see obvious differences in the mean precipitation from decade-to-decade. Overall, the points appear to be distributed randomly and uniformly above and below about 40 inches. Hence, we might say that the annual precipitation in Washington, DC is, on average, about 40 inches, with some decades above 40 inches and some decades below 40 inches. However, can we also say that the mean precipitation was unusually high in 1880–1889 and unusually low in 1960–1969? Caution must be taken when making such statements.

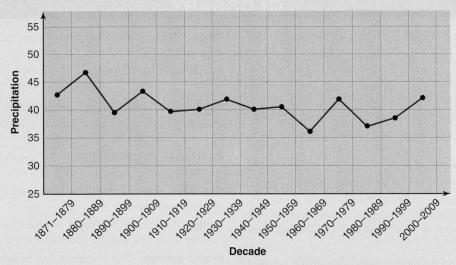

Figure 3.15   Mean precipitation by decade for Washington, DC from 1871 to 2009.

Even with the information presented in this graph, we are limited with regards to the interpretation of these data; there is important information that we are missing. The problem is that the mean, by itself, oversimplifies the data. By summarizing the data into a single numeric value, it is easy to be deceived into thinking that the mean tells us everything we need to know about the data. This is an incorrect assumption.

The mean of a data set, of and by itself, provides limited information. As an example, suppose that the average test score for a small class of three students was 50%. If the students' scores were 45%, 50%, and 55%, we might conclude that none of the students scored well on the exam. However, if the students' scores were 0%, 50%, and 100%, we conclude that one student was totally unprepared, one student needs to study more, and one student did extremely well on the test. To provide a complete interpretation of the data, we need to have more information to gain context. We need to know more than just the center of the distribution. We need a measure of the *variability* of the data.

In Figure 3.15, we can see some degree of variability in the mean precipitation from decade-to-decade, but what about the 10-year pattern of precipitation within each decade? This information is not provided in the graph. Now consider an error bar plot as shown in Figure 3.16. This graph shows the mean precipitation for each decade (represented by a solid dot) together with bars representing an interval of values spanning one standard deviation on either side of the mean.

We now have more information to obtain the necessary context. First, even though the distribution of the 10-year mean precipitation varies from decade-to-decade, the error bars indicate there is overlap in the variability for each decade. Second, even though the mean precipitation appears to be unusually high in 1880–1889 and unusually low in 1960–1969, the variation in the mean shows that, overall, the pattern of precipitation was not significantly different from that of the other decades. Third, we can see that some decades had a relatively consistent pattern of precipitation (1910–1919) while other decades experienced a wide variation in precipitation from year-to-year (2000–2009 among others).

The take-home message is clear. A measure of center is of limited use without an accompanying measure of variation. Failure to do so easily misleads readers into potentially drawing poor or misinformed conclusions. Additionally, we should be wary when we read results that report only the mean. Without context, we really do not know what the results "mean."

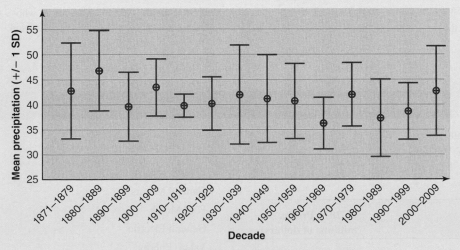

Figure 3.16   Mean precipitation in Washington, DC from 1871 to 2009.

GLOSSARY

Bimodal distribution   A distribution that has two modes.

Box-and-whisker plot   A plot that shows the center, spread, and skewness of a data set with a box and two whiskers using the median, the first quartile, the third quartile, and the smallest and the largest values in the data set between the lower and the upper inner fences.

Chebyshev's theorem   For any number k greater than 1, at least $(1 - 1/k^2)$ of the values for any distribution lie within k standard deviations of the mean.

Coefficient of variation   A measure of relative variability that expresses standard deviation as a percentage of the mean.

Empirical rule   For a specific bell-shaped distribution, about 68% of the observations fall in the interval $(\mu - \sigma)$ to $(\mu + \sigma)$, about 95% fall in the interval $(\mu - 2\sigma)$ to $(\mu + 2\sigma)$, and about 99.7% fall in the interval $(\mu - 3\sigma)$ to $(\mu + 3\sigma)$.

First quartile   The first of the three quartiles that divide a ranked data set into four equal parts. The value in a ranked data set such that about 25% of the measurements are smaller than this value and about 75% are larger. It is the median of the values that are smaller than the median of the whole data set.

Geometric mean   Calculated by taking the nth root of the product of all values in a data set.

Interquartile range (IQR)   The difference between the third and the first quartiles.

Lower inner fence   The value in a data set that is $1.5 \times$ IQR below the first quartile.

Lower outer fence   The value in a data set that is $3.0 \times$ IQR below the first quartile.

Mean   A measure of center calculated by dividing the sum of all values by the number of values in the data set.

Measures of center   Measures that describe the center of a distribution. The mean, median, and mode are three of the measures of center.

Measures of dispersion   Measures that give the spread of a distribution. The range, variance, and standard deviation are three such measures.

Measures of position   Measures that determine the position of a single value in relation to other values in a data set. Quartiles, percentiles, and percentile rank are examples of measures of position.

Median   The value that divides a ranked data set into two equal parts.

Mode   The value (or values) that occurs with highest frequency in a data set.

Multimodal distribution   A distribution that has more than two modes.

Parameter   A summary measure calculated for population data.

Percentile rank   The percentile rank of a value gives the percentage of values in the data set that are smaller than this value.

Percentiles   Ninety-nine values that divide a ranked data set into 100 equal parts.

Quartiles   Three summary measures that divide a ranked data set into four equal parts.

Range   A measure of spread obtained by taking the difference between the largest and the smallest values in a data set.

Second quartile   The middle or second of the three quartiles that divide a ranked data set into four equal parts. About 50% of the values in the data set are smaller and about 50% are larger than the second quartile. The second quartile is the same as the median.

Standard deviation   A measure of spread that is given by the positive square root of the variance.

Statistic   A summary measure calculated for sample data.

Third quartile   The third of the three quartiles that divide a ranked data set into four equal parts. About 75% of the values in a data set are smaller than the value of the third quartile and about 25% are larger. It is the median of the values that are greater than the median of the whole data set.

Trimmed mean   The k% trimmed mean is obtained by dropping k% of the smallest values and k% of the largest values from the given data and then calculating the mean of the remaining $(100 - 2k)$% of the values.

Unimodal distribution   A distribution that has only one mode.

Upper inner fence   The value in a data set that is $1.5 \times$ IQR above the third quartile.

Upper outer fence   The value in a data set that is $3.0 \times$ IQR above the third quartile.

Variance   A measure of spread.

Weighted mean   Mean of a data set whose values are assigned different weights before the mean is calculated.

SUPPLEMENTARY EXERCISES

3.82 The following data give the 2014 profits (in millions of dollars) of the top 10 companies listed in the 2014 *Fortune 500* (*source*: www.fortune.com).

Company	2014 Profits (millions of dollars)
Wal-Mart Stores	16,022
Exxon Mobil	32,580
Chevron	21,423
Berkshire Hathaway	19,476
Apple	37,037
Phillips 66	3726
General Motors	5346
Ford Motor	7155
General Electric	13,057
Valero Energy	2720

Find the mean and median for these data. Do these data have a mode? Explain.

3.83 The Belmont Stakes is the final race in the annual Triple Crown of thoroughbred horse racing. The race is 1.5 miles in length, and the record for the fastest time of 2 minutes, 24 seconds is held by Secretariat, the 1973 winner. We compared Secretariat's time from 1973 with the time of each winner of the Belmont Stakes for the years 1999–2011. The following data represent the differences (in seconds) between each winner's time for the years 1999–2011 and Secretariat's time in 1973. For example, the 1999 winner took 3.80 seconds longer than Secretariat to finish the race.

3.80	7.20	2.80	5.71	4.26	3.50	4.75
3.81	4.74	5.65	3.54	7.57	6.88	

   a. Calculate the mean and median. Do these data have a mode? Why or why not?

   b. Compute the range, variance, and standard deviation for these data.

3.84 An electronics store sold 4830 televisions last year. The following table lists the number of different models of televisions sold and the prices for which they were sold.

Television Model	Number of TVs Sold	Price (dollars)
RRV5023	485	1630
TSU7831	1324	625
WEV4920	856	899
HEG7609	633	1178
YOX6938	394	1727
LXT2384	1138	927

Calculate the weighted mean that represents the average price for which these 4830 televisions were sold.

3.85 A company makes five different models of cameras. Last month they sold 13,891 cameras of all five models. The following table lists the number of different models of cameras sold during the last month and the prices for which they were sold.

Camera Model	Number of Cameras Sold	Price (dollars)
B110	3216	425
Z310	1835	1299
N450	4036	361
J150	3142	681
L601	1662	1999

Calculate the weighted mean that represents the average price for which these 13,891 cameras were sold.

3.86 The following table gives the frequency distribution of the times (in minutes) that 50 commuter students at a large university spent looking for parking spaces on the first day of classes in the Fall semester of 2012.

Time	Number of Students
0 to less than 4	1
4 to less than 8	7
8 to less than 12	15
12 to less than 16	18
16 to less than 20	6
20 to less than 24	3

Find the mean, variance, and standard deviation. Are the values of these summary measures population parameters or sample statistics?

3.87 The mean time taken to learn the basics of a software program by all students is 200 minutes with a standard deviation of 20 minutes.

   a. Using Chebyshev's theorem, find the minimum percentage of students who learn the basics of this software program in

      i. 160 to 240 minutes   **ii.** 140 to 260 minutes

  *b. Using Chebyshev's theorem, find the interval that contains the times taken by at least 84% of all students to learn this software program.

3.88 According to the American Time Use Survey conducted by the Bureau of Labor Statistics (www.bls.gov/atus/), Americans spent an average of 985.50 hours watching television in 2010. Suppose that the standard deviation of the distribution of times that Americans spent watching television in 2010 is 285.20 hours.

   a. Using Chebyshev's theorem, find at least what percentage of Americans watched television in 2010 for

      i. 272.50 to 1698.50 hours   **ii.** 129.90 to 1841.10 hours

  *b. Using Chebyshev's theorem, find the interval that contains the time (in hours) that at least 75% of Americans spent watching television in 2010.

3.89 The annual earnings of all employees with CPA certification and 6 years of experience and working for large firms have a bell-shaped distribution with a mean of $134,000 and a standard deviation of $12,000.

   a. Using the empirical rule, find the percentage of all such employees whose annual earnings are between

      i. $98,000 and $170,000   **ii.** $110,000 and $158,000

  *b. Using the empirical rule, find the interval that contains the annual earnings of 68% of all such employees.

3.90 The following data give the number of new cars sold at a dealership during a 20-day period.

8	5	12	3	9	10	6	12	8	8
4	16	10	11	7	7	3	5	9	11

   a. Calculate the values of the three quartiles and the interquartile range. Where does the value of 4 lie in relation to these quartiles?

   b. Find the (approximate) value of the 25th percentile. Give a brief interpretation of this percentile.

   c. Find the percentile rank of 10. Give a brief interpretation of this percentile rank.

3.91 A student washes her clothes at a laundromat once a week. The data below give the time (in minutes) she spent in the laundromat for each of 15 randomly selected weeks. Here, time spent in the laundromat includes the time spent waiting for a machine to become available.

75	62	84	73	107	81	93	72
135	77	85	67	90	83	112	

Prepare a box-and-whisker plot. Is the data set skewed in any direction? If yes, is it skewed to the right or to the left? Does this data set contain any outliers?

3.92 The following data give the lengths of time (in weeks) taken to find a full-time job by 18 computer science majors who graduated in 2016 from a small college.

30	43	32	21	65	8	4	18	16
38	9	44	33	23	24	81	42	55

Make a box-and-whisker plot. Comment on the skewness of this data set. Does this data set contain any outliers?

ADVANCED EXERCISES

3.93 The heights of five starting players on a basketball team have a mean of 76 inches, a median of 78 inches, and a range of 11 inches.

    a. If the tallest of these five players is replaced by a substitute who is 2 inches taller, find the new mean, median, and range.

    b. If the tallest player is replaced by a substitute who is 4 inches shorter, which of the new values (mean, median, range) could you determine, and what would their new values be?

3.94 During the 2011–12 winter season, a homeowner received four deliveries of heating oil, as shown in the following table.

Gallons Purchased	Price per Gallon ($)
209	2.60
182	2.40
157	2.78
149	2.74

The homeowner claimed that the mean price he paid for oil during the season was $(2.60 + 2.40 + 2.78 + 2.74)/4 = \$2.63$ per gallon. Do you agree with this claim? If not, explain why this method of calculating the mean is not appropriate in this case and find the correct value of the mean price.

3.95 In the Olympic Games, when events require a subjective judgment of an athlete's performance, the highest and lowest of the judges' scores may be dropped. Consider a gymnast whose performance is judged by seven judges and the highest and the lowest of the seven scores are dropped.

    a. Gymnast A's scores in this event are 9.4, 9.7, 9.5, 9.5, 9.4, 9.6, and 9.5. Find this gymnast's mean score after dropping the highest and the lowest scores.

    b. The answer to part a is an example of (approximately) what percentage of trimmed mean?

    c. Write another set of scores for a gymnast B so that gymnast A has a higher mean score than gymnast B based on the trimmed mean, but gymnast B would win if all seven scores were counted. Do not use any scores lower than 9.0.

3.96 In a study of distances traveled to a college by commuting students, data from 100 commuters yielded a mean of 8.73 miles. After the mean was calculated, data came in late from three students, with respective distances of 11.5, 7.6, and 10.0 miles. Calculate the mean distance for all 103 students.

3.97 How much does the typical American family spend to go away on vacation each year? Twenty-five randomly selected households reported the following vacation expenditures (rounded to the nearest hundred dollars) during the past year:

2500	500	800	0	100
0	200	2200	0	200
0	1000	900	321,500	400
500	100	0	8200	900
0	1700	1100	600	3400

    a. Using both graphical and numerical methods, organize and interpret these data.

    b. What measure of center best answers the original question?

3.98 Suppose that there are 150 freshmen engineering majors at a college and each of them will take the same five courses next semester. Four of these courses will be taught in small sections of 25 students each, whereas the fifth course will be taught in one section containing all 150 freshmen. To accommodate all 150 students, there must be six sections of each of the four courses taught in 25-student sections. Thus, there are 24 classes of 25 students each and one class of 150 students.

    a. Find the mean size of these 25 classes.

    b. Find the mean class size from a student's point of view, noting that each student has five classes containing 25, 25, 25, 25, and 150 students, respectively.

Are the means in parts a and b equal? If not, why not?

3.99 The test scores for a large statistics class have a bell-shaped distribution with a mean of 70 points.

    a. If 16% of all students in the class scored above 85, what is the standard deviation of the scores?

    b. If 95% of the scores are between 60 and 80, what is the standard deviation?

3.100 The distribution of the lengths of fish in a certain lake is not known, but it is definitely not bell shaped. It is estimated that the mean length is 6 inches with a standard deviation of 2 inches.

    a. At least what proportion of fish in the lake are between 3 inches and 9 inches long?

    b. What is the smallest interval that will contain the lengths of at least 84% of the fish?

    c. Find an interval so that fewer than 36% of the fish have lengths outside this interval.

3.101 A local golf club has men's and women's summer leagues. The following data give the scores for a round of 18 holes of golf for 17 men and 15 women randomly selected from their respective leagues.

Men	87	68	92	79	83	67	71	92	112
	75	77	102	79	78	85	75	72	
Women	101	100	87	95	98	81	117	107	103
	97	90	100	99	94	94			

    a. Make a box-and-whisker plot for each of the data sets and use them to discuss the similarities and differences between the scores of the men and women golfers.

    b. Compute the various descriptive measures you have learned for each sample. How do they compare?

3.102 Answer the following questions.

    a. The total weight of all pieces of luggage loaded onto an airplane is 14,124 pounds, which works out to be an average of 51.36 pounds per piece. How many pieces of luggage are on the plane?

    b. A group of seven friends, having just gotten back a chemistry exam, discuss their scores. Six of the students reveal that they received grades of 81, 75, 97, 88, 82, and 83, respectively, but the seventh student is reluctant to say what grade she received. After some calculation she announces that the group averaged 86 on the exam. What is her score?

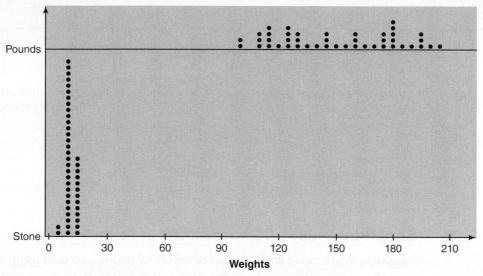

Figure 3.17 Stacked dotplot of weights in stones and pounds.

3.103 The following data give the weights (in pounds) of a random sample of 44 college students. (Here F and M indicate female and male, respectively.)

123 F	195 M	138 M	115 F	179 M	119 F
148 F	147 F	180 M	146 F	179 M	189 M
175 M	108 F	193 M	114 F	179 M	147 M
108 F	128 F	164 F	174 M	128 F	159 M
193 M	204 M	125 F	133 F	115 F	168 M
123 F	183 M	116 F	182 M	174 M	102 F
123 F	99 F	161 M	162 M	155 F	202 M
110 F	132 M				

Compute the mean, median, and standard deviation for the weights of all students, of men only, and of women only. Of the mean and median, which is the more informative measure of center? Write a brief note comparing the three measures for all students, men only, and women only.

3.104 Refer to the data in Problem 3.103. Two individuals, one from Canada and one from England, are interested in your analysis of these data but they need your results in different units. The Canadian individual wants the results in grams (1 pound = 435.59 grams), while the English individual wants the results in stones (1 stone = 14 pounds).

   a. Convert the data on weights from pounds to grams, and then recalculate the mean, median, and standard deviation of weight for males and females separately. Repeat the procedure, changing the unit from pounds to stones.

   b. Convert your answers from Problem 3.103 to grams and stones. What do you notice about these answers and your answers from part a?

   c. What happens to the values of the mean, median, and standard deviation when you convert from a larger unit to a smaller unit (e.g., from pounds to grams)? Does the same thing happen if you convert from a smaller unit (e.g., pounds) to a larger unit (e.g., stones)?

   d. Figure 3.17 gives a stacked dotplot of these weights in pounds and stones. Which of these two distributions has more variability? Use your results from parts a to c to explain why this is the case.

   e. Now consider the weights in pounds and grams. Make a stacked dotplot for these data and answer part d.

3.105 Refer to the data in Problem 3.101. It turns out that 117 was mistakenly entered. Although this person still had the highest score among the 15 women, her score was not a mild or extreme outlier according to the box-and-whisker plot, nor was she tied for the highest score. What are the possible scores that she could have shot?

3.106 The following stem-and-leaf diagram gives the distances (in thousands of miles) driven during the past year by a sample of drivers in a city.

```
0 | 3 6 9
1 | 2 8 5 1 0 5
2 | 5 1 6
3 | 8
4 | 1
5 |
6 | 2
```

   a. Compute the sample mean, median, and mode for the data on distances driven.

   b. Compute the range, variance, and standard deviation for these data.

   c. Compute the first and third quartiles.

   d. Compute the interquartile range. Describe the properties of the interquartile range. When would the IQR be preferable to using the standard deviation when measuring variation?

APPENDIX 3.1

A3.1.1 BASIC FORMULAS FOR THE VARIANCE AND STANDARD DEVIATION FOR UNGROUPED DATA

Example 3–28 below illustrates how to use the basic formulas to calculate the variance and standard deviation for ungrouped data. From Section 3.2.2, the basic formulas for variance for ungrouped data are

$$\sigma^2 = \frac{\sum (x - \mu)^2}{N} \quad \text{and} \quad s^2 = \frac{\sum (x - \bar{x})^2}{n - 1}$$

where σ^2 is the population variance and s^2 is the sample variance.

In either case, the standard deviation is obtained by taking the square root of the variance.

EXAMPLE 3–28 | Compensations of Female CEOs

Calculating the variance and standard deviation for ungrouped data using basic formulas.

Refer to Example 3–14, where short-cut formula was used to compute the variance and standard deviation for the data on the 2014 compensations of 11 female CEOs of American companies. Calculate the variance and standard deviation for that data set using the basic formula.

Solution Let x denote the 2014 compensations (in millions of dollars) of female CEOs of American companies. Table 3.16 shows all the required calculations to find the variance and standard deviation.

Table 3.16

x	$x - \bar{x}$	$(x - \bar{x})^2$
19.3	$19.3 - 23.77 = -4.47$	19.9809
16.2	$16.2 - 23.77 = -7.57$	57.3049
19.6	$19.6 - 23.77 = -4.17$	17.3889
19.3	$19.3 - 23.77 = -4.47$	19.9809
33.7	$33.7 - 23.77 = 9.93$	98.6049
21.0	$21.0 - 23.77 = -2.77$	7.6729
22.5	$22.5 - 23.77 = -1.27$	1.6129
16.9	$16.9 - 23.77 = -6.87$	47.1969
28.7	$28.7 - 23.77 = 4.93$	24.3049
42.1	$42.1 - 23.77 = 18.33$	335.9889
22.2	$22.2 - 23.77 = -1.57$	2.4649
$\sum x = 261.5$		$\sum (x - \bar{x})^2 = 632.5019$

The following steps are performed to compute the variance and standard deviation.

Step 1. Find the mean as follows:

$$\bar{x} = \frac{\sum x}{n} = \frac{261.5}{11} = 23.77$$

Step 2. Calculate $x - \bar{x}$, the deviation of each value of x from the mean. The results are shown in the second column of Table 3.16.

Step 3. Square each of the deviations of x from $\bar{x}$; that is, calculate each of the $(x - \bar{x})^2$ values. These values are called the *squared deviations*, and they are recorded in the third column of Table 3.16.

Step 4. Add all the squared deviations to obtain $\sum (x - \bar{x})^2$; that is, sum all the values given in the third column of Table 3.16. This gives

$$\sum (x - \bar{x})^2 = 632.5019$$

Step 5. Obtain the sample variance by dividing the sum of the squared deviations by $n - 1$. Thus

$$s^2 = \frac{\sum(x - \bar{x})^2}{n - 1} = \frac{632.5019}{11 - 1} = 63.25019$$

Step 6. Obtain the sample standard deviation by taking the positive square root of the variance. Hence,

$$s = \sqrt{63.25019} = 7.95 = \textbf{\$7.95 million}$$

A3.1.2 BASIC FORMULAS FOR THE VARIANCE AND STANDARD DEVIATION FOR GROUPED DATA

Example 3–29 demonstrates how to use the basic formulas to calculate the variance and standard deviation for grouped data. From Section 3.3.2, the basic formulas for these calculations are

$$\sigma^2 = \frac{\sum f(m - \mu)^2}{N} \quad \text{and} \quad s^2 = \frac{\sum f(m - \bar{x})^2}{n - 1}$$

where σ^2 is the population variance, s^2 is the sample variance, m is the midpoint of a class, and f is the frequency of a class.

In either case, the standard deviation is obtained by taking the square root of the variance.

EXAMPLE 3–29

Calculating the variance and standard deviation for grouped data using basic formulas.

In Example 3–20, we used the short-cut formula to compute the variance and standard deviation for the data on the number of orders received each day during the past 50 days at the office of a mail-order company. Calculate the variance and standard deviation for those data using the basic formula.

Solution   All the required calculations to find the variance and standard deviation appear in Table 3.17.

Table 3.17

Number of Orders	f	m	mf	$m - \bar{x}$	$(m - \bar{x})^2$	$f(m - \bar{x})^2$
10–12	4	11	44	−5.64	31.8096	127.2384
13–15	12	14	168	−2.64	6.9696	83.6352
16–18	20	17	340	.36	.1296	2.5920
19–21	14	20	280	3.36	11.2896	158.0544
	$n = 50$		$\sum mf = 832$			$\sum f(m - \bar{x})^2$ $= 371.5200$

The following steps are performed to compute the variance and standard deviation using the basic formula.

Step 1. Find the midpoint of each class. Multiply the corresponding values of m and f. Find $\sum mf$. From Table 3.17, $\sum mf = 832$.

Step 2. Find the mean as follows:

$$\bar{x} = \frac{\sum mf}{n} = \frac{832}{50} = 16.64$$

Step 3. Calculate $m - \bar{x}$, the deviation of each value of m from the mean. These calculations are shown in the fifth column of Table 3.17.

Step 4. Square each of the deviations $m - \bar{x}$; that is, calculate each of the $(m - \bar{x})^2$ values. These are called *squared deviations*, and they are recorded in the sixth column.

Step 5. Multiply the squared deviations by the corresponding frequencies (see the seventh column of Table 3.17). Adding the values of the seventh column, we obtain

$$\sum f(m - \bar{x})^2 = 371.5200$$

Step 6. Obtain the sample variance by dividing $\Sigma f(m - \bar{x})^2$ by $n - 1$. Thus,

$$s^2 = \frac{\Sigma f(m - \bar{x})^2}{n - 1} = \frac{371.5200}{50 - 1} = \mathbf{7.5820}$$

Step 7. Obtain the standard deviation by taking the positive square root of the variance.

$$s = \sqrt{s^2} = \sqrt{7.5820} = \mathbf{2.75\ orders}$$

SELF-REVIEW TEST

1. The value of the middle term in a ranked data set is called the
    a. mean   **b.** median   **c.** mode

2. Which of the following summary measures is/are influenced by extreme values?
    a. mean   **b.** median   **c.** mode   **d.** range

3. Which of the following summary measures can be calculated for qualitative data?
    a. mean   **b.** median   **c.** mode

4. Which of the following can have more than one value?
    a. mean   **b.** median   **c.** mode

5. Which of the following is obtained by taking the difference between the largest and the smallest values of a data set?
    a. variance   **b.** range   **c.** mean

6. Which of the following is the mean of the squared deviations of x values from the mean?
    a. standard deviation
    b. population variance
    c. sample variance

7. The values of the variance and standard deviation are
    a. never negative
    b. always positive
    c. never zero

8. A summary measure calculated for the population data is called
    a. a population parameter
    b. a sample statistic
    c. an outlier

9. A summary measure calculated for the sample data is called a
    a. population parameter
    b. sample statistic
    c. box-and-whisker plot

10. Chebyshev's theorem can be applied to
    a. any distribution
    b. bell-shaped distributions only
    c. skewed distributions only

11. The empirical rule can be applied to
    a. any distribution
    b. bell-shaped distributions only
    c. skewed distributions only

12. The first quartile is a value in a ranked data set such that about
    a. 75% of the values are smaller and about 25% are larger than this value
    b. 50% of the values are smaller and about 50% are larger than this value
    c. 25% of the values are smaller and about 75% are larger than this value

13. The third quartile is a value in a ranked data set such that about
    a. 75% of the values are smaller and about 25% are larger than this value
    b. 50% of the values are smaller and about 50% are larger than this value
    c. 25% of the values are smaller and about 75% are larger than this value

14. The 75th percentile is a value in a ranked data set such that about
    a. 75% of the values are smaller and about 25% are larger than this value
    b. 25% of the values are smaller and about 75% are larger than this value

15. Twenty randomly selected persons were asked to keep record of the number of times they used their debit cards during October 2015. The following data show their responses.

5 32 41 14 21 72 19 6 26 7 8 9 18 5 91 14 1 10 8 13

    a. Calculate the mean, median, and mode for these data.
    b. Compute the 10% trimmed mean for these data.
    c. Calculate the range, variance, and standard deviation for these data.
    d. Compute the coefficient of variation.
    e. Are the values of these summary measure calculated above population parameters or sample statistics? Explain.

16. A company makes five different models of refrigerators. Last year they sold 16,652 refrigerators of all five models. The following table lists the number of different models of refrigerators sold last year and the prices for which they were sold.

Refrigerator Model	Number of Refrigerators Sold	Price (dollars)
VX569	2842	2055
AX255	4364	1165
YU490	3946	1459
MT810	1629	2734
PR920	3871	1672

Calculate the weighted mean that represents the average price for which these 16,652 refrigerators were sold.

17. The mean, as a measure of center, has the disadvantage of being influenced by extreme values. Illustrate this point with an example.

18. The range, as a measure of spread, has the disadvantage of being influenced by extreme values. Illustrate this point with an example.

19. When is the value of the standard deviation for a data set zero? Give one example of such a data set. Calculate the standard deviation for that data set to show that it is zero.

20. The following table gives the frequency distribution of the numbers of computers sold during each of the past 25 weeks at an electronics store.

Computers Sold	Frequency
4 to 9	2
10 to 15	4
16 to 21	10
22 to 27	6
28 to 33	3

    a. What does the frequency column in the table represent?
    b. Calculate the mean, variance, and standard deviation.

21. The members of a very large health club were observed on a randomly selected day. The distribution of times they spent that day at the health club was found to have a mean of 91.8 minutes and a standard deviation of 16.2 minutes. Suppose these values of the mean and standard deviation hold true for all members of this club.
    a. Using Chebyshev's theorem, find the minimum percentage of this health club's members who spend time at this health club between
      i. 59.4 and 124.2 minutes   **ii.** 51.3 and 132.3 minutes
   *b. Using Chebyshev's theorem, find the interval that contains the times spent at this health club by at least 89% of members.

22. The ages of cars owned by all people living in a city have a bell-shaped distribution with a mean of 7.3 years and a standard deviation of 2.2 years.
    a. Using the empirical rule, find the (approximate) percentage of cars in this city that are
      i. 5.1 to 9.5 years old   **ii.** .7 to 13.9 years old
   *b. Using the empirical rule, find the interval that contains the ages of (approximate) 95% of the cars owned by all people in this city.

23. The following data give the number of hours worked last week by 18 randomly selected managers working for various Wall Street financial companies.

45	54	63	79	48	50	49	52	74
61	55	56	58	56	77	66	64	70

    a. Calculate the three quartiles and the interquartile range. Where does the value of 54 lie in relation to these quartiles?
    b. Find the (approximate) value of the 60th percentile. Give a brief interpretation of this value.
    c. Calculate the percentile rank of 64. Give a brief interpretation of this value.

24. Make a box-and-whisker plot for the data given in Problem 23. Comment on the skewness of this data set.

*25. The mean weekly wages of a sample of 15 employees of a company are $1035. The mean weekly wages of a sample of 20 employees of another company are $1090. Find the combined mean for these 35 employees.

*26. The mean GPA of five students is 3.21. The GPAs of four of these five students are, respectively, 3.85, 2.67, 3.45, and 2.91. Find the GPA of the fifth student.

*27. Consider the following two data sets.

Data Set I:	8	16	20	35
Data Set II:	5	13	17	32

Note that each value of the second data set is obtained by subtracting 3 from the corresponding value of the first data set.
    a. Calculate the mean for each of these two data sets. Comment on the relationship between the two means.
    b. Calculate the standard deviation for each of these two data sets. Comment on the relationship between the two standard deviations.

MINI-PROJECTS

Note: The Mini-Projects are located on the text's Web site, www.wiley.com/college/mann.

DECIDE FOR YOURSELF

Note: The Decide for Yourself feature is located on the text's Web site, www.wiley.com/college/mann.

TECHNOLOGY INSTRUCTIONS   CHAPTER 3

Note: Complete TI-84, Minitab, and Excel manuals are available for download at the textbook's Web site, www.wiley.com/college/mann.

TI-84 COLOR / TI-84

The TI-84 Color Technology Instructions feature of this text is written for the TI-84 Plus C color graphing calculator running the 4.0 operating system. Some screens, menus, and functions will be slightly different in older operating systems. The TI-84+ can perform all of the same functions but does not have the "Color" option referenced in some of the menus.

The TI-84 calculator does not have functions to calculate the coefficient of variation and trimmed mean or to make a dotplot.

Calculating Summary Statistics for Example 3–2 of the Text

Screen 3.1

1. Enter the data from Example 3–2 of the text into L1.

2. Select **STAT > CALC > 1-Var Stats**.

3. Use the following settings in the **1-Var Stats** menu:

   • At the **List** prompt, select the name of the list L1 by pressing 2nd > **STAT** and scroll through the names until you get to your list name.

   • Leave the **FreqList** prompt blank.

4. Highlight **Calculate** and press **ENTER**.

5. The calculator will display the output (see **Screen 3.1** and **3.2**).

6. To see all of the output, scroll down by pressing the down arrow. The output will be displayed in the following order:

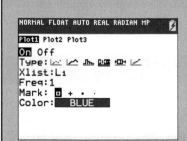

Screen 3.2

The mean, the sum of the data values, the sum of the squares of the data values, the sample standard deviation, the population standard deviation (used if your data are from a census), the number of data values (also called the sample size or population size), the minimum data value, the first quartile, the median, the third quartile, and the maximum data value.

Coefficient of Variation

There is no function to calculate the coefficient of variation on the TI. First calculate the sample mean and sample standard deviation as shown above. Then divide the sample standard deviation by the sample mean on the home screen to calculate the coefficient of variation.

Trimmed Mean

There is no function to calculate the trimmed mean on the TI. To calculate the trimmed mean, eliminate the most extreme data values from the data set and then follow the instructions given above to compute the mean from the resulting (trimmed) data.

Weighted Mean

Enter the data into L1. Enter the corresponding weights into L2. Follow the instructions given above, but in step 3, enter L2 at the **FreqList** prompt.

Creating a Box-and-Whisker Plot for Example 3–27 of the Text

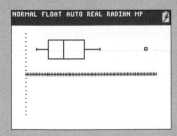

Screen 3.3

1. Enter the data from Example 3–27 of the text into L1.

2. Select **2nd > Y=** (the **STAT PLOT** menu).

3. Select **Plot 1**.

4. Use the following settings in the **Plot 1** menu (see **Screen 3.3**):

   • Select **On** to turn the plot on.

   • At the **Type** prompt, select either the box-and-whisker plot (the fifth choice) or the modified box-and-whisker plot (the fourth choice) which plots outliers as separate points.

   • At the **Xlist** prompt, access the name of your list by pressing 2nd > **STAT** and scroll through the names until you get to your list name.

   • At the **Freq** prompt, type 1.

   • At the **Color** prompt, select **BLUE**.

   (Skip this step if you do not have the TI-84 Color calculator.)

Screen 3.4

5. Select **ZOOM > ZoomStat** to display the boxplot. This function will automatically choose window settings for the boxplot. (See **Screen 3.4**.)

Minitab

The Minitab Technology Instructions feature of this text is written for Minitab version 17. Some screens, menus, and functions will be slightly different in older versions of Minitab.

Calculating Summary Statistics for Example 3–2 of the Text

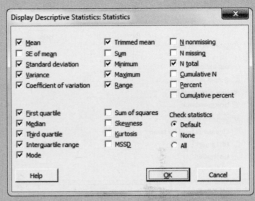
Screen 3.5

1. Enter the data from Example 3–2 of the text into column C1. Click below C1 in the blank column heading and name it *Ages*.

2. Select **Stat > Basic Statistics > Display Descriptive Statistics**.

3. In the dialog box that appears on screen, click inside the **Variables** box. Then either type C1 in this column or double click on C1 Ages on the left.

4. Click the **Statistics** button and check the boxes next to the summary statistics you wish to calculate. You can uncheck the boxes next to statistics that you do not want. (See **Screen 3.5**.)

5. Click **OK** in both dialog boxes. The output will appear in the **Session** window. (See **Screen 3.6**.)

Screen 3.6

In the above summary measures, you can obtain the values of various summary measures including the coefficient of variation and the trimmed mean.

Weighted Mean

Enter the data values in C1 and the corresponding weights in C2. Select **Calc > Calculator**. Type C3 in the **Store result in variable** box. Type SUM(C1*C2)/SUM(C2) in the **Expression** box. Click **OK**.

Creating a Box-and-Whisker Plot for Example 3–27 of the Text

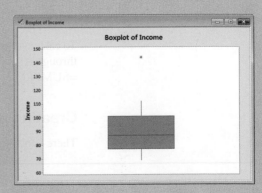
Screen 3.7

1. Enter the data from Example 3–27 of the text into column C1. Name the column *Income*.

2. Select **Graph > Boxplot > Simple** and click **OK**.

3. In the dialog box that appears on screen, click inside the **Graph variables** box. Then either type C1 in this column or double click on C1 Income on the left.

4. Click **OK** and the box-and-whisker plot will appear in a new window. (See **Screen 3.7**.)

Excel

The Excel Technology Instructions feature of this text is written for Excel 2013. Some screens, menus, and functions will be slightly different in older versions of Excel. To enable some of the advanced Excel functions, you must enable the Data Analysis Add-In for Excel.

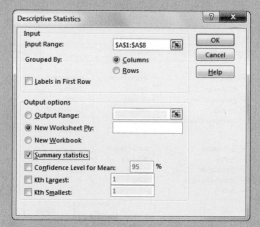

Screen 3.8

	A	B
1	*Column1*	
2		
3	Mean	45.25
4	Standard Error	4.24579624
5	Median	46.5
6	Mode	#N/A
7	Standard Deviation	12.00892525
8	Sample Variance	144.2142857
9	Kurtosis	-1.146585006
10	Skewness	-0.286315031
11	Range	34
12	Minimum	27
13	Maximum	61
14	Sum	362

Sheet2   Sheet1

Screen 3.9

Calculating Summary Statistics for Example 3–2 of the Text

1. Enter the data from Example 3–2 of the text into cells A1 through A8.

2. Click the **Data** tab and then click **Data Analysis** in the **Analysis** group.

3. In the **Data Analysis** dialog box, select **Descriptive Statistics**. Click **OK**.

4. In the **Descriptive Statistics** dialog box, click in the **Input Range** box. Select the cells where your data are located.

   Note: The easiest way to do this is to click and drag from cell A1 to cell A8.

5. Select **Columns** since these data are grouped in columns.

6. Select **New Worksheet Ply**. (See **Screen 3.8**.)

7. Check **Summary Statistics**. Click **OK**.

8. Excel will place the output, including the mean, median, mode, sample standard deviation, sample variance, range, minimum, and maximum in a new worksheet ply. You may need to make one or both columns wider to reveal all of the output. (See **Screen 3.9**.)

9. The summary statistics do not include the first and third quartiles. To find these quartiles, use the following procedure.

   - Click on an empty cell next to the original data.

   - Type =**quartile(**

   - Select the data by clicking and dragging from cell A1 to cell A8.

   - Type a comma, followed by a 1 for the first quartile (or a 3 for the third quartile).

   - Type a right parenthesis, and then press **Enter**.

To calculate the **coefficient of variation**, enter the data from Example 3–2 of the text into cells A1 through A8. Click in cell B1 and type =STDEV.S(A1:A8)/AVERAGE(A1:A8). Press **Enter**.

To calculate the **trimmed mean**, enter the data from Example 3–2 of the text into cells A1 through A8. Click in cell B1 and type =TRIMMEAN(A1:A8,0.2). Press **Enter**. *Note*: The 0.2 indicates that 20% of the total data set will be trimmed, 10% from each tail. To calculate the **weighted mean**, enter the data from Example 3–2 of the text into cells A1 through A8 and the desired weights in cells B1 through B8. Click in cell C1 and type =SUMPRODUCT(A1:A8,B1:B8)/SUM(B1:B8). Press **Enter**.

Creating a Box-and-Whisker Plot

There is no simple way to make a box-and-whisker plot with Excel.

TECHNOLOGY ASSIGNMENTS

TA3.1 Refer to City Data (Data Set I) on the prices of various products in different cities across the country. Select a subsample of the prices of regular unleaded gas for 40 cities.

  a. Find the mean, median, and standard deviation for the data of this subsample.

  b. Make a box-and-whisker plot for those data.

TA3.2 Refer to Data Set III on the National Football League. Calculate the mean, median, standard deviation, and interquartile range for the players' ages separately for each of the position groups. Is there a position group that tends to have younger players, on average, than the other position groups? Is there a position group that tends to have less variability in the ages of the players?

TA3.3 Refer to Data Set IX which contains information on Subway's menu items and accompanies this text (see Appendix A).

  a. Calculate the mean, median, range, standard deviation, quartiles, and the interquartile range for the data on sodium content listed in column 11.

  b. Create a box-and-whisker plot for the data on sodium content.

  c. Create a dotplot for the data on sodium content.

  d. If the technology that you are using lets you calculate trimmed mean and coefficient of variation, find these summary measures for the data on sodium content.

TA3.4 The following data give the total food expenditures (in dollars) for the past one month for a sample of 20 families.

| 1125 | 530 | 1234 | 595 | 427 | 872 | 1480 | 699 | 1274 | 1187 |
| 933 | 1127 | 716 | 1065 | 934 | 1630 | 1046 | 2199 | 1353 | 441 |

  a. Calculate the mean, median, standard deviation, quartiles, and the interquartile range for these data.

  b. Create a box-and-whisker plot for these data.

  c. Create a dotplot for these data.

  d. If the technology that you are using lets you calculate trimmed mean and coefficient of variation, find these summary measures for these data.

TA3.5 Using the data set Billboard (Data Set XIV), calculate the mean, median, range, standard deviation, and interquartile range, and create a boxplot of the number of weeks spent on the charts for the songs in the Billboard Hot 100 for the week of July 9, 2011. Discuss the features of the graph. Identify any outliers, and specify whether they are mild or extreme outliers. Now repeat the process for the songs ranked 1 through 50 and the songs ranked 51 through 100. Explain the differences and similarities between the two groups.

TA3.6 The following data give the odometer mileage (rounded to the nearest thousand miles) for all 20 cars that are for sale at a dealership.

| 62 | 86 | 58 | 84 | 72 | 40 | 27 | 38 | 50 | 43 |
| 27 | 40 | 90 | 43 | 94 | 36 | 28 | 48 | 86 | 77 |

  a. Calculate the mean, median, standard deviation, quartiles, and the interquartile range for these data.

  b. Create a box-and-whisker plot for these data.

  c. Create a dotplot for these data.

  d. If the technology that you are using lets you calculate trimmed mean and coefficient of variation, find these summary measures for these data.

TA3.7 Refer to Data Set VI on the stocks included in the Standard & Poor's 500 Index. Calculate the mean, median, standard deviation, and interquartile range for the data on the highest prices for the stocks in each of the market sectors. Compare the values of the various statistics for different sectors. Create a stacked dotplot of the highest prices for various sectors with each sector's data as one set of data. Explain how the results of your comparisons can be seen in the dotplot.

TA3.8 Refer to City Data (Data Set I) on the prices of various products in different cities across the country. Make a box-and-whisker plot for the data on the monthly rent.

CHAPTER 4

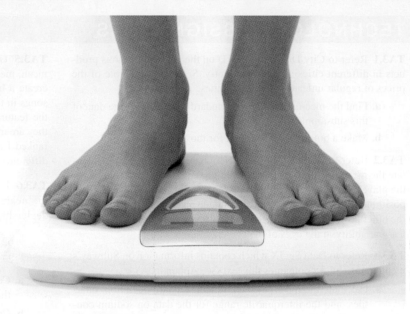

mediaphotos/Getty Images

Probability

Do you worry about your weight? According to a Gallup poll, 45% of American adults worry all or some of the time about their weight. The poll showed that more women than men worry about their weight all or some of the time. In this poll, 55% of adult women and 35% of adult men said that they worry all or some of the time about their weight. (See Case Study 4–1.)

We often make statements about probability. For example, a weather forecaster may predict that there is an 80% chance of rain tomorrow. A study may predict that a female, compared to a male, has a higher probability of being in favor of gun control. A college student may ask an instructor about the chances of passing a course or getting an A if he or she did not do well on the midterm examination.

Probability, which measures the likelihood that an event will occur, is an important part of statistics. It is the basis of inferential statistics, which will be introduced in later chapters. In inferential statistics, we make decisions under conditions of uncertainty. Probability theory is used to evaluate the uncertainty involved in those decisions. For example, estimating next year's sales for a company is based on many assumptions, some of which may happen to be true and others may not. Probability theory will help us make decisions under such conditions of imperfect information and uncertainty. Combining probability and probability distributions (which are discussed in Chapters 5 through 7) with descriptive statistics will help us make decisions about populations based on information obtained from samples. This chapter presents the basic concepts of probability and the rules for computing probability.

4.1 | Experiment, Outcome, and Sample Space

There are only a few things, if any, in life that have definite, certain, and sure outcomes. Most things in life have uncertain outcomes. For example, will you get an A grade in the statistics class that you are taking this semester? Suppose you recently opened a new business; will it be successful? Your friend will be getting married next month. Will she be happily married for the rest of her life? You just bought a lottery ticket. Will it be a winning ticket? You are playing black jack at a casino. Will you be a winner after 20 plays of the game? In all these situations, the outcomes are random. You know the various outcomes for each of these games/statistical experiments, but you do not know which one of these outcomes will actually occur. As another example, suppose you are working as a quality control manager at a factory where they are making hockey pucks. You randomly select a few pucks from the production line and inspect them for being good or defective. This act of inspecting a puck is an example of a statistical **experiment**. The result of this inspection will be that the puck is either "good" or "defective." Each of these two observations is called an **outcome** (also called a **basic** or **final outcome**) of the experiment, and these outcomes taken together constitute the **sample space** for this experiment. The elements of a sample space are called **sample points**.

> An **experiment** is a process that, when performed, results in one and only one of many observations. These observations are called the **outcomes** of the experiment. The collection of all outcomes for an experiment is called a **sample space**.

Table 4.1 lists some examples of experiments, their outcomes, and their sample spaces.

Table 4.1   Examples of Experiments, Outcomes, and Sample Spaces

Experiment	Outcomes	Sample Space
Toss a coin once	Head, Tail	{Head, Tail}
Roll a die once	1, 2, 3, 4, 5, 6	{1, 2, 3, 4, 5, 6}
Toss a coin twice	HH, HT, TH, TT	{HH, HT, TH, TT}
Play lottery	Win, Lose	{Win, Lose}
Take a test	Pass, Fail	{Pass, Fail}
Select a worker	Male, Female	{Male, Female}

The sample space for an experiment can also be illustrated by drawing a tree diagram. In a **tree diagram**, each outcome is represented by a branch of the tree. Tree diagrams help us understand probability concepts by presenting them visually. Examples 4–1 through 4–3 describe how to draw the tree diagrams for statistical experiments.

EXAMPLE 4–1   One Toss of a Coin

Drawing the tree diagram: one toss of a coin.

Draw the tree diagram for the experiment of tossing a coin once.

Solution   This experiment has two possible outcomes: head and tail. Consequently, the sample space is given by

$$\text{Sample space} = \{H, T\}, \quad \text{where } H = \text{Head} \quad \text{and} \quad T = \text{Tail}$$

To draw a tree diagram, we draw two branches starting at the same point, one representing the head and the second representing the tail. The two final outcomes are listed at the ends of the branches as shown in Figure 4.1.

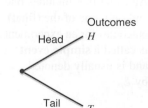

Figure 4.1   Tree diagram for one toss of a coin.

EXAMPLE 4–2   Two Tosses of a Coin

Drawing the tree diagram: two tosses of a coin.

Draw the tree diagram for the experiment of tossing a coin twice.

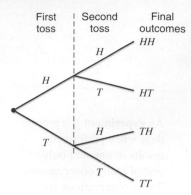

First toss | Second toss | Final outcomes

Figure 4.2 Tree diagram for two tosses of a coin.

Solution This experiment can be split into two parts: the first toss and the second toss. Suppose that the first time the coin is tossed, we obtain a head. Then, on the second toss, we can still obtain a head or a tail. This gives us two outcomes: *HH* (head on both tosses) and *HT* (head on the first toss and tail on the second toss). Now suppose that we observe a tail on the first toss. Again, either a head or a tail can occur on the second toss, giving the remaining two outcomes: *TH* (tail on the first toss and head on the second toss) and *TT* (tail on both tosses). Thus, the sample space for two tosses of a coin is

$$\text{Sample space} = \{HH, HT, TH, TT\}$$

The tree diagram is shown in Figure 4.2. This diagram shows the sample space for this experiment.

EXAMPLE 4–3   Selecting Two Workers

Drawing the tree diagram: two selections.

Suppose we randomly select two workers from a company and observe whether the worker selected each time is a man or a woman. Write all the outcomes for this experiment. Draw the tree diagram for this experiment.

Solution Let us denote the selection of a man by *M* and that of a woman by *W*. We can compare the selection of two workers to two tosses of a coin. Just as each toss of a coin can result in one of two outcomes, head or tail, each selection from the workers of this company can result in one of two outcomes, man or woman. As we can see from the tree diagram of Figure 4.3, there are four final outcomes: *MM, MW, WM, WW*. Hence, the sample space is written as

$$\text{Sample space} = \{MM, MW, WM, WW\}$$

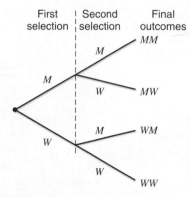

First selection | Second selection | Final outcomes

Figure 4.3 Tree diagram for selecting two workers.

An **event** is a collection of one or more of the outcomes of an experiment.

An event that includes one and only one of the (final) outcomes for an experiment is called a **simple event** and is usually denoted by E_i.

4.1.1   Simple and Compound Events

An **event** consists of one or more of the outcomes of an experiment.

An event may be a **simple event** or a **compound event**. A simple event is also called an **elementary event**, and a compound event is also called a **composite event**.

Simple Event

Each of the final outcomes for an experiment is called a **simple event**. In other words, a simple event includes one and only one outcome. Usually, simple events are denoted by E_1, E_2, E_3, and so forth. However, we can denote them by any other letter—that is, by *A, B, C,* and so forth. Many times we denote events by the same letter and use subscripts to distinguish them, as in $A_1, A_2, A_3, \ldots$.

Example 4–4 describes simple events.

EXAMPLE 4–4   Selecting Two Workers

Illustrating simple events.

Reconsider Example 4–3 on selecting two workers from a company and observing whether the worker selected each time is a man or a woman. Each of the final four outcomes (*MM, MW, WM,* and *WW*) for this experiment is a simple event. These four events can be denoted by E_1, E_2, E_3, and E_4, respectively. Thus,

$$E_1 = \{MM\}, \quad E_2 = \{MW\}, \quad E_3 = \{WM\}, \quad \text{and} \quad E_4 = \{WW\}$$

Compound Event

A **compound event** consists of more than one outcome.

Compound events are denoted by A, B, C, D, ..., or by A_1, A_2, A_3, ..., or by B_1, B_2, B_3, ..., and so forth. Examples 4–5 and 4–6 describe compound events.

> A **compound event** is a collection of more than one outcome for an experiment.

EXAMPLE 4–5	**Selecting Two Workers**

Illustrating a compound event: two selections.

Reconsider Example 4–3 on selecting two workers from a company and observing whether the worker selected each time is a man or a woman. Let A be the event that at most one man is selected. Is event A a simple or a compound event?

Solution   Here *at most one man* means one or no man is selected. Thus, event A will occur if either no man or one man is selected. Hence, the event A is given by

$$A = \text{at most one man is selected} = \{MW, WM, WW\}$$

Because event A contains more than one outcome, it is a compound event. The diagram in Figure 4.4 gives a graphic presentation of compound event A.

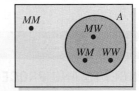

Figure 4.4   Diagram showing event A.

EXAMPLE 4–6	**Preference for Ice Tea**

Illustrating simple and compound events: two selections.

In a group of college students, some like ice tea and others do not. There is no student in this group who is indifferent or has no opinion. Two students are randomly selected from this group.

   (a) How many outcomes are possible? List all the possible outcomes.

   (b) Consider the following events. List all the outcomes included in each of these events. Mention whether each of these events is a simple or a compound event.

      (i) Both students like ice tea.

     (ii) At most one student likes ice tea.

    (iii) At least one student likes ice tea.

    (iv) Neither student likes ice tea.

Solution   Let L denote the event that a student likes ice tea and N denote the event that a student does not like ice tea.

   (a) This experiment has four outcomes, which are listed below and shown in Figure 4.5.

       LL = Both students like ice tea

       LN = The first student likes ice tea but the second student does not

       NL = The first student does not like ice tea but the second student does

       NN = Both students do not like ice tea

   (b) **(i)** The event *both students like ice tea* will occur if LL happens. Thus,

$$\text{Both students like ice tea} = \{LL\}$$

       Since this event includes only one of the four outcomes, it is a **simple** event.

     (ii) The event *at most one student likes ice tea* will occur if one or none of the two students likes ice tea, which will include the events LN, NL, and NN. Thus,

$$\text{At most one student likes ice tea} = \{LN, NL, NN\}$$

       Since this event includes three outcomes, it is a **compound** event.

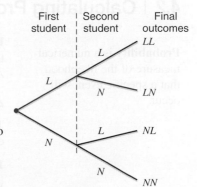

Figure 4.5   Tree diagram.

(iii) The event *at least one student likes ice tea* will occur if one or two of the two students like ice tea, which will include the events *LN, NL,* and *LL*. Thus,

At least one student likes ice tea = {*LN, NL, LL*}

Since this event includes three outcomes, it is a **compound** event.

(iv) The event *neither student likes ice tea* will occur if neither of the two students likes ice tea, which will include the event *NN*. Thus,

Neither student likes ice tea = {*NN*}

Since this event includes one outcome, it is a **simple** event.

EXERCISES

CONCEPTS AND PROCEDURES

4.1   Briefly explain the terms *experiment, outcome, sample space, simple event* and *compound event.*

4.2   What are the simple events for each of the following statistical experiments in a sample space *S.*
    a. One roll of a die
    b. Three tosses of a coin
    c. One toss of a coin and one roll of a die

4.3   There are three items labeled X, Y, and Z in a box. Two items are selected at random (without replacement) from this box. List all the possible outcomes for this experiment. Write the sample space *S.*

APPLICATIONS

4.4   Three students are randomly selected from a statistics class, and it is observed whether or not they suffer from math anxiety. How many total outcomes are possible? Draw a tree diagram for this experiment.

4.5   In a group of adults, some are computer literate, and the others are computer illiterate. If two adults are randomly selected from this group, how many total outcomes are possible? Draw a tree diagram for this experiment.

4.6   An automated teller machine at a local bank is stocked with $50 and $100 bills. When a customer withdraws $200 from the machine, it dispenses either two $100 bills or four $50 bills. If two customers withdraw $200 each, how many outcomes are possible? Draw a tree diagram for this experiment.

4.7   A box contains a certain number of computer parts, a few of which are defective. Two parts are selected at random from this box and inspected to determine if they are good or defective. How many total outcomes are possible? Draw a tree diagram for this experiment.

4.8   Draw a tree diagram for three tosses of a coin. List all outcomes for this experiment in a sample space *S.*

4.9   A test contains two multiple-choice questions. If a student makes a random guess to answer each question, how many outcomes are possible? Depict all these outcomes in a Venn diagram. Also draw a tree diagram for this experiment. (*Hint:* Consider two outcomes for each question—either the answer is correct or it is wrong.)

4.10   Refer to Exercise 4.6. List all of the outcomes in each of the following events and mention which of these are simple and which are compound events.
    a. Exactly one customer receives $100 bills.
    b. Both customers receive $50 bills.
    c. At most one customer receives $100 bills.
    d. The first customer receives $50 bills and the second receives $100 bills.

4.2 | Calculating Probability

> **Probability** is a numerical measure of the likelihood that a specific event will occur.

Probability, which gives the likelihood of occurrence of an event, is denoted by *P*. The probability that a simple event E_i will occur is denoted by $P(E_i)$, and the probability that a compound event *A* will occur is denoted by $P(A)$.

4.2.1   Two Properties of Probability

There are two important properties of probability that we should always remember. These properties are mentioned below.

1.   The probability of an event always lies in the range 0 to 1.

Whether it is a simple or a compound event, the probability of an event is never less than 0 or greater than 1. We can write this property as follows.

First Property of Probability

$$0 \le P(E_i) \le 1$$
$$0 \le P(A) \le 1$$

An event that cannot occur has zero probability and is called an **impossible (or null) event**. An event that is certain to occur has a probability equal to 1 and is called a **sure (or certain) event**. In the following examples, the first event is an impossible event and the second one is a sure event.

$$P(\text{a tossed coin will stand on its edge}) = 0$$

$$P(\text{a child born today will eventually die}) = 1.0$$

There are very few events in real life that have probability equal to either zero or 1.0. Most of the events in real life have probabilities that are between zero and 1.0. In other words, these probabilities are greater than zero but less than 1.0. A higher probability such as .82 indicates that the event is more likely to occur. On the other hand, an event with a lower probability such as .12 is less likely to occur. Sometime events with very low (.05 or lower) probabilities are also called **rare events**.

2. **The sum of the probabilities of all simple events (or final outcomes) for an experiment, denoted by $\sum P(E_i)$, is always 1.**

Second Property of Probability   For an experiment with outcomes $E_1, E_2, E_3, \ldots$,
$$\sum P(E_i) = P(E_1) + P(E_2) + P(E_3) + \cdots = 1.0$$

For example, if you buy a lottery ticket, you may either win or lose. The probabilities of these two events must add to 1.0, that is:

$$P(\text{you will win}) + P(\text{you will lose}) = 1.0$$

Similarly, for the experiment of one toss of a coin,

$$P(\text{Head}) + P(\text{Tail}) = 1.0$$

For the experiment of two tosses of a coin,

$$P(HH) + P(HT) + P(TH) + P(TT) = 1.0$$

For one game of football by a professional team,

$$P(\text{win}) + P(\text{loss}) + P(\text{tie}) = 1.0$$

4.2.2   Three Conceptual Approaches to Probability

How do we assign probabilities to events? For example, we may say that the probability of obtaining a head in one toss of a coin is .50, or that the probability that a randomly selected family owns a home is .68, or that the Los Angeles Dodgers will win the Major League Baseball championship next year is .14. How do we obtain these probabilities? We will learn the procedures that are used to obtain such probabilities in this section. There are three conceptual approaches to probability: (1) classical probability, (2) the relative frequency concept of probability, and (3) the subjective probability concept. These three concepts are explained next.

Classical Probability

Many times, various outcomes for an experiment may have the same probability of occurrence. Such outcomes are called **equally likely outcomes**. The classical probability rule is applied to compute the probabilities of events for an experiment for which all outcomes are equally likely.

Two or more outcomes that have the same probability of occurrence are said to be **equally likely outcomes**.

For example, head and tail are two equally likely outcomes when a fair coin is tossed once. Each of these two outcomes has the same chance of occurrence.

Earlier in this section, we learned that the total probability for all simple outcomes of an experiment is 1.0. This total probability is distributed over all the outcomes of the experiment. When all the outcomes of an experiment are equally likely, this total probability will be equally distributed over various outcomes. For example, for a fair coin, there are two equally likely outcomes—a head and a tail. Thus, if we distribute the total probability of 1.0 equally among these two outcomes, then each of these outcomes will have a probability of .50 of occurrence.

According to the **classical probability rule**, to find the probability of a simple event, we divide 1.0 by the total number of outcomes for the experiment. On the other hand, to find the probability of a compound event A, we divide the number of outcomes favorable to event A by the total number of outcomes for the experiment.

Classical Probability Rule to Find Probability   Suppose E_i is a simple event and A is a compound event for an experiment with equally likely outcomes. Then, applying the classical approach, the probabilities of E_i and A are:

$$P(E_i) = \frac{1}{\text{Total number of outcomes for the experiment}}$$

$$P(A) = \frac{\text{Number of outcomes favorable to } A}{\text{Total number of outcomes for the experiment}}$$

Examples 4–7 through 4–9 illustrate how probabilities of events are calculated using the classical probability rule.

EXAMPLE 4–7   One Toss of a Coin

Calculating the probability of a simple event.

Find the probability of obtaining a head and the probability of obtaining a tail for one toss of a coin.

Solution   If we toss a fair coin once, the two outcomes, head and tail, are equally likely outcomes. Hence, this is an example of a classical experiment. Therefore,[1]

$$P(\text{head}) = \frac{1}{\text{Total number of outcomes}} = \frac{1}{2} = .50$$

Similarly,

$$P(\text{tail}) = \frac{1}{2} = .50$$

Since this experiment has only two outcomes, a head and a tail, their probabilities add to 1.0.

EXAMPLE 4–8   One Roll of a Die

Calculating the probability of a compound event.

Find the probability of obtaining an even number in one roll of a die.

Solution   This experiment of rolling a die once has a total of six outcomes: 1, 2, 3, 4, 5, and 6. Given that the die is fair, these outcomes are equally likely. Let A be an event that an even number is observed on the die. Event A includes three outcomes: 2, 4, and 6; that is,

$$A = \text{an even number is obtained} = \{2, 4, 6\}$$

---

[1]If the final answer for the probability of an event does not terminate within four decimal places, it will be rounded to four decimal places.

If any one of these three numbers is obtained, event A is said to occur. Since three out of six outcomes are included in the event that an even number is obtained, its probability is:

$$P(A) = \frac{\text{Number of outcomes included in } A}{\text{Total number of outcomes}} = \frac{3}{6} = .50$$

EXAMPLE 4–9   Selecting One Out of Five Homes

Calculating the probability of a compound event.

Jim and Kim have been looking for a house to buy in New Jersey. They like five of the homes they have looked at recently and two of those are in West Orange. They cannot decide which of the five homes they should pick to make an offer. They put five balls (of the same size) marked 1 through 5 (each number representing a home) in a box and asked their daughter to select one of these balls. Assuming their daughter's selection is random, what is the probability that the selected home is in West Orange?

Solution   With random selection, each home has the same probability of being selected and the five outcomes (one for each home) are equally likely. Two of the five homes are in West Orange. Hence,

$$P(\text{selected home is in West Orange}) = \frac{2}{5} = .40$$

Relative Frequency Concept of Probability

Suppose we want to calculate the following probabilities:

1. The probability that the next car that comes out of an auto factory is a "lemon"
2. The probability that a randomly selected family owns a home
3. The probability that a randomly selected woman is an excellent driver
4. The probability that an 80-year-old person will live for at least 1 more year
5. The probability that a randomly selected adult is in favor of increasing taxes to reduce the national debt
6. The probability that a randomly selected person owns a sport-utility vehicle (SUV)

These probabilities cannot be computed using the classical probability rule because the various outcomes for the corresponding experiments are not equally likely. For example, the next car manufactured at an auto factory may or may not be a lemon. The two outcomes, "the car is a lemon" and "the car is not a lemon," are not equally likely. If they were, then (approximately) half the cars manufactured by this company would be lemons, and this might prove disastrous to the survival of the factory.

Although the various outcomes for each of these experiments are not equally likely, each of these experiments can be performed again and again to generate data. In such cases, to calculate probabilities, we either use past data or generate new data by performing the experiment a large number of times. Using these data, we calculate the frequencies and relative frequencies for various outcomes. The relative frequency of an event is used as an approximation for the probability of that event. This method of assigning a probability to an event is called the **relative frequency concept of probability**. Because relative frequencies are determined by performing an experiment, the probabilities calculated using relative frequencies may change when an experiment is repeated. For example, every time a new sample of 500 cars is selected from the production line of an auto factory, the number of lemons in those 500 cars is expected to be different. However, the variation in the percentage of lemons will be small if the sample size is large. Note that if we are considering the population, the relative frequency will give an exact probability.

Using Relative Frequency as an Approximation of Probability   If an experiment is repeated n times and an event A is observed f times where f is the frequency, then, according to the relative frequency concept of probability:

$$P(A) = \frac{f}{n} = \frac{\text{Frequency of } A}{\text{Sample size}}$$

Examples 4–10 and 4–11 illustrate how the probabilities of events are approximated using the relative frequencies.

EXAMPLE 4–10   Lemons in Car Production

Approximating probability by relative frequency: sample data.

Ten of the 500 randomly selected cars manufactured at a certain auto factory are found to be lemons. Assuming that lemons are manufactured randomly, what is the probability that the next car manufactured at this auto factory is a lemon?

Solution   Let n denote the total number of cars in the sample and f the number of lemons in n. Then, from the given information:

$$n = 500 \quad \text{and} \quad f = 10$$

Using the relative frequency concept of probability, we obtain

$$P(\text{next car is a lemon}) = \frac{f}{n} = \frac{10}{500} = .02$$

This probability is actually the relative frequency of lemons in 500 cars. Table 4.2 lists the frequency and relative frequency distributions for this example.

Table 4.2   **Frequency and Relative Frequency Distributions for the Sample of Cars**

Car	f	Relative Frequency
Good	490	490/500 = .98
Lemon	10	10/500 = .02
	$n = 500$	Sum = 1.00

The column of relative frequencies in Table 4.2 is used as the column of approximate probabilities. Thus, from the relative frequency column,

$$P(\text{next car is a lemon}) \quad = .02$$
$$P(\text{next car is a good car}) = .98$$

Law of Large Numbers ▶   **Note that relative frequencies are not exact probabilities but are approximate probabilities unless they are based on a census.** However, if the experiment is repeated again and again, this approximate probability of an outcome obtained from the relative frequency will approach the actual probability of that outcome. This is called the **Law of Large Numbers**.

We used Minitab to simulate the tossing of a coin with a different number of tosses. Table 4.3 lists the results of these simulations.

Table 4.3   Simulating the Tosses of a Coin

Number of Tosses	Number of Heads	Number of Tails	P(H)	P(T)
3	0	3	.00	1.00
8	6	2	.75	.25
25	9	16	.36	.64
100	61	39	.61	.39
1000	522	478	.522	.478
10,000	4962	5038	.4962	.5038
1,000,000	500,313	499,687	.5003	.4997

> **Law of Large Numbers:**
> If an experiment is repeated again and again, the probability of an event obtained from the relative frequency approaches the actual (or theoretical) probability.

As we can observe from this table, when the coin was tossed three times, based on relative frequencies, the probability of a head obtained from this simulation was .00 and that of a tail was 1.00. When the coin was tossed eight times, the probability of a head obtained from this simulation was .75 and that of a tail was .25. Notice how these probabilities change. As the number of tosses increases, the probabilities of both a head and a tail converge toward .50. When the simulated tosses of the coin are 1,000,000, the relative frequencies of head and tail are almost equal. This is what is meant by the **Law of Large Numbers**. If the experiment is repeated only a few times, the probabilities obtained may not be close to the actual probabilities. As the number of repetitions increases, the probabilities of outcomes obtained become very close to the actual probabilities. Note that in the example of tossing a fair coin, the actual probability of head and tail is .50 each.

EXAMPLE 4–11 | Owning a Home

Approximating probability by relative frequency.

Allison wants to determine the probability that a randomly selected family from New York State owns a home. How can she determine this probability?

Solution   There are two outcomes for a randomly selected family from New York State: "This family owns a home" and "This family does not own a home." These two outcomes are not equally likely. (Note that these two outcomes will be equally likely if exactly half of the families in New York State own homes and exactly half do not own homes.) Hence, the classical probability rule cannot be applied. However, we can repeat this experiment again and again. In other words, we can select a sample of families from New York State and observe whether or not each of them owns a home. Hence, we will use the relative frequency approach to probability.

Suppose Allison selects a random sample of 1000 families from New York State and observes that 730 of them own homes and 270 do not own homes. Then,

$$n = \text{sample size} = 1000$$
$$f = \text{number of families who own homes} = 730$$

Consequently,

$$P(\text{a randomly selected family owns a home}) = \frac{f}{n} = \frac{730}{1000} = \mathbf{.730}$$

Again, note that .730 is just an approximation of the probability that a randomly selected family from New York State owns a home. Every time Allison repeats this experiment she may obtain a different probability for this event. However, because the sample size ($n = 1000$) in this example is large, the variation is expected to be relatively small.

Subjective Probability

Many times we face experiments that neither have equally likely outcomes nor can be repeated to generate data. In such cases, we cannot compute the probabilities of events using the classical probability rule or the relative frequency concept. For example, consider the following probabilities of events:

1. The probability that Carol, who is taking a statistics course, will earn an A in the course
2. The probability that the Dow Jones Industrial Average will be higher at the end of the next trading day
3. The probability that the New York Giants will win the Super Bowl next season
4. The probability that Joe will lose the lawsuit he has filed against his landlord

> **Subjective probability** is the probability assigned to an event based on subjective judgment, experience, information, and belief. There are no definite rules to assign such probabilities.

Neither the classical probability rule nor the relative frequency concept of probability can be applied to calculate probabilities for these examples. All these examples belong to experiments that have neither equally likely outcomes nor the potential of being repeated. For example, Carol, who is taking statistics, will take the test (or tests) only once, and based on that she will either earn an A or not. The two events "she will earn an A" and "she will not earn an A" are not equally likely. Also, she cannot take the test (or tests) again and again to calculate the relative frequency of getting or not getting an A grade. She will take the test (or tests) only once. The probability assigned to an event in such cases is called **subjective probability**. It is based on the individual's judgment, experience, information, and belief. Carol may be very confident and assign a higher probability to the event that she will earn an A in statistics, whereas her instructor may be more cautious and assign a lower probability to the same event.

Subjective probability is assigned arbitrarily. It is usually influenced by the biases, preferences, and experience of the person assigning the probability.

EXERCISES

CONCEPTS AND PROCEDURES

4.11 What are the two basic properties of probability? Explain them briefly.

4.12 Differentiate between an impossible event and a sure event. What is the probability of the occurrence of each of these two events?

4.13 With the help of a suitable example for each, describe briefly the three approaches to probability.

4.14 For what kind of experiments in calculating probabilities of events do we use:
  (a) classical approach
  (b) relative frequency approach

4.15 Which of the following values cannot be probabilities of events and why?

.46   2/3   −.09   1.42   .96   9/4   −1/4   .02

APPLICATIONS

4.16 Suppose a randomly selected passenger is about to go through the metal detector at JFK Airport in New York City. Consider the following two outcomes: The passenger sets off the metal detector, and the passenger does not set off the metal detector. Are these two outcomes equally likely? Explain why or why not. If you are to find the probability of these two outcomes, would you use the classical approach or the relative frequency approach? Explain why.

4.17 Thirty-five persons have applied for a security guard position with a company. Of them, 9 have previous experience in this area and

26 do not. Suppose one applicant is selected at random. Consider the following two events: This applicant has previous experience, and this applicant does not have previous experience. If you are to find the probabilities of these two events, would you use the classical approach or the relative frequency approach? Explain why.

4.18 The president of a company has a hunch that there is a .75 probability that the company will be successful in marketing a new brand of ice cream. Is this a case of classical, relative frequency, or subjective probability? Explain why.

4.19 A hat contains 50 marbles. Of them, 24 are red and 26 are green. If one marble is randomly selected out of this hat, what is the probability that this marble is
  a. red?   **b.** green?

4.20 A die is rolled once. What is the probability that
  a. a number more than 2 is obtained?
  b. a number 3 to 6 is obtained?

4.21 A random sample of 3000 adults showed that 1980 of them have shopped at least once on the Internet. What is the (approximate) probability that a randomly selected adult has shopped on the Internet?

4.22 In a statistics class of 54 students, 34 have volunteered for community service in the past. Find the probability that a randomly selected student from this class has volunteered for community service in the past.

4.23 There are 1320 eligible voters in a town, and 1012 of them are registered to vote. If one eligible voter is selected at random from this town, what is the probability that this voter is
  a. registered?   **b.** not registered?

Do these two probabilities add up to 1.0? If yes, why?

4.24 A multiple-choice question on a test has four answers. If Dianne chooses one answer based on "pure guess," what is the probability that her answer is
    a. correct?   **b.** wrong?

Do these two probabilities add up to 1.0? If yes, why?

4.25 In a sample of 300 adults, 123 like chocolate ice cream and 84 like vanilla ice cream. One adult is randomly selected from these adults.
    a. What is the probability that this adult likes chocolate ice cream?
    b. What is the probability that this adult likes vanilla ice cream?

Do these two probabilities add to 1.0? Why or why not? Explain.

4.26 According to an article in *The Sacramento Bee* (www.sacbee.com/2011/08/04/3816872/medicareprescription-premiums.html), approximately 10% of Medicare beneficiaries lack a prescription drug care plan. Suppose that a town in Florida has 2384 residents who are Medicare beneficiaries, and 216 of them do not have a prescription drug care plan. If one of the Medicare beneficiaries is chosen at random from this town, what is the probability that this person has a prescription drug care plan? What is the probability that this person does not have a prescription drug care plan? Do these probabilities add up to 1.0? If yes, why? If no, why not?

4.27 A sample of 900 adults showed that 90 of them had no credit cards, 124 had one card each, 103 had two cards each, 77 had three cards each, 51 had four cards each, and 455 had five or more cards each. Write the frequency distribution table for the number of credit cards an adult possesses. Calculate the relative frequencies for all categories. Suppose one adult is randomly selected from these 900 adults. Find the probability that this adult has
    a. three credit cards   **b.** five or more credit cards

4.28 Suppose you want to find the (approximate) probability that a randomly selected family from Los Angeles earns more than $175,000 a year. How would you find this probability? What procedure would you use? Explain briefly.

4.3 | Marginal Probability, Conditional Probability, and Related Probability Concepts

In this section first we discuss marginal and conditional probabilities, and then we discuss the concepts (in that order) of mutually exclusive events, independent and dependent events, and complementary events.

4.3.1 Marginal and Conditional Probabilities

Suppose all 100 employees of a company were asked whether they are in favor of or against paying high salaries to CEOs of U.S. companies. Table 4.4 gives a two-way classification of the responses of these 100 employees. Assume that every employee responds either *in favor* or *against*.

Table 4.4 Two-Way Classification of Employee Responses

	In Favor	Against
Male	15	45
Female	4	36

Table 4.4 shows the distribution of 100 employees based on two variables or characteristics: gender (male or female) and opinion (in favor or against). Such a table is called a *contingency table* or a *two-way table*. In Table 4.4, each box that contains a number is called a *cell*. Notice that there are four cells. Each cell gives the frequency for two characteristics. For example, 15 employees in this group possess two characteristics: "male" and "in favor of paying high salaries to CEOs." We can interpret the numbers in other cells the same way.

By adding the row totals and the column totals to Table 4.4, we write Table 4.5.

Table 4.5 Two-Way Classification of Employee Responses with Totals

	In Favor	Against	Total
Male	15	45	60
Female	4	36	40
Total	19	81	100

Marginal probability is the probability of a single event without consideration of any other event.

Suppose one employee is selected at random from these 100 employees. This employee may be classified either on the basis of gender alone or on the basis of opinion alone. If only one characteristic is considered at a time, the employee selected can be a male, a female, in favor, or against. The probability of each of these four characteristics or events is called **marginal probability**. These probabilities are called marginal probabilities because they are calculated by dividing the corresponding row margins (totals for the rows) or column margins (totals for the columns) by the grand total.

For Table 4.5, the four marginal probabilities are calculated as follows:

$$P(\text{male}) = \frac{\text{Number of males}}{\text{Total number of employees}} = \frac{60}{100} = .60$$

As we can observe, the probability that a male will be selected is obtained by dividing the total of the row labeled "Male" (60) by the grand total (100). Similarly,

$$P(\text{female}) = 40/100 = .40$$

$$P(\text{in favor}) = 19/100 = .19$$

$$P(\text{against}) = 81/100 = .81$$

These four marginal probabilities are shown along the right side and along the bottom of Table 4.6.

Table 4.6   Listing the Marginal Probabilities

	In Favor (A)	Against (B)	Total	
Male (M)	15	45	60	$P(M) = 60/100 = .60$
Female (F)	4	36	40	$P(F) = 40/100 = .40$
Total	19	81	100	

$$P(A) = 19/100 \qquad P(B) = 81/100$$
$$= .19 \qquad\qquad = .81$$

Conditional probability is the probability that an event will occur given that another event has already occurred. If A and B are two events, then the conditional probability of A given B is written as

$$P(A \mid B)$$

and read as "the probability of A given that B has already occurred."

Now suppose that one employee is selected at random from these 100 employees. Furthermore, assume it is known that this (selected) employee is a male. In other words, the event that the employee selected is a male has already occurred. Given that this selected employee is a male, he can be in favor or against. What is the probability that the employee selected is in favor of paying high salaries to CEOs? This probability is written as follows:

Read as "given"

$$P(\text{in favor} \mid \text{male})$$

The event whose probability is to be determined

This event has already occurred

This probability, $P(\text{in favor} \mid \text{male})$, is called the **conditional probability** of "in favor" given that the event "male" has already happened. It is read as "the probability that the employee selected is in favor given that this employee is a male."

EXAMPLE 4–12   Opinions of Employees

Calculating the conditional probability: two-way table.

Compute the conditional probability $P(\text{in favor} \mid \text{male})$ for the data on 100 employees given in Table 4.5.

Solution   The probability $P(\text{in favor} \mid \text{male})$ is the conditional probability that a randomly selected employee is in favor given that this employee is a male. It is known that the event "male" has already occurred. Based on the information that the employee selected is a male, we can infer that the employee selected must be one of the 60 males and, hence, must belong to the first row of Table 4.5. Therefore, we are concerned only with the first row of that table.

	In Favor	Against	Total
Male	15	45	60

Males who are in favor ↑ (under In Favor 15)

Total number of males ↑ (under Total 60)

There are 60 males, and 15 of them are in favor. Hence, the required conditional probability is calculated as follows:

$$P(\text{in favor} \mid \text{male}) = \frac{\text{Number of males who are in favor}}{\text{Total number of males}} = \frac{15}{60} = .25$$

As we can observe from this computation of conditional probability, the total number of males (the event that has already occurred) is written in the denominator and the number of males who are in favor (the event whose probability we are to find) is written in the numerator. Note that we are considering the row of the event that has already occurred.

CASE STUDY 4–1   Do You Worry About Your Weight?

DO YOU WORRY ABOUT YOUR WEIGHT?

35% Men   45% All adults   55% Women

Data source: Gallup poll of 1013 adults aged 18 and older conducted July 7–10, 2014

A Gallup poll of 1013 American adults of age 18 years and older conducted July 7–10, 2014, asked them, "How often do you worry about your weight?" The accompanying chart shows the percentage of adults included in the poll who said that they worry all or some of the time about their weight. According to this information, 45% of the adults in the sample said that they worry all or some of the time about their weight. When broken down based on gender, this percentage is 35% for men and 55% for women.

Assume that these percentages are true for the current population of American adults. Suppose we randomly select one American adult. Based on the overall percentage, the probability that this adult worries all or some of the time about his/her weight is

P(a randomly selected adult worries all or some of the time about his/her weight) = .45

This is a marginal probability because there is no condition imposed here.

Now suppose we randomly select one American adult. Then, given that this adult is a man, the probability is .35 that he worries all or some of the time about his weight. If the selected adult is a woman, the probability is .55 that she worries all or some of the time about her weight. These are two conditional probabilities, which can be written as follows:

P(a randomly selected adult worries all or some of the time about his weight | man) = .35

P(a randomly selected adult worries all or some of the time about her weight | woman) = .55

Note that these are approximate probabilities because the percentages given in the chart are based on a sample survey of 1013 adults.

Source: http://www.gallup.com.

EXAMPLE 4–13   Opinions of Employees

Calculating the conditional probability: two-way table.

For the data of Table 4.5, calculate the conditional probability that a randomly selected employee is a female given that this employee is in favor of paying high salaries to CEOs.

Solution   Here one employee is selected and this employee happens to be *in favor*. Given that this employee is *in favor*, the employee can be either a male or a female. We are to find the probability of a *female* given that the selected employee is *in favor*, that is, we are to compute the probability P(female | in favor). Because it is known that the employee selected is in favor of paying high salaries to CEOs, this employee must belong to the first column (the column labeled "in favor") and must be one of the 19 employees who are in favor. Of these 19 employees who are in favor, four are females.

In Favor
15
4 ◄—— Females who are in favor
19 ◄—— Total number of employees who are in favor

Hence, the required probability is

$$P(\text{female | in favor}) = \frac{\text{Number of females who are in favor}}{\text{Total number of employees who are in favor}} = \frac{4}{19} = .2105$$

4.3.2 Mutually Exclusive Events

Events that cannot occur together are said to be **mutually exclusive events**.

Suppose you toss a coin once. You will obtain either a head or a tail, but not both, because, in one toss, the events head and tail cannot occur together. Such events are called **mutually exclusive events**. Such events do not have any common outcomes. If two or more events are mutually exclusive, then at most one of them will occur every time we repeat the experiment. Thus the occurrence of one event excludes the occurrence of the other event or events.

For any experiment, the final outcomes are always mutually exclusive because one and only one of these outcomes is expected to occur in one repetition of the experiment. For example, consider tossing a coin twice. This experiment has four outcomes: *HH*, *HT*, *TH*, and *TT*. These outcomes are mutually exclusive because one and only one of them will occur when we toss this coin twice. As another example, suppose one student is selected at random from a statistics class and the gender of this student is observed. This student can be a *male* or a *female*. These are two mutually exclusive events. These events cannot happen together if only one student is selected. But now, if you observe whether the selected student is a math major or a business major, then these two events will not be mutually exclusive if there is at least one student in the class who is a double major in math and business. However, if there is no student with both majors, then these two events will be mutually exclusive events.

| **EXAMPLE 4-14** | **One Roll of a Die** |

Illustrating mutually exclusive and mutually nonexclusive events.

Consider the following events for one roll of a die:

$$A = \text{an even number is observed} = \{2, 4, 6\}$$
$$B = \text{an odd number is observed} = \{1, 3, 5\}$$
$$C = \text{a number less than 5 is observed} = \{1, 2, 3, 4\}$$

Are events A and B mutually exclusive? Are events A and C mutually exclusive?

Solution   Figures 4.6 and 4.7 show the diagrams of events A and B and of events A and C, respectively.

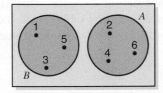

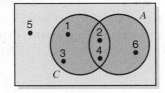

Figure 4.6   Mutually exclusive events A and B.

Figure 4.7   Mutually nonexclusive events A and C.

As we can observe from the definitions of events A and B and from Figure 4.6, events A and B have no common element. For one roll of a die, only one of the two events A and B can happen. Hence, these are two mutually exclusive events.

We can observe from the definitions of events A and C and from Figure 4.7 that events A and C have two common outcomes: 2-spot and 4-spot. Thus, if we roll a die and obtain either a 2-spot or a 4-spot, then A and C happen at the same time. Hence, events A and C are not mutually exclusive.

| **EXAMPLE 4-15** | **Shopping on the Internet** |

Illustrating mutually exclusive events.

Consider the following two events for a randomly selected adult:

$$Y = \text{this adult has shopped on the Internet at least once}$$
$$N = \text{this adult has never shopped on the Internet}$$

Are events Y and N mutually exclusive?

Solution   Note that event Y consists of all adults who have shopped on the Internet at least once, and event N includes all adults who have never shopped on the Internet. These two events are illustrated in the diagram in Figure 4.8.

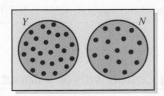

Figure 4.8   Mutually exclusive events Y and N.

As we can observe from the definitions of events Y and N and from Figure 4.8, events Y and N have no common outcome. They represent two distinct sets of adults: the ones who have shopped on the Internet at least once and the ones who have never shopped on the Internet. Hence, these two events are mutually exclusive.

4.3.3   Independent versus Dependent Events

> Two events are said to be *independent* if the occurrence of one event does not affect the probability of the occurrence of the other event. In other words, *A* and *B* are **independent events** if
>
> either $P(A \mid B) = P(A)$
>
> or
>
> $P(B \mid A) = P(B)$

Consider two tosses of a coin. Does the outcome of the first toss affect the outcome of the second toss? In other words, whether we obtain a head or a tail in the first toss, will it change the probability of obtaining a head or a tail in the second toss? The answer is: No. The outcome of the first toss does not affect the outcome of the second toss. Whether we get a head or a tail in the first toss, the probability of obtaining a head in the second toss is still .50 and that of a tail is also .50. This is an example of a statistical experiment where the outcomes of two tosses are independent. In the case of two **independent events**, the occurrence of one event does not change the probability of the occurrence of the other event.

Now consider one toss of a coin. Suppose you toss this coin and obtain a head. Did obtaining a head affect the probability of obtaining a tail? Yes, it did. Before you tossed the coin, the probability of obtaining a tail was .50. But once you tossed the coin and obtained a head, the probability of obtaining a tail in that toss becomes zero. In other words, it is obvious that you did not obtain a tail. This is an example of two dependent events or outcomes. Thus, if the occurrence of one event affects the probability of the occurrence of the other event, then the two events are said to be **dependent events**. In probability notation, the two events are dependent if either $P(A \mid B) \neq P(A)$ or $P(B \mid A) \neq P(B)$.

EXAMPLE 4–16	Opinions of Employees

Illustrating two dependent events: two-way table.

Refer to the information on 100 employees given in Table 4.5 in Section 4.3.1. Are events *in favor* and *female* independent?

Solution   To check whether events *in favor* and *female* are independent, we find the (marginal) probability of *in favor*, and then the conditional probability of *in favor* given that *female* has already happened. If these two probabilities are equal then these two events are independent, otherwise they are dependent.

From Table 4.5, there are 19 employees in favor out of 100 total employees. Hence,

$$P(\text{in favor}) = \frac{19}{100} = .19$$

Now we find the conditional probability of *in favor* given that the event *female* has already occurred. We know that the person selected is a *female* because this event has already happened. There are a total of 40 employees who are *female* and 4 of them are *in favor*. Hence,

$$P(\text{in favor} \mid \text{female}) = \frac{4}{40} = .10$$

Because these two probabilities are not equal, the two events are dependent. Here, dependence of events means that the percentage of female employees who are in favor of paying high salaries to CEOs is different from the percentage of male employees who are in favor of this issue. This percentage of female employees who are in favor of paying high salaries to CEOs is also different from the overall percentage of employees who are in favor of paying high salaries to CEOs.

EXAMPLE 4–17	Gender and Drinking Coffee with or without Sugar

Illustrating two independent events.

In a survey, 500 randomly selected adults who drink coffee were asked whether they usually drink coffee with or without sugar. Of these 500 adults, 240 are men, and 175 drink coffee without sugar. Of the 240 men, 84 drink coffee without sugar. Are the events drinking coffee *without sugar* and *man* independent?

Solution   To check if the events drinking coffee *without sugar* and *man* are independent, first we find the probability that a randomly selected adult from these 500 adults drinks coffee *without sugar* and then find the conditional probability that this selected adult drinks coffee *without sugar* given that this adult is a *man*. From the given information:

$$P(\text{drinks coffee without sugar}) = \frac{175}{500} = .35$$

Now to find the conditional probability that a randomly selected adult from these 500 adults drinks coffee *without sugar* given that this adult is a *man*, we divide the number of men in the adults who drink coffee without sugar by the total number of adults who drink coffee without sugar. There are a total of 175 adults who drink coffee without sugar and 84 of them are men. Also, there are a total of 240 men. Hence,

$$P(\text{drinks coffee without sugar} \mid \text{man}) = \frac{84}{240} = .35$$

Since the two probabilities are equal, the two events drinking coffee *without sugar* and *man* are independent. In this example, independence of these events means that men and women have the same preferences in regard to drinking coffee with or without sugar. Table 4.7 has been written using the given information. The numbers in the shaded cells are given to us. If we calculate the percentages of men and women who drink coffee with and without sugar, they are the same. Sixty-five percent of both, men and women, drink coffee with sugar and 35% of both drink coffee without sugar. If the percentages of men and women who drink coffee with and without sugar are different, then these events will be dependent.

Table 4.7   Gender and Drinking Coffee with or without Sugar

	With Sugar	Without Sugar	Total
Men	156	84	240
Women	169	91	260
Total	325	175	500

We can make the following two important observations about mutually exclusive, independent, and dependent events.

◀ *Two Important Observations*

1. Two events are either mutually exclusive or independent.[2]
    a. Mutually exclusive events are always dependent.
    b. Independent events are never mutually exclusive.
2. Dependent events may or may not be mutually exclusive.

4.3.4   Complementary Events

Consider one roll of a fair six-sided die. Let A be the event that an even number is obtained and B be the event that an odd number is obtained. Then event A includes outcomes 2, 4, and 6, and event B includes outcomes 1, 3, and 5. As we can notice, events A and B are mutually exclusive events as they do not contain any common outcomes. Taken together, events A and B include all six outcomes of this experiment. In this case, events A and B are called complementary events. Thus, we can state that two mutually exclusive events that taken together include all the outcomes for an experiment are called **complementary events**. Note that two complementary events are always mutually exclusive.

> The **complement of event** A, denoted by $\bar{A}$ and read as "*A* bar" or "*A* complement," is the event that includes all the outcomes for an experiment that are not in A. Event A and $\bar{A}$ are called **complementary events**.

[2]The exception to this rule occurs when at least one of the two events has a zero probability.

Events A and $\overline{A}$ are complements of each other. The diagram in Figure 4.9 shows the complementary events A and $\overline{A}$.

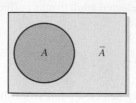

Figure 4.9   Two complementary events.

Because two complementary events, taken together, include all the outcomes for an experiment and because the sum of the probabilities of all outcomes is 1, it is obvious that

$$P(A) + P(\overline{A}) = 1.0$$

From this equation, we can deduce that

$$P(A) = 1 - P(\overline{A}) \quad \text{and} \quad P(\overline{A}) = 1 - P(A)$$

Thus, if we know the probability of an event, we can find the probability of its complementary event by subtracting the given probability from 1.

EXAMPLE 4–18   **Taxpayers Audited by IRS**

Calculating probability of complementary events.

In a group of 2000 taxpayers, 400 have been audited by the IRS at least once. If one taxpayer is randomly selected from this group, what are the two complementary events for this experiment, and what are their probabilities?

Solution   The two complementary events for this experiment are

A = the selected taxpayer has been audited by the IRS at least once
$\overline{A}$ = the selected taxpayer has never been audited by the IRS

Note that here event A includes the 400 taxpayers who have been audited by the IRS at least once, and $\overline{A}$ includes the 1600 taxpayers who have never been audited by the IRS. Hence, the probabilities of events A and $\overline{A}$ are

$$P(A) = 400/2000 = \mathbf{.20} \quad \text{and} \quad P(\overline{A}) = 1600/2000 = \mathbf{.80}$$

As we can observe, the sum of these two probabilities is 1. Figure 4.10 shows the two complementary events A and $\overline{A}$ for this example.

Figure 4.10   Complementary events A and $\overline{A}$.

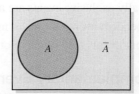

EXAMPLE 4–19   **Gender and Drinking Coffee with or without Sugar**

Calculating probabilities of complementary events.

Refer to the information given in Table 4.7. Of the 500 adults, 325 drink coffee with sugar. Suppose one adult is randomly selected from these 500 adults, and let A be the event that this adult drinks coffee with sugar. What is the complement of event A? What are the probabilities of the two events?

Solution   For this experiment, the two complementary events are:

$$A = \text{the selected adult drinks coffee with sugar}$$
$$\overline{A} = \text{the selected adult drinks coffee without sugar}$$

Note that here $\overline{A}$ includes 175 adults who drink coffee without sugar. The two events A and $\overline{A}$ are complements of each other. The probabilities of the two events are:

$$P(A) = P(\text{selected adult drinks coffee with sugar}) = \frac{325}{500} = .65$$

$$P(\overline{A}) = P(\text{selected adult drinks coffee without sugar}) = \frac{175}{500} = .35$$

As we can observe, the sum of these two probabilities is 1, which should be the case with two complementary events. Also, once we find $P(A)$, we can find $P(\overline{A})$ as:

$$P(\overline{A}) = 1 - P(A) = 1 - .65 = .35$$

Figure 4.11 shows these two complementary events.

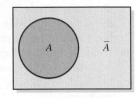

Figure 4.11 Complementary events drinking coffee with or without sugar.

EXERCISES

CONCEPTS AND PROCEDURES

4.29 Differentiate between the marginal and conditional probabilities of events. Give one example of each.

4.30 What is meant by two mutually exclusive events? Give one example of two mutually exclusive events and another example of two events that are not mutually exclusive.

4.31 Briefly explain the difference between independent and dependent events. Suppose X and Y are two events. What formula will you use to prove whether X and Y are independent or dependent?

4.32 What do you understand by the complement of an event? What is the sum of the probabilities of two complementary events?

4.33 A statistical experiment has 10 equally likely outcomes that are denoted by 1, 2, 3, 4, 5, 6, 7, 8, 9, and 10. Let event $A = \{3, 4, 6, 9\}$ and event $B = \{1, 2, 5\}$.

    a. Are events A and B mutually exclusive events?
    b. Are events A and B independent events?
    c. What are the complements of events A and B, respectively, and their probabilities?

APPLICATIONS

4.34 Two thousand randomly selected adults were asked whether or not they have ever shopped on the Internet. The following table gives a two-way classification of the responses.

	Have Shopped	Have Never Shopped
Male	400	800
Female	350	450

    a. If one adult is selected at random from these 2000 adults, find the probability that this adult

      i. has never shopped on the Internet
      ii. is a male
      iii. has shopped on the Internet given that this adult is a female
      iv. is a male given that this adult has never shopped on the Internet

    b. Are the events "male" and "female" mutually exclusive? What about the events "have shopped" and "male?" Why or why not?

    c. Are the events "female" and "have shopped" independent? Why or why not?

4.35 Five hundred employees were selected from a city's large private companies, and they were asked whether or not they have any retirement benefits provided by their companies. Based on this information, the following two-way classification table was prepared.

	Have Retirement Benefits	
	Yes	No
Men	240	60
Women	135	65

    a. If one employee is selected at random from these 500 employees, find the probability that this employee

      i. is a woman
      ii. has retirement benefits
      iii. has retirement benefits given the employee is a man
      iv. is a woman given that she does not have retirement benefits

    b. Are the events "man" and "yes" mutually exclusive? What about the events "yes" and "no?" Why or why not?

    c. Are the events "woman" and "yes" independent? Why or why not?

4.36 Two thousand randomly selected adults were asked if they are in favor of or against cloning. The following table gives the responses.

	In Favor	Against	No Opinion
Male	380	410	50
Female	400	680	80

a. If one person is selected at random from these 2000 adults, find the probability that this person is
  i. in favor of cloning
  ii. against cloning
  iii. in favor of cloning given the person is a female
  iv. a male given the person has no opinion
b. Are the events "male" and "in favor" mutually exclusive? What about the events "in favor" and "against?" Why or why not?
c. Are the events "female" and "no opinion" independent? Why or why not?

4.37   There are a total of 150 practicing physicians in a city. Of them, 70 are female and 20 are pediatricians. Of the 70 females, 10 are pediatricians. Are the events "female" and "pediatrician" independent? Are they mutually exclusive? Explain why or why not.

4.38   In a survey, 500 randomly selected adults who drink coffee were asked whether they usually drink coffee with or without sugar. Of these 500 adults, 290 are men and 200 drink coffee without sugar. Of the 200 who drink coffee without sugar, 130 are men. Are the events *man* and drinking coffee *without sugar* independent? (*Note:* Compare this exercise to Example 4–17.)

4.39   A July 21, 2009 (just a reminder that July 21 is National Junk Food Day) survey on www.HuffingtonPost.com asked people to choose their favorite junk food from a list of choices. Of the 8002 people who responded to the survey, 2049 answered chocolate, 345 said sugary candy, 1271 mentioned ice cream, 775 indicated fast food, 650 said cookies, 1107 mentioned chips, 490 said cake, and 1315 indicated pizza. Although the results were not broken down by gender, suppose that the following table represents the results for the 8002 people who responded, assuming that there were 4801 females and 3201 males included in the survey.

Favorite Junk Food	Female	Male
Chocolate	1518	531
Sugary candy	218	127
Ice cream	685	586
Fast food	312	463
Cookies	431	219
Chips	458	649
Cake	387	103
Pizza	792	523

a. If one person is selected at random from this sample of 8002 respondents, find the probability that this person
  i. is a female
  ii. responded *chips*
  iii. responded *chips* given that this person is a *female*
  iv. responded *chocolate* given that this person is a *male*
b. Are the events *chips* and *cake* mutually exclusive? What about the events *chips* and *female*? Why or why not?
c. Are the events *chips* and *female* independent? Why or why not?

4.40   Let *A* be the event that a number greater than 2 is obtained if we roll a die once. What is the probability of *A*? What is the complementary event of *A*, and what is its probability?

4.41   Thirty-five percent of last year's graduates from a university received job offers during their last semester in school. What are the two complementary events here and what are their probabilities?

4.42   The probability that a randomly selected college student attended at least one major league baseball game last year is .12. What is the complementary event? What is the probability of this complementary event?

4.43   There are 142 people participating in a local 5K road race. Sixty-five of these runners are females. Of the female runners, 19 are participating in their first 5K road race. Of the male runners, 28 are participating in their first 5K road race. Are the events *female* and *participating in their first 5K road race* independent? Are they mutually exclusive? Explain why or why not.

4.4 | Intersection of Events and the Multiplication Rule

This section discusses the intersection of two events and the application of the multiplication rule to compute the probability of the intersection of events.

4.4.1   Intersection of Events

Refer to the information given in Table 4.7. Suppose one adult is selected at random from this group of 500 adults. Consider the events a *woman* and drinking coffee *without sugar*. Then, 91 women in this group who drink coffee without sugar give the intersection of the events *woman* and drinking coffee *without sugar*. Note that the **intersection of two events** is given by the outcomes that are common to both events.

> **Intersection of Events**   Let *A* and *B* be two events defined in a sample space. The **intersection** of *A* and *B* represents the collection of all outcomes that are common to both *A* and *B* and is denoted by
>
> $$(A \text{ and } B)$$

The intersection of events A and B is also denoted by either $A \cap B$ or AB. Let

A = event that a woman is selected from the 500 adults in Table 4.7

B = event that an adult who drinks coffee without sugar is selected

Figure 4.12 illustrates the intersection of events A and B. The shaded area in this figure gives the intersection of events A and B, and it includes all the women who drink coffee without sugar.

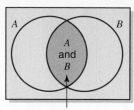

Intersection of A and B

Figure 4.12 Intersection of events A and B.

4.4.2   Multiplication Rule for Independent Events

Sometime we may need to find the probability that two or more events happen together, and this probability is called the **joint probability** of these events.

If two events are independent, then they do not affect each other. To find the probability that two independent events happen together, we multiply their marginal probabilities. For example, to find the probability of obtaining two heads in two tosses of a coin, we multiply the probability of obtaining a head in the first toss by the probability of obtaining a head in the second toss, since the outcomes of the two tosses are independent of each other.

> The probability of the intersection of two events is called their **joint probability**. It is written as
>
> $$P(A \text{ and } B)$$

> **Multiplication Rule to Calculate the Probability of Independent Events**   The probability of the intersection of two independent events A and B is
>
> $$P(A \text{ and } B) = P(A) \times P(B)$$

EXAMPLE 4–20   Two Fire Detectors

Calculating the joint probability of two independent events.

An office building has two fire detectors. The probability is .02 that any fire detector of this type will fail to go off during a fire. Find the probability that both of these fire detectors will fail to go off in case of a fire. Assume that these two fire detectors are independent of each other.

Solution   In this example, the two fire detectors are independent because whether or not one fire detector goes off during a fire has no effect on the second fire detector. We define the following two events:

A = the first fire detector fails to go off during a fire

B = the second fire detector fails to go off during a fire

Then, the joint probability of A and B is

$$P(A \text{ and } B) = P(A) \times P(B) = (.02) \times (.02) = \mathbf{.0004}$$

The multiplication rule can be extended to calculate the joint probability of more than two events. Example 4–21 illustrates such a case for independent events.

EXAMPLE 4–21   Three Students Getting Jobs

Calculating the joint probability of three independent events.

The probability that a college graduate will find a full-time job within three months after graduation from college is .27. Three college students, who will be graduating soon, are randomly selected. What is the probability that all three of them will find full-time jobs within three months after graduation from college?

Solution   Let:

A = the event that the first student finds a full-time job within three months after graduation from college

B = the event that the second student finds a full-time job within three months after graduation from college

C = the event that the third student finds a full-time job within three months after graduation from college

These three events, A, B, and C, are independent events because one student getting a full-time job within three months after graduation from college does not affect the probabilities of the other two students getting full-time jobs within three months after graduation from college. This is a reasonable assumption since the population of students graduating from colleges is very large and the students are randomly selected. Hence, the probability that all three students will find full-time jobs within three months after graduation from college is:

$$P(A \text{ and } B \text{ and } C) = P(A) \times P(B) \times P(C) = (.27) \times (.27) \times (.27) = \textbf{.0197}$$

4.4.3   Multiplication Rule for Dependent Events

We know that when the occurrence of one event affects the occurrence of another event, these events are said to be dependent events. For example, in the two-way Table 4.5, events *female* and *in favor* are dependent. Suppose we randomly select one person from the 100 persons included in Table 4.5 and we want to find the joint probability of *female* and *in favor*. To obtain this probability, we will not multiply the marginal probabilities of these two events. Here, to find the joint probability of these two events, we will multiply the marginal probability of *female* by the conditional probability of *in favor* given that a *female* has been selected.

Thus, the probability of the intersection of two dependent events is obtained by multiplying the marginal probability of one event by the conditional probability of the second event. This rule is called the **multiplication rule**.

> **Multiplication Rule to Find Joint Probability of Two Dependent Events**   The probability of the intersection of two dependent events A and B is
>
> $$P(A \text{ and } B) = P(A) \times P(B \mid A) \text{ or } P(B) \times P(A \mid B)$$

The joint probability of events A and B can also be denoted by $P(A \cap B)$ or $P(AB)$.

EXAMPLE 4–22   Gender and College Degree

Calculating the joint probability of two events: two-way table.

Table 4.8 gives the classification of all employees of a company by gender and college degree.

Table 4.8   Classification of Employees by Gender and Education

	College Graduate	Not a College Graduate	Total
Male	7	20	27
Female	4	9	13
Total	11	29	40

If one of these employees is selected at random for membership on the employee–management committee, what is the probability that this employee is a female and a college graduate?

Solution   We are to calculate the probability of the intersection of the events *female* and *college graduate*. This probability may be computed using the formula

$$P(\textit{female and college graduate}) = P(\textit{female}) \times P(\textit{college graduate} \mid \textit{female})$$

The shaded area in Figure 4.13 shows the intersection of the events *female* and *college graduate*. There are four females who are college graduates in 40 employees.

Notice that there are 13 females among 40 employees. Hence, the probability that a female is selected is

$$P(\textit{female}) = \frac{13}{40}$$

To calculate the probability P(college graduate | female), we know that the event *female* has already occurred. Consequently, the employee selected is one of the 13 females. In the table, there are 4 college graduates among 13 female employees. Hence, the conditional probability of the event *college graduate* given that the event *female* has already happened is

$$P(\textit{college graduate} \mid \textit{female}) = \frac{4}{13}$$

The joint probability of the events *female* and *college graduate* is

$$P(\textit{female and college graduate}) = P(\textit{female}) \times P(\textit{college graduate} \mid \textit{female})$$

$$= \frac{13}{40} \times \frac{4}{13} = .10$$

Thus, the probability is .10 that a randomly selected employee is a *female* and a *college graduate*.

The probability in this example can also be calculated without using the above multiplication rule. As we notice from Table 4.8:

Total employees = 40

Number of employees who are *females* and *college graduates* = 4

Hence, the joint probability of the events *female* and *college graduate* is obtained by dividing 4 by 40:

$$P(\textit{female and college graduate}) = \frac{4}{40} = .10$$

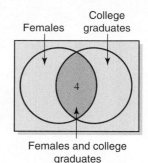

Females   College graduates

Females and college graduates

Figure 4.13   Intersection of events *female* and *college graduate*.

EXAMPLE 4–23   Liking Ice Tea

Calculating the joint probability of two events.

In a group of 20 college students, 7 like ice tea and others do not. Two students are randomly selected from this group.

(a) Find the probability that both of the selected students like ice tea.

(b) Find the probability that the first student selected likes ice tea and the second does not.

Solution   In these 20 students, 7 like ice tea and 13 do not. We find the required probabilities as follows.

(a)   There are 7 students in 20 who like ice tea. Hence, the probability that the first student likes ice tea is:

$$P(\textit{first student likes ice tea}) = \frac{7}{20} = .35$$

Now we have 19 students left in the group, of whom 6 like ice tea. Hence,

$$P(\textit{second student likes ice tea} \mid \textit{first student likes ice tea}) = \frac{6}{19} = .3158$$

Thus, the joint probability of the two events is:

P(both students like ice tea) =

P(first student likes ice tea) $\times$ P(second student likes ice tea | first student likes ice tea)

$$= (.35) \times (.3158) = \textbf{.1105}$$

(b) We find the probability that the first student selected likes ice tea and the second does not as follows.

$$P(first\ student\ likes\ ice\ tea) = \frac{7}{20} = .35$$

Now we have 19 students left in the group, of whom 13 do not like ice tea. Hence,

$$P(second\ student\ does\ not\ like\ ice\ tea\ |\ first\ student\ likes\ ice\ tea) = \frac{13}{19} = .6842$$

Thus, the joint probability of the two events is:

P(first student likes ice tea and second student does not like ice tea) =

P(first student likes ice tea) $\times$ P(second student does not like ice tea | first student likes ice tea)

$$= (.35) \times (.6842) = \textbf{.2395}$$

Conditional probability was discussed in Section 4.3.1. It is obvious from the formula for joint probability that if we know the probability of an event A and the joint probability of events A and B, then we can calculate the conditional probability of B given A.

Calculating Conditional Probability   If A and B are two events, then,

$$P(B \mid A) = \frac{P(A\ and\ B)}{P(A)} \quad \text{and} \quad P(A \mid B) = \frac{P(A\ and\ B)}{P(B)}$$

given that $P(A) \neq 0$ and $P(B) \neq 0$.

EXAMPLE 4–24   Major and Status of a Student

Calculating the conditional probability of an event.

The probability that a randomly selected student from a college is a senior is .20, and the joint probability that the student is a computer science major and a senior is .03. Find the conditional probability that a student selected at random is a computer science major given that the student is a senior.

Solution   From the given information, the probability that a randomly selected student is a senior is:

$$P(senior) = .20$$

The probability that the selected student is a senior and computer science major is:

$$P(senior\ and\ computer\ science\ major) = .03$$

Hence, the conditional probability that a randomly selected student is a computer science major given that he or she is a senior is:

P(computer science major | senior) $= P$(senior and computer science major)$/P$(senior)

$$= \frac{.03}{.20} = \textbf{.15}$$

Thus, the (conditional) probability is .15 that a student selected at random is a computer science major given that he or she is a senior.

4.4.4   Joint Probability of Mutually Exclusive Events

We know from an earlier discussion that two mutually exclusive events cannot happen together. Consequently, their joint probability is zero.

Joint Probability of Mutually Exclusive Events   The joint probability of two mutually exclusive events is always zero. If A and B are two mutually exclusive events, then,

$$P(A \text{ and } B) = 0$$

EXAMPLE 4–25	Car Loan Application

Illustrating the joint probability of two mutually exclusive events.

Consider the following two events for an application filed by a person to obtain a car loan:

$$A = \text{event that the loan application is approved}$$
$$R = \text{event that the loan application is rejected}$$

What is the joint probability of A and R?

Solution   The two events A and R are mutually exclusive. Either the loan application will be approved or it will be rejected. Hence,

$$P(A \text{ and } R) = \mathbf{0}$$

EXERCISES

CONCEPTS AND PROCEDURES

4.44   What is meant by the intersection of two events? Give one example.

4.45   Briefly explain the meaning of the joint probability of two or more events. Give one example.

4.46   How is the multiplication rule of probability for two dependent events different from the rule for two independent events?

4.47   Briefly explain the meaning of the joint probability of two mutually exclusive events.

4.48   Find the joint probability of A and B for the following.

    a. $P(A) = .40$ and $P(B \mid A) = .25$
    b. $P(B) = .65$ and $P(A \mid B) = .36$

4.49   Given that A and B are two independent events, find their joint probability for the following.

    a. $P(A) = .61$ and $P(B) = .27$
    b. $P(A) = .39$ and $P(B) = .63$

4.50   Given that A, B, and C are three independent events, find their joint probability for the following.

    a. $P(A) = .20$, $P(B) = .46$,   and $P(C) = .25$
    b. $P(A) = .44$, $P(B) = .27$,   and $P(C) = .43$

4.51   Given that $P(B) = .65$ and $P(A \text{ and } B) = .45$, find $P(A \mid B)$.

4.52   Given that $P(A \mid B) = .40$ and $P(A \text{ and } B) = .36$, find $P(B)$.

4.53   Given that $P(B \mid A) = .80$ and $P(A \text{ and } B) = .58$, find $P(A)$.

APPLICATIONS

4.54   Five hundred employees were selected from a city's large private companies and asked whether or not they have any retirement benefits provided by their companies. Based on this information, the following two-way classification table was prepared.

	Have Retirement Benefits	
	Yes	**No**
Men	240	60
Women	135	65

    a. Suppose one employee is selected at random from these 500 employees. Find the following probabilities.
        i. Probability of the intersection of events "woman" and "yes"
        ii. Probability of the intersection of events "no" and "man"

b. Mention what other joint probabilities you can calculate for this table and then find them. You may draw a tree diagram to find these probabilities.

4.55 In a sample survey, 1800 senior citizens were asked whether or not they have ever been victimized by a dishonest telemarketer. The following table gives the responses by age group (in years).

		Have Been Victimized	Have Never Been Victimized
	60–69 (A)	106	698
Age	70–79 (B)	145	447
	80 or over (C)	61	343

a. Suppose one person is randomly selected from these senior citizens. Find the following probabilities.
  i. *P*(have been victimized *and* C)
  ii. *P*(have never been victimized *and* A)
b. Find *P*(B and C). Is this probability zero? Explain why or why not.

4.56 A consumer agency randomly selected 1800 flights for two major airlines, A and B. The following table gives the two-way classification of these flights based on airline and arrival time. Note that "less than 30 minutes late" includes flights that arrived early or on time.

	Less Than 30 Minutes Late	30 Minutes to 1 Hour Late	More Than 1 Hour Late
Airline A	412	401	87
Airline B	385	362	153

a. Suppose one flight is selected at random from these 1800 flights. Find the following probabilities.
  i. *P*(more than 1 hour late *and* airline A)
  ii. *P*(airline B *and* less than 30 minutes late)
b. Find the joint probability of events "30 minutes to 1 hour late" and "more than 1 hour late." Is this probability zero? Explain why or why not.

4.57 In a statistics class of 54 students, 34 have volunteered for community service in the past. If two students are selected at random from this class, what is the probability that both of them have volunteered for community service in the past?

4.58 A company is to hire two new employees. They have prepared a final list of ten candidates, all of whom are equally qualified. Of these ten candidates, seven are women. If the company decides to select two persons randomly from these ten candidates, what is the probability that both of them are women?

4.59 In a political science class of 35 students, 21 favor abolishing the electoral college and thus electing the President of the United States by popular vote. If two students are selected at random from this class, what is the probability that both of them favor abolition of the electoral college?

4.60 In a group of 10 persons, 4 have a type A personality and 6 have a type B personality. If two persons are selected at random from this group, what is the probability that the first of them has a type A personality and the second has a type B personality?

4.61 Eight percent of all items sold by a mail-order company are returned by customers for a refund. Find the probability that, of two items sold during a given hour by this company,
  a. both will be returned for a refund
  b. neither will be returned for a refund

4.62 According to the Recording Industry Association of America, only 37% of music files downloaded from Web sites in 2009 were paid for. Suppose that this percentage holds true for such files downloaded this year. Three downloaded music files are selected at random. What is the probability that all three were paid for? What is the probability that none were paid for? Assume independence of events.

4.63 The probability that a student graduating from Suburban State University has student loans to pay off after graduation is .70. The probability that a student graduating from this university has student loans to pay off after graduation and is a male is .38. Find the conditional probability that a randomly selected student from this university is a male given that this student has student loans to pay off after graduation.

4.64 The probability that a farmer is in debt is .80. What is the probability that three randomly selected farmers are all in debt? Assume independence of events.

4.65 A telephone poll conducted of 1000 adult Americans for the *Washington Post* in March 2009 asked about current events in the United States. Suppose that of the 1000 respondents, 629 stated that they were cutting back on their daily spending. Suppose that 322 of the 629 people who stated that they were cutting back on their daily spending said that they were cutting back "somewhat" and 97 stated that they were cutting back "somewhat" and delaying the purchase of a new car by at least 6 months. If one of the 629 people who are cutting back on their spending is selected at random, what is the probability that he/she is delaying the purchase of a new car by at least 6 months given that he/she is cutting back on spending "somewhat?"

4.66 Suppose that 20% of all adults in a small town live alone, and 8% of the adults live alone and have at least one pet. What is the probability that a randomly selected adult from this town has at least one pet given that this adult lives alone?

4.5 | Union of Events and the Addition Rule

This section discusses the union of events and the addition rule that is applied to compute the probability of the union of events.

4.5.1 Union of Events

Suppose one student is randomly selected from a statistics class. Consider the event that this student is a *female* or *junior*. Here, a *female* or *junior* means that this student is either a female or junior or both. This is called the union of events *female* and *junior* and is written as:

(*female* or *junior*)

Thus, the **union of two events** A and B includes all outcomes that are either in A or in B or in both A and B.

The union of events A and B is also denoted by $A \cup B$. Example 4–26 illustrates the union of events A and B.

> Let A and B be two events defined in a sample space. The **union of events** A and B is the collection of all outcomes that belong either to A or to B or to both A and B and is denoted by
>
> $(A \text{ or } B)$

| EXAMPLE 4–26 | Gender and Hot Drinks |

Illustrating the union of two events.

All 440 employees of a company were asked what hot drink they like the most. Their responses are listed in Table 4.9.

Table 4.9 Hot Drinks Employees Like

	Regular Coffee	Regular Tea	Other	Total
Male	110	70	110	290
Female	60	50	40	150
Total	170	120	150	440

Describe the union of events *female* and likes *regular coffee*.

Solution The union of events *female* and likes *regular coffee* includes those employees who are either female or like regular coffee or both. As we can observe from Table 4.9, there are a total of 150 females in these employees, and 170 of all employees like regular coffee. However, 60 females are counted in each of these two numbers and, hence, are counted twice. To find the exact number of employees who are either female, or like regular coffee, or are both, we add 150 and 170 and then subtract 60 from this sum to avoid double counting. Thus, the number of employees who are either female or like regular coffee or both is:

$$150 + 170 - 60 = 260$$

These 260 employees give the union of events *female* and likes *regular coffee*, which is shown in Figure 4.14.

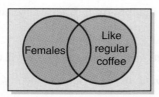

Figure 4.14 Union of events *female* and likes *regular coffee*.

Area shaded in red gives the union of events *female* and likes *regular coffee*, and includes 260 employees who are either female or like regular coffee or both.

4.5.2 Addition Rule for Mutually Nonexclusive Events

When two events are not mutually exclusive, which means that they can occur together, then the probability of the union of these events is given by the sum of their marginal probabilities minus their joint probability. We subtract the joint probability of the two events from the sum of their marginal probabilities due to the problem of double counting. For example, refer to Table 4.9 in Example 4–26. Suppose one employee is randomly selected from these 440 employees, and we want to find the probability of the union of events *female* and likes *regular coffee*. In other words, we are to find the probability $P(\text{female or regular coffee})$, which means that the selected employee is a female or likes regular coffee or both. If we add the number of females and the number of employees who like regular coffee, then the employees who are female and like regular coffee (which is 60 in Table 4.9) are counted twice, once in *females* and then in like *regular coffee*. To avoid this double counting, we subtract the joint probability of the two events from the sum of the two marginal probabilities. This method of calculating the probability of the union of events is called the **addition rule**.

Addition Rule to Find the Probability of Union of Two Mutually Nonexclusive Events   The probability of the union of two mutually nonexclusive events A and B is

$$P(A \text{ or } B) = P(A) + P(B) - P(A \text{ and } B)$$

Thus, to calculate the probability of the union of two events A and B, we add their marginal probabilities and subtract their joint probability from this sum. We must subtract the joint probability of A and B from the sum of their marginal probabilities to avoid double counting because of common outcomes in A and B. This is the case where events A and B are not mutually exclusive.

EXAMPLE 4–27   Taking a Course in Ethics

Calculating the probability of the union of two events: two-way table.

A university president proposed that all students must take a course in ethics as a requirement for graduation. Three hundred faculty members and students from this university were asked about their opinions on this issue. Table 4.10 gives a two-way classification of the responses of these faculty members and students.

Table 4.10   Two-Way Classification of Responses

	Favor	Oppose	Neutral	Total
Faculty	45	15	10	70
Student	90	110	30	230
Total	135	125	40	300

Find the probability that if one person is selected at random from these 300 persons, this person is a faculty member or is in favor of this proposal.

Solution   To find the probability of the union of events *faculty* and *in favor*, we will use the formula:

$$P(faculty \text{ or } in\,favor) = P(faculty) + P(in\,favor) - P(faculty \text{ and } in\,favor)$$

As we can observe from Table 4.10, there are a total of 70 faculty members and a total of 135 in favor of the proposal out of 300 total persons. Hence, the marginal probabilities of the events *faculty* and *in favor* are:

$$P(faculty) = \frac{70}{300} = .2333$$

$$P(in\,favor) = \frac{135}{300} = .4500$$

We also know from the table that there are a total of 45 faculty members who are in favor out of a total of 300 persons. (Note that these 45 faculty members are counted twice.) Hence, the joint probability of the two events *faculty* and *in favor* is:

$$P(faculty \text{ and } in\,favor) = \frac{45}{300} = .1500$$

The required probability is:

$$P(faculty \text{ or } in\,favor) = P(faculty) + P(in\,favor) - P(faculty \text{ and } in\,favor)$$
$$= .2333 + .4500 - .1500 = \mathbf{.5333}$$

Thus, the probability that a randomly selected person from these 300 persons is a faculty member or is in favor of this proposal is .5333.

The probability in this example can also be calculated without using the addition rule. The total number of persons in Table 4.10 who are either faculty members or in favor of this proposal or both is

$$45 + 15 + 10 + 90 = 160$$

Hence, the required probability is

$$P(faculty \text{ or } in \text{ } favor) = 160/300 = \textbf{.5333}$$

EXAMPLE 4–28 Gender and Vegetarian

Calculating the probability of the union of two events.

In a group of 2500 persons, 1400 are female, 600 are vegetarian, and 400 are female and vegetarian. What is the probability that a randomly selected person from this group is a male or vegetarian?

Solution From the given information, we know that there are a total of 2500 persons in the group, 1400 of them are female, 600 are vegetarian, and 400 are female and vegetarian. Using this information, we can write Table 4.11. In this table, the numbers in the shaded cells are given to us. The remaining numbers are calculated by doing some arithmetic manipulations.

Table 4.11 Two-Way Classification Table

	Vegetarian	Nonvegetarian	Total
Female	400	1000	1400
Male	200	900	1100
Total	600	1900	2500

We are to find the probability $P(male \text{ or } vegetarian)$. To find this probability, we will use the following addition rule formula:

$$P(male \text{ or } vegetarian) = P(male) + P(vegetarian) - P(male \text{ and } vegetarian)$$

We find the marginal probabilities of the two events and their joint probability as follows:

$$P(male) = \frac{1100}{2500} = .44$$

$$P(vegetarian) = \frac{600}{2500} = .24$$

$$P(male \text{ and } vegetarian) = \frac{200}{2500} = .08$$

Thus, the required probability is:

$$P(male \text{ or } vegetarian) = P(male) + P(vegetarian) - P(male \text{ and } vegetarian)$$
$$= .44 + .24 - .08 = \textbf{.60}$$

We can also calculate this probability as follows. From Table 4.11:

Total number of persons who are male or vegetarian or both = $1100 + 600 - 200 = 1500$

Hence, the probability of the union of events *male* and *vegetarian* is:

$$P(male \text{ or } vegetarian) = \frac{1500}{2500} = \textbf{.60}$$

4.5.3 Addition Rule for Mutually Exclusive Events

We know from an earlier discussion that the joint probability of two mutually exclusive events is zero. When A and B are mutually exclusive events, the term $P(A$ and $B)$ in the addition rule becomes zero and is dropped from the formula. Thus, the probability of the union of two mutually exclusive events is given by the sum of their marginal probabilities.

Addition Rule to Find the Probability of the Union of Mutually Exclusive Events   The probability of the union of two mutually exclusive events A and B is

$$P(A \text{ or } B) = P(A) + P(B)$$

EXAMPLE 4–29   Taking a Course in Ethics

Calculating the probability of the union of two mutually exclusive events: two-way table.

Refer to Example 4–27. A university president proposed that all students must take a course in ethics as a requirement for graduation. Three hundred faculty members and students from this university were asked about their opinion on this issue. The following table, reproduced from Table 4.10 in Example 4–27, gives a two-way classification of the responses of these faculty members and students.

	Favor	Oppose	Neutral	Total
Faculty	45	15	10	70
Student	90	110	30	230
Total	135	125	40	300

What is the probability that a randomly selected person from these 300 faculty members and students is in favor of the proposal or is neutral?

Solution   As is obvious from the information given in the table, the events *in favor* and *neutral* are mutually exclusive, as they cannot happen together. Hence, their joint probability is zero. There are a total of 135 persons in favor and 40 neutral in 300 persons. The marginal probabilities of the events *in favor* and *neutral* are:

$$P(in\ favor) = \frac{135}{300} = .4500$$

$$P(neutral) = \frac{40}{300} = .1333$$

The probability of the union of events *in favor* and *neutral* is:

$$P(in\ favor \text{ or } neutral) = .4500 + .1333 = \mathbf{.5833}$$

Also, from the information given in the table, we observe that there are a total of $135 + 40 = 175$ persons who are in favor or neutral to this proposal. Hence, the required probability is:

$$P(in\ favor \text{ or } neutral) = \frac{175}{300} = \mathbf{.5833}$$

EXAMPLE 4–30   Rolling a Die

Calculating the probability of the union of two mutually exclusive events.

Consider the experiment of rolling a die once. What is the probability that a number less than 3 or a number greater than 4 is obtained?

Solution   Here, the event *a number less than 3 is obtained* happens if either 1 or 2 is rolled on the die, and the event *a number greater than 4 is obtained* happens if either 5 or 6 is rolled on the die. Thus, these two events are mutually exclusive as they do not have any common outcome and cannot happen together. Hence, their joint probability is zero. The marginal probabilities of these two events are:

$$P(a \ number \ less \ than \ 3 \ is \ obtained) = \frac{2}{6}$$

$$P(a \ number \ greater \ than \ 4 \ is \ obtained) = \frac{2}{6}$$

The probability of the union of these two events is:

$$P(a \ number \ less \ than \ 3 \ is \ obtained \ or \ a \ number \ greater \ than \ 4 \ is \ obtained) = \frac{2}{6} + \frac{2}{6} = .6667$$

EXERCISES

CONCEPTS AND PROCEDURES

4.67   What is meant by the union of two events? Give one example.

4.68   What is the difference between the addition rule of probability for two mutually exclusive events and two events which are not mutually exclusive?

4.69   Consider the following addition rule to find the probability of the union of two events X and Y:

$$P(X \ or \ Y) = P(X) + P(Y) - P(X \ and \ Y)$$

When and why is the term $P(X \ and \ Y)$ subtracted from the sum of $P(X)$ and $P(Y)$? Give one example where you might use this formula.

4.70   When is the following addition rule used to find the probability of the union of two events X and Y?

$$P(X \ or \ Y) = P(X) + P(Y)$$

Give one example where you might use this formula.

4.71   Find $P(A \ or \ B)$ for the following.
   a. $P(A) = .66$, $P(B) = .47$,   and $P(A \ and \ B) = .33$
   b. $P(A) = .92$, $P(B) = .78$,   and $P(A \ and \ B) = .65$

4.72   Given that A and B are two mutually exclusive events, find $P(A \ or \ B)$ for the following.
   a. $P(A) = .25$ and $P(B) = .27$
   b. $P(A) = .58$ and $P(B) = .09$

APPLICATIONS

4.73   A consumer agency randomly selected 1800 flights for two major airlines, A and B. The following table gives the two-way classification of these flights based on airline and arrival time. Note that "less than 30 minutes late" includes flights that arrived early or on time.

	Less Than 30 Minutes Late	30 Minutes to 1 Hour Late	More Than 1 Hour Late
Airline A	412	401	87
Airline B	385	362	153

If one flight is selected at random from these 1800 flights, find the following probabilities.

   a. P(more than 1 hour late *or* airline A)
   b. P(airline B *or* less than 30 minutes late)
   c. P(airline A *or* airline B)

4.74   Two thousand randomly selected adults were asked whether or not they have ever shopped on the Internet. The following table gives a two-way classification of the responses.

	Have Shopped	Have Never Shopped
Male	400	800
Female	350	450

Suppose one adult is selected at random from these 2000 adults. Find the following probabilities.

   a. P(has never shopped on the Internet *or* is a female)
   b. P(is a male *or* has shopped on the Internet)
   c. P(has shopped on the Internet *or* has never shopped on the Internet)

4.75   Five hundred employees were selected from a city's large private companies, and they were asked whether or not they have any retirement benefits provided by their companies. Based on this information, the following two-way classification table was prepared.

	Have Retirement Benefits	
	Yes	No
Men	225	75
Women	150	50

Suppose one employee is selected at random from these 500 employees. Find the following probabilities.

   a. The probability of the union of events "woman" and "yes"
   b. The probability of the union of events "no" and "man"

4.76   There is an area of free (but illegal) parking near an inner-city sports arena. The probability that a car parked in this area will be ticketed by police is .50, that the car will be vandalized is .30, and that it will be ticketed and vandalized is .15. Find the probability that a car parked in this area will be ticketed or vandalized.

4.77   Jason and Lisa are planning an outdoor reception following their wedding. They estimate that the probability of bad weather is .28, that of a disruptive incident (a fight breaks out, the limousine is late, etc.) is .16, and that bad weather and a disruptive incident will occur is .09. Assuming these estimates are correct, find the probability that their reception will suffer bad weather or a disruptive incident.

4.78 The probability that a randomly selected elementary or secondary school teacher from a city is a female is .66, holds a second job is .33, and is a female and holds a second job is .27. Find the probability that an elementary or secondary school teacher selected at random from this city is a female or holds a second job.

4.79 According to a survey of 2000 home owners, 900 of them own homes with three bedrooms, and 700 of them own homes with four bedrooms. If one home owner is selected at random from these 2000 home owners, find the probability that this home owner owns a house that has three *or* four bedrooms. Explain why this probability is not equal to 1.0.

4.80 According to the U.S. Census Bureau's data on the marital status of the 242 million Americans aged 15 years and older, 124.2 million are currently married and 74.5 million have never married. If one person from these 242 million persons is selected at random, find the probability that this person is currently married *or*

has never been married. Explain why this probability is not equal to 1.0.

4.81 According to an Automobile Association of America report, 9.6% of Americans traveled by car over the 2011 Memorial Day weekend and 88.09% stayed home. What is the probability that a randomly selected American stayed home or traveled by car over the 2011 Memorial Day weekend? Explain why this probability does not equal 1.0.

4.82 Amy is trying to purchase concert tickets online for two of her favorite bands, the Leather Recliners and Double Latte No Foam. She estimates that her probability of being able to get tickets for the Leather Recliners concert is .14, the probability of being able to get tickets for the Double Latte No Foam concert is .23, and the probability of being able to get tickets for both concerts is .026. What is the probability that she will be able to get tickets for at least one of the two concerts?

4.83 The probability that an open-heart operation is successful is .85. What is the probability that in two randomly selected open-heart operations at least one will be successful?

4.6 | Counting Rule, Factorials, Combinations, and Permutations

In this section, first we discuss the counting rule that helps us calculate the total number of outcomes for experiments, and then we learn about factorials, combinations, and permutations, respectively.

4.6.1 Counting Rule

The experiments dealt with so far in this chapter have had only a few outcomes, which were easy to list. However, for experiments with a large number of outcomes, it may not be easy to list all outcomes. In such cases, we may use the **counting rule** to find the total number of outcomes.

Counting Rule to Find Total Outcomes If an experiment consists of three steps, and if the first step can result in m outcomes, the second step in n outcomes, and the third step in k outcomes, then

$$\text{Total outcomes for the experiment} = m \cdot n \cdot k$$

The counting rule can easily be extended to apply to an experiment that has fewer or more than three steps.

| EXAMPLE 4–31 | Three Tosses of a Coin |

Applying the counting rule: 3 steps.

Consider three tosses of a coin. How many total outcomes this experiment has?

Solution This experiment of tossing a coin three times has three steps: the first toss, the second toss, and the third toss. Each step has two outcomes: a head and a tail. Thus,

$$\text{Total outcomes for three tosses of a coin} = 2 \times 2 \times 2 = \textbf{8}$$

The eight outcomes for this experiment are *HHH, HHT, HTH, HTT, THH, THT, TTH,* and *TTT.*

| EXAMPLE 4–32 | Taking a Car Loan |

Applying the counting rule: 2 steps.

A prospective car buyer can choose between a fixed and a variable interest rate and can also choose a payment period of 36 months, 48 months, or 60 months. How many total outcomes are possible?

Solution   This experiment is made up of two steps: choosing an interest rate and selecting a loan payment period. There are two outcomes (a fixed or a variable interest rate) for the first step and three outcomes (a payment period of 36 months, 48 months, or 60 months) for the second step. Hence,

$$\text{Total outcomes} = 2 \times 3 = \mathbf{6}$$

EXAMPLE 4–33	Outcomes of a Football Game

Applying the counting rule: 16 steps.

A National Football League team will play 16 games during a regular season. Each game can result in one of three outcomes: a win, a loss, or a tie. How many total outcomes are possible?

Solution   The total possible outcomes for 16 games are calculated as follows:

$$\text{Total outcomes} = 3 \cdot 3 \cdot 3 \cdot 3 \cdot 3 \cdot 3 \cdot 3 \cdot 3 \cdot 3 \cdot 3 \cdot 3 \cdot 3 \cdot 3 \cdot 3 \cdot 3 \cdot 3$$

$$= 3^{16} = \mathbf{43{,}046{,}721}$$

One of the 43,046,721 possible outcomes is all 16 wins.

PhotoDisc, Inc./Getty Images

4.6.2   Factorials

The symbol ! (read as *factorial*) is used to denote **factorials**. The value of the factorial of a number is obtained by multiplying all the integers from that number to 1. For example, 7! is read as "seven factorial" and is evaluated by multiplying all the integers from 7 to 1.

> **Factorials**   The symbol $n!$, read as "n factorial," represents the product of all the integers from n to 1. In other words,
>
> $$n! = n(n - 1)(n - 2)(n - 3) \cdots 3 \cdot 2 \cdot 1$$
>
> By definition,
>
> $$0! = 1$$
>
> Note that some calculators use $r!$ instead of $n!$ on the factorial key.
>
> Thus, the factorial of a number is given by the product of all integers from that number to 1.

EXAMPLE 4–34	Factorial of a Number

Evaluating a factorial.

Evaluate 7!.

Solution   To evaluate 7!, we multiply all the integers from 7 to 1.

$$7! = 7 \cdot 6 \cdot 5 \cdot 4 \cdot 3 \cdot 2 \cdot 1 = \mathbf{5040}$$

Thus, the value of 7! is 5040.

EXAMPLE 4–35	Factorial of a Number

Evaluating a factorial.

Evaluate 10!.

Solution   The value of 10! is given by the product of all the integers from 10 to 1. Thus,

$$10! = 10 \cdot 9 \cdot 8 \cdot 7 \cdot 6 \cdot 5 \cdot 4 \cdot 3 \cdot 2 \cdot 1 = \mathbf{3{,}628{,}800}$$

EXAMPLE 4–36	Factorial of the Difference of Two Numbers

Evaluating a factorial of the difference between two numbers.

Evaluate (12 − 4)!.

Solution   The value of (12 − 4)! is

$$(12 - 4)! = 8! = 8 \cdot 7 \cdot 6 \cdot 5 \cdot 4 \cdot 3 \cdot 2 \cdot 1 = \mathbf{40{,}320}$$

EXAMPLE 4–37   Factorial of Zero

Evaluating a factorial of zero.

Evaluate $(5 - 5)!$.

Solution   The value of $(5 - 5)!$ is 1.

$$(5 - 5)! = 0! = \mathbf{1}$$

Note that $0!$ is always equal to 1.

Statistical software and most calculators can be used to find the values of factorials. Check if your calculator can evaluate factorials.

4.6.3   Combinations

Quite often we face the problem of selecting a few elements from a group of distinct elements. For example, a student may be required to attempt any two questions out of four in an examination. As another example, the faculty in a department may need to select 3 professors from 20 to form a committee, or a lottery player may have to pick 6 numbers from 49. The question arises: In how many ways can we make the selections in each of these examples? For instance, how many possible selections exist for the student who is to choose any two questions out of four? The answer is six. Suppose the four questions are denoted by the numbers 1, 2, 3, and 4. Then the six selections are

(1 and 2)   (1 and 3)   (1 and 4)   (2 and 3)   (2 and 4)   (3 and 4)

The student can choose questions 1 and 2, or 1 and 3, or 1 and 4, and so on. Note that in combinations, all selections are made without replacement.

Each of the possible selections in the above list is called a **combination**. All six combinations are distinct; that is, each combination contains a different set of questions. It is important to remember that the order in which the selections are made is not important in the case of combinations. Thus, whether we write (1 and 2) or (2 and 1), both of these arrangements represent only one combination.

Combinations Notation   *Combinations* give the number of ways x elements can be selected from n elements. The notation used to denote the total number of combinations is

$$_nC_x$$

which is read as "the number of combinations of n elements selected x at a time."

Note that some calculators use r instead of x, so that the combinations notation then reads $_nC_r$.

Suppose there are a total of n elements from which we want to select x elements. Then,

$$_nC_x = \text{the number of combinations of } n \text{ elements selected } x \text{ at a time}$$

┌— n denotes the total number of elements

└— x denotes the number of elements selected per selection

Number of Combinations   The **number of combinations** for selecting x from n distinct elements is given by the formula

$$_nC_x = \frac{n!}{x!(n - x)!}$$

where $n!$, $x!$, and $(n - x)!$ are read as "n factorial," "x factorial," and "n minus x factorial," respectively.

In the combinations formula, n represents the total items and x is the number of items selected out of n. Note that in combinations, n is always greater than or equal to x. If n is less than x, then we cannot select x distinct elements from n.

EXAMPLE 4–38 | **Choosing Ice Cream Flavors**

Finding the number of combinations using the formula.

An ice cream parlor has six flavors of ice cream. Kristen wants to buy two flavors of ice cream. If she randomly selects two flavors out of six, how many combinations are possible?

Solution   For this example,

$$n = \text{total number of ice cream flavors} = 6$$
$$x = \text{number of ice cream flavors to be selected} = 2$$

Therefore, the number of ways in which Kristen can select two flavors of ice cream out of six is

$$_6C_2 = \frac{6!}{2!(6-2)!} = \frac{6!}{2!\,4!} = \frac{6\cdot5\cdot4\cdot3\cdot2\cdot1}{2\cdot1\cdot4\cdot3\cdot2\cdot1} = \mathbf{15}$$

Thus, there are 15 ways for Kristen to select two ice cream flavors out of six.

EXAMPLE 4–39 | **Selecting a Committee**

Finding the number of combinations and listing them.

Three members of a committee will be randomly selected from five people. How many different combinations are possible?

Solution   There are a total of five persons, and we are to select three of them. Hence,

$$n = 5 \quad \text{and} \quad x = 3$$

Applying the combinations formula, we get

$$_5C_3 = \frac{5!}{3!(5-3)!} = \frac{5!}{3!\,2!} = \frac{120}{6\cdot2} = \mathbf{10}$$

If we assume that the five persons are A, B, C, D, and E, then the 10 possible combinations for the selection of three members of the committee are

ABC   ABD   ABE   ACD   ACE   ADE   BCD   BCE   BDE   CDE

Note that the number of combinations of *n* items selecting *x* at a time gives the number of samples of size *x* that can be selected from a population with *n* members. For example, in Example 4–39, the population is made of five people, and we are to select a sample of three to form a committee. There are 10 possible samples of three people each that can be selected from this population of five people.

EXAMPLE 4–40 | **Selecting Candidates for an Interview**

Using the combinations formula.

Marv & Sons advertised to hire a financial analyst. The company has received applications from 10 candidates who seem to be equally qualified. The company manager has decided to call only 3 of these candidates for an interview. If she randomly selects 3 candidates from the 10, how many total selections are possible?

Solution   The total number of ways to select 3 applicants from 10 is given by $_{10}C_3$. Here, $n = 10$ and $x = 3$. We find the number of combinations as follows:

$$_{10}C_3 = \frac{10!}{3!\,(10-3)!} = \frac{10!}{3!\,7!} = \frac{3,628,800}{(6)(5040)} = \mathbf{120}$$

Thus, the company manager can select 3 applicants from 10 in 120 ways.

Case Study 4–2 describes the probability of winning a Mega Million Lottery jackpot.

Probability of Winning a Mega Millions Lottery Jackpot

Large jackpot lotteries became popular in the United State during the 1970s and 1980s. The introduction of the Powerball lottery in 1992 resulted in the growth of multi-jurisdictional lotteries, as well as the development of a second extensive multi-jurisdictional game called Mega Millions in 2002. Both of these games are offered in many states, the District of Columbia, and the U.S. Virgin Islands, which has resulted in multiple jackpots of more than $300 million.

Both games operate on a similar premise. There are two bins—one containing white balls, and one containing red (Powerball) or gold (Mega Millions) balls. When a player fills out a ticket, he or she selects five numbers from the set of white balls (1–69 for Powerball, 1–75 for Mega Millions) and one number from the set of red (1–26) or gold (1–15) balls, depending on the game. Prizes awarded to players are based on how many balls of each color are matched. If all five white ball numbers and the colored ball number are matched, the player wins the jackpot. If more than one player matches all of the numbers, then the jackpot is divided among them. The following table lists the various prizes for the Mega Millions lottery.

Number of white balls matched	Number of gold balls matched	Prize	Number of white balls matched	Number of gold balls matched	Prize
5	1	Jackpot	3	0	$5
5	0	$1,000,000	2	1	$5
4	1	$5,000	1	1	$2
4	0	$500	0	1	$1
3	1	$50			

The probability of winning each of the various prizes listed in the table for the Mega Millions lottery can be calculated using combinations. To find the probability of winning the jackpot, first we need to calculate the number of ways to draw five white ball numbers from 75 and one gold ball number from 15. These combinations are, respectively,

$$_{75}C_5 = 17,259,390 \quad \text{and} \quad _{15}C_1 = 15$$

To obtain the total number of ways to select six numbers (five white ball numbers and one gold ball number), we multiply the two numbers obtained above, which gives us $17,259,390 \times 15 = 258,890,850$. Thus, there are 258,890,850 different sets of five white ball numbers and one gold ball number that can be drawn. Then, the probability that a player with one ticket wins the jackpot is

$$P(\text{winning the jackpot}) = 1/258,890,850 = .00000000386$$

To calculate the probability of winning each of the other prizes, we calculate the number of ways any prize can be won and divide it by 258,890,850. For example, to win a prize of $5,000, a player needs to match four white ball numbers and the gold ball number. As shown below, there are 350 ways to match four white ball numbers and one gold ball number.

$$_5C_4 \times _{70}C_1 \times _1C_1 \times _{14}C_0 = 5 \times 70 \times 1 \times 1 = 350$$

Here $_5C_4$ gives the number of ways to match four of the five winning white ball numbers, $_{70}C_1$ gives the number of ways to match one of the 70 nonwinning white ball numbers, $_1C_1$ gives the number of ways to match the winning gold ball number, and $_{14}C_0$ gives the number of ways to match none of the 14 nonwinning gold ball numbers. Then, the probability of winning a prize of $5,000 is

$$P(\text{winning a 5,000 prize}) = 350/258,890,850 = .00000135$$

We can calculate the probabilities of winning the other prizes listed in the table in the same way.

Statistical software and many calculators can be used to find combinations. Check to see whether your calculator can do so.

If the total number of elements and the number of elements to be selected are the same, then there is only one combination. In other words,

◄ *Remember*

$$_nC_n = 1$$

Also, the number of combinations for selecting zero items from n is 1; that is,

$$_nC_0 = 1$$

For example,

$$_5C_5 = \frac{5!}{5!(5-5)!} = \frac{5!}{5!\,0!} = \frac{120}{(120)(1)} = 1$$

$$_8C_0 = \frac{8!}{0!(8-0)!} = \frac{8!}{0!\,8!} = \frac{40,320}{(1)(40,320)} = 1$$

Permutations give the total number of selections of x elements from n (different) elements in such a way that the order of selection is important. The notation used to denote the permutations is

$$_nP_x$$

which is read as "the number of permutations of selecting x elements from n elements." Permutations are also called **arrangements**.

4.6.4 Permutations

The concept of permutations is very similar to that of combinations but with one major difference—here the order of selection is important. Suppose there are three marbles in a jar—red, green, and purple—and we select two marbles from these three. When the order of selection is not important, as we know from the previous section, there are three ways (combinations) to do so. Those three ways are RG, RP, and GP, where R represents that a red marble is selected, G means a green marble is selected, and P indicates a purple marble is selected. In these three combinations, the order of selection is not important, and, thus, RG and GR represent the same selection. However, if the order of selection is important, then RG and GR are not the same selections, but they are two different selections. Similarly, RP and PR are two different selections, and GP and PG are two different selections. Thus, if the order in which the marbles are selected is important, then there are six selections—RG, GR, RP, PR, GP, and PG. These are called six **permutations** or **arrangements**.

Number of Permutations   The following formula is used to find the number of permutations or arrangements of selecting x items out of n items. Note that here, the n items must all be different.

$$_nP_x = \frac{n!}{(n-x)!}$$

Example 4–41 shows how to apply this formula.

EXAMPLE 4–41   Selecting Office Holders of a Club

Finding the number of permutations using the formula.

A club has 20 members. They are to select three office holders—president, secretary, and treasurer—for next year. They always select these office holders by drawing 3 names randomly from the names of all members. The first person selected becomes the president, the second is the secretary, and the third one takes over as treasurer. Thus, the order in which 3 names are selected from the 20 names is important. Find the total number of arrangements of 3 names from these 20.

Solution   For this example,

$$n = \text{total members of the club} = 20$$
$$x = \text{number of names to be selected} = 3$$

Since the order of selections is important, we find the number of permutations or arrangements using the following formula:

$$_{20}P_3 = \frac{20!}{(20-3)!} = \frac{20!}{17!} = \mathbf{6840}$$

Thus, there are 6840 permutations or arrangements for selecting 3 names out of 20.

Statistical software and many calculators can find permutations. Check to see whether your calculator can do it.

EXERCISES

CONCEPTS AND PROCEDURES

4.84 How many different outcomes are possible for four rolls of a die?

4.85 How many different outcomes are possible for 10 tosses of a coin?

4.86 Determine the value of each of the following using the appropriate formula.

3!	$(9-3)!$	9!	$(14-12)!$	$_5C_3$
$_7C_4$	$_9C_3$	$_4C_0$	$_3C_3$	$_6P_2$ $_8P_4$

APPLICATIONS

4.87 A small ice cream shop has 12 flavors of ice cream and 6 kinds of toppings for its sundaes. How many different selections of one flavor of ice cream and one kind of topping are possible?

4.88 A restaurant menu has three kinds of soups, seven kinds of main courses, four kinds of desserts, and five kinds of drinks. If a customer randomly selects one item from each of these four categories, how many different outcomes are possible?

4.89 A ski patrol unit has eight members available for duty, and two of them are to be sent to rescue an injured skier. In how many ways can two of these eight members be selected? Now suppose the order of selection is important. How many arrangements are possible in this case?

4.90 An ice cream shop offers 25 flavors of ice cream. How many ways are there to select 3 different flavors from these 25 flavors? How many permutations are possible?

4.91 A man just bought 4 suits, 8 shirts, and 12 ties. All of these suits, shirts, and ties coordinate with each other. If he is to randomly select one suit, one shirt, and one tie to wear on a certain day, how many different outcomes (selections) are possible?

4.92 A student is to select three classes for next semester. If this student decides to randomly select one course from each of eight economics classes, six mathematics classes, and five computer classes, how many different outcomes are possible?

4.93 An environmental agency will randomly select 4 houses from a block containing 25 houses for a radon check. How many total selections are possible? How many permutations are possible?

4.94 In how many ways can a sample (without replacement) of 9 items be selected from a population of 20 items?

USES AND MISUSES...

ODDS AND PROBABILITY

One of the first things we learn in probability is that the sum of the probabilities of all outcomes for an experiment must equal 1.0. We also learn about the probabilities that are developed from relative frequencies and about subjective probabilities. In the latter case, many of the probabilities involve personal opinions of experts in the field. Still, both scenarios (probabilities obtained from relative frequencies and subjective probabilities) require that all probabilities must be non-negative and the sum of the probabilities of all (simple) outcomes for an experiment must equal 1.0.

Although probabilities and probability models are all around us — in weather prediction, medicine, financial markets, and so forth — they

are most obvious in the world of gaming and gambling. Sports betting agencies publish odds of each team winning a specific game or championship. The following table gives the odds, as of June 12, 2015, of each National Football League team winning Super Bowl 50 to be held on February 7, 2016. These odds were obtained from the Web site www.vegas.com.

Note that the odds listed in this table are called the odds in favor of winning the Super Bowl. For example, the defending champion New England Patriots had 1:8 (which is read as 1 to 8) odds of winning Super Bowl 50. If we switch the numbers around, we can state that the odds were 8:1 (or 8 to 1) against the Patriots winning Super Bowl 50.

Team	Odds	Team	Odds
Arizona Cardinals	1:30	Miami Dolphins	1:40
Atlanta Falcons	1:40	Minnesota Vikings	1:55
Baltimore Ravens	1:25	New York Giants	1:32
Buffalo Bills	1:40	New York Jets	1:50
Carolina Panthers	1:40	New England Patriots	1:8
Chicago Bears	1:55	New Orleans Saints	1:40
Cincinnati Bengals	1:35	Oakland Raiders	1:100
Cleveland Browns	1:100	Philadelphia Eagles	1:25
Dallas Cowboys	1:12	Pittsburgh Steelers	1:30
Denver Broncos	1:12	San Diego Chargers	1:45
Detroit Lions	1:40	San Francisco 49ers	1:45
Green Bay Packers	1:6.5	Seattle Seahawks	1:6
Houston Texans	1:50	St. Louis Rams	1:40
Indianapolis Colts	1:8.5	Tampa Bay Buccaneers	1:90
Jacksonville Jaguars	1:50	Tennessee Titans	1:200
Kansas City Chiefs	1:38	Washington Redskins	1:125

How do we convert these odds into probabilities? Let us consider the New England Patriots. Odds of 1:8 imply that out of 9 chances, there is 1 chance that the Patriots would win Super Bowl 50 and 8 chances that the Patriots would not win Super Bowl 50. Thus, the probability that the Patriots would win Super Bowl 50 is $\frac{1}{1+8} = \frac{1}{9} = .1111$ and the probability that Patriots would not win Super Bowl 50 is $\frac{8}{1+8} = \frac{8}{9} = .8889$. Similarly, for the Chicago Bears, the probability of winning Super Bowl 50 was $\frac{1}{1+55} = \frac{1}{56} = .0179$ and the probability of not winning Super Bowl 50 is $\frac{55}{1+55} = \frac{55}{56} = .9821$. We can calculate these probabilities for all teams listed in the table by using this procedure.

Note that here the 32 outcomes (that each team would win Super Bowl 50) are mutually exclusive events because it is impossible for two or more teams to win the Super Bowl during the same year. Hence, if we add the probabilities of winning Super Bowl 50 for all teams, we should obtain a value of 1.0 according to the second property of probability. However, if you calculate the probability of winning Super Bowl 50 for each of the 32 teams using the odds given in the table and then add all these probabilities, the sum is 1.588759461. So, what happened? Did these odds makers flunk their statistics and probability courses? Probably not.

Casinos and odds makers, who are in the business of making money, are interested in encouraging people to gamble. These probabilities, which seem to violate the basic rule of probability theory, still obey the primary rule for the casinos, which is that, on average, a casino is going to make a profit.

Note: When casinos create odds for sports betting, they recognize that many people will bet on one of their favorite teams, such as the Dallas Cowboys or the Pittsburgh Steelers. To meet the rule that the sum of all of the probabilities is 1.0, the probabilities for the teams more likely to win would have to be lowered. Lowering a probability corresponds to lowering the odds. For example, if the odds for the New England Patriots had been lowered from 1:8 to 1:20, the probability for them to win would have decreased from .1111 to .0476. If the Patriots had remained as one of the favorites, many people would have bet on them. However, if they had won, the casino would have paid $20, instead of $8, for every $1 bet. The casinos do not want to do this and, hence, they ignore the probability rule in order to make more money. However, the casinos cannot do this with their traditional games, which are bound by the standard rules. From a mathematical standpoint, it is not acceptable to ignore the rule that the probabilities of all final outcomes for an experiment must add to 1.0. (The reader should check who won Super Bowl 50 on February 7, 2016.)

GLOSSARY

Classical probability rule   The method of assigning probabilities to outcomes or events of an experiment with equally likely outcomes.

Combinations   The number of ways x elements can be selected from n elements. Here order of selection is not important.

Complementary events   Two events that taken together include all the outcomes for an experiment but do not contain any common outcome.

Compound event   An event that contains more than one outcome of an experiment. It is also called a *composite event*.

Conditional probability   The probability of an event subject to the condition that another event has already occurred.

Dependent events   Two events for which the occurrence of one changes the probability of the occurrence of the other.

Equally likely outcomes   Two (or more) outcomes or events that have the same probability of occurrence.

Event   A collection of one or more outcomes of an experiment.

Experiment   A process with well-defined outcomes that, when performed, results in one and only one of the outcomes per repetition.

Factorial   Denoted by the symbol !. The product of all the integers from a given number to 1. For example, $n!$ (read as "n factorial") represents the product of all the integers from n to 1.

Impossible event   An event that cannot occur.

Independent events   Two events for which the occurrence of one does not change the probability of the occurrence of the other.

Intersection of events   The intersection of events is given by the outcomes that are common to two (or more) events.

Joint probability   The probability that two (or more) events occur together.

Law of Large Numbers   If an experiment is repeated again and again, the probability of an event obtained from the relative frequency approaches the actual or theoretical probability.

Marginal probability   The probability of one event or characteristic without consideration of any other event.

Mutually exclusive events   Two or more events that do not contain any common outcome and, hence, cannot occur together.

Outcome   The result of the performance of an experiment.

Permutations   Number of arrangements of x items selected from n items. Here order of selection is important.

Probability   A numerical measure of the likelihood that a specific event will occur.

Relative frequency as an approximation of probability   Probability assigned to an event based on the results of an experiment or based on historical data.

Sample point   An outcome of an experiment.

Sample space   The collection of all sample points or outcomes of an experiment.

Simple event   An event that contains one and only one outcome of an experiment. It is also called an *elementary event*.

Subjective probability   The probability assigned to an event based on the information and judgment of a person.

Sure event   An event that is certain to occur.

Tree diagram   A diagram in which each outcome of an experiment is represented by a branch of a tree.

Union of two events   Given by the outcomes that belong either to one or to both events.

SUPPLEMENTARY EXERCISES

4.95 A car rental agency currently has 52 cars available, 38 of which have a GPS navigation system. One of the 52 cars is selected at random. Find the probability that this car
  a. has a GPS navigation system
  b. does not have a GPS navigation system

4.96 A random sample of 250 adults was taken, and they were asked whether they prefer watching sports or opera on television. The following table gives the two-way classification of these adults.

	Prefer Watching Sports	Prefer Watching Opera
Male	104	18
Female	55	73

  a. If one adult is selected at random from this group, find the probability that this adult
    i. prefers watching opera
    ii. is a male
    iii. prefers watching sports given that the adult is a female
    iv. is a male given that he prefers watching sports
    v. is a female *and* prefers watching opera
    vi. prefers watching sports *or* is a male
  b. Are the events "female" and "prefers watching sports" independent? Are they mutually exclusive? Explain why or why not.

4.97 A random sample of 100 lawyers was taken, and they were asked if they are in favor of or against capital punishment. The following table gives the two-way classification of their responses.

	Favors Capital Punishment	Opposes Capital Punishment
Male	44	38
Female	11	7

  a. If one lawyer is randomly selected from this group, find the probability that this lawyer
    i. favors capital punishment
    ii. is a female
    iii. opposes capital punishment given that the lawyer is a female
    iv. is a male given that he favors capital punishment
    v. is a female *and* favors capital punishment
    vi. opposes capital punishment *or* is a male
  b. Are the events "female" and "opposes capital punishment" independent? Are they mutually exclusive? Explain why or why not.

4.98 A random sample of 400 college students was asked if college athletes should be paid. The following table gives a two-way classification of the responses.

	Should Be Paid	Should Not Be Paid
Student athlete	110	20
Student nonathlete	170	100

  a. If one student is randomly selected from these 400 students, find the probability that this student

i. is in favor of paying college athletes

ii. favors paying college athletes given that the student selected is a nonathlete

iii. is an athlete *and* favors paying student athletes

iv. is a nonathlete *or* is against paying student athletes

b. Are the events "student athlete" and "should be paid" independent? Are they mutually exclusive? Explain why or why not.

4.99 An appliance repair company that makes service calls to customers' homes has found that 10% of the time there is nothing wrong with the appliance and the problem is due to customer error (appliance unplugged, controls improperly set, etc.). Two service calls are selected at random, and it is observed whether or not the problem is due to customer error. Draw a tree diagram. Find the probability that in this sample of two service calls

a. both problems are due to customer error

b. at least one problem is not due to customer error

4.100 According to the National Science Foundation, during the Fall 2008 semester 30.13% of all science and engineering graduate students enrolled in doctorate-granting colleges in the United States were temporary visa holders (www.nsf.gov/statistics/nsf11311/pdf/tab46.pdf). Assume that this percentage holds true for current such students. Suppose that two students from the aforementioned group are selected and their status (U.S. citizens/permanent residents or temporary visa holders) is observed. Draw a tree diagram for this problem using C to denote U.S. citizens/permanent residents and V to denote temporary visa holders. Find the probability that in this sample of two graduate students

a. at least one student is a temporary visa holder

b. both students are U.S. citizens/permanent residents

4.101 A company has installed a generator to back up the power in case there is a power failure. The probability that there will be a power failure during a snowstorm is .40. The probability that the generator will stop working during a snowstorm is .08. What is the probability that during a snowstorm the company will lose both sources of power? Assume that the two sources of power are independent.

4.102 Terry & Sons makes bearings for autos. The production system involves two independent processing machines so that each bearing passes through these two processes. The probability that the first processing machine is not working properly at any time is .09, and the probability that the second machine is not working properly at any time is .06. Find the probability that both machines will not be working properly at any given time.

ADVANCED EXERCISES

4.103 A player plays a roulette game in a casino by betting on a single number each time. Because the wheel has 38 numbers, the probability that the player will win in a single play is 1/38. Note that each play of the game is independent of all previous plays.

a. Find the probability that the player will win for the first time on the 5th play.

b. Find the probability that it takes the player more than 80 plays to win for the first time.

c. The gambler claims that because he has 1 chance in 38 of winning each time he plays, he is certain to win at least once if he plays 38 times. Does this sound reasonable to you? Find the probability that he will win at least once in 38 plays.

4.104 A trimotor plane has three engines—a central engine and an engine on each wing. The plane will crash only if the central engine fails *and* at least one of the two wing engines fails. The probability of failure during any given flight is .005 for the central engine and .008 for each of the wing engines. Assuming that the three engines operate independently, what is the probability that the plane will crash during a flight?

4.105 Powerball is a game of chance that has generated intense interest because of its large jackpots. To play this game, a player selects five different numbers from 1 through 59, and then picks a Powerball number from 1 through 39. The lottery organization randomly draws 5 different white balls from 59 balls numbered 1 through 59, and then randomly picks a Powerball number from 1 through 39. Note that it is possible for the Powerball number to be the same as one of the first five numbers.

a. If a player's first five numbers match the numbers on the five white balls drawn by the lottery organization and the player's Powerball number matches the Powerball number drawn by the lottery organization, the player wins the jackpot. Find the probability that a player who buys one ticket will win the jackpot. (Note that the order in which the five white balls are drawn is unimportant.)

b. If a player's first five numbers match the numbers on the five white balls drawn by the lottery organization, the player wins about $200,000. Find the probability that a player who buys one ticket will win this prize.

4.106 A box contains 12 red marbles and 12 green marbles.

a. Sampling at random from the box six times with replacement, you have drawn a red marble all six times. What is the probability of drawing a red marble the seventh time?

b. Sampling at random from the box six times without replacement, you have drawn a red marble all six times. Without replacing any of the marbles, what is the probability of drawing a red marble the seventh time?

c. You have tossed a fair coin five times and have obtained heads all five times. A friend argues that according to the law of averages, a tail is due to occur and, hence, the probability of obtaining a head on the sixth toss is less than .50. Is he right? Is coin tossing mathematically equivalent to the procedure mentioned in part a or the procedure mentioned in part b? Explain.

4.107 A gambler has given you two jars and 20 marbles. Of these 20 marbles, 10 are red and 10 are green. You must put all 20 marbles in these two jars in such a way that each jar must have at least one marble in it. Then a friend of yours, who is blindfolded, will select one of the two jars at random and then will randomly select a marble from this jar. If the selected marble is red, you and your friend win $100.

a. If you put 5 red marbles and 5 green marbles in each jar, what is the probability that your friend selects a red marble?

b. If you put 2 red marbles and 2 green marbles in one jar and the remaining marbles in the other jar, what is the probability that your friend selects a red marble?

c. How should these 20 marbles be distributed among the two jars in order to give your friend the highest possible probability of selecting a red marble?

4.108 An insurance company has information that 93% of its auto policy holders carry collision coverage or uninsured motorist coverage on their policies. Eighty percent of the policy holders carry collision coverage, and 60% have uninsured motorist coverage.

a. What percentage of these policy holders carry both collision and uninsured motorist coverage?

b. What percentage of these policy holders carry neither collision nor uninsured motorist coverage?

c. What percentage of these policy holders carry collision but not uninsured motorist coverage?

4.109 A screening test for a certain disease is prone to giving false positives or false negatives. If a patient being tested has the disease, the probability that the test indicates a (false) negative is .14. If the patient does not have the disease, the probability that the test indicates a (false) positive is .08. Assume that 3% of the patients being tested actually have the disease. Suppose that one patient is chosen at random and tested. Find the probability that

a. this patient has the disease and tests positive

b. this patient does not have the disease and tests positive

c. this patient tests positive

d. this patient has the disease given that he or she tests positive

(*Hint:* A tree diagram may be helpful in part c.)

4.110 A pizza parlor has 15 different toppings available for its pizzas, and 2 of these toppings are pepperoni and anchovies. If a customer picks 2 toppings at random, find the probability that

a. neither topping is anchovies

b. pepperoni is one of the toppings

4.111 Many states have a lottery game, usually called a Pick-4, in which you pick a four-digit number such as 7359. During the lottery drawing, there are four bins, each containing balls numbered 0 through 9. One ball is drawn from each bin to form the four-digit winning number.

a. You purchase one ticket with one four-digit number. What is the probability that you will win this lottery game?

b. There are many variations of this game. The primary variation allows you to win if the four digits in your number are selected in any order as long as they are the same four digits as obtained

by the lottery agency. For example, if you pick four digits making the number 1265, then you will win if 1265, 2615, 5216, 6521, and so forth, are drawn. The variations of the lottery game depend on how many unique digits are in your number. Consider the following four different versions of this game.

i. All four digits are unique (e.g., 1234)

ii. Exactly one of the digits appears twice (e.g., 1223 or 9095)

iii. Two digits each appear twice (e.g., 2121 or 5588)

iv. One digit appears three times (e.g., 3335 or 2722)

Find the probability that you will win this lottery in each of these four situations.

4.112 The Big Six Wheel (or Wheel of Fortune) is a casino and carnival game that is well known for being a big money maker for the casinos. The wheel has 54 equally likely slots (outcomes) on it. The slot that pays the largest amount of money is called the *joker*. If a player bets on the *joker*, the probability of winning is 1/54. The outcome of any given play of this game (a spin of the wheel) is independent of the outcomes of previous plays.

a. Find the probability that a player who always bets on *joker* wins for the first time on the 15th play of the game.

b. Find the probability that it takes a player who always bets on *joker* more than 70 plays to win for the first time.

4.113 A large university has 12,600 male students. Of these students, 5312 are members of so-called "Greek" social organizations (fraternities or sororities), 2844 are members of Greek service organizations, and the others are not members of either of these two types of Greek organizations. Similarly, the female students are members of Greek social organizations, Greek service organizations, or neither. Assuming that gender and membership are independent events, find the probabilities of the events in parts a to c:

a. A student is a member of a Greek social organization given that the student is a female.

b. A student is a member of a Greek service organization given that the student is a female.

c. A student is not a member of either of these two types of Greek organizations given that the student is a female.

d. If the university has 14,325 female students, is it possible that

$$P \text{ (Greek social organization | male)} = P \text{ (Greek social organization | female)}?$$

Explain why/why not.

SELF-REVIEW TEST

1. The collection of all outcomes for an experiment is called
a. a sample space
b. the intersection of events
c. joint probability

2. A final outcome of an experiment is called
a. a compound event
b. a simple event
c. a complementary event

3. A compound event includes
a. all final outcomes
b. exactly two outcomes
c. more than one outcome for an experiment

4. Two equally likely events
a. have the same probability of occurrence
b. cannot occur together
c. have no effect on the occurrence of each other

5. Which of the following probability approaches can be applied only to experiments with equally likely outcomes?
  a. Classical probability
  b. Empirical probability
  c. Subjective probability

6. Two mutually exclusive events
  a. have the same probability
  b. cannot occur together
  c. have no effect on the occurrence of each other

7. Two independent events
  a. have the same probability
  b. cannot occur together
  c. have no effect on the occurrence of each other

8. The probability of an event is always
  a. less than 0
  b. in the range 0 to 1.0
  c. greater than 1.0

9. The sum of the probabilities of all final outcomes of an experiment is always
  a. 100   **b.** 1.0   **c.** 0

10. The joint probability of two mutually exclusive events is always
  a. 1.0   **b.** between 0 and 1   **c.** 0

11. Two independent events are
  a. always mutually exclusive
  b. never mutually exclusive
  c. always complementary

12. A couple is planning their wedding reception. The bride's parents have given them a choice of four reception facilities, three caterers, five DJs, and two limo services. If the couple randomly selects one reception facility, one caterer, one DJ, and one limo service, how many different outcomes are possible?

13. Lucia graduated this year with an accounting degree from a university. She has received job offers from an accounting firm, an insurance company, and an airline. She cannot decide which of the three job offers she should accept. Suppose she decides to randomly select one of these three job offers. Find the probability that the job offer selected is
  a. from the insurance company
  b. not from the accounting firm

14. There are 200 students in a particular graduate program at a state university. Of them, 110 are females and 125 are out-of-state students. Of the 110 females, 70 are out-of-state students.
  a. Are the events *female* and *out-of-state student* independent? Are they mutually exclusive? Explain why or why not.

b. If one of these 200 students is selected at random, what is the probability that the student selected is
  i. a male?
  ii. an out-of-state student given that this student is a female?

15. Reconsider Problem 14. If one of these 200 students is selected at random, what is the probability that the selected student is a *female* or *an out-of-state student*?

16. Reconsider Problem 14. If two of these 200 students are selected at random, what is the probability that both of them are out-of-state students?

17. The probability that an adult has ever experienced a migraine headache is .35. If two adults are randomly selected, what is the probability that neither of them has ever experienced a migraine headache?

18. A hat contains five green, eight red, and seven blue marbles. Let *A* be the event that a red marble is drawn if we randomly select one marble out of this hat. What is the probability of *A*? What is the complementary event of *A*, and what is its probability?

19. The probability that a randomly selected student from a college is a female is .55 and the probability that a student works for more than 10 hours per week is .62. If these two events are independent, find the probability that a randomly selected student is a
  a. *male* and *works for more than 10 hours per week*
  b. *female* or *works for more than 10 hours per week*

20. A sample was selected of 506 workers who currently receive two weeks of paid vacation per year. These workers were asked if they were willing to accept a small pay cut to get an additional week of paid vacation a year. The following table shows the responses of these workers.

	Yes	No	No Response
Men	77	140	32
Women	104	119	34

  a. If one person is selected at random from these 506 workers, find the following probabilities.
    i. $P(\text{yes})$
    ii. $P(\text{yes} \mid \text{woman})$
    iii. $P(\text{woman and no})$
    iv. $P(\text{no response or man})$
  b. Are the events *woman* and *yes* independent? Are they mutually exclusive? Explain why or why not.

MINI-PROJECTS

Note: The Mini-Projects are located on the text's Web site, www.wiley.com/college/mann.

DECIDE FOR YOURSELF

Note: The Decide for Yourself feature is located on the text's Web site, www.wiley.com/college/mann.

TECHNOLOGY INSTRUCTIONS   **CHAPTER 4**

Note: Complete TI-84, Minitab, and Excel manuals are available for download at the textbook's Web site, www.wiley.com/college/mann.

TI-84 Color/TI-84

The TI-84 Color Technology Instructions feature of this text is written for the TI-84 Plus C color graphing calculator running the 4.0 operating system. Some screens, menus, and functions will be slightly different in older operating systems. The TI-84 and TI-84 Plus can perform all of the same functions but will not have the "Color" option referenced in some of the menus.

Generating Random Numbers

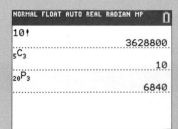

NORMAL FLOAT AUTO REAL RADIAN MP

```
randInt(1,50,10)
{11 35 28 42 33 3 20 33 6▶
Ans→L1
{11 35 28 42 33 3 20 33 6▶
```

Screen 4.1

To Generate Random Integers Uniformly Distributed between 1 and 50

1. Select **MATH > PROB > randInt**.

2. Use the following settings in the **randInt** menu:

   • Type 1 at the **lower** prompt.

   • Type 50 at the **upper** prompt.

   • Type 10 at the **n** prompt.

3. Highlight **Paste** and press **ENTER** twice. (See **Screen 4.1**.)

4. The output is 10 randomly chosen integers between 1 and 50. (See **Screen 4.1**.)

5. You may use any lower bound, upper bound, and number of integers.

6. To store these numbers in a list (see **Screen 4.1**):
   • Press **STO**.
   • Press **2ⁿᵈ > STAT** and select the name of the list.
   • Press **ENTER** twice.

Simulating Tosses of a Coin

Follow the instructions shown above, but generate random integers between 0 and 1. Assign 0 to be a tail and 1 to be a head (you can switch these if you prefer). To simulate 500 tosses, type 500 at the **n** prompt. To see a graph of the resulting distribution, store these values in a list and make a histogram of that list.

Simulating Tosses of a Die

NORMAL FLOAT AUTO REAL RADIAN MP

```
10!
                3628800
₅C₃
                      10
₂₀P₃
                    6840
```

Screen 4.2

Follow the instructions shown above, but generate random integers between 1 and 6. To simulate 500 tosses, type 500 at the **n** prompt. To see a graph of the resulting distribution, store these values in a list and make a histogram of that list.

Calculating 10! for Example 4–35 of the Text

1. Type 10.
2. Select **MATH > PROB > !**.
3. Press **ENTER**. (See **Screen 4.2**.)

Calculating ₅C₃ for Example 4–39 of the Text

1. Type 5.
2. Select **MATH > PROB ▷ nCr**.

3. Type 3.

4. Press **ENTER**. (See **Screen 4.2.**)

Calculating $_{20}P_3$ for Example 4–41 of the Text

1. Type 20.

2. Select **MATH > PROB > nPr**.

3. Type 3.

4. Press **ENTER**. (See **Screen 4.2.**)

Minitab

The Minitab Technology Instructions feature of this text is written for Minitab version 17. Some screens, menus, and functions will be slightly different in older versions of Minitab.

Generating Random Numbers

Generating 10 random Integers Uniformly Distributed between 1 and 50

1. Select **Calc > Random Data > Integer**.

2. Use the following settings in the dialog box that appears on screen:

- Type 10 in the **Number of rows of data to generate** box.

- Type C1 in the **Store in column(s)** box.

- Type 1 in the **Minimum value** box.

- Type 50 in the **Maximum value** box.

3. Click **OK**. Minitab will list 10 random integers between 1 and 50 in column C1.

Simulating Tosses of a Coin

Follow the instructions shown above, but generate random integers between 0 and 1. Assign 0 to be a tail and 1 to be a head (you can switch these if you prefer). To simulate 500 tosses, type 500 in the **Number of rows of data to generate** box. To see a graph of the resulting distribution, store these values in C1 and make a histogram of C1.

Simulating Tosses of a Die

Follow the instructions shown above, but generate random integers between 1 and 6. To simulate 500 tosses, type 500 in the **Number of rows of data to generate** box. To see a graph of the resulting distribution, store these values in C1 and make a histogram of C1.

Calculating 10! for Example 4–35 of the Text

1. Select **CALC > Calculator**.

2. Use the following settings in the dialog box that appears on screen (see **Screen 4.3**):

- In the **Store result in variable** box, type C1.

- In the **Expression** box, type factorial(10).

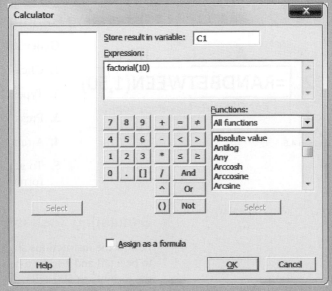

Screen 4.3

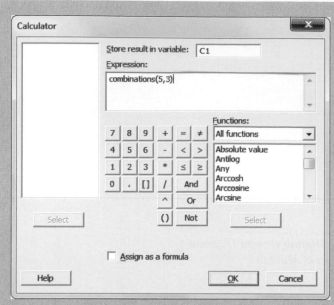

Screen 4.4

3. Click **OK**. Minitab will list the value of 10! in row 1 of column C1.

Calculating $_5C_3$ for Example 4–39 of the Text

1. Select **CALC > Calculator**.

2. Use the following settings in the dialog box that appears on screen (see **Screen 4.4**):

   - In the **Store result in variable** box, type C1.

   - In the **Expression** box, type combinations(5,3).

3. Click **OK**. Minitab will list the value of $_5C_3$ in row 1 of column C1.

Calculating $_{20}P_3$ for Example 4–41 of the Text

1. Select **CALC > Calculator**.

2. Use the following settings in the dialog box that appears on screen:

   - In the **Store result in variable** box, type C1.

   - In the **Expression** box, type permutations(20,3).

3. Click **OK**. Minitab will list the value of $_{20}P_3$ in row 1 of column C1.

Excel

The Excel Technology Instructions feature of this text is written for Excel 2013. Some screens, menus, and functions will be slightly different in older versions of Excel. To enable some of the advanced Excel functions, you must enable the Data Analysis Add-In for Excel.

Generating Random Numbers

Screen 4.5

Generating Random Integers Uniformly Distributed between 1 and 50

1. Click on cell A1.

2. Type =**randbetween(1,50)**. (See **Screen 4.5**.)

3. Press **ENTER**.

4. A randomly chosen integer between 1 and 50 will appear in cell A1.

5. To generate more than one such random number, copy and paste the formula into as many cells as you require.

Simulating Tosses of a Coin

Follow the instructions shown above, but generate random integers between 0 and 1. Assign 0 to be a tail and 1 to be a head (you can switch these if you prefer). To simulate 500 tosses, copy the formula to 500 cells. To see a graph of the resulting distribution, make a histogram of these 500 values.

Simulating Tosses of a Die

Follow the instructions shown above, but generate random integers between 1 and 6. To simulate 500 tosses, copy the formula to 500 cells. To see a graph of the resulting distribution, make a histogram of these values.

Calculating 10! for Example 4–35 of the Text

1. Click on cell A1.

2. Type =**FACT(10)**.

3. Press **ENTER**.

Calculating $_5C_3$ for Example 4–39 of the Text

1. Click on cell A1.

2. Type =**COMBIN(5,3)**.

3. Press **ENTER**. (See **Screen 4.6**.)

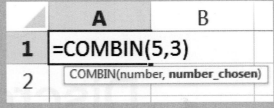

Screen 4.6

Calculating $_{20}P_3$ for Example 4–41 of the Text

1. Click on cell A1.

2. Type =**PERMUT(20,3)**.

3. Press **ENTER**.

TECHNOLOGY ASSIGNMENTS

TA4.1 Random number generators can be used to simulate the behavior of many different types of events, including those that have an infinite set of possibilities.

    a. Generate a set of 200 random numbers on the interval 0 to 1 and save them to a column or list in the technology you are using.

    b. Generate a second set of 200 random numbers, but on the interval 12.3 to 13.3 and save them to a different column or list in the technology you are using.

    c. Create histograms of the simulated data in each of the two columns for parts a and b. Compare the shapes of the histograms.

TA4.2 The data set *Simulated* (that is on the Web site of this text) contains four data sets named data1, data2, data3, and data4. Each of these four data sets consists of 1000 simulated values.

    a. Create a histogram and calculate the mean, median, and standard deviation for each of these four data sets.

    b. For data1, calculate the endpoints of the interval $\mu \pm 1\sigma$, that is, the interval from $\mu - 1\sigma$ to $\mu + 1\sigma$. Calculate the probability that a randomly selected value from data1 falls in this interval. (Note: To find this probability, you can sort data1 and then count the number of values that fall within this interval.)

    c. Now repeat part b for data1 to find intervals $\mu \pm 2\sigma$ and $\mu \pm 3\sigma$, respectively. Calculate the probability mentioned in part b for each of these intervals.

    d. Now repeat parts b and c for each of the other three data sets—data2, data3, and data4.

    e. How do the three probabilities (converted to percentages) for each data set compare with the probabilities (percentages) predicted by Chebyshev's Theorem and the Empirical Rule of Section 3.4? Do any of the four data sets appear to fit the percentages given by the Empirical Rule? If so, which data sets? Use the relevant histogram(s) that you created in part a to explain why this makes sense.

    f. All four of the data sets were simulated from four different populations that have the same mean. Based on the summary statistics, what does the value of the common mean of the four data sets appear to be?

TA4.3 Using a technology of your choice, find the number of samples of size 9 each that can be selected from a population of 36 cars.

TA4.4 Using a technology of your choice, find the number of permutations of 24 items selecting 9 at a time.

TA4.5 Using a technology of your choice, find 6!, 9!, and 15!.

TA4.6 Using a technology of your choice, find the number of combinations of 16 items selecting 7 at a time.

TA4.7 Using a technology of your choice, find the number of ways to select 9 items out of 24.

TA4.8 Using a technology of your choice, find the number of permutations of 16 items selecting 7 at a time.

CHAPTER **5**

Evan Lorne/Shutterstock

Discrete Random Variables and Their Probability Distributions

Now that you know a little about probability, do you feel lucky enough to play the lottery? If you have $20 to spend on lunch today, are you willing to spend it all on four $5 lottery tickets to increase your chance of winning? Do you think you will profit, on average, if you continue buying lottery tickets over time? Can lottery players beat the state, on average? Not a chance! (See Case Study 5–1 for answers.)

Chapter 4 discussed the concepts and rules of probability. This chapter extends the concept of probability to explain probability distributions. As we saw in Chapter 4, any given statistical experiment has more than one outcome. It is impossible to predict with certainty which of the many possible outcomes will occur if an experiment is performed. Consequently, decisions are made under uncertain conditions. For example, a lottery player does not know in advance whether or not he is going to win that lottery. If he knows that he is not going to win, he will definitely not play. It is the uncertainty about winning (some positive probability of winning) that makes him play. This chapter shows that if the outcomes and their probabilities for a statistical experiment are known, we can find out what will happen, on average, if that experiment is performed many times. For the lottery example, we can find out how much a lottery player can expect to win (or lose), on average, if he continues playing this lottery again and again.

In this chapter, random variables and types of random variables are explained. Then, the concept of a probability distribution and its mean and standard deviation for a discrete random variable are discussed. Finally, three special probability distributions for a discrete random variable—the binomial probability distribution, the hypergeometric probability distribution, and the Poisson probability distribution—are developed.

5.1 | Random Variables

Randomness plays an important part in our lives. Most things happen randomly. For example, consider the following events: winning a lottery, your car breaking down, getting sick, getting involved in an accident, losing a job, making money in the stock market, and so on. We cannot predict when, where, and to whom these things can or will happen. Who will win a lottery and when? Who will get sick and when? Where, when, and who will get involved in an accident? All these events are uncertain, and they happen randomly to people. In other words, there is no definite time, place, and person for these events to happen. Similarly, consider the following events: how many customers will visit a bank, a grocery store, or a gas station on a given day? How many cars will pass a bridge on a given day? How many students will be absent from a class on a given day? In all these examples, the *number* of customers, cars, and students are random; that is, each of these can assume any value within a certain interval.

Suppose Table 5.1 gives the frequency and relative frequency distributions of the number of vehicles owned by all 2000 families living in a small town.

Table 5.1   **Frequency and Relative Frequency Distributions of the Number of Vehicles Owned by Families**

Number of Vehicles Owned	Frequency	Relative Frequency
0	30	30/2000 = .015
1	320	320/2000 = .160
2	910	910/2000 = .455
3	580	580/2000 = .290
4	160	160/2000 = .080
	$N = 2000$	Sum = 1.000

Suppose one family is randomly selected from this population. The process of randomly selecting a family is called a *random* or *chance experiment*. Let x denote the number of vehicles owned by the selected family. Then x can assume any of the five possible values (0, 1, 2, 3, and 4) listed in the first column of Table 5.1. The value assumed by x depends on which family is selected. Thus, this value depends on the outcome of a random experiment. Consequently, x is called a **random variable** or a **chance variable**. In general, a random variable is denoted by x or y.

A **random variable** is a variable whose value is determined by the outcome of a random experiment.

As will be explained next, a random variable can be discrete or continuous.

5.1.1   Discrete Random Variable

A **discrete random variable** assumes values that can be counted. In other words, the consecutive values of a discrete random variable are separated by a certain gap.

In Table 5.1, *the number of vehicles owned by a family* is an example of a discrete random variable because the values of the random variable x are countable: 0, 1, 2, 3, and 4. Here are a few other examples of discrete random variables:

A random variable that assumes countable values is called a **discrete random variable**.

1. The number of cars sold at a dealership during a given month
2. The number of houses in a certain block
3. The number of fish caught on a fishing trip

4. The number of complaints received at the office of an airline on a given day
5. The number of customers who visit a bank during any given hour
6. The number of heads obtained in three tosses of a coin

5.1.2 Continuous Random Variable

A random variable that can assume any value contained in one or more intervals is called a **continuous random variable**.

A random variable whose values are not countable is called a **continuous random variable**. A continuous random variable can assume any value over an interval or intervals.

Because the number of values contained in any interval is infinite, the possible number of values that a continuous random variable can assume is also infinite. Moreover, we cannot count these values. Consider the life of a battery. We can measure it as precisely as we want. For instance, the life of this battery may be 40 hours, or 40.25 hours, or 40.247 hours. Assume that the maximum life of a battery is 200 hours. Let x denote the life of a randomly selected battery of this kind. Then, x can assume any value in the interval 0 to 200. Consequently, x is a continuous random variable. As shown in the diagram below, every point on the line representing the interval 0 to 200 gives a possible value of x.

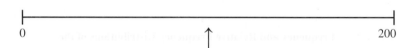

Every point on this line represents a possible value of x that denotes the life of a battery. There is an infinite number of points on this line. The values represented by points on this line are uncountable.

The following are a few other examples of continuous random variables:

1. The length of a room
2. The time taken to commute from home to work
3. The amount of milk in a gallon (note that we do not expect "a gallon" to contain exactly one gallon of milk but either slightly more or slightly less than one gallon).
4. The weight of a letter
5. The price of a house

Note that money is often treated as a continuous random variable, specifically when there are a large number of unique values.

This chapter is limited to the discussion of discrete random variables and their probability distributions. Continuous random variables will be discussed in Chapter 6.

EXERCISES

CONCEPTS AND PROCEDURES

5.1 Define the following terms in statistics: a random variable, a discrete random variable, and a continuous random variable. Give one example each of a discrete random variable and a continuous random variable.

5.2 Indicate which of the following random variables are discrete and which are continuous.

    a. The amount of rainfall in a city during a specific month
    b. The number of students on a waitlist to register for a class
    c. The price of one ounce of gold at the close of trading on a given day

    d. The number of vacation trips taken by a family during a given year
    e. The amount of gasoline in your car's gas tank at a given time
    f. The distance you walked to class this morning

APPLICATIONS

5.3 One of the four gas stations located at an intersection of two major roads is a Texaco station. Suppose the next six cars that stop at any of these four gas stations make their selections randomly and independently. Let x be the number of cars in these six that stop at the Texaco station. Is x a discrete or a continuous random variable? Explain. What are the possible values that x can assume?

5.2 | Probability Distribution of a Discrete Random Variable

Let x be a discrete random variable. The **probability distribution** of x describes how the probabilities are distributed over all the possible values of x.

We learned in Chapter 4 about the classical and the relative frequency approaches to probability. Depending on the type of the statistical experiment, we can use one of these two approaches to write the probability distribution of a random variable. Example 5–1 uses the relative frequency approach to write the probability distribution.

> The **probability distribution of a discrete random variable** lists all the possible values that the random variable can assume and their corresponding probabilities.

EXAMPLE 5–1   Number of Vehicles Owned by Families

Writing the probability distribution of a discrete random variable.

Recall the frequency and relative frequency distributions of the number of vehicles owned by families given in Table 5.1. That table is reproduced here as Table 5.2. Let x be the number of vehicles owned by a randomly selected family. Write the probability distribution of x and make a histogram for this probability distribution.

Table 5.2   **Frequency and Relative Frequency Distributions of the Number of Vehicles Owned by Families**

Number of Vehicles Owned	Frequency	Relative Frequency
0	30	.015
1	320	.160
2	910	.455
3	580	.290
4	160	.080
	$N = 2000$	Sum = 1.000

Solution   In Chapter 4, we learned that the relative frequencies obtained from an experiment or a sample can be used as approximate probabilities. However, when the relative frequencies represent the population, as in Table 5.2, they give the actual (theoretical) probabilities of outcomes. Using the relative frequencies of Table 5.2, we can write the *probability distribution* of the discrete random variable x in Table 5.3. Note that the values of x listed in Table 5.3 are pairwise mutually exclusive events.

Table 5.3   **Probability Distribution of the Number of Vehicles Owned by Families**

Number of Vehicles Owned x	Probability $P(x)$
0	.015
1	.160
2	.455
3	.290
4	.080
	$\Sigma P(x) = 1.000$

Using the probability distribution of Table 5.3, we make the histogram given in Figure 5.1. Such a histogram can show whether the probability distribution is symmetric or skewed, and if skewed then whether it is skewed to the left or right.

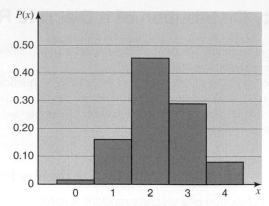

Figure 5.1   Histogram for the probability distribution of Table 5.3.

From the probability distribution listed in Table 5.3, we observe two characteristics:

1.   Each probability listed in the $P(x)$ column is in the range zero to 1.0.
2.   The sum of all the probabilities listed in the $P(x)$ column is 1.0.

These two characteristics are always true for a valid probability distribution. These two characteristics are the same that we discussed in Chapter 4, and can be written as in the following box.

Two Characteristics of a Probability Distribution   The probability distribution of a discrete random variable possesses the following two characteristics.

1. $0 \leq P(x) \leq 1$ for each value of x
2. $\Sigma P(x) = 1$

These two characteristics are also called the *two conditions* that a probability distribution must satisfy. Notice that in Table 5.3 each probability listed in the column labeled $P(x)$ is between 0 and 1. Also, $\Sigma P(x) = 1.0$. Because both conditions are satisfied, Table 5.3 represents the probability distribution of x.

Using the probability distribution listed in Table 5.3, we can find the probability of any outcome or event. Example 5–2 illustrates this.

EXAMPLE 5–2   **Number of Vehicles Owned by Families**

Finding probabilities of events using Table 5.3.

Using the probability distribution listed in Table 5.3 of Example 5–1, find the following probabilities:

  (a)   The probability that a randomly selected family owns two vehicles
  (b)   The probability that a randomly selected family owns at least two vehicles
  (c)   The probability that a randomly selected family owns at most one vehicle
  (d)   The probability that a randomly selected family owns three or more vehicles

Solution   Using the probabilities listed in Table 5.3, we find the required probabilities as follows.

  (a) P(selected family owns two vehicles) = $P(2) = .455$
  (b) P(selected family owns at least two vehicles) = $P(2$ or 3 or $4) = P(2) + P(3) + P(4)$

$$= .455 + .290 + .080 = \mathbf{.825}$$

  (c) P(selected family owns at most one vehicle) = $P(0$ or $1) = P(0) + P(1)$

$$= .015 + .160 = \mathbf{.175}$$

(d) P(selected family owns three or more vehicles) = P(3 or 4) = P(3) + P(4)

$$= .290 + .080 = \mathbf{.370}$$

EXAMPLE 5–3

Verifying the conditions of a probability distribution.

Each of the following tables lists certain values of x and their probabilities. Determine whether or not each table represents a valid probability distribution.

(a)

x	$P(x)$
0	.08
1	.11
2	.39
3	.27

(b)

x	$P(x)$
2	.25
3	.34
4	.28
5	.13

(c)

x	$P(x)$
7	.70
8	.50
9	−.20

Solution

(a) Because each probability listed in this table is in the range 0 to 1, it satisfies the first condition of a probability distribution. However, the sum of all probabilities is not equal to 1.0 because $\Sigma P(x) = .08 + .11 + .39 + .27 = .85$. Therefore, the second condition is not satisfied. Consequently, this table does not represent a valid probability distribution.

(b) Each probability listed in this table is in the range 0 to 1. Also, $\Sigma P(x) = .25 + .34 + .28 + .13 = 1.0$. Consequently, this table represents a valid probability distribution.

(c) Although the sum of all probabilities listed in this table is equal to 1.0, one of the probabilities is negative. This violates the first condition of a probability distribution. Therefore, this table does not represent a valid probability distribution.

EXAMPLE 5–4   Number of Breakdowns for a Machine

Finding the probabilities of events for a discrete random variable.

The following table lists the probability distribution of the number of breakdowns per week for a machine based on past data.

Breakdowns per week	0	1	2	3
Probability	.15	.20	.35	.30

Find the probability that the number of breakdowns for this machine during a given week is

(a) exactly 2   (b) 0 to 2

(c) more than 1   (d) at most 1

Solution   Let x denote the number of breakdowns for this machine during a given week. Table 5.4 lists the probability distribution of x.

Table 5.4   **Probability Distribution of the Number of Breakdowns**

x	$P(x)$
0	.15
1	.20
2	.35
3	.30
	$\Sigma P(x) = 1.00$

Using Table 5.4, we can calculate the required probabilities as follows.

(a) The probability of exactly two breakdowns is

$$P(\text{exactly 2 breakdowns}) = P(x = 2) = P(2) = \mathbf{.35}$$

(b) The probability of 0 to 2 breakdowns is given by the sum of the probabilities of 0, 1, and 2 breakdowns:

$$P(0 \text{ to } 2 \text{ breakdowns}) = P(0 \leq x \leq 2) = P(0 \text{ or } 1 \text{ or } 2)$$
$$= P(0) + P(1) + P(2)$$
$$= .15 + .20 + .35 = \mathbf{.70}$$

(c) The probability of more than 1 breakdown is obtained by adding the probabilities of 2 and 3 breakdowns:

$$P(\text{more than } 1 \text{ breakdown}) = P(x > 1) = P(2 \text{ or } 3)$$
$$= P(2) + P(3)$$
$$= .35 + .30 = \mathbf{.65}$$

(d) The probability of at most 1 breakdown is given by the sum of the probabilities of 0 and 1 breakdown:

$$P(\text{at most } 1 \text{ breakdown}) = P(x \leq 1) = P(0 \text{ or } 1)$$
$$= P(0) + P(1)$$
$$= .15 + .20 = \mathbf{.35}$$

EXAMPLE 5–5　Students Suffering from Math Anxiety

Constructing a probability distribution.

According to a survey, 60% of all students at a large university suffer from math anxiety. Two students are randomly selected from this university. Let x denote the number of students in this sample who suffer from math anxiety. Construct the probability distribution of x.

Solution　Let us define the following two events:

$$N = \text{the student selected does not suffer from math anxiety}$$
$$M = \text{the student selected suffers from math anxiety}$$

As we can observe from the tree diagram of Figure 5.2, there are four possible outcomes for this experiment: NN (neither of the students suffers from math anxiety), NM (the first student does not suffer from math anxiety and the second does), MN (the first student suffers from math anxiety and the second does not), and MM (both students suffer from math anxiety). The probabilities of these four outcomes are listed in the tree diagram. Because 60% of the students suffer from math anxiety and 40% do not, the probability is .60 that any student selected suffers from math anxiety and .40 that he or she does not.

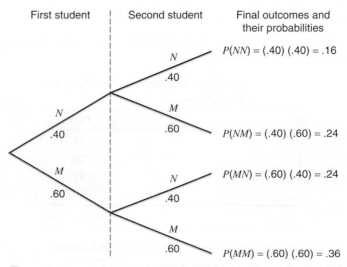

Figure 5.2　Tree diagram.

In a sample of two students, the number who suffer from math anxiety can be 0 (given by *NN*), 1 (given by *NM* or *MN*), or 2 (given by *MM*). Thus, *x* can assume any of three possible values: 0, 1, or 2. The probabilities of these three outcomes are calculated as follows:

$$P(0) = P(NN) = .16$$
$$P(1) = P(NM \text{ or } MN) = P(NM) + P(MN) = .24 + .24 = .48$$
$$P(2) = P(MM) = .36$$

Using these probabilities, we can write the probability distribution of *x* as in Table 5.5.

Table 5.5 Probability Distribution of the Number of Students with Math Anxiety

x	P(x)
0	.16
1	.48
2	.36
	$\Sigma P(x) = 1.00$

EXERCISES

CONCEPTS AND PROCEDURES

5.4 What do you understand by the probability distribution of a discrete random variable? Give one example of such a probability distribution.

5.5 Describe briefly the two characteristics (conditions) of the probability distribution of a discrete random variable.

5.6 Each of the following tables lists certain values of *x* and their probabilities. Verify whether or not each represents a valid probability distribution.

a.

x	P(x)
0	.24
1	.31
2	.40
3	.10

b.

x	P(x)
2	.15
3	.25
4	.40
5	.20

c.

x	P(x)
7	.65
8	.55
9	−.15

5.7 The following table gives the probability distribution of a discrete random variable *x*.

x	0	1	2	3	4	5
P(x)	.05	.19	.28	.31	.12	.05

Find the following probabilities.

a. $P(x = 1)$ **b.** $P(x \leq 1)$ **c.** $P(x \geq 3)$ **d.** $P(0 \leq x \leq 2)$
e. Probability that *x* assumes a value less than 3
f. Probability that *x* assumes a value greater than 3
g. Probability that *x* assumes a value in the interval 2 to 4

APPLICATIONS

5.8 A review of emergency room records at rural Millard Fellmore Memorial Hospital was performed to determine the probability distribution of the number of patients entering the emergency room during a 1-hour period. The following table lists this probability distribution.

Patients per hour	0	1	2	3	4	5	6
Probability	.2165	.4183	.2112	.0998	.0314	.0138	.0090

a. Graph the probability distribution.
b. Determine the probability that the number of patients entering the emergency room during a randomly selected 1-hour period is
    i. 2 or more   **ii.** exactly 5   **iii.** fewer than 3   **iv.** at most 1

5.9 The H2 Hummer limousine has eight tires on it. A fleet of 1300 H2 limos was fit with a batch of tires that mistakenly passed quality testing. The following table lists the frequency distribution of the number of defective tires on the 1300 H2 limos.

Number of defective tires	0	1	2	3	4	5	6	7	8
Number of H2 limos	59	224	369	347	204	76	18	2	1

a. Construct a probability distribution table for the numbers of defective tires on these limos. Draw a bar graph for this probability distribution.
b. Are the probabilities listed in the table of part a exact or approximate probabilities of the various outcomes? Explain.
c. Let *x* denote the number of defective tires on a randomly selected H2 limo. Find the following probabilities.
    i. $P(x = 0)$ **ii.** $P(x < 4)$ **iii.** $P(3 \leq x < 7)$ **iv.** $P(x \geq 2)$

5.10 Nathan Cheboygan, a singing gambler from northern Michigan, is famous for his loaded dice. The following table shows the probability distribution for the sum, denoted by *x*, of the faces on a pair of Nathan's dice.

x	2	3	4	5	6	7	8	9	10	11	12
P(x)	.065	.065	.080	.095	.110	.170	.110	.095	.080	.065	.065

a. Draw a bar graph for this probability distribution.
b. Determine the probability that the sum of the faces on a single roll of Nathan's dice is
    i. an even number   **ii.** 7 or 11   **iii.** 4 to 6   **iv.** no less than 9

5.11 Suppose that currently 16.1% of motorists in the United States are uninsured. Suppose that two motorists are selected at random. Let

x denote the number of motorists in this sample of two who are uninsured. Construct the probability distribution table of x. Draw a tree diagram for this problem.

5.12 In a 10K run, 27.4% of the 5248 participants finished the race in 49 minutes 42 seconds (49:42) or faster, which is equivalent to a pace of less than 8 minutes per mile. Suppose that this result holds true for all people who would participate in and finish a 10K race. Suppose that two 10K runners are selected at random. Let x denote the number of runners in these two who would finish a 10K race in 49:42 or less. Construct the probability distribution table of x. Draw a tree diagram for this problem.

5.13 Suppose that currently 3.7% of Americans with Alzheimer's disease are younger than the age of 65 years (which means that they were diagnosed with early onset of Alzheimer's). Suppose that two Americans with Alzheimer's disease are selected at random. Let x denote the number in this sample of two Americans with Alzheimer's disease who are younger than the age of 65 years. Construct the probability distribution table of x. Draw a tree diagram for this problem.

*5.14 In a group of 20 athletes, 6 have used performance-enhancing drugs that are illegal. Suppose that 2 athletes are randomly selected from this group. Let x denote the number of athletes in this sample who have used such illegal drugs. Write the probability distribution of x. You may draw a tree diagram and use that to write the probability distribution. (*Hint:* Note that the selections are made without replacement from a small population. Hence, the probabilities of outcomes do not remain constant for each selection.)

5.3 | Mean and Standard Deviation of a Discrete Random Variable

In this section, we will learn how to calculate the mean and standard deviation of a discrete random variable and how to interpret them.

5.3.1 Mean of a Discrete Random Variable

The **mean of a discrete random variable**, denoted by μ, is actually the mean of its probability distribution. The mean of a discrete random variable x is also called its *expected value* and is denoted by $E(x)$. The mean (or expected value) of a discrete random variable is the value that we expect to observe per repetition, on average, if we perform an experiment a large number of times. For example, we may expect a car salesperson to sell, on average, 2.4 cars per week. This does not mean that every week this salesperson will sell exactly 2.4 cars. (Obviously one cannot sell exactly 2.4 cars.) This simply means that if we observe for many weeks, this salesperson will sell a different number of cars during different weeks; however, the average for all these weeks will be 2.4 cars per week.

To calculate the mean of a discrete random variable x, we multiply each value of x by the corresponding probability and sum the resulting products. This sum gives the mean (or expected value) of the discrete random variable x.

Mean of a Discrete Random Variable The **mean of a discrete random variable** x is the value that is expected to occur per repetition, on average, if an experiment is repeated a large number of times. It is denoted by μ and calculated as

$$\mu = \Sigma xP(x)$$

The mean of a discrete random variable x is also called its expected value and is denoted by $E(x)$; that is,

$$E(x) = \Sigma xP(x)$$

Example 5–6 illustrates the calculation of the mean of a discrete random variable.

| EXAMPLE 5–6 | Machine Breakdowns per Week |

Calculating and interpreting the mean of a discrete random variable.

The following table lists the number of breakdowns per week and their probabilities for a machine based on past data.

Breakdowns	Probability
0	.15
1	.20
2	.35
3	.30

Find the mean number of breakdowns per week for this machine.

Solution   Let x represent the number of breakdowns per week and $P(x)$ be the probability of x breakdowns. To find the mean number of breakdowns per week for this machine, we multiply each value of x by its probability and add these products. This sum gives the mean of the probability distribution of x. The products $xP(x)$ are listed in the third column of Table 5.6. The sum of these products gives $\Sigma xP(x)$, which is the mean of x.

Table 5.6   **Calculating the Mean for the Probability Distribution of Breakdowns**

x	$P(x)$	$xP(x)$
0	.15	$0(.15) = .00$
1	.20	$1(.20) = .20$
2	.35	$2(.35) = .70$
3	.30	$3(.30) = .90$
		$\Sigma xP(x) = 1.80$

The mean is

$$\mu = \Sigma xP(x) = \mathbf{1.80}$$

Thus, on average, this machine is expected to break down 1.80 times per week over a period of time. In other words, if this machine is used for many weeks, then for certain weeks we will observe no breakdowns; for some other weeks we will observe one breakdown per week; and for still other weeks we will observe two or three breakdowns per week. The mean number of breakdowns is expected to be 1.80 per week for the entire period.

Note that $\mu = 1.80$ is also the **expected value** of x. It can also be written as

$$E(x) = 1.80$$

Again, this expected value of x means that we will expect an average of 1.80 breakdowns per week for this machine.

Case Study 5–1 illustrates the calculation of the mean amount that an instant lottery player is expected to win.

5.3.2   Standard Deviation of a Discrete Random Variable

The **standard deviation of a discrete random variable**, denoted by σ, measures the spread of its probability distribution. A higher value for the standard deviation of a discrete random variable indicates that x can assume values over a larger range about the mean. In contrast, a smaller value for the standard deviation indicates that most of the values that x can assume are clustered closely about the mean. The basic formula to compute the standard deviation of a discrete random variable is

$$\sigma = \sqrt{\Sigma[(x - \mu)^2 \cdot P(x)]}$$

However, it is more convenient to use the following shortcut formula to compute the standard deviation of a discrete random variable.

Standard Deviation of a Discrete Random Variable   The **standard deviation of a discrete random variable** x measures the spread of its probability distribution and is computed as

$$\sigma = \sqrt{\Sigma x^2 P(x) - \mu^2}$$

All State Lottery

A hypothetical country has an instant lottery game called All State Lottery. The cost of each ticket for this lottery game is $5. A player can instantly win $500,000, $10,000, $1000, $100, $50, $20, and $10. Each ticket has seven spots that are covered by latex coating. The top one of these spots, labeled Prize, contains the winning prize. The remaining six spots labeled *Player's Spots* belong to the player. A player will win if any of the *Player's Spots* contains the prize amount that matches the winning prize in the top spot. If none of the six spots contains a dollar amount that matches the winning prize, the player loses.

Initially a total of 20,000,000 tickets are printed for this lottery. The following table lists the prizes and the number of tickets with those prizes. As is obvious from this table, out of a total of 20,000,000 tickets, 18,000,000 are non-winning tickets (the ones with a prize of $0 in this table). Of the remaining 2,000,000 tickets with prizes, 1,620,000 tickets have a prize of $10 each, 364,000 tickets contain a prize of $20 each, and so forth.

Prize (dollars)	Number of Tickets
0	18,000,000
10	1,620,000
20	364,000
50	10,000
100	5000
1000	730
10,000	200
500,000	70
	Total = 20,000,000

The net amount a player wins for each of the winning tickets is equal to the amount of the prize minus $5, which is the cost of the ticket. Thus, the net gain for each of the non-winning tickets is –$5, which is the cost of the ticket. Let

x = the net amount a player wins by playing this lottery game

The following table shows the probability distribution of x, and all the calculations required to compute the mean of x for this probability distribution. The probability of an outcome (net winnings) is calculated by dividing the number of tickets with that outcome (prize) by the total number of tickets.

x (dollars)	$P(x)$	$x P(x)$
–5	18,000,000/20,000,000 = .9000000	–4.5000000
5	1,620,000/20,000,000 = .0810000	.4050000
15	364,000/20,000,000 = .0182000	.2730000
45	10,000/20,000,000 = .0005000	.0225000
95	5000/20,000,000 = .0002500	.0237500
995	730/20,000,000 = .0000365	.0363175
9995	200/20,000,000 = .0000100	.0999500
499,995	70/20,000,000 = .0000035	1.7499825
		$\Sigma x P(x) = -1.8895000$

Hence, the mean or expected value of x is

$$\mu = \Sigma x P(x) = -\$1.8895000 \approx -\$1.89$$

This value of the mean gives the expected value of the random variable x, that is,

$$E(x) = \Sigma x P(x) = -\$1.89$$

Thus, the mean of net winnings for this lottery game is −$1.89. In other words, all players taken together will lose an average of $1.89 per ticket. That is, out of every $5 (the price of a ticket), $3.11 will be returned to players in the form of prizes and $1.89 will go to the government, which will cover the costs of operating the lottery, the commission paid to agents, and profit to the government. Note that $1.89 is 37.8% of $5. Thus, we can also state that 37.8% of the total money spent by players on this lottery game will go to the government and 100 − 37.8 = 62.2% will be returned to the players in the form of prizes.

Note that the variance σ^2 of a discrete random variable is obtained by squaring its standard deviation.

Example 5–7 illustrates how to use the shortcut formula to compute the standard deviation of a discrete random variable.

EXAMPLE 5–7 | Defective Computer Parts

Calculating the standard deviation of a discrete random variable.

Baier's Electronics manufactures computer parts that are supplied to many computer companies. Despite the fact that two quality control inspectors at Baier's Electronics check every part for defects before it is shipped to another company, a few defective parts do pass through these inspections undetected. Let x denote the number of defective computer parts in a shipment of 400. The following table gives the probability distribution of x.

x	0	1	2	3	4	5
$P(x)$	.02	.20	.30	.30	.10	.08

Compute the standard deviation of x.

Solution   Table 5.7 shows all the calculations required for the computation of the standard deviation of x.

Table 5.7   Computations to Find the Standard Deviation

x	$P(x)$	$xP(x)$	x^2	$x^2P(x)$
0	.02	.00	0	.00
1	.20	.20	1	.20
2	.30	.60	4	1.20
3	.30	.90	9	2.70
4	.10	.40	16	1.60
5	.08	.40	25	2.00
		$\Sigma x P(x) = 2.50$		$\Sigma x^2 P(x) = 7.70$

We perform the steps listed next to compute the standard deviation of x.

Step 1.   Compute the mean of the discrete random variable.

The sum of the products $xP(x)$, recorded in the third column of Table 5.7, gives the mean of x.

$$\mu = \Sigma xP(x) = 2.50 \text{ defective computer parts in } 400$$

Step 2.   Compute the value of $\Sigma x^2 P(x)$.

First we square each value of x and record it in the fourth column of Table 5.7. Then we multiply these values of x^2 by the corresponding values of $P(x)$. The resulting values of $x^2 P(x)$ are recorded in the fifth column of Table 5.7. The sum of this column is

$$\Sigma x^2 P(x) = 7.70$$

Step 3.   Substitute the values of μ and $\Sigma x^2 P(x)$ in the formula for the standard deviation of x and simplify.

By performing this step, we obtain

$$\sigma = \sqrt{\Sigma x^2 P(x) - \mu^2} = \sqrt{7.70 - (2.50)^2} = \sqrt{1.45}$$
$$= \mathbf{1.204} \text{ defective computer parts}$$

Thus, a given shipment of 400 computer parts is expected to contain an average of 2.50 defective parts with a standard deviation of 1.204 defective parts.

Remember ➤   Because the standard deviation of a discrete random variable is obtained by taking the positive square root, its value is never negative.

EXAMPLE 5–8   Profits from New Makeup Product

Loraine Corporation is planning to market a new makeup product. According to the analysis made by the financial department of the company, it will earn an annual profit of $4.5 million if this product has high sales, it will earn an annual profit of $1.2 million if the sales are mediocre, and it will lose $2.3 million a year if the sales are low. The probabilities of these three scenarios are .32, .51, and .17, respectively.

(a) Let x be the profits (in millions of dollars) earned per annum from this product by the company. Write the probability distribution of x.

(b) Calculate the mean and standard deviation of x.

Solution

Writing the probability distribution of a discrete random variable.

(a) The table below lists the probability distribution of x. Note that because x denotes profits earned by the company, the loss is written as a *negative profit* in the table.

x	$P(x)$
4.5	.32
1.2	.51
−2.3	.17

Calculating the mean and standard deviation of a discrete random variable.

(b) Table 5.8 shows all the calculations that are required for the computation of the mean and standard deviation of x.

Table 5.8   Computations to Find the Mean and Standard Deviation

x	$P(x)$	$xP(x)$	x^2	$x^2P(x)$
4.5	.32	1.440	20.25	6.4800
1.2	.51	.612	1.44	.7344
−2.3	.17	−.391	5.29	.8993
		$\Sigma xP(x) = 1.661$		$\Sigma x^2P(x) = 8.1137$

The mean of x is

$$\mu = \sum xP(x) = \$1.661 \text{ million}$$

The standard deviation of x is

$$\sigma = \sqrt{\sum x^2 P(x) - \mu^2} = \sqrt{8.1137 - (1.661)^2} = \$2.314 \text{ million}$$

Thus, it is expected that Loraine Corporation will earn an average of \$1.661 million in profits per year from the new product, with a standard deviation of \$2.314 million.

Interpretation of the Standard Deviation

The standard deviation of a discrete random variable can be interpreted or used the same way as the standard deviation of a data set in Section 3.4 of Chapter 3. In that section, we learned that according to Chebyshev's theorem, at least $[1 - (1/k^2)] \times 100\%$ of the total area under a curve lies within k standard deviations of the mean, where k is any number greater than 1. Thus, if $k = 2$, then at least 75% of the area under a curve lies between $\mu - 2\sigma$ and $\mu + 2\sigma$. In Example 5–7,

$$\mu = 2.50 \quad \text{and} \quad \sigma = 1.204$$

Hence,

$$\mu - 2\sigma = 2.50 - 2(1.204) = .092$$
$$\mu + 2\sigma = 2.50 + 2(1.204) = 4.908$$

Using Chebyshev's theorem, we can state that at least 75% of the shipments (each containing 400 computer parts) are expected to contain .092 to 4.908 defective computer parts each.

EXERCISES

CONCEPTS AND PROCEDURES

5.15 Describe the concept of the mean and standard deviation of a discrete random variable.

5.16 Find the mean and standard deviation for each of the following probability distributions.

a.

x	$P(x)$
3	.09
4	.36
5	.28
6	.15
7	.12

b.

x	$P(x)$
0	.53
1	.26
2	.17
3	.04

APPLICATIONS

5.17 Let x be the number of cars that a randomly selected auto mechanic repairs on a given day. The following table lists the probability distribution of x.

x	2	3	4	5	6
$P(x)$	.05	.22	.40	.23	.10

Find the mean and standard deviation of x. Give a brief interpretation of the value of the mean.

5.18 Let x be the number of emergency root canal surgeries performed by Dr. Sharp on a given Monday. The following table lists the probability distribution of x.

x	0	1	2	3	4	5
$P(x)$	.13	.28	.30	.17	.08	.04

Calculate the mean and standard deviation of x. Give a brief interpretation of the value of the mean.

5.19 Let x be the number of potential weapons detected by a metal detector at an airport on a given day. The following table lists the probability distribution of x.

x	0	1	2	3	4	5
$P(x)$	.14	.28	.22	.18	.12	.06

Calculate the mean and standard deviation for this probability distribution and give a brief interpretation of the value of the mean.

5.20 Let x be the number of heads obtained in two tosses of a coin. The following table lists the probability distribution of x.

x	0	1	2
$P(x)$	.25	.50	.25

Calculate the mean and standard deviation of x. Give a brief interpretation of the value of the mean.

5.21 Let x be the number of magazines a person reads every week. Based on a sample survey of adults, the following probability distribution table was prepared.

x	0	1	2	3	4	5
$P(x)$	.38	.27	.15	.09	.06	.05

Find the mean and standard deviation of x.

5.22   The following table lists the probability distribution of the number of exercise machines sold per day at Elmo's Sporting Goods store.

Machines sold per day	4	5	6	7	8	9	10
Probability	.08	.11	.14	.19	.20	.16	.12

Calculate the mean and standard deviation for this probability distribution. Give a brief interpretation of the value of the mean.

5.23   A contractor has submitted bids on three state jobs: an office building, a theater, and a parking garage. State rules do not allow a contractor to be offered more than one of these jobs. If this contractor is awarded any of these jobs, the profits earned from these contracts are $10 million from the office building, $5 million from the theater, and $2 million from the parking garage. His profit is zero if he gets no contract. The contractor estimates that the probabilities of getting the office building contract, the theater contract, the parking garage contract, or nothing are .15, .30, .45, and .10, respectively. Let x be the random variable that represents the contractor's profits in millions of dollars. Write the probability distribution of x. Find the mean and

standard deviation of x. Give a brief interpretation of the values of the mean and standard deviation.

5.24   An instant lottery ticket costs $2. Out of a total of 20,000 tickets printed for this lottery, 2000 tickets contain a prize of $5 each, 200 tickets have a prize of $10 each, 10 tickets have a prize of $1000 each, and 1 ticket has a prize of $10,000. Let x be the random variable that denotes the net amount a player wins by playing this lottery. Write the probability distribution of x. Determine the mean and standard deviation of x. How will you interpret the values of the mean and standard deviation of x?

5.25   In a group of 20 athletes, 6 have used performance-enhancing drugs that are illegal. Suppose that 2 athletes are randomly selected from this group. Let x denote the number of athletes in this sample who have used such illegal drugs. The following table lists the probability distribution of x. Calculate the mean and standard deviation of x for this distribution.

x	0	1	2
$P(x)$	.4789	.4422	.0789

5.4 | The Binomial Probability Distribution

The **binomial probability distribution** is one of the most widely used discrete probability distributions. It is applied to find the probability that an outcome will occur x times in n performances of an experiment. For example, given that 75% of students at a college use Instagram, we may want to find the probability that in a random sample of five students at this college, exactly three use Instagram. As a second example, we may be interested in finding the probability that a baseball player with a batting average of .250 will have no hits in 10 trips to the plate.

To apply the binomial probability distribution, the random variable x must be a discrete dichotomous random variable. In other words, the variable must be a discrete random variable, and each repetition of the experiment must result in one of two possible outcomes. The binomial distribution is applied to experiments that satisfy the four conditions of a *binomial experiment*. (These conditions are discussed next.) Each repetition of a binomial experiment is called a **trial** or a **Bernoulli trial** (after Jacob Bernoulli). For example, if an experiment is defined as one toss of a coin and this experiment is repeated 10 times, then each repetition (toss) is called a trial. Consequently, there are 10 total trials for this experiment.

5.4.1   The Binomial Experiment

An experiment that satisfies the following four conditions is called a **binomial experiment**.

1. There are n **identical trials**. In other words, the given experiment is repeated n times, where n is a positive integer. All of these repetitions are performed under identical conditions.

2. Each trial has two and only two outcomes. These outcomes are usually called a *success* and a *failure,* respectively. In case there are more than two outcomes for an experiment, we can combine outcomes into two events and then apply binomial probability distribution.

3. The probability of success is denoted by p and that of failure by q, and $p + q = 1$. The probabilities p and q remain constant for each trial.

4. The **trials are independent**. In other words, the outcome of one trial does not affect the outcome of another trial.

Conditions of a Binomial Experiment   A binomial experiment must satisfy the following four conditions.

1. There are n identical trials.
2. Each trial has only two possible outcomes (or events). In other words, the outcomes of a trial are divided into two mutually exclusive events.
3. The probabilities of the two outcomes (or events) remain constant.
4. The trials are independent.

Note that one of the two outcomes (or events) of a trial is called a *success* and the other a *failure*. Notice that a success does not mean that the corresponding outcome is considered favorable or desirable. Similarly, a failure does not necessarily refer to an unfavorable or undesirable outcome. Success and failure are simply the names used to denote the two possible outcomes of a trial. The outcome to which the question refers is usually called a success; the outcome to which it does not refer is called a failure. For example, if we are to find the probability of four heads in 10 tosses of a coin, we will call the head a *success* and the tail a *failure*. But if the question asks to find the probability of six tails in 10 tosses of a coin, we will call the tail a *success* and the head a *failure*.

EXAMPLE 5–9   Ten Tosses of a Coin

Verifying the conditions of a binomial experiment.

Consider the experiment consisting of 10 tosses of a coin. Determine whether or not it is a binomial experiment.

Solution   The experiment consisting of 10 tosses of a coin satisfies all four conditions of a binomial experiment as explained below.

1. There are a total of 10 trials (tosses), and they are all identical. All 10 tosses are performed under identical conditions. Here, $n = 10$.
2. Each trial (toss) has only two possible outcomes: a head and a tail. Let a head be called a success and a tail be called a failure.
3. The probability of obtaining a head (a success) is $1/2$ and that of a tail (a failure) is $1/2$ for any toss. That is,

$$p = P(H) = 1/2 \quad \text{and} \quad q = P(T) = 1/2$$

The sum of these two probabilities is 1.0. Also, these probabilities remain the same for each toss.
4. The trials (tosses) are independent. The result of any preceding toss has no bearing on the result of any succeeding toss.

Since this experiment satisfies all four conditions, it is a binomial experiment.

EXAMPLE 5–10   Students Using Instagram

Verifying the conditions of a binomial experiment.

(a) Seventy five percent of students at a college with a large student population use Instagram. A sample of five students from this college is selected, and these students are asked whether or not they use Instagram. Is this experiment a binomial experiment?

(b) In a group of 12 students at a college, 9 use Instagram. Five students are selected from this group of 12 and are asked whether or not they use Instagram. Is this experiment a binomial experiment?

Solution

(a) Below we check whether all four conditions of a binomial experiment are satisfied.

    1. This example consists of five identical trials. A trial represents the selection of a student.

2. Each trial has two outcomes: a student uses Instagram or a student does not use Instagram. Let us call the outcome a *success* if a student uses Instagram and a *failure* if he/she does not use Instagram.

3. Seventy five percent of students at this college use Instagram. Hence, the probability p that a student uses Instagram is .75, and the probability q that a student does not use Instagram is .25. These probabilities, p and q, remain constant for each selection of a student. Also, these two probabilities add to 1.0.

4. Each trial (student) is independent. In other words, if one student uses Instagram, it does not affect the outcome of another student using or not using Instagram. This is so because the size of the population is very large compared to the sample size.

   Since all four conditions are satisfied, this is an example of a binomial experiment.

(b) Below we check whether all four conditions of a binomial experiment are satisfied.

1. This example consists of five identical trials. A trial represents the selection of a student.

2. Each trial has two outcomes: a student uses Instagram or a student does not use Instagram. Let us call the outcome a *success* if a student uses Instagram and a *failure* if he/she does not use Instagram.

3. There are a total of 12 students, and 9 of them use Instagram. Let p be the probability that a student uses Instagram and q be the probability that a student does not use Instagram. These two probabilities, p and q, do not remain constant for each selection of a student because of the limited number (12) of students. The probability of each outcome changes with each selection depending on what happened in the previous selections.

4. Because p and q do not remain constant for each selection, the trials are not independent. The outcome of the first selection affects the outcome of the second selection, and so on.

   Since the third and fourth conditions of a binomial experiment are not satisfied, this is not an example of a binomial experiment.

5.4.2 The Binomial Probability Distribution and Binomial Formula

The random variable x that represents the number of successes in n trials for a binomial experiment is called a *binomial random variable*. The probability distribution of x in such experiments is called the **binomial probability distribution** or simply the *binomial distribution*. Thus, the binomial probability distribution is applied to find the probability of x successes in n trials for a binomial experiment. The number of successes x in such an experiment is a discrete random variable. Consider Example 5–10a. Let x be the number of college students in a sample of five who use Instagram. Because the number of students who use Instagram in a sample of five can be any number from zero to five, x can assume any of the values 0, 1, 2, 3, 4, and 5. Since the values of x are countable, x is a discrete random variable.

Binomial Formula   For a binomial experiment, the probability of exactly x successes in n trials is given by the binomial formula

$$P(x) = {}_nC_x\, p^x q^{n-x}$$

where

$$n = \text{total number of trials}$$
$$p = \text{probability of success}$$
$$q = 1 - p = \text{probability of failure}$$
$$x = \text{number of successes in } n \text{ trials}$$
$$n - x = \text{number of failures in } n \text{ trials}$$

In the binomial formula, n is the total number of trials and x is the total number of successes. The difference between the total number of trials and the total number of successes, $n - x$, gives the total number of failures in n trials. The value of $_nC_x$ gives the number of ways to obtain x successes in n trials. As mentioned earlier, p and q are the probabilities of success and failure, respectively. Again, although it does not matter which of the two outcomes is called a success and which a failure, usually the outcome to which the question refers is called a success.

To solve a binomial problem, we determine the values of n, x, $n - x$, p, and q and then substitute these values in the binomial formula. To find the value of $_nC_x$, we can use either the combinations formula from Section 4.6.3 or a calculator.

To find the probability of x successes in n trials for a binomial experiment, the only values needed are those of n and p. These are called the *parameters of the binomial probability distribution* or simply the **binomial parameters**. The value of q is obtained by subtracting the value of p from 1.0. Thus, $q = 1 - p$.

Next we solve a binomial problem by using the binomial formula.

EXAMPLE 5–11	**Students Using Instagram**

Calculating the probability of one success in three trials using the binomial formula.

Seventy five percent of students at a college with a large student population use the social media site Instagram. Three students are randomly selected from this college. What is the probability that exactly two of these three students use Instagram?

Solution   Here, there are three trials (students selected), and we are to find the probability of two successes (two students using Instagram) in these three trials. We can find the probability of two successes in three trials using the procedures learned in Chapter 4, but that will be a time-consuming and complex process. First, we will have to write all the possible outcomes that give two successes in three trials, and then we will have to find the probability that any one of those outcomes occurs using the union of events. But using the binomial probability distribution to find this probability is very easy, as described below. Here, we are given that:

$$n = \text{total number of trials} = \text{number of students selected} = 3$$

$$x = \text{number of successes} = \text{number of students in three who use Instagram} = 2$$

$$p = \text{probability of success} = \text{probability that a student uses Instagram} = .75$$

From this given information, we can calculate the values of $n - x$ and q as follows:

$$n - x = \text{number of failures} = \text{number of students not using Instagram} = 3 - 2 = 1$$

$$q = \text{probability of failure} = \text{probability that a student does not use Instagram} = 1 - .75 = .25$$

The probability of two successes is denoted by $P(x = 2)$ or simply by $P(2)$. By substituting all of the values in the binomial formula, we obtain

$$P(2) = {_3C_2} \, (.75)^2 \, (.25)^1 = (3)(.5625)(.25) = \mathbf{.4219}$$

Number of ways to obtain 2 success in 3 trials · Number of successes · Number of failures · Probability of success · Probability of failure

Note that the value of $_3C_2$ in the formula can either be obtained from a calculator or be computed as follows:

$$_3C_2 = \frac{3!}{2! \, (3 - 2)!} = \frac{3 \cdot 2 \cdot 1}{2 \cdot 1 \cdot 1} = 3$$

In the above computation, $_3C_2 = 3$ gives the number of ways to select two students who use Instagram in a random sample of three students. Suppose I denotes a student who uses Instagram and N denotes a student who does not use Instagram. The three ways to select two students who use Instagram are *IIN*, *INI*, and *NII*.

| EXAMPLE 5–12 | Home Delivery Service |

Calculating the probability using the binomial formula.

At the Express House Delivery Service, providing high-quality service to customers is the top priority of the management. The company guarantees a refund of all charges if a package it is delivering does not arrive at its destination by the specified time. It is known from past data that despite all efforts, 2% of the packages mailed through this company do not arrive at their destinations within the specified time. Suppose a corporation mails 10 packages through Express House Delivery Service on a certain day.

(a) Find the probability that exactly one of these 10 packages will not arrive at its destination within the specified time.

(b) Find the probability that at most one of these 10 packages will not arrive at its destination within the specified time.

Solution Let us call an event a success if a package does not arrive at its destination within the specified time and a failure if a package does arrive within the specified time. Note that because we are to find the probability that one package *will not arrive* within the specified time, we will call *not arriving* a *success*. Then,

$$n = \text{total number of packages mailed} = 10$$
$$p = P(\text{success}) = .02$$
$$q = P(\text{failure}) = 1 - .02 = .98$$

(a) For this part,

$$x = \text{number of successes} = 1$$
$$n - x = \text{number of failures} = 10 - 1 = 9$$

Substituting all values in the binomial formula, we obtain

$$P(1) = {}_{10}C_1 \, (.02)^1 \, (.98)^9 = \frac{10!}{1! \, (10 - 1)!} \, (.02)^1 \, (.98)^9$$
$$= (10) \, (.02) \, (.83374776) = \mathbf{.1667}$$

Thus, there is a .1667 probability that exactly one of the 10 packages mailed will not arrive at its destination within the specified time.

(b) The probability that at most one of the 10 packages will not arrive at its destination within the specified time is given by the sum of the probabilities of $x = 0$ and $x = 1$. Thus,

$$P(x \leq 1) = P(0) + P(1)$$
$$= {}_{10}C_0 \, (.02)^0 \, (.98)^{10} + {}_{10}C_1 \, (.02)^1 \, (.98)^9$$
$$= (1) \, (1) \, (.81707281) + (10) \, (.02) \, (.83374776)$$
$$= .8171 + .1667 = \mathbf{.9838}$$

Thus, the probability that at most one of the 10 packages will not arrive at its destination within the specified time is .9838.

| EXAMPLE 5–13 | Employees Changing Their Jobs |

Constructing a binomial probability distribution and its graph.

According to a survey, 33% of American employees do not plan to change their jobs in the near future. Let x denote the number of employees in a random sample of three American employees who do not plan to change their jobs in the near future. Write the probability distribution of x and draw a histogram for this probability distribution.

Solution Let x be the number of employees who do not plan to change their jobs in the near future in a random sample of three American employees. Then, $n - x$ is the number of American employees who do plan to change their jobs in the near future. (We have included the employees who are uncertain or have no opinion in this later group of $n - x$ employees.) From the given information,

$n =$ total employees in the sample $= 3$

$p = P$(an employee does not plan to change his/her job in the near future) $= .33$

$q = P$(an employee does plan to change his/her job in the near future) $= 1 - .33 = .67$

The possible values that x can assume are 0, 1, 2, and 3. In other words, the number of employees in this sample of three who do not plan to change their jobs in the near future can be 0, 1, 2, or 3. The probability of each of these four outcomes is calculated as follows.

If $x = 0$, then $n - x = 3$. Using the binomial formula, we obtain the probability of $x = 0$ as follows:

$$P(0) = {}_3C_0 (.33)^0 (.67)^3 = (1)(1)(.300763) = .3008$$

Note that ${}_3C_0$ is equal to 1 by definition and $(.33)^0$ is equal to 1 because any (nonzero) number raised to the power zero is always 1.

If $x = 1$, then $n - x = 2$. Using the binomial formula, the probability of $x = 1$ is

$$P(1) = {}_3C_1 (.33)^1 (.67)^2 = (3)(.33)(.4489) = .4444$$

Similarly, if $x = 2$, then $n - x = 1$, and if $x = 3$, then $n - x = 0$. The probabilities of $x = 2$ and $x = 3$ are, respectively,

$$P(2) = {}_3C_2 (.33)^2 (.67)^1 = (3)(.1089)(.67) = .2189$$

$$P(3) = {}_3C_3 (.33)^3 (.67)^0 = (1)(.035937)(1) = .0359$$

These probabilities are written in Table 5.9. Figure 5.3 shows the histogram for the probability distribution of Table 5.9.

Table 5.9 Probability Distribution of x

x	$P(x)$
0	.3008
1	.4444
2	.2189
3	.0359

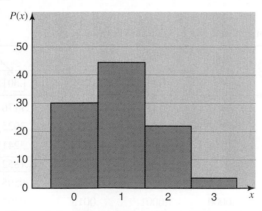

Figure 5.3   Histogram for the probability distribution of Table 5.9.

5.4.3   Using the Table of Binomial Probabilities

The probabilities for a binomial experiment can also be read from Table I, the table of binomial probabilities, in Appendix B. This table lists the probabilities of x for $n = 1$ to $n = 25$ and for selected values of p. Example 5–14 illustrates how to read Table I.

EXAMPLE 5–14 | Time Spent on Facebook by College Students

Using the binomial table to find probabilities and to construct the probability distribution and graph.

According to a survey, 30% of college students said that they spend too much time on Facebook. (The remaining 70% said that they do not spend too much time on Facebook or had no opinion.) Suppose this result holds true for the current population of all college students. A random sample of six college students is selected. Using Table I of Appendix B, answer the following.

(a) Find the probability that exactly three of these six college students will say that they spend too much time on Facebook.

(b) Find the probability that at most two of these six college students will say that they spend too much time on Facebook.

(c) Find the probability that at least three of these six college students will say that they spend too much time on Facebook.

(d) Find the probability that one to three of these six college students will say that they spend too much time on Facebook.

(e) Let x denote the number in a random sample of six college students who will say that they spend too much time on Facebook. Write the probability distribution of x and draw a histogram for this probability distribution.

Solution

(a) To read the required probability from Table I of Appendix B, we first determine the values of n, x, and p. For this example,

n = number of students in the sample = 6

x = number of students in this sample who spend too much time on Facebook = 3

p = P(a student spends too much time on Facebook) = .30

Then we locate $n = 6$ in the column labeled n in Table I of Appendix B. The relevant portion of Table I with $n = 6$ is reproduced as Table 5.10 here. Next, we locate 3 in the column for x in the portion of the table for $n = 6$ and locate $p = .30$ in the row for p at the top of the table. The entry at the intersection of the row for $x = 3$ and the column for $p = .30$ gives the probability of three successes in six trials when the probability of success is .30. From Table I of Appendix B or Table 5.10,

$$P(3) = \mathbf{.1852}$$

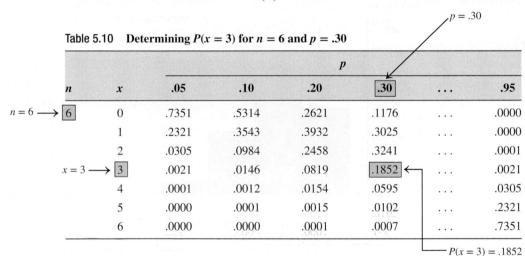

Table 5.10   Determining $P(x = 3)$ for $n = 6$ and $p = .30$

					p		
n	x	.05	.10	.20	.30	...	.95
$n = 6$ → 6	0	.7351	.5314	.2621	.1176	...	.0000
	1	.2321	.3543	.3932	.3025	...	.0000
	2	.0305	.0984	.2458	.3241	...	.0001
$x = 3$ → 3		.0021	.0146	.0819	.1852	...	.0021
	4	.0001	.0012	.0154	.0595	...	.0305
	5	.0000	.0001	.0015	.0102	...	.2321
	6	.0000	.0000	.0001	.0007	...	.7351

$P(x = 3) = .1852$

Table 5.11   Portion of Table I for $n = 6$ and $p = .30$

n	x	p .30
6	0	.1176
	1	.3025
	2	.3241
	3	.1852
	4	.0595
	5	.0102
	6	.0007

Using Table I of Appendix B or Table 5.10, we write Table 5.11, which can be used to answer the remaining parts of this example.

(b) The event that at most two students will say that they spend too much time on Facebook will occur if x is equal to 0, 1, or 2. From Table I of Appendix B or from Table 5.11, the required probability is

$$P(\text{at most 2}) = P(0 \text{ or } 1 \text{ or } 2) = P(0) + P(1) + P(2)$$
$$= .1176 + .3025 + .3241 = \mathbf{.7442}$$

(c) The probability that at least three students will say that they spend too much time on Facebook is given by the sum of the probabilities of 3, 4, 5, or 6. Using Table I of Appendix B or from Table 5.11, we obtain

$$P(\text{at least 3}) = P(3 \text{ or } 4 \text{ or } 5 \text{ or } 6)$$
$$= P(3) + P(4) + P(5) + P(6)$$
$$= .1852 + .0595 + .0102 + .0007 = \mathbf{.2556}$$

(d) The probability that one to three students will say that they spend too much time on Facebook is given by the sum of the probabilities of $x = 1$, 2, and 3. Using Table I of Appendix B or from Table 5.11, we obtain

$$P(1 \text{ to } 3) = P(1) + P(2) + P(3)$$
$$= .3025 + .3241 + .1852 = \mathbf{.8118}$$

(e) Using Table I of Appendix B or from Table 5.11, we list the probability distribution of x for $n = 6$ and $p = .30$ in Table 5.12. Figure 5.4 shows the histogram of the probability distribution of x.

Table 5.12 **Probability Distribution of x for $n = 6$ and $p = .30$**

x	$P(x)$
0	.1176
1	.3025
2	.3241
3	.1852
4	.0595
5	.0102
6	.0007

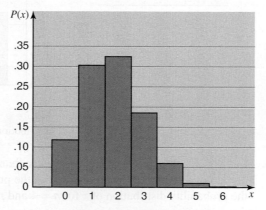

Figure 5.4 Histogram for the probability distribution of Table 5.12.

5.4.4 Probability of Success and the Shape of the Binomial Distribution

For any number of trials n:

1. The binomial probability distribution is symmetric if $p = .50$.
2. The binomial probability distribution is skewed to the right if p is less than .50.
3. The binomial probability distribution is skewed to the left if p is greater than .50.

These three cases are illustrated next with examples and graphs.

1. Let $n = 4$ and $p = .50$. Using Table I of Appendix B, we have written the probability distribution of x in Table 5.13 and plotted the distribution in Figure 5.5. As we can observe from Table 5.13 and Figure 5.5, the probability distribution of x is symmetric.

Table 5.13 **Probability Distribution of x for $n = 4$ and $p = .50$**

x	$P(x)$
0	.0625
1	.2500
2	.3750
3	.2500
4	.0625

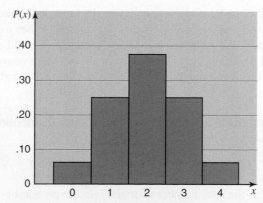

Figure 5.5 Histogram for the probability distribution of Table 5.13.

2. Let $n = 4$ and $p = .30$ (which is less than .50). Table 5.14, which is written by using Table I of Appendix B, and the graph of the probability distribution in Figure 5.6 show that the probability distribution of x for $n = 4$ and $p = .30$ is skewed to the right.

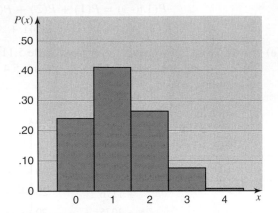

Table 5.14	Probability Distribution of x for $n = 4$ and $p = .30$
x	$P(x)$
0	.2401
1	.4116
2	.2646
3	.0756
4	.0081

Figure 5.6   Histogram for the probability distribution of Table 5.14.

3. Let $n = 4$ and $p = .80$ (which is greater than .50). Table 5.15, which is written by using Table I of Appendix B, and the graph of the probability distribution in Figure 5.7 show that the probability distribution of x for $n = 4$ and $p = .80$ is skewed to the left.

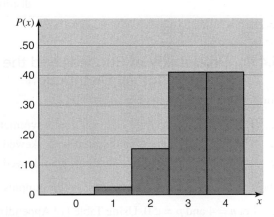

Table 5.15	Probability Distribution of x for $n = 4$ and $p = .80$
x	$P(x)$
0	.0016
1	.0256
2	.1536
3	.4096
4	.4096

Figure 5.7   Histogram for the probability distribution of Table 5.15.

5.4.5   Mean and Standard Deviation of the Binomial Distribution

Section 5.3 explained how to compute the mean and standard deviation, respectively, for a probability distribution of a discrete random variable. When a discrete random variable has a binomial distribution, the formulas learned in Section 5.3 could still be used to compute its mean and standard deviation. However, it is simpler and more convenient to use the following formulas to find the mean and standard deviation of a binomial distribution.

Mean and Standard Deviation of a Binomial Distribution   The *mean* and *standard deviation of a binomial distribution* are, respectively,

$$\mu = np \quad \text{and} \quad \sigma = \sqrt{npq}$$

where n is the total number of trials, p is the probability of success, and q is the probability of failure.

Example 5–15 describes the calculation of the mean and standard deviation of a binomial distribution.

EXAMPLE 5–15   U.S. Adults with No Religious Affiliation

Calculating the mean and standard deviation of a binomial distribution.

According to a Pew Research Center survey released on May 12, 2015, 22.8% of U.S. adults do not have a religious affiliation (*Time*, May 25, 2015). Assume that this result is true for the current population of U.S. adults. A sample of 50 U.S. adults is randomly selected. Let x be the number of adults in this sample who do not have a religious affiliation. Find the mean and standard deviation of the probability distribution of x.

Solution   This is a binomial experiment with a total of 50 trials (U.S. adults). Each trial has two outcomes: (1) the selected adult does not have a religious affiliation, (2) the selected adult has a religious affiliation or does not want to answer the question. The probabilities p and q for these two outcomes are .228 and .772, respectively. Thus,

$$n = 50, \quad p = .228 \quad \text{and} \quad q = .772$$

Using the formulas for the mean and standard deviation of the binomial distribution, we obtain

$$\mu = np = 50(.228) = \mathbf{11.4}$$

$$\sigma = \sqrt{npq} = \sqrt{(50)(.228)(.772)} = \mathbf{2.9666}$$

Thus, the mean of the probability distribution of x is 11.4, and the standard deviation is 2.9666. The value of the mean is what we expect to obtain, on average, per repetition of the experiment. In this example, if we select many samples of 50 U.S. adults each, we expect that each sample will contain an average of 11.4 adults, with a standard deviation of 2.9666, who will not have a religious affiliation.

EXERCISES

CONCEPTS AND PROCEDURES

5.26   Briefly explain the following:
  a. A binomial experiment
  b. A trial
  c. A binomial random variable.

5.27   Briefly explain about the parameters of the binomial probability distribution.

5.28   Which of the following are binomial experiments? Explain why.
  a. Rolling a die many times and observing the number of spots
  b. Rolling a die many times and observing whether the number obtained is even or odd
  c. Selecting a few voters from a very large population of voters and observing whether or not each of them favors a certain proposition in an election when 54% of all voters are known to be in favor of this proposition.

5.29   Which of the following are binomial experiments? Explain why.
  a. Drawing 3 balls with replacement from a box that contains 12 balls, 8 of which are red and 4 are blue, and observing the colors of the drawn balls
  b. Drawing 3 balls without replacement from a box that contains 10 balls, 6 of which are red and 4 are blue, and observing the colors of the drawn balls
  c. Selecting a few households from New York City and observing whether or not they own stocks when it is known that 32% of all households in New York City own stocks

5.30   Let x be a discrete random variable that possesses a binomial distribution. Using the binomial formula, find the following probabilities.
  a. $P(x = 0)$ for $n = 4$ and $p = .01$
  b. $P(x = 2)$ for $n = 6$ and $p = .30$
  c. $P(x = 7)$ for $n = 8$ and $p = .60$

Verify your answers by using Table I of Appendix B.

5.31   Let x be a discrete random variable that possesses a binomial distribution.
  a. Using Table I of Appendix B, write the probability distribution of x for $n = 7$ and $p = .30$ and graph it.
  b. What are the mean and standard deviation of the probability distribution developed in part a?

5.32   The binomial probability distribution is symmetric for $p = .50$, skewed to the right for $p < .50$, and skewed to the left for $p > .50$. Illustrate each of these three cases by writing a probability distribution table and drawing a graph. Choose any values of n (equal to 4 or higher) and p and use the table of binomial probabilities (Table I of Appendix B) to write the probability distribution tables.

APPLICATIONS

5.33   The most recent data from the Department of Education show that 34.8% of students who submitted otherwise valid applications for a Title IV Pell Grant in 2005–2006 were ineligible to receive such a grant (www2.ed.gov/finaid/prof/resources/data/pell-2005-06/eoy-05-06.pdf). Suppose that this result is true for the current population of students who submitted otherwise valid applications for this grant.

a. Let *x* be a binomial random variable that denotes the number of students in a random sample of 20 who submitted otherwise valid applications for a Title IV Pell Grant but were ineligible to receive one. What are the possible values that *x* can assume?

b. Find the probability that exactly 6 students are ineligible to receive a Title IV Pell Grant in a random sample of 20 who submitted otherwise valid applications for this grant. Use the binomial probability distribution formula.

5.34 According to a 2011 poll, 55% of Americans do not know that GOP stands for Grand Old Party (*Time*, October 17, 2011). Suppose that this result is true for the current population of Americans.

a. Let *x* be a binomial random variable that denotes the number of people in a random sample of 17 Americans who do not know that GOP stands for Grand Old Party. What are the possible values that *x* can assume?

b. Find the probability that exactly 8 people in a random sample of 17 Americans do not know that GOP stands for Grand Old Party. Use the binomial probability distribution formula.

5.35 In a poll, men and women were asked, "When someone yelled or snapped at you at work, how did you want to respond?" Twenty percent of the women in the survey said that they felt like crying (*Time*, April 4, 2011). Suppose that this result is true for the current population of women employees. A random sample of 24 women employees is selected. Use the binomial probabilities table (Table I of Appendix B) or technology to find the probability that the number of women employees in this sample of 24 who will hold the above opinion in response to the said question is

    a. at least 5   **b.** 1 to 3   **c.** at most 6

5.36 According to a Wakefield Research survey of adult women, 50% of the women said that they had tried five or more diets in their lifetime (*USA TODAY*, June 21, 2011). Suppose that this result is true for the current population of adult women. A random sample of 13 adult women is selected. Use the binomial probabilities table (Table I of Appendix B) or technology to find the probability that the number of women in this sample of 13 who had tried five or more diets in their lifetime is

    a. at most 7   **b.** 5 to 8   **c.** at least 7

5.37 Magnetic resonance imaging (MRI) is a process that produces internal body images using a strong magnetic field. Some

patients become claustrophobic and require sedation because they are required to lie within a small, enclosed space during the MRI test. Suppose that 25% of all patients undergoing MRI testing require sedation due to claustrophobia. If five patients are selected at random, find the probability that the number of patients in these five who require sedation is

    a. exactly 2   **b.** all 5   **c.** exactly 4

5.38 A professional basketball player makes 85% of the free throws he tries. Assuming this percentage will hold true for future attempts, find the probability that in the next six tries, the number of free throws he will make is

    a. exactly 6   **b.** exactly 4

5.39 A fast food chain store conducted a taste survey before marketing a new hamburger. The results of the survey showed that 70% of the people who tried this hamburger liked it. Encouraged by this result, the company decided to market the new hamburger. Assume that 70% of all people like this hamburger. On a certain day, eight customers bought it for the first time.

a. Let *x* denote the number of customers in this sample of eight who will like this hamburger. Using the binomial probabilities table, obtain the probability distribution of *x* and draw a graph of the probability distribution. Determine the mean and standard deviation of *x*.

b. Using the probability distribution of part a, find the probability that exactly three of the eight customers will like this hamburger.

5.40 Johnson Electronics makes calculators. Consumer satisfaction is one of the top priorities of the company's management. The company guarantees a refund or a replacement for any calculator that malfunctions within 2 years from the date of purchase. It is known from past data that despite all efforts, 5% of the calculators manufactured by the company malfunction within a 2-year period. The company mailed a package of 20 randomly selected calculators to a store.

a. Let *x* denote the number of calculators in this package of 20 that will be returned for refund or replacement within a 2-year period. Using the binomial probabilities table, obtain the probability distribution of *x* and draw a graph of the probability distribution. Determine the mean and standard deviation of *x*.

b. Using the probability distribution of part a, find the probability that exactly 4 of the 20 calculators will be returned for refund or replacement within a 2-year period.

5.5 | The Hypergeometric Probability Distribution

In Section 5.4, we learned that one of the conditions required to apply the binomial probability distribution is that the trials are independent, so that the probabilities of the two outcomes or events (success and failure) remain constant. If the trials are not independent, we cannot apply the binomial probability distribution to find the probability of *x* successes in *n* trials. In such cases we replace the binomial probability distribution by the **hypergeometric probability distribution**. Such a case occurs when a sample is drawn without replacement from a finite population.

As an example, suppose 20% of all auto parts manufactured at a company are defective. Four auto parts are selected at random. What is the probability that three of these four parts are good? Note that we are to find the probability that three of the four auto parts are good and one is defective. In this case, the population is very large and the probability of the first, second, third, and fourth auto parts being defective remains the same at .20. Similarly, the probability of any of the parts being good remains unchanged at .80. Consequently, we will apply the binomial probability distribution to find the probability of three good parts in four.

Now suppose this company shipped 25 auto parts to a dealer. Later, it finds out that 5 of those parts were defective. By the time the company manager contacts the dealer, 4 auto parts from that shipment have already been sold. What is the probability that 3 of those 4 parts were good parts and 1 was defective? Here, because the 4 parts were selected without replacement from a small population, the probability of a part being good changes from the first selection to the second selection, to the third selection, and to the fourth selection. In this case we cannot apply the binomial probability distribution. In such instances, we use the hypergeometric probability distribution to find the required probability.

Hypergeometric Probability Distribution

Let

$$N = \text{total number of elements in the population}$$
$$r = \text{number of successes in the population}$$
$$N - r = \text{number of failures in the population}$$
$$n = \text{number of trials (sample size)}$$
$$x = \text{number of successes in } n \text{ trials}$$
$$n - x = \text{number of failures in } n \text{ trials}$$

The probability of x successes in n trials is given by

$$P(x) = \frac{_rC_x \times _{N-r}C_{n-x}}{_NC_n}$$

Examples 5–16 and 5–17 provide applications of the hypergeometric probability distribution.

EXAMPLE 5–16   Defective Auto Parts in a Shipment

Calculating the probability using the hypergeometric distribution formula.

Brown Manufacturing makes auto parts that are sold to auto dealers. Last week the company shipped 25 auto parts to a dealer. Later, it found out that 5 of those parts were defective. By the time the company manager contacted the dealer, 4 auto parts from that shipment had already been sold. What is the probability that 3 of those 4 parts were good parts and 1 was defective?

Solution   Let a good part be called a success and a defective part be called a failure. From the given information,

$$N = \text{total number of elements (auto parts) in the population} = 25$$
$$r = \text{number of successes (good parts) in the population} = 20$$
$$N - r = \text{number of failures (defective parts) in the population} = 5$$
$$n = \text{number of trials (sample size)} = 4$$
$$x = \text{number of successes in four trials} = 3$$
$$n - x = \text{number of failures in four trials} = 1$$

Using the hypergeometric formula, we calculate the required probability as follows:

$$P(3) = \frac{_rC_x \times _{N-r}C_{n-x}}{_NC_n} = \frac{_{20}C_3 \times _5C_1}{_{25}C_4} = \frac{\dfrac{20!}{3!\,(20-3)!} \cdot \dfrac{5!}{1!\,(5-1)!}}{\dfrac{25!}{4!\,(25-4)!}}$$

$$= \frac{(1140)\,(5)}{12,650} = .4506$$

Thus, the probability that 3 of the 4 parts sold are good and 1 is defective is .4506.

In the above calculations, the values of combinations can either be calculated using the formula learned in Section 4.6.3 (as done here) or by using a calculator.

EXAMPLE 5–17	Selecting Employees from a Group

Calculating the probability using the hypergeometric distribution formula.

Dawn Corporation has 12 employees who hold managerial positions. Of them, 7 are females and 5 are males. The company is planning to send 3 of these 12 managers to a conference. If 3 managers are randomly selected out of 12,

    (a) find the probability that all 3 of them are females

    (b) find the probability that at most 1 of them is a female

Solution   Let the selection of a female be called a success and the selection of a male be called a failure.

    (a) From the given information,

$$N = \text{total number of managers in the population} = 12$$
$$r = \text{number of successes (females) in the population} = 7$$
$$N - r = \text{number of failures (males) in the population} = 5$$
$$n = \text{number of selections (sample size)} = 3$$
$$x = \text{number of successes (females) in three selections} = 3$$
$$n - x = \text{number of failures (males) in three selections} = 0$$

Using the hypergeometric formula, we calculate the required probability as follows:

$$P(3) = \frac{{}_rC_x \times {}_{N-r}C_{n-x}}{{}_NC_n} = \frac{{}_7C_3 \times {}_5C_0}{{}_{12}C_3} = \frac{(35)\,(1)}{220} = .1591$$

Thus, the probability that all 3 of the managers selected are females is .1591.

    (b) The probability that at most 1 of them is a female is given by the sum of the probabilities that either none or 1 of the selected managers is a female.

        To find the probability that none of the selected managers is a female, we use

$$N = \text{total number of managers in the population} = 12$$
$$r = \text{number of successes (females) in the population} = 7$$
$$N - r = \text{number of failures (males) in the population} = 5$$
$$n = \text{number of selections (sample size)} = 3$$
$$x = \text{number of successes (females) in three selections} = 0$$
$$n - x = \text{number of failures (males) in three selections} = 3$$

Using the hypergeometric formula, we calculate the required probability as follows:

$$P(0) = \frac{{}_rC_x \times {}_{N-r}C_{n-x}}{{}_NC_n} = \frac{{}_7C_0 \times {}_5C_3}{{}_{12}C_3} = \frac{(1)\,(10)}{220} = .0455$$

To find the probability that 1 of the selected managers is a female, we use

$$N = \text{total number of managers in the population} = 12$$
$$r = \text{number of successes (females) in the population} = 7$$
$$N - r = \text{number of failures (males) in the population} = 5$$
$$n = \text{number of selections (sample size)} = 3$$
$$x = \text{number of successes (females) in three selections} = 1$$
$$n - x = \text{number of failures (males) in three selections} = 2$$

Using the hypergeometric formula, we obtain the required probability as follows:

$$P(1) = \frac{{}_rC_x \times {}_{N-r}C_{n-x}}{{}_NC_n} = \frac{{}_7C_1 \times {}_5C_2}{{}_{12}C_3} = \frac{(7)\,(10)}{220} = .3182$$

The probability that at most 1 of the 3 managers selected is a female is

$$P(x \leq 1) = P(0) + P(1) = .0455 + .3182 = \mathbf{.3637}$$

EXERCISES

CONCEPTS AND PROCEDURES

5.41   Explain the hypergeometric probability distribution. Under what conditions is this probability distribution applied to find the probability of a discrete random variable x? Give one example of an application of the hypergeometric probability distribution.

5.42   Let $N = 12$, $r = 6$, and $n = 5$. Using the hypergeometric probability distribution formula, find
    a. $P(x = 4)$ **b.** $P(x = 5)$ **c.** $P(x \leq 1)$

5.43   Let $N = 16$, $r = 8$, and $n = 4$. Using the hypergeometric probability distribution formula, find
    a. $P(x = 4)$ **b.** $P(x = 0)$ **c.** $P(x \leq 1)$

APPLICATIONS

5.44   An average of 1.4 private airplanes arrive per hour at an airport.
    a. Find the probability that during a given hour no private airplane will arrive at this airport.

b. Let x denote the number of private airplanes that will arrive at this airport during a given hour. Write the probability distribution of x. Use the appropriate probabilities table from Appendix B.

5.45   In a list of 15 households, 9 own homes and 6 do not own homes. Four households are randomly selected from these 15 households. Find the probability that the number of households in these 4 who own homes is
    a. exactly 3   **b.** at most 1   **c.** exactly 4

5.46   A high school boys' basketball team averages 1.2 technical fouls per game.
    a. Using the appropriate formula, find the probability that in a given basketball game this team will commit exactly 3 technical fouls.
    b. Let x denote the number of technical fouls that this team will commit during a given basketball game. Using the appropriate probabilities table from Appendix B, write the probability distribution of x.

5.6 | The Poisson Probability Distribution

The **Poisson probability distribution**, named after the French mathematician Siméon-Denis Poisson, is another important probability distribution of a discrete random variable that has a large number of applications. Suppose a washing machine in a laundromat breaks down an average of three times a month. We may want to find the probability of exactly two breakdowns during the next month. This is an example of a Poisson probability distribution problem. Each breakdown is called an **occurrence** in Poisson probability distribution terminology. The Poisson probability distribution is applied to experiments with random and independent occurrences. The occurrences are random in the sense that they do not follow any pattern, and, hence, they are unpredictable. Independence of occurrences means that one occurrence (or nonoccurrence) of an event does not influence the successive occurrences or nonoccurrences of that event. The occurrences are always considered with respect to an interval. In the example of the washing machine, the interval is one month. The interval may be a time interval, a space interval, or a volume interval. The actual number of occurrences within an interval is random and independent. If the average number of occurrences for a given interval is known, then by using the Poisson probability distribution, we can compute the probability of a certain number of occurrences, x, in that interval. Note that the number of actual occurrences in an interval is denoted by x.

Conditions to Apply the Poisson Probability Distribution   The following three conditions must be satisfied to apply the Poisson probability distribution.

1. x is a discrete random variable.
2. The occurrences are random.
3. The occurrences are independent.

The following are three examples of discrete random variables for which the occurrences are random and independent. Hence, these are examples to which the Poisson probability distribution can be applied.

1. Consider the number of telemarketing phone calls received by a household during a given day. In this example, the receiving of a telemarketing phone call by a household is called an occurrence, the interval is one day (an interval of time), and the occurrences are random (that is, there is no specified time for such a phone call to come in) and discrete. The total number of telemarketing phone calls received by a household during a given day may be 0, 1, 2, 3, 4, and so forth. The independence of occurrences in this example means that the telemarketing phone calls are received individually and none of two (or more) of these phone calls are related.

2. Consider the number of defective items in the next 100 items manufactured on a machine. In this case, the interval is a volume interval (100 items). The occurrences (number of defective items) are random and discrete because there may be 0, 1, 2, 3, . . . , 100 defective items in 100 items. We can assume the occurrence of defective items to be independent of one another.

3. Consider the number of defects in a 5-foot-long iron rod. The interval, in this example, is a space interval (5 feet). The occurrences (defects) are random because there may be any number of defects in a 5-foot iron rod. We can assume that these defects are independent of one another.

The following examples also qualify for the application of the Poisson probability distribution.

1. The number of accidents that occur on a given highway during a 1-week period
2. The number of customers entering a grocery store during a 1-hour interval
3. The number of television sets sold at a department store during a given week

In contrast, consider the arrival of patients at a physician's office. These arrivals are nonrandom if the patients have to make appointments to see the doctor. The arrival of commercial airplanes at an airport is nonrandom because all planes are scheduled to arrive at certain times, and airport authorities know the exact number of arrivals for any period (although this number may change slightly because of late or early arrivals and cancellations). The Poisson probability distribution cannot be applied to these examples.

In the Poisson probability distribution terminology, the average number of occurrences in an interval is denoted by λ (Greek letter *lambda*). The actual number of occurrences in that interval is denoted by x. Then, using the Poisson probability distribution, we find the probability of x occurrences during an interval given that the mean number of occurrences during that interval is λ.

Poisson Probability Distribution Formula According to the *Poisson probability distribution*, the probability of x occurrences in an interval is

$$P(x) = \frac{\lambda^x e^{-\lambda}}{x!}$$

where λ (pronounced *lambda*) is the mean number of occurrences in that interval and the value of e is approximately 2.71828.

The mean number of occurrences in an interval, denoted by λ, is called the *parameter of the Poisson probability distribution* or the **Poisson parameter**. As is obvious from the Poisson probability distribution formula, we need to know only the value of λ to compute the probability of any given value of x. We can read the value of $e^{-\lambda}$ for a given λ from Table II of Appendix B. Examples 5–18 through 5–20 illustrate the use of the Poisson probability distribution formula.

EXAMPLE 5–18 | Telemarketing Phone Calls Received

Using the Poisson formula: x equals a specific value.

On average, a household receives 9.5 telemarketing phone calls per week. Using the Poisson probability distribution formula, find the probability that a randomly selected household receives exactly 6 telemarketing phone calls during a given week.

Solution   Let λ be the mean number of telemarketing phone calls received by a household per week. Then, $\lambda = 9.5$. Let x be the number of telemarketing phone calls received by a household during a given week. We are to find the probability of $x = 6$. Substituting all of the values in the Poisson formula, we obtain

$$P(6) = \frac{\lambda^x\, e^{-\lambda}}{x!} = \frac{(9.5)^6\, e^{-9.5}}{6!} = \frac{(735,091.8906)\,(.00007485)}{720} = .0764$$

To do these calculations, we can find the value of 6! either by using the factorial key on a calculator or by multiplying all integers from 1 to 6, and we can find the value of $e^{-9.5}$ by using the e^x key on a calculator or from Table II in Appendix B.

EXAMPLE 5–19	Washing Machine Breakdowns

Calculating probabilities using the Poisson formula.

A washing machine in a laundromat breaks down an average of three times per month. Using the Poisson probability distribution formula, find the probability that during the next month this machine will have

(a) exactly two breakdowns

(b) at most one breakdown

Solution   Let λ be the mean number of breakdowns per month, and let x be the actual number of breakdowns observed during the next month for this machine. Then,

$$\lambda = 3$$

(a) The probability that exactly two breakdowns will be observed during the next month is

$$P(2) = \frac{\lambda^x\, e^{-\lambda}}{x!} = \frac{(3)^2\, e^{-3}}{2!} = \frac{(9)\,(.04978707)}{2} = .2240$$

(b) The probability that at most one breakdown will be observed during the next month is given by the sum of the probabilities of zero and one breakdown. Thus,

$$P(\text{at most 1 breakdown}) = P(0 \text{ or } 1 \text{ breakdown}) = P(0) + P(1)$$

$$= \frac{(3)^0\, e^{-3}}{0!} + \frac{(3)^1\, e^{-3}}{1!}$$

$$= \frac{(1)\,(.04978707)}{1} + \frac{(3)\,(.04978707)}{1}$$

$$= .0498 + .1494 = .1992$$

One important point about the Poisson probability distribution is that *the intervals for λ and x must be equal*. If they are not, the mean λ should be redefined to make them equal. Example 5–20 illustrates this point.

◀ *Remember*

EXAMPLE 5–20	Returning the Purchased Items

Calculating probability using the Poisson formula.

Cynthia's Mail Order Company provides free examination of its products for 7 days. If not completely satisfied, a customer can return the product within that period and get a full refund. According to past records of the company, an average of 2 of every 10 products sold by this company are returned for a refund. Using the Poisson probability distribution formula, find the probability that exactly 6 of the 40 products sold by this company on a given day will be returned for a refund.

Solution   Let x denote the number of products in 40 that will be returned for a refund. We are to find $P(x = 6)$, which is usually written as $P(6)$. The given mean is defined per 10 products, but

x is defined for 40 products. As a result, we should first find the mean for 40 products. Because, on average, 2 out of 10 products are returned, the mean number of products returned out of 40 will be 8. Thus, $\lambda = 8$. Substituting $x = 6$ and $\lambda = 8$ in the Poisson probability distribution formula, we obtain

$$P(6) = \frac{\lambda^x e^{-\lambda}}{x!} = \frac{(8)^6 e^{-8}}{6!} = \frac{(262{,}144)\,(.00033546)}{720} = \textbf{.1221}$$

Thus, the probability is .1221 that exactly 6 products out of 40 sold on a given day will be returned.

Note that Example 5–20 is actually a binomial problem with $p = 2/10 = .20$, $n = 40$, and $x = 6$. In other words, the probability of success (that is, the probability that a product is returned) is .20 and the number of trials (products sold) is 40. We are to find the probability of six successes (returns). However, we used the Poisson distribution to solve this problem. This is referred to as *using the Poisson distribution as an approximation to the binomial distribution*. We can also use the binomial distribution to find this probability as follows:

$$P(6) = {}_{40}C_6\,(.20)^6\,(.80)^{34} = \frac{40!}{6!(40-6)!}\,(.20)^6\,(.80)^{34}$$

$$= (3{,}838{,}380)\,(.000064)\,(.00050706) = \textbf{.1246}$$

Thus the probability $P(6)$ is .1246 when we use the binomial distribution.

As we can observe, simplifying the above calculations for the binomial formula is a little complicated when n is large. It is much easier to solve this problem using the Poisson probability distribution. As a general rule, if it is a binomial problem with $n > 25$ but $\mu \leq 25$, then we can use the Poisson probability distribution as an approximation to the binomial distribution. However, if $n > 25$ and $\mu > 25$, we prefer to use the normal distribution as an approximation to the binomial distribution. The latter case will be discussed in Chapter 6. However, if you are using technology, it does not matter how large n is. You can always use the binomial probability distribution if it is a binomial problem.

Case Study 5–2 presents an application of the Poisson probability distribution.

5.6.1 Using the Table of Poisson Probabilities

The probabilities for a Poisson distribution can also be read from Table III in Appendix B, the table of Poisson probabilities. The following example describes how to read that table.

EXAMPLE 5–21	**New Bank Accounts Opened per Day**

Using the table of Poisson probabilities.

On average, two new accounts are opened per day at an Imperial Savings Bank branch. Using Table III of Appendix B, find the probability that on a given day the number of new accounts opened at this bank will be

(a) exactly 6   (b) at most 3   (c) at least 7

Solution   Let

λ = mean number of new accounts opened per day at this bank

x = number of new accounts opened at this bank on a given day

(a) The values of λ and x are

$$\lambda = 2 \quad \text{and} \quad x = 6$$

In Table III of Appendix B, we first locate the column that corresponds to $\lambda = 2$. In this column, we then read the value for $x = 6$. The relevant portion of that table is shown here

Global Birth and Death Rates

GLOBAL BIRTH AND DEATH RATES

BIRTH RATE
4.3 per second

Congratulations, It's a boy!

Congratulations, It's a girl!

DEATH RATE
1.8 per second

Data source: The International Data Base and U.S. Census Bureau

The accompanying graph shows the average global birth and death rates. According to this information, on average, 4.3 children are born per second and 1.8 persons die per second in the world. If we assume that the global birth and death rates follow the Poisson probability distribution, we can find the probability of any given number of global births or deaths for a given time interval. For example, if x is the actual number of global births during a given 1-second interval, then x can assume any (nonnegative integer) value, such as 0, 1, 2, 3, ... The same is true for the number of global deaths during a given 1-second interval. For example, if y is the actual number of global deaths during a given 1-second interval, then y can assume any (nonnegative integer) value, such as 0, 1, 2, 3, ... Here x and y are both discrete random variables.

Using the Poisson formula or Table III of Appendix B, we can find the probability of any values of x and y. For example, if we want to find the probability of at most three global births during any given 1-second interval, then, using $\lambda = 4.3$, we find this probability from Table III as

$$P(x \leq 3) = P(0) + P(1) + P(2) + P(3) = .0136 + .0583 + .1254 + .1798 = .3771$$

Now suppose we want to find the probability of exactly six global births using the Poisson formula. This probability is

$$P(6) = \frac{\lambda^x e^{-\lambda}}{x!} = \frac{4.3^6 e^{-4.3}}{6!} = .1191$$

As mentioned earlier, let y be the number of global deaths in a given 1-second interval. If we want to find the probability of at most two global deaths during any given 1-second interval, then, using $\lambda = 1.8$, we find this probability from Table III as

$$P(y \leq 2) = P(0) + P(1) + P(2) = .1653 + .2975 + .2678 = .7306$$

Now suppose we want to find the probability of exactly three global deaths in a given 1-second interval using the Poisson formula. This probability is

$$P(3) = \frac{\lambda^y e^{-\lambda}}{y!} = \frac{1.8^3 e^{-1.8}}{3!} = .1607$$

Using Table III of Appendix B, we can prepare the probability distributions of x and y.

as Table 5.16. The probability that exactly 6 new accounts will be opened on a given day is .0120. Therefore,

$$P(6) = .0120$$

Table 5.16 Portion of Table III for $\lambda = 2.0$

x	1.1	1.2	λ ...	2.0	
					← $\lambda = 2.0$
0				.1353	
1				.2707	
2				.2707	
3				.1804	
4				.0902	
5				.0361	
$x = 6 \longrightarrow$ 6				.0120	← $P(6)$
7				.0034	
8				.0009	
9				.0002	

Actually, Table 5.16 gives the probability distribution of x for $\lambda = 2.0$. Note that the sum of the 10 probabilities given in Table 5.16 is .9999 and not 1.0. This is so for two reasons. First, these probabilities are rounded to four decimal places. Second, on a given day more than 9 new accounts might be opened at this bank. However, the probabilities of 10, 11, 12, ... new accounts are very small, and they are not listed in the table.

(b) The probability that at most three new accounts are opened on a given day is obtained by adding the probabilities of 0, 1, 2, and 3 new accounts. Thus, using Table III of Appendix B or Table 5.16, we obtain

$$P(\text{at most } 3) = P(0) + P(1) + P(2) + P(3)$$
$$= .1353 + .2707 + .2707 + .1804 = \mathbf{.8571}$$

(c) The probability that at least 7 new accounts are opened on a given day is obtained by adding the probabilities of 7, 8, and 9 new accounts. Note that 9 is the last value of x for $\lambda = 2.0$ in Table III of Appendix B or Table 5.16. Hence, 9 is the last value of x whose probability is included in the sum. However, this does not mean that on a given day more than 9 new accounts cannot be opened. It simply means that the probability of 10 or more accounts is close to zero. Thus,

$$P(\text{at least } 7) = P(7) + P(8) + P(9)$$
$$= .0034 + .0009 + .0002 = \mathbf{.0045}$$

EXAMPLE 5–22 Cars Sold per Day by a Salesman

Constructing a Poisson probability distribution and graphing it.

An auto salesperson sells an average of .9 car per day. Let x be the number of cars sold by this salesperson on any given day. Using the Poisson probability distribution table, write the probability distribution of x. Draw a graph of the probability distribution.

Solution Let λ be the mean number of cars sold per day by this salesperson. Hence, $\lambda = .9$. Using the portion of Table III of Appendix B that corresponds to $\lambda = .9$, we write the probability distribution of x in Table 5.17. Figure 5.8 shows the histogram for the probability distribution of Table 5.17.

Table 5.17	Probability Distribution of x for $\lambda = .9$
x	$P(x)$
0	.4066
1	.3659
2	.1647
3	.0494
4	.0111
5	.0020
6	.0003

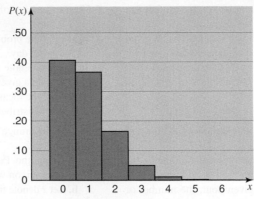

Figure 5.8 Histogram for the probability distribution of Table 5.17.

Note that 6 is the largest value of x for $\lambda = .9$ listed in Table III for which the probability is greater than zero. However, this does not mean that this salesperson cannot sell more than six cars on a given day. What this means is that the probability of selling seven or more cars is very small. Actually, the probability of $x = 7$ for $\lambda = .9$ calculated by using the Poisson formula is .000039. When rounded to four decimal places, this probability is .0000, as listed in Table III.

5.6.2   Mean and Standard Deviation of the Poisson Probability Distribution

For the Poisson probability distribution, the mean and variance both are equal to λ, and the standard deviation is equal to $\sqrt{\lambda}$. That is, for the Poisson probability distribution,

$$\mu = \lambda, \quad \sigma^2 = \lambda, \quad \text{and} \quad \sigma = \sqrt{\lambda}$$

For Example 5–22, $\lambda = .9$. Therefore, for the probability distribution of x in Table 5.17, the mean, variance, and standard deviation are, respectively,

$$\mu = \lambda = .9 \text{ car}$$
$$\sigma^2 = \lambda = .9$$
$$\sigma = \sqrt{\lambda} = \sqrt{.9} = .949 \text{ car}$$

EXERCISES

CONCEPTS AND PROCEDURES

5.47 State the conditions that must be satisfied to apply the Poisson probability distribution.

5.48 What is parameter of the Poisson probability distribution and what does it mean?

5.49 Using the Poisson formula, find the following probabilities.
    a. $P(x < 2)$ for $\lambda = 3$ **b.** $P(x = 8)$ for $\lambda = 6.5$
Verify these probabilities using Table III of Appendix B.

5.50 Let x be a Poisson random variable. Using the Poisson probabilities table, write the probability distribution of x for each of the following. Find the mean, variance, and standard deviation for each of these probability distributions. Draw a graph for each of these probability distributions.
    a. $\lambda = .6$ **b.** $\lambda = 1.8$

APPLICATIONS

5.51 A household receives an average of 1.8 pieces of junk mail per day. Find the probability that this household will receive exactly 3 pieces of junk mail on a certain day. Use the Poisson probability distribution formula.

5.52 On average, 12.5 rooms stay vacant per day at a large hotel in a city. Find the probability that on a given day exactly 4 rooms will be vacant. Use the Poisson formula.

5.53 A commuter airline receives an average of 9.7 complaints per day from its passengers. Using the Poisson formula, find the probability that on a certain day this airline will receive exactly 5 complaints.

5.54 Despite all efforts by the quality control department, the fabric made at Benton Corporation always contains a few defects. A certain type of fabric made at this corporation contains an average of .5 defect per 500 yards.

    a. Using the Poisson formula, find the probability that a given piece of 500 yards of this fabric will contain exactly 1 defect.

    b. Using the Poisson probabilities table, find the probability that the number of defects in a given 500-yard piece of this fabric will be

      i. 2 to 4   **ii.** more than 3   **iii.** less than 2

5.55 An average of 4.8 customers come to Columbia Savings and Loan every half hour.

    a. Find the probability that exactly 2 customers will come to this savings and loan during a given hour.

    b. Find the probability that during a given hour, the number of customers who will come to this savings and loan is

      i. 2 or fewer   **ii.** 10 or more

5.56 According to a study performed by the NCAA, the average rate of injuries occurring in collegiate women's soccer is 8.6 per 1000 participants (www.fastsports.com/tips/tip12/).

    a. Using the Poisson formula, find the probability that the number of injuries in a sample of 1000 women's soccer participants is

      i. exactly 12   **ii.** exactly 5

    b. Using the Poisson probabilities table, find the probability that the number of injuries in a sample of 1000 women's soccer participants is

      i. more than 3   **ii.** less than 10   **iii.** 8 to 13

5.57 An average of .6 accident occur per day in a particular large city.

    a. Find the probability that no accident will occur in this city on a given day.

    b. Let x denote the number of accidents that will occur in this city on a given day. Write the probability distribution of x.

    c. Find the mean, variance, and standard deviation of the probability distribution developed in part b.

5.58 An insurance salesperson sells an average of 1.4 policies per day.

    a. Using the Poisson formula, find the probability that this salesperson will sell no insurance policy on a certain day.

    b. Let x denote the number of insurance policies that this salesperson will sell on a given day. Using the Poisson probabilities table from Appendix B, write the probability distribution of x.

    c. Find the mean, variance, and standard deviation of the probability distribution developed in part b.

*5.59 On average, 30 households in 50 own answering machines.

    a. Using the Poisson formula, find the probability that in a random sample of 50 households, exactly 25 will own answering machines.

    b. Using the Poisson probabilities table, find the probability that the number of households in 50 who own answering machines is

      i. at most 12   **ii.** 13 to 17   **iii.** at least 40

USES AND MISUSES...

BATTER UP!

In all of sport, there is no other sports that has the history and reputation for the use of statistics than baseball. The book and movie adaption of *Moneyball: The Art of Winning an Unfair Game* by Michael Lewis brought to light the importance of statistics in baseball to the general public. Of course, this is nothing new to true fans of the game who participate in "Fantasy Baseball" and have become experts in *sabermetrics*, the term coined by Bill James for the statistical analysis of baseball in-game activity—the term derives from the acronym SABR (Society for American Baseball Research). In fact, there is quite a following for data published by the Player Empirical Comparison and Optimization Test Algorithm (PECOTA) for projecting baseball player performance.

One of the most commonly reported baseball statistics is the batting average. Even though this is often considered to be a somewhat weak measure of player ability overall because of its singular dependence on one facet of the game, there is still keen interest in this statistic. In baseball, the battling average is defined as the number of hits divided by the number of times at bat. For example, if a batter goes to the plate 10 times and gets a hit on 3 of those occasions, then the batting average is 3/10 = .3. In baseball parlance, this

is translated to three decimal places, so we would report this batting average to be .300.

One would think that most players, inasmuch as batting is such an important part of the game, would have very high batting averages. However, Ty Cobb, who holds the record for highest career batting average, only hit .366; the lowest career average was recorded for Bill Bergen (.170). The highest single season average in the modern era was .426, which was achieved in 1901 by Napoleon Lajoie. Hitting a .400 is a goal of many players, though it is extraordinarily rare even for short periods of time, let alone an entire season or over a career.

To see an example of this, it is important to consider that it is not uncommon for an average player who plays regularly to have over 600 at-bats in a given season. Hence, such a player would have to get 240 hits in order to reach the elusive .400. We can perform some quick calculations to see just how rare it would be for a player of a given batting average to hit .400 or higher. For example, if a player normally bats a .350, how likely would he be able to maintain a batting average of .400 or higher over several at-bats? We can use the binomial distribution to answer this question. For example, we could calculate the probability of hitting .400 or higher over 10 chances at the plate. Out of 10 at-bats, this would be equivalent to calculating $P(x \geq 4) = 1 - P(x \leq 3)$. Using the binomial formula from this chapter, this probability works out to be .4862, or about a 48.62% chance that a regular .350 hitter will bat .400 or higher in 10 batting opportunities. If we repeat this calculation for a .300 average batter, a .250 average batter, and a .200 average batter, we can complete the following table:

Regular Season Batting Average	Probability of batting at least .400 in 10 at-bats
.350	.4862
.300	.3504
.250	.2241
.200	.1209

As can be seen, the probability of hitting a .400 or better for the average batter, even over a relatively small number of bats, is very low. We can easily repeat this procedure for any number of batting opportunities.

In the table below, we examine up to 100 at-bats (or about 1/6 of a standard season) for players with a batting average ranging from .200 to .350.

Probability of Batting at Least .400

Number of at-bats	Regular Season Batting Average			
	.350	.300	.250	.200
5	.572	.472	.367	.263
10	.486	.350	.224	.121
15	.436	.278	.148	.061
20	.399	.228	.102	.032
25	.370	.189	.071	.017
30	.345	.159	.051	.009
35	.324	.135	.036	.005
40	.305	.115	.026	.003
45	.289	.099	.019	.002
50	.274	.085	.014	.001
55	.260	.073	.010	.001
60	.247	.063	.007	<.001
65	.235	.055	.006	<.001
70	.225	.048	.004	<.001
75	.214	.041	.003	<.001
80	.205	.036	.002	<.001
85	.196	.031	.002	<.001
90	.188	.027	.001	<.001
95	.180	.024	.001	<.001
100	.172	.021	.001	<.001

It is clear that reaching .400 happens with very low probability. Even for a very good hitter (.350 average), maintaining a .400 or higher average declines very rapidly as the number of at-bats increases.

Here we have seen that the binomial distribution can be used to make projections of reaching the nearly unattainable .400 batting average. Of course, there are players who manage to exceed a batting average of .400 but just not for very long. There are many other factors that could play into high batting average (e.g., who is pitching, a player having a good batting day just by chance, etc.). But can a player maintain a high batting average for long? We should not bet on it.

GLOSSARY

Bernoulli trial   One repetition of a binomial experiment. Also called a *trial*.

Binomial experiment   An experiment that contains n identical trials such that each of these n trials has only two possible outcomes (or events), the probabilities of these two outcomes (or events) remain constant for each trial, and the trials are independent.

Binomial parameters   The total trials n and the probability of success p for the binomial probability distribution.

Binomial probability distribution   The probability distribution that gives the probability of x successes in n trials when the probability of success is p for each trial of a binomial experiment.

Continuous random variable   A random variable that can assume any value in one or more intervals.

Discrete random variable   A random variable whose values are countable.

Hypergeometric probability distribution   The probability distribution that is applied to determine the probability of x successes in n trials when the trials are not independent.

Mean of a discrete random variable   The mean of a discrete random variable x is the value that is expected to occur per repetition, on average, if an experiment is repeated a large number of times. The mean of a discrete random variable is also called its *expected value*.

Poisson parameter   The average occurrences, denoted by λ, during an interval for a Poisson probability distribution.

Poisson probability distribution   The probability distribution that gives the probability of x occurrences in an interval when the average number of occurrences in that interval is λ.

Probability distribution of a discrete random variable   A list of all the possible values that a discrete random variable can assume and their corresponding probabilities.

Random variable   A variable, denoted by x, whose value is determined by the outcome of a random experiment. Also called a *chance variable*.

Standard deviation of a discrete random variable   A measure of spread for the probability distribution of a discrete random variable.

SUPPLEMENTARY EXERCISES

5.60 Let x be the number of errors that appear on a randomly selected page of a book. The following table lists the probability distribution of x.

x	0	1	2	3	4
$P(x)$	.73	.16	.06	.04	.01

Find the mean and standard deviation of x.

5.61 The following table gives the probability distribution of the number of camcorders sold on a given day at an electronics store.

Camcorders sold	0	1	2	3	4	5	6
Probability	.05	.12	.19	.30	.20	.10	.04

Calculate the mean and standard deviation of this probability distribution. Give a brief interpretation of the value of the mean.

5.62 GESCO Insurance Company charges a $350 premium per annum for a $100,000 life insurance policy for a 40-year-old female. The probability that a 40-year-old female will die within 1 year is .002.

    a. Let x be a random variable that denotes the gain of the company for next year from a $100,000 life insurance policy sold to a 40-year-old female. Write the probability distribution of x.

    b. Find the mean and standard deviation of the probability distribution of part a. Give a brief interpretation of the value of the mean.

5.63 The student health center at a university treats an average of seven cases of mononucleosis per day during the week of final examinations.

    a. Using the appropriate formula, find the probability that on a given day during the finals week exactly four cases of mononucleosis will be treated at this health center.

    b. Using the appropriate probabilities table from Appendix B, find the probability that on a given day during the finals week the number of cases of mononucleosis treated at this health center will be

      i. at least 7   **ii.** at most 3   **iii.** 2 to 5

5.64 One of the toys made by Dillon Corporation is called Speaking Joe, which is sold only by mail. Consumer satisfaction is one of the top priorities of the company's management. The company guarantees a refund or a replacement for any Speaking Joe toy if the chip that is installed inside becomes defective within 1 year from the date of purchase. It is known from past data that 10% of these chips become defective within a 1-year period. The company sold 15 Speaking Joes on a given day.

    a. Let x denote the number of Speaking Joes in these 15 that will be returned for a refund or a replacement within a 1-year period. Using the appropriate probabilities table from Appendix B, obtain the probability distribution of x and draw a graph of the probability distribution. Determine the mean and standard deviation of x.

    b. Using the probability distribution constructed in part a, find the probability that exactly 5 of the 15 Speaking Joes will be returned for a refund or a replacement within a 1-year period.

5.65 Twenty corporations were asked whether or not they provide retirement benefits to their employees. Fourteen of the corporations said they do provide retirement benefits to their employees, and 6 said they do not. Five corporations are randomly selected from these 20. Find the probability that

    a. exactly 2 of them provide retirement benefits to their employees.

    b. none of them provides retirement benefits to their employees.

    c. at most one of them provides retirement benefits to employees.

5.66 A really bad carton of 18 eggs contains 7 spoiled eggs. An unsuspecting chef picks 4 eggs at random for his "Mega-Omelet Surprise." Find the probability that the number of *unspoiled* eggs among the 4 selected is

    a. exactly 4

    b. 2 or fewer

    c. more than 1

5.67 Alison Bender works for an accounting firm. To make sure her work does not contain errors, her manager randomly checks on her work. Alison recently filled out 12 income tax returns for the company's clients. Unknown to anyone, two of these 12 returns have minor errors. Alison's manager randomly selects three returns from these 12 returns. Find the probability that

    a. exactly 1 of them contains errors

    b. none of them contains errors

    c. exactly 2 of them contain errors

5.68 Uniroyal Electronics Company buys certain parts for its refrigerators from Bob's Corporation. The parts are received in shipments of 400 boxes, each box containing 16 parts. The quality control department at Uniroyal Electronics first randomly selects 1 box from each shipment and then randomly selects 4 parts from that box. The shipment is accepted if at most 1 of the 4 parts is defective. The quality control inspector at Uniroyal Electronics selected a box from a recently received shipment of such parts. Unknown to the inspector, this box contains 3 defective parts.

    a. What is the probability that this shipment will be accepted?

    b. What is the probability that this shipment will not be accepted?

5.69 Six jurors are to be selected from a pool of 20 potential candidates to hear a civil case involving a lawsuit between two families. Unknown to the judge or any of the attorneys, 4 of the 20 prospective jurors are potentially prejudiced by being acquainted with one or more of the litigants. They will not disclose this during the jury selection process. If 6 jurors are selected at random from this group of 20, find the probability that the number of potentially prejudiced jurors among the 6 selected jurors is

    a. exactly 1   **b.** none   **c.** at most 2

5.70 Bender Electronics buys keyboards for its computers from another company. The keyboards are received in shipments of 100 boxes, each box containing 20 keyboards. The quality control department at Bender Electronics first randomly selects one box from each shipment and then randomly selects 5 keyboards from that box. The shipment is accepted if not more than 1 of the 5 keyboards is defective. The quality control inspector at Bender Electronics selected a box from a recently received shipment of keyboards. Unknown to the inspector, this box contains 6 defective keyboards.

    a. What is the probability that this shipment will be accepted?

    b. What is the probability that this shipment will not be accepted?

ADVANCED EXERCISES

5.71 Scott offers you the following game: You will roll two fair dice. If the sum of the two numbers obtained is 2, 3, 4, 9, 10, 11, or 12, Scott will pay you $20. However, if the sum of the two numbers is 5, 6, 7, or 8, you will pay Scott $20. Scott points out that you have seven winning numbers and only four losing numbers. Is this game fair to you? Should you accept this offer? Support your conclusion with appropriate calculations.

5.72 Two teams, A and B, will play a best-of-seven series, which will end as soon as one of the teams wins four games. Thus, the series may end in four, five, six, or seven games. Assume that each team has an equal chance of winning each game and that all games are independent of one another. Find the following probabilities.
  a. Team A wins the series in four games.
  b. Team A wins the series in five games.
  c. Seven games are required for a team to win the series.

5.73 York Steel Corporation produces a special bearing that must meet rigid specifications. When the production process is running properly, 10% of the bearings fail to meet the required specifications. Sometimes problems develop with the production process that cause the rejection rate to exceed 10%. To guard against this higher rejection rate, samples of 15 bearings are taken periodically and carefully inspected. If more than 2 bearings in a sample of 15 fail to meet the required specifications, production is suspended for necessary adjustments.
  a. If the true rate of rejection is 10% (that is, the production process is working properly), what is the probability that the production will be suspended based on a sample of 15 bearings?
  b. What assumptions did you make in part a?

5.74 Suppose that a certain casino has the "money wheel" game. The money wheel is divided into 50 sections, and the wheel has an equal probability of stopping on each of the 50 sections when it is spun. Twenty-two of the sections on this wheel show a $1 bill, 14 show a $2 bill, 7 show a $5 bill, 3 show a $10 bill, 2 show a $20 bill, 1 shows a flag, and 1 shows a joker. A gambler may place a bet on any of the seven possible outcomes. If the wheel stops on the outcome that the gambler bet on, he or she wins. The net payoffs for these outcomes for $1 bets are as follows.

Symbol bet on	$1	$2	$5	$10	$20	Flag	Joker
Payoff (dollars)	1	2	5	10	20	40	40

  a. If the gambler bets on the $1 outcome, what is the expected net payoff?
  b. Calculate the expected net payoffs for each of the other six outcomes.
  c. Which bet(s) is (are) best in terms of expected net payoff? Which is (are) worst?

5.75 A high school history teacher gives a 50-question multiple-choice examination in which each question has four choices. The scoring includes a penalty for guessing. Each correct answer is worth 1 point, and each wrong answer costs 1/2 point. For example, if a student answers 35 questions correctly, 8 questions incorrectly, and does not answer 7 questions, the total score for this student will be $35 - (1/2)(8) = 31$.
  a. What is the expected score of a student who answers 38 questions correctly and guesses on the other 12 questions? Assume

that the student randomly chooses one of the four answers for each of the 12 guessed questions.
  b. Does a student increase his expected score by guessing on a question if he has no idea what the correct answer is? Explain.
  c. Does a student increase her expected score by guessing on a question for which she can eliminate one of the wrong answers? Explain.

5.76 A history teacher has given her class a list of seven essay questions to study before the next test. The teacher announced that she will choose four of the seven questions to give on the test, and each student will have to answer three of those four questions.
  a. In how many ways can the teacher choose four questions from the set of seven?
  b. Suppose that a student has enough time to study only five questions. In how many ways can the teacher choose four questions from the set of seven so that the four selected questions include both questions that the student did not study?
  c. What is the probability that the student in part b will have to answer a question that he or she did not study? That is, what is the probability that the four questions on the test will include both questions that the student did not study?

5.77 Brad Henry is a stone products salesman. Let x be the number of contacts he visits on a particular day. The following table gives the probability distribution of x.

x	1	2	3	4
$P(x)$	.12	.25	.56	.07

Let y be the total number of contacts Brad visits on two randomly selected days. Write the probability distribution for y.

5.78 The number of calls that come into a small mail-order company follows a Poisson distribution. Currently, these calls are serviced by a single operator. The manager knows from past experience that an additional operator will be needed if the rate of calls exceeds 20 per hour. The manager observes that 9 calls came into the mail-order company during a randomly selected 15-minute period.
  a. If the rate of calls is actually 20 per hour, what is the probability that 9 or more calls will come in during a given 15-minute period?
  b. If the rate of calls is really 30 per hour, what is the probability that 9 or more calls will come in during a given 15-minute period?
  c. Based on the calculations in parts a and b, do you think that the rate of incoming calls is more likely to be 20 or 30 per hour?
  d. Would you advise the manager to hire a second operator? Explain.

5.79 Customers arrive at the checkout counter of a supermarket at an average rate of 10 per hour, and these arrivals follow a Poisson distribution. Using each of the following two methods, find the probability that exactly 4 customers will arrive at this checkout counter during a 2-hour period.
  a. Use the arrivals in each of the two nonoverlapping 1-hour periods and then add these. (Note that the numbers of arrivals in two nonoverlapping periods are independent of each other.)
  b. Use the arrivals in a single 2-hour period.

SELF-REVIEW TEST

1. Briefly explain the meaning of a random variable, a discrete random variable, and a continuous random variable. Give one example each of a discrete and a continuous random variable.

2. What name is given to a table that lists all of the values that a discrete random variable x can assume and their corresponding probabilities?

3. For the probability distribution of a discrete random variable, the probability of any single value of x is always
    a. in the range 0 to 1   **b.** 1.0   **c.** less than zero

4. For the probability distribution of a discrete random variable, the sum of the probabilities of all possible values of x is always
    a. greater than 1   **b.** 1.0   **c.** less than 1.0

5. State the four conditions of a binomial experiment. Give one example of such an experiment.

6. The parameters of the binomial probability distribution are
    a. n, p, and q **b.** n and p **c.** n, p, and x

7. The mean and standard deviation of a binomial probability distribution with $n = 25$ and $p = .20$ are
    a. 5 and 2   **b.** 8 and 4   **c.** 4 and 3

8. The binomial probability distribution is symmetric if
    a. $p < .5$ **b.** $p = .5$ **c.** $p > .5$

9. The binomial probability distribution is skewed to the right if
    a. $p < .5$ **b.** $p = .5$ **c.** $p > .5$

10. The binomial probability distribution is skewed to the left if
    a. $p < .5$ **b.** $p = .5$ **c.** $p > .5$

11. Briefly explain when a hypergeometric probability distribution is used. Give one example of a hypergeometric probability distribution.

12. The parameter/parameters of the Poisson probability distribution is/are
    a. λ **b.** λ and x **c.** λ and e

13. Describe the three conditions that must be satisfied to apply the Poisson probability distribution.

14. Let x be the number of homes sold per week by all four real estate agents who work at a realty office. The following table lists the probability distribution of x.

x	0	1	2	3	4	5
$P(x)$	.15	.24	.29	.14	.10	.08

Calculate the mean and standard deviation of x. Give a brief interpretation of the value of the mean.

15. According to a survey, 60% of adults believe that all college students should be required to perform a specified number of hours of community service to graduate. Assume that this percentage is true for the current population of adults.
    a. Find the probability that the number of adults in a random sample of 12 who hold this view is
        i. exactly 8 (use the appropriate formula)
        ii. at least 6 (use the appropriate table from Appendix B)
        iii. less than 4 (use the appropriate table from Appendix B)
    b. Let x be the number of adults in a random sample of 12 who believe that all college students should be required to perform a specified number of hours of community service to graduate. Using the appropriate table from Appendix B, write the probability distribution of x. Find the mean and standard deviation of x.

16. The Red Cross honors and recognizes its best volunteers from time to time. One of the Red Cross offices has received 12 nominations for the next group of 4 volunteers to be recognized. Eight of these 12 nominated volunteers are females. If the Red Cross office decides to randomly select 4 names out of these 12 nominated volunteers, find the probability that of these 4 volunteers
    a. exactly 3 are females
    b. exactly 1 is female
    c. at most 1 is female

17. The police department in a large city has installed a traffic camera at a busy intersection. Any car that runs a red light will be photographed with its license plate visible, and the driver will receive a citation. Suppose that during the morning rush hour of weekdays, an average of 10 drivers are caught running the red light per day by this system.
    a. Find the probability that during the morning rush hour on a given weekday this system will catch
        i. exactly 14 drivers (use the appropriate formula)
        ii. at most 7 drivers (use the appropriate table from Appendix B)
        iii. 13 to 18 drivers (use the appropriate table from Appendix B)
    b. Let x be the number of drivers caught by this system during the morning rush hour on a given weekday. Write the probability distribution of x. Use the appropriate table from Appendix B.

18. The binomial probability distribution is symmetric when $p = .50$, it is skewed to the right when $p < .50$, and it is skewed to the left when $p > .50$. Illustrate these three cases by writing three probability distributions and graphing them. Choose any values of n (4 or higher) and p and use the table of binomial probabilities (Table I of Appendix B).

MINI-PROJECTS

Note: The Mini-Projects are located on the text's Web site, www.wiley.com/college/mann.

DECIDE FOR YOURSELF

Note: The Decide for Yourself feature is located on the text's Web site, www.wiley.com/college/mann.

TECHNOLOGY INSTRUCTIONS   **CHAPTER 5**

Note: Complete TI-84, Minitab, and Excel manuals are available for download at the textbook's Web site, www.wiley.com/college/mann.

TI-84 Color/TI-84

The TI-84 Color Technology Instructions feature of this text is written for the TI-84 Plus C color graphing calculator running the 4.0 operating system. Some screens, menus, and functions will be slightly different in older operating systems. The TI-84 and TI-84 Plus can perform all of the same functions but will not have the "Color" option referenced in some of the menus.

Calculating a Binomial Probability for Example 5–14(a) of the Text

1. Select **2ⁿᵈ > VARS > binompdf(**.

2. Use the following settings in the **binompdf** menu (see **Screen 5.1**):

   • At the **trials** prompt, type 6.

   • At the **p** prompt, type 0.3.

   • At the **x value** prompt, type 3.

3. Highlight **Paste** and press **ENTER** twice.

4. The output $P(3) = .18522$ will appear on the screen.

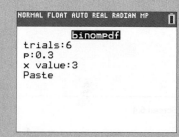

Screen 5.1

Calculating a Cumulative Binomial Probability for Example 5–14(b) of the Text

1. Select **2ⁿᵈ > VARS > binomcdf(**.

2. Use the following settings in the **binomcdf** menu:

   • At the **trials** prompt, type 6.

   • At the **p** prompt, type 0.3.

   • At the **x value** prompt, type 2.

3. Highlight **Paste** and press **ENTER** twice.

4. The output $P(x \leq 2) = .74431$ will appear on the screen.

   Note: The difference between **binompdf** and **binomcdf** is that **binompdf** calculates the probability $P(2)$ while **binomcdf** calculates the cumulative probability $P(x \leq 2)$.

Preparing a Binomial Probability Distribution for Example 5–14(e) of the Text

1. Select **STAT > EDIT > Edit**.

2. Type 0, 1, 2, 3, 4, 5, and 6 into **L1**. These are all possible values of x for a binomial probability distribution when $n = 6$. (See **Screen 5.3**.)

3. Select **2ⁿᵈ > MODE** to return to the home screen.

4. Select **2ⁿᵈ > VARS > binompdf(**.

5. Use the following settings in the **binompdf** menu:

   • At the **trials** prompt, type 6.

   • At the **p** prompt, type 0.3.

   • At the **x value** prompt, select **2ⁿᵈ > STAT > L1**. (See **Screen 5.2**.)

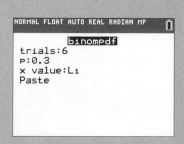

Screen 5.2

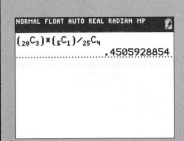

Screen 5.3

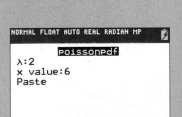

Screen 5.4

6. Highlight **Paste** and press **ENTER**.

7. Press **STO**.

8. Select 2nd > **STAT** > **L2**.

9. Press **ENTER**.

10. Select **STAT** > **EDIT** > **Edit...** to see the probability distribution. (See **Screen 5.3**.)

Calculating a Hypergeometric Probability for Example 5–17(a) of the Text

1. Access the Home screen.

2. Enter $_{20}C_3$ by following these steps (see **Screen 5.4**):

   • Type (20.

   • Select **MATH** > **PROB** > **nCr**.

   • Type 3).

3. Press ×.

4. Enter $_5C_1$ by following these steps (see **Screen 5.4**):

   • Type (5.

   • Select **MATH** > **PROB** > **nCr**.

   • Type 1).

5. Press ÷.

6. Enter $_{25}C_4$ by following these steps (see **Screen 5.4**):

   • Type 25.

   • Select **MATH** > **PROB** > **nCr**.

   • Type 4.

7. Press **ENTER**. $P(3) = .4505928854$ will appear on the screen.

Calculating a Poisson Probability for Example 5–21(a) of the Text

1. Select 2nd > **VARS** > **poissonpdf(**.

2. Use the following settings in the **poissonpdf** menu (see **Screen 5.5**):

   • At the λ prompt, type 2.

   • At the **x value** prompt, type 6.

3. Highlight **Paste** and press **ENTER** twice.

4. The output $P(6) = .012029803$ will appear on the screen.

Calculating a Cumulative Poisson Probability for Example 5–21(b) of the Text

1. Select 2nd > **VARS** > **poissoncdf(**.

2. Use the following settings in the **poissoncdf** menu:

   • At the λ prompt, type 2.

   • At the **x value** prompt, type 3.

Screen 5.5

3. Highlight **Paste** and press **ENTER** twice.

4. The output $P(x \leq 3) = .8571234606$ will appear on the screen.

Note: The difference between **poissonpdf** and **poissoncdf** is that **poissonpdf** calculates the probability $P(x = 3)$ while **poissoncdf** calculates the cumulative probability $P(x \leq 3)$.

Minitab

The Minitab Technology Instructions feature of this text is written for Minitab version 17. Some screens, menus, and functions will be slightly different in older versions of Minitab.

Calculating a Binomial Probability for Example 5–14(a) of the Text

1. Select **Calc > Probability Distributions > Binomial**.

2. Use the following settings in the dialog box that appears on screen (see **Screen 5.6**):

- Select **Probability**.

- In the **Number of trials** box, type 6.

- In the **Event probability** box, type 0.3

- Select **Input constant** and type 3 in the **input column** box.

3. Click **OK**. The output $P(3) = .18522$ will appear in the **Session** window. (See **Screen 5.7**.)

Screen 5.6

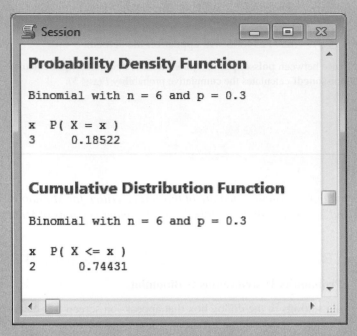

Screen 5.7

Calculating a Cumulative Binomial Probability for Example 5–14(b) of the Text

1. Select **Calc > Probability Distributions > Binomial**.

2. Use the following settings in the dialog box that appears on screen:

   • Select **Cumulative probability**.

   • Type 6 in the **Number of trials** box.

   • Type 0.3 in the **Event probability** box.

   • Select **Input constant** and type 2 in the **Input constant** box.

3. Click **OK**. The output $P(x \leq 2) = .74431$ will appear in the **Session** window. (See **Screen 5.7**.)

Preparing a Binomial Probability Distribution for Example 5–14(e) of the Text

1. Type 0, 1, 2, 3, 4, 5, and 6 into column C1. These are all possible values of x for a binomial probability distribution when $n = 6$.

2. Select **Calc > Probability Distributions > Binomial**.

3. Use the following settings in the dialog box that appears on screen (see **Screen 5.8**):

   • Select **Probability**.

   • Type 6 in the **Number of trials** box.

   • Type 0.3 in the **Event probability** box.

   • Select **Input column** and type C1 in the **Input column** box.

   • Type C2 in the **Optional storage** box.

4. Click **OK**. The output will appear in the **Worksheet** window.

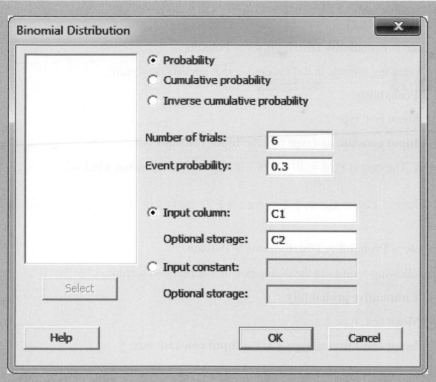

Screen 5.8

Calculating a Hypergeometric Probability for Example 5–17(a) of the Text

1. Select **Calc > Probability Distributions > Hypergeometric**.

2. Use the following settings in the dialog box that appears on screen:

   - Select **Probability**.

   - In the **Population size** box, type 12.

   - In the **Event count in population** box, type 7.

   - In the **Sample size** box, type 3.

   - Select **Input constant** and type 3 in the **Input constant** box.

3. Click **OK**. The output $P(3) = .159091$ will appear in the **Session** window.

Calculating a Cumulative Hypergeometric Probability for Example 5–17(b) of the Text

1. Select **Calc > Probability Distributions > Hypergeometric**.

2. Use the following settings in the dialog box that appears on screen:

   - Select **Cumulative probability**.

   - In the **Population size** box, type 12.

   - In the **Event count in population** box, type 7.

   - In the **Sample size** box, type 3.

   - Select **Input constant** and type 1 in the **Input constant** box.

3. Click **OK**. The output $P(x \leq 1) = .363636$ will appear in the **Session** window.

Calculating a Poisson Probability for Example 5–21(a) of the Text

1. Select **Calc > Probability Distributions > Poisson**.

2. Use the following settings in the dialog box that appears on screen:

   - Select **Probability**.

   - In the **Mean** box, type 2.

   - Select **Input constant** and type 6 in the **Input constant** box.

3. Click **OK**. The output $P(6) = .0120298$ will appear in the **Session** window.

Calculating a Cumulative Poisson Probability for Example 5–21(b) of the Text

1. Select **Calc > Probability Distributions > Poisson**.

2. Use the following settings in the dialog box that appears on screen:

   - Select **Cumulative probability**.

   - In the **Mean** box, type 2.

   - Select **Input constant** and type 3 in the **Input constant** box.

3. Click **OK**. The output $P(x \leq 3) = .857123$ will appear in the **Session** window.

Excel

The Excel Technology Instructions feature of this text is written for Excel 2013. Some screens, menus, and functions will be slightly different in older versions of Excel. To enable some of the advanced Excel functions, you must enable the Data Analysis Add-In for Excel. For Excel 2007 and older versions of Excel, replace the function **BINOM.DIST** *with the function* **BINOMDIST**.

Calculating a Binomial Probability for Example 5–14(a) of the Text

1. Click on cell A1.

2. Type **=BINOM.DIST(3, 6, 0.3, 0)**. (See **Screen 5.9**.)

   Note: Entering a 0 as the fourth value in this function causes Excel to calculate the probability $P(0)$.

3. Press **ENTER**. The probability $P(3) = .18522$ will appear in cell A1.

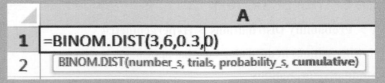

Screen 5.9

Calculating a Cumulative Binomial Probability for Example 5–14(b) of the Text

1. Click on cell A1.

2. Type **=BINOM.DIST(2, 6, 0.3, 1)**.

Note: Entering a 1 as the fourth value in this function causes Excel to calculate the cumulative probability $P(x \leq 2)$.

3. Press **ENTER**. The probability $P(x \leq 2) = .74431$ will appear in cell A1.

Preparing a Binomial Probability Distribution for Example 5–14(e) of the Text

1. Type 0, 1, 2, 3, 4, 5, and 6 into cells A1 to A7. These are all possible values of x for a binomial probability distribution when $n = 6$.

2. Click on cell B1.

3. Type =**BINOM.DIST(A1, 6, 0.3, 0)**. (See **Screen 5.10**.)

4. Press **ENTER**.

5. Copy and paste the formula from cell B1 into cells B2 to B7.

	A	B
1	0	=BINOM.DIST(A1,6,0.3,0)
2	1	BINOM.DIST(number_s, trials, probability_s, **cumulative**)
3	2	
4	3	
5	4	
6	5	
7	6	

Screen 5.10

Calculating a Hypergeometric Probability for Example 5–17(a) of the Text

1. Click on cell A1.

2. Type =**HYPGEOM.DIST(3, 3, 7, 12, 0)**.

Note: Entering a 1 as the fifth value in this function causes Excel to calculate the probability $P(3)$.

3. Press **ENTER**. The probability $P(3) = .159091$ will appear in cell A1.

Calculating a Cumulative Hypergeometric Probability for Example 5–17(b) of the Text

1. Click on cell A1.

2. Type =**HYPGEOM.DIST(1, 3, 7, 12, 1)**.

Note: Entering a 1 as the fifth value in this function causes Excel to calculate the cumulative probability $P(x \leq 1)$.

3. Press **ENTER**. The probability $P(x \leq 1) = .363636$ will appear in cell A1.

Calculating a Poisson Probability for Example 5–21(a) of the Text

1. Click on cell A1.

2. Type =**POISSON.DIST(6,2,0)**.

3. Press **ENTER**. The probability $P(6) = .01203$ will appear in cell A1.

Calculating a Cumulative Poisson Probability for Example 5–21(b) of the Text

1. Click on cell A1.

2. Type =**POISSON.DIST(3,2,1)**.

   Note: Entering a 1 as the third value in this function causes Excel to calculate the cumulative probability $P(x \leq 1)$.

3. Press **ENTER**. The probability $P(x \leq 1) = .857123$ will appear in cell A1.

TECHNOLOGY ASSIGNMENTS

TA5.1 According to an NPD Group poll, 63% of 18 to 24-year-old women said they shop only at their favorite stores (*Forbes*, November 1, 2004). Assume that this percentage is true for the current population of such women.

  a. Find the probability that in a random sample of 20 such women, exactly 14 will say that they shop only at their favorite stores.

  b. Find the probability that in a random sample of 30 such women, exactly 18 will say that they shop only at their favorite stores.

  c. Find the probability that in a random sample of 25 such women, at most 15 will say that they shop only at their favorite stores.

  d. Find the probability that in a random sample of 40 such women, at least 30 will say that they shop only at their favorite stores.

TA5.2 According to an October 2010 *Consumer Reports* survey, 39% of car owners in the United States are considering a hybrid or a plug-in for their next car purchase (news.consumerreports.org/cars/2010/10/consumer-reports-shares-preliminary-green-car-survey-findings-at-gridweek-conference.html).

  a. Find the probability that in a random sample of 70 car owners in the United States, exactly 32 would be considering a hybrid or a plug-in for their next car purchase.

  b. Find the probability that in a random sample of 70 car owners in the United States, 31 or more would be considering a hybrid or a plug-in for their next car purchase.

  c. Find the probability that in a random sample of 700 car owners in the United States, 301 or more would be considering a hybrid or a plug-in for their next car purchase.

  d. A reporter states that he believes that the sample results in parts b and c imply that the percentage of car owners in the United States who are considering a hybrid or a plug-in for their next car purchase is higher than 39%. Using the probabilities from parts b and c, state whether you believe that the reporter's inference is reasonable and explain why.

TA5.3 A mail-order company receives an average of 45 orders per day.

  a. Find the probability that it will receive exactly 55 orders on a certain day.

  b. Find the probability that it will receive at most 30 orders on a certain day.

  c. Let x denote the number of orders received by this company on a given day. Obtain the probability distribution of x.

  d. Using the probability distribution obtained in part c, find the following probabilities.

   i. $P(x \geq 50)$ **ii.** $P(x < 35)$ **iii.** $P(36 < x < 52)$

TA5.4 A commuter airline receives an average of 14 complaints per week from its passengers. Let x denote the number of complaints received by this airline during a given week.

  a. Find $P(x = 0)$. If your answer is zero, does it mean that this cannot happen? Explain.

  b. Find $P(x \leq 10)$.

  c. Obtain the probability distribution of x.

  d. Using the probability distribution obtained in part c, find the following probabilities.

   i. $P(x > 19)$ **ii.** $P(x \leq 9)$ **iii.** $P(10 \leq x \leq 17)$

TA5.5 The purpose of this assignment is to investigate some of the properties of the binomial probability distribution.

  a. Suppose that for a binomial experiment with $n = 10$ trials, the probability of success is $p = .10$, and x is the number of successes. Obtain the probability distribution of x. Make a histogram for the probability distribution and comment on its shape. Find the mean and standard deviation of x.

  b. Suppose that for a binomial experiment with $n = 10$ trials, the probability of success is $p = .50$, and x is the number of successes. Obtain the probability distribution of x. Make a histogram for the probability distribution and comment on its shape. Find the mean and standard deviation of x.

  c. Suppose that for a binomial experiment with $n = 10$ trials, the probability of success is $p = .80$, and x is the number of successes. Obtain the probability distribution of x. Make a histogram for the probability distribution and comment on its shape. Find the mean and standard deviation of x.

  d. Which of the distributions in parts a–c had the greatest variability in the number of successes? Why does this make sense?

CHAPTER 6

Zoran Karapancev/Shutterstock

Continuous Random Variables and the Normal Distribution

Have you ever participated in a road race? If you have, where did you stand in comparison to the other runners? Do you think the time taken to finish a road race varies as much among runners as the runners themselves? See Case Study 6–1 for the distribution of times for runners who completed the Manchester (Connecticut) Road Race in 2014.

Discrete random variables and their probability distributions were presented in Chapter 5. Section 5.1 defined a continuous random variable as a variable that can assume any value in one or more intervals.

The possible values that a continuous random variable can assume are infinite and uncountable. For example, the variable that represents the time taken by a worker to commute from home to work is a continuous random variable. Suppose 5 minutes is the minimum time and 130 minutes is the maximum time taken by all workers to commute from home to work. Let x be a continuous random variable that denotes the time taken to commute from home to work by a randomly selected worker. Then x can assume any value in the interval 5 to 130 minutes. This interval contains an infinite number of values that are uncountable.

A continuous random variable can possess one of many probability distributions. In this chapter, we discuss the normal probability distribution and the normal distribution as an approximation to the binomial distribution.

6.1 | Continuous Probability Distribution and the Normal Probability Distribution

In this section we will learn about the continuous probability distribution and its properties and then discuss the normal probability distribution.

In Chapter 5, we defined a **continuous random variable** as a random variable whose values are not countable. A continuous random variable can assume any value over an interval or intervals. Because the number of values contained in any interval is infinite, the possible number of values that a continuous random variable can assume is also infinite. Moreover, we cannot count these values. In Chapter 5, it was stated that the life of a battery, heights of people, time taken to complete an examination, amount of milk in a gallon container, weights of babies, and prices of houses are all examples of continuous random variables. Note that although money can be counted, variables involving money are often represented by continuous random variables. This is so because a variable involving money often has a very large number of outcomes.

6.1.1   Continuous Probability Distribution

Suppose 5000 female students are enrolled at a university, and x is the continuous random variable that represents the height of a randomly selected female student. Table 6.1 lists the frequency and relative frequency distributions of x.

Table 6.1   Frequency and Relative Frequency Distributions of Heights of Female Students

Height of a Female Student (inches) x	f	Relative Frequency
60 to less than 61	90	.018
61 to less than 62	170	.034
62 to less than 63	460	.092
63 to less than 64	750	.150
64 to less than 65	970	.194
65 to less than 66	760	.152
66 to less than 67	640	.128
67 to less than 68	440	.088
68 to less than 69	320	.064
69 to less than 70	220	.044
70 to less than 71	180	.036
	$N = 5000$	Sum = 1.0

The relative frequencies given in Table 6.1 can be used as the probabilities of the respective classes. Note that these are exact probabilities because we are considering the population of all female students enrolled at the university.

Figure 6.1 displays the histogram and polygon for the relative frequency distribution of Table 6.1. Figure 6.2 shows the smoothed polygon for the data of Table 6.1. The smoothed polygon is an approximation of the *probability distribution curve* of the continuous random variable x. Note that each class in Table 6.1 has a width equal to 1 inch. If the width of classes is more (or less) than 1 unit, we first obtain the *relative frequency densities* and then graph these relative frequency densities to obtain the distribution curve. The relative frequency density of a class is obtained by dividing the relative frequency of that class by the class width. The relative frequency densities are calculated to make the sum of the areas of all rectangles in the histogram equal to 1.0. Case Study 6–1, which appears later in this section, illustrates this procedure. The probability distribution curve of a continuous random variable is also called its *probability density function.*

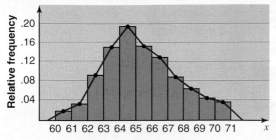

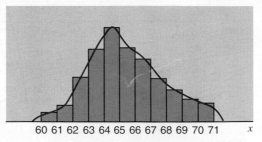

Figure 6.1   Histogram and polygon for Table 6.1.

Figure 6.2   Probability distribution curve for heights.

The probability distribution of a continuous random variable possesses the following *two* *characteristics*.

1. The probability that x assumes a value in any interval lies in the range 0 to 1.

2. The total probability of all the (mutually exclusive) intervals within which x can assume a value is 1.0.

The first characteristic states that the area under the probability distribution curve of a continuous random variable between any two points is between 0 and 1, as shown in Figure 6.3. The second characteristic indicates that the total area under the probability distribution curve of a continuous random variable is always 1.0, or 100%, as shown in Figure 6.4.

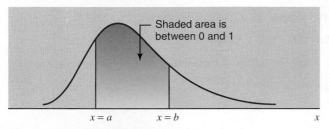

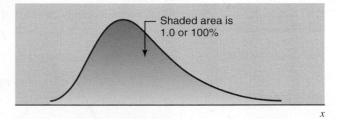

Figure 6.3   Area under a curve between two points.

Figure 6.4   Total area under a probability distribution curve.

The probability that a continuous random variable x assumes a value within a certain interval is given by the area under the curve between the two limits of the interval, as shown in Figure 6.5. The shaded area under the curve from a to b in this figure gives the probability that x falls in the interval a to b; that is,

$$P(a \leq x \leq b) = \text{Area under the curve from } a \text{ to } b$$

Note that the interval $a \leq x \leq b$ states that x is greater than or equal to a but less than or equal to b.

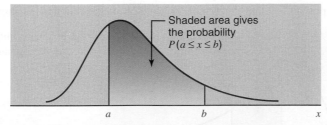

Figure 6.5   Area under the curve as probability.

Reconsider the example on the heights of all female students at a university. The probability that the height of a randomly selected female student from this university lies in the interval 65 to 68 inches is given by the area under the distribution curve of the heights of all female students from $x = 65$ to $x = 68$, as shown in Figure 6.6. This probability is written as

$$P(65 \leq x \leq 68)$$

which states that x is greater than or equal to 65 but less than or equal to 68.

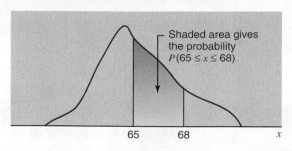

Figure 6.6   Probability that x lies in the interval 65 to 68.

For a continuous probability distribution, the probability is always calculated for an interval. For example, in Figure 6.6, the interval representing the shaded area is from 65 to 68. Consequently, the shaded area in that figure gives the probability for the interval $65 \leq x \leq 68$. In other words, this shaded area gives the probability that the height of a randomly selected female student from this university is in the interval 65 to 68 inches.

The probability that a continuous random variable x assumes a single value is always zero. This is so because the area of a line, which represents a single point, is zero. For example, if x is the height of a randomly selected female student from that university, then the probability that this student is exactly 66.8 inches tall is zero; that is,

$$P(x = 66.8) = 0$$

This probability is shown in Figure 6.7. Similarly, the probability for x to assume any other single value is zero.

In general, if a and b are two of the values that x can assume, then

$$P(a) = 0 \quad \text{and} \quad P(b) = 0$$

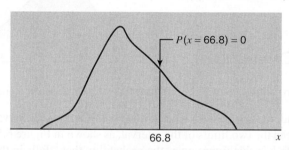

Figure 6.7   The probability of a single value of x is zero.

From this we can deduce that for a continuous random variable,

$$P(a \leq x \leq b) = P(a < x \leq b) = P(a \leq x < b) = P(a < x < b)$$

In other words, the probability that x assumes a value in the interval a to b is the same whether or not the values a and b are included in the interval. For the example on the heights of female students, the probability that a randomly selected female student is between 65 and 68 inches tall is the same as the probability that this female is 65 to 68 inches tall. This is shown in Figure 6.8.

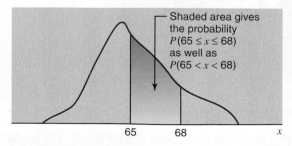

Figure 6.8   Probability "from 65 to 68" and "between 65 and 68."

Note that the interval "between 65 and 68" represents "$65 < x < 68$" and it does not include 65 and 68. On the other hand, the interval "from 65 to 68" represents "$65 \leq x \leq 68$" and it does include 65 and 68. However, as mentioned previously, in the case of a continuous random variable, both of these intervals contain the same probability or area under the curve.

Case Study 6–1 on the next page describes how we obtain the probability distribution curve of a continuous random variable.

Distribution of Time Taken to Run a Road Race

The following table gives the frequency and relative frequency distributions for the time (in minutes) taken to complete the Manchester Road Race (held on November 27, 2014) by a total of 11,682 participants who finished that race. This event is held every year on Thanksgiving Day in Manchester, Connecticut. The total distance of the course is 4.748 miles. The relative frequencies in the following table are used to construct the histogram and polygon in Figure 6.9.

Class	Frequency	Relative Frequency
20 to less than 25	53	.0045
25 to less than 30	246	.0211
30 to less than 35	763	.0653
35 to less than 40	1443	.1235
40 to less than 45	1633	.1398
45 to less than 50	1906	.1632
50 to less than 55	2164	.1852
55 to less than 60	1418	.1214
60 to less than 65	672	.0575
65 to less than 70	380	.0325
70 to less than 75	230	.0197
75 to less than 80	176	.0151
80 to less than 85	186	.0159
85 to less than 90	130	.0111
90 to less than 95	113	.0097
95 to less than 100	90	.0077
100 to less than 105	33	.0028
105 to less than 110	36	.0031
110 to less than 115	9	.0008
115 to less than 120	1	.0001
	$\Sigma f = 11{,}682$	Sum = 1.000

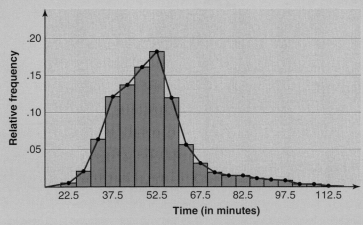

Figure 6.9 Histogram and polygon for the road race data.

To derive the probability distribution curve for these data, we calculate the relative frequency densities by dividing the relative frequencies by the class widths. The width of each class in the above table is 5. By dividing the relative frequencies by 5, we obtain the relative frequency densities, which are recorded in the next table. Using the relative frequency densities, we draw a histogram and a smoothed polygon, as shown in Figure 6.10. The curve in this figure is the probability distribution curve for the road race data.

Note that the areas of the rectangles in Figure 6.9 do not give probabilities (which are approximated by relative frequencies). Rather here the heights of the rectangles give the probabilities. This is so because the base of each rectangle is 5 in this histogram. Consequently, the area of any rectangle is given by its height multiplied by 5. Thus the total area of all the rectangle in Figure 6.9 is 5.0, not 1.0. However, in Figure 6.10, it is the areas, not the heights, of rectangles that give the probabilities of the respective classes. Thus, if we add the areas of all the rectangles in Figure 6.10, we obtain the sum of all probabilities equal to 1.0. Consequently the total area under the curve is equal to 1.0.

Class	Relative Frequency Density
20 to less than 25	.00090
25 to less than 30	.00422
30 to less than 35	.01306
35 to less than 40	.02470
40 to less than 45	.02796
45 to less than 50	.03264
50 to less than 55	.03704
55 to less than 60	.02428
60 to less than 65	.01150
65 to less than 70	.00650
70 to less than 75	.00394
75 to less than 80	.00302
80 to less than 85	.00318
85 to less than 90	.00222
90 to less than 95	.00194
95 to less than 100	.00154
100 to less than 105	.00056
105 to less than 110	.00062
110 to less than 115	.00016
115 to less than 120	.00002

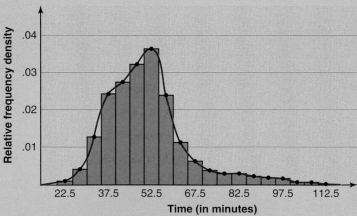

Figure 6.10 Probability distribution curve for the road race data.

The probability distribution of a continuous random variable has a mean and a standard deviation, which are denoted by μ and σ, respectively. The mean and standard deviation of the probability distribution given in the above table and Figure 6.10 are 50.873 and 14.011 minutes, respectively. These values of μ and σ are calculated by using the data on all 11,682 participants.

Source: This case study is based on data published on the official Web site of the Manchester Road Race.

6.1.2 The Normal Distribution

The normal distribution is one of the many probability distributions that a continuous random variable can possess. The normal distribution is the most important and most widely used of all probability distributions. A large number of phenomena in the real world are approximately normally distributed. The continuous random variables representing heights and weights of people, scores on an examination, weights of packages (e.g., cereal boxes, boxes of cookies), amount of milk in a gallon, life of an item (such as a light-bulb or a television set), and time taken to complete a certain job have all been observed to have an approximate normal distribution.

A normal probability distribution or a *normal curve* is a bell-shaped (symmetric) curve. Such a curve is shown in Figure 6.11. Its mean is denoted by μ and its standard deviation by σ. A continuous random variable x that has a normal distribution is called a *normal random variable*. Note that not all bell-shaped curves represent a normal distribution curve. Only a specific kind of bell-shaped curve represents a normal curve.

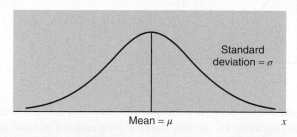

Figure 6.11   Normal distribution with mean μ and standard deviation σ.

Standard deviation $= \sigma$

Mean $= \mu$

x

Normal Probability Distribution   A *normal probability distribution*, when plotted, gives a bell-shaped curve such that:

1. The total area under the curve is 1.0.
2. The curve is symmetric about the mean.
3. The two tails of the curve extend indefinitely.

A normal distribution possesses the following three characteristics:

1. The total area under a normal distribution curve is 1.0, or 100%, as shown in Figure 6.12.

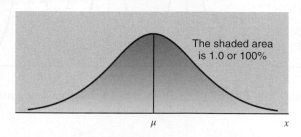

Figure 6.12   Total area under a normal curve.

The shaded area is 1.0 or 100%

μ

x

2. A normal distribution curve is symmetric about the mean, as shown in Figure 6.13. Consequently, 50% of the total area under a normal distribution curve lies on the left side of the mean, and 50% lies on the right side of the mean.

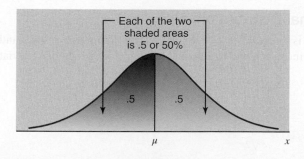

Figure 6.13   A normal curve is symmetric about the mean

Each of the two shaded areas is .5 or 50%

.5   .5

μ

x

3. The tails of a normal distribution curve extend indefinitely in both directions without touching or crossing the horizontal axis. Although a normal distribution curve never meets the horizontal axis, beyond the points represented by $\mu - 3\sigma$ and $\mu + 3\sigma$ it becomes so close to

this axis that the area under the curve beyond these points in both directions is very small and can be taken as very close to zero (but not zero). The actual area in each tail of the standard normal distribution curve beyond three standard deviations of the mean is .0013. These areas are shown in Figure 6.14.

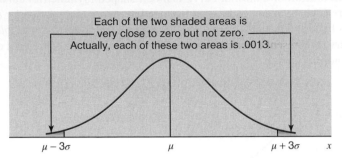

Figure 6.14   Areas of the normal curve beyond $\mu \pm 3\sigma$.

The mean, μ, and the standard deviation, σ, are the **parameters** of the normal distribution. Given the values of these two parameters, we can find the area under a normal distribution curve for any interval. Remember, there is not just one normal distribution curve but a *family* of normal distribution curves. Each different set of values of μ and σ gives a normal distribution curve with different height and/or spread. The value of μ determines the center of a normal distribution curve on the horizontal axis, and the value of σ gives the spread of the normal distribution curve. The three normal distribution curves shown in Figure 6.15 have the same mean but different standard deviations. By contrast, the three normal distribution curves in Figure 6.16 have different means but the same standard deviation.

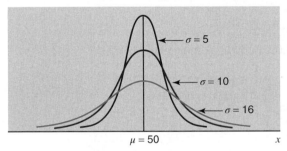

Figure 6.15   Three normal distribution curves with the same mean but different standard deviations.

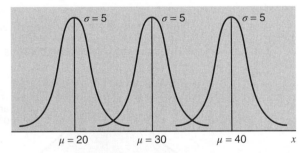

Figure 6.16   Three normal distribution curves with different means but the same standard deviation.

Like the binomial and Poisson probability distributions discussed in Chapter 5, the normal probability distribution can also be expressed by a mathematical equation.[1] However, we will not use this equation to find the area under a normal distribution curve. Instead, we will use Table IV of Appendix B.

6.1.3   The Standard Normal Distribution

> The normal distribution with $\mu = 0$ and $\sigma = 1$ is called the **standard normal distribution**.

The **standard normal distribution** is a special case of the normal distribution. For the standard normal distribution, the value of the mean is equal to zero and the value of the standard deviation is equal to 1.

[1]The equation of the normal distribution is

$$f(x) = \frac{1}{\sigma \sqrt{2\pi}}\, e^{-(1/2)[(x-\mu)/\sigma]^2}$$

where $e = 2.71828$ and $\pi = 3.14159$ approximately; $f(x)$, called the probability density function, gives the vertical distance between the horizontal axis and the curve at point x. For the information of those who are familiar with integral calculus, the definite integral of this equation from a to b gives the probability that x assumes a value between a and b.

Figure 6.17 displays the standard normal distribution curve. The random variable that possesses the standard normal distribution is denoted by z. In other words, the units for the standard normal distribution curve are denoted by z and are called the **z values** or **z scores**. They are also called *standard units* or *standard scores*.

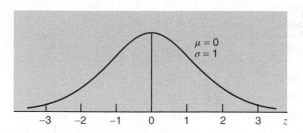

Figure 6.17   The standard normal distribution curve.

$\mu = 0$
$\sigma = 1$

-3 -2 -1 0 1 2 3 z

The units marked on the horizontal axis of the standard normal curve are denoted by z and are called the **z values** or **z scores**. A specific value of z gives the distance between the mean and the point represented by z in terms of the standard deviation.

In Figure 6.17, the horizontal axis is labeled z. The z values on the right side of the mean are positive and those on the left side are negative. **The z value for a point on the horizontal axis gives the distance between the mean and that point in terms of the standard deviation**. For example, a point with a value of $z = 2$ is two standard deviations to the right of the mean. Similarly, a point with a value of $z = -2$ is two standard deviations to the left of the mean.

The standard normal distribution table, Table IV of Appendix B, lists the areas under the standard normal curve to the left of z values from −3.49 to 3.49. To read the standard normal distribution table, we look for the given z value in the table and record the area corresponding to that z value. As shown in Figure 6.18, Table IV gives what is called the cumulative probability to the left of any z value.

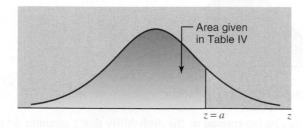

Figure 6.18   Area under the standard normal curve.

Area given in Table IV

$z = a$ z

Although the values of z on the left side of the mean are negative, the area under the curve is always positive.

◀ *Remember*

The area under the standard normal curve between any two points can be interpreted as the probability that z assumes a value within that interval. Examples 6–1 through 6–4 describe how to read Table IV of Appendix B to find areas under the standard normal curve.

EXAMPLE 6–1

Finding the area to the left of a positive z.

Find the area under the standard normal curve to the left of $z = 1.95$.

Solution   We divide the given number 1.95 into two portions: 1.9 (the digit before the decimal and one digit after the decimal) and .05 (the second digit after the decimal). (Note that $1.95 = 1.9 + .05$.) To find the required area under the standard normal curve, we locate 1.9 in the column for z on the left side of Table IV of Appendix B and .05 in the row for z at the top of

Table IV. The entry where the row for 1.9 and the column for .05 intersect gives the area under the standard normal curve to the left of $z = 1.95$. The relevant portion of Table IV is reproduced here as Table 6.2. From Table IV or Table 6.2, the entry where the row for 1.9 and the column for .05 cross is .9744. Consequently, the area under the standard normal curve to the left of $z = 1.95$ is .9744. This area is shown in Figure 6.19. (It is always helpful to sketch the curve and mark the area you are determining.)

Table 6.2   Area Under the Standard Normal Curve to the Left of $z = 1.95$

z	.00	.01	...	.05	...	.09
−3.4	.0003	.0003	...	.0003	...	.0002
−3.3	.0005	.0005	...	.0004	...	.0003
−3.2	.0007	.0007	...	.0006	...	.0005
.	.	.	...	.	...	.
.	.	.	...	.	...	.
.	.	.	...	.	...	.
1.9	.9713	.9719	...	.9744	...	.9767
.	.	.	...	.	...	.
.	.	.	...	.	...	.
.	.	.	...	.	...	.
3.4	.9997	.9997	...	.9997	...	.9998

Required area

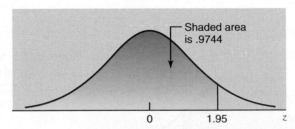

Figure 6.19   Area to the left of $z = 1.95$.

The area to the left of $z = 1.95$ can be interpreted as the probability that z assumes a value less than 1.95; that is,

$$\text{Area to the left of } 1.95 = P(z < 1.95) = \mathbf{.9744}$$

As mentioned in Section 6.1, the probability that a continuous random variable assumes a single value is zero. Therefore, the probability that z is exactly equal to 1.95 is zero, that is:

$$P(z = 1.95) = 0$$

Hence,

$$P(z < 1.95) = P(z \leq 1.95) = \mathbf{.9744}$$

EXAMPLE 6–2

Finding the area between a negative z and z = 0.

Find the area under the standard normal curve from $z = -2.17$ to $z = 0$.

Solution   To find the area from $z = -2.17$ to $z = 0$, which is the shaded area in Figure 6.20, first we find the areas to the left of $z = 0$ and to the left of $z = -2.17$ in the standard normal distribution table (Table IV). As shown in Table 6.3, these two areas are .5000 and .0150, respectively. Next we subtract .0150 from .5000 to find the required area.

Table 6.3   Area Under the Standard Normal Curve

z	.00	.01	...	.07	...	.09
−3.4	.0003	.0003	...	.0003	...	.0002
−3.3	.0005	.0005	...	.0004	...	.0003
−3.2	.0007	.0007	...	.0005	...	.0005
.	.	.	...	.	...	.
.	.	.	...	.	...	.
.	.	.	...	.	...	.
−2.1	.0179	.0174	...	.0150 ←	...	.0143
.	.	.	...	.	...	.
.	.	.	...	.	...	.
.	.	.	...	.	...	.
0.0	.5000 ←	.5040	...	.5279	...	.5359
.	.	.	...	.	...	.
.	.	.	...	.	...	.
3.4	.9997	.9997	...	.9997	...	.9998

Area to the left of $z = 0$ Area to the left of $z = -2.17$

The area from $z = -2.17$ to $z = 0$ gives the probability that z lies in the interval -2.17 to 0 (see Figure 6.20); that is,

Area from -2.17 to $0 = P(-2.17 \leq z \leq 0) = .5000 - .0150 = .4850$

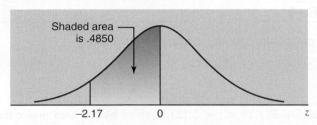

Figure 6.20   Area from $z = -2.17$ to $z = 0$.

EXAMPLE 6–3

Finding the area in the right and left tails.

Find the following areas under the standard normal curve.

(a) Area to the right of $z = 2.32$

(b) Area to the left of $z = -1.54$

Solution

(a) Here we are to find the area that is shaded in Figure 6.21. As mentioned earlier, the normal distribution table gives the area to the left of a z value. To find the area to the right of $z = 2.32$, first we find the area to the left of $z = 2.32$. Then we subtract this area from 1.0, which is the total area under the curve. From Table IV, the area to the left of $z = 2.32$ is .9898. Consequently, the required area is $1.0 - .9898 = .0102$, as shown in Figure 6.21.

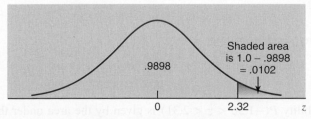

Figure 6.21   Area to the right of $z = 2.32$.

Note that the area to the right of $z = 2.32$ gives the probability that z is greater than 2.32. Thus,

$$\text{Area to the right of } 2.32 = P(z > 2.32) = 1.0 - .9898 = \textbf{.0102}$$

(b) To find the area under the standard normal curve to the left of $z = -1.54$, we find the area in Table IV that corresponds to -1.5 in the z column and .04 in the top row. This area is .0618. This area is shown in Figure 6.22.

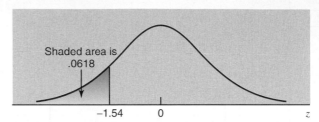

Figure 6.22   Area to the left of $z = -1.54$.

The area to the left of $z = -1.54$ gives the probability that z is less than -1.54. Thus,

$$\text{Area to the left of } -1.54 = P(z < -1.54) = \textbf{.0618}$$

EXAMPLE 6–4

Find the following probabilities for the standard normal curve.

(a) $P(1.19 < z < 2.12)$ **(b)** $P(-1.56 < z < 2.31)$ **(c)** $P(z > -.75)$

Solution

Finding the area between two positive values of z.

(a) The probability $P(1.19 < z < 2.12)$ is given by the area under the standard normal curve between $z = 1.19$ and $z = 2.12$, which is the shaded area in Figure 6.23.

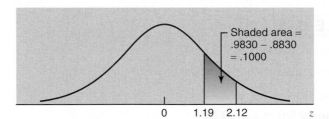

Figure 6.23   Finding $P(1.19 < z < 2.12)$.

To find the area between $z = 1.19$ and $z = 2.12$, first we find the areas to the left of $z = 1.19$ and $z = 2.12$. Then we subtract the smaller area (to the left of $z = 1.19$) from the larger area (to the left of $z = 2.12$).

From Table IV for the standard normal distribution, we find

$$\text{Area to the left of } 1.19 = .8830$$
$$\text{Area to the left of } 2.12 = .9830$$

Then, the required probability is

$$P(1.19 < z < 2.12) = \text{Area between } 1.19 \text{ and } 2.12$$
$$= .9830 - .8830 = \textbf{.1000}$$

Finding the area between a positive and a negative value of z.

(b) The probability $P(-1.56 < z < 2.31)$ is given by the area under the standard normal curve between $z = -1.56$ and $z = 2.31$, which is the shaded area in Figure 6.24.

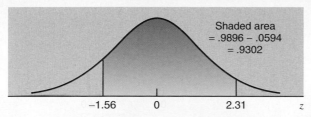

Figure 6.24   Finding $P(-1.56 < z < 2.31)$.

From Table IV for the standard normal distribution, we have

$$\text{Area to the left of } -1.56 = .0594$$
$$\text{Area to the left of } 2.31 = .9896$$

The required probability is

$$P(-1.56 < z < 2.31) = \text{Area between } -1.56 \text{ and } 2.31$$
$$= .9896 - .0594 = \mathbf{.9302}$$

(c)   The probability $P(z > -.75)$ is given by the area under the standard normal curve to the right of $z = -.75$, which is the shaded area in Figure 6.25.

Finding the area to the right of a negative value of z.

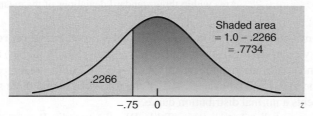

Figure 6.25   Finding $P(z > -.75)$.

From Table IV for the standard normal distribution,

$$\text{Area to the left of } -.75 = .2266$$

The required probability is

$$P(z > -.75) = \text{Area to the right of } -.75 = 1.0 - .2266 = \mathbf{.7734}$$

In the discussion in Section 3.4 of Chapter 3 on the use of the standard deviation, we discussed the empirical rule for a bell-shaped curve. That empirical rule is based on the standard normal distribution. By using the normal distribution table, we can now verify the empirical rule as follows.

1.   The total area within one standard deviation of the mean is 68.26%. This area is given by the difference between the area to the left of $z = 1.0$ and the area to the left of $z = -1.0$. As shown in Figure 6.26, this area is $.8413 - .1587 = .6826$, or 68.26%.

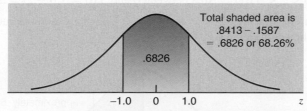

Figure 6.26   Area within one standard deviation of the mean.

2.   The total area within two standard deviations of the mean is 95.44%. This area is given by the difference between the area to the left of $z = 2.0$ and the area to the left of $z = -2.0$. As shown in Figure 6.27, this area is $.9772 - .0228 = .9544$, or 95.44%.

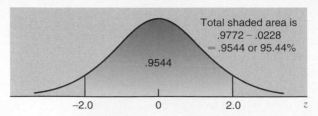

Figure 6.27   Area within two standard deviations of the mean.

3.   The total area within three standard deviations of the mean is 99.74%. This area is given by the difference between the area to the left of $z = 3.0$ and the area to the left of $z = -3.0$. As shown in Figure 6.28, this area is $.9987 - .0013 = .9974$, or 99.74%.

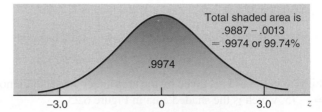

Figure 6.28   Area within three standard deviations of the mean.

Again, as mentioned earlier, only a specific bell-shaped curve represents the normal distribution. Now we can state that a bell-shaped curve that contains (about) 68.26% of the total area within one standard deviation of the mean, (about) 95.44% of the total area within two standard deviations of the mean, and (about) 99.74% of the total area within three standard deviations of the mean represents a normal distribution curve.

The standard normal distribution table, Table IV of Appendix B, goes from $z = -3.49$ to $z = 3.49$. Consequently, if we need to find the area to the left of $z = -3.50$ or a smaller value of z, we can assume it to be approximately 0.0. If we need to find the area to the left of $z = 3.50$ or a larger number, we can assume it to be approximately 1.0. Example 6–5 illustrates this procedure.

EXAMPLE 6–5

Find the following probabilities for the standard normal curve.

    (a) $P(0 < z < 5.67)$ **(b)** $P(z < -5.35)$

Solution

Finding the area between z = 0 and a value of z greater than 3.49.

    (a) The probability $P(0 < z < 5.67)$ is given by the area under the standard normal curve between $z = 0$ and $z = 5.67$. Because $z = 5.67$ is greater than 3.49 and is not in Table IV, the area under the standard normal curve to the left of $z = 5.67$ can be approximated by 1.0. Also, the area to the left of $z = 0$ is .5. Hence, the required probability is

$$P(0 < z < 5.67) = \text{Area between 0 and 5.67} = 1.0 - .5 = \textbf{.5} \text{ approximately}$$

Note that the area between $z = 0$ and $z = 5.67$ is not exactly .5 but very close to .5. This area is shown in Figure 6.29.

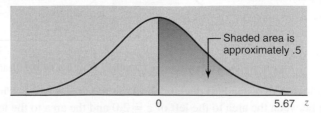

Figure 6.29   Area between $z = 0$ and $z = 5.67$.

(b) The probability $P(z < -5.35)$ represents the area under the standard normal curve to the left of $z = -5.35$. Since $z = -5.35$ is not in the table, we can assume that this area is approximately .00. This is shown in Figure 6.30.

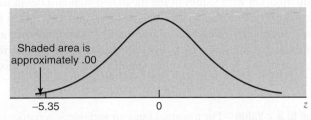

Shaded area is approximately .00

−5.35   0   z

Figure 6.30   Area to the left of $z = -5.35$.

Finding the area to the left of a z that is less than −3.49.

The required probability is

$$P(z < -5.35) = \text{Area to the left of } -5.35 = .00 \text{ approximately}$$

Again, note that the area to the left of $z = -5.35$ is not exactly .00 but very close to .00. We can find the exact areas for parts (a) and (b) of this example by using technology. The reader should do that.

EXERCISES

CONCEPTS AND PROCEDURES

6.1   What is the difference between the probability distribution of a discrete random variable and that of a continuous random variable? Explain.

6.2   Let x be a continuous random variable. What is the probability that x assumes a single value, such as a?

6.3   For a continuous probability distribution, explain why the following holds true.

$$P(a < x < b) = P(a < x \leq b) = P(a \leq x < b) = P(a \leq x \leq b)$$

6.4   Briefly explain the main characteristics of a normal distribution. Illustrate with the help of graphs.

6.5   Briefly describe the standard normal distribution curve.

6.6   What are the parameters of the normal distribution?

6.7   How do the width and height of a normal distribution change when its mean remains the same but its standard deviation decreases?

6.8   Do the width and/or height of a normal distribution change when its standard deviation remains the same but its mean increases?

6.9   For the standard normal distribution, what does z represent?

6.10   For the standard normal distribution, find the area within 1.5 standard deviations of the mean—that is, the area between $\mu - 1.5\sigma$ and $\mu + 1.5\sigma$.

6.11   For the standard normal distribution, what is the area within two standard deviations of the mean?

6.12   For the standard normal distribution, what is the area within 2.5 standard deviations of the mean?

6.13   Find the area under the standard normal curve
  a. from $z = 0$ to $z = 1.85$
  b. between $z = 0$ and $z = -1.43$
  c from $z = .84$ to $z = 1.95$
  d. between $z = -.40$ and $z = -2.49$
  e. between $z = -2.15$ and $z = 1.34$

6.14   Find the area under the standard normal curve
  a. to the right of $z = 1.36$
  b. to the left of $z = -1.76$
  c. to the right of $z = -2.48$
  d. to the left of $z = 1.89$

6.15   Obtain the area under the standard normal curve
  a. to the right of $z = 1.24$ **b.** to the left of $z = -1.85$
  c. to the right of $z = -.45$ **d.** to the left of $z = .73$

6.16   Determine the following probabilities for the standard normal distribution.
  a. $P(-2.46 \leq z \leq 1.88)$ **b.** $P(0 \leq z \leq 1.85)$
  c. $P(-2.58 \leq z \leq 0)$ **d.** $P(z \geq .67)$

6.17   Find the following probabilities for the standard normal distribution.
  a. $P(z < -1.48)$ **b.** $P(.67 \leq z \leq 2.34)$
  c. $P(-1.08 \leq z \leq -.93)$ **d.** $P(z < 2.15)$

6.18   Obtain the following probabilities for the standard normal distribution.
  a. $P(z > -1.86)$ **b.** $P(-.68 \leq z \leq 1.85)$
  c. $P(0 \leq z \leq 3.85)$ **d.** $P(-4.34 \leq z \leq 0)$
  e. $P(z > 3.96)$ **f.** $P(z < -5.76)$

6.2 | Standardizing a Normal Distribution

As was shown in the previous section, Table IV of Appendix B can be used to find areas under the standard normal curve. However, in real-world applications, a (continuous) random variable may have a normal distribution with values of the mean and standard deviation that are different from 0 and 1, respectively. The first step in such a case is to convert the given normal distribution to the standard normal distribution. This procedure is called **standardizing a normal distribution**. The units of a normal distribution (which is not the standard normal distribution) are denoted by x. We know from Section 6.1.3 that units of the standard normal distribution are denoted by z.

> **Converting an *x* Value to a *z* Value**   For a normal random variable x, a particular value of x can be converted to its corresponding z value by using the formula
>
> $$z = \frac{x - \mu}{\sigma}$$
>
> where μ and σ are the mean and standard deviation of the normal distribution of x, respectively. When x follows a normal distribution, z follows the standard normal distribution.

Thus, to find the z value for an x value, we calculate the difference between the given x value and the mean, μ, and divide this difference by the standard deviation, σ. If the value of x is equal to μ, then its z value is equal to zero. Note that we will always round z values to two decimal places.

Remember ➤   The z value for the mean of a normal distribution is always zero. The z value for an x value greater than the mean is positive and the z value for an x value smaller than the mean is negative.

Examples 6–6 through 6–10 describe how to convert x values to the corresponding z values and how to find areas under a normal distribution curve.

EXAMPLE 6–6

Converting x values to the corresponding z values.

Let x be a continuous random variable that has a normal distribution with a mean of 50 and a standard deviation of 10. Convert the following x values to z values and find the probability to the left of these points.

(a) $x = 55$ (b) $x = 35$

Solution   For the given normal distribution, $\mu = 50$ and $\sigma = 10$.

(a) The z value for $x = 55$ is computed as follows:

$$z = \frac{x - \mu}{\sigma} = \frac{55 - 50}{10} = .50$$

Thus, the z value for $x = 55$ is .50. The z values for $\mu = 50$ and $x = 55$ are shown in Figure 6.31. Note that the z value for $\mu = 50$ is zero. The value $z = .50$ for $x = 55$ indicates that the distance between $\mu = 50$ and $x = 55$ is 1/2 of the standard deviation, which is 10. Consequently, we can state that the z value represents the distance between μ and x in terms of the standard deviation. Because $x = 55$ is greater than $\mu = 50$, its z value is positive.

From this point on, we will usually show only the z axis below the x axis and not the standard normal curve itself.

To find the probability to the left of $x = 55$, we find the area to the left of $z = .50$ from Table IV. This area is .6915. Therefore,

$$P(x < 55) = P(z < .50) = \textbf{.6915}$$

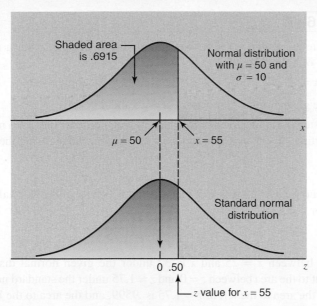

Figure 6.31 Area to the left of $x = 55$.

(b) The z value for $x = 35$ is computed as follows and is shown in Figure 6.32:

$$z = \frac{x - \mu}{\sigma} = \frac{35 - 50}{10} = -1.50$$

Because $x = 35$ is on the left side of the mean (i.e., 35 is less than $\mu = 50$), its z value is negative. As a general rule, whenever an x value is less than the value of μ, its z value is negative.

To find the probability to the left of $x = 35$, we find the area under the normal curve to the left of $z = -1.50$. This area from Table IV is .0668. Hence,

$$P(x < 35) = P(z < -1.50) = \mathbf{.0668}$$

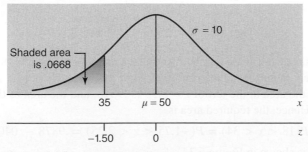

Figure 6.32 Area to the left of $x = 35$.

The z value for an x value that is greater than μ is positive, the z value for an x value that is equal to μ is zero, and the z value for an x value that is less than μ is negative.

◄ *Remember*

To find the area between two values of x for a normal distribution, we first convert both values of x to their respective z values. Then we find the area under the standard normal curve between those two z values. The area between the two z values gives the area between the corresponding x values. Example 6–7 illustrates this case.

EXAMPLE 6–7

Let x be a continuous random variable that is normally distributed with a mean of 25 and a standard deviation of 4. Find the area

(a) between $x = 25$ and $x = 32$ (b) between $x = 18$ and $x = 34$

Solution   For the given normal distribution, $\mu = 25$ and $\sigma = 4$.

(a) The first step in finding the required area is to standardize the given normal distribution by converting $x = 25$ and $x = 32$ to their respective z values using the formula

$$z = \frac{x - \mu}{\sigma}$$

The z value for $x = 25$ is zero because it is the mean of the normal distribution. The z value for $x = 32$ is

$$z = \frac{32 - 25}{4} = 1.75$$

The area between $x = 25$ and $x = 32$ under the given normal distribution curve is equivalent to the area between $z = 0$ and $z = 1.75$ under the standard normal curve. From Table IV, the area to the left of $z = 1.75$ is .9599, and the area to the left of $z = 0$ is .50. Hence, the required area is $.9599 - .50 = .4599$, which is shown in Figure 6.33.

   The area between $x = 25$ and $x = 32$ under the normal curve gives the probability that x assumes a value between 25 and 32. This probability can be written as

$$P(25 < x < 32) = P(0 < z < 1.75) = .9599 - .50 = \mathbf{.4599}$$

Figure 6.33   Area between $x = 25$ and $x = 32$.

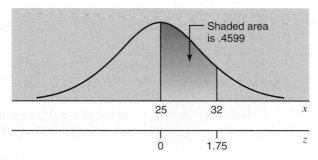

Shaded area is .4599

25   32 x

0   1.75 z

(b) First, we calculate the z values for $x = 18$ and $x = 34$ as follows:

$$\text{For } x = 18: \quad z = \frac{18 - 25}{4} = -1.75$$

$$\text{For } x = 34: \quad z = \frac{34 - 25}{4} = 2.25$$

The area under the given normal distribution curve between $x = 18$ and $x = 34$ is given by the area under the standard normal curve between $z = -1.75$ and $z = 2.25$. From Table IV, the area to the left of $z = 2.25$ is .9878, and the area to the left of $z = -1.75$ is .0401. Hence, the required area is

$$P(18 < x < 34) = P(-1.75 < z < 2.25) = .9878 - .0401 = \mathbf{.9477}$$

This area is shown in Figure 6.34.

Figure 6.34   Area between $x = 18$ and $x = 34$.

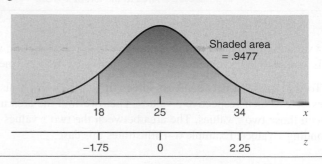

Shaded area = .9477

18   25   34 x

−1.75   0   2.25 z

EXAMPLE 6–8

Let x be a normal random variable with its mean equal to 40 and standard deviation equal to 5. Find the following probabilities for this normal distribution.

    (a) $P(x > 55)$ **(b)** $P(x < 49)$

Solution   For the given normal distribution, $\mu = 40$ and $\sigma = 5$.

  (a) The probability that x assumes a value greater than 55 is given by the area under the normal distribution curve to the right of $x = 55$, as shown in Figure 6.35. This area is calculated by subtracting the area to the left of $x = 55$ from 1.0, which is the total area under the curve.

Calculating the probability to the right of a value of x.

$$\text{For } x = 55: \quad z = \frac{55 - 40}{5} = 3.00$$

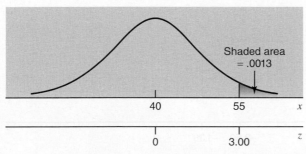

Figure 6.35   Finding $P(x > 55)$.

The required probability is given by the area to the right of $z = 3.00$. To find this area, first we find the area to the left of $z = 3.00$, which is .9987. Then we subtract this area from 1.0. Thus,

$$P(x > 55) = P(z > 3.00) = 1.0 - .9987 = \mathbf{.0013}$$

  (b) The probability that x will assume a value less than 49 is given by the area under the normal distribution curve to the left of 49, which is the shaded area in Figure 6.36. This area is obtained from Table IV as follows.

Calculating the probability to the left of a value of x.

$$\text{For } x = 49: \quad z = \frac{49 - 40}{5} = 1.80$$

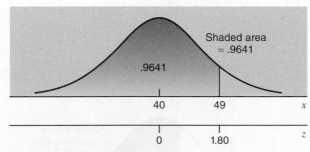

Figure 6.36   Finding $P(x < 49)$.

The required probability is given by the area to the left of $z = 1.80$. This area from Table IV is .9641. Therefore, the required probability is

$$P(x < 49) = P(z < 1.80) = \mathbf{.9641}$$

EXAMPLE 6-9

Finding the area between two x values that are less than the mean.

Let x be a continuous random variable that has a normal distribution with $\mu = 50$ and $\sigma = 8$. Find the probability $P(30 \leq x \leq 39)$.

Solution   For this normal distribution, $\mu = 50$ and $\sigma = 8$. The probability $P(30 \leq x \leq 39)$ is given by the area from $x = 30$ to $x = 39$ under the normal distribution curve. As shown in Figure 6.37, this area is given by the difference between the area to the left of $x = 30$ and the area to the left of $x = 39$.

$$\text{For } x = 30: \quad z = \frac{30 - 50}{8} = -2.50$$

$$\text{For } x = 39: \quad z = \frac{39 - 50}{8} = -1.38$$

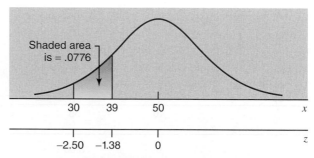

Figure 6.37   Finding $P(30 \leq x \leq 39)$.

To find the required area, we first find the area to the left of $z = -2.50$, which is .0062. Then, we find the area to the left of $z = -1.38$, which is .0838. The difference between these two areas gives the required probability, which is

$$P(30 \leq x \leq 39) = P(-2.50 \leq z \leq -1.38) = .0838 - .0062 = \mathbf{.0776}$$

EXAMPLE 6-10

Let x be a continuous random variable that has a normal distribution with a mean of 80 and a standard deviation of 12. Find the area under the normal distribution curve

    (a) from $x = 70$ to $x = 135$ **(b)** to the left of 27

Solution   For the given normal distribution, $\mu = 80$ and $\sigma = 12$.

Finding the area between two x values that are on different sides of the mean.

    (a) The z values for $x = 70$ and $x = 135$ are:

$$\text{For } x = 70: \quad z = \frac{70 - 80}{12} = -.83$$

$$\text{For } x = 135: \quad z = \frac{135 - 80}{12} = 4.58$$

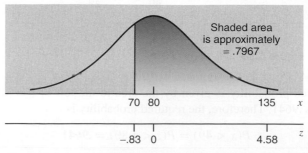

Figure 6.38   Area between $x = 70$ and $x = 135$.

Thus, to find the required area we find the areas to the left of $z = -.83$ and to the left of $z = 4.58$ under the standard normal curve. From Table IV, the area to the left of $z = -.83$ is .2033 and the area to the left of $z = 4.58$ is approximately 1.0. Note that $z = 4.58$ is not in Table IV.

Hence,

$$P(70 \leq x \leq 135) = P(-.83 \leq z \leq 4.58)$$
$$= 1.0 - .2033 = .7967 \text{ approximately}$$

Figure 6.38 shows this area.

(b) First we find the z value for $x = 27$.

Finding an area in the left tail.

$$\text{For } x = 27: \quad z = \frac{27 - 80}{12} = -4.42$$

As shown in Figure 6.39, the required area is given by the area under the standard normal distribution curve to the left of $z = -4.42$. This area is approximately zero.

$$P(x < 27) = P(z < -4.42) = .00 \text{ approximately}$$

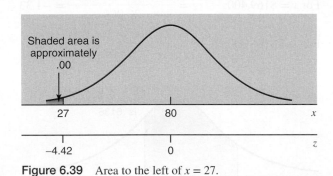

Figure 6.39   Area to the left of $x = 27$.

EXERCISES

CONCEPTS AND PROCEDURES

6.19   Determine the z value for each of the following x values for a normal distribution with $\mu = 16$ and $\sigma = 3$.

    a. $x = 15$ **b.** $x = 24$ **c.** $x = 18$ **d.** $x = 10$

6.20   Find the following areas under a normal distribution curve with $\mu = 14$ and $\sigma = 2$.

    a. Area between $x = 7.76$ and $x = 14$
    b. Area between $x = 14.48$ and $x = 16.54$
    c. Area from $x = 8.22$ to $x = 10.06$

6.21   Find the area under a normal distribution curve with $\mu = 18.3$ and $\sigma = 3.6$

    a. to the left of $x = 10.9$ **b.** to the right of $x = 14$
    c. to the left of $x = 22.7$ **d.** to the right of $x = 29.2$

6.22   Let x be a continuous random variable that is normally distributed with a mean of 25 and a standard deviation of 6. Find the probability that x assumes a value

    a. between 28 and 34   **b.** between 20 and 35

6.23   Let x be a continuous random variable that has a normal distribution with a mean of 30 and a standard deviation of 2. Find the probability that x assumes a value

    a. between 29 and 35   **b.** from 34 to 50

6.24   Let x be a continuous random variable that is normally distributed with a mean of 80 and a standard deviation of 12. Find the probability that x assumes a value

    a. greater than 69   **b.** less than 73
    c. greater than 101   **d.** less than 87

6.3 | Applications of the Normal Distribution

Sections 6.1 and 6.2 discussed the normal distribution, how to convert a normal distribution to the standard normal distribution, and how to find areas under a normal distribution curve. This section presents examples that illustrate the applications of the normal distribution.

EXAMPLE 6–11 **Earnings of Internal Medicine Physicians**

Using the normal distribution: the area between two points on different sides of the mean.

According to the 2015 Physician Compensation Report by Medscape (a subsidiary of WebMD), American internal medicine physicians earned an average of $196,000 in 2014. Suppose that the 2014 earnings of all American internal medicine physicians are normally distributed with a mean of $196,000 and a standard deviation of $20,000. Find the probability that the 2014 earnings of a randomly selected American internal medicine physician are between $169,400 and $206,800.

Solution   Let x denote the 2014 earnings of a randomly selected American internal medicine physician. Then, x is normally distributed with

$$\mu = \$196,000 \quad \text{and} \quad \sigma = \$20,000$$

The probability that the 2014 earnings of a randomly selected American internal medicine physician are between $169,400 and $206,800 is given by the area under the normal distribution curve of x that falls between $x = \$169,400$ and $x = \$206,800$ as shown in Figure 6.40. To find this area, first we find the areas to the left of $x = \$169,400$ and $x = \$206,800$, respectively, and then take the difference between these two areas. We find the z values for the two x values as follows.

$$\text{For } x = \$169,400: \quad z = \frac{169,400 - 196,000}{20,000} = -1.33$$

$$\text{For } x = \$206,800: \quad z = \frac{206,800 - 196,000}{20,000} = .54$$

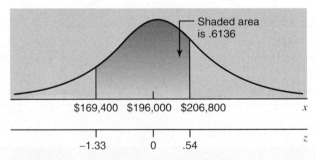

Figure 6.40   Area between $x = \$169,400$ and $x = \$206,800$.

Thus, the required probability is given by the difference between the areas under the standard normal curve to the left of $z = -1.33$ and to the left of $z = .54$. From Table IV in Appendix B, the area to the left of $z = -1.33$ is .0918, and the area to the left of $z = .54$ is .7054. Hence, the required probability is

$$P(\$169,400 < x < \$206,800) = P(-1.33 < z < .54) = .7054 - .0918 = \mathbf{.6136}$$

Thus, the probability is .6136 that the 2014 earnings of a randomly selected American internal medicine physician are between $169,400 and $206,800. Converting this probability into a percentage, we can also state that (about) 61.36% of American internal medicine physicians earned between $169,400 and $206,800 in 2014.

EXAMPLE 6–12 **Time Taken to Assemble a Toy**

Using the normal distribution: probability that x is less than a value that is to the right of the mean.

A racing car is one of the many toys manufactured by the Mack Corporation. The assembly times for this toy follow a normal distribution with a mean of 55 minutes and a standard deviation of 4 minutes. The company closes at 5 P.M. every day. If one worker starts to assemble a racing car at 4 P.M., what is the probability that she will finish this job before the company closes for the day?

Solution Let x denote the time this worker takes to assemble a racing car. Then, x is normally distributed with

$$\mu = 55 \text{ minutes} \quad \text{and} \quad \sigma = 4 \text{ minutes}$$

We are to find the probability that this worker can assemble this car in 60 minutes or less (between 4 and 5 P.M.). This probability is given by the area under the normal curve to the left of $x = 60$ minutes as shown in Figure 6.41. First we find the z value for $x = 60$ as follows.

$$\text{For } x = 60: \quad z = \frac{60 - 55}{4} = 1.25$$

The required probability is given by the area under the standard normal curve to the left of $z = 1.25$, which is .8944 from Table IV of Appendix B. Thus, the required probability is

$$P(x \le 60) = P(z \le 1.25) = \textbf{.8944}$$

Thus, the probability is .8944 that this worker will finish assembling this racing car before the company closes for the day.

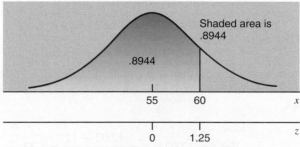

Figure 6.41 Area to the left of $x = 60$.

EXAMPLE 6–13 | Amount of Soda in a Can

Using the normal distribution.

Hupper Corporation produces many types of soft drinks, including Orange Cola. The filling machines are adjusted to pour 12 ounces of soda into each 12-ounce can of Orange Cola. However, the actual amount of soda poured into each can is not exactly 12 ounces; it varies from can to can. It has been observed that the net amount of soda in such a can has a normal distribution with a mean of 12 ounces and a standard deviation of .015 ounce.

(a) What is the probability that a randomly selected can of Orange Cola contains 11.97 to 11.99 ounces of soda?

(b) What percentage of the Orange Cola cans contain 12.02 to 12.07 ounces of soda?

Solution Let x be the net amount of soda in a can of Orange Cola. Then, x has a normal distribution with $\mu = 12$ ounces and $\sigma = .015$ ounce.

(a) The probability that a randomly selected can contains 11.97 to 11.99 ounces of soda is given by the area under the normal distribution curve from $x = 11.97$ to $x = 11.99$. This area is shown in Figure 6.42. First we convert the two values of x to the corresponding z values.

Calculating the probability between two points that are to the left of the mean.

$$\text{For } x = 11.97: \quad z = \frac{11.97 - 12}{.015} = -2.00$$

$$\text{For } x = 11.99: \quad z = \frac{11.99 - 12}{.015} = -.67$$

The required probability is given by the area under the standard normal curve between $z = -2.00$ and $z = -.67$. From Table IV of Appendix B, the area to the left of $z = -2.00$ is .0228, and the area to the left of $z = -.67$ is .2514. Hence, the required probability is

$$P(11.97 \leq x \leq 11.99) = P(-2.00 \leq z \leq -.67) = .2514 - .0228 = \textbf{.2286}$$

Thus, the probability is .2286 that a randomly selected can of Orange Cola will contain 11.97 to 11.99 ounces of soda. We can also state that about 22.86% of Orange Cola cans contain 11.97 to 11.99 ounces of soda.

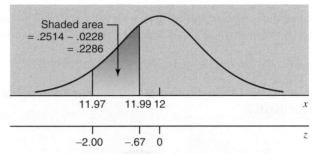

Figure 6.42   Area between $x = 11.97$ and $x = 11.99$.

Calculating the probability between two points that are to the right of the mean.

(b) The percentage of Orange Cola cans that contain 12.02 to 12.07 ounces of soda is given by the area under the normal distribution curve from $x = 12.02$ to $x = 12.07$, as shown in Figure 6.43.

$$\text{For } x = 12.02: \quad z = \frac{12.02 - 12}{.015} = 1.33$$

$$\text{For } x = 12.07: \quad z = \frac{12.07 - 12}{.015} = 4.67$$

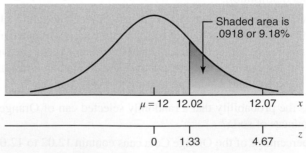

Figure 6.43   Area from $x = 12.02$ to $x = 12.07$.

The required probability is given by the area under the standard normal curve between $z = 1.33$ and $z = 4.67$. From Table IV of Appendix B, the area to the left of $z = 1.33$ is .9082, and the area to the left of $z = 4.67$ is approximately 1.0. Hence, the required probability is

$$P(12.02 \leq x \leq 12.07) = P(1.33 \leq z \leq 4.67) = 1.0 - .9082 = \textbf{.0918}$$

Converting this probability to a percentage, we can state that approximately 9.18% of all Orange Cola cans are expected to contain 12.02 to 12.07 ounces of soda.

EXAMPLE 6–14   Life Span of a Calculator

Finding the area to the left of x that is less than the mean.

Suppose the life span of a calculator manufactured by the Calculators Corporation has a normal distribution with a mean of 54 months and a standard deviation of 8 months. The company guarantees that any calculator that starts malfunctioning within 36 months of the purchase will be replaced by a new one. About what percentage of calculators made by this company are expected to be replaced?

Solution   Let x be the life span of such a calculator. Then x has a normal distribution with $\mu = 54$ and $\sigma = 8$ months. The probability that a randomly selected calculator will start to malfunction within 36 months is given by the area under the normal distribution curve to the left of $x = 36$, as shown in Figure 6.44.

$$\text{For } x = 36: \quad z = \frac{36 - 54}{8} = -2.25$$

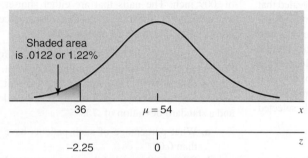

Figure 6.44   Area to the left of $x = 36$.

The required percentage is given by the area under the standard normal curve to the left of $z = -2.25$. From Table IV of Appendix B, this area is .0122. Hence, the required probability is

$$P(x < 36) = P(z < -2.25) = .0122$$

The probability that any randomly selected calculator manufactured by the Calculators Corporation will start to malfunction within 36 months is .0122. Converting this probability to a percentage, we can state that approximately 1.22% of all calculators manufactured by this company are expected to start malfunctioning within 36 months. Hence, 1.22% of the calculators are expected to be replaced.

EXERCISES

APPLICATIONS

6.25   Let x denote the time taken to run a road race. Suppose x is approximately normally distributed with a mean of 180 minutes and a standard deviation of 14 minutes. If one runner is selected at random, what is the probability that this runner will complete this road race

    a. in less than 160 minutes?　　**b.** in 215 to 245 minutes?

6.26   Tommy Wait, a minor league baseball pitcher, is notorious for taking an excessive amount of time between pitches. In fact, his times between pitches are normally distributed with a mean of 36 seconds and a standard deviation of 2.5 seconds. What percentage of his times between pitches are

    a. longer than 40 seconds?　　**b.** between 25 and 30 seconds?

6.27   A construction zone on a highway has a posted speed limit of 40 miles per hour. The speeds of vehicles passing through this construction zone are normally distributed with a mean of 45 miles per

hour and a standard deviation of 4 miles per hour. Find the percentage of vehicles passing through this construction zone that are

    a. exceeding the posted speed limit
    b. traveling at speeds between 55 and 60 miles per hour

6.28   The Bank of Connecticut issues Visa and MasterCard credit cards. It is estimated that the balances on all Visa credit cards issued by the Bank of Connecticut have a mean of $920 and a standard deviation of $270. Assume that the balances on all these Visa cards follow a normal distribution.

    a. What is the probability that a randomly selected Visa card issued by this bank has a balance between $1000 and $1440?
    b. What percentage of the Visa cards issued by this bank have a balance of $730 or more?

6.29   According to the U.S. Employment and Training Administration, the average weekly unemployment benefit paid out in 2008 was $297 (http://www.ows.doleta.gov/unemploy/hb394.asp). Suppose that the current distribution of weekly unemployment benefits paid out is approximately normally distributed with a mean of $297 and a standard deviation of $74.42. Find the probability that a randomly selected American who is receiving unemployment benefits is receiving

    a. more than $450 per week
    b. between $200 and $360 per week

6.30 The management of a supermarket wants to adopt a new promotional policy of giving a free gift to every customer who spends more than a certain amount per visit at this supermarket. The expectation of the management is that after this promotional policy is advertised, the expenditures for all customers at this supermarket will be normally distributed with a mean of $110 and a standard deviation of $20. If the management decides to give free gifts to all those customers who spend more than $150 at this supermarket during a visit, what percentage of the customers are expected to get free gifts?

6.31 A 2011 analysis performed by ReadWrite Mobile revealed that the average number of apps downloaded per day per iOS device (such as iPhone, iPod, and iPad) exceeds 60 (www.readwriteweb.com/mobile/2011/01/more-than-60-apps-downloaded-per-ios-device.php). Suppose that the current distribution of apps downloaded per day per iOS device is approximately normal with a mean of 65 and a standard deviation of 19.4. Find the probability that the number of apps downloaded on a randomly selected day by a randomly selected owner of an iOS device is

    a. 100 or more   **b.** 45 or fewer

6.32 According to the U.S. Department of Agriculture, the average American consumed 54.3 pounds (approximately seven gallons) of salad and cooking oils in 2008 (www.ers.usda.gov/data/foodconsumption). Suppose that the current distribution of salad and cooking oil consumption is approximately normally distributed with a mean of 54.3 pounds and a standard deviation of 14.5 pounds. What percentage of Americans' annual salad and cooking oil consumption is

    a. less than 10 pounds   **b.** between 40 and 60 pounds
    c. more than 90 pounds   **d.** between 50 and 70 pounds

6.33 According to the National Retail Federation's recent Back to College Consumer Intentions and Actions survey, families of college students spend an average of $616.13 on new apparel, furniture for dorms or apartments, school supplies, and electronics (www.nrf.com/modules.php?name=News&op=viewlive&sp_id=966). Suppose that the expenses on such Back to College items for the current year are approximately normally distributed with a mean of $616.13 and a standard deviation of $120. Find the probability that the amount of money spent on such items by a randomly selected family of a college student is

    a. less than $450   **b.** between $500 and $750

6.34 The lengths of 3-inch nails manufactured on a machine are normally distributed with a mean of 3.0 inches and a standard deviation of .009 inch. The nails that are either shorter than 2.97 inches or longer than 3.03 inches are unusable. What percentage of all the nails produced by this machine are unusable?

6.35 A psychologist has devised a stress test for dental patients sitting in the waiting rooms. According to this test, the stress scores (on a scale of 1 to 10) for patients waiting for root canal treatments are found to be approximately normally distributed with a mean of 7.59 and a standard deviation of .73.

    a. What percentage of such patients have a stress score lower than 6.0?
    b. What is the probability that a randomly selected root canal patient sitting in the waiting room has a stress score between 7.0 and 8.0?
    c. The psychologist suggests that any patient with a stress score of 9.0 or higher should be given a sedative prior to treatment. What percentage of patients waiting for root canal treatments would need a sedative if this suggestion is accepted?

6.4 | Determining the *z* and *x* Values When an Area Under the Normal Distribution Curve Is Known

So far in this chapter we have discussed how to find the area under a normal distribution curve for an interval of *z* or *x*. Now we invert this procedure and learn how to find the corresponding value of *z* or *x* when an area under a normal distribution curve is known. Examples 6–15 through 6–17 describe this procedure for finding the *z* value.

EXAMPLE 6–15

Finding z when the area to the left of z is known.

Find the value of *z* such that the area under the standard normal curve to the left of *z* is .9251.

Solution As shown in Figure 6.45, we are to find the *z* value such that the area to the left of *z* is .9251. Since this area is greater than .50, *z* is positive and lies to the right of zero.

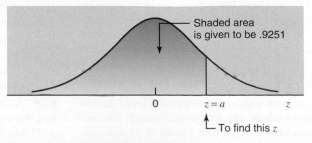

Figure 6.45 Finding the *z* value.

To find the required value of z, we locate .9251 in the body of the normal distribution table, Table IV of Appendix B. The relevant portion of that table is reproduced as Table 6.4 here. Next we read the numbers in the column and row for z that correspond to .9251. As shown in Table 6.4, these numbers are 1.4 and .04, respectively. Combining these two numbers, we obtain the required value of $z = 1.44$. Thus, the area to the left of $z = 1.44$ is .9251.

Table 6.4   Finding the z Value When Area Is Known

z	.00	.01	...	.04	...	.09
−3.4	.0003	.0003	...		...	.0002
−3.3	.0005	.0005	...		...	.0003
−3.2	.0007	.0007	...		...	.0005
.	.	.	...		...	
.	.	.	...		...	
.	.	.	...		...	.
1.4			...	.9251	...	...
.	.	.	...	.	...	.
.	.	.	...	.	...	
.	.	.	...		...	.
3.4	.9997	.9997	...	.9997	...	.9998

We locate this value in Table IV of Appendix B

EXAMPLE 6–16

Finding z when the area in the right tail is known.

Find the value of z such that the area under the standard normal curve in the right tail is .0050.

Solution   To find the required value of z, we first find the area to the left of z by subtracting .0050 from 1.0. Hence,

$$\text{Area to the left of } z = 1.0 - .0050 = .9950$$

This area is shown in Figure 6.46.

Now we look for .9950 in the body of the normal distribution table. Table IV does not contain .9950. So we find the value closest to .9950, which is either .9949 or .9951. We can use either of these two values. If we choose .9951, the corresponding z value is 2.58. Hence, the required value of z is **2.58**, and the area to the right of $z = 2.58$ is approximately .0050. Note that there is no apparent reason to choose .9951 and not to choose .9949. We can use either of the two values. If we choose .9949, the corresponding z value will be 2.57.

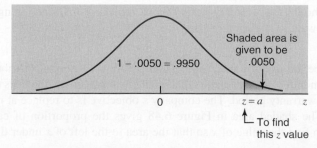

Figure 6.46   Finding the z value.

EXAMPLE 6–17

Finding z when the area in the left tail is known.

Find the value of z such that the area under the standard normal curve in the left tail is .05.

Solution   Because .05 is less than .5 and it is the area in the left tail, the value of z is negative. This area is shown in Figure 6.47.

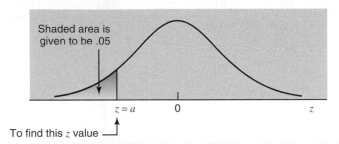

Shaded area is given to be .05

$z = a$ 0 z

To find this z value

Figure 6.47   Finding the z value.

Next, we look for .0500 in the body of the normal distribution table. The value closest to .0500 in the normal distribution table is either .0505 or .0495. Suppose we use the value .0495. The corresponding z value is −1.65. Thus, the required value of z is **−1.65** and the area to the left of $z = -1.65$ is approximately .05.

To find an x value when an area under a normal distribution curve is given, first we find the z value corresponding to that x value from the normal distribution table. Then, to find the x value, we substitute the values of μ, σ, and z in the following formula, which is obtained from $z = (x - \mu)/\sigma$ by doing some algebraic manipulations. Also, if we know the values of x, z, and σ, we can find μ using this same formula.

> **Finding an *x* Value for a Normal Distribution**   For a normal curve with known values of μ and σ and for a given area under the curve to the left of x, the x value is calculated as
>
> $$x = \mu + z\sigma$$

Examples 6–18 and 6–19 illustrate how to find an x value when an area under a normal distribution curve is known.

EXAMPLE 6–18   Life Span of a Calculator

Finding x when the area in the left tail is known.

Recall Example 6–14. It is known that the life of a calculator manufactured by Calculators Corporation has a normal distribution with a mean of 54 months and a standard deviation of 8 months. What should the warranty period be to replace a malfunctioning calculator if the company does not want to replace more than 1% of all the calculators sold?

Solution   Let x be the life of a calculator. Then, x follows a normal distribution with $\mu = 54$ months and $\sigma = 8$ months. The calculators that would be replaced are the ones that start malfunctioning during the warranty period. The company's objective is to replace at most 1% of all the calculators sold. The shaded area in Figure 6.48 gives the proportion of calculators that are replaced. We are to find the value of x so that the area to the left of x under the normal curve is 1%, or .01.

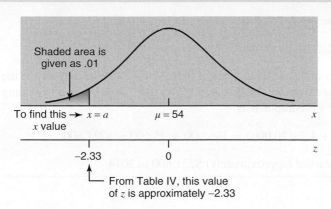

Figure 6.48   Finding an x value.

In the first step, we find the z value that corresponds to the required x value.

We find the z value from the normal distribution table for .0100. Table IV of Appendix B does not contain a value that is exactly .0100. The value closest to .0100 in the table is .0099, and the z value for .0099 is −2.33. Hence,

$$z = -2.33$$

Substituting the values of μ, σ, and z in the formula $x = \mu + z\sigma$, we obtain

$$x = \mu + z\sigma = 54 + (-2.33)(8) = 54 - 18.64 = \mathbf{35.36}$$

Thus, the company should replace all the calculators that start to malfunction within 35.36 months (which can be rounded to 35 months) of the date of purchase so that they will not have to replace more than 1% of the calculators.

EXAMPLE 6–19   Earnings of Internal Medicine Physicians

Finding x when the area in the right tail is known.

According to the 2015 Physician Compensation Report by Medscape (a subsidiary of WebMD), American internal medicine physicians earned an average of $196,000 in 2014. Suppose that the 2014 earnings of all American internal medicine physicians are normally distributed with a mean of $196,000 and a standard deviation of $20,000. Dr. Susan Garcia practices internal medicine in New Jersey. What were Dr. Garcia's 2014 earnings if 10% of all American internal medicine physicians earned more than Dr. Susan Garcia in 2014?

Solution   Let x denote the 2014 earnings of a randomly selected American internal medicine physician. Then, x is normally distributed with

$$\mu = \$196{,}000 \quad \text{and} \quad \sigma = \$20{,}000$$

We are to find the value of x such that the area under the normal distribution curve to the right of x is 10%, as shown in Figure 6.49.

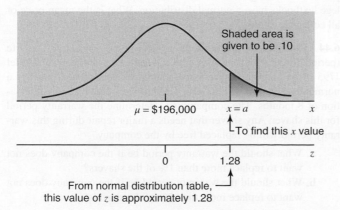

Figure 6.49   Finding an x value.

First, we find the area under the normal distribution curve to the left of the x value.

$$\text{Area to the left of the } x \text{ value} = 1.0 - .10 = .9000$$

To find the z value that corresponds to the required x value, we look for .9000 in the body of the normal distribution table. The value closest to .9000 in Table IV is .8997, and the corresponding z value is 1.28. Hence, the value of x is computed as

$$x = \mu + z\sigma = 196,000 + 1.28(20,000) = 196,000 + 25,600 = \$221,600$$

Thus, we can state that Dr. Garcia earned (approximately) \$221,600 in 2014.

EXERCISES

CONCEPTS AND PROCEDURES

6.36 Find the value of z so that the area under the standard normal curve

a. from 0 to z is (approximately) .1965 and z is positive
b. between 0 and z is (approximately) .2740 and z is negative
c. in the left tail is (approximately) .2050
d. in the right tail is (approximately) .1053

6.37 Calculate the value of z so that the area under the standard normal curve

a. in the right tail is .0250
b. in the left tail is .0500
c. in the left tail is .0010
d. in the right tail is .0100

6.38 Calculate the value of z so that the area under the standard normal curve

a. in the right tail is .0500
b. in the left tail is .0250
c. in the left tail is .0100
d. in the right tail is .0050

6.39 Let x be a continuous random variable that follows a normal distribution with a mean of 200 and a standard deviation of 25.

a. Find the value of x so that the area under the normal curve to the left of x is approximately .6330.
b. Find the value of x so that the area under the normal curve to the right of x is approximately .05.
c. Find the value of x so that the area under the normal curve to the right of x is .8051.
d. Find the value of x so that the area under the normal curve to the left of x is .0150.
e. Find the value of x so that the area under the normal curve between μ and x is .4525 and the value of x is less than μ.
f. Find the value of x so that the area under the normal curve between μ and x is approximately .4800 and the value of x is greater than μ.

APPLICATIONS

6.40 According to the records of an electric company serving the Boston area, the mean electricity consumption during winter for all households is 1650 kilowatt-hours per month. Assume that the monthly electric consumptions during winter by all households in this area have a normal distribution with a mean of 1650 kilowatt-hours and a standard deviation of 320 kilowatt-hours. The company sent a notice to Bill Johnson informing him that about 90% of the households use less electricity per month than he does. What is Bill Johnson's monthly electricity consumption?

6.41 The management of a supermarket wants to adopt a new promotional policy of giving a free gift to every customer who spends more than a certain amount per visit at this supermarket. The expectation of the management is that after this promotional policy is advertised, the expenditures for all customers at this supermarket will be normally distributed with a mean of \$110 and a standard deviation of \$20. If the management wants to give free gifts to at most 10% of the customers, what should the amount be above which a customer would receive a free gift?

6.42 Fast Auto Service provides oil and lube service for cars. It is known that the mean time taken for oil and lube service at this garage is 15 minutes per car and the standard deviation is 2.4 minutes. The management wants to promote the business by guaranteeing a maximum waiting time for its customers. If a customer's car is not serviced within that period, the customer will receive a 50% discount on the charges. The company wants to limit this discount to at most 5% of the customers. What should the maximum guaranteed waiting time be? Assume that the times taken for oil and lube service for all cars have a normal distribution.

*6.43 A study has shown that 20% of all college textbooks have a price of \$100 or higher. It is known that the standard deviation of the prices of all college textbooks is \$7.50. Suppose the prices of all college textbooks have a normal distribution. What is the mean price of all college textbooks?

6.44 Rockingham Corporation makes electric shavers. The life (period during which a shaver does not need a major repair) of Model J795 of an electric shaver manufactured by this corporation has a normal distribution with a mean of 70 months and a standard deviation of 8 months. The company is to determine the warranty period for this shaver. Any shaver that needs a major repair during this warranty period will be replaced free by the company.

a. What should the warranty period be if the company does not want to replace more than 1% of the shavers?
b. What should the warranty period be if the company does not want to replace more than 5% of the shavers?

6.5 | The Normal Approximation to the Binomial Distribution

Recall from Chapter 5 that:

1. The binomial distribution is applied to a discrete random variable.
2. Each repetition, called a trial, of a binomial experiment results in one of two possible outcomes (or events), either a success or a failure.
3. The probabilities of the two (possible) outcomes (or events) remain the same for each repetition of the experiment.
4. The trials are independent.

The binomial formula, which gives the probability of x successes in n trials, is

$$P(x) = {}_nC_x \, p^x q^{n-x}$$

The use of the binomial formula becomes very tedious when n is large. In such cases, the normal distribution can be used to approximate the binomial probability. Note that for a binomial problem, the exact probability is obtained by using the binomial formula. If we apply the normal distribution to solve a binomial problem, the probability that we obtain is an approximation to the exact probability. The approximation obtained by using the normal distribution is very close to the exact probability when n is large and p is very close to .50. However, this does not mean that we should not use the normal approximation when p is not close to .50. The reason the approximation is closer to the exact probability when p is close to .50 is that the binomial distribution is symmetric when $p = .50$. The normal distribution is always symmetric. Hence, the two distributions are very close to each other when n is large and p is close to .50. However, this does not mean that whenever $p = .50$, the binomial distribution is the same as the normal distribution because not every symmetric bell-shaped curve is a normal distribution curve.

> **Normal Distribution as an Approximation to Binomial Distribution**   Usually, the normal distribution is used as an approximation to the binomial distribution when np and nq are both greater than 5, that is, when
>
> $$np > 5 \quad \text{and} \quad nq > 5$$

Table 6.5 gives the binomial probability distribution of x for $n = 12$ and $p = .50$. This table is written using Table I of Appendix B. Figure 6.50 shows the histogram and the smoothed polygon for the probability distribution of Table 6.5. As we can observe, the histogram in Figure 6.50 is symmetric, and the curve obtained by joining the upper midpoints of the rectangles is approximately bell shaped.

Table 6.5	The Binomial Probability Distribution for $n = 12$ and $p = .50$
x	**$P(x)$**
0	.0002
1	.0029
2	.0161
3	.0537
4	.1208
5	.1934
6	.2256
7	.1934
8	.1208
9	.0537
10	.0161
11	.0029
12	.0002

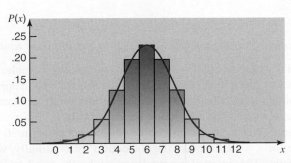

Figure 6.50   Histogram for the probability distribution of Table 6.5.

Examples 6–20 through 6–22 illustrate the application of the normal distribution as an approximation to the binomial distribution.

EXAMPLE 6–20   Number of Credit Cards People Have

Using the normal approximation to the binomial distribution: x equals a specific value.

According to an estimate, 50% of people in the United States have at least one credit card. If a random sample of 30 persons is selected, what is the probability that 19 of them will have at least one credit card?

Solution   Let n be the total number of persons in the sample, x be the number of persons in the sample who have at least one credit card, and p be the probability that a person has at least one credit card. Then, this is a binomial problem with

$$n = 30, \quad p = .50, \quad q = 1 - p = .50,$$
$$x = 19, \quad n - x = 30 - 19 = 11$$

Using the binomial formula, the exact probability that 19 persons in a sample of 30 have at least one credit card is

$$P(19) = {}_{30}C_{19}\,(.50)^{19}\,(.50)^{11} = \mathbf{.0509}$$

Now let us solve this problem using the normal distribution as an approximation to the binomial distribution. For this example,

$$np = 30\,(.50) = 15 \quad \text{and} \quad nq = 30\,(.50) = 15$$

Because np and nq are both greater than 5, we can use the normal distribution as an approximation to solve this binomial problem. We perform the following three steps.

Step 1.   *Compute μ and σ for the binomial distribution.*

To use the normal distribution, we need to know the mean and standard deviation of the distribution. Hence, the first step in using the normal approximation to the binomial distribution is to compute the mean and standard deviation of the binomial distribution. As we know from Chapter 5, the mean and standard deviation of a binomial distribution are given by np and $\sqrt{npq}$, respectively. Using these formulas, we obtain

$$\mu = np = 30(.50) = 15$$
$$\sigma = \sqrt{npq} = \sqrt{30(.50)(.50)} = 2.73861279$$

Step 2.   *Convert the discrete random variable into a continuous random variable.*

The normal distribution applies to a continuous random variable, whereas the binomial distribution applies to a discrete random variable. The second step in applying the normal approximation to the binomial distribution is to convert the discrete random variable to a continuous random variable by making the **correction for continuity**.

As shown in Figure 6.51, the probability of 19 successes in 30 trials is given by the area of the rectangle for $x = 19$. To make the correction for continuity, we use the interval 18.5 to 19.5 for 19 persons. This interval is actually given by the two boundaries of the rectangle for $x = 19$, which are obtained by subtracting .5 from 19 and by adding .5 to 19. Thus, $P(x = 19)$ for the binomial problem will be approximately equal to $P(18.5 \leq x \leq 19.5)$ for the normal distribution.

> The addition of .5 and/or subtraction of .5 from the value(s) of x when the normal distribution is used as an approximation to the binomial distribution, where x is the number of successes in n trials, is called the **continuity correction factor**.

The area contained by the rectangle for $x = 19$ is approximated by the area under the normal curve between 18.5 and 19.5.

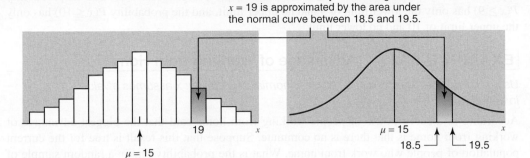

Step 3. *Compute the required probability using the normal distribution.*

As shown in Figure 6.52, the area under the normal distribution curve between $x = 18.5$ and $x = 19.5$ will give us the (approximate) probability that 19 persons have at least one credit card. We calculate this probability as follows:

$$\text{For } x = 18.5: \quad z = \frac{18.5 - 15}{2.73861279} = 1.28$$

$$\text{For } x = 19.5: \quad z = \frac{19.5 - 15}{2.73861279} = 1.64$$

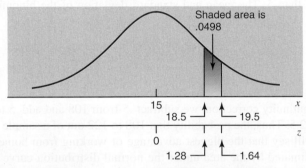

Figure 6.52   Area between $x = 18.5$ and $x = 19.5$.

The required probability is given by the area under the standard normal curve between $z = 1.28$ and $z = 1.64$. This area is obtained by subtracting the area to the left of $z = 1.28$ from the area to the left of $z = 1.64$. From Table IV of Appendix B, the area to the left of $z = 1.28$ is .8997 and the area to the left of $z = 1.64$ is .9495. Hence, the required probability is

$$P(18.5 \leq x \leq 19.5) = P(1.28 \leq z \leq 1.64) = .9495 - .8997 = \textbf{.0498}$$

Thus, based on the normal approximation, the probability that 19 persons in a sample of 30 will have at least one credit card is approximately .0498. Earlier, using the binomial formula, we obtained the exact probability .0509. The error due to using the normal approximation is .0509 − .0498 = .0011. Thus, the exact probability is underestimated by .0011 if the normal approximation is used.

When applying the normal distribution as an approximation to the binomial distribution, always make a *correction for continuity*. The continuity correction is made by subtracting .5 from the lower limit of the interval and/or by adding .5 to the upper limit of the interval. For example, the binomial probability $P(7 \leq x \leq 12)$ will be approximated by the probability $P(6.5 \leq x \leq 12.5)$ for the normal distribution; the binomial probability $P(x \geq 9)$ will be approximated by the probability $P(x \geq 8.5)$ for the normal distribution; and the binomial probability $P(x \leq 10)$ will be

◀ *Remember*

approximated by the probability $P(x \le 10.5)$ for the normal distribution. Note that the probability $P(x \ge 9)$ has only the lower limit of 9 and no upper limit, and the probability $P(x \le 10)$ has only the upper limit of 10 and no lower limit.

EXAMPLE 6–21   Advantage of Working from Home

Using the normal approximation to the binomial probability: x assumes a value in an interval.

According to a survey, 32% of people working from home said that the biggest advantage of working from home is that there is no commute. Suppose that this result is true for the current population of people who work from home. What is the probability that in a random sample of 400 people who work from home, 108 to 122 will say that the biggest advantage of working from home is that there is no commute?

Solution   Let n be the number of people in the sample who work from home, x be the number of people in the sample who say that the biggest advantage of working from home is that there is no commute, and p be the probability that a person who works from home says that the biggest advantage of working from home is that there is no commute. Then, this is a binomial problem with

$$n = 400, \quad p = .32, \quad \text{and} \quad q = 1 - .32 = .68$$

We are to find the probability of 108 to 122 successes in 400 trials. Because n is large, it is easier to apply the normal approximation than to use the binomial formula. We can check that np and nq are both greater than 5. The mean and standard deviation of the binomial distribution are, respectively,

$$\mu = np = 400\,(.32) = 128$$
$$\sigma = \sqrt{npq} = \sqrt{400(.32)(.68)} = 9.32952303$$

To make the continuity correction, we subtract .5 from 108 and add .5 to 122 to obtain the interval 107.5 to 122.5. Thus, the probability that 108 to 122 out of a sample of 400 people who work from home will say that the biggest advantage of working from home is that there is no commute is approximated by the area under the normal distribution curve from $x = 107.5$ to $x = 122.5$. This area is shown in Figure 6.53. The z values for $x = 107.5$ and $x = 122.5$ are calculated as follows:

$$\text{For } x = 107.5: \qquad z = \frac{107.5 - 128}{9.32952303} = -2.20$$

$$\text{For } x = 122.5: \qquad z = \frac{122.5 - 128}{9.32952303} = -.59$$

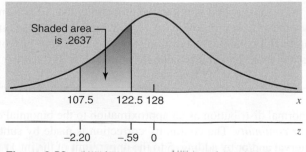

Figure 6.53   Area between $x = 107.5$ and $x = 122.5$.

The required probability is given by the area under the standard normal curve between $z = -2.20$ and $z = -.59$. This area is obtained by taking the difference between the areas under the standard normal curve to the left of $z = -2.20$ and to the left of $z = -.59$. From Table IV of Appendix B, the area to the left of $z = -2.20$ is .0139, and the area to the left of $z = -.59$ is .2776. Hence, the required probability is

$$P(107.5 \leq x \leq 122.5) = P(-2.20 \leq z \leq -.59) = .2776 - .0139 = \mathbf{.2637}$$

Thus, the probability that 108 to 122 people in a sample of 400 who work from home will say that the biggest advantage of working from home is that there is no commute is approximately .2637.

EXAMPLE 6–22	Women Supporting Red Light Cameras at Intersections

Using the normal approximation to the binomial probability: x is greater than or equal to a value.

According to a FindLaw.com survey of American adult women, 61% support red light cameras at intersections (*USA TODAY*, January 7, 2015). Assume that this percentage is true for the current population of adult American women. What is the probability that 500 or more such women in a random sample of 800 women will support red light cameras at intersections?

Solution   Let n be the sample size, x be the number of women in the sample who support red light cameras at intersections, and p be the probability that a randomly selected American adult woman supports red light cameras at intersections. Then, this is a binomial problem with

$$n = 800, \quad p = .61, \quad \text{and} \quad q = 1 - .61 = .39$$

We are to find the probability of 500 or more successes in 800 trials. The mean and standard deviation of the binomial distribution are

$$\mu = np = 800\,(.61) = 488$$
$$\sigma = \sqrt{npq} = \sqrt{800(.61)(.39)} = 13.79565149$$

For the continuity correction, we subtract .5 from 500, which gives 499.5. Thus, the probability that 500 or more women in a random sample of 800 support red light cameras at intersections is approximated by the area under the normal distribution curve to the right of $x = 499.5$, as shown in Figure 6.54. The z-value for $x = 499.5$ is calculated as follows.

$$\text{For } x = 499.5: \qquad z = \frac{499.5 - 488}{13.79565149} = .83$$

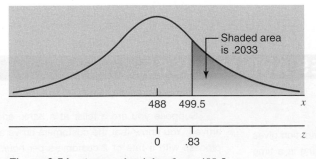

Figure 6.54   Area to the right of $x = 499.5$.

To find the required probability, we find the area to the left of $z = .83$ and subtract this area from 1.0. From Table IV of Appendix B, the area to the left of $z = .83$ is .7967. Hence,

$$P(x \geq 499.5) = P(z \geq .83) = 1.0 - .7967 = \mathbf{.2033}$$

Thus, the probability that 500 or more adult American women in a random sample of 800 will support red light cameras at intersections is approximately .2033.

EXERCISES

CONCEPTS AND PROCEDURES

6.45 What are the conditions under which the normal distribution is used as an approximation to the binomial distribution?

6.46 For a binomial probability distribution, $n = 20$ and $p = .60$.

    a. Find the probability $P(x = 12)$ by using the table of binomial probabilities (Table I of Appendix B).

    b. Find the probability $P(x = 12)$ by using the normal distribution as an approximation to the binomial distribution. What is the difference between this approximation and the exact probability calculated in part a?

6.47 For a binomial probability distribution, $n = 100$ and $p = .50$. Let x be the number of successes in 100 trials.

    a. Find the mean and standard deviation of this binomial distribution.

    b. Find $P(x \geq 50)$ using the normal approximation.

    c. Find $P(51 \leq x \leq 59)$ using the normal approximation.

6.48 Find the following binomial probabilities using the normal approximation.

    a. $n = 70,$ $p = .30,$ $P(x = 18)$
    b. $n = 200,$ $p = .70,$ $P(133 \leq x \leq 145)$
    c. $n = 85,$ $p = .40,$ $P(x \geq 30)$
    d. $n = 150,$ $p = .38,$ $P(x \leq 62)$

APPLICATIONS

6.49 According to an Allstate/National Journal poll, 39% of the U.S. adults polled said that it is *extremely or very likely* that "there will be a female president within 10–15 years" in the United States (*USA Today*, March 28, 2012). Suppose that this percentage is true for the current population of U.S. adults. Find the probability that in a random sample of 800 U.S. adults, more than 330 would hold the foregoing belief.

6.50 The percentage of women in the work force has increased tremendously during the past few decades. Whereas only 35% of all employees in the United States in 1970 were women, this percentage was 49% in 2012 (*Bloomberg Businessweek*, January 9–15, 2012). Suppose that currently 49% of all employees in the United States are women. Find the probability that in a random sample of 400 employees, the number of women is

    a. exactly 205   **b.** less than 190   **c.** 210 to 220

6.51 According to a November 8, 2010 report on www.teleread. com, 7% of U.S. adults with online services currently read e-books. Assume that this percentage is true for the current population of U.S. adults with online services. Find the probability that in a random sample of 600 U.S. adults with online services, the number who read e-books is

    a. exactly 45   **b.** at most 53   **c.** 30 to 50

6.52 An office supply company conducted a survey before marketing a new paper shredder designed for home use. In the survey, 80% of the people who tried the shredder were satisfied with it. Because of this high satisfaction rate, the company decided to market the new shredder. Assume that 90% of all people are satisfied with this shredder. During a certain month, 100 customers bought this shredder. Find the probability that of these 100 customers, the number who are satisfied is

    a. exactly 85   **b.** 80 or fewer   **c.** 81 to 95

6.53 Stress on the job is a major concern of a large number of people who go into managerial positions. It is estimated that 80% of the managers of all companies suffer from job-related stress.

    a. What is the probability that in a sample of 200 managers of companies, exactly 150 suffer from job-related stress?

    b. Find the probability that in a sample of 200 managers of companies, at least 170 suffer from job-related stress.

    c. What is the probability that in a sample of 200 managers of companies, 165 or fewer suffer from job-related stress?

    d. Find the probability that in a sample of 200 managers of companies, 164 to 172 suffer from job-related stress.

USES AND MISUSES...

(1) DON'T LOSE YOUR MEMORY

As discussed in the previous chapter, the Poisson distribution gives the probability of a specified number of events occurring in a time interval. The Poisson distribution provides a model for the number of emails a server might receive during a certain time period or the number of people arriving in line at a bank during lunch hour. These are nice to know for planning purposes, but sometimes we want to know the specific times at which emails or customers arrive. These times are governed by a special continuous probability distribution with certain unusual properties. This distribution is called the *exponential distribution*, and it is derived from the Poisson probability distribution.

Suppose you are a teller at a bank, and a customer has just arrived. You know that the customers arrive according to a Poisson process with a rate of λ customers per hour. Your boss might care how many customers arrive on average during a given time interval to ensure there are enough tellers available to handle the customers efficiently; you are more concerned with the time when the next customer will arrive. Remember that the probability that x customers arrive in an interval of length t is

$$P(x) = \frac{(\lambda t)^x e^{-\lambda t}}{x!}$$

The probability that at least one customer arrives within time t is 1 minus the probability that no customer arrives within time t. Hence,

$$P \text{ (at least one customer arrives within time } t) = 1 - P(0)$$
$$= 1 - \frac{(\lambda t)^0 e^{-\lambda t}}{0!}$$
$$= 1 - e^{-\lambda t}$$

If the bank receives an average of 15 customers per hour—an average of one every 4 minutes—and a customer has just arrived, the probability that a customer arrives within 4 minutes is $1 - e^{-\lambda t} = 1 - e^{-(15/60)4} = .6321$. In the same way, the probability that a customer arrives within 8 minutes is .8647.

Let us say that a customer arrived and went to your co-worker's window. No additional customer arrived within the next 2 minutes—an event with probability .6065—and you dozed off for 2 more minutes. When you open your eyes, you see that a customer has not arrived yet. What is the probability that a customer arrives within the next 4 minutes? From the calculation above, you might say that the answer is .8647. After all, you know that a customer arrived 8 minutes earlier. But .8647 is not the correct answer.

The exponential distribution, which governs the time between arrivals of a Poisson process, has a property called the *memoryless* property. For you as a bank teller, this means that if you know a customer has not arrived during the past 4 minutes, then the clock is reset to zero, as if the previous customer had just arrived. So even after your nap, the probability that a customer arrives within 4 minutes is .6321. This interesting property reminds us again that we should be careful when we use mathematics to model real-world phenomena.

(2) QUALITY IS JOB 1

During the early 1980s, Ford Motor Company adopted a new marketing slogan: "Quality Is Job 1." The new slogan coincided with Ford's release of the Taurus and the Mercury Sable, but the groundwork that resulted in Ford's sudden need to improve quality was laid some 30 years earlier by an American statistician—not in Detroit, but in Japan.

W. Edwards Deming is one of the most famous statisticians, if not *the* most famous statistican, in the field of statistical process control, a field that debunked the myth that you could not improve quality and lower costs simultaneously. After World War II, Deming was asked by the U.S. Armed Forces to assist with planning for the 1951 census in Japan. While there, he taught statistical process control and quality management to the managers and engineers at many of Japan's largest companies. After adopting Deming's principles, the quality of and the demand for Japanese products, including Japanese automobiles, increased tremendously.

Ford's interest in Japanese quality resulted from the fact that Ford was having a specific transmission produced simultaneously in Japan and the United States. Ford's U.S. customers were requesting cars with Japanese-produced transmissions, even if it required them to wait longer for the car. Despite the fact that the transmissions were made to the same specifications in the two countries, the parts used in the Japanese transmissions were much closer to the desired size than those used in the transmissions made in America. Knowing that Deming had done a great deal of work with Japanese companies, Ford hired him as a consultant. The result of Deming's work in statistical process control and proper management methods, along with Ford's willingness to implement his recommendations, was the production of Ford's Taurus and Mercury Sable lines of automobiles, which resulted in Ford earning a profit after numerous years of losses.

W. Edwards Deming died in 1993 at the age of 93 years, but he left a legacy. Japan introduced the Deming Prize in 1950. The Deming Prize is awarded annually to individuals and companies whose work has advanced knowledge in the statistical process control area. More information about Deming and The W. Edwards Deming Institute is available at www.deming.org.

Source: The W. Edwards Deming Institute (www.deming.org), and en.wikipedia.org/wiki/W._Edwards_Deming.

GLOSSARY

Continuity correction factor   Addition of .5 and/or subtraction of .5 from the value(s) of x when the normal distribution is used as an approximation to the binomial distribution, where x is the number of successes in n trials.

Continuous random variable   A random variable that can assume any value in one or more intervals.

Normal probability distribution   The probability distribution of a continuous random variable that, when plotted, gives a specific bell-shaped curve. The parameters of the normal distribution are the mean μ and the standard deviation σ.

Standard normal distribution   The normal distribution with $\mu = 0$ and $\sigma = 1$. The units of the standard normal distribution are denoted by z.

z value or **z score**   The units of the standard normal distribution that are denoted by z.

SUPPLEMENTARY EXERCISES

6.54 The management at Ohio National Bank does not want its customers to wait in line for service for too long. The manager of a branch of this bank estimated that the customers currently have to wait an average of 8 minutes for service. Assume that the waiting times for all customers at this branch have a normal distribution with a mean of 8 minutes and a standard deviation of 2 minutes.

   a. Find the probability that a randomly selected customer will have to wait for less than 3 minutes.

b. What percentage of the customers have to wait for 10 to 13 minutes?

c. What percentage of the customers have to wait for 6 to 12 minutes?

d. Is it possible that a customer may have to wait longer than 16 minutes for service? Explain.

6.55 The print on the package of Sylvania CFL 65W replacement bulbs that use only 16W claims that these bulbs have an average life of 8000 hours. Assume that the distribution of lives of all such bulbs is normal with a mean of 8000 hours and a standard deviation of 400 hours. Let x be the life of a randomly selected such light bulb.

a. Find x so that about 22.5% of such light bulbs have lives longer than this value.

b. Find x so that about 63% of such light bulbs have lives shorter than this value.

6.56 According to an article on Yahoo.com on February 19, 2012, the average salary of actuaries in the U.S. was $98,620 a year (http://education. yahoo.net/articles/careers_for_shy_people_2.htm?kid=1KWO3). Suppose that currently the distribution of annual salaries of all actuaries in the U.S. is approximately normal with a mean of $98,620 and a standard deviation of $18,000. How much would an actuary have to be paid in order to be in the highest-paid 10% of all actuaries?

6.57 Jenn Bard, who lives in the San Francisco Bay area, commutes by car from home to work. She knows that it takes her an average of 28 minutes for this commute in the morning. However, due to the variability in the traffic situation every morning, the standard deviation of these commutes is 5 minutes. Suppose the population of her morning commute times has a normal distribution with a mean of 28 minutes and a standard deviation of 5 minutes. Jenn has to be at work by 8:30 A.M. every morning. By what time must she leave home in the morning so that she is late for work at most 1% of the time?

6.58 Hurbert Corporation makes font cartridges for laser printers that it sells to Alpha Electronics Inc. The cartridges are shipped to Alpha Electronics in large volumes. The quality control department at Alpha Electronics randomly selects 100 cartridges from each shipment and inspects them for being good or defective. If this sample contains 7 or more defective cartridges, the entire shipment is rejected. Hurbert Corporation promises that of all the cartridges, only 5% are defective.

a. Find the probability that a given shipment of cartridges received by Alpha Electronics will be accepted.

b. Find the probability that a given shipment of cartridges received by Alpha Electronics will not be accepted.

6.59 The amount of time taken by a bank teller to serve a randomly selected customer has a normal distribution with a mean of 2 minutes and a standard deviation of .5 minute.

a. What is the probability that both of two randomly selected customers will take less than 1 minute each to be served?

b. What is the probability that at least one of four randomly selected customers will need more than 2.25 minutes to be served?

6.60 Johnson Electronics makes calculators. Consumer satisfaction is one of the top priorities of the company's management. The company guarantees the refund of money or a replacement for any calculator that malfunctions within two years from the date of purchase. It is known from past data that despite all efforts, 5% of the calculators manufactured by this company malfunction within a 2-year period. The company recently mailed 500 such calculators to its customers.

a. Find the probability that exactly 29 of the 500 calculators will be returned for refund or replacement within a 2-year period.

b. What is the probability that 27 or more of the 500 calculators will be returned for refund or replacement within a 2-year period?

c. What is the probability that 15 to 22 of the 500 calculators will be returned for refund or replacement within a 2-year period?

ADVANCED EXERCISES

6.61 It is known that 15% of all homeowners pay a monthly mortgage of more than $2500 and that the standard deviation of the monthly mortgage payments of all homeowners is $350. Suppose that the monthly mortgage payments of all homeowners have a normal distribution. What is the mean monthly mortgage paid by all homeowners?

6.62 Two companies, A and B, drill wells in a rural area. Company A charges a flat fee of $3500 to drill a well regardless of its depth. Company B charges $1000 plus $12 per foot to drill a well. The depths of wells drilled in this area have a normal distribution with a mean of 250 feet and a standard deviation of 40 feet.

a. What is the probability that Company B would charge more than Company A to drill a well?

b. Find the mean amount charged by Company B to drill a well.

6.63 Otto is trying out for the javelin throw to compete in the Olympics. The lengths of his javelin throws are normally distributed with a mean of 253 feet and a standard deviation of 8.4 feet. What is the probability that the longest of three of his throws is 270 feet or more?

6.64 The Jen and Perry Ice Cream company makes a gourmet ice cream. Although the law allows ice cream to contain up to 50% air, this product is designed to contain only 20% air. Because of variability inherent in the manufacturing process, management is satisfied if each pint contains between 18% and 22% air. Currently two of Jen and Perry's plants are making gourmet ice cream. At Plant A, the mean amount of air per pint is 20% with a standard deviation of 2%. At Plant B, the mean amount of air per pint is 19% with a standard deviation of 1%. Assuming the amount of air is normally distributed at both plants, which plant is producing the greater proportion of pints that contain between 18% and 22% air?

6.65 The highway police in a certain state are using aerial surveillance to control speeding on a highway with a posted speed limit of 55 miles per hour. Police officers watch cars from helicopters above a straight segment of this highway that has large marks painted on the pavement at 1-mile intervals. After the police officers observe how long a car takes to cover the mile, a computer estimates that car's speed. Assume that the errors of these estimates are normally distributed with a mean of 0 and a standard deviation of 2 miles per hour.

a. The state police chief has directed his officers not to issue a speeding citation unless the aerial unit's estimate of speed is at least 65 miles per hour. What is the probability that a car traveling at 60 miles per hour or slower will be cited for speeding?

b. Suppose the chief does not want his officers to cite a car for speeding unless they are 99% sure that it is traveling at 60 miles per hour or faster. What is the minimum estimate of speed at which a car should be cited for speeding?

6.66 A soft-drink vending machine is supposed to pour 8 ounces of the drink into a paper cup. However, the actual amount poured into a cup varies. The amount poured into a cup follows a normal distribution with a mean that can be set to any desired amount by adjusting the machine. The standard deviation of the amount poured is always .07 ounce regardless of the mean amount. If the owner of the machine wants to be 99% sure that the amount in each cup is 8 ounces or more, to what level should she set the mean?

6.67 According to a College Board, the mean SAT mathematics score for all college-bound seniors was 511 in 2016. Suppose that this is true for the current population of college-bound seniors. Furthermore, assume that 17% of college-bound seniors scored below 410 in this test. Assume that the distribution of SAT mathematics scores for college-bound seniors is approximately normal.

a. Find the standard deviation of the mathematics SAT scores for college-bound seniors.

b. Find the percentage of college-bound seniors whose mathematics SAT scores were above 660.

6.68 A charter bus company is advertising a singles outing on a bus that holds 60 passengers. The company has found that, on average, 10% of ticket holders do not show up for such trips; hence, the company routinely overbooks such trips. Assume that passengers act independently of one another.

a. If the company sells 65 tickets, what is the probability that the bus can hold all the ticket holders who actually show up? In other words, find the probability that 60 or fewer passengers show up.

b. What is the largest number of tickets the company can sell and still be at least 95% sure that the bus can hold all the ticket holders who actually show up?

6.69 Refer to Exercise 6.66. In that exercise, suppose the mean is set to be 8 ounces, but the standard deviation is unknown. The cups used in the machine can hold up to 8.2 ounces, but these cups will overflow if more than 8.2 ounces is dispensed by the machine. What is the smallest possible standard deviation that will result in overflows occurring 3% of the time?

6.70 A variation of a roulette wheel has slots that are not of equal size. Instead, the width of any slot is proportional to the probability that a standard normal random variable z takes on a value between a and $(a + .1)$, where $a = -3.0, -2.9, -2.8, \ldots, 2.9, 3.0$. In other words, there are slots for the intervals $(-3.0, -2.9)$, $(-2.9, -2.8)$, $(-2.8, -2.7)$ through $(2.9, 3.0)$. There is one more slot that represents the probability that z falls outside the interval $(-3.0, 3.0)$. Find the following probabilities.

a. The ball lands in the slot representing $(.3, .4)$.

b. The ball lands in any of the slots representing $(-.1, .4)$.

c. In at least one out of five games, the ball lands in the slot representing $(-.1, .4)$.

d. In at least 100 out of 500 games, the ball lands in the slot representing $(.4, .5)$.

APPENDIX 6.1

NORMAL QUANTILE PLOTS

Many of the methods that are used in statistics require that the sampled data come from a normal distribution. While it is impossible to determine if this holds true without taking a census (i.e., looking at the population data), there are statistical tools that can be used to determine if this is a reasonable assumption. One of the simplest tools to use is called a normal quantile plot. The idea of the plot is to compare the values in a data set with the corresponding values one would predict for a standard normal distribution.

Although normal quantile plots are typically created using technology, it is helpful to see an example to understand how they are created and what the various numbers represent. To demonstrate, consider the data in the following table, which contains the 2001 salaries of the mayors of 10 large cities.

City	Mayor's Salary ($)	City	Mayor's Salary ($)
Chicago, IL	170,000	Newark, NJ	147,000
New York, NY	165,000	San Francisco, CA	146,891
Houston, TX	160,500	Jacksonville, FL	127,230
Detroit, MI	157,300	Baltimore, MD	125,000
Los Angeles, CA	147,390	Boston, MA	125,000

Each data point represents 1/10 of the distribution, with the smallest value representing the smallest 10%, the next representing the 10%–20% interval, and so on. In each case, we estimate that the data points fall in the middle of their respective intervals. For these 10 data points, these midpoints would be at the 5%, 15%, 25%, and so on, locations, while if we have 20 data points, these locations would be at 2.5%, 7.5%, 12.5%, and so on. Next we determine the z scores for these locations. The following table shows the z scores for the 10–data point scenario and for the 20–data point scenario.

Ten data points

Location (%)	5	15	25	35	45	55	65	75	85	95
z Score	−1.645	−1.036	−0.674	−0.385	−0.126	0.126	0.385	0.674	1.036	1.645

Twenty data points

Location (%)	2.5	7.5	12.5	17.5	22.5	27.5	32.5	37.5	42.5	47.5
z Score	−1.960	−1.440	−1.150	−0.935	−0.755	−0.598	−0.454	−0.319	−0.189	−0.063
Location (%)	52.5	57.5	62.5	67.5	72.5	77.5	82.5	87.5	92.5	97.5
z Score	0.063	0.189	0.319	0.454	0.598	0.755	0.935	1.150	1.440	1.960

Next we make a two-dimensional plot that places the data on the horizontal axis and the z scores on the vertical axis.

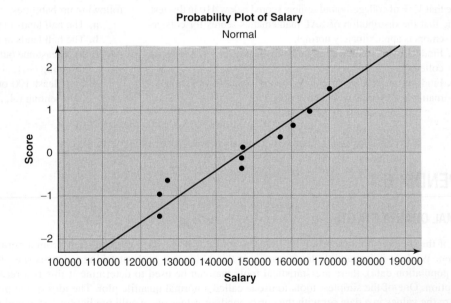

Probability Plot of Salary
Normal

If the data are in complete agreement with a normal distribution, the points will lie on the line displayed in the graph. As the likelihood that the given data come from a normal distribution decreases, the plot of data points will become less linear.

So, how do we interpret the plot of 10 salaries in the graph? There are a couple of features that we can point out. There are two groups of points that are stacked almost vertically (near $125,000 and $147,000). Depending on the software, multiple data points of the same value will be stacked or will appear as one point. In addition, there is a fairly big gap between these two groups. This is not unusual with small data sets, even if the data come from a normal distribution. Most times, in order to state that a very small data set does not come from a normal distribution, many people will be able to see that the data are very strongly skewed or have an outlier simply by looking at a sorted list of the data.

To understand the correspondence between the shape of a data set and its normal quantile plot, it is useful to look at a dotplot or histogram side by side with the normal quantile plot. We will consider a few common cases here.

(1). First, following are the two graphs for 20 data points randomly selected from a normal distribution.

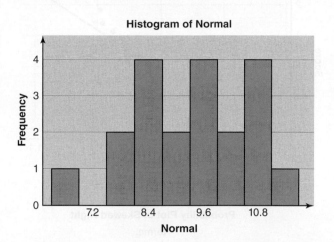

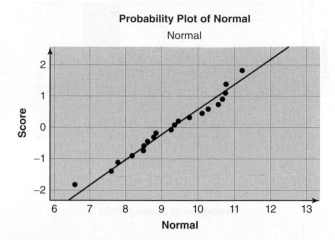

As you can see, just because data come from a normal distribution does not imply that they will be perfectly linear. In general, the points are close to the line, but small patterns such as in the upper right or the gap in the lower left can occur without invalidating the normality assumption.

(2). In the next case, we consider data that come from a distribution that is *heavy tailed*, which means that the distribution has higher percentages of values in its tails than one would expect in a normal distribution. For example, the Empirical Rule states that a normal distribution has approximately 2.5% of the observations below $\mu - 2\sigma$ and 2.5% of the observations above $\mu + 2\sigma$. If a data set has 10% of the observations below $\mu - 2\sigma$ and 10% of the observations above $\mu + 2\sigma$, it would be classified as being heavy-tailed. The resulting shape of the normal quantile plot will look somewhat like a playground slide. As the tails get heavier, the ends of the plot become steeper and the middle gets flatter.

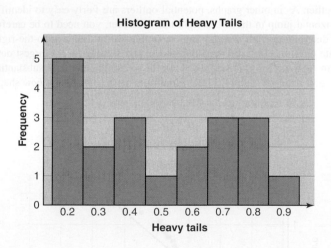

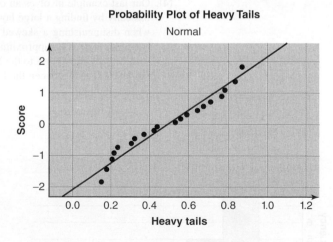

(3). Skewed distributions have normal quantile plots that are shaped somewhat like a boomerang, which has a rounded V shape, with one end of the boomerang stretched out more than the other side. Just like all other graphs of skewed distributions, the side that is stretched out identifies the direction of the skew. Again, as the distribution becomes more skewed, the bend in the quantile plot will become more severe.

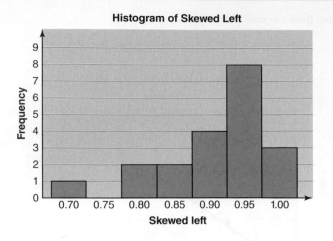

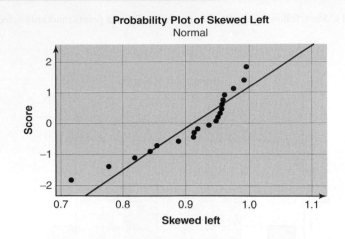

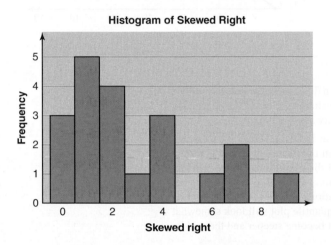

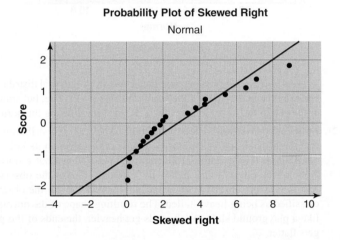

(4). Our last example involves an outlier. As in other graphs, potential outliers are fairly easy to identify, basically by finding a large horizontal jump in the left or right tail. However, you need to be careful when distinguishing a skewed distribution from one that has an outlier. In our skewed-to-the-right example, there is an approximate difference of 2 between the two largest values, yet the largest data value is still fairly close to the line in a vertical direction. In our outlier example, there is a substantial vertical distance between the largest data value and the line. Moreover, we do not see the *bow* shape in the latter plot.

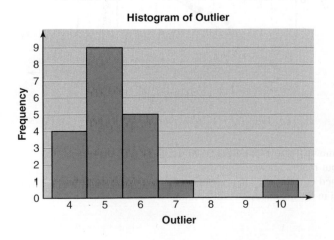

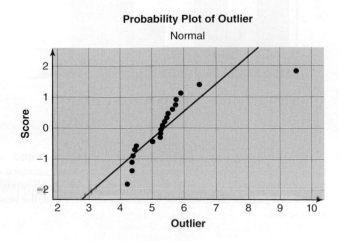

It is important to remember that these plots contain examples of a variety of common features. However, it is also important to remember that some of these features are not mutually exclusive. As one example, it is possible for a distribution to have heavy tails and an outlier. Identifying issues that would reject the notion of normality will be important in determining the types of inference procedures that can be used, which we will begin examining in Chapter 8.

SELF-REVIEW TEST

1. The normal probability distribution is applied to
  a. a continuous random variable
  b. a discrete random variable
  c. any random variable

2. For a continuous random variable, the probability of a single value of x is always
  a. zero   **b.** 1.0   **c.** between 0 and 1

3. Which of the following is not a characteristic of the normal distribution?
  a. The total area under the curve is 1.0.
  b. The curve is symmetric about the mean.
  c. The two tails of the curve extend indefinitely.
  d. The value of the mean is always greater than the value of the standard deviation.

4. The parameters of a normal distribution are
  a. μ, z, and σ **b.** μ and σ **c.** μ, x, and σ

5. For the standard normal distribution,
  a. $\mu = 0$ and $\sigma = 1$
  b. $\mu = 1$ and $\sigma = 0$
  c. $\mu = 100$ and $\sigma = 10$

6. The z value for μ for a normal distribution curve is always
  a. positive   **b.** negative   **c.** 0

7. For a normal distribution curve, the z value for an x value that is less than μ is always
  a. positive   **b.** negative   **c.** 0

8. Usually the normal distribution is used as an approximation to the binomial distribution when
  a. $n \geq 30$ **b.** $np > 5$ and $nq > 5$ **c.** $n > 20$ and $p = .50$

9. Find the following probabilities for the standard normal distribution.
  a. $P(.85 \leq z \leq 2.33)$ **b.** $P(-2.97 \leq z \leq 1.49)$
  c. $P(z \leq -1.29)$ **d.** $P(z > -.74)$

10. Find the value of z for the standard normal curve such that the area
  a. in the left tail is .1000
  b. between 0 and z is .2291 and z is positive
  c. in the right tail is .0500
  d. between 0 and z is .3571 and z is negative

11. In a National Highway Traffic Safety Administration (NHTSA) report, data provided to the NHTSA by Goodyear stated that the average tread life of properly inflated automobile tires is 45,000 miles (*Source*: http://www.nhtsa.dot.gov/cars/rules/rulings/TPMS_FMVSS_No138/part5.5.html). Suppose that the current distribution of tread life of properly inflated automobile tires is normally distributed with a mean of 45,000 miles and a standard deviation of 2360 miles.
  a. Find the probability that a randomly selected automobile tire has a tread life between 42,000 and 46,000 miles.
  b. What is the probability that a randomly selected automobile tire has a tread life of less than 38,000 miles?
  c. What is the probability that a randomly selected automobile tire has a tread life of more than 50,000 miles?
  d. Find the probability that a randomly selected automobile tire has a tread life between 46,500 and 47,500 miles.

12. Refer to Problem 11.
  a. Suppose that 6% of all automobile tires with the longest tread life have a tread life of at least x miles. Find the value of x.
  b. Suppose that 2% of all automobile tires with the shortest tread life have a tread life of at most x miles. Find the value of x.

13. Gluten sensitivity, which is also known as wheat intolerance, affects approximately 15% of people. The condition involves great difficulty in digesting wheat, but is not the same as wheat allergy, which has much more severe reactions (*Source*: http://www.foodintol.com/wheat.asp). A random sample of 800 individuals is selected.
  a. Find the probability that the number of individuals in this sample who have wheat intolerance is
    i. exactly 115   **ii.** 103 to 142   **iii.** at least 107
    iv. at most 100   **v.** between 111 and 123
  b. Find the probability that at least 675 of the individuals in this sample do *not* have wheat intolerance.
  c. Find the probability that 682 to 697 of the individuals in this sample do *not* have wheat intolerance.

MINI-PROJECTS

Note: The Mini-Projects are located on the text's Web site, www.wiley.com/college/mann.

DECIDE FOR YOURSELF

Note: The Decide for Yourself feature is located on the text's Web site, www.wiley.com/college/mann.

TECHNOLOGY INSTRUCTIONS   CHAPTER 6

Note: Complete TI-84, Minitab, and Excel manuals are available for download at the textbook's Web site, www.wiley.com/college/mann.

TI-84 Color/TI-84

The TI-84 Color Technology Instructions feature of this text is written for the TI-84 Plus C color graphing calculator running the 4.0 operating system. Some screens, menus, and functions will be slightly different in older operating systems. The TI-84+ can perform all of the same functions but does not have the "Color" option referenced in some of the menus.

Calculating a Left-Tail Probability for Example 6–6(b) of the Text

1. Select 2nd > **VARS** > **normalcdf(**.

NORMAL FLOAT AUTO REAL RADIAN MP
        normalcdf
lower: -1E99
upper:35
μ:50
σ:10
Paste
```
Screen 6.1

2. Use the following settings in the **normalcdf** menu (see **Screen 6.1**):
   - At the **lower** prompt, type -1E99. (The "E" symbol means scientific notation. Press $2^{nd}$ followed by the comma key to get the "E" symbol. -1E99 is calculator notation for the number $-1 \times 10^{99}$, which indicates that there is no lower limit of the interval.)
   - At the **upper** prompt , type 35.
   - At the $\mu$ prompt, type 50.
   - At the $\sigma$ prompt, type 10.

3. Highlight **Paste** and press **ENTER** twice.

4. The output gives the probability of .0668072287 that $x$ is less than 35 for a normal distribution with $\mu = 50$ and $\sigma = 10$ (see **Screen 6.2**).

### Calculating a Probability Between Two Values for Example 6–7(b) of the Text

NORMAL FLOAT AUTO REAL RADIAN MP
normalcdf( -1E99,35,50,10)
                .0668072287
normalcdf(18,34,25,4)
                .9477164531
normalcdf(55,1E99,40,5)
                .0013499672
invNorm(.05,0,1)
                -1.644853626
```
Screen 6.2

1. Select 2nd > **VARS** > **normalcdf(**.
2. Use the following settings in the **normalcdf** menu:
   - At the **lower** prompt, type 18.
   - At the **upper** prompt, type 34.
   - At the μ prompt, type 25.
   - At the σ prompt, type 4.

3. Highlight **Paste** and press **ENTER** twice.

4. The output gives the probability of .9477164531 that x is between 18 and 34 for a normal distribution with $\mu = 25$ and $\sigma = 4$. (See **Screen 6.2**.)

Calculating a Right-Tail Probability for Example 6–8(a) of the Text

1. Select 2nd > **VARS** > **normalcdf(**.
2. Use the following settings in the **normalcdf** menu:
   - At the **lower** prompt, type 55.
   - At the **upper** prompt, type 1E99. (The "E" symbol means scientific notation. Press 2nd followed by the comma key to get the "E" symbol. 1E99 is calculator notation for the number 1×10^{99}, which indicates that there is no upper limit of the interval.)
   - At the μ prompt, type 40.
   - At the σ prompt, type 5.

3. Highlight **Paste** and press **ENTER** twice.

4. The output gives the probability of .0013499672 that *x* is greater than 55 for a normal distribution with $\mu = 40$ and $\sigma = 5$. (See **Screen 6.2**.)

Determining *z* When a Probability Is Known for Example 6–17 of the Text

1. Select 2nd > **VARS** > **invNorm(**.

2. Use the following settings in the **invNorm** menu (see **Screen 6.3**):

   - At the **area** prompt, type 0.05.

   - At the μ prompt, type 0.

   - At the σ prompt, type 1.

3. Highlight **Paste** and press **ENTER** twice.

4. The output is -1.644853626, the value of *z* such that the area under the standard normal curve in the left tail is 0.05. (See **Screen 6.2**.)

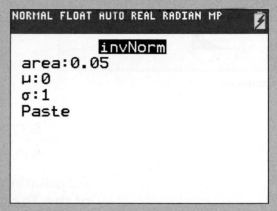

Screen 6.3

Minitab

The Minitab Technology Instructions feature of this text is written for Minitab version 17. Some screens, menus, and functions will be slightly different in older versions of Minitab.

Calculating a Left-Tail Probability for Example 6–6(b) of the Text

1. Select **Graph** > **Probability Distribution Plot**.

2. In the dialog box that appears on screen, select **View Probability** and click **OK**.

3. Use the following settings in the dialog box that appears on screen (see **Screen 6.4**):

   - Select Normal in the **Distribution** box.

   - Type 50 in the **Mean** box.

   - Type 10 in the **Standard deviation** box.

4. Click on the **Shaded Area** tab, and then use the following settings (see **Screen 6.5**):

   - Select X Value under **Define Shaded Area By**.

   - Select Left Tail.

   - Type 35 in the **X value** box.

5. Click **OK**.

6. A new window will appear showing the shaded area of 0.06681 (see **Screen 6.6**).

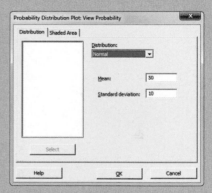

Screen 6.4

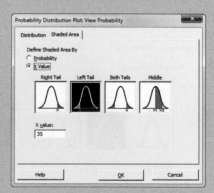

Screen 6.5

Calculating a Probability Between Two Values for Example 6–7(b) of the Text

1. Select **Graph** > **Probability Distribution Plot**.

2. In the dialog box that appears on screen, select **View Probability** and click **OK**.

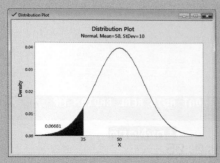

Screen 6.6

3. Use the following settings in the dialog box that appears on screen:
   • Select Normal in the **Distribution** box.
   • Type 25 in the **Mean** box.
   • Type 4 in the **Standard deviation** box.

4. Click on the **Shaded Area** tab, and then use the following settings:
   • Select X Value under **Define Shaded Area By**.
   • Select Middle.
   • Type 18 in the **X value 1** box.
   • Type 34 in the **X value 2** box.

5. Click **OK**.

6. A new window will appear showing the shaded area of 0.9477.

Calculating a Right-Tail Probability for Example 6–8(a) of the Text

1. Select **Graph > Probability Distribution Plot**.

2. In the dialog box that appears on screen, select **View Probability** and click **OK**.

3. Use the following settings in the dialog box that appears on screen:
   • Select Normal in the **Distribution** box.
   • Type 40 in the **Mean** box.
   • Type 5 in the **Standard deviation** box.

4. Click on the **Shaded Area** tab, and then use the following settings:
   • Select X Value under **Define Shaded Area By**.
   • Select Right Tail.
   • Type 55 in the **X value** box.

5. Click **OK**.

6. A new window will appear showing the shaded area of 0.001350.

Determining *z* When a Probability Is Known for Example 6–17 of the Text

1. Select **Graph > Probability Distribution Plot**.

2. In the dialog box that appears on screen, select **View Probability** and click **OK**.

3. Use the following settings in the dialog box that appears on screen:
   • Select Normal in the **Distribution** box.
   • Type 0.0 in the **Mean** box.
   • Type 1.0 in the **Standard deviation** box.

4. Click on the **Shaded Area** tab, then use the following settings (see **Screen 6.7**):
   • Select Probability under **Define Shaded Area By**.
   • Select Left Tail.
   • Type 0.05 in the **Probability** box.

Screen 6.7

5. Click **OK**.

6. A new window will appear showing the *z* value of −1.645 (see **Screen 6.8**).

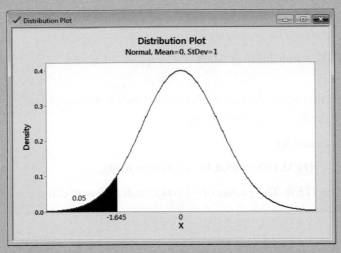

Screen 6.8

Excel

The Excel Technology Instructions feature of this text is written for Excel 2013. Some screens, menus, and functions will be slightly different in older versions of Excel. To enable some of the advanced Excel functions, you must enable the Data Analysis Add-In for Excel.

Calculating a Left-Tail Probability for Example 6–6(b) of the Text

1. Click on cell A1.

2. Type **=NORM.DIST(35,50,10,1)** (see **Screen 6.9**).

> *Note*: Entering a 1 as the fourth value in this function will cause Excel to calculate the cumulative probability, which we will always do for the normal probability distribution.

	A
1	=NORM.DIST(35,50,10,1)
2	NORM.DIST(x, mean, standard_dev, **cumulative**)

Screen 6.9

3. Press **ENTER**. The probability of 0.066807 will appear in cell A1.

Calculating a Probability Between Two Values for Example 6–7(b) of the Text

1. Click on cell A1.

2. Type **=NORM.DIST(34,25,4,1)−NORM.DIST(18,25,4,1)**.

> *Note*: Entering a 1 as the fourth value in these functions will cause Excel to calculate the cumulative probability, which we will always do for the normal probability distribution.

3. Press **ENTER**. The probability of 0.947716 will appear in cell A1.

Calculating a Right-Tail Probability for Example 6–8(a) of the Text

1. Click on cell A1.

2. Type =**1–NORM.DIST(55,40,5,1)**.

   Note: Entering a 1 as the fourth value in this function will cause Excel to calculate the cumulative probability, which we will always do for the normal probability distribution.

3. Press **ENTER**. The probability of 0.00135 will appear in cell A1.

Determining z When a Probability Is Known for Example 6–17 of the Text

	A
1	=NORM.INV(0.05,0,1)
2	NORM.INV(probability, mean, **standard_dev**)

Screen 6.10

1. Click on cell A1.

2. Type =**NORM.INV(0.05,0,1)** (see **Screen 6.10**).

3. Press **ENTER**. The *z* value of −1.64485 will appear in cell A1.

TECHNOLOGY ASSIGNMENTS

TA6.1 Find the area under the standard normal curve.

   a. between $z = 0$ and $z = 1.95$

   b. between $z = 0$ and $z = -2.05$

   c. to the right of $z = 1.36$

   d. to the left of $z = -1.97$

   e. to the right of $z = -2.05$

   f. to the left of $z = 1.76$

TA6.2 Determine the area under a normal distribution curve with

   A. $\mu = 55$ and $\sigma = 7$

      a. to the right of $x = 58$

      b. to the right of $x = 43$

      c. to the left of $x = 68$

      d. to the left of $x = 22$

   B. $\mu = 20$ and $\sigma = 4$

      a. Area between $x = 20$ and $x = 27$

      b. Area between $x = 9.5$ and $x = 17$

TA6.3 The transmission on a particular model of car has a warranty for 40,000 miles. It is known that the life of such a transmission has a normal distribution with a mean of 72,000 miles and a standard deviation of 12,000 miles. Answer the following questions.

   a. What percentage of the transmissions will fail before the end of the warranty period?

   b. What percentage of the transmissions will be good for more than 100,000 miles?

   c. What percentage of the transmissions will be good for 80,000 to 100,000 miles?

TA6.4 According to the *New York Times*, working men spend an average of 48 minutes per day caring for their families (*Time*, September 27, 2004). Assume that the times that working men currently spend per day caring for their families are normally distributed with a mean of 48 minutes and a standard deviation of 11 minutes.

   a. Find the probability that a randomly selected working man spends more than 68 minutes per day caring for his family.

   b. What percentage of working men spend between 30 and 73 minutes per day caring for their families?

TA6.5 According to the Kaiser Family Foundation, the average amount of time spent watching TV by 8-to-18-year-olds is 231 minutes per day (*USA TODAY*, March 10, 2005). Suppose currently the times spent watching TV by all 8-to-18-year-olds have a normal distribution with a mean of 231 minutes and a standard deviation of 45 minutes. What percentage of the 8-to-18-year-olds watch TV for

   a. more than 290 minutes per day?

   b. less than 150 minutes per day?

   c. 180 to 320 minutes per day?

   d. 270 to 350 minutes per day?

TA6.6 Refer to Exercise 6.29. Assume that the distribution of weekly unemployment benefits in the United States is approximately normal with a mean of $297 and a standard deviation of $74.42.

a. Find the probability that the weekly unemployment benefit received by a randomly selected person who is currently receiving an unemployment benefit is

   i. more than $200

   ii. $275 to $375

   iii. $0 or less (theoretically the normal distribution extends from negative infinity to positive infinity; realistically, unemployment benefits must be positive, so this answer provides an idea of the level of approximation used in modeling this variable)

   iv. more than $689, which is the maximum weekly unemployment benefit that a person can receive in the United States (this is similar to part c, except that we are looking at the upper tail of the distribution)

b. What is the amount of the weekly unemployment benefit that will place someone in the highest 6.5% of all weekly unemployment benefits received?

TA6.7 According to the CDC (http:\\www.cdc.gov/nchs/data/series/sr_11/sr11_252.pdf), American women age 20 and over have a mean height of 63.8 inches. The standard deviation of their heights is about 2.3 inches. Suppose these heights have an approximate normal distribution.

a. What percentage of women age 20 and over are shorter than 62 inches?

b. What proportion of women age 20 and over are taller than 70 inches?

c. What proportion of women age 20 and over are between 59 and 65 inches tall?

d. The tallest 10% of women age 20 and over are at least how tall?

CHAPTER 7

Sampling Distributions

garajstock/Shutterstock

You read about opinion polls in newspapers, magazines, and on the Web every day. These polls are based on sample surveys. Have you heard of sampling and nonsampling errors? It is good to be aware of such errors while reading these opinion poll results. Sound sampling methods are essential for opinion poll results to be valid and to minimize effects of such errors.

Chapters 5 and 6 discussed probability distributions of discrete and continuous random variables. This chapter extends the concept of probability distribution to that of a sample statistic. As we discussed in Chapter 3, a sample statistic is a numerical summary measure calculated for sample data. The mean, median, mode, and standard deviation calculated for sample data are called *sample statistics*. On the other hand, the same numerical summary measures calculated for population data are called *population parameters*. A population parameter is always a constant (at a given point in time), whereas a sample statistic is always a random variable. Because every random variable must possess a probability distribution, each sample statistic possesses a probability distribution. The probability distribution of a sample statistic is more commonly called its *sampling distribution*. This chapter discusses the sampling distributions of the sample mean and the sample proportion. The concepts covered in this chapter are the foundation of the inferential statistics discussed in succeeding chapters.

7.1 | Sampling Distribution, Sampling Error, and Nonsampling Errors

This section introduces the concepts of sampling distribution, sampling error, and nonsampling errors. Before we discuss these concepts, we will briefly describe the concept of a population distribution.

If we have information on all members of a population and we use this information to prepare a probability distribution, this will be called the population probability distribution.

Suppose there are only five students in an advanced statistics class and the midterm scores of these five students are

> The **population probability distribution** is the probability distribution of the population data.

70   78   80   80   95

Let x denote the score of a student. Using single-valued classes (because there are only five data values, and there is no need to group them), we can write the frequency distribution of scores as in Table 7.1 along with the relative frequencies of the classes, which are obtained by dividing the frequencies of the classes by the population size. Table 7.2, which lists the probabilities of various x values, presents the probability distribution of the population. Note that these probabilities are the same as the relative frequencies.

Table 7.1   **Population Frequency and Relative Frequency Distributions**

x	f	Relative Frequency
70	1	$1/5 = .20$
78	1	$1/5 = .20$
80	2	$2/5 = .40$
95	1	$1/5 = .20$
	$N = 5$	Sum $= 1.00$

Table 7.2   **Population Probability Distribution**

x	$P(x)$
70	.20
78	.20
80	.40
95	.20
	$\Sigma P(x) = 1.00$

The values of the mean and standard deviation calculated for the probability distribution of Table 7.2 give the values of the population parameters μ and σ. These values are $\mu = 80.60$ and $\sigma = 8.09$. The values of μ and σ for the probability distribution of Table 7.2 can be calculated using the formulas given in Section 5.3 of Chapter 5. We can also obtain these values of μ and σ by using the five scores and the appropriate formulas from Chapter 3.

7.1.1   Sampling Distribution

As mentioned at the beginning of this chapter, the value of a population parameter is always constant. For example, for any population data set, there is only one value of the population mean, μ. However, we cannot say the same about the sample mean, $\bar{x}$. We would expect different samples of the same size drawn from the same population to yield different values of the sample mean, $\bar{x}$. The value of the sample mean for any one sample will depend on the elements included in that sample. Consequently, **the sample mean, $\bar{x}$, is a random variable**. Therefore, like other random variables, the sample mean possesses a probability distribution, which is more commonly called the **sampling distribution of $\bar{x}$**. Other sample statistics, such as the median, mode, and standard deviation, also possess sampling distributions.

Remember that randomness does not play any role in the calculation of a population parameter, but randomness plays a significant role in obtaining the values of a sample statistic. For example, if we calculate the mean μ of a population data set, there will be no randomness involved, as all the data values in the population are included in the calculation of mean μ. However, if we select a sample from this population and calculate the sample mean $\bar{x}$, it will assume one of the numerous (in most cases millions) possible values. Assuming we select this

> The probability distribution of $\bar{x}$ is called its sampling distribution. It lists the various values that $\bar{x}$ can assume and the probability of each value of $\bar{x}$.
>
> In general, the probability distribution of a sample statistic is called its **sampling distribution**.

sample randomly, the value of $\bar{x}$ depends on which values from the population are included in the sample.

Reconsider the population of midterm scores of five students given in Table 7.1. Consider all possible samples of three scores each that can be selected, without replacement, from that population. The total number of possible samples, given by the combinations formula discussed in Chapter 4, is 10; that is,

$$\text{Total number of samples} = {}_5C_3 = \frac{5!}{3!(5-3)!} = \frac{5 \cdot 4 \cdot 3 \cdot 2 \cdot 1}{3 \cdot 2 \cdot 1 \cdot 2 \cdot 1} = 10$$

Suppose we assign the letters A, B, C, D, and E to the scores of the five students, so that

$$A = 70, \quad B = 78, \quad C = 80, \quad D = 80, \quad E = 95$$

Then, the 10 possible samples of three scores each are

$$\text{ABC, ABD, ABE, ACD, ACE, ADE, BCD, BCE, BDE, CDE}$$

These 10 samples and their respective means are listed in Table 7.3. Note that the first two samples have the same three scores. The reason for this is that two of the students (C and D) have the same score of 80, and, hence, the samples ABC and ABD contain the same values. The mean of each sample is obtained by dividing the sum of the three scores included in that sample by 3. For instance, the mean of the first sample is $(70 + 78 + 80)/3 = 76$. Note that the values of the means of the samples in Table 7.3 are rounded to two decimal places.

By using the values of $\bar{x}$ given in Table 7.3, we record the frequency distribution of $\bar{x}$ in Table 7.4. By dividing the frequencies of the various values of $\bar{x}$ by the sum of all frequencies, we obtain the relative frequencies of the classes, which are listed in the third column of Table 7.4. These relative frequencies are used as probabilities and listed in Table 7.5. This table gives the sampling distribution of $\bar{x}$.

If we select just one sample of three scores from the population of five scores, we may draw any of the 10 possible samples. Hence, the sample mean, $\bar{x}$, can assume any of the values listed in Table 7.5 with the corresponding probability. For instance, the probability that the mean of a randomly selected sample of three scores is 81.67 is .20. This probability can be written as

$$P(\bar{x} = 81.67) = .20$$

Table 7.3 All Possible Samples and Their Means When the Sample Size is 3

Sample	Scores in the Sample	$\bar{x}$
ABC	70, 78, 80	76.00
ABD	70, 78, 80	76.00
ABE	70, 78, 95	81.00
ACD	70, 80, 80	76.67
ACE	70, 80, 95	81.67
ADE	70, 80, 95	81.67
BCD	78, 80, 80	79.33
BCE	78, 80, 95	84.33
BDE	78, 80, 95	84.33
CDE	80, 80, 95	85.00

Table 7.4 Frequency and Relative Frequency Distributions of $\bar{x}$ When the Sample Size is 3

$\bar{x}$	f	Relative Frequency
76.00	2	$2/10 = .20$
76.67	1	$1/10 = .10$
79.33	1	$1/10 = .10$
81.00	1	$1/10 = .10$
81.67	2	$2/10 = .20$
84.33	2	$2/10 = .20$
85.00	1	$1/10 = .10$
	$\Sigma f = 10$	Sum $= 1.00$

Table 7.5 Sampling Distribution of $\bar{x}$ When the Sample Size is 3

$\bar{x}$	$P(x)$
76.00	.20
76.67	.10
79.33	.10
81.00	.10
81.67	.20
84.33	.20
85.00	.10
	$\Sigma P(\bar{x}) = 1.00$

7.1.2   Sampling and Nonsampling Errors

Usually, different samples selected from the same population will give different results because they contain different elements. This is obvious from Table 7.3, which shows that the mean of a sample of three scores depends on which three of the five scores are included in the sample. The result obtained from any one sample will generally be different from the result obtained from the corresponding population. The difference between the value of a sample statistic obtained from a sample and the value of the corresponding population parameter obtained from the population is called the **sampling error**. Note that this difference represents the sampling error only if the sample is random and no nonsampling error has been made. Otherwise, only a part of this difference will be due to the sampling error.

However, in the real world, it is not possible to find the sampling error because μ is not known. If μ is known then we do not need to find $\bar{x}$. But the concept of sampling error is very important in making inferences in later chapters.

It is important to remember that **a sampling error occurs because of chance**. The errors that occur for other reasons, such as errors made during collection, recording, and tabulation of data, are called **nonsampling errors**. These errors occur because of human mistakes, and not chance. Note that there is only one kind of sampling error—the error that occurs due to chance. However, there is not just one nonsampling error, but there are many nonsampling errors that may occur for different reasons.

The following paragraph, reproduced from the *Current Population Reports* of the U.S. Bureau of the Census, explains how nonsampling errors can occur.

> Nonsampling errors can be attributed to many sources, e.g., inability to obtain information about all cases in the sample, definitional difficulties, differences in the interpretation of questions, inability or unwillingness on the part of the respondents to provide correct information, inability to recall information, errors made in collection such as in recording or coding the data, errors made in processing the data, errors made in estimating values for missing data, biases resulting from the differing recall periods caused by the interviewing pattern used, and failure of all units in the universe to have some probability of being selected for the sample (undercoverage).

The following are the main reasons for the occurrence of nonsampling errors.

1. If a sample is nonrandom (and, hence, most likely nonrepresentative), the sample results may be too different from the census results. Even a randomly selected sample can become nonrandom if some of the members included in the sample cannot be contacted. A very good example of this comes from an article published in a magazine in 1988. As reported in a July 11, 1988, article in *U.S. News & World Report* ("The Numbers Racket: How Polls and Statistics Lie"), during the 1984 presidential election a test poll was conducted in which the only subjects interviewed were those who could be reached on the first try. The results of this poll indicated that the Republican incumbant Ronald Reagan had a 3 percentage point lead over the Democratic challenger Walter Mondale. However, when interviewers made an effort to contact everyone on their lists (calling some households up to 30 times before reaching someone), Reagan's lead increased to 13%. It turned out that this 13% lead was much closer to the actual election results. Apparently, people who planned to vote Republican spent less time at home.

2. The questions may be phrased in such a way that they are not fully understood by the members of the sample or population. As a result, the answers obtained are not accurate.

3. The respondents may intentionally give false information in response to some sensitive questions. For example, people may not tell the truth about their drinking habits, incomes, or opinions about minorities. Sometimes the respondents may give wrong answers because of ignorance. For example, a person may not remember the exact amount he or she spent on clothes last year. If asked in a survey, he or she may give an inaccurate answer.

4. The poll taker may make a mistake and enter a wrong number in the records or make an error while entering the data on a computer.

Sampling error is the difference between the value of a sample statistic and the value of the corresponding population parameter. In the case of the mean,

$$\text{Sampling error} = \bar{x} - \mu$$

assuming that the sample is random and no nonsampling error has been made. The sampling error occurs only in a sample survey, and not in a census.

The errors that occur in the collection, recording, and tabulation of data are called **nonsampling errors**. The nonsampling errors can occur both in a sample survey and in a census.

As mentioned earlier, the nonsampling errors can occur both in a sample survey and in a census, whereas sampling error occurs only when a sample survey is conducted. Nonsampling errors can be minimized by preparing the survey questionnaire carefully and handling the data cautiously. However, it is impossible to avoid sampling error.

Example 7–1 illustrates the sampling and nonsampling errors using the mean.

EXAMPLE 7–1   Scores of Students

Illustrating sampling and nonsampling errors.

Reconsider the population of five scores given in Table 7.1. Suppose one sample of three scores is selected from this population, and this sample includes the scores 70, 80, and 95. Find the sampling error.

Solution   The scores of the five students are 70, 78, 80, 80, and 95. The population mean is

$$\mu = \frac{70 + 78 + 80 + 80 + 95}{5} = 80.60$$

Now a random sample of three scores from this population is taken and this sample includes the scores 70, 80, and 95. The mean for this sample is

$$\bar{x} = \frac{70 + 80 + 95}{3} = 81.67$$

Consequently,

$$\text{Sampling error} = \bar{x} - \mu = 81.67 - 80.60 = \mathbf{1.07}$$

That is, the mean score estimated from the sample is 1.07 higher than the mean score of the population. Note that this difference occurred due to chance—that is, because we used a sample instead of the population.

Now suppose, when we select the sample of three scores, we mistakenly record the second score as 82 instead of 80. As a result, we calculate the sample mean as

$$\bar{x} = \frac{70 + 82 + 95}{3} = 82.33$$

Consequently, the difference between this sample mean and the population mean is

$$\bar{x} - \mu = 82.33 - 80.60 = 1.73$$

However, this difference between the sample mean and the population mean does not entirely represent the sampling error. As we calculated earlier, only 1.07 of this difference is due to the sampling error. The remaining portion, which is equal to $1.73 - 1.07 = .66$, represents the nonsampling error because it occurred due to the error we made in recording the second score in the sample. Thus, in this case,

$$\text{Sampling error} = \mathbf{1.07}$$

$$\text{Nonsampling error} = \mathbf{.66}$$

Figure 7.1 shows the sampling and nonsampling errors for these calculations.

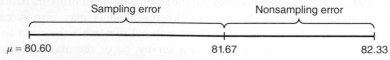

Figure 7.1   Sampling and nonsampling errors.

Thus, the sampling error is the difference between the correct value of $\bar{x}$ and the value of μ, where the correct value of $\bar{x}$ is the value of $\bar{x}$ that does not contain any nonsampling errors. In contrast, the nonsampling error(s) is (are) obtained by subtracting the correct value of $\bar{x}$ from the incorrect value of $\bar{x}$, where the incorrect value of $\bar{x}$ is the value that contains the nonsampling error(s). For our example,

$$\text{Sampling error} = \bar{x} - \mu = 81.67 - 80.60 = 1.07$$

$$\text{Nonsampling error} = \text{Incorrect } \bar{x} - \text{Correct } \bar{x} = 82.33 - 81.67 = .66$$

As mentioned earlier, in the real world we do not know the mean of a population. Hence, we select a sample to use the sample mean as an estimate of the population mean. Consequently, we never know the size of the sampling error.

EXERCISES

CONCEPTS AND PROCEDURES

7.1 Describe in short the meaning of a sampling distribution and a population probability distribution. Give an example of each.

7.2 What is meant by a sampling error? Give an example. Does such an error occur only in a sample survey, or can it occur in both a sample survey and a census?

7.3 Explain briefly the meaning of nonsampling errors. Give an example. Do such errors occur only in a sample survey, or can they occur in both a sample survey and a census?

7.4 Consider the following population of six numbers.

$$16 \quad 13 \quad 7 \quad 20 \quad 9 \quad 12$$

a. Find the population mean.
b. Liza selected one sample of four numbers from this population. The sample included the numbers 13, 7, 9, and 12. Calculate the sample mean and sampling error for this sample.
c. Refer to part b. When Liza calculated the sample mean, she mistakenly used the numbers 13, 7, 6, and 12 to calculate the sample mean. Find the sampling and nonsampling errors in this case.

d. List all samples of four numbers (without replacement) that can be selected from this population. Calculate the sample mean and sampling error for each of these samples.

APPLICATIONS

7.5 Using the formulas of Section 5.3 of Chapter 5 for the mean and standard deviation of a discrete random variable, verify that the mean and standard deviation for the population probability distribution of Table 7.2 are 80.60 and 8.09, respectively.

7.6 The following data give the years of teaching experience for all six faculty members of a department at a university.

$$7 \quad 8 \quad 14 \quad 7 \quad 20 \quad 12$$

a. Let x denote the years of teaching experience for a faculty member of this department. Write the population distribution of x.
b. List all the possible samples of size five (without replacement) that can be selected from this population. Calculate the mean for each of these samples. Write the sampling distribution of $\bar{x}$.
c. Calculate the mean for the population data. Select one random sample of size five and calculate the sample mean $\bar{x}$. Compute the sampling error.

7.2 | Mean and Standard Deviation of $\bar{x}$

The mean and standard deviation calculated for the sampling distribution of $\bar{x}$ are called the **mean** and **standard deviation of $\bar{x}$**. Actually, the mean and standard deviation of $\bar{x}$ are, respectively, the mean and standard deviation of the means of all samples of the same size selected from a population. The standard deviation of $\bar{x}$ is also called the **standard error of $\bar{x}$**.

If we calculate the mean and standard deviation of the 10 values of $\bar{x}$ listed in Table 7.3, we obtain the mean, $\mu_{\bar{x}}$ and the standard deviation, $\sigma_{\bar{x}}$ of $\bar{x}$. Alternatively, we can calculate the mean and standard deviation of the sampling distribution of $\bar{x}$ listed in Table 7.5. These will also be the values of $\mu_{\bar{x}}$ and $\sigma_{\bar{x}}$. From these calculations, we will obtain $\mu_{\bar{x}} = 80.60$ and $\sigma_{\bar{x}} = 3.30$.

The mean of the sampling distribution of $\bar{x}$ is always equal to the mean of the population.

> The mean and standard deviation of the sampling distribution of $\bar{x}$ are called the **mean and standard deviation of $\bar{x}$** and are denoted by $\mu_{\bar{x}}$ and $\sigma_{\bar{x}}$, respectively.

Mean of the Sampling Distribution of $\bar{x}$ The **mean of the sampling distribution of $\bar{x}$** is always equal to the mean of the population. Thus,

$$\mu_{\bar{x}} = \mu$$

Thus, if we select all possible samples (of the same size) from a population and calculate their means, the mean of all these sample means ($\mu_{\bar{x}}$) will be the same as the mean of the population (μ). If we calculate the mean for the population probability distribution of Table 7.2 and the mean for the sampling distribution of Table 7.5 by using the formula learned in Section 5.3 of Chapter 5, we get the same value of 80.60 for μ and $\mu_{\bar{x}}$.

The sample mean, $\bar{x}$, is called an **estimator** of the population mean, μ. When the expected value (or mean) of a sample statistic is equal to the value of the corresponding population parameter, that sample statistic is said to be an **unbiased estimator**. For the sample mean $\bar{x}$, $\mu_{\bar{x}} = \mu$. Hence, $\bar{x}$ is an unbiased estimator of μ. This is a very important property that an estimator should possess.

However, the standard deviation of $\bar{x}$, $\sigma_{\bar{x}}$, is not equal to the standard deviation of the population probability distribution, σ. The standard deviation of $\bar{x}$ is equal to the standard deviation of the population divided by the square root of the sample size; that is,

$$\sigma_{\bar{x}} = \frac{\sigma}{\sqrt{n}}$$

This formula for the standard deviation of $\bar{x}$ holds true only when the sampling is done either with replacement from a finite population or with or without replacement from an infinite population. These two conditions can be replaced by the condition that the above formula holds true if the sample size is small in comparison to the population size. The sample size is considered to be small compared to the population size if the sample size is equal to or less than 5% of the population size; that is, if

$$\frac{n}{N} \leq .05$$

If this condition is not satisfied, we use the following formula to calculate $\sigma_{\bar{x}}$:

$$\sigma_{\bar{x}} = \frac{\sigma}{\sqrt{n}} \sqrt{\frac{N-n}{N-1}}$$

where the factor $\sqrt{\dfrac{N-n}{N-1}}$ is called the **finite population correction factor**.

In most practical applications, the sample size is small compared to the population size. Consequently, in most cases, the formula used to calculate $\sigma_{\bar{x}}$ is $\sigma_{\bar{x}} = \sigma/\sqrt{n}$.

Standard Deviation of the Sampling Distribution of $\bar{x}$ The **standard deviation of the sampling distribution of $\bar{x}$** is

$$\sigma_{\bar{x}} = \frac{\sigma}{\sqrt{n}}$$

where σ is the standard deviation of the population and n is the sample size. This formula is used when $n/N \leq .05$, where N is the population size.

Following are two important observations regarding the sampling distribution of $\bar{x}$.

1. *The spread of the sampling distribution of $\bar{x}$ is smaller than the spread of the corresponding population distribution.* In other words, $\sigma_{\bar{x}} < \sigma$. This is obvious from the formula for $\sigma_{\bar{x}}$. When n is greater than 1, which is usually true, the denominator in $\sigma/\sqrt{n}$ is greater than 1. Hence, $\sigma_{\bar{x}}$ is smaller than σ.

2. *The standard deviation of the sampling distribution of $\bar{x}$ decreases as the sample size increases.* This feature of the sampling distribution of $\bar{x}$ is also obvious from the formula

$$\sigma_{\bar{x}} = \frac{\sigma}{\sqrt{n}}$$

If the standard deviation of a sample statistic decreases as the sample size is increased, that statistic is said to be a **consistent estimator**. This is another important property that an estimator should possess. It is obvious from the above formula for $\sigma_{\bar{x}}$ that as n increases, the value of $\sqrt{n}$ also increases and, consequently, the value of $\sigma/\sqrt{n}$ decreases. Thus, the sample mean $\bar{x}$ is a consistent estimator of the population mean μ. Example 7–2 illustrates this feature.

EXAMPLE 7–2   Wages of Employees

Finding the mean and standard deviation of $\bar{x}$.

The mean wage per hour for all 5000 employees who work at a large company is $27.50, and the standard deviation is $3.70. Let $\bar{x}$ be the mean wage per hour for a random sample of certain employees selected from this company. Find the mean and standard deviation of $\bar{x}$ for a sample size of

(a) 30   **(b)** 75   **(c)** 200

Solution   From the given information, for the population of all employees,

$$N = 5000, \quad \mu = \$27.50, \quad \text{and} \quad \sigma = \$3.70$$

(a) The mean of the sampling distribution of $\bar{x}$, $\mu_{\bar{x}}$, is

$$\mu_{\bar{x}} = \mu = \$27.50$$

In this case, $n = 30$, $N = 5000$, and $n/N = 30/5000 = .006$. Because n/N is less than .05, the standard deviation of $\bar{x}$ is obtained by using the formula $\sigma/\sqrt{n}$. Hence,

Image Source/GettyImages, Inc.

$$\sigma_{\bar{x}} = \frac{\sigma}{\sqrt{n}} = \frac{3.70}{\sqrt{30}} = \$.676$$

Thus, we can state that if we take all possible samples of size 30 from the population of all employees of this company and prepare the sampling distribution of $\bar{x}$, the mean and standard deviation of this sampling distribution of $\bar{x}$ will be $27.50 and $.676, respectively.

(b) In this case, $n = 75$ and $n/N = 75/5000 = .015$, which is less than .05. The mean and standard deviation of $\bar{x}$ are

$$\mu_{\bar{x}} = \mu = \$27.50 \quad \text{and} \quad \sigma_{\bar{x}} = \frac{\sigma}{\sqrt{n}} = \frac{3.70}{\sqrt{75}} = \$.427$$

(c) In this case, $n = 200$ and $n/N = 200/5000 = .04$, which is less than .05. Therefore, the mean and standard deviation of $\bar{x}$ are

$$\mu_{\bar{x}} = \mu = \$27.50 \quad \text{and} \quad \sigma_{\bar{x}} = \frac{\sigma}{\sqrt{n}} = \frac{3.70}{\sqrt{200}} = \$.262$$

From the preceding calculations we observe that the mean of the sampling distribution of $\bar{x}$ is always equal to the mean of the population whatever the size of the sample. However, the value of the standard deviation of $\bar{x}$ decreases from $.676 to $.427 and then to $.262 as the sample size increases from 30 to 75 and then to 200.

Note that in Example 7–2, $\mu_{\bar{x}} = \mu$ in all three parts indicates that $\bar{x}$ is an unbiased estimator of μ. We also observe that $\sigma_{\bar{x}}$ decreases in this example as the sample size increases. This demonstrates that $\bar{x}$ is a consistent estimator of μ.

EXERCISES

CONCEPTS AND PROCEDURES

7.7   Let $\bar{x}$ be the mean of a sample selected from a population.

    a. What is the mean of the sampling distribution of $\bar{x}$ equal to?

    b. What is the standard deviation of the sampling distribution of $\bar{x}$ equal to? Assume $n/N \leq .05$.

7.8   What is an estimator? When is an estimator unbiased? Is the sample mean, $\bar{x}$, an unbiased estimator of μ? Explain.

7.9   When is an estimator said to be consistent? Is the sample mean, $\bar{x}$, a consistent estimator of μ? Explain.

7.10   How does the value of $\sigma_{\bar{x}}$ change as the sample size increases? Explain.

7.11   Consider a large population with $\mu = 90$ and $\sigma = 18$. Assuming $n/N \leq .05$, find the mean and standard deviation of the sample mean, $\bar{x}$, for a sample size of

    a. 15   **b.** 40

7.12   A population of $N = 100,000$ has $\sigma = 50$. In each of the following cases, which formula will you use to calculate $\sigma_{\bar{x}}$ and why? Using the appropriate formula, calculate $\sigma_{\bar{x}}$ for each of these cases.

    a. $n = 2500$ **b.** $n = 7000$

*7.13   For a population, $\mu = 46$ and $\sigma = 10$.

    a. For a sample selected from this population, $\mu_{\bar{x}} = 46$ and $\sigma_{\bar{x}} = 2.0$. Find the sample size. Assume $n/N \leq .05$.

    b. For a sample selected from this population, $\mu_{\bar{x}} = 46$ and $\sigma_{\bar{x}} = 1.6$. Find the sample size. Assume $n/N \leq .05$.

APPLICATIONS

7.14   The mean monthly out-of-pocket cost of prescription drugs for all senior citizens in a particular city is $520 with a standard deviation of $72. Let $\bar{x}$ be the mean of such costs for a random sample of 25 senior citizens from this city. Find the mean and standard deviation of the sampling distribution of $\bar{x}$.

7.15   According to the American Automobile Association's 2012 annual report *Your Driving Costs*, the cost of owning and operating a four-wheel drive SUV is $11,350 per year (*USA TODAY*, April 27, 2012). Note that this cost includes expenses for gasoline, maintenance, insurance, and financing for a vehicle that is driven 15,000 miles a year. Suppose that the distribution of such costs of owning and operating all four-wheel drive SUVs has a mean of $11,350 with a standard deviation of $2390. Let $\bar{x}$ be the average of such costs of owning and operating a four-wheel drive SUV based on a random sample of 400 four-wheel drive SUVs. Find the mean and standard deviation of the sampling distribution of $\bar{x}$.

*7.16   Suppose the standard deviation of recruiting costs per player for all female basketball players recruited by all public universities in the Midwest is $2000. Let $\bar{x}$ be the mean recruiting cost for a sample of a certain number of such players. What sample size will give the standard deviation of $\bar{x}$ equal to $125? Assume $n/N \leq .05$.

*7.17   Consider the sampling distribution of $\bar{x}$ given in Table 7.5.

    a. Calculate the value of $\mu_{\bar{x}}$ using the formula $\mu_{\bar{x}} = \Sigma \bar{x} P(\bar{x})$. Is the value of $\mu = 80.60$ calculated earlier in Exercise 7.5 the same as the value of $\mu_{\bar{x}}$ calculated here?

    b. Calculate the value of $\sigma_{\bar{x}}$ by using the formula

$$\sigma_{\bar{x}} = \sqrt{\Sigma \bar{x}^2 P(\bar{x}) - (\mu_{\bar{x}})^2}$$

    c. For the five scores, $\sigma = 8.09$. Also, our sample size is 3, so that $n = 3$. Therefore, $\sigma/\sqrt{n} = 8.09/\sqrt{3} = 4.67$. From part b, you should get $\sigma_{\bar{x}} = 3.30$. Why does $\sigma/\sqrt{n}$ not equal $\sigma_{\bar{x}}$ in this case?

    d. In our example (given in the beginning of Section 7.1) on scores, $N = 5$ and $n = 3$. Hence, $n/N = 3/5 = .60$. Because n/N is greater than .05, the appropriate formula to find $\sigma_{\bar{x}}$ is

$$\sigma_{\bar{x}} = \frac{\sigma}{\sqrt{n}} \sqrt{\frac{N - n}{N - 1}}$$

Show that the value of $\sigma_{\bar{x}}$ calculated by using this formula gives the same value as the one calculated in part b above.

7.3 | Shape of the Sampling Distribution of $\bar{x}$

The shape of the sampling distribution of $\bar{x}$ relates to the following two cases:

1.   The population from which samples are drawn has a normal distribution.
2.   The population from which samples are drawn does not have a normal distribution.

7.3.1   Sampling from a Normally Distributed Population

When the population from which samples are drawn is normally distributed with its mean equal to μ and standard deviation equal to σ, then:

1.   The mean of $\bar{x}$, $\mu_{\bar{x}}$, is equal to the mean of the population, μ.
2.   The standard deviation of $\bar{x}$, $\sigma_{\bar{x}}$, is equal to $\sigma/\sqrt{n}$, assuming $n/N \leq .05$.
3.   The shape of the sampling distribution of $\bar{x}$ is normal, whatever the value of n.

Sampling Distribution of $\bar{x}$ When the Population Has a Normal Distribution   If the population from which the samples are drawn is normally distributed with mean μ and standard deviation σ, then the sampling distribution of the sample mean, $\bar{x}$, will also be normally distributed with the following mean and standard deviation, regardless of the sample size:

$$\mu_{\bar{x}} = \mu \quad \text{and} \quad \sigma_{\bar{x}} = \frac{\sigma}{\sqrt{n}}$$

For $\sigma_{\bar{x}} = \sigma/\sqrt{n}$ to be true, n/N must be less than or equal to .05.

◄ *Remember*

Figure 7.2*a* shows the probability distribution curve for a population. The distribution curves in Figure 7.2*b* through Figure 7.2*e* show the sampling distributions of $\bar{x}$ for different sample sizes taken from the population of Figure 7.2*a*. As we can observe, the population has a normal distribution. Because of this, the sampling distribution of $\bar{x}$ is normal for each of the four cases illustrated in Figure 7.2*b* through Figure 7.2*e*. Also notice from Figure 7.2*b* through Figure 7.2*e* that the spread of the sampling distribution of $\bar{x}$ decreases as the sample size increases.

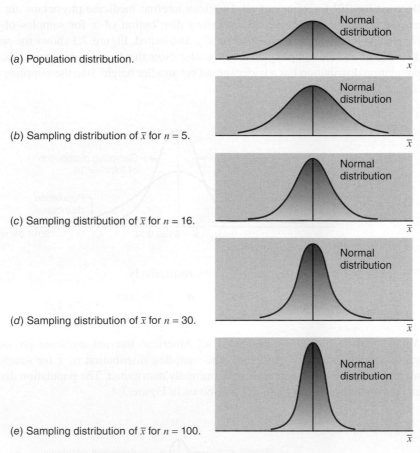

(a) Population distribution.

(b) Sampling distribution of $\bar{x}$ for $n = 5$.

(c) Sampling distribution of $\bar{x}$ for $n = 16$.

(d) Sampling distribution of $\bar{x}$ for $n = 30$.

(e) Sampling distribution of $\bar{x}$ for $n = 100$.

Figure 7.2   Population distribution and sampling distributions of $\bar{x}$.

Example 7–3 illustrates the calculation of the mean and standard deviation of $\bar{x}$ and the description of the shape of its sampling distribution.

EXAMPLE 7–3   Earnings of Internal Medicine Physicians

Finding the mean, standard deviation, and sampling distribution of $\bar{x}$: normally distributed population.

According to the 2015 Physician Compensation Report by Medscape (a subsidiary of WebMD), American internal medicine physicians earned an average of $196,000 in 2014. Suppose that the 2014 earnings of all American internal medicine physicians are approximately normally distributed with a mean of $196,000 and a standard deviation of $20,000. Let $\bar{x}$ be the mean 2014 earnings of a random sample of American internal medicine physicians. Calculate the mean and standard deviation of $\bar{x}$ and describe the shape of its sampling distribution when the sample size is

    (a) 16   **(b)** 50   **(c)** 1000

Solution   Let μ and σ be the mean and standard deviation of the 2014 earnings of all American internal medicine physicians, and $\mu_{\bar{x}}$ and $\sigma_{\bar{x}}$ be the mean and standard deviation of the sampling distribution of $\bar{x}$, respectively. Then, from the given information,

$$\mu = \$196{,}000 \quad \text{and} \quad \sigma = \$20{,}000$$

(a) The mean and standard deviation of $\bar{x}$ are, respectively,

$$\mu_{\bar{x}} = \mu = \mathbf{\$196{,}000} \quad \text{and} \quad \sigma_{\bar{x}} = \frac{\sigma}{\sqrt{n}} = \frac{\$20{,}000}{\sqrt{16}} = \mathbf{\$5000}$$

Because the 2014 earnings of all American internal medicine physicians are approximately normally distributed, the sampling distribution of $\bar{x}$ for samples of 16 such physicians is also approximately normally distributed. Figure 7.3 shows the population distribution and the sampling distribution of $\bar{x}$. Note that because σ is greater than $\sigma_{\bar{x}}$, the population distribution has a wider spread but smaller height than the sampling distribution of $\bar{x}$ in Figure 7.3.

Figure 7.3

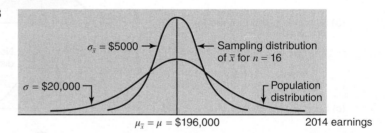

(b) The mean and standard deviation of $\bar{x}$ are, respectively,

$$\mu_{\bar{x}} = \mu = \mathbf{\$196{,}000} \quad \text{and} \quad \sigma_{\bar{x}} = \frac{\sigma}{\sqrt{n}} = \frac{\$20{,}000}{\sqrt{50}} = \mathbf{\$2828.427}$$

Again, because the 2014 earnings of all American internal medicine physicians are approximately normally distributed, the sampling distribution of $\bar{x}$ for samples of 50 such physicians is also approximately normally distributed. The population distribution and the sampling distribution of $\bar{x}$ are shown in Figure 7.4.

Figure 7.4

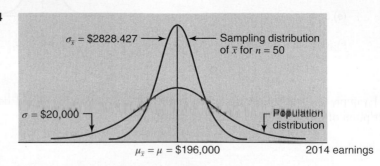

(c) The mean and standard deviation of $\bar{x}$ are, respectively,

$$\mu_{\bar{x}} = \mu = \mathbf{\$196{,}000} \quad \text{and} \quad \sigma_{\bar{x}} = \frac{\sigma}{\sqrt{n}} = \frac{\$20{,}000}{\sqrt{1000}} = \mathbf{\$632.456}$$

Again, because the 2014 earnings of all American internal medicine physicians are approximately normally distributed, the sampling distribution of $\bar{x}$ for samples of 1000 such physicians is also approximately normally distributed. The two distributions are shown in Figure 7.5.

Figure 7.5

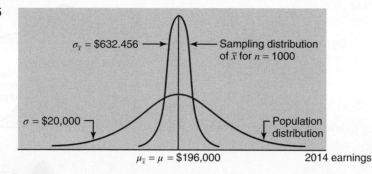

Thus, whatever the sample size, the sampling distribution of $\bar{x}$ is normal when the population from which the samples are drawn is normally distributed.

7.3.2   Sampling from a Population That Is Not Normally Distributed

Most of the time the population from which the samples are selected is not normally distributed. In such cases, the shape of the sampling distribution of $\bar{x}$ is inferred from a very important theorem called the **central limit theorem**.

Central Limit Theorem   According to the **central limit theorem**, for a large sample size, the sampling distribution of $\bar{x}$ is approximately normal, irrespective of the shape of the population distribution. The mean and standard deviation of the sampling distribution of $\bar{x}$ are, respectively,

$$\mu_{\bar{x}} = \mu \quad \text{and} \quad \sigma_{\bar{x}} = \frac{\sigma}{\sqrt{n}}$$

In case of the mean, the sample size is usually considered to be large if $n \geq 30$.

Note that when the population does not have a normal distribution, the shape of the sampling distribution is not exactly normal, but it is approximately normal for a large sample size. The approximation becomes more accurate as the sample size increases. Another point to remember is that the central limit theorem applies to *large* samples only. Usually, in case of the mean, if the sample size is 30 or larger, it is considered sufficiently large so that the central limit theorem can be applied to the sampling distribution of $\bar{x}$. Thus:

1. When $n \geq 30$, the shape of the sampling distribution of $\bar{x}$ is approximately normal irrespective of the shape of the population distribution. This is so due to the central limit theorem.
2. The mean of $\bar{x}$, $\mu_{\bar{x}}$, is equal to the mean of the population, μ.
3. The standard deviation of $\bar{x}$, $\sigma_{\bar{x}}$, is equal to $\sigma/\sqrt{n}$ if $n/N \leq .05$.

Again, remember that for $\sigma_{\bar{x}} = \sigma/\sqrt{n}$ to apply, n/N must be less than or equal to .05, otherwise we multiply $\sigma/\sqrt{n}$ by the finite population correction factor explained earlier in this chapter.

Figure 7.6a shows the probability distribution curve for a population. The distribution curves in Figure 7.6b through Figure 7.6e show the sampling distributions of $\bar{x}$ for different sample sizes taken from the population of Figure 7.6a. As we can observe, the population is not normally

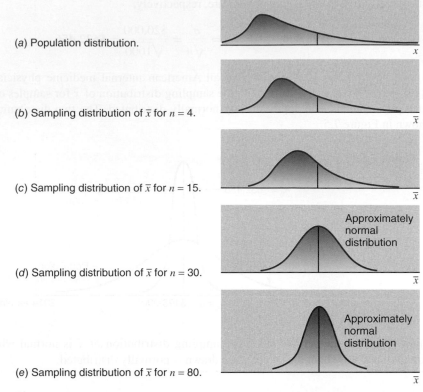

(a) Population distribution.

(b) Sampling distribution of $\bar{x}$ for $n = 4$.

(c) Sampling distribution of $\bar{x}$ for $n = 15$.

(d) Sampling distribution of $\bar{x}$ for $n = 30$.

Approximately normal distribution

Approximately normal distribution

(e) Sampling distribution of $\bar{x}$ for $n = 80$.

Figure 7.6 Population distribution and sampling distributions of $\bar{x}$.

distributed. The sampling distributions of $\bar{x}$ shown in parts b and c, when $n < 30$, are also not normal. However, the sampling distributions of $\bar{x}$ shown in parts d and e, when $n \geq 30$, are (approximately) normal. Also notice that the spread of the sampling distribution of $\bar{x}$ decreases as the sample size increases.

Example 7–4 illustrates the calculation of the mean and standard deviation of $\bar{x}$ and describes the shape of the sampling distribution of $\bar{x}$ when the sample size is large.

EXAMPLE 7–4 Rents Paid by Tenants in a City

Finding the mean, standard deviation, and sampling distribution of $\bar{x}$:
population not normally distributed.

The mean rent paid by all tenants in a small city is $1550 with a standard deviation of $225. However, the population distribution of rents for all tenants in this city is skewed to the right. Calculate the mean and standard deviation of $\bar{x}$ and describe the shape of its sampling distribution when the sample size is

(a) 30   **(b)** 100

Solution Although the population distribution of rents paid by all tenants is not normal, in each case the sample size is large ($n \geq 30$). Hence, the central limit theorem can be applied to infer the shape of the sampling distribution of $\bar{x}$.

(a) Let $\bar{x}$ be the mean rent paid by a sample of 30 tenants. Then, the sampling distribution of $\bar{x}$ is approximately normal with the values of the mean and standard deviation given as

$$\mu_{\bar{x}} = \mu = \$1550 \quad \text{and} \quad \sigma_{\bar{x}} = \frac{\sigma}{\sqrt{n}} = \frac{225}{\sqrt{30}} = \$41.079$$

Figure 7.7 shows the population distribution and the sampling distribution of $\bar{x}$.

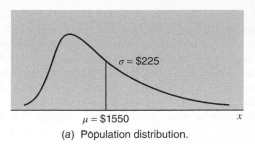

(a) Population distribution.

(b) Sampling distribution of $\bar{x}$ for $n = 30$.

Figure 7.7

(b) Let $\bar{x}$ be the mean rent paid by a sample of 100 tenants. Then, the sampling distribution of $\bar{x}$ is approximately normal with the values of the mean and standard deviation given as

$$\mu_{\bar{x}} = \mu = \mathbf{\$1550} \quad \text{and} \quad \sigma_{\bar{x}} = \frac{\sigma}{\sqrt{n}} = \frac{225}{\sqrt{100}} = \mathbf{\$22.50}$$

Figure 7.8 shows the population distribution and the sampling distribution of $\bar{x}$.

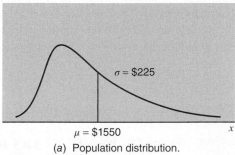

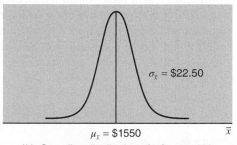

(a) Population distribution.

(b) Sampling distribution of $\bar{x}$ for $n = 100$.

Figure 7.8

EXERCISES

CONCEPTS AND PROCEDURES

7.18 Write the condition or conditions that must hold true for the sampling distribution of the sample mean to be normal when the sample size is less than 30.

7.19 Explain the central limit theorem.

7.20 A population has a distribution that is skewed to the left. Indicate in which of the following cases the central limit theorem will apply to describe the sampling distribution of the sample mean.

    a. $n = 500$ **b.** $n = 22$ **c.** $n = 34$

7.21 A population has a distribution that is skewed to the right. A sample of size n is selected from this population. Describe the shape of the sampling distribution of the sample mean for each of the following cases.

    a. $n = 25$ **b.** $n = 90$ **c.** $n = 29$

7.22 A population has a normal distribution. A sample of size n is selected from this population. Describe the shape of the sampling distribution of the sample mean for each of the following cases.

    a. $n = 23$ **b.** $n = 450$

APPLICATIONS

7.23 The delivery times for all food orders at a fast-food restaurant during the lunch hour are normally distributed with a mean of 8.4 minutes and a standard deviation of 1.8 minutes. Let $\bar{x}$ be the mean delivery time for a random sample of 16 orders at this restaurant. Calculate the mean and standard deviation of $\bar{x}$, and describe the shape of its sampling distribution.

7.24 Among college students who hold part-time jobs during the school year, the distribution of the time spent working per week is approximately normally distributed with a mean of 20.20 hours and a standard deviation of 2.60 hours. Let $\bar{x}$ be the average time spent working per week for 18 randomly selected college students who hold part-time jobs during the school year. Calculate the mean and the standard deviation of the sampling distribution of $\bar{x}$, and describe the shape of this sampling distribution.

7.25 The weights of all people living in a particular town have a distribution that is skewed to the right with a mean of 142 pounds and a standard deviation of 31 pounds. Let $\bar{x}$ be the mean weight of a random sample of 45 persons selected from this town. Find the mean and standard deviation of $\bar{x}$ and comment on the shape of its sampling distribution.

7.26 According to an estimate, the average age at first marriage for men in the United States was 28.2 years in 2010 (*Time*, March 21, 2011). Suppose that currently the mean age for all U.S. men at the time of first marriage is 28.2 years with a standard deviation of 6 years and that this distribution is strongly skewed to the right. Let $\bar{x}$ be the average age at the time of first marriage for 25 randomly selected U.S. men. Find the mean and the standard deviation of the sampling distribution of $\bar{x}$. What if the sample size is 100? How do the shapes of the sampling distributions differ for the two sample sizes?

7.27 Suppose the annual per capita (average per person) chewing gum consumption in the United States is 200 pieces. Suppose that the standard deviation of per capita consumption is 145 pieces per year. Let $\bar{x}$ be the average annual chewing gum consumption of 84 randomly selected Americans. Find the mean and the standard deviation of the sampling distribution of $\bar{x}$. What is the shape of the sampling distribution of $\bar{x}$? Do you need to know the shape of the population distribution to make this conclusion? Explain why or why not.

7.4 | Applications of the Sampling Distribution of $\bar{x}$

From the central limit theorem, for large samples, the sampling distribution of $\bar{x}$ is approximately normal with mean $\mu_{\bar{x}} = \mu$ and standard deviation $\sigma_{\bar{x}} = \sigma/\sqrt{n}$. Based on this result, we can make the following statements about $\bar{x}$ for large samples. Note that if the population is normally distributed, then sample does not have to be large. The areas under the curve of $\bar{x}$ mentioned in these statements are found from the normal distribution table.

1. *If we take all possible samples of the same (large) size from a population and calculate the mean for each of these samples, then about 68.26% of the sample means will be within one standard deviation* ($\sigma_{\bar{x}}$) *of the population mean.* Alternatively, we can state that if we take one sample (of $n \geq 30$) from a population and calculate the mean for this sample, the probability that this sample mean will be within one standard deviation ($\sigma_{\bar{x}}$) of the population mean is .6826. That is,

$$P(\mu - 1\sigma_{\bar{x}} \leq \bar{x} \leq \mu + 1\sigma_{\bar{x}}) = .8413 - .1587 = .6826$$

This probability is shown in Figure 7.9.

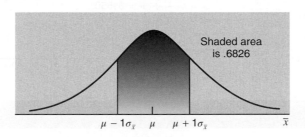

Figure 7.9 $P(\mu - 1\sigma_{\bar{x}} \leq \bar{x} \leq \mu + 1\sigma_{\bar{x}})$.

Shaded area is .6826

$\mu - 1\sigma_{\bar{x}}$ μ $\mu + 1\sigma_{\bar{x}}$ $\bar{x}$

2. *If we take all possible samples of the same (large) size from a population and calculate the mean for each of these samples, then about 95.44% of the sample means will be within two standard deviations* ($\sigma_{\bar{x}}$) *of the population mean.* Alternatively, we can state that if we take one sample (of $n \geq 30$) from a population and calculate the mean for this sample, the probability that this sample mean will be within two standard deviations ($\sigma_{\bar{x}}$) of the population mean is .9544. That is,

$$P(\mu - 2\sigma_{\bar{x}} \leq \bar{x} \leq \mu + 2\sigma_{\bar{x}}) = .9772 - .0228 = .9544$$

This probability is shown in Figure 7.10.

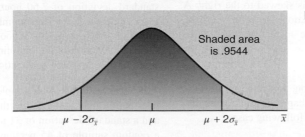

Figure 7.10 $P(\mu - 2\sigma_{\bar{x}} \leq \bar{x} \leq \mu + 2\sigma_{\bar{x}})$.

Shaded area is .9544

$\mu - 2\sigma_{\bar{x}}$ μ $\mu + 2\sigma_{\bar{x}}$ $\bar{x}$

3. *If we take all possible samples of the same (large) size from a population and calculate the mean for each of these samples, then about 99.74% of the sample means will be within three standard deviations ($\sigma_{\bar{x}}$) of the population mean.* Alternatively, we can state that if we take one sample (of $n \geq 30$) from a population and calculate the mean for this sample, the probability that this sample mean will be within three standard deviations ($\sigma_{\bar{x}}$) of the population mean is .9974. That is,

$$P(\mu - 3\sigma_{\bar{x}} \leq \bar{x} \leq \mu + 3\sigma_{\bar{x}}) = .9987 - .0013 = .9974$$

This probability is shown in Figure 7.11.

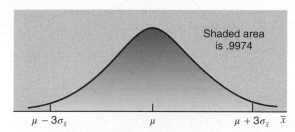

Figure 7.11 $P(\mu - 3\sigma_{\bar{x}} \leq \bar{x} \leq \mu + 3\sigma_{\bar{x}})$.

When conducting a survey, we usually select one sample and compute the value of $\bar{x}$ based on that sample. We never select all possible samples of the same size and then prepare the sampling distribution of $\bar{x}$. Rather, we are more interested in finding the probability that the value of $\bar{x}$ computed from one sample falls within a given interval. Examples 7–5 and 7–6 illustrate this procedure.

EXAMPLE 7-5 | **Weights of Packages of Cookies**

Calculating the probability of $\bar{x}$ in an interval: normal population.

Assume that the weights of all packages of a certain brand of cookies are approximately normally distributed with a mean of 32 ounces and a standard deviation of .3 ounce. Find the probability that the mean weight, $\bar{x}$, of a random sample of 20 packages of this brand of cookies will be between 31.8 and 31.9 ounces.

Solution   Although the sample size is small ($n < 30$), the shape of the sampling distribution of $\bar{x}$ is approximately normal because the population is approximately normally distributed. The mean and standard deviation of $\bar{x}$ are, respectively,

$$\mu_{\bar{x}} = \mu = 32 \text{ ounces} \quad \text{and} \quad \sigma_{\bar{x}} = \frac{\sigma}{\sqrt{n}} = \frac{.3}{\sqrt{20}} = .06708204 \text{ ounce}$$

We are to compute the probability that the value of $\bar{x}$ calculated for one randomly drawn sample of 20 packages is between 31.8 and 31.9 ounces; that is,

$$P(31.8 < \bar{x} < 31.9)$$

This probability is given by the area under the normal distribution curve for $\bar{x}$ between the points $\bar{x} = 31.8$ and $\bar{x} = 31.9$. The first step in finding this area is to convert the two $\bar{x}$ values to their respective z values.

***z* Value for a Value of $\bar{X}$**   The *z value for a value of $\bar{x}$* is calculated as

$$z = \frac{\bar{x} - \mu}{\sigma_{\bar{x}}}$$

The z values for $\bar{x} = 31.8$ and $\bar{x} = 31.9$ are computed below, and they are shown on the z scale below the normal distribution curve for $\bar{x}$ in Figure 7.12.

$$\text{For } \bar{x} = 31.8: \quad z = \frac{31.8 - 32}{.06708204} = -2.98$$

$$\text{For } \bar{x} = 31.9: \quad z = \frac{31.9 - 32}{.06708204} = -1.49$$

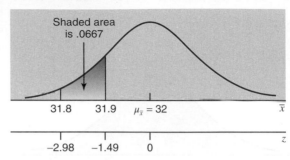

Figure 7.12 $P(31.8 < \bar{x} < 31.9)$.

The probability that $\bar{x}$ is between 31.8 and 31.9 is given by the area under the standard normal curve between $z = -2.98$ and $z = -1.49$, which is obtained by subtracting the area to the left of $z = -2.98$ from the area to the left of $z = -1.49$. Thus, the required probability is

$$P(31.8 < \bar{x} < 31.9) = P(-2.98 < z < -1.49)$$
$$= P(z < -1.49) - P(z < -2.98)$$
$$= .0681 - .0014 = \textbf{.0667}$$

Therefore, the probability that the mean weight of a sample of 20 packages will be between 31.8 and 31.9 ounces is .0667.

EXAMPLE 7–6   Cost of a Checking Account at U.S. Banks

Calculating the probability of $\bar{x}$ in an interval: $n \geq 30$.

According to Moebs Services Inc., an individual checking account at major U.S. banks costs the banks between \$350 and \$450 per year. Suppose that the current average cost of all checking accounts at major U.S. banks is \$400 per year with a standard deviation of \$30. Let $\bar{x}$ be the current average annual cost of a random sample of 225 individual checking accounts at major banks in America.

   (a) What is the probability that the average annual cost of the checking accounts in this sample is within \$4 of the population mean?

   (b) What is the probability that the average annual cost of the checking accounts in this sample is less than the population mean by \$2.70 or more?

Solution   From the given information, for the annual costs of all individual checking accounts at major banks in America,

$$\mu = \$400 \qquad \text{and} \qquad \sigma = \$30$$

Although the shape of the probability distribution of the population (annual costs of all individual checking accounts at major U.S. banks) is unknown, the sampling distribution of $\bar{x}$ is approximately normal because the sample is large ($n \geq 30$). Remember that when the sample is large, the central limit theorem applies. The mean and standard deviation of the sampling distribution of $\bar{x}$ are, respectively,

$$\mu_{\bar{x}} = \mu = \$400 \qquad \text{and} \qquad \sigma_{\bar{x}} = \frac{\sigma}{\sqrt{n}} = \frac{30}{\sqrt{225}} = \$2.00$$

(a) The probability that the mean of the annual costs of checking accounts in this sample is within \$4 of the population mean is written as

$$P(396 \leq \bar{x} \leq 404)$$

This probability is given by the area under the normal distribution curve for $\bar{x}$ between $\bar{x} = \$396$ and $\bar{x} = \$404$, as shown in Figure 7.13. We find this area as follows:

For $\bar{x} = \$396$: $z = \dfrac{\bar{x} - \mu}{\sigma_{\bar{x}}} = \dfrac{396 - 400}{2.00} = -2.00$

For $\bar{x} = \$404$: $z = \dfrac{\bar{x} - \mu}{\sigma_{\bar{x}}} = \dfrac{404 - 400}{2.00} = 2.00$

Figure 7.13 $P(\$396 \leq \bar{x} \leq \$400)$.

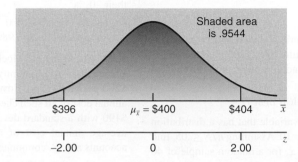

Hence, the required probability is

$$P(\$396 \leq \bar{x} \leq \$404) = P(-2.00 \leq z \leq 2.00)$$
$$= P(z \leq 2.00) - P(z \leq -2.00)$$
$$= .9772 - .0228 = \mathbf{.9544}$$

Therefore, the probability that the average annual cost of the 225 checking accounts in this sample is within \$4 of the population mean is .9544.

(b) The probability that the average annual cost of the checking accounts in this sample is less than the population mean by \$2.70 or more is written as

$$P(\bar{x} \leq 397.30)$$

This probability is given by the area under the normal curve for $\bar{x}$ to the left of $\bar{x} = \$397.30$, as shown in Figure 7.14. We find this area as follows:

For $\bar{x} = \$397.30$: $z = \dfrac{\bar{x} - \mu}{\sigma_{\bar{x}}} = \dfrac{397.30 - 400}{2.00} = -1.35$

Figure 7.14 $P(\bar{x} \leq \$397.30)$.

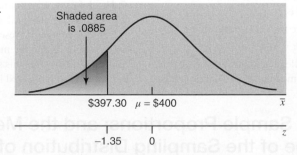

Hence, the required probability is

$$P(\bar{x} \leq 397.30) = P(z \leq -1.35) = \mathbf{.0885}$$

Thus, the probability that the average annual cost of the checking accounts in this sample is less than the population mean by \$2.70 or more is .0885.

EXERCISES

CONCEPTS AND PROCEDURES

7.28 If all possible samples of the same (large) size are selected from a population, what percentage of all the sample means will be within 1.5 standard deviations ($\sigma_{\bar{x}}$) of the population mean?

7.29 If all possible samples of the same (large) size are selected from a population, what percentage of all the sample means will be within 2.5 standard deviations ($\sigma_{\bar{x}}$) of the population mean?

7.30 For a population, $N = 205,000$, $\mu = 66$, and $\sigma = 14$. Find the z value for each of the following for $n = 49$.

  a. $\bar{x} = 66.44$ **b.** $\bar{x} = 58.75$
  c. $\bar{x} = 62.35$ **d.** $\bar{x} = 72.34$

7.31 Let x be a continuous random variable that has a normal distribution with $\mu = 75$ and $\sigma = 12$. Assuming $n/N \leq .05$, find the probability that the sample mean, $\bar{x}$, for a random sample of 20 taken from this population will be

  a. between 68.5 and 77.3   **b.** less than 72.4

7.32 Let x be a continuous random variable that has a distribution skewed to the right with $\mu = 60$ and $\sigma = 10$. Assuming $n/N \leq .05$, find the probability that the sample mean, $\bar{x}$, for a random sample of 40 taken from this population will be

  a. less than 62.40   **b.** between 62.2 and 64.2

7.33 Let x be a continuous random variable that follows a distribution skewed to the left with $\mu = 90$ and $\sigma = 18$. Assuming $n/N \leq .05$, find the probability that the sample mean, $\bar{x}$, for a random sample of 64 taken from this population will be

  a. less than 82.3   **b.** greater than 86.7

APPLICATIONS

7.34 The GPAs of all students enrolled at a large university have an approximately normal distribution with a mean of 3.13 and a standard deviation of .52. Find the probability that the mean GPA of a random sample of 20 students selected from this university is

  a. 3.20 or higher   **b.** 3.00 or lower   **c.** 3.05 to 3.21

7.35 According to an article by Laroche et al. (*The Journal of the American Board of Family Medicine* 2007;20:9–15), the average daily fat intake of U.S. adults with children in the household is 91.4 grams, with a standard deviation of 93.25 grams. Find the probability that the average daily fat intake of a random sample of 75 U.S. adults with children in the household is

  a. less than 80 grams   **b.** more than 100 grams
  c. 95 to 102 grams

7.36 The credit card debts of all college students have a distribution that is skewed to the right with a mean of $2840 and a standard deviation of $672. Find the probability that the mean credit card debt for a random sample of 36 college students would be

  a. between $2600 and $2950   **b.** less than $3060

7.37 According to a PNC Financial Independence Survey released in March 2012, U.S. adults in their 20s "hold an average debt of about $45,000, which includes everything from cars to credit cards to student loans to mortgages" (*USA TODAY*, April 24, 2012). Suppose that the current distribution of debts of all U.S. adults in their 20s has a mean of $45,000 and a standard deviation of $12,720. Find the probability that the average debt of a random sample of 144 U.S. adults in their 20s is

  a. less than $42,600   **b.** more than $46,240
  c. $43,190 to $46,980

7.38 According to Moebs Services Inc., an individual checking account at U.S. community banks costs these banks between $175 and $200 per year (*Time*, November 21, 2011). Suppose that the average annual cost of all such checking accounts at U.S. community banks is $190 with a standard deviation of $20. Find the probability that the average annual cost of a random sample of 100 such checking accounts at U.S. community banks is

  a. less than $179   **b.** more than $191.5
  c. $191.70 to $194.5

7.39 The delivery times for all food orders at a fast-food restaurant during the lunch hour are normally distributed with a mean of 8.4 minutes and a standard deviation of 1.8 minutes. Find the probability that the mean delivery time for a random sample of 16 such orders at this restaurant is

  a. between 8 and 9 minutes
  b. within 1 minute of the population mean
  c. less than the population mean by 1 minute or more

7.40 Among college students who hold part-time jobs during the school year, the distribution of the time spent working per week is approximately normally distributed with a mean of 20.20 hours and a standard deviation of 2.60 hours. Find the probability that the average time spent working per week for 18 randomly selected college students who hold part-time jobs during the school year is

  a. not within 1 hour of the population mean
  b. 20 to 20.50 hours
  c. at least 22 hours
  d. no more than 21 hours

7.41 Johnson Electronics Corporation makes electric tubes. It is known that the standard deviation of the lives of these tubes is 150 hours. The company's research department takes a sample of 100 such tubes and finds that the mean life of these tubes is 2250 hours. What is the probability that this sample mean is within 25 hours of the mean life of all tubes produced by this company?

7.5 | Population and Sample Proportions; and the Mean, Standard Deviation, and Shape of the Sampling Distribution of $\hat{p}$

The concept of proportion is the same as the concept of relative frequency discussed in Chapter 2 and the concept of probability of success in a binomial experiment. The relative frequency of a category or class gives the proportion of the sample or population that belongs to that category or class. Similarly, the probability of success in a binomial experiment represents the proportion of the sample or population that possesses a given characteristic.

In this section, first we will learn about the population and sample proportions. Then we will discuss the mean, standard deviation, and shape of the sampling distribution of $\hat{p}$.

7.5.1   Population and Sample Proportions

The **population proportion**, denoted by p, is obtained by taking the ratio of the number of elements in a population with a specific characteristic to the total number of elements in the population. The **sample proportion**, denoted by $\hat{p}$ (pronounced p *hat*), gives a similar ratio for a sample.

Population and Sample Proportions   The **population and sample proportions**, denoted by p and $\hat{p}$, respectively, are calculated as

$$p = \frac{X}{N} \quad \text{and} \quad \hat{p} = \frac{x}{n}$$

where

N = total number of elements in the population
n = total number of elements in the sample
X = number of elements in the population that possess a specific characteristic
x = number of elements in the sample that possess the same specific characteristic

Example 7–7 illustrates the calculation of the population and sample proportions.

EXAMPLE 7–7   Families Owning Homes

Calculating the population and sample proportions.

Suppose a total of 789,654 families live in a particular city and 563,282 of them own homes. A sample of 240 families is selected from this city, and 158 of them own homes. Find the proportion of families who own homes in the population and in the sample.

Solution   For the population of this city,

N = population size = 789,654

X = number of families in the population who own homes = 563,282

The population proportion of families in this city who own homes is

$$p = \frac{X}{N} = \frac{563,282}{789,654} = .71$$

Now, a sample of 240 families is taken from this city, and 158 of them are home-owners. Then,

n = sample size = 240

x = number of families in the sample who own homes = 158

The sample proportion is

$$\hat{p} = \frac{x}{n} = \frac{158}{240} = .66$$

As in the case of the mean, the difference between the sample proportion and the corresponding population proportion gives the **sampling error**, assuming that the sample is random and no nonsampling error has been made. Thus, in the case of the proportion,

$$\text{Sampling error} = \hat{p} - p$$

For instance, for Example 7–7,

$$\text{Sampling error} = \hat{p} - p = .66 - .71 = -.05$$

Here, the sampling error being equal to $-.05$ means that the sample proportion underestimates the population proportion by .05.

7.5.2 Sampling Distribution of $\hat{p}$

The probability distribution of the sample proportion, $\hat{p}$, is called its **sampling distribution**. It gives the various values that $\hat{p}$ can assume and their probabilities.

Just like the sample mean $\bar{x}$, the sample proportion $\hat{p}$ is a random variable. In other words, the population proportion p is a contant as it assumes one and only one value. However, the sample proportion $\hat{p}$ can assume one of a large number of possible values depending on which sample is selected. Hence, $\hat{p}$ is a random variable and it possesses a probability distribution, which is called its **sampling distribution**.

The value of $\hat{p}$ calculated for a particular sample depends on what elements of the population are included in that sample. Example 7–8 illustrates the concept of the sampling distribution of $\hat{p}$.

EXAMPLE 7–8 | **Employees' Knowledge of Statistics**

Illustrating the sampling distribution of $\hat{p}$.

Boe Consultant Associates has five employees. Table 7.6 gives the names of these five employees and information concerning their knowledge of statistics.

Table 7.6 **Information on the Five Employees of Boe Consultant Associates**

Name	Knows Statistics
Ally	Yes
John	No
Susan	No
Lee	Yes
Tom	Yes

If we define the population proportion, p, as the proportion of employees who know statistics, then

$$p = 3/5 = .60$$

Note that this population proportion, $p = .60$, is a constant. As long as the population does not change, this value of p will not change.

Now, suppose we draw all possible samples of three employees each and compute the proportion of employees, for each sample, who know statistics. The total number of samples of size three that can be drawn from the population of five employees is

$$\text{Total number of samples} = {}_5C_3 = \frac{5!}{3!(5-3)!} = \frac{5 \cdot 4 \cdot 3 \cdot 2 \cdot 1}{3 \cdot 2 \cdot 1 \cdot 2 \cdot 1} = 10$$

Table 7.7 lists these 10 possible samples and the proportion of employees who know statistics for each of those samples. Note that we have rounded the values of $\hat{p}$ to two decimal places.

Table 7.7   All Possible Samples of Size 3 and the Value of $\hat{p}$ for Each Sample

Sample	Proportion Who Know Statistics $\hat{p}$
Ally, John, Susan	$1/3 = .33$
Ally, John, Lee	$2/3 = .67$
Ally, John, Tom	$2/3 = .67$
Ally, Susan, Lee	$2/3 = .67$
Ally, Susan, Tom	$2/3 = .67$
Ally, Lee, Tom	$3/3 = 1.00$
John, Susan, Lee	$1/3 = .33$
John, Susan, Tom	$1/3 = .33$
John, Lee, Tom	$2/3 = .67$
Susan, Lee, Tom	$2/3 = .67$

Using Table 7.7, we prepare the frequency distribution of $\hat{p}$ as recorded in Table 7.8, along with the relative frequencies of classes, which are obtained by dividing the frequencies of classes by the total number of possible samples. The relative frequencies are used as probabilities and listed in Table 7.9. This table gives the sampling distribution of $\hat{p}$.

Table 7.8   Frequency and Relative Frequency Distributions of $\hat{p}$ When the Sample Size is 3

$\hat{p}$	f	Relative Frequency
.33	3	$3/10 = .30$
.67	6	$6/10 = .60$
1.00	1	$1/10 = .10$
	$\Sigma f = 10$	Sum $= 1.00$

Table 7.9   Sampling Distribution of $\hat{p}$ When the Sample Size is 3

$\hat{p}$	$P(\hat{p})$
.33	.30
.67	.60
1.00	.10
	$\Sigma P(\hat{p}) = 1.00$

7.5.3   Mean and Standard Deviation of $\hat{p}$

The **mean of $\hat{p}$**, which is the same as the mean of the sampling distribution of $\hat{p}$, is always equal to the population proportion, p, just as the mean of the sampling distribution of $\bar{x}$ is always equal to the population mean, μ.

In Example 7–8, $p = .60$. If we calculate the mean of the 10 values of $\hat{p}$ listed in Table 7.7, it will give us $\mu_{\hat{p}} = .60$.

Mean of the Sample Proportion   The **mean of the sample proportion**, $\hat{p}$, is denoted by $\mu_{\hat{p}}$ and is equal to the population proportion, p. Thus,

$$\mu_{\hat{p}} = p$$

The sample proportion, $\hat{p}$, is called an **estimator** of the population proportion, p. As mentioned earlier, when the expected value (or mean) of a sample statistic is equal to the value of the corresponding population parameter, that sample statistic is said to be an **unbiased estimator**. Since for the sample proportion $\mu_{\hat{p}} = p$, $\hat{p}$ is an unbiased estimator of p.

The **standard deviation of** $\hat{p}$, denoted by $\sigma_{\hat{p}}$, is given by the following formula. This formula is true only when the sample size is small compared to the population size. As we know from Section 7.2, the sample size is said to be small compared to the population size if $n/N \leq .05$.

Standard Deviation of the Sample Proportion The **standard deviation of the sample proportion** $\hat{p}$ is denoted by $\sigma_{\hat{p}}$ and is given by the formula

$$\sigma_{\hat{p}} = \sqrt{\frac{pq}{n}}$$

where p is the population proportion, $q = 1 - p$, and n is the sample size. This formula is used when $n/N \leq .05$, where N is the population size.

However, if n/N is greater than .05, then $\sigma_{\hat{p}}$ is calculated as follows:

$$\sigma_{\hat{p}} = \sqrt{\frac{pq}{n}} \ \sqrt{\frac{N - n}{N - 1}}$$

where the factor

$$\sqrt{\frac{N - n}{N - 1}}$$

is called the **finite population correction factor**.

In almost all cases, the sample size is small compared to the population size and, consequently, the formula used to calculate $\sigma_{\hat{p}}$ is $\sqrt{pq/n}$.

As mentioned earlier, if the standard deviation of a sample statistic decreases as the sample size is increased, that statistic is said to be a **consistent estimator**. It is obvious from the above formula for $\sigma_{\hat{p}}$ that as n increases, the value of $\sqrt{pq/n}$ decreases. Thus, the sample proportion, $\hat{p}$, is a consistent estimator of the population proportion, p.

7.5.4 Shape of the Sampling Distribution of $\hat{p}$

The shape of the sampling distribution of $\hat{p}$ is inferred from the central limit theorem.

Central Limit Theorem for Sample Proportion According to the central limit theorem, the **sampling distribution of** $\hat{p}$ is approximately normal for a sufficiently large sample size. In the case of proportion, the sample size is considered to be sufficiently large if np and nq are both greater than 5; that is, if

$$np > 5 \quad \text{and} \quad nq > 5$$

Note that the sampling distribution of $\hat{p}$ will be approximately normal if $np > 5$ and $nq > 5$. This is the same condition that was required for the application of the normal approximation to the binomial probability distribution in Chapter 6.

Example 7–9 shows the calculation of the mean and standard deviation of $\hat{p}$ and describes the shape of its sampling distribution.

EXAMPLE 7–9 Owning a Home: an American Dream

Finding the mean and standard deviation, and describing the shape of the sampling distribution of $\hat{p}$.

According to a recent *New York Times*/CBS News poll, 55% of adults polled said that owning a home is a *very* important part of the American Dream. Assume that this result is true for the current population of American adults. Let $\hat{p}$ be the proportion of American adults in a random sample of 2000 who will say that owning a home is a *very* important part of the American Dream. Find the mean and standard deviation of $\hat{p}$ and describe the shape of its sampling distribution.

Solution Let p be the proportion of all American adults who will say that owning a home is a *very* important part of the American Dream. Then,

$$p = .55, \qquad q = 1 - p = 1 - .55 = .45, \qquad \text{and} \qquad n = 2000$$

The mean of the sampling distribution of $\hat{p}$ is

$$\mu_{\hat{p}} = p = \textbf{.55}$$

The standard deviation of $\hat{p}$ is

$$\sigma_{\hat{p}} = \sqrt{\frac{pq}{n}} = \sqrt{\frac{(.55)(.45)}{2000}} = \textbf{.0111}$$

The values of np and nq are

$$np = 2000(.55) = 1100 \qquad \text{and} \qquad nq = 2000(.45) = 900$$

Because np and nq are both greater than 5, we can apply the central limit theorem to make an inference about the shape of the sampling distribution of $\hat{p}$. Thus, the sampling distribution of $\hat{p}$ is approximately normal with a mean of .55 and a standard deviation of .0111, as shown in Figure 7.15.

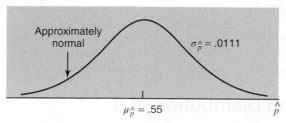

Figure 7.15 Sampling distribution of $\hat{p}$.

EXERCISES

CONCEPTS AND PROCEDURES

7.42 In a population of 1000 subjects, 580 possess a certain characteristic. A sample of 50 subjects selected from this population has 32 subjects who possess the same characteristic. What are the values of the population and sample proportions?

7.43 In a population of 5000 subjects, 800 possess a certain characteristic. A sample of 150 subjects selected from this population contains 20 subjects who possess the same characteristic. What are the values of the population and sample proportions?

7.44 In a population of 16,400 subjects, 30% possess a certain characteristic. In a sample of 200 subjects selected from this population, 25% possess the same characteristic. How many subjects in the population and sample, respectively, possess this characteristic?

7.45 Let $\hat{p}$ be the proportion of elements in a sample that possess a characteristic.

  a. What is the mean of $\hat{p}$?
  b. What is the formula to calculate the standard deviation of $\hat{p}$? Assume $n/N \leq .05$.
  c. What condition(s) must hold true for the sampling distribution of $\hat{p}$ to be approximately normal?

7.46 For a population, $N = 10,000$ and $p = .68$. A random sample of 900 elements selected from this population gave $\hat{p} = .64$. Find the sampling error.

7.47 For a population, $N = 2800$ and $p = .28$. A random sample of 100 elements selected from this population gave $\hat{p} = .33$. Find the sampling error.

7.48 Is the sample proportion a consistent estimator of the population proportion? Explain why or why not.

7.49 How does the value of $\sigma_{\hat{p}}$ change as the sample size increases? Explain. Assume $n/N \leq .05$.

7.50 What is the estimator of the population proportion? Is this estimator an unbiased estimator of p? Explain why or why not.

7.51 Consider a large population with $p = .34$. Assuming $n/N \leq .05$, find the mean and standard deviation of the sample proportion $\hat{p}$ for a sample size of

  a. 300   **b.** 850

7.52 A population of $N = 4000$ has a population proportion equal to .12. In each of the following cases, which formula will you use to calculate $\sigma_{\hat{p}}$ and why? Using the appropriate formula, calculate $\sigma_{\hat{p}}$ for each of these cases.

  a. $n = 800$ **b.** $n = 30$

7.53 According to the central limit theorem, the sampling distribution of $\hat{p}$ is approximately normal when the sample is large. What is considered a large sample in the case of the proportion? Briefly explain.

7.54 Indicate in which of the following cases the central limit theorem will apply to describe the sampling distribution of the sample proportion.

    a. $n = 400$ and $p = .28$ **b.** $n = 80$ and $p = .05$
    c. $n = 350$ and $p = .01$ **d.** $n = 200$ and $p = .022$

7.55 Indicate in which of the following cases the central limit theorem will apply to describe the sampling distribution of the sample proportion.

    a. $n = 20$ and $p = .45$ **b.** $n = 75$ and $p = .22$
    c. $n = 60$ and $p = .12$ **d.** $n = 100$ and $p = .035$

APPLICATIONS

7.56 A company manufactured five television sets on a given day, and these TV sets were inspected for being good or defective. The results of the inspection follow.

    Good   Good   Defective   Defective   Good

    a. What proportion of these TV sets are good?
    b. How many total samples (without replacement) of size four can be selected from this population?
    c. List all the possible samples of size four that can be selected from this population and calculate the sample proportion, $\hat{p}$, of television sets that are good for each sample. Prepare the sampling distribution of $\hat{p}$.
    d. For each sample listed in part c, calculate the sampling error.

7.57 A 2009 nonscientific poll on the Web site of the *Daily Gazette* of Schenectady, New York, asked readers the following question: "Are you less inclined to buy a General Motors or Chrysler vehicle now that they have filed for bankruptcy?" Of the readers who responded, 56.1% answered *Yes* (http://www.dailygazette.com/polls/2009/jun/Bankruptcy/). Assume that this result is true for the current population of vehicle owners in the United States. Let $\hat{p}$ be the proportion in a random sample of 340 U.S. vehicle owners who are less inclined to buy a General Motors or Chrysler vehicle after these corporations filed for bankruptcy. Find the mean and standard deviation of the sampling distribution of $\hat{p}$, and describe its shape.

7.58 According to a poll, 55% of Americans do not know that GOP stands for Grand Old Party (*Time*, October 17, 2011). Assume that this percentage is true for the current population of Americans. Let $\hat{p}$ be the proportion in a random sample of 800 Americans who do not know that GOP stands for Grand Old Party. Find the mean and standard deviation of the sampling distribution of $\hat{p}$ and describe its shape.

7.59 In a *Time* Magazine/Aspen poll of American adults conducted by the strategic research firm Penn Schoen Berland, these adults were asked, "In your opinion, what is more important for the U.S. to focus on in the next decade?" Eighty-three percent of the adults polled said *domestic issues* (*Time*, July 11, 2011). Assume that this percentage is true for the current population of American adults. Let $\hat{p}$ be the proportion in a random sample of 1000 American adults who hold the above opinion. Find the mean and standard deviation of the sampling distribution of $\hat{p}$ and describe its shape.

7.6 | Applications of the Sampling Distribution of $\hat{p}$

As mentioned in Section 7.4, when we conduct a study, we usually take only one sample and make all decisions or inferences on the basis of the results of that one sample. We use the concepts of the mean, standard deviation, and shape of the sampling distribution of $\hat{p}$ to determine the probability that the value of $\hat{p}$ computed from one sample falls within a given interval. Examples 7–10 and 7–11 illustrate this application.

EXAMPLE 7–10   **College Education Too Expensive**

Calculating the probability that $\hat{p}$ is in an interval.

In a recent Pew Research Center nationwide telephone survey of American adults, 75% of adults said that college education has become too expensive for most people. Suppose that this result is true for the current population of American adults. Let $\hat{p}$ be the proportion in a random sample of 1400 adult Americans who will hold the said opinion. Find the probability that 76.5% to 78% of adults in this sample will hold this opinion.

Solution   From the given information,

$$n = 1400, \qquad p = .75, \qquad \text{and} \qquad q = 1 - p = 1 - .75 = .25$$

where p is the proportion of all adult Americans who hold the said opinion.

The mean of the sample proportion $\hat{p}$ is

$$\mu_{\hat{p}} = p = .75$$

The standard deviation of $\hat{p}$ is

$$\sigma_{\hat{p}} = \sqrt{\frac{pq}{n}} = \sqrt{\frac{(.75)(.25)}{1400}} = .01157275$$

The values of np and nq are

$$np = 1400\,(.75) = 1050 \qquad \text{and} \qquad nq = 1400\,(.25) = 350$$

Because *np* and *nq* are both greater than 5, we can infer from the central limit theorem that the sampling distribution of $\hat{p}$ is approximately normal. The probability that $\hat{p}$ is between .765 and .78 is given by the area under the normal curve for $\hat{p}$ between $\hat{p} = .765$ and $\hat{p} = .78$, as shown in Figure 7.16.

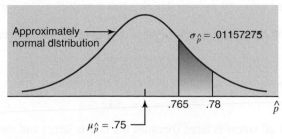

Figure 7.16 $P(.765 < \hat{p} < .78)$.

The first step in finding the area under the normal distribution curve between $\hat{p} = .765$ and $\hat{p} = .78$ is to convert these two values to their respective *z* values. The *z* value for $\hat{p}$ is computed using the following formula

---

z Value for a Value of $\hat{p}$ The *z* **value for a value of** $\hat{p}$ is calculated as

$$z = \frac{\hat{p} - p}{\sigma_{\hat{p}}}$$

The two values of $\hat{p}$ are converted to their respective *z* values, and then the area under the normal curve between these two points is found using the normal distribution table.

$$\text{For } \hat{p} = .765: \quad z = \frac{.765 - .75}{.01157275} = 1.30$$

$$\text{For } \hat{p} = .78: \quad z = \frac{.78 - .75}{.01157275} = 2.59$$

Thus, the probability that $\hat{p}$ is between .765 and .78 is given by the area under the standard normal curve between $z = 1.30$ and $z = 2.59$. This area is shown in Figure 7.17. The required probability is

$$P(.765 < \hat{p} < .78) = P(1.30 < z < 2.59) = P(z < 2.59) - P(z < 1.30)$$
$$= .9952 - .9032 = \mathbf{.0920}$$

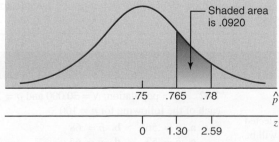

Figure 7.17 $P(.765 < \hat{p} < .78)$.

Thus, the probability that 76.5% to 78% of American adults in a random sample of 1400 will say that college education has become too expensive for most people is .0920.

EXAMPLE 7–11   Voting in an Election

Calculating the probability that $\hat{p}$ is less than a certain value.

Maureen Webster, who is running for mayor in a large city, claims that she is favored by 53% of all eligible voters of that city. Assume that this claim is true. What is the probability that in a random sample of 400 registered voters taken from this city, less than 49% will favor Maureen Webster?

Solution   Let p be the proportion of all eligible voters who favor Maureen Webster. Then,

$$p = .53 \quad \text{and} \quad q = 1 - p = 1 - .53 = .47$$

The mean of the sampling distribution of the sample proportion $\hat{p}$ is

$$\mu_{\hat{p}} = p = .53$$

The population of all voters is large (because the city is large) and the sample size is small compared to the population. Consequently, we can assume that $n/N \leq .05$. Hence, the standard deviation of $\hat{p}$ is calculated as

$$\sigma_{\hat{p}} = \sqrt{\frac{pq}{n}} = \sqrt{\frac{(.53)(.47)}{400}} = .02495496$$

From the central limit theorem, the shape of the sampling distribution of $\hat{p}$ is approximately normal. (The reader should check that $np > 5$ and $nq > 5$ and, hence, the sample size is large.) The probability that $\hat{p}$ is less than .49 is given by the area under the normal distribution curve for $\hat{p}$ to the left of $\hat{p} = .49$, as shown in Figure 7.18. The z value for $\hat{p} = .49$ is

$$z = \frac{\hat{p} - p}{\sigma_{\hat{p}}} = \frac{.49 - .53}{.02495496} = -1.60$$

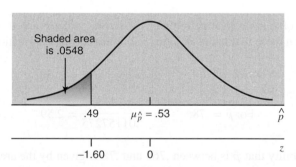

Figure 7.18 $P(\hat{p} < .49)$.

Thus, the required probability from Table IV is

$$P(\hat{p} < .49) = P(z < -1.60)$$

$$= \mathbf{.0548}$$

Hence, the probability that less than 49% of the voters in a random sample of 400 will favor Maureen Webster is .0548.

EXERCISES

CONCEPTS AND PROCEDURES

7.60   If all possible samples of the same (large) size are selected from a population, what percentage of all sample proportions will be within 3.0 standard deviations ($\sigma_{\hat{p}}$) of the population proportion?

7.61   If all possible samples of the same (large) size are selected from a population, what percentage of all sample proportions will be within 2.0 standard deviations ($\sigma_{\hat{p}}$) of the population proportion?

7.62   For a population, $N = 30,000$ and $p = .62$. Find the z value for each of the following for $n = 100$.

    a. $\hat{p} = .56$ **b.** $\hat{p} = .68$
    c. $\hat{p} = .53$ **d.** $\hat{p} = .65$

APPLICATIONS

7.63   According to a 2008 survey by the Royal Society of Chemistry, 30% of adults in Great Britain stated that they typically run the water for a period of 6 to 10 minutes before they take a shower. Let $\hat{p}$ be the proportion in a random sample of 180 adults from Great Britain who typically run the water for a period of 6 to 10 minutes before they take a shower. Find the probability that the value of $\hat{p}$ will be

  a. greater than .35 **b.** between .22 and .27

7.64 Beginning in the second half of 2011, there were widespread protests in many American cities that were primarily against Wall Street corruption and the gap between the rich and the poor in America. According to a *Time* Magazine/ABT SRBI poll conducted by telephone during October 9–10, 2011, 86% of adults who were familiar with those protests agreed that Wall Street and lobbyists have too much influence in Washington (*The New York Times*, October 22, 2011). Assume that this percentage is true for the current population of American adults. Let $\hat{p}$ be the proportion in a random sample of 400 American adults who hold the opinion that Wall Street and lobbyists have too much influence in Washington. Find the probability that the value of $\hat{p}$ will be

  a. greater than .88 **b.** between .82 and .84

7.65 Brooklyn Corporation manufactures DVDs. The machine that is used to make these DVDs is known to produce 4% defective DVDs.

The quality control inspector selects a sample of 150 DVDs every week and inspects them for being good or defective. If 8% or more of the DVDs in the sample are defective, the process is stopped and the machine is readjusted. What is the probability that based on a sample of 150 DVDs, the process will be stopped to readjust the machine?

7.66 According to the American Time Use Survey results released by the Bureau of Labor Statistics on June 24, 2015, on a typical day, 65% of American men age 15 and over spent some time doing household activities such as housework, cooking, lawn care, or financial and other household management. Assume that this percentage is true for the current population of all American men age 15 and over. A random sample of 600 American men age 15 and over is selected.

  a. Find the probability that the sample proportion is
   i. less than .68 **ii.** between .63 and .69
  b. What is the probability that the sample proportion is within .025 of the population proportion?
  c. What is the probability that the sample proportion is greater than the population proportion by .03 or more?

USES AND MISUSES...

BEWARE OF BIAS

Mathematics tells us that the sample mean, $\bar{x}$, is an unbiased and consistent estimator for the population mean, μ. This is great news because it allows us to estimate properties of a population based on those of a sample; this is the essence of statistics. But statistics always makes a number of assumptions about the sample from which the mean and standard deviation are calculated. Failure to respect these assumptions can introduce bias in your calculations. In statistics, *bias* means a deviation of the expected value of a statistical estimator from the parameter it is intended to estimate.

Let's say you are a quality control manager for a refrigerator parts company. One of the parts that you manufacture has a specification that the length of the part be 2.0 centimeters plus or minus .025 centimeter. The manufacturer expects that the parts it receives have a mean length of 2.0 centimeters and a small variation around that mean. The manufacturing process is to mold the part to something a little bit bigger than necessary—say, 2.1 centimeters—and finish the process by hand. Because the action of cutting material is

irreversible, the machinists tend to miss their target by approximately .01 centimeter, so the mean length of the parts is not 2.0 centimeters, but rather 2.01 centimeters. It is your job to catch this.

One of your quality control procedures is to select completed parts randomly and test them against specification. Unfortunately, your measurement device is also subject to variation and might consistently underestimate the length of the parts. If your measurements are consistently .01 centimeter too short, your sample mean will not catch the manufacturing error in the population of parts.

The solution to the manufacturing problem is relatively straightforward: Be certain to calibrate your measurement instrument. Calibration becomes very difficult when working with people. It is known that people tend to overestimate the number of times that they vote and underestimate the time it takes to complete a project. Basing statistical results on this type of data can result in distorted estimates of the properties of your population. It is very important to be careful to weed out bias in your data because once it gets into your calculations, it is very hard to get it out.

GLOSSARY

Central limit theorem The theorem from which it is inferred that for a large sample size ($n \geq 30$), the shape of the sampling distribution of $\bar{x}$ is approximately normal. Also, by the same theorem, the shape of the sampling distribution of $\hat{p}$ is approximately normal for a sample for which $np > 5$ and $nq > 5$.

Consistent estimator A sample statistic with a standard deviation that decreases as the sample size increases.

Estimator The sample statistic that is used to estimate a population parameter.

Mean of $\hat{p}$ The mean of the sampling distribution of $\hat{p}$, denoted by $\mu_{\hat{p}}$ is equal to the population proportion p.

Mean of $\bar{x}$ The mean of the sampling distribution of $\bar{x}$, denoted by $\mu_{\bar{x}}$, is equal to the population mean μ.

Nonsampling errors The errors that occur during the collection, recording, and tabulation of data.

Population probability distribution The probability distribution of the population data.

Population proportion p   The ratio of the number of elements in a population with a specific characteristic to the total number of elements in the population.

Sample proportion $\hat{p}$ The ratio of the number of elements in a sample with a specific characteristic to the total number of elements in that sample.

Sampling distribution of $\hat{p}$ The probability distribution of all the values of $\hat{p}$ calculated from all possible samples of the same size selected from a population.

Sampling distribution of $\bar{x}$ The probability distribution of all the values of $\bar{x}$ calculated from all possible samples of the same size selected from a population.

Sampling error   The difference between the value of a sample statistic calculated from a random sample and the value of the corresponding population parameter. This type of error occurs due to chance.

Standard deviation of $\hat{p}$ The standard deviation of the sampling distribution of $\hat{p}$, denoted by $\sigma_{\hat{p}}$, is equal to $\sqrt{pq/n}$ when $n/N \le .05$.

Standard deviation of $\bar{x}$ The standard deviation of the sampling distribution of $\bar{x}$, denoted by $\sigma_{\bar{x}}$, is equal to $\sigma/\sqrt{n}$ when $n/N \le .05$.

Unbiased estimator   An estimator with an expected value (or mean) that is equal to the value of the corresponding population parameter.

SUPPLEMENTARY EXERCISES

7.67 The print on the package of 100-watt General Electric soft-white lightbulbs claims that these bulbs have an average life of 750 hours. Assume that the lives of all such bulbs have a normal distribution with a mean of 750 hours and a standard deviation of 55 hours. Let $\bar{x}$ be the mean life of a random sample of 25 such bulbs. Find the mean and standard deviation of $\bar{x}$, and describe the shape of its sampling distribution.

7.68 The package of Sylvania CFL 65-watt replacement bulbs that use only 16 watts claims that these bulbs have an average life of 8000 hours. Assume that the lives of all such bulbs have a normal distribution with a mean of 8000 hours and a standard deviation of 400 hours. Find the probability that the mean life of a random sample of 25 such bulbs is

    a. less than 7890 hours
    b. between 7850 and 7910 hours
    c. within 130 hours of the population mean
    d. less than the population mean by 150 hours or more

7.69 According to a *Time* Magazine/ABT SRBI poll conducted by telephone during October 9–10, 2011, 73% of adults age 18 years and older said that they are in favor of raising taxes on those with annual incomes of $1 million or more to help cut the federal deficit (*Time*, October 24, 2011). Assume that this percentage is true for the current population of all American adults age 18 years and older. Let $\hat{p}$ be the proportion of American adults age 18 years and older in a random sample of 900 who will hold the above opinion. Find the mean and standard deviation of the sampling distribution of $\hat{p}$ and describe its shape.

7.70 Suppose that 88% of the cases of car burglar alarms that go off are false. Let $\hat{p}$ be the proportion of false alarms in a random sample of

80 cases of car burglar alarms that go off. Calculate the mean and standard deviation of $\hat{p}$, and describe the shape of its sampling distribution.

7.71 Seventy percent of adults favor some kind of government control on the prices of medicines. Assume that this percentage is true for the current population of all adults. Let $\hat{p}$ be the proportion of adults in a random sample of 400 who favor government control on the prices of medicines. Calculate the mean and standard deviation of $\hat{p}$ and describe the shape of its sampling distribution.

7.72 Gluten sensitivity, which is also known as wheat intolerance, affects approximately 15% of people in the United States (*Source*: http://www.foodintol.com/wheat.asp). Let $\hat{p}$ be the proportion in a random sample of 800 individuals who have gluten sensitivity. Find the probability that the value of $\hat{p}$ is

    a. within .02 of the population proportion
    b. not within .02 of the population proportion
    c. greater than the population proportion by .025 or more
    d. less than the population proportion by .03 or more

7.73 Mong Corporation makes auto batteries. The company claims that 80% of its LL70 batteries are good for 70 months or longer. Assume that this claim is true. Let $\hat{p}$ be the proportion in a sample of 100 such batteries that are good for 70 months or longer.

    a. What is the probability that this sample proportion is within .05 of the population proportion?
    b. What is the probability that this sample proportion is less than the population proportion by .06 or more?
    c. What is the probability that this sample proportion is greater than the population proportion by .07 or more?

ADVANCED EXERCISES

7.74 Let μ be the mean annual salary of Major League Baseball players for 2012. Assume that the standard deviation of the salaries of these players is $2,845,000. What is the probability that the 2012 mean salary of a random sample of 32 baseball players was within $500,000 of the population mean, μ? Assume that $n/N \le .05$.

7.75 A chemist has a 10-gallon sample of river water taken just downstream from the outflow of a chemical plant. He is concerned about the concentration, c (in parts per million), of a certain toxic

substance in the water. He wants to take several measurements, find the mean concentration of the toxic substance for this sample, and have a 95% chance of being within .5 part per million of the true mean value of c. If the concentration of the toxic substance in all measurements is normally distributed with $\sigma = 0.8$ part per million, how many measurements are necessary to achieve this goal?

7.76 A Census Bureau report revealed that 43.7% of Americans who moved between 2009 and 2010 did so for housing-related reasons, such as the desire to live in a new or better home or apartment (http://www.census.gov/newsroom/releases/archives/mobility_of_the_population/cb11-91.html). Suppose that this percentage is true for the current population of Americans.

a. Suppose that 49% of the people in a random sample of 100 Americans who moved recently did so for housing-related reasons. How likely is it for the sample proportion in a sample of 100 to be .49 or more when the population proportion is .437?

b. Refer to part a. How likely is it for the sample proportion in a random sample of 200 to be .49 or more when the population proportion is .437?

c. What is the smallest sample size that will produce a sample proportion of .49 or more is no more than 5% of all sample surveys of that size?

7.77 Refer to the sampling distribution discussed in Section 7.1. Calculate and replace the sample means in Table 7.3 with the sample medians, and then calculate the average of these sample medians. Does this average of the medians equal the population mean? If yes, why does this make sense? If no, how could you change exactly two of the five data values in this example so that the average of the sample medians equals the population mean?

SELF-REVIEW TEST

1. A sampling distribution is the probability distribution of
   a. a population parameter
   b. a sample statistic
   c. any random variable

2. Nonsampling errors are
   a. the errors that occur because the sample size is too large in relation to the population size
   b. the errors made while collecting, recording, and tabulating data
   c. the errors that occur because an untrained person conducts the survey

3. A sampling error is
   a. the difference between the value of a sample statistic based on a random sample and the value of the corresponding population parameter
   b. the error made while collecting, recording, and tabulating data
   c. the error that occurs because the sample is too small

4. The mean of the sampling distribution of $\bar{x}$ is always equal to
   a. μ **b.** $\mu - 5$ **c.** $\sigma/\sqrt{n}$

5. The condition for the standard deviation of the sample mean to be $\sigma/\sqrt{n}$ is that
   a. $np > 5$ **b.** $n/N \leq .05$ **c.** $n > 30$

6. The standard deviation of the sampling distribution of the sample mean decreases when
   a. x increases   **b.** n increases   **c.** n decreases

7. When samples are selected from a normally distributed population, the sampling distribution of the sample mean has a normal distribution
   a. if $n \geq 30$ **b.** if $n/N \leq .05$ **c.** all the time

8. When samples are selected from a nonnormally distributed population, the sampling distribution of the sample mean has an approximately normal distribution
   a. if $n \geq 30$ **b.** if $n/N \leq .05$ **c.** always

9. In a sample of 200 customers of a mail-order company, 174 are found to be satisfied with the service they receive from the company. The proportion of customers in this sample who are satisfied with the company's service is
   a. .87   **b.** .174   **c.** .148

10. The mean of the sampling distribution of $\hat{p}$ is always equal to
   a. p **b.** μ **c.** $\hat{p}$

11. The condition for the standard deviation of the sampling distribution of the sample proportion to be $\sqrt{pq/n}$ is
   a. $np > 5$ and $nq > 5$ **b.** $n > 30$ **c.** $n/N \leq .05$

12. The sampling distribution of $\hat{p}$ is approximately normal if
   a. $np > 5$ and $nq > 5$ **b.** $n > 30$ **c.** $n/N \leq .05$

13. Briefly state and explain the central limit theorem.

14. According to a 2014 Kaiser Family Foundation Health Benefits Survey released in 2015, the total mean cost of employer-sponsored family health coverage was $16,834 per family per year, of which workers were paying an average of $4823. Suppose that currently the distribution of premiums paid by all workers who have employer-sponsored family health coverage is approximately normal with the mean and standard deviation of $4823 and $700, respectively. Let $\bar{x}$ be the average premium paid by a random sample of certain workers who have employer-sponsored family health coverage. Calculate the mean and standard deviation of $\bar{x}$ and describe the shape of its sampling distribution when the sample size is
   a. 20   **b.** 100   **c.** 800

15. According to a Bureau of Labor Statistics release of March 25, 2015, statisticians earn an average of $84,010 a year. Suppose that the current annual earnings of all statisticians have the mean and standard deviation of $84,010 and $20,000, respectively, and the shape of this distribution is skewed to the right. Let $\bar{x}$ be the average earnings of a random sample of a certain number of statisticians. Calculate the mean and standard deviation of $\bar{x}$ and describe the shape of its sampling distribution when the sample size is
   a. 20   **b.** 100   **c.** 800

16. According to a Bureau of Labor Statistics release of March 25, 2015, statisticians earn an average of $84,010 a year. Suppose that the current annual earnings of all statisticians have the mean and standard deviation of $84,010 and $20,000, respectively, and the shape of this distribution is skewed to the right. Find the probability that the mean earnings of a random sample of 256 statisticians is
   a. between $81,400 and $82,600
   b. within $1500 of the population mean
   c. $85,300 or more
   d. not within $1600 of the population mean
   e. less than $83,000
   f. less than $85,000
   g. more than $86,000
   h. between $83,000 and 85,500

17. At Jen and Perry Ice Cream Company, the machine that fills 1-pound cartons of Top Flavor ice cream is set to dispense 16 ounces of ice cream into every carton. However, some cartons contain slightly less than and some contain slightly more than 16 ounces of ice cream. The amounts of ice cream in all such cartons have an approximate normal distribution with a mean of 16 ounces and a standard deviation of .18 ounce.

    a. Find the probability that the mean amount of ice cream in a random sample of 16 such cartons will be

      i. between 15.90 and 15.95 ounces

      ii. less than 15.95 ounces

      iii. more than 15.97 ounces

    b. What is the probability that the mean amount of ice cream in a random sample of 16 such cartons will be within .10 ounce of the population mean?

    c. What is the probability that the mean amount of ice cream in a random sample of 16 such cartons will be less than the population mean by .135 ounce or more?

18. According to a Pew Research survey conducted in February 2014, 24% of American adults said they trust the government in Washington, D.C., always or most of the time. Suppose that this result is true for the current population of American adults. Let $\hat{p}$ be the proportion of American adults in a random sample who hold the aforementioned opinion. Find the mean and standard deviation of the sampling distribution of $\hat{p}$ and describe its shape when the sample size is:

    a. 30   **b.** 300   **c.** 3000

19. According to a Gallup poll conducted August 7–10, 2014, 48% of American workers said that they were completely satisfied with their jobs. Assume that this result is true for the current population of American workers.

    a. Find the probability that in a random sample of 1150 American workers, the proportion who will say they are completely satisfied with their jobs is

      i. greater than .50   **ii.** between .46 and .51

      iii. less than .45   **iv.** between .495 and .515

      v. less than .49   **vi.** more than .47

    b. What is the probability that in a random sample of 1150 American workers, the proportion who will say they are completely satisfied with their jobs is within .025 of the population proportion?

    c. What is the probability that in a random sample of 1150 American workers, the proportion who will say they are completely satisfied with their jobs is not within .03 of the population proportion?

    d. What is the probability that in a random sample of 1150 American workers, the proportion who will say they are completely satisfied with their jobs is greater than the population proportion by .02 or more?

MINI-PROJECTS

Note: The Mini-Projects are located on the text's Web site, www.wiley.com/college/mann.

DECIDE FOR YOURSELF

Note: The Decide for Yourself feature is located on the text's Web site, www.wiley.com/college/mann.

TECHNOLOGY INSTRUCTIONS   CHAPTER 7

Note: Complete TI-84, Minitab, and Excel manuals are available for download at the textbook's Web site, www.wiley.com/college/mann.

TI-84 Color/TI-84

The TI-84 Color Technology Instructions feature of this text is written for the TI-84 Plus C color graphing calculator running the 4.0 operating system. Some screens, menus, and functions will be slightly different in older operating systems. The TI-84+ can perform all of the same functions but does not have the "Color" option referenced in some of the menus.

```
NORMAL FLOAT AUTO REAL RADIAN MP
          seq
Expr:mean(randNorm(196000
Variable:X
start:1
end:100
step:1
Paste
```

Screen 7.1

Approximating the Sampling Distribution of $\bar{x}$ for Example 7–3(a) of the Text

1. Select **2ⁿᵈ** > **STAT** > **OPS** > **seq(**.

2. Use the following settings in the **seq** menu (see **Screen 7.1** and **Screen 7.2**):

    • Use the following settings at the **Expr** prompt:

      • Select **2ⁿᵈ** > **STAT** > **MATH** > **mean(**.

- Select **MATH > PROB > randNorm(**.
- Type 196000, 20000, 16)). Be sure to include two right parentheses in this step.

Screen 7.2

- At the **Variable** prompt, type X.
- At the **start** prompt, type 1.
- At the **end** prompt, type 100.
- At the **step** prompt, type 1.
- Highlight **Paste** and press **ENTER**.

3. Select **STO**.

4. Select **2ⁿᵈ > 1**.

5. Press **ENTER**.

   Note: This process will take a little while to complete. Be patient.

6. There should now be 100 sample means stored in list 1 of your calculator. You can create a histogram of the values in list 1 and compute statistics as described in previous chapters.

   Note: You have only created an approximate sampling distribution. To create the complete sampling distribution, you would need to have each sample mean from every possible sample of 16 American internal medicine physicians.

   Note: You can complete parts (b) and (c) of this example by changing the number 16 in step 2 to 50 and 1000, respectively.

Approximating the Sampling Distribution of $\hat{p}$ for Example 7–9 of the Text

1. Select **2ⁿᵈ > STAT > OPS > seq(**.

2. Use the following settings in the **seq** menu (see **Screen 7.3**):

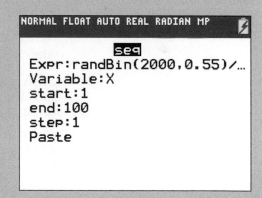

Screen 7.3

   - Use the following settings at the **Expr** prompt:
     - Select **MATH > PROB > randBin(**.
     - Type 2000,0.55)/2000.
   - At the **Variable** prompt, type X.
   - At the **start** prompt, type 1.
   - At the **end** prompt, type 100.
   - At the **step** prompt, type 1.
   - Highlight **Paste** and press **ENTER**.

3. Select **STO**.

4. Select **2ⁿᵈ > 1**.

5. Press **ENTER**.

   Note: This process will take a little while to complete. Be patient.

6. There should now be 100 sample proportions stored in list 1 of your calculator. You can create a histogram of the values in list 1 and compute statistics as described in previous chapters.

   Note: You have only created an approximate sampling distribution. To create the complete sampling distribution, you would need to have each sample proportion from every possible sample of size 2000 from this population.

Minitab

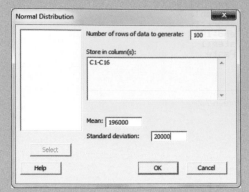

Screen 7.4

The Minitab Technology Instructions feature of this text is written for Minitab version 17. Some screens, menus, and functions will be slightly different in older versions of Minitab.

Approximating the Sampling Distribution of $\bar{x}$ for Example 7–3(a) of the Text

1. Select **Calc > Random Data > Normal**.

2. Use the following settings in the dialog box that appears on screen (see **Screen 7.4**):

   • In the **Number of rows of data to generate** box, type 100.

   • In the **Store in column(s)** box, type C1-C16.

   • In the **Mean** box, type 196000.

   • In the **Standard deviation** box, type 20000.

3. Click **OK**.

4. Select **Calc > Row Statistics**.

5. Use the following settings in the dialog box that appears on screen (see **Screen 7.5**):

   • In the **Statistic** box, select Mean.

   • In the **Input variables** box, type C1-C16.

   • In the **Store result in** box, type C17.

6. Click **OK**.

7. There should now be 100 sample means stored in C17. You can create a histogram of the values in C17 and compute statistics as described in previous chapters.

Note: You have only created an approximate sampling distribution. To create the complete sampling distribution, you would need to have each sample mean from every possible sample of 16 American internal medicine physicians.

Note: You can complete parts (b) and (c) of this example by changing the number 16 in step 2 to 50 and 1000, respectively, and the number 17 in step 5 to 51 and 1001, respectively.

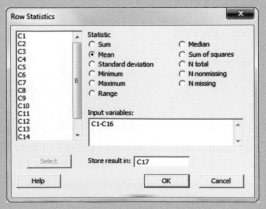

Screen 7.5

Approximating the Sampling Distribution of $\hat{p}$ for Example 7–9 of the Text

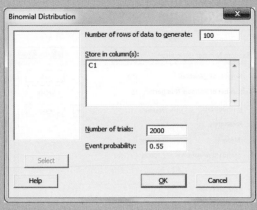

Screen 7.6

1. Select **Calc > Random Data > Binomial**.

2. Use the following settings in the dialog box that appears on screen (see **Screen 7.6**):

   - In the **Number of rows of data to generate** box, type 100.

   - In the **Store in column(s)** box, type C1.

   - In the **Number of trials** box, type 2000.

   - In the **Event probability** box, type 0.55.

3. Click **OK**.

4. Each row in C1 should now contain a number of successes from a sample of size 2000. We need to convert these numbers of successes into sample proportions of successes.

5. Select **Calc > Calculator**.

6. Use the following settings in the dialog box that appears on screen (see **Screen 7.7**):

   - In the **Store result in variable** box, type C2.

   - In the **Expression** box, type C1/2000.

7. Click **OK**.

8. There should now be 100 sample proportions stored in C2. You can create a histogram of the values in C2 and compute statistics as described in previous chapters. (See **Screen 7.8**.)

   Note: You have only created an approximate sampling distribution. To create the complete sampling distribution, you would need to have each sample proportion from every possible sample of size 2000 from this population.

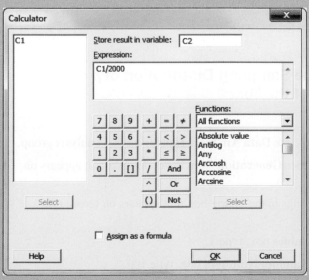

Screen 7.7

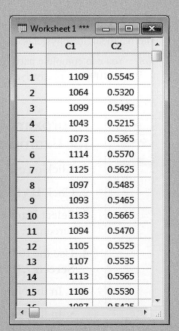

Screen 7.8

Excel

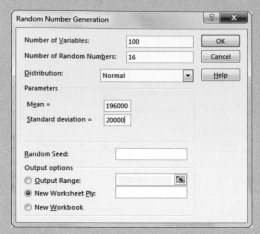

Screen 7.9

The Excel Technology Instructions feature of this text is written for Excel 2013. Some screens, menus, and functions will be slightly different in older versions of Excel. To enable some of the advanced Excel functions, you must enable the Data Analysis Add-In for Excel.

Approximating the Sampling Distribution of $\bar{x}$ for Example 7–3(a) of the Text

1. Click on cell A1.

2. Click **DATA** and then click **Data Analysis Tools** from the **Analysis** group.

3. Select **Random Number Generation** from the dialog box that appears on screen.

4. Use the following settings in the dialog box that appears on screen (see **Screen 7.9**):

   • In the **Number variables** box, type 100.

   • In the **Number of Random Numbers** box, type 16.

   • From the **Distribution** dropdown menu, select Normal.

   • In the **Mean** box, type 196000.

   • In the **Standard deviation** box, type 20000.

5. Click **OK**.

6. Click on cell A17 and type =AVERAGE(A1:A16).

7. Copy the formula from cell A17 to cells B17 through CV17.

8. There should now be 100 sample means stored in row 17 of the spreadsheet. You can create a histogram of the values in row 17 and compute statistics as described in previous chapters.

Note: You have only created an approximate sampling distribution. To create the complete sampling distribution, you would need to have every sample mean from every possible sample of 16 American internal medicine physicians.

Note: You can complete parts (b) and (c) of this example by changing the number 16 in steps 4 and 6 to 50 and 1000, respectively, and changing the number 17 in steps 6 and 7 to 51 and 1001, respectively.

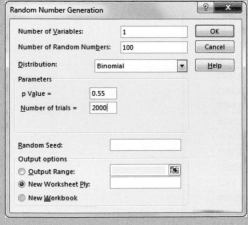

Screen 7.10

Approximating the Sampling Distribution of $\hat{p}$ for Example 7–9 of the Text

1. Click on cell A1.

2. Click **DATA** and then click **Data Analysis Tools** from the **Analysis** group.

3. Select **Random Number Generation** from the dialog box that appears on screen.

4. Use the following settings in the dialog box that appears on screen (see **Screen 7.10**):

   • In the **Number variables** box, type 1.

   • In the **Number of Random Numbers** box, type 100.

- From the **Distribution** drop-down menu, select Binomial.
- In the **p Value** box, type 0.55.
- In the **Number of trials** box, type 2000.

5. Click **OK**.

6. Each cell in column A should now contain a number of successes from a sample of size 2000. We need to convert these numbers of successes into sample proportions of successes.

7. Click on cell B1 and type = A1/2000.

8. Copy the formula from cell B1 to cells B2 through B100.

9. There should now be 100 sample proportions stored in column B. You can create a histogram of the values in column B and compute statistics as described in previous chapters.

Note: You have only created an approximate sampling distribution. To create the complete sampling distribution, you would need to have every sample proportion from every possible sample of size 2000 from this population.

TECHNOLOGY ASSIGNMENTS

TA7.1 Create 200 samples, each containing the results of 30 rolls of a die. Calculate the means of these 200 samples. Construct the histogram, and calculate the mean and standard deviation of these 200 sample means.

TA7.2 Create 150 samples each containing the results of selecting 35 numbers from 1 through 100. Calculate the means of these 150 samples. Construct the histogram, and calculate the mean and standard deviation of these 150 sample means.

TA7.3 It has been conjectured that 10% of all people are left-handed. Use technology to simulate 200 binomial experiments with 20 trials each and a probability of success (observing a left-handed person) of .10. Calculate the sample proportion of successes for each of the 200 experiments.

a. Find the mean of these 200 sample proportions.

b. Find the standard deviation of the 200 sample proportions.

c. Create a histogram of the 200 sample proportions. Describe the shape revealed by the histogram.

CHAPTER 8

Monkey Business Images/Shutterstock

Estimation of the Mean and Proportion

Do you plan to become a registered nurse? If you do, do you know how much registered nurses earn a year? According to the U.S. Bureau of Labor Statistics, registered nurses earned an average of $69,790 in 2014. The earnings of registered nurses varied greatly from state to state. Whereas the 2014 average earnings of registered nurses was $98,400 in California, it was $62,720 in Florida. (See Case Study 8–1.)

Now we are entering that part of statistics called *inferential statistics*. In Chapter 1 inferential statistics was defined as the part of statistics that helps us make decisions about some characteristics of a population based on sample information. In other words, inferential statistics uses the sample results to make decisions and draw conclusions about the population from which the sample is drawn. Estimation is the first topic to be considered in our discussion of inferential statistics. Estimation and hypothesis testing (discussed in Chapter 9) taken together are usually referred to as inference making. This chapter explains how to estimate the population mean and population proportion for a single population.

8.1 | Estimation, Point Estimate, and Interval Estimate

In this section, first we discuss the concept of estimation and then the concepts of point and interval estimates.

8.1.1 Estimation: An Introduction

Estimation is a procedure by which a numerical value or values are assigned to a population parameter based on the information collected from a sample.

In inferential statistics, μ is called the **true population mean** and p is called the **true population proportion**. There are many other population parameters, such as the median, mode, variance, and standard deviation.

The following are a few examples of estimation: an auto company may want to estimate the mean fuel consumption for a particular model of a car; a manager may want to estimate the average time taken by new employees to learn a job; the U.S. Census Bureau may want to find the mean housing expenditure per month incurred by households; and a polling agency may want to find the proportion or percentage of adults who are in favor of raising taxes on rich people to reduce the budget deficit.

The examples about estimating the mean fuel consumption, estimating the average time taken to learn a job by new employees, and estimating the mean housing expenditure per month incurred by households are illustrations of estimating the *true population mean*, μ. The example about estimating the proportion (or percentage) of all adults who are in favor of raising taxes on rich people is an illustration of estimating the *true population proportion*, p.

If we can conduct a *census* (a survey that includes the entire population) each time we want to find the value of a population parameter, then the estimation procedures explained in this and subsequent chapters are not needed. For example, if the U.S. Census Bureau can contact every household in the United States to find the mean housing expenditure incurred by households, the result of the survey (which is actually a census) will give the value of μ, and the procedures learned in this chapter will not be needed. However, it is too expensive, very time consuming, or virtually impossible to contact every member of a population to collect information to find the true value of a population parameter. Therefore, we usually take a sample from the population and calculate the value of the appropriate sample statistic. Then we assign a value or values to the corresponding population parameter based on the value of the sample statistic. This chapter (and subsequent chapters) explains how to assign values to population parameters based on the values of sample statistics.

For example, to estimate the mean time taken to learn a certain job by new employees, the manager will take a sample of new employees and record the time taken by each of these employees to learn the job. Using this information, he or she will calculate the sample mean, $\bar{x}$. Then, based on the value of $\bar{x}$, he or she will assign certain values to μ. As another example, to estimate the mean housing expenditure per month incurred by all households in the United States, the Census Bureau will take a sample of certain households, collect the information on the housing expenditure that each of these households incurs per month, and compute the value of the sample mean, $\bar{x}$. Based on this value of $\bar{x}$, the bureau will then assign values to the population mean, μ. Similarly, the polling agency who wants to find the proportion or percentage of adults who are in favor of raising taxes on rich people to reduce the budget deficit will take a sample of adults and determine the value of the sample proportion, $\hat{p}$, which represents the proportion of adults in the sample who are in favor of raising taxes on rich people to reduce the budget deficit. Using this value of the sample proportion, $\hat{p}$, the agency will assign values to the population proportion, p.

The value(s) assigned to a population parameter based on the value of a sample statistic is called an **estimate** of the population parameter. For example, suppose the manager takes a sample of 40 new employees and finds that the mean time, $\bar{x}$, taken to learn this job for these employees is 5.5 hours. If he or she assigns this value to the population mean, then 5.5 hours is called an estimate of μ. The sample statistic used to estimate a population parameter is called an **estimator**. Thus, the sample mean, $\bar{x}$, is an estimator of the population mean, μ; and the sample proportion, $\hat{p}$, is an estimator of the population proportion, p.

The estimation procedure involves the following steps.

1. Select a sample.
2. Collect the required information from the members of the sample.

> The assignment of value(s) to a population parameter based on a value of the corresponding sample statistic is called **estimation**.

> The value(s) assigned to a population parameter based on the value of a sample statistic is called an **estimate**. The sample statistic used to estimate a population parameter is called an **estimator**.

3. Calculate the value of the sample statistic.
4. Assign value(s) to the corresponding population parameter.

Remember, **the procedures to be learned in this chapter assume that the sample taken is a simple random sample**. If the sample is not a simple random sample, then the procedures to be used to estimate a population mean or proportion become more complex. These procedures are outside the scope of this book.

8.1.2 Point and Interval Estimates

An estimate may be a point estimate or an interval estimate. These two types of estimates are described in this section.

A Point Estimate

> The value of a sample statistic that is used to estimate a population parameter is called a **point estimate**.

If we select a sample and compute the value of a sample statistic for this sample, then this value gives the **point estimate** of the corresponding population parameter.

Thus, the value computed for the sample mean, $\bar{x}$, from a sample is a point estimate of the corresponding population mean, μ. For the example mentioned earlier, suppose the Census Bureau takes a random sample of 10,000 households and determines that the mean housing expenditure per month, $\bar{x}$, for this sample is $2970. Then, using $\bar{x}$ as a point estimate of μ, the Bureau can state that the mean housing expenditure per month, μ, for all households is about $2970. Thus,

Point estimate of a population parameter = Value of the corresponding sample statistic

Each sample selected from a population is expected to yield a different value of the sample statistic. Thus, the value assigned to a population mean, μ, based on a point estimate depends on which of the samples is drawn. Consequently, the point estimate assigns a value to μ that almost always differs from the true value of the population mean.

An Interval Estimate

> In **interval estimation**, an interval is constructed around the point estimate, and it is stated that this interval contains the corresponding population parameter with a certain confidence level.

In the case of **interval estimation**, instead of assigning a single value to a population parameter, an interval is constructed around the point estimate, and then a probabilistic statement that this interval contains the corresponding population parameter is made.

For the example about the mean housing expenditure, instead of saying that the mean housing expenditure per month for all households is $2970, we may obtain an interval by subtracting a number from $2970 and adding the same number to $2970. Then we state that this interval contains the population mean, μ. For purposes of illustration, suppose we subtract $340 from $2970 and add $340 to $2970. Consequently, we obtain the interval ($2970 − $340) to ($2970 + $340), or $2630 to $3310. Then we state that the interval $2630 to $3310 is likely to contain the population mean, μ, and that the mean housing expenditure per month for all households in the United States is between $2630 and $3310. This procedure is called **interval estimation**. The value $2630 is called the **lower limit** of the interval, and $3310 is called the **upper limit** of the interval. The number we add to and subtract from the point estimate is called the **margin of error** or the **maximum error of the estimate**. Figure 8.1 illustrates the concept of interval estimation.

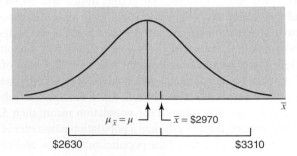

Figure 8.1   Interval estimation.

The question arises: What number should we subtract from and add to a point estimate to obtain an interval estimate? The answer to this question depends on two considerations:

1. The standard deviation $\sigma_{\bar{x}}$ of the sample mean, $\bar{x}$
2. The level of confidence to be attached to the interval

First, the larger the standard deviation of $\bar{x}$, the greater is the number subtracted from and added to the point estimate. Thus, it is obvious that if the range over which $\bar{x}$ can assume values is larger, then the interval constructed around $\bar{x}$ must be wider to include μ.

Second, the quantity subtracted and added must be larger if we want to have a higher confidence in our interval. We always attach a probabilistic statement to the interval estimation. This probabilistic statement is given by the **confidence level**. An interval constructed based on this confidence level is called a **confidence interval**.

> Each interval is constructed with regard to a given **confidence level** and is called a **confidence interval**. The confidence interval is given as
>
> $$\text{Point estimate} \pm \text{Margin of error}$$
>
> The confidence level associated with a confidence interval states how much confidence we have that this interval contains the true population parameter. The confidence level is denoted by $(1 - \alpha)100\%$.

As mentioned above, the confidence level is denoted by $(1 - \alpha)100\%$, where α is the Greek letter **alpha**. When expressed as probability, it is called the *confidence coefficient* and is denoted by $1 - \alpha$. In passing, note that α is called the *significance level*, which will be explained in detail in Chapter 9.

Although any value of the confidence level can be chosen to construct a confidence interval, the more common values are 90%, 95%, and 99%. The corresponding confidence coefficients are .90, .95, and .99, respectively. The next section describes how to construct a confidence interval for the population mean when the population standard deviation, σ, is known.

Sections 8.2 and 8.3 discuss the procedures that are used to estimate a population mean μ. In Section 8.2 we assume that the population standard deviation σ is known, and in Section 8.3 we do not assume that the population standard deviation σ is known. In the latter situation, we use the sample standard deviation s instead of σ. In the real world, the population standard deviation σ is almost never known. Consequently, we (almost) always use the sample standard deviation s.

EXERCISES

CONCEPTS AND PROCEDURES

8.1 Briefly explain the meaning of an estimator and an estimate.

8.2 Explain the meaning of a point estimate and an interval estimate.

8.2 | Estimation of a Population Mean: σ Known

This section explains how to construct a confidence interval for the population mean μ when the population standard deviation σ is known. Here, there are three possible cases, as follows.

Case I. If the following three conditions are fulfilled:

1. The population standard deviation σ is known
2. The sample size is small (i.e., $n < 30$)
3. The population from which the sample is selected is approximately normally distributed,

then we use the normal distribution to make the confidence interval for μ because from Section 7.3.1 of Chapter 7 the sampling distribution of $\bar{x}$ is normal with its mean equal to μ and the standard deviation equal to $\sigma_{\bar{x}} = \sigma/\sqrt{n}$, assuming that $n/N \leq .05$.

Case II. If the following two conditions are fulfilled:

1. The population standard deviation σ is known
2. The sample size is large (i.e., $n \geq 30$),

then, again, we use the normal distribution to make the confidence interval for μ because from Section 7.3.2 of Chapter 7, due to the central limit theorem, the sampling distribution of $\bar{x}$ is (approximately) normal with its mean equal to μ and the standard deviation equal to $\sigma_{\bar{x}} = \sigma/\sqrt{n}$, assuming that $n/N \leq .05$.

Case III. If the following three conditions are fulfilled:

1. The population standard deviation σ is known
2. The sample size is small (i.e., $n < 30$)
3. The population from which the sample is selected is not normally distributed (or its distribution is unknown),

then we use a nonparametric method to make the confidence interval for μ. Such procedures are covered in Chapter 15, which is on the Web site of the text.

This section will cover the first two cases. The procedure for making a confidence interval for μ is the same in both these cases. Note that in Case I, the population does not have to be exactly normally distributed. As long as it is close to the normal distribution without any outliers, we can use the normal distribution procedure. In Case II, although 30 is considered a large sample, if the population distribution is very different from the normal distribution, then 30 may not be a large enough sample size for the sampling distribution of $\bar{x}$ to be normal and, hence, to use the normal distribution.

The following chart summarizes the above three cases.

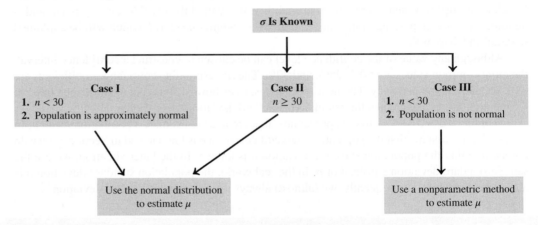

Confidence Interval for μ   The $(1 - \alpha)100\%$ **confidence interval for μ** under Cases I and II is

$$\bar{x} \pm z\sigma_{\bar{x}}$$

where

$$\sigma_{\bar{x}} = \frac{\sigma}{\sqrt{n}}$$

The value of z used here is obtained from the standard normal distribution table (Table IV of Appendix B) for the given confidence level.

The **margin of error** for the estimate of μ, denoted by E, is the quantity that is subtracted from and added to the value of $\bar{x}$ to obtain a confidence interval for μ. Thus,

$$E = z\sigma_{\bar{x}}$$

The quantity $z\sigma_{\bar{x}}$ in the confidence interval formula is called the **margin of error** and is denoted by E.

The value of z in the confidence interval formula is obtained from the standard normal distribution table (Table IV of Appendix B) for the given confidence level. To illustrate, suppose we want to construct a 95% confidence interval for μ. A 95% confidence level means that the total area under the standard normal curve between two points (at the same distance) on different sides of μ is 95%, or .95, as shown in Figure 8.2. Note that we have denoted these two points by $-z$ and z in Figure 8.2. To find the value of z for a 95% confidence level, we first find the areas to the left of these two points, $-z$ and z. Then we find the z values for these two areas from the normal

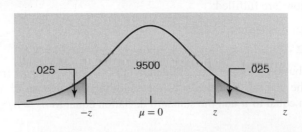

Figure 8.2   Finding z for a 95% confidence level.

distribution table. Note that these two values of z will be the same but with opposite signs. To find these values of z, we perform the following two steps:

1. The first step is to find the areas to the left of $-z$ and z, respectively. Note that the area between $-z$ and z is denoted by $1 - \alpha$. Hence, the total area in the two tails is α because the total area under the curve is 1.0. Therefore, the area in each tail, as shown in Figure 8.3, is $\alpha/2$. In our example, $1 - \alpha = .95$. Hence, the total area in both tails is $\alpha = 1 - .95 = .05$. Consequently, the area in each tail is $\alpha/2 = .05/2 = .025$. Then, the area to the left of $-z$ is .0250, and the area to the left of z is $.0250 + .95 = .9750$.

2. Now find the z values from Table IV of Appendix B such that the areas to the left of $-z$ and z are .0250 and .9750, respectively. These z values are -1.96 and 1.96, respectively.

Thus, for a confidence level of 95%, we will use $z = 1.96$ in the confidence interval formula.

Figure 8.3   Area in the tails.

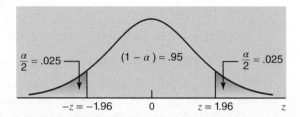

$$\frac{\alpha}{2} = .025 \qquad (1 - \alpha) = .95 \qquad \frac{\alpha}{2} = .025$$

$$-z = -1.96 \qquad 0 \qquad z = 1.96 \qquad z$$

Table 8.1 lists the z values for some of the most commonly used confidence levels. **Note that we always use the positive value of z in the formula**.

Table 8.1 z Values for Commonly Used Confidence Levels

Confidence Level	Areas to Look for in Table IV	z Value
90%	.0500 and .9500	1.64 or 1.65
95%	.0250 and .9750	1.96
96%	.0200 and .9800	2.05
97%	.0150 and .9850	2.17
98%	.0100 and .9900	2.33
99%	.0050 and .9950	2.57 or 2.58

Example 8–1 describes the procedure used to construct a confidence interval for μ when σ is known, the sample size is small, but the population from which the sample is drawn is approximately normally distributed.

EXAMPLE 8–1   Prices of Textbooks

Finding the point estimate and confidence interval for μ: σ known, $n < 30$, and population normal.

A publishing company has just published a new college textbook. Before the company decides the price at which to sell this textbook, it wants to know the average price of all such textbooks in the market. The research department at the company took a random sample of 25 comparable textbooks and collected information on their prices. This information produced a mean price of $145 for this sample. It is known that the standard deviation of the prices of all such textbooks is $35 and the population distribution of such prices is approximately normal.

(a) What is the point estimate of the mean price of all such college textbooks?

(b) Construct a 90% confidence interval for the mean price of all such college textbooks.

Solution   Here, σ is known and, although $n < 30$, the population is approximately normally distributed. Hence, we can use the normal distribution. From the given information,

$$n = 25, \quad \bar{x} = \$145, \quad \text{and} \quad \sigma = \$35$$

The standard deviation of $\bar{x}$ is

$$\sigma_{\bar{x}} = \frac{\sigma}{\sqrt{n}} = \frac{35}{\sqrt{25}} = \$7.00$$

(a) The point estimate of the mean price of all such college textbooks is $145; that is,

$$\text{Point estimate of } \mu = \bar{x} = \mathbf{\$145}$$

(b) The confidence level is 90%, or .90. First we find the z value for a 90% confidence level. Here, the area in each tail of the normal distribution curve is $\alpha/2 = (1 - .90)/2 = .05$. Now in Table IV of Appendix B, look for the areas .0500 and .9500 and find the corresponding values of z. These values are (approximately) $z = -1.65$ and $z = 1.65$.[1]

Next, we substitute all the values in the confidence interval formula for μ. The 90% confidence interval for μ is

$$\bar{x} \pm z\sigma_{\bar{x}} = 145 \pm 1.65(7.00) = 145 \pm 11.55$$

$$= (145 - 11.55) \text{ to } (145 + 11.55) = \mathbf{\$133.45 \text{ to } \$156.55}$$

Thus, we are 90% confident that the mean price of all such college textbooks is between $133.45 and $156.55. Note that we cannot say for sure whether the interval $133.45 to $156.55 contains the true population mean or not. Since μ is a constant, we cannot say that the probability is .90 that this interval contains μ because either it contains μ or it does not. Consequently, the probability that this interval contains μ is either 1.0 or 0. All we can say is that we are 90% confident that the mean price of all such college textbooks is between $133.45 and $156.55.

In the above estimate, $11.55 is called the margin of error or give-and-take figure.

How do we interpret a 90% confidence level? In terms of Example 8–1, if we take all possible samples of 25 such college textbooks each and construct a 90% confidence interval for μ around each sample mean, we can expect that 90% of these intervals will include μ and 10% will not. In Figure 8.4 we show means $\bar{x}_1$, $\bar{x}_2$, and $\bar{x}_3$ of three different samples of the same size drawn from the same population. Also shown in this figure are the 90% confidence intervals constructed around these three sample means. As we observe, the 90% confidence intervals constructed around $\bar{x}_1$ and $\bar{x}_2$ include μ, but the one constructed around $\bar{x}_3$ does not. We can state for a 90% confidence level that if we take many samples of the same size from a population and construct 90% confidence intervals around the means of these samples, then we expect 90% of these confidence intervals will be like the ones around $\bar{x}_1$ and $\bar{x}_2$ in Figure 8.4, which include μ, and 10% will be like the one around $\bar{x}_3$, which does not include μ.

Figure 8.4   Confidence intervals.

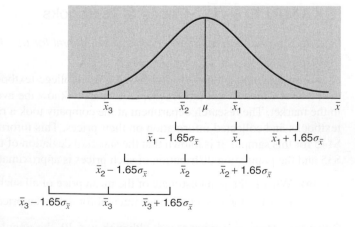

[1]Note that there is no apparent reason for choosing .0495 and .9505 and not choosing .0505 and .9495 in Table IV. If we choose .0505 and .9495, the z values will be -1.64 and 1.64. An alternative is to use the average of 1.64 and 1.65, which is 1.645, which we will not do in this text.

CASE STUDY 8–1

Annual Salaries of Registered Nurses, 2014

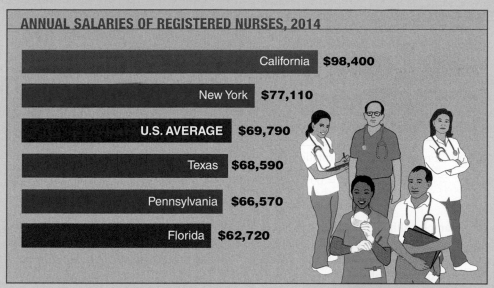

ANNUAL SALARIES OF REGISTERED NURSES, 2014

California	$98,400
New York	$77,110
U.S. AVERAGE	$69,790
Texas	$68,590
Pennsylvania	$66,570
Florida	$62,720

Data source: U.S. Bureau of Labor Statistics, May 2014

As shown in the accompanying chart, according to the May 2014 Occupational Employment and Wages report by the U.S. Bureau of Labor Statistics, registered nurses in the United States earned an average of $69,790 in 2014. The average earnings of registered nurses varied greatly from state to state. Whereas the 2014 average earnings of registered nurses was $98,400 in California, it was only $53,970 in South Dakota (which is not shown in the graph). The 2014 earnings of registered nurses also varied greatly among different metropolitan areas. Whereas such average earnings was $128,190 in the San Francisco–San Mateo–Redwood City (California) metropolitan area, it was $94,140 in the Los Angeles–Long Beach–Glendale metropolitan division. This average was $63,130 for the Greenville, North Carolina, metropolitan area. (Note that these numbers for metropolitan areas are not shown in the accompanying graph.) As we know, such estimates are based on sample surveys. If we know the sample size and the population standard deviation for any state or metropolitan area, we can find a confidence interval for the 2014 average earnings of registered nurses for that state or metropolitan area. For example, if we know the sample size and the population standard deviation of the 2014 earnings of registered nurses in Texas, we can make a confidence interval for the 2014 average earnings of all registered nurses in Texas using the following formula:

$$\bar{x} \pm z\sigma_{\bar{x}}$$

In this formula, we can substitute the values of $\bar{x}$, z, and $\sigma_{\bar{x}}$ to obtain the confidence interval. Remember that $\sigma_{\bar{x}} = \sigma/\sqrt{n}$. Suppose we want to find a 98% confidence interval for the 2014 average earnings of all registered nurses in Texas. Suppose that the 2014 average earnings of registered nurses in Texas (given in the graph) is based on a random sample of 1600 registered nurses and that the population standard deviation for such 2014 earnings is $6240. Then the 98% confidence interval for the corresponding population mean is calculated as follows:

$$\sigma_{\bar{x}} = \frac{\sigma}{\sqrt{n}} = \frac{6240}{\sqrt{1600}} = \$156.00$$

$$\bar{x} \pm z\sigma_{\bar{x}} = 68,590 \pm 2.33\,(156.00) = 68,590 \pm 363.48 = \$68,226.52 \text{ to } \$68,953.48$$

Thus, we can state with 98% confidence that the 2014 average earnings of all registered nurses in Texas is in the interval $68,226.52 to $68,953.48. We can find the confidence intervals for the other states mentioned in the graph the same way. Note that the sample means given in the graph are the point estimates of the corresponding population means.

In practice, we typically do not know the value of the population standard deviation, but we do know the value of the sample standard deviation, which is calculated from the sample data. In this case, we will find a confidence interval for the population mean using the *t* distribution procedure, which is explained in the next section.

Source: U.S. Bureau of Labor Statistics, May 2014; http://www.bls.gov/oes/current/oes291141.htm.

Example 8–2 illustrates how to obtain a confidence interval for μ when σ is known and the sample size is large ($n \geq 30$).

EXAMPLE 8–2   Cost of a Checking Account to Banks

Constructing a confidence interval for μ: σ known and $n \geq 30$.

According to a 2013 study by Moebs Services Inc., an individual checking account at major U.S. banks costs the banks more than \$380 per year. A recent random sample of 600 such checking accounts produced a mean annual cost of \$500 to major U.S. banks. Assume that the standard deviation of annual costs to major U.S. banks of all such checking accounts is \$40. Make a 99% confidence interval for the current mean annual cost to major U.S. banks of all such checking accounts.

Solution   From the given information,

$$n = 600, \qquad \bar{x} = \$500, \qquad \sigma = \$40,$$

$$\text{Confidence level} = 99\% \text{ or } .99$$

In this example, although the shape of the population distribution is unknown, the population standard deviation is known, and the sample size is large ($n \geq 30$). Hence, we can use the normal distribution to make a confidence interval for μ. To make this confidence interval, first we find the standard deviation of $\bar{x}$. The value of $\sigma_{\bar{x}}$ is

$$\sigma_{\bar{x}} = \frac{\sigma}{\sqrt{n}} = \frac{40}{\sqrt{600}} = 1.63299316$$

To find z for a 99% confidence level, first we find the area in each of the two tails of the normal distribution curve, which is $(1 - .99)/2 = .0050$. Then, we look for areas .0050 and $.0050 + .99 = .9950$ in the normal distribution table to find the two z values. These two z values are (approximately) -2.58 and 2.58. Thus, we will use $z = 2.58$ in the confidence interval formula. Substituting all the values in the formula, we obtain the 99% confidence interval for μ,

$$\bar{x} \pm z\sigma_{\bar{x}} = 500 \pm 2.58\,(1.63299316) = 500 \pm 4.21 = \textbf{\$495.79 to \$504.21}$$

Thus, we can state with 99% confidence that the current mean annual cost to major U.S. banks of all individual checking accounts is between \$495.79 and \$504.21.

The **width of a confidence interval** depends on the size of the margin of error, $z\sigma_{\bar{x}}$, which depends on the values of z, σ, and n because $\sigma_{\bar{x}} = \sigma/\sqrt{n}$. However, the value of σ is not under the control of the investigator. Hence, the width of a confidence interval can be controlled using

1.   The value of z, which depends on the confidence level
2.   The sample size n

The confidence level determines the value of z, which in turn determines the size of the margin of error. The value of z increases as the confidence level increases, and it decreases as the confidence level decreases. For example, the value of z is approximately 1.65 for a 90% confidence level, 1.96 for a 95% confidence level, and approximately 2.58 for a 99% confidence level. Hence, the higher the confidence level, the larger the width of the confidence interval as long as σ and n remain the same.

For the same confidence level and the same value of σ, an increase in the sample size decreases the value of $\sigma_{\bar{x}}$, which in turn decreases the size of the margin of error. Therefore, an increase in the sample size decreases the width of the confidence interval.

Thus, if we want to decrease the width of a confidence interval, we have two choices:

1.   Lower the confidence level
2.   Increase the sample size

Lowering the confidence level is not a good choice because a lower confidence level may give less reliable results. Therefore, we should always prefer to increase the sample size if we want to decrease the width of a confidence interval. Next we illustrate, using Example 8–2, how either a decrease in the confidence level or an increase in the sample size decreases the width of the confidence interval.

❶ Confidence Level and the Width of the Confidence Interval

Reconsider Example 8–2. Suppose all the information given in that example remains the same. First, let us decrease the confidence level to 95%. From the normal distribution table, $z = 1.96$ for a 95% confidence level. Then, using $z = 1.96$ in the confidence interval for Example 8–2, we obtain

$$\bar{x} \pm z\sigma_{\bar{x}} = 500 \pm 1.96\ (1.63299316) = 500 \pm 3.20 = \textbf{\$496.80 to \$503.20}$$

Comparing this confidence interval to the one obtained in Example 8–2, we observe that the width of the confidence interval for a 95% confidence level is smaller than the one for a 99% confidence level.

❷ Sample Size and the Width of the Confidence Interval

Consider Example 8–2 again. Now suppose the information given in that example is based on a sample size of 1000. Further assume that all other information given in that example, including the confidence level, remains the same. First, we calculate the standard deviation of the sample mean using $n = 1000$:

$$\sigma_{\bar{x}} = \frac{\sigma}{\sqrt{n}} = \frac{40}{\sqrt{1000}} = 1.26491106$$

Then, the 99% confidence interval for μ is

$$\bar{x} \pm z\sigma_{\bar{x}} = 500 \pm 2.58\ (1.26491106) = 500 \pm 3.26 = \textbf{\$496.74 to \$503.26}$$

Comparing this confidence interval to the one obtained in Example 8–2, we observe that the width of the 99% confidence interval for $n = 1000$ is smaller than the 99% confidence interval for $n = 600$.

8.2.1   Determining the Sample Size for the Estimation of Mean

One reason we usually conduct a sample survey and not a census is that almost always we have limited resources at our disposal. In light of this, if a smaller sample can serve our purpose, then we will be wasting our resources by taking a larger sample. For instance, suppose we want to estimate the mean life of a certain auto battery. If a sample of 40 batteries can give us the desired margin of error, then we will be wasting money and time if we take a sample of a much larger size—say, 500 batteries. In such cases, if we know the confidence level and the margin of error that we want, then we can find the (approximate) size of the sample that will produce the required result.

From earlier discussion, we learned that $E = z\sigma_{\bar{x}}$ is called the margin of error of the interval estimate for μ. As we know, the standard deviation of the sample mean is equal to $\sigma/\sqrt{n}$. Therefore, we can write the margin of error of estimate for μ as

$$E = z \cdot \frac{\sigma}{\sqrt{n}}$$

Suppose we predetermine the size of the margin of error, E, and want to find the size of the sample that will yield this margin of error. From the above expression, the following formula is obtained that determines the required sample size n.

Determining the Sample Size for the Estimation of μ   Given the confidence level and the standard deviation of the population, the sample size that will produce a predetermined margin of error E of the confidence interval estimate of μ is

$$n = \frac{z^2\sigma^2}{E^2}$$

If we do not know σ, we can take a preliminary sample (of any arbitrarily determined size) and find the sample standard deviation, s. Then we can use s for σ in the formula. However, note that using s for σ may give a sample size that eventually may produce an error much larger (or smaller) than the predetermined margin of error. This will depend on how close s and σ are.

Example 8–3 illustrates how we determine the sample size that will produce the margin of error of estimate for μ to within a certain limit.

EXAMPLE 8–3	Student Loan Debt of College Graduates

Determining the sample size for the estimation of μ.

An alumni association wants to estimate the mean debt of this year's college graduates. It is known that the population standard deviation of the debts of this year's college graduates is $11,800. How large a sample should be selected so that a 99% confidence interval of the estimate is within $800 of the population mean?

Solution   The alumni association wants the 99% confidence interval for the mean debt of this year's college graduates to be

$$\bar{x} \pm 800$$

Hence, the maximum size of the margin of error of estimate is to be $800; that is,

$$E = \$800$$

The value of z for a 99% confidence level is 2.58. The value of σ is given to be $11,800. Therefore, substituting all values in the formula and simplifying, we obtain

$$n = \frac{z^2 \sigma^2}{E^2} = \frac{(2.58)^2(11,800)^2}{(800)^2} = 1448.18 \approx \mathbf{1449}$$

Thus, the minimum required sample size is 1449. If the alumni association takes a random sample of 1449 of this year's college graduates, computes the mean debt for this sample, and then makes a 99% confidence interval around this sample mean, the margin of error of estimate will be approximately $800. Note that we have rounded the final answer for the sample size to the next higher integer. This is always the case when determining the sample size because rounding down to 1448 will result in a margin of error slightly greater than $800.

EXERCISES

CONCEPTS AND PROCEDURES

8.3   What is the point estimator of the population mean, μ? How would you calculate the margin of error for an estimate of μ?

8.4   Explain the various alternatives for decreasing the width of a confidence interval. Which is the best alternative?

8.5   Briefly explain how the width of a confidence interval decreases with an increase in the sample size. Give an example.

8.6   Briefly explain how the width of a confidence interval decreases with a decrease in the confidence level. Give an example.

8.7   Briefly explain the difference between a confidence level and a confidence interval.

8.8   What is the margin of error of estimate for μ when σ is known? How is it calculated?

8.9   How will you interpret a 99% confidence interval for μ? Explain.

8.10   Find z for each of the following confidence levels.

   a. 90%   **b.** 95%   **c.** 96%
   d. 97%   **e.** 98%   **f.** 99%

8.11   For a data set obtained from a random sample, $n = 20$ and $\bar{x} = 20.4$. It is known that $\sigma = 2.3$. The population is normally distributed.

   a. What is the point estimate of μ?
   b. Make a 99% confidence interval for μ.
   c. What is the margin of error of estimate for part b?

8.12   The standard deviation for a population is $\sigma = 12.4$. A random sample of 25 observations selected from this population gave a mean equal to 143.72. The population is known to have a normal distribution.

   a. Make a 99% confidence interval for μ.
   b. Construct a 95% confidence interval for μ.
   c. Determine a 90% confidence interval for μ.
   d. Does the width of the confidence intervals constructed in parts a through c decrease as the confidence level decreases? Explain your answer.

8.13   The standard deviation for a population is $\sigma = 15.3$. A random sample of 36 observations selected from this population gave a mean equal to 74.8.

   a. Make a 90% confidence interval for μ.
   b. Construct a 95% confidence interval for μ.
   c. Determine a 99% confidence interval for μ.
   d. Does the width of the confidence intervals constructed in parts a through c increase as the confidence level increases? Explain your answer.

8.14   The standard deviation for a population is $\sigma = 6.35$. A random sample selected from this population gave a mean equal to 81.90. The population is known to be normally distributed.

   a. Make a 99% confidence interval for μ assuming $n = 16$.
   b. Construct a 99% confidence interval for μ assuming $n = 20$.
   c. Determine a 99% confidence interval for μ assuming $n = 25$.
   d. Does the width of the confidence intervals constructed in parts a through c decrease as the sample size increases? Explain.

8.15 For a population, the value of the standard deviation is 2.35. A sample of 35 observations taken from this population produced the following data.

42	51	42	31	28	36	45
24	46	37	32	29	33	41
47	41	28	46	34	39	48
26	30	37	38	46	48	39
29	31	44	41	37	38	46

  a. What is the point estimate of μ?
  b. Make a 98% confidence interval for μ.
  c. What is the margin of error of estimate for part b?

8.16 For a population data set, $\sigma = 12.5$.

  a. How large a sample should be selected so that the margin of error of estimate for a 99% confidence interval for μ is 2.50?
  b. How large a sample should be selected so that the margin of error of estimate for a 96% confidence interval for μ is 3.20?

8.17 For a population data set, $\sigma = 18.50$.

  a. What should the sample size be for a 98% confidence interval for μ to have a margin of error of estimate equal to 5.50?
  b. What should the sample size be for a 95% confidence interval for μ to have a margin of error of estimate equal to 4.00?

8.18 Determine the sample size for the estimate of μ for the following.

  a. $E = 2.3$, $\sigma = 15.40$,   confidence level = 99%
  b. $E = 4.1$, $\sigma = 23.45$,   confidence level = 95%
  c. $E = 25.9$, $\sigma = 122.25$,   confidence level = 90%

8.19 Determine the sample size for the estimate of μ for the following.

  a. $E = .17$, $\sigma = .80$,   confidence level = 99%
  b. $E = 1.45$, $\sigma = 5.82$,   confidence level = 95%
  c. $E = 5.65$, $\sigma = 16.40$,   confidence level = 90%

APPLICATIONS

8.20 A sample of 1500 homes sold recently in a state gave the mean price of homes equal to $287,350. The population standard deviation of the prices of homes in this state is $53,850. Construct a 99% confidence interval for the mean price of all homes in this state.

8.21 Yunan Corporation produces bolts that are supplied to other companies. These bolts are supposed to be 4 inches long. The machine that makes these bolts does not produce each bolt exactly 4 inches long but the length of each bolt varies slightly. It is known that when the machine is working properly, the mean length of the bolts made on this machine is 4 inches. The standard deviation of the lengths of all bolts produced on this machine is always equal to .04 inch. The quality control department takes a random sample of 20 such bolts every week, calculates the mean length of these bolts, and makes a 98% confidence interval for the population mean. If either the upper limit of this confidence interval is greater than 4.02 inches or the lower limit of this confidence interval is less than 3.98 inches, the machine is stopped and adjusted. A recent such sample of 20 bolts produced a mean length of 3.99 inches. Based on this sample, will you conclude that the machine needs an adjustment? Assume that the population distribution is approximately normal.

8.22 York Steel Corporation produces iron rings that are supplied to other companies. These rings are supposed to have a diameter of 24 inches. The machine that makes these rings does not produce each ring with a diameter of exactly 24 inches. The diameter of each of the rings varies slightly. It is known that when the machine is working properly, the rings made on this machine have a mean diameter of 24 inches. The standard deviation of the diameters of all rings produced on this machine is always equal to .06 inch. The quality control department takes a random sample of 25 such rings every week, calculates the mean of the diameters for these rings, and makes a 99% confidence interval for the population mean. If either the lower limit of this confidence interval is less than 23.975 inches or the upper limit of this confidence interval is greater than 24.025 inches, the machine is stopped and adjusted. A recent such sample of 25 rings produced a mean diameter of 24.015 inches. Based on this sample, can you conclude that the machine needs an adjustment? Explain. Assume that the population distribution is approximately normal.

8.23 Lazurus Steel Corporation produces iron rods that are supposed to be 36 inches long. The machine that makes these rods does not produce each rod exactly 36 inches long. The lengths of the rods vary slightly. It is known that when the machine is working properly, the mean length of the rods made on this machine is 36 inches. The standard deviation of the lengths of all rods produced on this machine is always equal to .10 inch. The quality control department takes a sample of 30 such rods every week, calculates the mean length of these rods, and makes a 99% confidence interval for the population mean. If either the upper limit of this confidence interval is greater than 36.05 inches or the lower limit of this confidence interval is less than 35.95 inches, the machine is stopped and adjusted. A recent sample of 20 rods produced a mean length of 36.02 inches. Based on this sample, will you conclude that the machine needs an adjustment? Assume that the lengths of all such rods have a normal distribution.

8.24 A bank manager wants to know the mean amount owed on credit card accounts that become delinquent. A random sample of 100 delinquent credit card accounts taken by the manager produced a mean amount owed on these accounts equal to $2640. The population standard deviation was $578.

  a. What is the point estimate of the mean amount owed on all delinquent credit card accounts at this bank?
  b. Construct a 97% confidence interval for the mean amount owed on all delinquent credit card accounts for this bank.

8.25 A marketing researcher wants to find a 95% confidence interval for the mean amount that visitors to a theme park spend per person per day. She knows that the standard deviation of the amounts spent per person per day by all visitors to this park is $25. How large a sample should the researcher select so that the estimate will be within $3 of the population mean?

8.26 A department store manager wants to estimate at a 98% confidence level the mean amount spent by all customers at this store. The manager knows that the standard deviation of amounts spent by all customers at this store is $41. What sample size should he choose so that the estimate is within $4 of the population mean?

8.27 The principal of a large high school is concerned about the amount of time that his students spend on jobs to pay for their cars, to buy clothes, and so on. He would like to estimate the mean number of hours worked per week by these students. He knows that the standard deviation of the times spent per week on such jobs by all students is 2.5 hours. What sample size should he choose so that the estimate is within .75 hours of the population mean? The principal wants to use a 98% confidence level.

8.28 Computer Action Company sells computers and computer parts by mail. The company assures its customers that products are mailed as soon as possible after an order is placed with the company. A sample of 25 recent orders showed that the mean time taken to mail products for these orders was 70 hours. Suppose the population standard deviation is 16 hours and the population distribution is normal.

  a. Construct a 95% confidence interval for the mean time taken to mail products for all orders received at the office of this company.

b. Explain why we need to make the confidence interval. Why can we not say that the mean time taken to mail products for all orders received at the office of this company is 70 hours?

*8.29 You are interested in estimating the mean commuting time from home to school for all commuter students at your school. Briefly explain the procedure you will follow to conduct this study. Collect the required data from a sample of 30 or more such students and then estimate the population mean at a 99% confidence level. Assume that the population standard deviation for all such times is 5.5 minutes.

*8.30 You are interested in estimating the mean age of cars owned by all people in the United States. Briefly explain the procedure you will follow to conduct this study. Collect the required data on a sample of 30 or more cars and then estimate the population mean at a 95% confidence level. Assume that the population standard deviation is 2.4 years.

8.3 | Estimation of a Population Mean: σ Not Known

This section explains how to construct a confidence interval for the population mean μ when the population standard deviation σ is not known. Here, again, there are three possible cases:

Case I. If the following three conditions are fulfilled:

1. The population standard deviation σ is not known
2. The sample size is small (i.e., $n < 30$)
3. The population from which the sample is selected is approximately normally distributed,

then we use the t distribution (explained in Section 8.3.1) to make the confidence interval for μ.

Case II. If the following two conditions are fulfilled:

1. The population standard deviation σ is not known
2. The sample size is large (i.e., $n \geq 30$),

then again we use the t distribution to make the confidence interval for μ.

Case III. If the following three conditions are fulfilled:

1. The population standard deviation σ is not known
2. The sample size is small (i.e., $n < 30$)
3. The population from which the sample is selected is not normally distributed (or its distribution is unknown),

then we use a nonparametric method to make the confidence interval for μ. Such procedures are covered in Chapter 15, which is on the Web site for this text.

The following chart summarizes the above three cases.

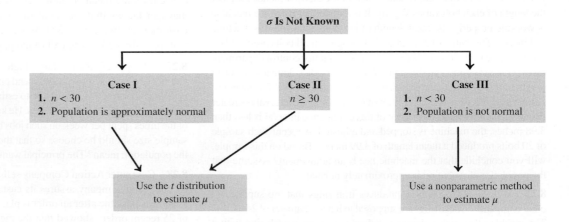

In the next subsection, we discuss the t distribution, and then in Section 8.3.2 we show how to use the t distribution to make a confidence interval for μ when σ is not known and conditions of Cases I or II are satisfied.

8.3.1   The *t* Distribution

The ***t* distribution** was developed by W. S. Gosset in 1908 and published under the pseudonym *Student*. As a result, the *t* distribution is also called *Student's t distribution*. The *t* distribution is similar to the normal distribution in some respects. Like the normal distribution curve, the *t* distribution curve is symmetric (bell shaped) about the mean and never meets the horizontal axis. The total area under a *t* distribution curve is 1.0, or 100%. However, the *t* distribution curve is flatter and wider than the standard normal distribution curve. In other words, the *t* distribution curve has a lower height and a greater spread (or, we can say, a larger standard deviation) than the standard normal distribution. However, as the sample size increases, the *t* distribution approaches the standard normal distribution. The units of a *t* distribution are denoted by *t*.

The shape of a particular *t* distribution curve depends on the number of **degrees of freedom (*df*)**. For the purpose of this chapter and Chapter 9, the number of degrees of freedom for a *t* distribution is equal to the sample size minus one, that is,

$$df = n - 1$$

The number of degrees of freedom is the only parameter of the *t* distribution. There is a different *t* distribution for each number of degrees of freedom. Like the standard normal distribution, the mean of the *t* distribution is 0. But unlike the standard normal distribution, whose standard deviation is 1, the standard deviation of a *t* distribution is $\sqrt{df/(df - 2)}$ for $df > 2$. Thus, the standard deviation of a *t* distribution is always greater than 1 and, hence, is larger than the standard deviation of the standard normal distribution.

Figure 8.5 shows the standard normal distribution and the *t* distribution for 9 degrees of freedom. The standard deviation of the standard normal distribution is 1.0, and the standard deviation of the *t* distribution is $\sqrt{df/(df - 2)} = \sqrt{9/(9 - 2)} = 1.134$.

> The ***t* distribution** is a specific type of bell-shaped distribution with a lower height and a greater spread than the standard normal distribution. As the sample size becomes larger, the *t* distribution approaches the standard normal distribution. The *t* distribution has only one parameter, called the degrees of freedom (*df*). The mean of the *t* distribution is equal to 0, and its standard deviation is $\sqrt{df/(df - 2)}$.

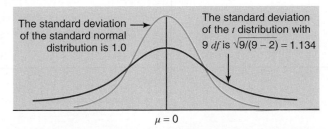

The standard deviation of the standard normal distribution is 1.0

The standard deviation of the *t* distribution with 9 *df* is $\sqrt{9/(9 - 2)} = 1.134$

$\mu = 0$

Figure 8.5   The *t* distribution for $df = 9$ and the standard normal distribution.

As stated earlier, the number of degrees of freedom for a *t* distribution for the purpose of this chapter is $n - 1$. *The number of degrees of freedom is defined as the number of observations that can be chosen freely.* As an example, suppose we know that the mean of four values is 20. Consequently, the sum of these four values is $20(4) = 80$. Now, how many values out of four can we choose freely so that the sum of these four values is 80? The answer is that we can freely choose $4 - 1 = 3$ values. Suppose we choose 27, 8, and 19 as the three values. Given these three values and the information that the sum of the four values is $20 \times 4 = 80$, the fourth value is $80 - 27 - 8 - 19 = 26$. Thus, once we have chosen three values, the fourth value is automatically determined. Consequently, the number of degrees of freedom for this example is

$$df = n - 1 = 4 - 1 = 3$$

We subtract 1 from *n* because we lose 1 degree of freedom to calculate the mean.

Table V of Appendix B lists the values of *t* for the given number of degrees of freedom and areas in the right tail of a *t* distribution. Because the *t* distribution is symmetric, these are also the values of $-t$ for the same number of degrees of freedom and the same areas in the left tail of the *t* distribution. Example 8–4 describes how to read Table V of Appendix B.

EXAMPLE 8–4

Reading the t distribution table.

Find the value of t for 16 degrees of freedom and .05 area in the right tail of a t distribution curve.

Solution   In Table V of Appendix B, we locate 16 in the column of degrees of freedom (labeled *df*) and .05 in the row of *Area in the right tail under the t distribution curve* at the top of the table. The entry at the intersection of the row of 16 and the column of .05, which is 1.746, gives the required value of *t*. The relevant portion of Table V of Appendix B is shown here as Table 8.2. The value of *t* read from the *t* distribution table is shown in Figure 8.6.

Table 8.2   **Determining *t* for 16 *df* and .05 Area in the Right Tail**

| | | | Area in the right tail | | |
| | | **Area in the Right Tail Under the *t* Distribution Curve** | | | |
df	**.10**	**.05**	**.025**	**...**	**.001**
1	3.078	6.314	12.706	...	318.309
2	1.886	2.920	4.303	...	22.327
3	1.638	2.353	3.182	...	10.215
.	...	...	...	...	...
.	...	...	...	...	...
.	...	...	...	...	...
16	1.337	1.746	2.120	...	3.686
.	...	...	...	...	...
.	...	...	...	...	...
.	...	...	...	...	...
75	1.293	1.665	1.992	...	3.202
∞	1.282	1.645	1.960	...	3.090

The required value of *t* for 16 *df* and .05 area in the right tail

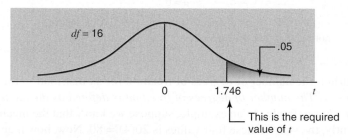

This is the required value of *t*

Figure 8.6   The value of *t* for 16 *df* and .05 area in the right tail.

Because of the symmetric shape of the *t* distribution curve, the value of *t* for 16 degrees of freedom and .05 area in the left tail is −1.746. Figure 8.7 illustrates this case.

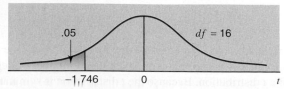

Figure 8.7   The value of *t* for 16 *df* and .05 area in the left tail.

8.3.2   Confidence Interval for μ Using the t Distribution

To reiterate, when the conditions mentioned under Cases I or II in the beginning of this section hold true, we use the t distribution to construct a confidence interval for the population mean μ.

When the population standard deviation σ is not known, then we replace it by the sample standard deviation s, which is its estimator. Consequently, for the standard deviation of $\bar{x}$, we use

$$s_{\bar{x}} = \frac{s}{\sqrt{n}}$$

for $\sigma_{\bar{x}} = \sigma/\sqrt{n}$. Note that the value of $s_{\bar{x}}$ is a point estimate of $\sigma_{\bar{x}}$.

Confidence Interval for μ Using the t Distribution   The $(1 - \alpha)100\%$ **confidence interval for μ** is

$$\bar{x} \pm t s_{\bar{x}}$$

where

$$s_{\bar{x}} = \frac{s}{\sqrt{n}}$$

The value of t is obtained from the t distribution table for $n - 1$ degrees of freedom and the given confidence level. Here $t s_{\bar{x}}$ is the margin of error or the maximum error of the estimate; that is,

$$E = t s_{\bar{x}}$$

Examples 8–5 and 8–6 describe the procedure of constructing a confidence interval for μ using the t distribution.

EXAMPLE 8–5   **Premium for Health Insurance Coverage**

Constructing a 95% confidence interval for μ using the t distribution.

According to a 2014 Kaiser Family Foundation Health Benefits Survey released in 2015, the total mean cost of employer-sponsored family health coverage was $16,834 per family per year, of which workers were paying an average of $4823. A recent random sample of 25 workers from New York City who have employer-provided health insurance coverage paid an average premium of $6600 for family health insurance coverage with a standard deviation of $800. Make a 95% confidence interval for the current average premium paid for family health insurance coverage by all workers in New York City who have employer-provided health insurance coverage. Assume that the distribution of premiums paid for family health insurance coverage by all workers in New York City who have employer-provided health insurance coverage is approximately normally distributed.

Solution   Here, σ is not known, $n < 30$, and the population is normally distributed. All conditions mentioned in Case I of the chart given in the beginning of this section are satisfied. Therefore, we will use the t distribution to make a confidence interval for μ. From the given information,

$$n = 25, \quad \bar{x} = \$6600, \quad s = \$800, \quad \text{Confidence level} = 95\% \text{ or } .95$$

The value of $s_{\bar{x}}$ is

$$s_{\bar{x}} = \frac{s}{\sqrt{n}} = \frac{800}{\sqrt{25}} = \$160$$

To find the value of t from Table V of Appendix B, we need to know the degrees of freedom and the area under the t distribution curve in each tail.

$$\text{Degrees of freedom} = n - 1 = 25 - 1 = 24$$

To find the area in each tail, we divide the confidence level by 2 and subtract the number obtained from .5. Thus,

$$\text{Area in each tail} = .5 - (.95/2) = .5 - .4750 = .025$$

From the t distribution table, Table V of Appendix B, the value of t for $df = 24$ and .025 area in the right tail is 2.064. The value of t is shown in Figure 8.8.

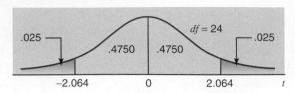

Figure 8.8   The value of t.

By substituting all values in the formula for the confidence interval for μ, we obtain the 95% confidence interval as

$$\bar{x} \pm t s_{\bar{x}} = 6600 \pm 2.064\,(160) = 6600 \pm 330.24 = \textbf{\$6269.76 to \$6930.24}$$

Thus, we can state with 95% confidence that the current average premium paid for family health insurance coverage by all workers in New York City who have employer-provided health insurance coverage is between \$6269.76 and \$6930.24.

Note that $\bar{x} = \$6600$ is a point estimate of μ in this example, and \$330.24 is the margin of error.

EXAMPLE 8–6   Annual Expenses on Books

Constructing a 99% confidence interval for μ using the t distribution.

Sixty-four randomly selected adults who buy books for general reading were asked how much they usually spend on books per year. This sample produced a mean of \$1450 and a standard deviation of \$300 for such annual expenses. Determine a 99% confidence interval for the corresponding population mean.

Solution   From the given information,

$$n = 64, \quad \bar{x} = \$1450, \quad s = \$300,$$

and   Confidence level = 99% or .99

Here σ is not known, but the sample size is large ($n \geq 30$). Hence, we will use the t distribution to make a confidence interval for μ. First we calculate the standard deviation of $\bar{x}$, the number of degrees of freedom, and the area in each tail of the t distribution.

$$s_{\bar{x}} = \frac{s}{\sqrt{n}} = \frac{300}{\sqrt{64}} = \$37.50$$

$$df = n - 1 = 64 - 1 = 63$$

$$\text{Area in each tail} = .5 - (.99/2) = .5 - .4950 = .005$$

From the t distribution table, $t = 2.656$ for 63 degrees of freedom and .005 area in the right tail. The 99% confidence interval for μ is

$$\bar{x} \pm t s_{\bar{x}} = \$1450 \pm 2.656(37.50)$$

$$= \$1450 \pm \$99.60 = \textbf{\$1350.40 to \$1549.60}$$

Thus, we can state with 99% confidence that based on this sample the mean annual expenditure on books by all adults who buy books for general reading is between \$1350.40 and \$1549.60.

Again, we can decrease the width of a confidence interval for μ either by lowering the confidence level or by increasing the sample size, as was done in Section 8.2. However, increasing the sample size is the better alternative.

Note: What If the Sample Size Is Large and the Number of *df* Is Not in the *t* Distribution Table?

In the above section, when σ is not known, we used the *t* distribution to make a confidence interval for μ in Cases I and II. Note that in Case II, the sample size must be large. If we have access to technology, it does not matter how large (greater than 30) the sample size is; we can use the *t* distribution. However, if we are using the *t* distribution table (Table V of Appendix B), this may pose a problem. Usually such a table goes only up to a certain number of degrees of freedom. For example, Table V in Appendix B goes only up to 75 degrees of freedom. Thus, if the sample size is larger than 76, we cannot use Table V to find the *t* value for the given confidence level to use in the confidence interval. In such a situation when *n* is large (for example, 500) and the number of *df* is not included in the *t* distribution table, there are two options:

1. Use the *t* value from the last row (the row of ∞) in Table V.
2. Use the normal distribution as an approximation to the *t* distribution.

Note that the *t* values you will obtain from the last row of the *t* distribution table are the same as obtained from the normal distribution table for the same confidence levels, the only difference being the decimal places. To use the normal distribution as an approximation to the *t* distribution to make a confidence interval for μ, the procedure is exactly like the one in Section 8.2, except that now we replace σ by s, and $\sigma_{\bar{x}}$ by $s_{\bar{x}}$.

Again, note that here we can use the normal distribution as a convenience and as an approximation, but if we can, we should use the *t* distribution by using technology.

EXERCISES

CONCEPTS AND PROCEDURES

8.31 Briefly explain the similarities and the differences between the standard normal distribution and the *t* distribution.

8.32 What are the parameters of a normal distribution and a *t* distribution? Explain.

8.33 Briefly explain the meaning of the degrees of freedom for a *t* distribution. Give one example.

8.34 What assumptions must hold true to use the *t* distribution to make a confidence interval for μ?

8.35 Find the value of *t* for the *t* distribution for each of the following.
  a. Area in the right tail = .05 and $df = 9$
  b. Area in the left tail = .025 and $n = 58$
  c. Area in the left tail = .001 and $df = 34$
  d. Area in the right tail = .005 and $n = 27$

8.36 a. Find the value of *t* for the *t* distribution with a sample size of 18 and area in the left tail equal to .10.
  b. Find the value of *t* for the *t* distribution with a sample size of 22 and area in the right tail equal to .025.
  c. Find the value of *t* for the *t* distribution with 45 degrees of freedom and .001 area in the right tail.
  d. Find the value of *t* for the *t* distribution with 37 degrees of freedom and .005 area in the left tail.

8.37 For each of the following, find the area in the appropriate tail of the *t* distribution.
  a. $t = 2.467$ and $df = 32$ **b.** $t = -1.762$ and $df = 62$
  c. $t = -2.670$ and $n = 55$ **d.** $t = 2.819$ and $n = 18$

8.38 Find the value of *t* from the *t* distribution table for each of the following.
  a. Confidence level = 99% and $df = 21$
  b. Confidence level = 95% and $n = 42$
  c. Confidence level = 90% and $df = 12$

8.39 A random sample of 18 observations taken from a normally distributed population produced the following data.

| 28.4 | 25.4 | 25.5 | 25.5 | 31.1 | 23.0 | 25.4 | 24.6 | 28.4 |
| 37.2 | 23.9 | 30.4 | 27.9 | 25.1 | 27.2 | 25.3 | 23.1 | 28.9 |

  a. What is the point estimate of μ?
  b. Make a 99% confidence interval for μ.
  c. What is the margin of error of estimate for μ in part b?

8.40 A random sample of 10 observations taken from a normally distributed population produced the following data.

| 44 | 52 | 31 | 48 | 46 | 39 | 47 | 36 | 41 | 57 |

  a. What is the point estimate of μ?
  b. Make a 95% confidence interval for μ.
  c. What is the margin of error of estimate for μ in part b?

8.41 Suppose, for a random sample selected from a population, $\bar{x} = 25.5$ and $s = 4.9$.
  a. Construct a 95% confidence interval for μ assuming $n = 47$.
  b. Construct a 99% confidence interval for μ assuming $n = 47$. Is the width of the 99% confidence interval larger than the width of the 95% confidence interval calculated in part a? If yes, explain why.

c. Find a 95% confidence interval for μ assuming $n = 32$. Is the width of the 95% confidence interval for μ with $n = 32$ larger than the width of the 95% confidence interval for μ with $n = 47$ calculated in part a? If so, why? Explain.

8.42 a. A random sample of 100 observations taken from a population produced a sample mean equal to 55.32 and a standard deviation equal to 8.4. Make a 90% confidence interval for μ.

b. Another sample of 100 observations taken from the same population produced a sample mean equal to 57.40 and a standard deviation equal to 7.5. Make a 90% confidence interval for μ.

c. A third sample of 100 observations taken from the same population produced a sample mean equal to 56.25 and a standard deviation equal to 7.9. Make a 90% confidence interval for μ.

d. The true population mean for this population is 55.80. Which of the confidence intervals constructed in parts a through c cover this population mean and which do not?

APPLICATIONS

8.43 A random sample of 20 acres gave a mean yield of wheat equal to 41.2 bushels per acre with a standard deviation of 3 bushels. Assuming that the yield of wheat per acre is normally distributed, construct a 90% confidence interval for the population mean μ.

8.44 Almost all employees working for financial companies in New York City receive large bonuses at the end of the year. A random sample of 65 employees selected from financial companies in New York City showed that they received an average bonus of $55,000 last year with a standard deviation of $12,000. Construct a 95% confidence interval for the average bonus that all employees working for financial companies in New York City received last year.

8.45 A random sample of 16 airline passengers at the Bay City airport showed that the mean time spent waiting in line to check in at the ticket counters was 28 minutes with a standard deviation of 4 minutes. Construct a 99% confidence interval for the mean time spent waiting in line by all passengers at this airport. Assume that such waiting times for all passengers are normally distributed.

8.46 A random sample of 36 mid-sized cars tested for fuel consumption gave a mean of 24.3 miles per gallon with a standard deviation of 3.1 miles per gallon.

a. Find a 99% confidence interval for the population mean, μ.

b. Suppose the confidence interval obtained in part a is too wide. How can the width of this interval be reduced? Describe all possible alternatives. Which alternative is the best and why?

8.47 Jack's Auto Insurance Company customers sometimes have to wait a long time to speak to a customer service representative when they call regarding disputed claims. A random sample of 25 such calls yielded a mean waiting time of 22 minutes with a standard deviation of 6 minutes. Construct a 99% confidence interval for the population mean of such waiting times. Assume that such waiting times for the population follow a normal distribution.

8.48 The following data give the speeds (in miles per hour), as measured by radar, of 10 cars traveling on Interstate I-15.

76   74   80   68   79   74   58   78   82   68

Assuming that the speeds of all cars traveling on this highway have a normal distribution, construct a 90% confidence interval for the mean speed of all cars traveling on this highway.

8.49 A dentist wants to find the average time taken by one of her hygienists to take X-rays and clean teeth for patients. She recorded the time to serve 24 randomly selected patients by this hygienist. The data (in minutes) are as follows:

36.80   39.80   38.60   38.30   35.80   32.60   38.70   34.50   37.00
32.70   40.90   33.80   37.10   31.00   35.10   38.20   36.60   38.80
39.60   39.70   35.10   38.20   32.70   40.50

Assume that such times for this hygienist for all patients are approximately normal.

a. What is the point estimate of the corresponding population mean.

b. Construct a 99% confidence interval for the average time taken by this hygienist to take X-rays and to clean teeth for all patients.

8.50 The following data are the times (in seconds) of eight finalists in the Girls' 100-meter dash at the North Carolina 1A High School Track and Field Championships (*Source:* prepinsiders.blogspot.com).

12.25   12.37   12.68   12.84   12.90   12.97   13.02   13.35

Assume that these times represent a random sample of times for girls who would qualify for the finals of this event, and that the population distribution of such times is normal. Determine the 98% confidence interval for the average time in the Girls' 100-meter dash finals using these data.

8.51 A random sample of 36 participants in a Zumba dance class had their heart rates measured before and after a moderate 10-minute workout. The following data correspond to the increase in each individual's heart rate (in beats per minute):

59   70   57   42   57   59   41   54   44   36   59   61
52   42   41   32   60   54   52   53   51   47   62   48
62   44   69   50   37   50   54   48   52   61   45   63

a. What is the point estimate of the corresponding population mean?

b. Make a 98% confidence interval for the average increase in a person's heart rate after a moderate 10-minute Zumba workout.

8.52 An article in the *Los Angeles Times* (latimesblogs.latimes.com/pardonourdust/) quoted from the National Association of Realtors that "we now sell our homes and move an average of every six years." Suppose that the average time spent living in a house prior to selling it for a random sample of 400 recent home sellers was 6.18 years and the sample standard deviation was 2.87 years.

a. What is the point estimate of the corresponding population mean?

b. Construct a 98% confidence interval for the average time spent living in a house prior to selling it for all home owners. What is the margin of error for this estimate?

*8.53 You are working for a supermarket. The manager has asked you to estimate the mean time taken by a cashier to serve customers at this supermarket. Briefly explain how you will conduct this study. Collect data on the time taken by any supermarket cashier to serve 40 customers. Then estimate the population mean. Choose your own confidence level.

*8.54 You are working for a bank. The bank manager wants to know the mean waiting time for all customers who visit this bank. She has asked you to estimate this mean by taking a sample. Briefly explain how you will conduct this study. Collect data on the waiting times for 45 customers who visit a bank. Then estimate the population mean. Choose your own confidence level.

8.4 | Estimation of a Population Proportion: Large Samples

Often we want to estimate the population proportion or percentage. (Recall that a percentage is obtained by multiplying the proportion by 100.) For example, the production manager of a company may want to estimate the proportion of defective items produced on a machine. A bank manager may want to find the percentage of customers who are satisfied with the service provided by the bank.

Again, if we can conduct a census each time we want to find the value of a population proportion, there is no need to learn the procedures discussed in this section. However, we usually derive our results from sample surveys. Hence, to take into account the variability in the results obtained from different sample surveys, we need to know the procedures for estimating a population proportion.

Recall from Chapter 7 that the population proportion is denoted by p, and the sample proportion is denoted by $\hat{p}$. This section explains how to estimate the population proportion, p, using the sample proportion, $\hat{p}$. The sample proportion, $\hat{p}$, is a sample statistic, and it possesses a sampling distribution. From Chapter 7, we know that:

1. The sampling distribution of the sample proportion $\hat{p}$ is approximately normal for a large sample.
2. The mean of the sampling distribution of $\hat{p}$, $\mu_{\hat{p}}$, is equal to the population proportion, p.
3. The standard deviation of the sampling distribution of $\hat{p}$ is $\sigma_{\hat{p}} = \sqrt{pq/n}$, where $q = 1 - p$, given that $\dfrac{n}{N} \leq .05$.

In the case of a proportion, a sample is considered to be large if np and nq are both greater than 5. If p and q are not known, then $n\hat{p}$ and $n\hat{q}$ should each be greater than 5 for the sample to be large.

◀ *Remember*

When estimating the value of a population proportion, we do not know the values of p and q. Consequently, we cannot compute $\sigma_{\hat{p}}$. Therefore, in the estimation of a population proportion, we use the value of $s_{\hat{p}}$ as an estimate of $\sigma_{\hat{p}}$. The value of $s_{\hat{p}}$ is calculated using the following formula.

Estimator of the Standard Deviation of $\hat{p}$ The value of $s_{\hat{p}}$, which gives a point estimate of $\sigma_{\hat{p}}$, is calculated as follows. Here, $s_{\hat{p}}$ is an estimator of $\sigma_{\hat{p}}$.

$$s_{\hat{p}} = \sqrt{\frac{\hat{p}\hat{q}}{n}}$$

Note that the condition $\dfrac{n}{N} \leq .05$ must hold true to use this formula.

The sample proportion, $\hat{p}$, is the point estimator of the corresponding population proportion p. Then to find the confidence interval for p, we add to and subtract from $\hat{p}$ a number that is called the **margin of error**, E.

Confidence Interval for the Population Proportion, p For a large sample, the $(1 - \alpha)100\%$ **confidence interval for the population proportion**, p, is

$$\hat{p} \pm z\, s_{\hat{p}}$$

The value of z is obtained from the standard normal distribution table for the given confidence level, and $s_{\hat{p}} = \sqrt{\hat{p}\,\hat{q}/n}$. The term $z\, s_{\hat{p}}$ is called the **margin of error**, or the maximum error of the estimate, and is denoted by E.

Examples 8–7 and 8–8 illustrate the procedure for constructing a confidence interval for p.

EXAMPLE 8–7   **Having Basic Needs Met as the American Dream**

Finding the point estimate and 99% confidence interval for p: large sample.

PolicyInteractive of Eugene, Oregon, conducted a study in April 2014 for the Center for a New American Dream that included a sample of 1821 American adults. Seventy five percent of the people included in this study said that having basic needs met is very or extremely important in their vision of the American dream (www.newdream.org).

(a) What is the point estimate of the corresponding population proportion?

(b) Find, with a 99% confidence level, the percentage of all American adults who will say that having basic needs met is very or extremely important in their vision of the American dream. What is the margin of error of this estimate?

Solution   Let p be the proportion of all American adults who will say that having basic needs met is very or extremely important in their vision of the American dream, and let $\hat{p}$ be the corresponding sample proportion. From the given information,

$$n = 1821, \qquad \hat{p} = .75, \qquad \text{and} \qquad \hat{q} = 1 - \hat{p} = 1 - .75 = .25$$

First, we calculate the value of the standard deviation of the sample proportion as follows:

$$s_{\hat{p}} = \sqrt{\frac{\hat{p}\hat{q}}{n}} = \sqrt{\frac{(.75)(.25)}{1821}} = .010147187$$

Note that $n\hat{p}$ and $n\hat{q}$ are both greater than 5. (The reader should check this condition.) Consequently, the sampling distribution of $\hat{p}$ is approximately normal and we will use the normal distribution to make a confidence interval about p.

(a) The point estimate of the proportion of all American adults who will say that having basic needs met is very or extremely important in their vision of the American dream is equal to .75; that is,

$$\text{Point estimate of } p = \hat{p} = \textbf{.75}$$

(b) The confidence level is 99%, or .99. To find z for a 99% confidence level, first we find the area in each of the two tails of the normal distribution curve, which is $(1 - .99)/2 = .0050$. Then, we look for .0050 and $.0050 + .99 = .9950$ areas in the normal distribution table to find the two values of z. These two z values are (approximately) -2.58 and 2.58. Thus, we will use $z = 2.58$ in the confidence interval formula. Substituting all the values in the confidence interval formula for p, we obtain

$$\hat{p} \pm z s_{\hat{p}} = .75 \pm 2.58 \, (.010147187) = .75 \pm .026$$

$$= \textbf{.724 to .776} \text{ or } \textbf{72.4\% to 77.6\%}$$

Thus, we can state with 99% confidence that .724 to .776 or 72.4% to 77.6% of all American adults will say that having basic needs met is very or extremely important in their vision of the American dream.

The margin of error associated with this estimate of p is .026 or 2.6%, that is,

$$\text{Margin of error} = z s_{\hat{p}} = .026 \text{ or } 2.6\%$$

EXAMPLE 8–8   **Excited About Learning in College**

Constructing a 97% confidence interval for p: large sample.

According to a Gallup-Purdue University study of college graduates conducted during February 4 to March 7, 2014, 63% of college graduates polled said that they had at least one college professor who made them feel excited about learning (www.gallup.com). Suppose that this study was based on a random sample of 2000 college graduates. Construct a 97% confidence interval for the corresponding population proportion.

Solution Let p be the proportion of all college graduates who would say that they had at least one college professor who made them feel excited about learning, and let $\hat{p}$ be the corresponding sample proportion. From the given information,

$$n = 2000, \qquad \hat{p} = .63, \qquad \hat{q} = 1 - \hat{p} = 1 - .63 = .37$$

and

$$\text{Confidence level} = 97\% \text{ or } .97$$

The standard deviation of the sample proportion is

$$s_{\hat{p}} = \sqrt{\frac{\hat{p}\hat{q}}{n}} = \sqrt{\frac{(.63)(.37)}{2000}} = .01079583$$

Note that if we check $n\hat{p}$ and $n\hat{q}$, both will be greater than 5. Consequently, we can use the normal distribution to make a confidence interval for p.

From the normal distribution table, the value of z for the 97% confidence level is 2.17. Note that to find this z value, you will look for the areas .0150 and .9850 in Table IV of Appendix B. Substituting all the values in the formula, the 97% confidence interval for p is

$$\hat{p} \pm z s_{\hat{p}} = .63 \pm 2.17(.01079583) = .63 \pm .023$$

$$= \textbf{.607 to .653} \text{ or } \textbf{60.7\% to 65.3\%}$$

Thus, we can state with 97% confidence that the proportion of all college graduates who would say that they had at least one college professor who made them feel excited about learning is between .607 and .653 or between 60.7% and 65.3%.

Again, we can decrease the width of a confidence interval for p either by lowering the confidence level or by increasing the sample size. However, lowering the confidence level is not a good choice because it simply decreases the likelihood that the confidence interval contains p. Hence, to decrease the width of a confidence interval for p, we should always choose to increase the sample size.

8.4.1   Determining the Sample Size for the Estimation of Proportion

Just as we did with the mean, we can also determine the minimum sample size for estimating the population proportion to within a predetermined margin of error, E. By knowing the sample size that can give us the required results, we can save our scarce resources by not taking an unnecessarily large sample. From earlier discussion in this section, the margin of error, E, of the interval estimate of the population proportion is

$$E = z\, s_{\hat{p}} = z \times \sqrt{\frac{\hat{p}\hat{q}}{n}}$$

By manipulating this expression algebraically, we obtain the following formula to find the required sample size given E, $\hat{p}$, $\hat{q}$, and z.

Determining the Sample Size for the Estimation of p Given the confidence level and the values of $\hat{p}$ and $\hat{q}$, the sample size that will produce a predetermined margin of error E of the confidence interval **estimate of p** is

$$n = \frac{z^2 \hat{p}\hat{q}}{E^2}$$

We can observe from this formula that to find n, we need to know the values of $\hat{p}$ and $\hat{q}$. However, the values of $\hat{p}$ and $\hat{q}$ are not known to us. In such a situation, we can choose one of the following two alternatives.

Americans' Efforts to Lose Weight Still Trail Desires

AMERICANS' EFFORTS TO LOSE WEIGHT STILL TRAIL DESIRES

Want to lose weight   51%

Seriously trying to lose weight   26%

Data source: www.gallup.com

The Gallup polling agency conducted a poll of 828 American adults aged 18 and over on the issue of weight. These adults were asked if they want to lose weight and if they were seriously trying to lose weight. As shown in the graph, 51% of these adults said that they want to lose weight but only 26% said that they were seriously trying to lose weight. The poll was conducted November 6 to 9, 2014. As these percentages are based on a sample, using the procedure learned in this section, we can make a confidence interval for each of the two population proportions as shown in the table below.

Category	Sample Proportion	Confidence Interval
Adults who want to lose weight	.51	$.51 \pm z\, s_{\hat{p}}$
Adults who are seriously trying to lose weight	.26	$.26 \pm z\, s_{\hat{p}}$

For each of the two confidence intervals listed in the table, we can substitute the value of z and the value of $s_{\hat{p}}$, which is calculated as $\sqrt{\dfrac{\hat{p}\hat{q}}{n}}$. For example, suppose we want to find a 96% confidence interval for the proportion of all adults who will say that they want to lose weight. Then this confidence interval is determined as follows.

$$s_{\hat{p}} = \sqrt{\frac{\hat{p}\hat{q}}{n}} = \sqrt{\frac{(.51)(.49)}{828}} = .01737273$$

$$\hat{p} \pm z s_{\hat{p}} = .51 \pm 2.05\,(.01737273) = .51 \pm .036 = .474 \text{ to } .546$$

Thus, we can state with 96% confidence that 47.4% to 54.6% of all adults will say that they want to lose weight. Here the margin of error is .036 or 3.6%.

In the same way, we can find the confidence interval for the population proportion of all adults who will say that they are seriously trying to lose weight.

Source: http://www.gallup.com/poll/179771/americans-effort-lose-weight-trails-desire.aspx.

1. We make the **most conservative estimate** of the sample size n by using $\hat{p} = .50$ and $\hat{q} = .50$. For a given E, these values of $\hat{p}$ and $\hat{q}$ will give us the largest sample size in comparison to any other pair of values of $\hat{p}$ and $\hat{q}$ because the product of $\hat{p} = .50$ and $\hat{q} = .50$ is greater than the product of any other pair of values for $\hat{p}$ and $\hat{q}$.

2. We take a **preliminary sample** (of arbitrarily determined size) and calculate $\hat{p}$ and $\hat{q}$ for this sample. Then, we use these values of $\hat{p}$ and $\hat{q}$ to find n.

Examples 8–9 and 8–10 illustrate how to determine the sample size that will produce the error of estimation for the population proportion within a predetermined margin of error value. Example 8–9 gives the most conservative estimate of n, and Example 8–10 uses the results from a preliminary sample to determine the required sample size.

EXAMPLE 8–9   Proportion of Parts That Are Defective

Determining the most conservative estimate of n for the estimation of p.

Lombard Electronics Company has just installed a new machine that makes a part that is used in clocks. The company wants to estimate the proportion of these parts produced by this machine that are defective. The company manager wants this estimate to be within .02 of the population proportion for a 95% confidence level. What is the most conservative estimate of the sample size that will limit the margin of error to within .02 of the population proportion?

Solution   The company manager wants the 95% confidence interval to be

$$\hat{p} \pm .02$$

Therefore,

$$E = .02$$

The value of z for a 95% confidence level is 1.96. For the most conservative estimate of the sample size, we will use $\hat{p} = .50$ and $\hat{q} = .50$. Hence, the required sample size is

$$n = \frac{z^2 \hat{p}\hat{q}}{E^2} = \frac{(1.96)^2(.50)(.50)}{(.02)^2} = \textbf{2401}$$

Thus, if the company takes a sample of 2401 parts, there is a 95% chance that the estimate of p will be within .02 of the population proportion.

EXAMPLE 8–10   Proportion of Parts That Are Defective

Determining n for the estimation of p using preliminary sample results.

Consider Example 8–9 again. Suppose a preliminary sample of 200 parts produced by this machine showed that 7% of them are defective. How large a sample should the company select so that the 95% confidence interval for p is within .02 of the population proportion?

Solution   Again, the company wants the 95% confidence interval for p to be

$$\hat{p} \pm .02$$

Hence,

$$E = .02$$

The value of z for a 95% confidence level is 1.96. From the preliminary sample,

$$\hat{p} = .07 \qquad \text{and} \qquad \hat{q} = 1 - .07 = .93$$

Using these values of $\hat{p}$ and $\hat{q}$, we obtain

$$n = \frac{z^2 \hat{p}\hat{q}}{E^2} = \frac{(1.96)^2(.07)(.93)}{(.02)^2} = \frac{(3.8416)(.07)(.93)}{.0004} = 625.22 \approx \textbf{626}$$

Note that if the value of n is not an integer, we always round it up.

Thus, if the company takes a sample of 626 items, there is a 95% chance that the estimate of p will be within .02 of the population proportion. However, we should note that this sample size will produce the margin of error within .02 only if $\hat{p}$ is .07 or less for the new sample. If $\hat{p}$ for the new sample happens to be much higher than .07, the margin of error will not be within .02. Therefore, to avoid such a situation, we may be more conservative and take a sample much larger than 626 items.

EXERCISES

CONCEPTS AND PROCEDURES

8.55 What assumption(s) must hold true to use the normal distribution to make a confidence interval for the population proportion, p?

8.56 What is the point estimator of the population proportion, p?

8.57 Check if the sample size is large enough to use the normal distribution to make a confidence interval for p for each of the following cases.

a. $n = 50$ and $\hat{p} = .25$ b. $n = 160$ and $\hat{p} = .04$
c. $n = 500$ and $\hat{p} = .65$ d. $n = 75$ and $\hat{p} = .05$

8.58 Check if the sample size is large enough to use the normal distribution to make a confidence interval for p for each of the following cases.

a. $n = 120$ and $\hat{p} = .04$ b. $n = 60$ and $\hat{p} = .08$
c. $n = 40$ and $\hat{p} = .50$ d. $n = 900$ and $\hat{p} = .15$

8.59 a. A random sample of 400 observations taken from a population produced a sample proportion of .63. Make a 95% confidence interval for p.

b. Another sample of 400 observations taken from the same population produced a sample proportion of .59. Make a 95% confidence interval for p.

c. A third sample of 400 observations taken from the same population produced a sample proportion of .67. Make a 95% confidence interval for p.

d. The true population proportion for this population is .65. Which of the confidence intervals constructed in parts a through c cover this population proportion and which do not?

8.60 A random sample of 500 observations selected from a population produced a sample proportion equal to .68.

a. Make a 90% confidence interval for p.
b. Construct a 95% confidence interval for p.
c. Make a 99% confidence interval for p.
d. Does the width of the confidence intervals constructed in parts a through c increase as the confidence level increases? If yes, explain why.

8.61 A random sample of 200 observations selected from a population gave a sample proportion equal to .27.

a. Make a 99% confidence interval for p.
b. Construct a 97% confidence interval for p.
c. Make a 90% confidence interval for p.
d. Does the width of the confidence intervals constructed in parts a through c decrease as the confidence level decreases? If yes, explain why.

8.62 a. How large a sample should be selected so that the margin of error of estimate for a 99% confidence interval for p is .035 when the value of the sample proportion obtained from a preliminary sample is .29?

b. Find the most conservative sample size that will produce the margin of error for a 99% confidence interval for p equal to .035.

8.63 a. How large a sample should be selected so that the margin of error of estimate for a 98% confidence interval for p is .045 when the value of the sample proportion obtained from a preliminary sample is .53?

b. Find the most conservative sample size that will produce the margin of error for a 98% confidence interval for p equal to .045.

8.64 Determine the most conservative sample size for the estimation of the population proportion for the following.

a. $E = .03$,   confidence level = 99%
b. $E = .04$,   confidence level = 95%
c. $E = .01$,   confidence level = 90%

8.65 Determine the sample size for the estimation of the population proportion for the following, where $\hat{p}$ is the sample proportion based on a preliminary sample.

a. $E = .03$, $\hat{p} = .32$,   confidence level = 99%
b. $E = .04$, $\hat{p} = .78$,   confidence level = 95%
c. $E = .02$, $\hat{p} = .64$,   confidence level = 90%

APPLICATIONS

8.66 According to a Pew Research Center nationwide telephone survey of adults conducted March 15 to April 24, 2011, 55% of college graduates said that their college education prepared them for a job (*Time*, May 30, 2011). Suppose that this survey included 1450 college graduates.

a. What is the point estimate of the corresponding population proportion?

b. Construct a 98% confidence interval for the proportion of all college graduates who will say that their college education prepared them for a job. What is the margin of error for this estimate?

8.67 According to an estimate, 11% of junk mail is thrown out without being opened (*Source:* www.uoregon.edu/~recycle/events_topics_junkmail_text.htm). Suppose that this percentage is based on a random sample of 1400 pieces of junk mail.

a. What is the point estimate of the corresponding population proportion?

b. Construct a 95% confidence interval for the proportion of all pieces of junk mail that is thrown out without being opened. What is the margin of error for this estimate?

8.68 According to data from Ipsos Global Express, 64% of Americans say there is never enough time in the day to get things done (*Business Week*, November 22, 2004). Suppose this percentage is based on a random sample of 900 Americans.

a. What is the point estimate of the corresponding population proportion?

b. Find a 90% confidence interval for the corresponding population proportion. What is the margin of error for this estimate?

8.69 In a random sample of 50 homeowners selected from a large suburban area, 19 said that they had serious problems with excessive noise from their neighbors.

    a. Make a 99% confidence interval for the percentage of all homeowners in this suburban area who have such problems.

    b. Suppose the confidence interval obtained in part a is too wide. How can the width of this interval be reduced? Discuss all possible alternatives. Which option is best?

8.70 An Accountemps survey asked workers to identify what behavior of coworkers irritates them the most. Forty-one percent of the workers surveyed said that *sloppy work* is the most irritating behavior. Suppose that this percentage is based on a random sample of 500 workers.

    a. Construct a 95% confidence interval for the proportion of all workers who will say that *sloppy work* is the most irritating behavior of their coworkers.

    b. Suppose the confidence interval obtained in part a is too wide. How can the width of this interval be reduced? Discuss all possible alternatives. Which alternative is the best?

8.71 A researcher wanted to know the percentage of judges who are in favor of the death penalty. He took a random sample of 16 judges and asked them whether or not they favor the death penalty. The responses of these judges are given here.

| Yes | No | Yes | Yes | Yes | No | No | Yes |
| Yes | Yes | Yes | Yes | Yes | No | Yes | Yes |

    a. What is the point estimate of the population proportion?

    b. Make a 95% confidence interval for the percentage of all judges who are in favor of the death penalty.

8.72 A random sample of 20 managers was taken, and they were asked whether or not they usually take work home. The responses of these managers are given below, where *yes* indicates they usually take work home and *no* means they do not.

| Yes | Yes | No | No | No | Yes | No | No | No | No |
| Yes | Yes | No | Yes | Yes | No | No | No | No | Yes |

Make a 99% confidence interval for the percentage of all managers who take work home.

8.73 Tony's Pizza guarantees all pizza deliveries within 30 minutes of the placement of orders. An agency wants to estimate the proportion of all pizzas delivered within 30 minutes by Tony's. What is the most conservative estimate of the sample size that would limit the margin of error to within .05 of the population proportion for a 99% confidence interval?

8.74 Refer to Exercise 8.73. Assume that a preliminary study has shown that 93% of all Tony's pizzas are delivered within 30 minutes. How large should the sample size be so that the 99% confidence interval for the population proportion has a margin of error of .05?

*8.75 You want to estimate the proportion of students at your college who hold off-campus (part-time or full-time) jobs. Briefly explain how you will make such an estimate. Collect data from 40 students at your college on whether or not they hold off-campus jobs. Then calculate the proportion of students in this sample who hold off-campus jobs. Using this information, estimate the population proportion. Select your own confidence level.

*8.76 You want to estimate the percentage of students at your college or university who are satisfied with the campus food services. Briefly explain how you will make such an estimate. Select a sample of 30 students and ask them whether or not they are satisfied with the campus food services. Then calculate the percentage of students in the sample who are satisfied. Using this information, find the confidence interval for the corresponding population percentage. Select your own confidence level.

USES AND MISUSES...

NATIONAL VERSUS LOCAL UNEMPLOYMENT RATE

Reading a newspaper article, you learn that the national unemployment rate is 5.1%. The next month you read another article that states that a recent survey in your area, based on a random sample of the labor force, estimates that the local unemployment rate is 4.7% with a margin of error of .5%. Thus, you conclude that the unemployment rate in your area is somewhere between 4.2% and 5.2%.

So, what does this say about the local unemployment picture in your area versus the national unemployment situation? Since a major portion of the interval for the local unemployment rate is below 5.1%, is it reasonable to conclude that the local unemployment rate is below the national unemployment rate? Not really. When looking at the confidence interval, you have some degree of confidence, usually between 90% and 99%. If we use z = 1.96 to calculate the margin of error, which is the z value for a 95% confidence level, we can state that we are 95% confident that the local unemployment rate falls in the interval we obtain by using the margin of error. However, since 5.1% is in the interval for the local unemployment rate, the one thing that you can say is that it appears reasonable to conclude that the local and national unemployment rates are not different. However, if the national rate was 5.3%, then a conclusion that the two rates differ is reasonable because we are confident that the local unemployment rate falls between 4.2% and 5.2%.

When making conclusions based on the types of confidence intervals you have learned and will learn in this course, you will only be able to conclude that either there is a difference or there is not a difference. However, the methods you will learn in Chapter 9 will also allow you to determine the validity of a conclusion that states that the local rate is lower (or higher) than the national rate.

GLOSSARY

Confidence interval   An interval constructed around the value of a sample statistic to estimate the corresponding population parameter.

Confidence level   Confidence level, denoted by $(1 - \alpha)100\%$, that states how much confidence we have that a confidence interval contains the true population parameter.

Degrees of freedom (df)   The number of observations that can be chosen freely. For the estimation of μ using the t distribution, the degrees of freedom is $n - 1$.

Estimate   The value of a sample statistic that is used to find the corresponding population parameter.

Estimation   A procedure by which a numerical value or values are assigned to a population parameter based on the information collected from a sample.

Estimator   The sample statistic that is used to estimate a population parameter.

Interval estimate   An interval constructed around the point estimate that is likely to contain the corresponding population parameter. Each interval estimate has a confidence level.

Margin of error   The quantity that is subtracted from and added to the value of a sample statistic to obtain a confidence interval for the corresponding population parameter.

Point estimate   The value of a sample statistic assigned to the corresponding population parameter.

t distribution   A continuous probability distribution with a specific type of bell-shaped curve with its mean equal to 0 and standard deviation equal to $\sqrt{df/(df - 2)}$ for $df > 2$.

SUPPLEMENTARY EXERCISES

8.77 A bank manager wants to know the mean amount of mortgage paid per month by homeowners in an area. A random sample of 120 homeowners selected from this area showed that they pay an average of $1575 per month for their mortgages. The population standard deviation of all such mortgages is $215.

   a. Find a 97% confidence interval for the mean amount of mortgage paid per month by all homeowners in this area.

   b. Suppose the confidence interval obtained in part a is too wide. How can the width of this interval be reduced? Discuss all possible alternatives. Which alternative is the best?

8.78 At Farmer's Dairy, a machine is set to fill 32-ounce milk cartons. However, this machine does not put exactly 32 ounces of milk into each carton; the amount varies slightly from carton to carton. It is known that when the machine is working properly, the mean net weight of these cartons is 32 ounces. The standard deviation of the amounts of milk in all such cartons is always equal to .15 ounce. The quality control department takes a random sample of 25 such cartons every week, calculates the mean net weight of these cartons, and makes a 99% confidence interval for the population mean. If either the upper limit of this confidence interval is greater than 32.15 ounces or the lower limit of this confidence interval is less than 31.85 ounces, the machine is stopped and adjusted. A recent sample of 25 such cartons produced a mean net weight of 31.94 ounces. Based on this sample, will you conclude that the machine needs an adjustment? Assume that the amounts of milk put in all such cartons have an approximate normal distribution.

8.79 Lazurus Steel Corporation produces iron rods that are supposed to be 36 inches long. The machine that makes these rods does not produce each rod exactly 36 inches long; the lengths of the rods vary slightly. It is known that when the machine is working properly, the mean length of the rods made on this machine is 36 inches. The standard deviation of the lengths of all rods produced on this machine is always equal to .10 inch. The quality control department takes a random sample of 20 such rods every week, calculates the mean length of these rods, and makes a 99% confidence interval for the population mean. If either the upper limit of this confidence interval is greater than 36.05 inches or the lower limit of this confidence interval is less than 35.95 inches, the machine is stopped and adjusted. A recent sample of 20 rods produced a mean length of 36.02 inches. Based on this

sample, will you conclude that the machine needs an adjustment? Assume that the lengths of all such rods have an approximate normal distribution.

8.80 A hospital administration wants to estimate the mean time spent by patients waiting for treatment at the emergency room. The waiting times (in minutes) recorded for a random sample of 32 such patients are given below.

110	42	88	19	35	76	10	151
2	44	27	77	53	102	66	39
20	108	92	55	14	52	3	62
78	15	60	121	40	35	11	72

Construct a 98% confidence interval for the corresponding population mean. Use the t distribution.

8.81 A travel magazine wanted to estimate the mean amount of leisure time per week enjoyed by adults. The research department at the magazine took a sample of 36 adults and obtained the following data on the weekly leisure time (in hours).

15	12	18	23	11	21	16	13	9	19	26	14
7	18	11	15	23	26	10	8	17	21	12	7
19	21	11	13	21	16	14	9	15	12	10	14

Construct a 99% confidence interval for the mean leisure time per week enjoyed by all adults. Use the t distribution.

8.82 A drug that provides relief from headaches was tried on 18 randomly selected patients. The experiment showed that the mean time to get relief from headaches for these patients after taking this drug was 24 minutes with a standard deviation of 4.5 minutes. Assuming that the time taken to get relief from a headache after taking this drug is (approximately) normally distributed, determine a 95% confidence interval for the mean relief time for this drug for all patients.

8.83 A random sample of 300 female members of health clubs in Los Angeles showed that they spend, on average, 4.5 hours per week doing physical exercise with a standard deviation of .75 hour. Find a 98% confidence interval for the population mean.

8.84 A company that produces eight-ounce low-fat yogurt cups wanted to estimate the mean number of calories for such cups. A random sample of 10 such cups produced the following numbers of calories.

$$147 \quad 159 \quad 153 \quad 146 \quad 144 \quad 148 \quad 163 \quad 153 \quad 143 \quad 158$$

Construct a 99% confidence interval for the population mean. Assume that the numbers of calories for such cups of yogurt produced by this company have an approximately normal distribution.

8.85 An auto company wanted to know the percentage of people who prefer to own safer cars (that is, cars that possess more safety features) even if they have to pay a few thousand dollars more. A random sample of 500 persons showed that 44% of them will not mind paying a few thousand dollars more to have safer cars.

    a. What is the point estimate of the percentage of all people who will not mind paying a few thousand dollars more to have safer cars?

    b. Construct a 90% confidence interval for the percentage of all people who will not mind paying a few thousand dollars more to have safer cars.

8.86 Salaried workers at a large corporation receive 2 weeks' paid vacation per year. Sixteen randomly selected workers from this corporation were asked whether or not they would be willing to take a 3% reduction in their annual salaries in return for 2 additional weeks of paid vacation. The following are the responses of these workers.

No	Yes	No	No	Yes	No	No	Yes
Yes	No	No	No	Yes	No	No	No

Construct a 97% confidence interval for the percentage of all salaried workers at this corporation who would accept a 3% pay cut in return for 2 additional weeks of paid vacation.

8.87 A large city with chronic economic problems is considering legalizing casino gambling. The city council wants to estimate the proportion of all adults in the city who favor legalized casino gambling. Assume that a preliminary sample has shown that 63% of the adults in this city favor legalized casino gambling. How large should the sample size be so that the 95% confidence interval for the population proportion has a margin of error of .05?

8.88 A researcher wants to determine a 99% confidence interval for the mean number of hours that adults spend per week doing community service. How large a sample should the researcher select so that the estimate is within 1.2 hours of the population mean? Assume that the standard deviation for time spent per week doing community service by all adults is 3 hours.

8.89 An economist wants to find a 90% confidence interval for the mean sale price of houses in a state. How large a sample should she select so that the estimate is within $3500 of the population mean? Assume that the standard deviation for the sale prices of all houses in this state is $31,500.

8.90 A large city with chronic economic problems is considering legalizing casino gambling. The city council wants to estimate the proportion of all adults in the city who favor legalized casino gambling. What is the most conservative estimate of the minimum sample size that would limit the margin of error to be within .05 of the population proportion for a 95% confidence interval?

ADVANCED EXERCISES

8.91 Let μ be the hourly wage (excluding tips) for workers who provide hotel room service in a large city. A random sample of a number (more than 30) of such workers yielded a 95% confidence interval for μ of $8.46 to $9.86 using the normal distribution with a known population standard deviation.

    a. Find the value of $\bar{x}$ for this sample.

    b. Find a 99% confidence interval for μ based on this sample.

8.92 A group of veterinarians wants to test a new canine vaccine for Lyme disease. (Lyme disease is transmitted by the bite of an infected deer tick.) In an area that has a high incidence of Lyme disease, 100 dogs are randomly selected (with their owners' permission) to receive the vaccine. Over a 12-month period, these dogs are periodically examined by veterinarians for symptoms of Lyme disease. At the end of 12 months, 10 of these 100 dogs are diagnosed with the disease. During the same 12-month period, 18% of the unvaccinated dogs in the area have been found to have Lyme disease. Let p be the proportion of all potential vaccinated dogs who would contract Lyme disease in this area.

    a. Find a 95% confidence interval for p.

    b. Does 18% lie within your confidence interval of part a? Does this suggest the vaccine might or might not be effective to some degree?

    c. Write a brief critique of this experiment, pointing out anything that may have distorted the results or conclusions.

8.93 A couple considering the purchase of a new home would like to estimate the average number of cars that go past the location per day.

The couple guesses that the number of cars passing this location per day has a population standard deviation of 170.

    a. On how many randomly selected days should the number of cars passing the location be observed so that the couple can be 99% certain the estimate will be within 100 cars of the true average?

    b. Suppose the couple finds out that the population standard deviation of the number of cars passing the location per day is not 170 but is actually 272. If they have already taken a sample of the size computed in part a, what confidence does the couple have that their point estimate is within 100 cars of the true average?

    c. If the couple has already taken a sample of the size computed in part a and later finds out that the population standard deviation of the number of cars passing the location per day is actually 130, they can be 99% confident their point estimate is within how many cars of the true average?

8.94 In April 2012, N3L Optics conducted a telephone poll of 1080 adult Americans aged 18 years and older. One of the questions asked respondents to identify which outdoor activities and sports they favor for fitness. Respondents could choose more than one activity/sport. Of the respondents, 76% said walking, 35% mentioned hiking, and 27% said team sports (http://n3loptics.com/news_items/57). Using these results, find a 98% confidence interval for the population percentage that corresponds to each response. Write a one-page report to present your results to a group of college students who have not taken statistics. Your report should answer questions such as the following: (1) What is a confidence interval? (2) Why is a range of values (interval) more informative than a single percentage (point estimate)? (3) What does 98% confidence mean in this context? (4) What assumptions, if any, are you making when you construct each confidence interval?

8.95 The U.S. Senate just passed a bill by a vote of 55–45 (with all 100 senators voting). A student who took an elementary statistics course last semester says, "We can use these data to make a confidence interval about p. We have $n = 100$ and $\hat{p} = 55/100 = .55$." Hence, according to him, a 95% confidence interval for p is

$$\hat{p} \pm z\sigma_{\hat{p}} = .55 \pm 1.96\sqrt{\frac{(.55)(.45)}{100}} = .55 \pm .098 = .452 \text{ to } .648$$

Does this make sense? If not, what is wrong with the student's reasoning?

8.96 When calculating a confidence interval for the population mean μ with a known population standard deviation σ, describe the effects

of the following two changes on the confidence interval: (1) doubling the sample size, (2) quadrupling (multiplying by 4) the sample size. Give two reasons why this relationship does not hold true if you are calculating a confidence interval for the population mean μ with an unknown population standard deviation.

8.97 Calculating a confidence interval for the proportion requires a minimum sample size. Calculate a confidence interval, using any confidence level of 90% or higher, for the population proportion for each of the following.

  a. $n = 200$ and $\hat{p} = .01$
  b. $n = 160$ and $\hat{p} = .9875$

Explain why these confidence intervals reveal a problem when the conditions for using the normal approximation do not hold.

SELF-REVIEW TEST

1. Complete the following sentences using the terms *population parameter* and *sample statistic*.

  a. Estimation means assigning values to a _____ based on the value of a _____.
  b. An estimator is a _____ used to estimate a _____.
  c. The value of a _____ is called the point estimate of the corresponding _____.

2. A 95% confidence interval for μ can be interpreted to mean that if we take 100 samples of the same size and construct 100 such confidence intervals for μ, then

  a. 95 of them will not include μ
  b. 95 will include μ
  c. 95 will include $\bar{x}$

3. The confidence level is denoted by
  a. $(1 - \alpha)100\%$ **b.** $100\alpha\%$ **c.** α

4. The margin of error of estimate for μ is
  a. $z\sigma_{\bar{x}}$ (or $ts_{\bar{x}}$)   **b.** $\sigma/\sqrt{n}$ (or $s/\sqrt{n}$)   **c.** $\sigma_{\bar{x}}$ (or $s_{\bar{x}}$)

5. Which of the following assumptions is not required to use the t distribution to make a confidence interval for μ?

  a. Either the population from which the sample is taken is (approximately) normally distributed or $n \geq 30$.
  b. The population standard deviation, σ, is not known.
  c. The sample size is at least 10.

6. The parameter(s) of the t distribution is (are)
  a. n
  b. degrees of freedom
  c. μ and degrees of freedom

7. A random sample of 36 vacation homes built during the past 2 years in a coastal resort region gave a mean construction cost of $159,000 with a population standard deviation of $27,000.

  a. What is the point estimate of the corresponding population mean?
  b. Make a 99% confidence interval for the mean construction cost for all vacation homes built in this region during the past 2 years. What is the margin of error here?

8. A random sample of 25 malpractice lawsuits filed against doctors showed that the mean compensation awarded to the plaintiffs was $610,425 with a standard deviation of $94,820. Find a 95% confidence interval for the mean compensation awarded to plaintiffs of all

such lawsuits. Assume that the compensations awarded to plaintiffs of all such lawsuits are approximately normally distributed.

9. Harris Interactive conducted an online poll of 2097 American adults between July 17 and 21, 2014, on the topic "Who are we lying to?" In response to one of the questions, 37% of American adults said that they have lied to get out of work (www.harrisinteractive.com).

  a. What is the point estimate of the corresponding population proportion?
  b. Construct a 99% confidence interval for the proportion of all American adults who will say that they have lied to get out of work.

10. A company that makes toaster ovens has done extensive testing on the accuracy of its temperature-setting mechanism. For a previous toaster model of this company, the standard deviation of the temperatures when the mechanism is set for 350°F is 5.78°. Assume that this is the population standard deviation for a new toaster model that uses the same temperature mechanism. How large a sample must be taken so that the estimate of the mean temperature when the mechanism is set for 350°F is within 1.25° of the population mean temperature? Use a 95% confidence level.

11. A college registrar has received numerous complaints about the online registration procedure at her college, alleging that the system is slow, confusing, and error prone. She wants to estimate the proportion of all students at this college who are dissatisfied with the online registration procedure. What is the most conservative estimate of the minimum sample size that would limit the margin of error to be within .05 of the population proportion for a 90% confidence interval?

12. Refer to Problem 11. Assume that a preliminary study has shown that 70% of the students surveyed at this college are dissatisfied with the current online registration system. How large a sample should be taken in this case so that the margin of error is within .05 of the population proportion for a 90% confidence interval?

13. Dr. Garcia estimated the mean stress score before a statistics test for a random sample of 25 students. She found the mean and standard deviation for this sample to be 7.1 (on a scale of 1 to 10) and 1.2, respectively. She used a 97% confidence level. However, she thinks that the confidence interval is too wide. How can she reduce the width of the confidence interval? Describe all possible alternatives. Which alternative do you think is best and why?

*14. You want to estimate the mean number of hours that students at your college work per week. Briefly explain how you will conduct this study using a small sample. Take a sample of 12 students from your college who hold a job. Collect data on the number of hours that these students spent working last week. Then estimate the population mean. Choose your own confidence level. What assumptions will you make to estimate this population mean?

*15. You want to estimate the proportion of people who are happy with their current jobs. Briefly explain how you will conduct this study. Take a sample of 35 persons and collect data on whether or not they are happy with their current jobs. Then estimate the population proportion. Choose your own confidence level.

MINI-PROJECTS

Note: The Mini-Projects are located on the text's Web site, www.wiley.com/college/mann.

DECIDE FOR YOURSELF

Note: The Decide for Yourself feature is located on the text's Web site, www.wiley.com/college/mann.

TECHNOLOGY INSTRUCTIONS | CHAPTER 8

Note: Complete TI-84, Minitab, and Excel manuals are available for download at the textbook's Web site, www.wiley.com/college/mann.

TI-84 Color/TI-84

The TI-84 Color Technology Instructions feature of this text is written for the TI-84 Plus C color graphing calculator running the 4.0 operating system. Some screens, menus, and functions will be slightly different in older operating systems. The TI-84 and TI-84 Plus can perform all of the same functions but will not have the "Color" option referenced in some of the menus.

Estimating a Population Mean, σ Known for Example 8–1(b) of the Text

1. Select **STAT > TESTS > ZInterval**.

2. Use the following settings in the **Zinterval** menu (see **Screen 8.1**):

   • Select **Stats** at the **Inpt** prompt.

     Note: If you have the data in a list, select **Data** at the **Inpt** prompt.

   • At the σ prompt, type 35.

   • At the $\bar{x}$ prompt, type 145.

   • At the **n** prompt, type 25.

   • At the **C-Level** prompt, type 0.90.

3. Highlight **Calculate** and press **ENTER**.

4. The output includes the confidence interval. (See **Screen 8.2**.)

   Note: The confidence interval from your calculator may differ slightly from the one in the text since the calculator uses a more precise value for *z*.

Screen 8.1

Screen 8.2

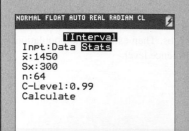

Screen 8.3

Screen 8.4

Estimating a Population Mean, σ Unknown for Example 8–6 of the Text

1. Select **STAT > TESTS > TInterval**.

2. Use the following settings in the **Tinterval** menu (see **Screen 8.3**):

   - At the **Inpt** prompt, select **Stats**.

     Note: If you have the data in a list, select **Data** at the **Inpt** prompt.

   - At the $\bar{x}$ prompt, type 1450.

   - At the **Sx** prompt, type 300.

   - At the **n** prompt, type 64.

   - At the **C-Level** prompt, type 0.99.

3. Highlight **Calculate** and press **ENTER**.

4. The output includes the confidence interval. (See **Screen 8.4**.)

   Note: The confidence interval from your calculator may differ slightly from the one in the text since the calculator uses a more precise value for *t*.

Estimating a Population Proportion for Example 8–7 of the Text

1. Select **STAT > TESTS > 1-PropZInt**.

2. Use the following settings in the **1-PropZInt** menu (see **Screen 8.5**):

   - At the **x** prompt, type 1366.

     Note: The value of **x** is the number of successes in the sample, and it must be a whole number or the calculator will return an error message. If **x** is not given, multiply *n* by $\hat{p}$ to obtain **x** and round the result to the nearest whole number.

   - At the **n** prompt, type 1821.

   - At the **C-Level** prompt, type 0.99.

3. Highlight **Calculate** and press **ENTER**.

4. The output includes the confidence interval. (See **Screen 8.6**.)

   Note: The confidence interval from your calculator may differ slightly from the one in the text since the calculator uses a more precise value for *z*.

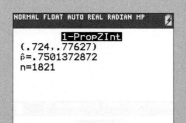

Screen 8.5

Screen 8.6

Minitab

The Minitab Technology Instructions feature of this text is written for Minitab version 17. Some screens, menus, and functions will be slightly different in older versions of Minitab.

Estimating a Population Mean, σ Known for Example 8–1(b) of the Text

1. Select **Stat > Basic Statistics > 1-Sample Z**.

2. Use the following settings in the dialog box that appears on screen (see **Screen 8.7**):

- From the dropdown box, select **Summarized Data**.

  Note: If you have the data in a column, select **One or more samples, each in a column**, type the column name(s) in the box, and move to step 3 below.

- In the **Sample size** box, type 25.

- In the **Sample mean** box, type 145.

- In the **Known standard deviation** box, type 35.

3. Select **Options** and type 90 in the **Confidence level** box when the dialog box appears on screen. (See **Screen 8.8**.)

4. Click **OK** in both dialog boxes.

5. The confidence interval will be displayed in the Session window. (See **Screen 8.9**.)

   Note: The confidence interval from Minitab may differ slightly from the one in the text since Minitab uses a more precise critical value for *z*.

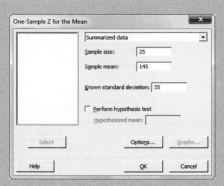

Screen 8.7

Screen 8.8

Estimating a Population Mean, σ Unknown for Example 8–6 of the Text

1. Select **Stat > Basic Statistics > 1-Sample t**.

2. Use the following settings in the dialog box that appears on screen (see **Screen 8.10**):

   - Select **Summarized Data**.

     Note: If you have the data in a column, select **Samples in columns**, type the column name(s) in the box, and move to step 3 below.

   - In the **Sample size** box, type 64.

   - In the **Sample mean** box, type 1450.

   - In the **Standard deviation** box, type 300.

3. Select **Options** and type 99 in the **Confidence level** box when the dialog box appears on screen.

4. Click **OK** in both dialog boxes.

5. The confidence interval will be displayed in the Session window. (See **Screen 8.9**.)

   Note: The confidence interval from Minitab may differ slightly from the one in the text since Minitab uses a more precise value for *t*.

Screen 8.9

Estimating a Population Proportion for Example 8–7 of the Text

1. Select **Stat > Basic Statistics > 1 Proportion**.

2. Use the following settings in the dialog box that appears on screen (see **Screen 8.11**):

   - From the drop-down box, select **Summarized Data**.

     Note: If you have the data in a column, select **One or more samples, each in a column**, type the column name(s) in the box, and move to step 3 below.

Screen 8.10

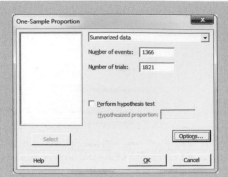

Screen 8.11

- In the **Number of events** box, type 1366.

  Note: The value 1366 is the number of successes in the sample, and it must be a whole number or the calculator will return an error message. If the number of successes is not given, multiply n by $\hat{p}$ to obtain the number of successes and round the result to the nearest whole number.

- In the **Number of trials** box, type 1821.

3. Select **Options**. Use the following settings in the dialog box that appears on screen:

- In the **Confidence level** box, type 99.

- In the Method box, select **Normal approximation** from the drop-down menu.

4. Click **OK** in both dialog boxes.

5. The confidence interval will be displayed in the Session window. (See **Screen 8.9**.)

  Note: The confidence interval from Minitab may differ slightly from the one in the text since Minitab uses a more precise value for z.

Excel

The Excel Technology Instructions feature of this text is written for Excel 2013. Some screens, menus, and functions will be slightly different in older versions of Excel. To enable some of the advanced Excel functions, you must enable the Data Analysis Add-In for Excel. For example, in Excel 2007 and older versions of Excel, replace the function **CONFIDENCE.NORM** *with the function* **CONFIDENCE** *and replace the function* **NORM.S.INV** *with the function* **NORMSINV***. For Excel 2007 and older versions of Excel, there is no function that corresponds to* **CONFIDENCE.T***.*

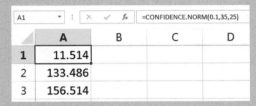

Screen 8.12

Estimating a Population Mean, σ Known for Example 8–1(b) of the Text

1. Click on cell A1.

2. Type =**CONFIDENCE.NORM(0.10,35,25)**. This function calculates the margin of error for the confidence interval. (See **Screen 8.12**.)

  Note: The first number in this function, denoted alpha, is calculated by subtracting the confidence level (as a decimal) from 1. The second number is the population standard deviation σ, and the third number is the sample size.

3. Click on cell A2 and type =**145 – A1** to calculate the lower bound for the confidence interval. (See **Screen 8.12**.)

4. Click on cell A3 and type =**145 + A1** to calculate the upper bound for the confidence interval. (See **Screen 8.12**.)

  Note: The confidence interval from Excel may differ slightly from the one in the text since Excel uses a more precise value for z.

Estimating a Population Mean, σ Unknown for Example 8–6 of the Text

1. Click on cell A1.

2. Type =**CONFIDENCE.T(0.01,300,64)**. This function calculates the margin of error for the confidence interval.

  Note: The first number in this function, denoted alpha, is calculated by subtracting the confidence level (as a decimal) from 1. The second number is the sample standard deviation s, and the third number is the sample size.

3. Click on cell A2 and type =**1450 − A1** to calculate the lower bound for the confidence interval.

4. Click on cell A3 and type =**1450 + A1** to calculate the upper bound for the confidence interval.

Note: The confidence interval from Excel may differ slightly from the one in the text since Excel uses a more precise value for *t*.

Estimating a Population Proportion for Example 8–7 of the Text

1. There is no native Excel function that can be used to create a confidence interval for a population proportion, so we must enter a few calculations in sequence. To make our work easier to understand, type the following text into the corresponding spreadsheet cells (see **Screen 8.13**):

   • Type "**p-hat**" in cell A1.

   • Type "**n**" in cell A2.

   • Type "**SE of p-hat**" in cell A3.

   • Type "**Critical z**" in cell A4.

   • Type "**Lower Bound**" in cell A5.

   • Type "**Upper Bound**" in cell A6.

2. Click on cell B1 and type **0.75**.

3. Click on cell B2 and type **1821**.

4. Click on cell B3 and type =**SQRT(B1*(1-B1)/B2)**.

5. Click on cell B4 and type =**NORM.S.INV(0.995)**.

   Note: The value 0.995 is calculated by adding the confidence level (as a decimal) to the area in the left tail (as a decimal).

6. Click on cell B5 and type =**B1–B4*B3**.

7. Click on cell B6 and type =**B1+B4*B3**.

8. The lower bound of the interval is in cell B5 and the upper bound of the interval is in cell B6. (See **Screen 8.13**.)

   Note: The confidence interval from Excel may differ slightly from the one in the text since Excel uses a more precise value for *z*.

B6		fx	=B1+B4*B3
	A	**B**	
1	p-hat	0.75	
2	n	1821	
3	SE of p-hat	0.01014719	
4	Critical z	2.5758293	
5	Lower Bound	0.72386258	
6	Upper Bound	0.77613742	

Screen 8.13

TECHNOLOGY ASSIGNMENTS

TA8.1 The following data give the annual incomes (in thousands of dollars) before taxes for a sample of 36 randomly selected families from a city:

21.6	33.0	25.6	37.9	50.0	148.1
50.1	21.5	70.0	72.8	58.2	85.4
91.2	57.0	72.2	45.0	95.0	27.8
92.8	79.4	45.3	76.0	48.6	69.3
40.6	69.0	75.5	57.5	49.7	75.1
96.3	44.5	84.0	43.0	61.7	126.0

Construct a 99% confidence interval for μ assuming that the population standard deviation is $23.75 thousand.

TA 8.2 Chicken eggs are graded AA, A, or B according to strict guidelines set by the USDA. A producer of chicken eggs selects a random sample of 500 eggs and finds that 463 of these are grade A. Construct a 92% confidence interval for the relevant population parameter.

TA8.3 The following data give the checking account balances (in dollars) on a certain day for a randomly selected sample of 30 households:

500	100	650	1917	2200	500	180	3000	1500	1300
319	1500	1102	405	124	1000	134	2000	150	800
200	750	300	2300	40	1200	500	900	20	160

Construct a 97% confidence interval for μ assuming that the population standard deviation is unknown.

TA8.4 Refer to Data Set I (that accompanies this text) on the prices of various products in different cities across the country. Using the data on monthly rent of an unfurnished two-bedroom apartment, make a 98% confidence interval for the population mean μ.

TA8.5 Refer to the Manchester Road Race data set (Data Set IV) for all participants that accompanies this text (see Appendix A). Take a sample of 100 observations from this data set.

    a. Using the sample data, make a 95% confidence interval for the mean time (listed in column 3) taken to complete this race by all participants.

    b. Now calculate the mean time taken to run this race by all participants. Does the confidence interval made in part a include this population mean?

TA8.6 Repeat Technology Assignment TA8.5 for a sample of 50 observations. Assume that the distribution of times taken to run this race by all participants is approximately normal.

TA8.7 The following data give the prices (in thousands of dollars) of 16 recently sold houses in an area.

| 341 | 163 | 327 | 204 | 197 | 203 | 313 | 279 |
| 456 | 228 | 383 | 289 | 533 | 399 | 271 | 381 |

Construct a 98% confidence interval for the mean price of all houses in this area. Assume that the distribution of prices of all houses in the given area is approximately normal.

TA8.8 A researcher wanted to estimate the mean contributions made to charitable causes by major companies. A random sample of 18 companies produced the following data on contributions (in millions of dollars) made by them.

| 1.8 | .6 | 1.2 | .3 | 2.6 | 1.9 | 3.4 | 2.6 | .2 |
| 2.4 | 1.4 | 2.5 | 3.1 | .9 | 1.2 | 2.0 | .8 | 1.1 |

Make a 99% confidence interval for the mean contributions made to charitable causes by all major companies. Assume that the contributions made to charitable causes by all major companies have an approximate normal distribution.

TA8.9 A mail-order company promises its customers that their orders will be processed and mailed within 72 hours after an order is placed. The quality control department at the company checks from time to time to see if this promise is kept. Recently the quality control department took a sample of 250 orders and found that 224 of them were processed and mailed within 72 hours of the placement of the orders. Make a 99% confidence interval for the corresponding population proportion.

TA8.10 One of the major problems faced by department stores is a high percentage of returns. The manager of a department store wanted to estimate the percentage of all sales that result in returns. A random sample of 750 sales showed that 143 of them had products returned within the time allowed for returns. Make a 98% confidence interval for the corresponding population proportion.

TA8.11 One of the major problems faced by auto insurance companies is the filing of fraudulent claims. An insurance company carefully investigated 2000 auto claims filed with it and found 215 of them to be fraudulent. Make a 96% confidence interval for the corresponding population proportion.

TA 8.12 A manufacturer of contact lenses must ensure that each contact lens is properly manufactured with no flaws, since flaws lead to poor vision or even eye damage. In a recent quality control check, 9 of 1000 lenses were found to have flaws.

    a. Construct a 98% confidence interval for the population proportion of lenses that have flaws.

    b. The company advertises that less than 1% of all its lenses have flaws. Based on the interval you constructed in part a, do you have reason to doubt this claim?

CHAPTER 9

Hypothesis Tests About the Mean and Proportion

Gary Alvis/iStockphoto

Will you graduate from college with debt? If yes, how much money do you think you will owe? Do you know that students who graduated from public and nonprofit colleges in 2013 with loans had an average debt of $28,400? The average debt of the class of 2013 varied a great deal from state to state, with the highest average debt for students who graduated from colleges in New Hampshire at $32,795 and the lowest average for students who graduated from colleges in New Mexico at $18,656. (See Case Study 9–1.)

This chapter introduces the second topic in inferential statistics: tests of hypotheses. In a test of hypothesis, we test a certain given theory or belief about a population parameter. We may want to find out, using some sample information, whether or not a given claim (or statement) about a population parameter is true. This chapter discusses how to make such tests of hypotheses about the population mean, μ, and the population proportion, p.

As an example, a soft-drink company may claim that, on average, its cans contain 12 ounces of soda. A government agency may want to test whether or not such cans do contain, on average, 12 ounces of soda. As another example, in a Gallup poll conducted April 9–12, 2015, American adults were asked, "Do you feel that the distribution of money and wealth in this country today is fair, or do you feel that the money and wealth in this country should be more evenly distributed among a larger percentage of the people?" Of the adults included in the poll, 63% said that money and wealth in this country should be more evenly distributed among a larger percentage of the people. A researcher wants to check if this percentage is still true. In the first of these two examples we are to test a hypothesis about the population mean, μ, and in the second example we are to test a hypothesis about the population proportion, p.

9.1 | Hypothesis Tests: An Introduction

Why do we need to perform a test of hypothesis? Reconsider the example about soft-drink cans. Suppose we take a sample of 100 cans of the soft drink under investigation. We then find out that the mean amount of soda in these 100 cans is 11.89 ounces. Based on this result, can we state that all such cans contain, on average, less than 12 ounces of soda and that the company is lying to the public? Not until we perform a test of hypothesis can we make such an accusation. The reason is that the mean, $\bar{x} = 11.89$ ounces, is obtained from a sample. The difference between 12 ounces (the required average amount for the population) and 11.89 ounces (the observed average amount for the sample) may have occurred only because of the sampling error (assuming that no nonsampling errors have been committed). Another sample of 100 cans may give us a mean of 12.04 ounces. Therefore, we perform a test of hypothesis to find out how large the difference between 12 ounces and 11.89 ounces is and whether or not this difference has occurred as a result of chance alone. Now, if 11.89 ounces is the mean for all cans and not for just 100 cans, then we do not need to make a test of hypothesis. Instead, we can immediately state that the mean amount of soda in all such cans is less than 12 ounces. We perform a test of hypothesis only when we are making a decision about a population parameter based on the value of a sample statistic.

9.1.1 Two Hypotheses

Consider as a nonstatistical example a person who has been indicted for committing a crime and is being tried in a court. Based on the available evidence, the judge or jury will make one of two possible decisions:

1. The person is not guilty.
2. The person is guilty.

At the outset of the trial, the person is presumed not guilty. The prosecutor's job is to prove that the person has committed the crime and, hence, is guilty.

> A **null hypothesis** is a claim (or statement) about a population parameter that is assumed to be true until it is declared false.

In statistics, *the person is not guilty* is called the **null hypothesis** and *the person is guilty* is called the **alternative hypothesis**. The null hypothesis is denoted by H_0, and the alternative hypothesis is denoted by H_1. In the beginning of the trial it is assumed that the person is not guilty. The null hypothesis is usually the hypothesis that is assumed to be true to begin with. The two hypotheses for the court case are written as follows (notice the colon after H_0 and H_1):

$$\text{Null hypothesis:} \qquad H_0\text{: The person is not guilty}$$

$$\text{Alternative hypothesis:} \qquad H_1\text{: The person is guilty}$$

> An **alternative hypothesis** is a claim about a population parameter that will be declared true if the null hypothesis is declared to be false.

In a statistics example, the null hypothesis states that a given claim (or statement) about a population parameter is true. Reconsider the example of the soft-drink company's claim that its cans contain, on average, 12 ounces of soda. In reality, this claim may or may not be true. However, we will initially assume that the company's claim is true (that is, the company is not guilty of cheating and lying). To test the claim of the soft-drink company, the null hypothesis will be that the company's claim is true. Let μ be the mean amount of soda in all cans. The company's claim will be true if $\mu = 12$ ounces. Thus, the null hypothesis will be written as

$$H_0\text{: } \mu = 12 \text{ ounces} \quad \text{(The company's claim is true)}$$

In this example, the null hypothesis can also be written as $\mu \geq 12$ ounces because the claim of the company will still be true if the cans contain, on average, more than 12 ounces of soda. The company will be accused of cheating the public only if the cans contain, on average, less than 12 ounces of soda. However, it will not affect the test whether we use an = or a ≥ sign in the null hypothesis as long as the alternative hypothesis has a < sign. Remember that in the null hypothesis (and in the alternative hypothesis also) we use a population parameter (such as μ or p) and not a sample statistic (such as $\bar{x}$ or $\hat{p}$).

The alternative hypothesis in our statistics example will be that the company's claim is false and its soft-drink cans contain, on average, less than 12 ounces of soda—that is, $\mu < 12$ ounces. The alternative hypothesis will be written as

$$H_1: \mu < 12 \text{ ounces} \quad \text{(The company's claim is false)}$$

Let us return to the example of the court trial. The trial begins with the assumption that the null hypothesis is true—that is, the person is not guilty. The prosecutor assembles all the possible evidence and presents it in court to prove that the null hypothesis is false and the alternative hypothesis is true (that is, the person is guilty). In the case of our statistics example, the information obtained from a sample will be used as evidence to decide whether or not the claim of the company is true. In the court case, the decision made by the judge (or jury) depends on the amount of evidence presented by the prosecutor. At the end of the trial, the judge (or jury) will consider whether or not the evidence presented by the prosecutor is sufficient to declare the person guilty. The amount of evidence that will be considered to be sufficient to declare the person guilty depends on the discretion of the judge (or jury).

9.1.2 Rejection and Nonrejection Regions

In Figure 9.1, which represents the court case, the point marked 0 indicates that there is no evidence against the person being tried. The farther we move toward the right on the horizontal axis, the more convincing the evidence is that the person has committed the crime. We have arbitrarily marked a point C on the horizontal axis. Let us assume that a judge (or jury) considers any amount of evidence from point C to the right of it to be sufficient and any amount of evidence to the left of C to be insufficient to declare the person guilty. Point C is called the **critical value** or **critical point** in statistics. If the amount of evidence presented by the prosecutor falls in the area to the left of point C, the verdict will reflect that there is not enough evidence to declare the person guilty. Consequently, the accused person will be declared *not guilty*. In statistics, this decision is stated as *do not reject H_0* or failing to reject H_0. It is equivalent to saying that there is not enough evidence to declare the null hypothesis false. The area to the left of point C is called the *nonrejection region*; that is, this is the region where the null hypothesis is not rejected. However, if the amount of evidence falls at point C or to the right of point C, the verdict will be that there is sufficient evidence to declare the person guilty. In statistics, this decision is stated as *reject H_0* or *the null hypothesis is false*. Rejecting H_0 is equivalent to saying that *the alternative hypothesis is true*. The area to the right of point C (including point C) is called the *rejection region*; that is, this is the region where the null hypothesis is rejected.

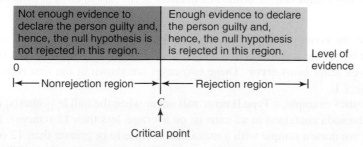

Figure 9.1   Nonrejection and rejection regions for the court case.

9.1.3 Two Types of Errors

We all know that a court's verdict is not always correct. If a person is declared guilty at the end of a trial, there are two possibilities.

1. The person has *not* committed the crime but is declared guilty (because of what may be false evidence).

2. The person *has* committed the crime and is rightfully declared guilty.

In the first case, the court has made an error by punishing an innocent person. In statistics, this kind of error is called a **Type I** or an α (*alpha*) **error**. In the second case, because the guilty person has been punished, the court has made the correct decision. The second row in the shaded portion of Table 9.1 shows these two cases. The two columns of Table 9.1, corresponding to *the person is not guilty* and *the person is guilty*, give the two actual situations. Which one of these is true is known only to the person being tried. The two rows in this table, corresponding to *the person is not guilty* and *the person is guilty*, show the two possible court decisions.

Table 9.1   Four Possible Outcomes for a Court Case

		Actual Situation	
		The Person Is Not Guilty	**The Person Is Guilty**
Court's decision	The person is not guilty	Correct decision	Type II or β error
	The person is guilty	Type I or α error	Correct decision

A Type I error occurs when a true null hypothesis is rejected. The value of α represents the probability of committing this type of error; that is,

$$\alpha = P(H_0 \text{ is rejected} \mid H_0 \text{ is true})$$

The value of α represents the **significance level** of the test.

In our statistics example, a Type I error will occur when H_0 is actually true (that is, the cans do contain, on average, 12 ounces of soda), but it just happens that we draw a sample with a mean that is much less than 12 ounces and we wrongfully reject the null hypothesis, H_0. The value of α, called the **significance level** of the test, represents the probability of making a Type I error. In other words, α is the probability of rejecting the null hypothesis, H_0, when in fact H_0 is true.

The size of the **rejection region** in a statistics problem of a test of hypothesis depends on the value assigned to α. In one approach to the test of hypothesis, we assign a value to α before making the test. Although any value can be assigned to α, commonly used values of α are .01, .025, .05, and .10. Usually the value assigned to α does not exceed .10 (or 10%).

Now, suppose that in the court trial case the person is declared not guilty at the end of the trial. Such a verdict does not indicate that the person has indeed *not* committed the crime. It is possible that the person is guilty but there is not enough evidence to prove the guilt. Consequently, in this situation there are again two possibilities.

1. The person has *not* committed the crime and is declared not guilty.
2. The person *has* committed the crime but, *because of the lack of enough evidence*, is declared not guilty.

In the first case, the court's decision is correct. In the second case, however, the court has committed an error by setting a guilty person free. In statistics, this type of error is called a **Type II** or a β (the Greek letter *beta*) **error**. These two cases are shown in the first row of the shaded portion of Table 9.1.

A Type II error occurs when a false null hypothesis is not rejected. The value of β represents the probability of committing a Type II error; that is,

$$\beta = P(H_0 \text{ is not rejected} \mid H_0 \text{ is false})$$

The value of $1 - \beta$ is called the **power of the test**. It represents the probability of rejecting H_0 when it is false.

In our statistics example, a Type II error will occur when the null hypothesis, H_0, is actually false (that is, the soda contained in all cans is, on average, less than 12 ounces), but it happens by chance that we draw a sample with a mean that is close to or greater than 12 ounces and we wrongfully conclude *do not reject H_0*. The value of β represents the probability of making a Type II error. It represents the probability that H_0 is not rejected when actually H_0 is false. The value of $1 - \beta$ is called the **power of the test**. It represents the probability of not making a Type II error. In other words, $1 - \beta$ is the probability that H_0 is rejected when it is false.

The two types of errors that occur in tests of hypotheses depend on each other. We cannot lower the values of α and β simultaneously for a test of hypothesis for a fixed sample size. Lowering the value of α will raise the value of β, and lowering the value of β will raise the value of α. However, we can decrease both α and β simultaneously by increasing the sample size. The explanation of how α and β are related and the computation of β are not within the scope of this text.

Table 9.2, which is similar to Table 9.1, is written for the statistics problem of a test of hypothesis. In Table 9.2 *the person is not guilty* is replaced by H_0 *is true, the person is guilty* by H_0 *is false*, and the *court's decision* by *decision*.

Table 9.2   **Four Possible Outcomes for a Test of Hypothesis**

		Actual Situation	
		H_0 Is True	H_0 Is False
Decision	Do not reject H_0	Correct decision	Type II or β error
	Reject H_0	Type I or α error	Correct decision

9.1.4   Tails of a Test

The statistical hypothesis-testing procedure is similar to the trial of a person in court but with two major differences. The first major difference is that in a statistical test of hypothesis, the partition of the total region into rejection and nonrejection regions is not arbitrary. Instead, it depends on the value assigned to α (Type I error). As mentioned earlier, α is also called the significance level of the test.

The second major difference relates to the rejection region. In the court case, the rejection region is at and to the right side of the critical point, as shown in Figure 9.1. However, in statistics, the rejection region for a hypothesis-testing problem can be on both sides, with the nonrejection region in the middle, or it can be on the left side or right side of the nonrejection region. These possibilities are explained in the next three parts of this section. A test with two rejection regions is called a **two-tailed test**, and a test with one rejection region is called a **one-tailed test**. The one-tailed test is called a **left-tailed test** if the rejection region is in the left tail of the distribution curve, and it is called a **right-tailed test** if the rejection region is in the right tail of the distribution curve.

> A **two-tailed test** has rejection regions in both tails, a **left-tailed test** has the rejection region in the left tail, and a **right-tailed test** has the rejection region in the right tail of the distribution curve.

A Two-Tailed Test

According to the U.S. Bureau of Labor Statistics, workers in the United States who were employed earned an average of \$47,230 a year in 2014. Suppose an economist wants to check whether this mean has changed since 2014. The key word here is *changed*. The mean annual earnings of employed Americans has changed if it has either increased or decreased since 2014. This is an example of a two-tailed test. Let μ be the mean annual earnings of employed Americans. The two possible decisions are as follows:

1. The mean annual earnings of employed Americans has not changed since 2014, that is, currently $\mu = \$47,230$.

2. The mean annual earnings of employed Americans has changed since 2014, that is, currently $\mu \neq \$47,230$.

We will write the null and alternative hypotheses for this test as follows:

H_0: $\mu = \$47,230$ (The mean annual earnings of employed Americans has not changed)

H_1: $\mu \neq \$47,230$ (The mean annual earnings of employed Americans has changed)

To determine whether a test is two-tailed or one-tailed, we look at the sign in the alternative hypothesis. If the alternative hypothesis has a *not equal to* ($\neq$) sign, as in this example, it is a two-tailed test. As shown in Figure 9.2, a two-tailed test has two rejection regions, one in each tail of the distribution curve. Figure 9.2 shows the sampling distribution of $\bar{x}$, assuming it has a normal distribution. Assuming H_0 is true, $\bar{x}$ has a normal distribution with its mean equal to \$47,230 (the value of μ in H_0). In Figure 9.2, the area of each of the two rejection regions is $\alpha/2$

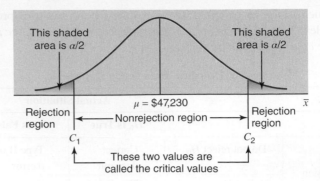

Figure 9.2 A two-tailed test.

and the total area of both rejection regions is α (the significance level). As shown in this figure, a two-tailed test of hypothesis has two critical values that separate the two rejection regions from the nonrejection region. We will reject H_0 if the value of $\bar{x}$ obtained from the sample falls in either of the two rejection regions. We will not reject H_0 if the value of $\bar{x}$ lies in the nonrejection region. By rejecting H_0, we are saying that the difference between the value of μ stated in H_0 and the value of $\bar{x}$ obtained from the sample is too large to have occurred because of the sampling error alone. Consequently, this difference appears to be real. By not rejecting H_0, we are saying that the difference between the value of μ stated in H_0 and the value of $\bar{x}$ obtained from the sample is small and may have occurred because of the sampling error alone.

A Left-Tailed Test

Reconsider the example of the mean amount of soda in all soft-drink cans produced by a company. The company claims that these cans contain, on average, 12 ounces of soda. However, if these cans contain less than the claimed amount of soda, then the company can be accused of underfilling the cans. Suppose a consumer agency wants to test whether the mean amount of soda per can is less than 12 ounces. Note that the key phrase this time is *less than*, which indicates a left-tailed test. Let μ be the mean amount of soda in all cans. The two possible decisions are as follows:

1. The mean amount of soda in all cans is equal to 12 ounces, that is, $\mu = 12$ ounces.
2. The mean amount of soda in all cans is less than 12 ounces, that is, $\mu < 12$ ounces.

The null and alternative hypotheses for this test are written as

$$H_0: \mu = 12 \text{ ounces} \quad \text{(The mean is equal to 12 ounces)}$$

$$H_1: \mu < 12 \text{ ounces} \quad \text{(The mean is less than 12 ounces)}$$

In this case, we can also write the null hypothesis as $H_0: \mu \geq 12$. This will not affect the result of the test as long as the sign in H_1 is *less than* ($<$).

When the alternative hypothesis has a *less than* ($<$) sign, as in this case, the test is always left-tailed. In a left-tailed test, the rejection region is in the left tail of the distribution curve, as shown in Figure 9.3, and the area of this rejection region is equal to α (the significance level). We can observe from this figure that there is only one critical value in a left-tailed test.

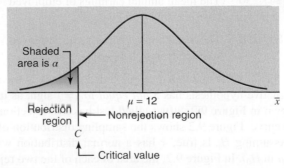

Figure 9.3 A left-tailed test.

Assuming H_0 is true, the sampling distribution of $\bar{x}$ has a mean equal to 12 ounces (the value of μ in H_0). We will reject H_0 if the value of $\bar{x}$ obtained from the sample falls in the rejection region; we will not reject H_0 otherwise.

A Right-Tailed Test

To illustrate the third case, according to www.city-data.com, the average price of detached homes in West Orange, New Jersey, was $471,257 in 2013. Suppose a real estate researcher wants to check whether the current mean price of homes in this town is higher than $471,257. The key phrase in this case is *higher than*, which indicates a right-tailed test. Let μ be the current mean price of homes in this town. The two possible decisions are as follows:

1. The current mean price of homes in this town is not higher than $471,257, that is, currently $\mu = \$471,257$.

2. The current mean price of homes in this town is higher than $471,257, that is, currently $\mu > \$471,257$.

We write the null and alternative hypotheses for this test as follows:

H_0: $\mu = \$471,257$ (The current mean price of homes in this town is not higher than $471,257)

H_1: $\mu > \$471,257$ (The current mean price of homes in this town is higher than $471,257)

Note that here we can also write the null hypothesis as H_1: $\mu \leq \$471,257$, which states that the current mean price of homes in this area is either equal to or less than $471,257. Again, the result of the test will not be affected by whether we use an *equal to* (=) or a *less than or equal to* ($\leq$) sign in H_0 as long as the alternative hypothesis has a *greater than* (>) sign.

When the alternative hypothesis has a *greater than* (>) sign, the test is always right-tailed. As shown in Figure 9.4, in a right-tailed test, the rejection region is in the right tail of the distribution curve. The area of this rejection region is equal to α, the significance level. Like a left-tailed test, a right-tailed test has only one critical value.

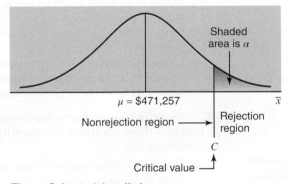

Shaded area is α

$\mu = \$471,257$ $\bar{x}$

Nonrejection region → Rejection region

C

Critical value

Figure 9.4   A right-tailed test.

Again, assuming H_0 is true, the sampling distribution of $\bar{x}$ has a mean equal to $471,257 (the value of μ in H_0). We will reject H_0 if the value of $\bar{x}$ obtained from the sample falls in the rejection region. Otherwise, we will not reject H_0.

Table 9.3 summarizes the foregoing discussion about the relationship between the signs in H_0 and H_1 and the tails of a test.

Table 9.3   Signs in H_0 and H_1 and Tails of a Test

	Two-Tailed Test	Left-Tailed Test	Right-Tailed Test
Sign in the null hypothesis H_0	=	= or $\geq$	= or $\leq$
Sign in the alternative hypothesis H_1	$\neq$	<	>
Rejection region	In both tails	In the left tail	In the right tail

Note that the null hypothesis always has an *equal to* (=) or a *greater than or equal to* (≥) or a *less than or equal to* (≤) sign, and the alternative hypothesis always has a *not equal to* (≠) or a *less than* (<) or a *greater than* (>) sign.

In this text we will use the following two procedures to make tests of hypothesis:

1. **The *p*-value approach.** Under this procedure, we calculate what is called the *p*-value for the observed value of the sample statistic. If we have a predetermined significance level, then we compare the *p*-value with this significance level and make a decision. Note that **here *p* stands for probability**.

2. **The critical-value approach.** In this approach, we find the critical value(s) from a table (such as the normal distribution table or the *t* distribution table) and find the value of the test statistic for the observed value of the sample statistic. Then we compare these two values and make a decision.

Remember ➤ Remember, the procedures to be learned in this chapter assume that data are obtained from a simple random sample. Also, remember that the critical point is included in the rejection region.

EXERCISES

CONCEPTS AND PROCEDURES

9.1 Briefly explain the meaning of each of the following terms.

   a. Null hypothesis   **b.** Alternative hypothesis
   c. Critical point(s)   **d.** Significance level
   e. Nonrejection region   **f.** Rejection region
   g. Tails of a test   **h.** Two types of errors

9.2 What are the four possible outcomes for a test of hypothesis? Show these outcomes by writing a table. Briefly describe the Type I and Type II errors.

9.3 Explain how the tails of a test depend on the sign in the alternative hypothesis. Describe the signs in the null and alternative hypotheses for a two-tailed, a left-tailed, and a right-tailed test, respectively.

9.4 Explain which of the following is a two-tailed test, a left-tailed test, or a right-tailed test.

   a. $H_0: \mu = 32$, $H_1: \mu > 32$
   b. $H_0: \mu = 23$, $H_1: \mu \neq 23$
   c. $H_0: \mu \geq 80$, $H_1: \mu < 80$

Show the rejection and nonrejection regions for each of these cases by drawing a sampling distribution curve for the sample mean, assuming that it is normally distributed.

9.5 Which of the two hypotheses (null and alternative) is initially assumed to be true in a test of hypothesis?

9.6 Consider $H_0: \mu = 60$ versus $H_1: \mu \neq 60$.

   a. What type of error would you make if the null hypothesis is actually false and you fail to reject it?

   b. What type of error would you make if the null hypothesis is actually true and you reject it?

APPLICATIONS

9.7 Write the null and alternative hypotheses for each of the following examples. Determine if each is a case of a two-tailed, a left-tailed, or a right-tailed test.

   a. To test if the mean number of hours spent working per week by college students who hold jobs is different from 20 hours

   b. To test whether or not a bank's ATM is out of service for an average of more than 8 hours per month

   c. To test if the mean length of experience of airport security guards is different from 5 years

   d. To test if the mean credit card debt of college seniors is less than $1000

   e. To test if the mean time a customer has to wait on the phone to speak to a representative of a mail-order company about unsatisfactory service is less than 15 minutes

9.8 Write the null and alternative hypotheses for each of the following examples. Determine if each is a case of a two-tailed, a left-tailed, or a right-tailed test.

   a. To test if the mean amount of time spent per week watching sports on television by all adult men is different from 6 hours

   b. To test if the mean amount of money spent by all customers at a supermarket is more than $80

   c. To test whether the mean starting salary of college graduates is higher than $39,000 per year

   d. To test if the mean waiting time at the drive-through window at a fast food restaurant during rush hour differs from 10 minutes

   e. To test if the mean time spent per week on house chores by all housewives is less than 30 hours

9.2 | Hypothesis Tests About μ: σ Known

This section explains how to perform a test of hypothesis for the population mean μ when the population standard deviation σ is known. As in Section 8.2 of Chapter 8, here also there are three possible cases as follows.

Case I. If the following three conditions are fulfilled:

1. The population standard deviation σ is known

2. The sample size is small (i.e., $n < 30$)
3. The population from which the sample is selected is approximately normally distributed,

then we use the normal distribution to perform a test of hypothesis about μ because from Section 7.3.1 of Chapter 7 the sampling distribution of $\bar{x}$ is normal with its mean equal to μ and the standard deviation equal to $\sigma_{\bar{x}} = \sigma/\sqrt{n}$, assuming that $n/N \leq .05$.

Case II. If the following two conditions are fulfilled:

1. The population standard deviation σ is known
2. The sample size is large (i.e., $n \geq 30$),

then, again, we use the normal distribution to perform a test of hypothesis about μ because from Section 7.3.2 of Chapter 7, due to the central limit theorem, the sampling distribution of $\bar{x}$ is (approximately) normal with its mean equal to μ and the standard deviation equal to $\sigma_{\bar{x}} = \sigma/\sqrt{n}$, assuming that $n/N \leq .05$.

Case III. If the following three conditions are fulfilled:

1. The population standard deviation σ is known
2. The sample size is small (i.e., $n < 30$)
3. The population from which the sample is selected is not normally distributed (or the shape of its distribution is unknown),

then we use a nonparametric method (explained in Chapter 15) to perform a test of hypothesis about μ.

This section will cover the first two cases. The procedure for performing a test of hypothesis about μ is the same in both these cases. Note that in Case I, the population does not have to be exactly normally distributed. As long as it is close to the normal distribution without any outliers, we can use the normal distribution procedure. In Case II, although 30 is considered a large sample, if the population distribution is very different from the normal distribution, then 30 may not be a large enough sample size for the sampling distribution of $\bar{x}$ to be normal and, hence, to use the normal distribution.

The following chart summarizes the above three cases.

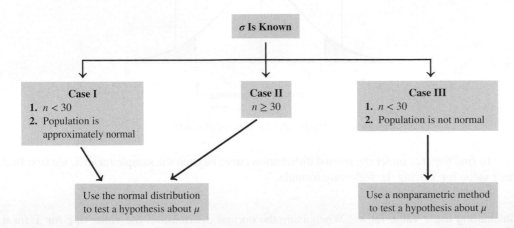

Below we explain two procedures, the p-value approach and the critical-value approach, to test hypotheses about μ under Cases I and II. We will use the normal distribution to perform such tests.

Note that the two approaches—the p-value approach and the critical-value approach—are not mutually exclusive. We do not need to use one or the other. We can use both at the same time.

9.2.1 The p-Value Approach

In this procedure, we find a probability value such that a given null hypothesis is rejected for any α (significance level) greater than or equal to this value and it is not rejected for any α less than this value. The **probability-value approach**, more commonly called the p-value approach, gives such a value. In this approach, we calculate the p-value for the test, which is defined as the smallest

Assuming that the null hypothesis is true, the *p*-value can be defined as the probability that a sample statistic (such as the sample mean) is at least as far away from the hypothesized value of the parameter in the direction of the alternative hypothesis as the one obtained from the sample data under consideration. Note that the **p-value** is the smallest significance level at which the null hypothesis is rejected.

level of significance at which the given null hypothesis is rejected. Using this *p*-value, we state the decision. If we have a predetermined value of α, then we compare the value of *p* with α and make a decision.

Using the *p*-value approach, we reject the null hypothesis if

$$p\text{-value} \leq \alpha \quad \text{or} \quad \alpha \geq p\text{-value}$$

and we do not reject the null hypothesis if

$$p\text{-value} > \alpha \quad \text{or} \quad \alpha < p\text{-value}$$

For a one-tailed test, the *p*-value is given by the area in the tail of the sampling distribution curve beyond the observed value of the sample statistic. Figure 9.5 shows the *p*-value for a right-tailed test about μ. For a left-tailed test, the *p*-value will be the area in the lower tail of the sampling distribution curve to the left of the observed value of $\bar{x}$.

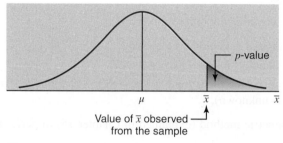

Figure 9.5 The *p*-value for a right-tailed test.

For a two-tailed test, the *p*-value is twice the area in the tail of the sampling distribution curve beyond the observed value of the sample statistic. Figure 9.6 shows the *p*-value for a two-tailed test. Each of the areas in the two tails gives one-half the *p*-value.

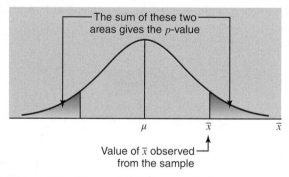

Figure 9.6 The *p*-value for a two-tailed test.

To find the area under the normal distribution curve beyond the sample mean $\bar{x}$, we first find the *z* value for $\bar{x}$ using the following formula.

Calculating the z Value for $\bar{x}$ When using the normal distribution, **the value of z for $\bar{x}$ for a test of hypothesis about μ** is computed as follows:

$$z = \frac{\bar{x} - \mu}{\sigma_{\bar{x}}} \quad \text{where} \quad \sigma_{\bar{x}} = \frac{\sigma}{\sqrt{n}}$$

The value of *z* calculated for $\bar{x}$ using this formula is also called the **observed value of z**.

Then we find the area under the tail of the normal distribution curve beyond this value of *z*. This area gives the *p*-value or one-half the *p*-value, depending on whether the test is a one-tailed test or a two-tailed test.

A test of hypothesis procedure that uses the *p*-value approach involves the following four steps.

Steps to Perform a Test of Hypothesis Using the *p*-Value Approach

1. State the null and alternative hypotheses.
2. Select the distribution to use.
3. Calculate the *p*-value.
4. Make a decision.

Examples 9–1 and 9–2 illustrate the calculation and use of the *p*-value to test a hypothesis using the normal distribution.

EXAMPLE 9–1   Learning a Food Processing Job

Performing a hypothesis test using the p-value approach for a two-tailed test with the normal distribution.

At Canon Food Corporation, it used to take an average of 90 minutes for new workers to learn a food processing job. Recently the company installed a new food processing machine. The supervisor at the company wants to find if the mean time taken by new workers to learn the food processing procedure on this new machine is different from 90 minutes. A random sample of 20 workers showed that it took, on average, 85 minutes for them to learn the food processing procedure on the new machine. It is known that the learning times for all new workers are approximately normally distributed with a population standard deviation of 7 minutes. Find the *p*-value for the test that the mean learning time for the food processing procedure on the new machine is different from 90 minutes. What will your conclusion be if $\alpha = .01$?

Solution   Let μ be the mean time (in minutes) taken to learn the food processing procedure on the new machine by all workers, and let $\bar{x}$ be the corresponding sample mean. From the given information,

$$n = 20, \quad \bar{x} = 85 \text{ minutes}, \quad \sigma = 7 \text{ minutes}, \quad \text{and} \quad \alpha = .01$$

To calculate the *p*-value and perform the test, we apply the following four steps.

Step 1.   *State the null and alternative hypotheses.*

$$H_0: \mu = 90 \text{ minutes}$$
$$H_1: \mu \neq 90 \text{ minutes}$$

Note that the null hypothesis states that the mean time for learning the food processing procedure on the new machine is 90 minutes, and the alternative hypothesis states that this time is different from 90 minutes.

Step 2.   *Select the distribution to use.*

Here, the population standard deviation σ is known, the sample size is small ($n < 30$), but the population distribution is approximately normal. Hence, the sampling distribution of $\bar{x}$ is approximately normal with its mean equal to μ and the standard deviation equal to $\sigma_{\bar{x}} = \sigma/\sqrt{n}$. Consequently, we will use the normal distribution to find the *p*-value and make the test.

Step 3.   *Calculate the p-value.*

The $\neq$ sign in the alternative hypothesis indicates that the test is two-tailed. The *p*-value is equal to twice the area in the tail of the sampling distribution curve of $\bar{x}$ to the left of $\bar{x} = 85$, as shown in Figure 9.7 on page 350. To find this area, we first find the *z* value for $\bar{x} = 85$ as follows:

$$\sigma_{\bar{x}} = \frac{\sigma}{\sqrt{n}} = \frac{7}{\sqrt{20}} = 1.56524758 \text{ minutes}$$

$$z = \frac{\bar{x} - \mu}{\sigma_{\bar{x}}} = \frac{85 - 90}{1.56524758} = -3.19$$

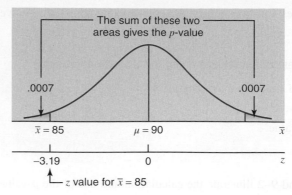

Figure 9.7   The p-value for a two-tailed test.

The area to the left of $\bar{x} = 85$ is equal to the area under the standard normal curve to the left of $z = -3.19$. From the normal distribution table, the area to the left of $z = -3.19$ is .0007.

Consequently, the p-value is

$$p\text{-value} = 2(.0007) = \mathbf{.0014}$$

Step 4.   *Make a decision.*

Thus, based on the p-value of .0014, we can state that for any α (significance level) greater than or equal to .0014, we will reject the null hypothesis stated in Step 1, and for any α less than .0014, we will not reject the null hypothesis.

Because $\alpha = .01$ is greater than the p-value of .0014, we reject the null hypothesis at this significance level. Therefore, we conclude that the mean time for learning the food processing procedure on the new machine is different from 90 minutes.

EXAMPLE 9–2   Losing Weight by Joining a Health Club

Performing a hypothesis test using the p-value approach for a one-tailed test with the normal distribution.

The management of Priority Health Club claims that its members lose an average of 10 pounds or more within the first month after joining the club. A consumer agency that wanted to check this claim took a random sample of 36 members of this health club and found that they lost an average of 9.2 pounds within the first month of membership. The population standard deviation is known to be 2.4 pounds. Find the p-value for this test. What will your decision be if $\alpha = .01$? What if $\alpha = .05$?

Solution   Let μ be the mean weight lost during the first month of membership by all members of this health club, and let $\bar{x}$ be the corresponding mean for the sample. From the given information,

$$n = 36, \quad \bar{x} = 9.2 \text{ pounds}, \quad \text{and} \quad \sigma = 2.4 \text{ pounds}$$

The claim of the club is that its members lose, on average, 10 pounds or more within the first month of membership. To perform the test using the p-value approach, we apply the following four steps.

Step 1.   *State the null and alternative hypotheses.*

$$H_0: \mu \geq 10 \quad \text{(The mean weight lost is 10 pounds or more.)}$$

© Christopher Futcher/iStockphoto

$$H_1: \mu < 10 \quad \text{(The mean weight lost is less than 10 pounds.)}$$

Step 2.   *Select the distribution to use.*

Here, the population standard deviation σ is known, and the sample size is large ($n \geq 30$). Hence, the sampling distribution of $\bar{x}$ is normal (due to the central limit theorem) with its mean equal to μ and the standard deviation equal to $\sigma_{\bar{x}} = \sigma/\sqrt{n}$. Consequently, we will use the normal distribution to find the p-value and perform the test.

Step 3.   *Calculate the p-value.*

The $<$ sign in the alternative hypothesis indicates that the test is left-tailed. The p-value is given by the area to the left of $\bar{x} = 9.2$ under the sampling distribution curve of $\bar{x}$, as shown in Figure 9.8. To find this area, we first find the z value for $\bar{x} = 9.2$ as follows:

$$\sigma_{\bar{x}} = \frac{\sigma}{\sqrt{n}} = \frac{2.4}{\sqrt{36}} = .40$$

$$z = \frac{\bar{x} - \mu}{\sigma_{\bar{x}}} = \frac{9.2 - 10}{.40} = -2.00$$

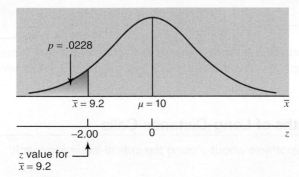

Figure 9.8   The p-value for a left-tailed test.

The area to the left of $\bar{x} = 9.2$ under the sampling distribution of $\bar{x}$ is equal to the area under the standard normal curve to the left of $z = -2.00$. From the normal distribution table, the area to the left of $z = -2.00$ is .0228. Consequently,

$$p\text{-value} = .0228$$

Step 4.   *Make a decision.*

Thus, based on the p-value of .0228, we can state that for any α (significance level) greater than or equal to .0228 we will reject the null hypothesis stated in Step 1, and for any α less than .0228 we will not reject the null hypothesis.

Since $\alpha = .01$ is less than the p-value of .0228, we do not reject the null hypothesis at this significance level. Consequently, we conclude that there is not significant evidence to reject the claim that the mean weight lost within the first month of membership by the members of this club is 10 pounds or more.

Now, because $\alpha = .05$ is greater than the p-value of .0228, we reject the null hypothesis at this significance level. In this case we conclude that the mean weight lost within the first month of membership by the members of this club is less than 10 pounds.

9.2.2   The Critical-Value Approach

In this procedure, we have a predetermined value of the significance level α. The value of α gives the total area of the rejection region(s). First we find the critical value(s) of z from the normal distribution table for the given significance level. Then we find the value of the test statistic z for the observed value of the sample statistic $\bar{x}$. Finally we compare these two values and make a decision. Remember, if the test is one-tailed, there is only one critical value of z, and it is obtained by using the value of α which gives the area in the left or right tail of the normal distribution curve depending on whether the test is left-tailed or right-tailed, respectively. However, if the test is two-tailed, there are two critical values of z and they are obtained by using $\alpha/2$ area in each tail of the normal distribution curve. The value of the test statistic is obtained as follows.

Test Statistic   In tests of hypotheses about μ using the normal distribution, the random variable

$$z = \frac{\bar{x} - \mu}{\sigma_{\bar{x}}} \quad \text{where} \quad \sigma_{\bar{x}} = \frac{\sigma}{\sqrt{n}}$$

is called the **test statistic**. The test statistic can be defined as a rule or criterion that is used to make the decision on whether or not to reject the null hypothesis.

A test of hypothesis procedure that uses the critical-value approach involves the following five steps.

> ### Steps to Perform a Test of Hypothesis with the Critical-Value Approach
>
> **1.** State the null and alternative hypotheses.
> **2.** Select the distribution to use.
> **3.** Determine the rejection and nonrejection regions.
> **4.** Calculate the value of the test statistic.
> **5.** Make a decision.

Examples 9–3 and 9–4 illustrate the use of these five steps to perform tests of hypotheses about the population mean μ. Example 9–3 is concerned with a two-tailed test, and Example 9–4 describes a one-tailed test.

EXAMPLE 9–3 Lengths of Long-Distance Calls

Conducting a two-tailed test of hypothesis about μ using the critical-value approach: σ known and $n \geq 30$.

The TIV Telephone Company provides long-distance telephone service in an area. According to the company's records, the average length of all long-distance calls placed through this company in 2015 was 12.44 minutes. The company's management wanted to check if the mean length of the current long-distance calls is different from 12.44 minutes. A sample of 150 such calls placed through this company produced a mean length of 13.71 minutes. The standard deviation of all such calls is 2.65 minutes. Using a 2% significance level, can you conclude that the mean length of all current long-distance calls is different from 12.44 minutes?

Solution Let μ be the mean length of all current long-distance calls placed through this company and $\bar{x}$ be the corresponding mean for the sample. From the given information,

$$n = 150, \quad \bar{x} = 13.71 \text{ minutes}, \quad \text{and} \quad \sigma = 2.65 \text{ minutes}$$

We are to test whether or not the mean length of all current long-distance calls is different from 12.44 minutes. The significance level α is .02; that is, the probability of rejecting the null hypothesis when it actually is true should not exceed .02. This is the probability of making a Type I error. We perform the test of hypothesis using the five steps as follows.

Step 1. *State the null and alternative hypotheses.*

Notice that we are testing to find whether or not the mean length of all current long-distance calls is different from 12.44 minutes. We write the null and alternative hypotheses as follows.

H_0: $\mu = 12.44$ (The mean length of all current long-distance calls is 12.44 minutes.)

H_1: $\mu \neq 12.44$ (The mean length of all current long-distance calls is different from 12.44 minutes.)

Step 2. *Select the distribution to use.*

Here, the population standard deviation σ is known, and the sample size is large ($n \geq 30$). Hence, the sampling distribution of $\bar{x}$ is (approximately) normal (due to the central limit theorem) with its mean equal to μ and the standard deviation equal to $\sigma_{\bar{x}} = \sigma/\sqrt{n}$. Consequently, we will use the normal distribution to perform the test.

Step 3. *Determine the rejection and nonrejection regions.*

The significance level is .02. The $\neq$ sign in the alternative hypothesis indicates that the test is two-tailed with two rejection regions, one in each tail of the normal distribution curve of $\bar{x}$. Because the total area of both rejection regions is .02 (the significance level), the area of the rejection region in each tail is .01; that is,

$$\text{Area in each tail} = \alpha/2 = .02/2 = .01$$

These areas are shown in Figure 9.9. Two critical points in this figure separate the two rejection regions from the nonrejection region. Next, we find the z values for the two critical points using the area of the rejection region. To find the z values for these critical points, we look for .0100 and .9900 areas in the normal distribution table. From Table IV of Appendix B, the z values of the two critical points, as shown in Figure 9.9, are approximately -2.33 and 2.33.

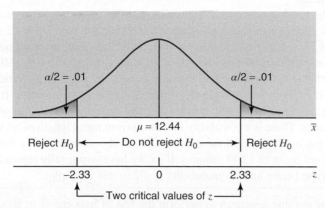

Figure 9.9   Rejection and nonrejection regions.

Step 4.   *Calculate the value of the test statistic.*

The decision to reject or not to reject the null hypothesis will depend on whether the evidence from the sample falls in the rejection or the nonrejection region. If the value of $\bar{x}$ falls in either of the two rejection regions, we reject H_0. Otherwise, we do not reject H_0. The value of $\bar{x}$ obtained from the sample is called the *observed value of* $\bar{x}$. To locate the position of $\bar{x} = 13.71$ on the sampling distribution curve of $\bar{x}$ in Figure 9.9, we first calculate the z value for $\bar{x} = 13.71$. This is called the *value of the test statistic*. Then, we compare the value of the test statistic with the two critical values of z, -2.33 and 2.33, shown in Figure 9.9. If the value of the test statistic is between -2.33 and 2.33, we do not reject H_0. If the value of the test statistic is either greater than or equal to 2.33 or less than or equal to -2.33, we reject H_0.

Calculating the Value of the Test Statistic   When using the normal distribution, **the value of the test statistic** z for $\bar{x}$ for a test of hypothesis about μ is computed as follows:

$$z = \frac{\bar{x} - \mu}{\sigma_{\bar{x}}}$$

where

$$\sigma_{\bar{x}} = \frac{\sigma}{\sqrt{n}}$$

This value of z for $\bar{x}$ is also called the **observed value of** z.

The value of $\bar{x}$ from the sample is 13.71. We calculate the z value as follows:

$$\sigma_{\bar{x}} = \frac{\sigma}{\sqrt{n}} = \frac{2.65}{\sqrt{150}} = .21637159$$

$$z = \frac{\bar{x} - \mu}{\sigma_{\bar{x}}} = \frac{13.71 - 12.44}{.21637159} = 5.87 \qquad \text{From } H_0$$

The value of μ in the calculation of the z value is substituted from the null hypothesis. The value of $z = 5.87$ calculated for $\bar{x}$ is called the *computed value of the test statistic* z. This is the value of z that corresponds to the value of $\bar{x}$ observed from the sample. It is also called the *observed value of z*.

Step 5.   *Make a decision.*

In the final step we make a decision based on the location of the value of the test statistic z computed for $\bar{x}$ in Step 4. This value of $z = 5.87$ is greater than the critical value of $z = 2.33$, and it falls in the rejection region in the right tail in Figure 9.9. Hence, we reject H_0 and conclude that based on the sample information, it appears that the current mean length of all such calls is not equal to 12.44 minutes.

By rejecting the null hypothesis, we are stating that the difference between the sample mean, $\bar{x} = 13.71$ minutes, and the hypothesized value of the population mean, $\mu = 12.44$ minutes, is too large and may not have occurred because of chance or sampling error alone. This difference seems to be real and, hence, the mean length of all such calls is currently different from 12.44 minutes. Note that the rejection of the null hypothesis does not necessarily indicate that the mean length of all such calls is currently definitely different from 12.44 minutes. It simply indicates that there is strong evidence (from the sample) that the current mean length of such calls is not equal to 12.44 minutes. There is a possibility that the current mean length of all such calls is equal to 12.44 minutes, but by the luck of the draw we selected a sample with a mean that is too far from the hypothesized mean of 12.44 minutes. If so, we have wrongfully rejected the null hypothesis H_0. This is a Type I error and its probability is .02 in this example.

We can use the *p*-value approach to perform the test of hypothesis in Example 9–3. In this example, the test is two-tailed. The *p*-value is equal to twice the area under the sampling distribution of $\bar{x}$ to the right of $\bar{x} = 13.71$. As calculated in Step 4 above, the z value for $\bar{x} = 13.71$ is 5.87. From the normal distribution table, the area to the right of $z = 5.87$ is (approximately) zero. Hence, the *p*-value is (approximately) zero. (If you use technology, you will obtain the *p*-value of .000000002.) As we know from earlier discussions, we will reject the null hypothesis for any α (significance level) that is greater than or equal to the *p*-value. Consequently, in this example, we will reject the null hypothesis for any $\alpha > 0$ approximately. Since $\alpha = .02$ here, which is greater than zero, we reject the null hypothesis.

EXAMPLE 9–4   Net Worth of Families in a City

Conducting a left-tailed test of hypothesis about μ using the critical-value approach: σ known, $n < 30$, and population normal.

The mayor of a large city claims that the average net worth of families living in this city is at least \$300,000. A random sample of 25 families selected from this city produced a mean net worth of \$288,000. Assume that the net worths of all families in this city have an approximate normal distribution with the population standard deviation of \$80,000. Using a 2.5% significance level, can you conclude that the mayor's claim is false?

Solution   Let μ be the mean net worth of families living in this city and $\bar{x}$ be the corresponding mean for the sample. From the given information,

$$n = 25, \quad \bar{x} = \$288{,}000, \quad \text{and} \quad \sigma = \$80{,}000$$

The significance level is $\alpha = .025$.

Step 1.   *State the null and alternative hypotheses.*

We are to test whether or not the mayor's claim is false. The mayor's claim is that the average net worth of families living in this city is at least \$300,000. Hence, the null and alternative hypotheses are as follows:

$$H_0: \mu \geq \$300{,}000 \quad \text{(The mayor's claim is true. The mean net worth is at least \$300,000.)}$$

$$H_1: \mu < \$300{,}000 \quad \text{(The mayor's claim is false. The mean net worth is less than \$300,000.)}$$

Step 2.   *Select the distribution to use.*

Here, the population standard deviation σ is known, the sample size is small ($n < 30$), but the population distribution is approximately normal. Hence, the sampling distribution of $\bar{x}$ is approximately normal with its mean equal to μ and the standard deviation equal to $\sigma_{\bar{x}} = \sigma/\sqrt{n}$. Consequently, we will use the normal distribution to perform the test.

Step 3.   *Determine the rejection and nonrejection regions.*

The significance level is .025. The < sign in the alternative hypothesis indicates that the test is left-tailed with the rejection region in the left tail of the sampling distribution curve of $\bar{x}$. The critical value of z, obtained from the normal table for .0250 area in the left tail, is −1.96, as shown in Figure 9.10.

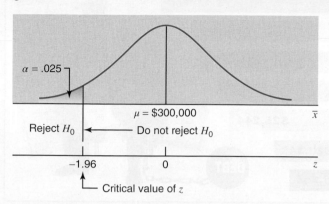

Figure 9.10   Rejection and nonrejection regions.

Step 4.   *Calculate the value of the test statistic.*

The value of the test statistic z for $\bar{x} = \$288{,}000$ is calculated as follows:

$$\sigma_{\bar{x}} = \frac{\sigma}{\sqrt{n}} = \frac{80{,}000}{\sqrt{25}} = \$16{,}000$$

$$z = \frac{\bar{x} - \mu}{\sigma_{\bar{x}}} = \frac{288{,}000 - 300{,}000}{16{,}000} = -.75$$

From H_0

Step 5.   *Make a decision.*

The value of the test statistic $z = -.75$ is greater than the critical value of $z = -1.96$, and it falls in the nonrejection region. As a result, we fail to reject H_0. Therefore, we can state that based on the sample information, it appears that the mean net worth of families in this city is not less than \$300,000. Note that we are not concluding that the mean net worth is definitely not less than \$300,000. By not rejecting the null hypothesis, we are saying that the information obtained from the sample is not strong enough to reject the null hypothesis and to conclude that the mayor's claim is false.

We can use the *p*-value approach to perform the test of hypothesis in Example 9–4. In this example, the test is left-tailed. The *p*-value is given by the area under the sampling distribution of $\bar{x}$ to the left of $\bar{x} = \$288{,}000$. As calculated in Step 4 above, the z value for $\bar{x} = \$288{,}000$ is −.75. From the normal distribution table, the area to the left of $z = -.75$ is .2266. Hence, the *p*-value is .2266. We will reject the null hypothesis for any α (significance level) that is greater than or equal to the *p*-value. Consequently, we will reject the null hypothesis in this example for any $\alpha \geq .2266$. Since in this example $\alpha = .025$, which is less than .2266, we fail to reject the null hypothesis.

In studies published in various journals, authors usually use the terms **significantly different** and **not significantly different** when deriving conclusions based on hypothesis tests. These terms are short versions of the terms *statistically significantly different* and *statistically not significantly different*. The expression *significantly different* means that the difference between the observed value of the sample mean $\bar{x}$ and the hypothesized value of the population mean μ is so large that it probably did not occur because of the sampling error alone. Consequently, the null hypothesis is rejected. In other words, the difference between $\bar{x}$ and μ is statistically significant. Thus, the statement *significantly different* is equivalent to saying that the *null hypothesis is rejected*. In Example 9–3, we can state as a conclusion that the observed value of $\bar{x} = 13.71$ minutes is significantly different from the hypothesized value of $\mu = 12.44$ minutes. That is, the mean length of all current long-distance calls is different from 12.44 minutes.

Average Student Loan Debt for the Class of 2013

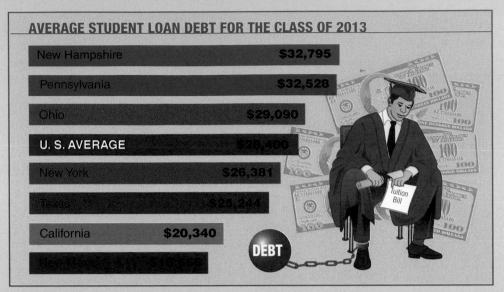

AVERAGE STUDENT LOAN DEBT FOR THE CLASS OF 2013

New Hampshire	$32,795
Pennsylvania	$32,528
Ohio	$29,090
U.S. AVERAGE	$28,400
New York	$26,381
Texas	$25,244
California	$20,340
New Mexico	$18,656

Data source: The Project on Student Debt, The Institute for College Access & Success

According to estimates by the Project on Student Debt study, 69% of college students who graduated in 2013 had loans to pay. The accompanying chart lists the U.S. average and the averages of such debts for a few selected states for students in the class of 2013 who graduated with loans. Remember that each of these averages is based on a survey of students who graduated in 2013 with debt. For example, U.S. college graduates in 2013 owed an average of $28,400. Students who graduated with loans from colleges in New Hampshire in 2013 had the highest average debt of $32,795, and those who graduated with loans from colleges in New Mexico had the lowest average debt of $18,656. Note that these averages are based on sample surveys. Suppose we want to find out if the average debt for students in Ohio was higher than the U.S. average of $28,400. Suppose that the average debt for Ohio college students in the class of 2013 is based on a random sample of 900 students who graduated with loans. Assume that the standard deviation of the loans for all Ohio students in the class of 2013 was $4800 and the significance level is 1%. The test is right-tailed because we are testing the hypothesis that the average debt for the class of 2013 in Ohio (which was $29,090) was higher than $28,400. The null and alternative hypotheses are

$$H_0: \mu = \$28,400$$

$$H_1: \mu > \$28,400$$

Here, $n = 900$, $\bar{x} = \$29,090$, $\sigma = \$4800$, and $\alpha = .01$. The population standard deviation is known, and the sample is large. Hence, we can use the normal distribution to perform this test. Using the normal distribution to perform the test, we find that the critical value of z is 2.33 for .01 area in the right tail of the normal curve. We find the observed value of z as follows.

$$\sigma_{\bar{x}} = \frac{\sigma}{\sqrt{n}} = \frac{4800}{\sqrt{900}} = \$160$$

$$z = \frac{\bar{x} - \mu}{\sigma_{\bar{x}}} = \frac{29,090 - 28,400}{160} = 4.31$$

The value of the test statistic $z = 4.31$ for $\bar{x}$ is larger than the critical value of $z = 2.33$, and it falls in the rejection region. Consequently, we reject H_0 and conclude that the average debt of students who graduated in 2013 from colleges in Ohio is higher than $28,400, which is the average for the United States.

To use the p-value approach, we find the area under the normal curve to the right of $z = 4.31$ from the normal distribution table. This area is (approximately) .0000. Therefore, the p-value is .0000. Since $\alpha = .01$ is larger than the p-value = .0000, we reject the null hypothesis.

Source: http://projectonstudentdebt.org/files/pub/classof2013.pdf.

On the other hand, the statement *not significantly different* means that the difference between the observed value of the sample mean $\bar{x}$ and the hypothesized value of the population mean μ is so small that it may have occurred just because of chance. Consequently, the null hypothesis is not rejected. Thus, the expression *not significantly different* is equivalent to saying that we *fail to reject the null hypothesis*. In Example 9–4, we can state as a conclusion that the observed value of $\bar{x} = \$288,000$ is not significantly less than the hypothesized value of $\mu = \$300,000$. In other words, the current mean net worth of households in this city is not less than $\$300,000$.

Note that here *significant* does not mean that results are of practical importance to us. Significant here is related to rejecting or not rejecting the null hypothesis.

EXERCISES

CONCEPTS AND PROCEDURES

9.9 What are the five steps of a test of hypothesis using the critical value approach? Explain briefly.

9.10 What does the level of significance represent in a test of hypothesis? Explain.

9.11 By rejecting the null hypothesis in a test of hypothesis example, are you stating that the alternative hypothesis is true?

9.12 What is the difference between the critical value of z and the observed value of z?

9.13 Briefly explain the procedure used to calculate the p-value for a two-tailed and for a one-tailed test, respectively.

9.14 Find the p-value for each of the following hypothesis tests.

   a. H_0: $\mu = 46$, H_1: $\mu \neq 46$, $n = 40$, $\bar{x} = 49.60$, $\sigma = 9.7$

   b. H_0: $\mu = 26$, H_1: $\mu < 26$, $n = 33$, $\bar{x} = 24.30$, $\sigma = 4.3$

   c. H_0: $\mu = 18$, H_1: $\mu > 18$, $n = 55$, $\bar{x} = 20.50$, $\sigma = 7.8$

9.15 Consider H_0: $\mu = 29$ versus H_1: $\mu \neq 29$. A random sample of 25 observations taken from this population produced a sample mean of 25.3. The population is normally distributed with $\sigma = 8$.

   a. Calculate the p-value.

   b. Considering the p-value of part a, would you reject the null hypothesis if the test were made at a significance level of .05?

   c. Considering the p-value of part a, would you reject the null hypothesis if the test were made at a significance level of .01?

9.16 For each of the following examples of tests of hypotheses about μ, show the rejection and nonrejection regions on the sampling distribution of the sample mean assuming it is normal.

   a. A two-tailed test with $\alpha = .01$ and $n = 100$

   b. A left-tailed test with $\alpha = .005$ and $n = 27$

   c. A right-tailed test with $\alpha = .025$ and $n = 36$

9.17 Consider the following null and alternative hypotheses:

$$H_0: \mu = 25 \quad \text{versus} \quad H_1: \mu \neq 25$$

Suppose you perform this test at $\alpha = .05$ and reject the null hypothesis. Would you state that the difference between the hypothesized value of the population mean and the observed value of the sample mean is "statistically significant" or would you state that this difference is "statistically not significant?" Explain.

9.18 Consider the following null and alternative hypotheses:

$$H_0: \mu = 30 \quad \text{versus} \quad H_1: \mu \neq 30$$

Suppose you perform this test at $\alpha = .05$ and reject the null hypothesis. Would you state that the difference between the hypothesized value of the population mean and the observed value of the sample mean is "statistically significant" or would you state that this difference is "statistically not significant"? Explain.

9.19 For each of the following significance levels, what is the probability of making a Type I error?

   a. $\alpha = .03$ **b.** $\alpha = .06$ **c.** $\alpha = .02$ **d.** $\alpha = .005$

9.20 A random sample of 80 observations produced a sample mean of 88.50. Find the critical and observed values of z for each of the following tests of hypothesis using $\alpha = .10$. The population standard deviation is known to be 6.40.

   a. H_0: $\mu = 91$ versus H_1: $\mu \neq 91$

   b. H_0: $\mu = 91$ versus H_1: $\mu < 91$

9.21 Consider the null hypothesis H_0: $\mu = 625$. Suppose that a random sample of 32 observations is taken from a normally distributed population with $\sigma = 29$. Using a significance level of .01, show the rejection and nonrejection regions on the sampling distribution curve of the sample mean and find the critical value(s) of z when the alternative hypothesis is as follows.

   a. H_1: $\mu \neq 625$ **b.** H_1: $\mu > 625$ **c.** H_1: $\mu < 625$

9.22 Consider H_0: $\mu = 100$ versus H_1: $\mu \neq 100$.

   a. A random sample of 64 observations produced a sample mean of 96. Using $\alpha = .01$, would you reject the null hypothesis? The population standard deviation is known to be 12.

   b. Another random sample of 64 observations taken from the same population produced a sample mean of 105. Using $\alpha = .01$, would you reject the null hypothesis? The population standard deviation is known to be 12.

Comment on the results of parts a and b.

9.23 Consider the null hypothesis H_0: $\mu = 8$. A random sample of 150 observations is taken from a population with $\sigma = 17$. Using $\alpha = .05$, show the rejection and nonrejection regions on the sampling distribution curve of the sample mean and find the critical value(s) of z for the following.

   a. a right-tailed test   **b.** a left-tailed test   **c.** a two-tailed test

9.24 Make the following tests of hypotheses.

   a. H_0: $\mu = 25$, H_1: $\mu \neq 25$, $n = 81$, $\bar{x} = 28.5$, $\sigma = 3$, $\alpha = .01$

   b. H_0: $\mu = 12$, H_1: $\mu < 12$, $n = 45$, $\bar{x} = 11.25$, $\sigma = 4.5$, $\alpha = .05$

   c. H_0: $\mu = 40$, H_1: $\mu > 40$, $n = 100$, $\bar{x} = 47$, $\sigma = 7$, $\alpha = .10$

APPLICATIONS

9.25 The manufacturer of a certain brand of auto batteries claims that the mean life of these batteries is 45 months. A consumer protection agency that wants to check this claim took a random sample of 20 such batteries and found that the mean life for this sample is 43.05 months. The lives of all such batteries have a normal distribution with the population standard deviation of 6.5 months.

a. Find the p-value for the test of hypothesis with the alternative hypothesis that the mean life of these batteries is less than 45 months. Will you reject the null hypothesis at $\alpha = .025$?

b. Test the hypothesis of part a using the critical-value approach and $\alpha = .025$.

9.26 According to the U.S. Bureau of Labor Statistics, all workers in America who had a bachelor's degree and were employed earned an average of \$1038 a week in 2010. A sample of 400 American workers who have a bachelor's degree showed that they earn an average of \$1060 per week. Suppose that the population standard deviation of such earnings is \$160.

a. Find the p-value for the test of hypothesis with the alternative hypothesis that the current mean weekly earning of American workers who have a bachelor's degree is higher than \$1038. Will you reject the null hypothesis at $\alpha = .025$?

b. Test the hypothesis of part a using the critical-value approach and $\alpha = .025$.

9.27 A study claims that all adults spend an average of 14 hours or more on chores during a weekend. A researcher wanted to check if this claim is true. A random sample of 200 adults taken by this researcher showed that these adults spend an average of 13.85 hours on chores during a weekend. The population standard deviation is known to be 3.0 hours.

a. Find the p-value for the hypothesis test with the alternative hypothesis that all adults spend more than 14 hours on chores during a weekend. Will you reject the null hypothesis at $\alpha = .01$?

b. Test the hypothesis of part a using the critical-value approach and $\alpha = .01$?

9.28 According to the U.S. Postal Service, the average weight of mail received by Americans in 2011 through the Postal Service was 57.2 pounds (*The New York Times*, December 4, 2011). One hundred randomly selected Americans were asked to keep all their mail for last year. It was found that they received an average of 56.1 pounds of mail last year. Suppose that the population standard deviation is 9.2 pounds.

a. Find the p-value for the test of hypothesis with the alternative hypothesis that the average weight of mail received by all Americans last year was less than 57.2 pounds. Will you reject the null hypothesis at $\alpha = .01$? Explain. What if $\alpha = .025$?

b. Test the hypothesis of part a using the critical-value approach. Will you reject the null hypothesis at $\alpha = .01$? What if $\alpha = .025$?

9.29 Records in a three-county area show that in the last few years, Girl Scouts sell an average of 45.38 boxes of cookies per year, with a population standard deviation of 9.15 boxes per year. Fifty randomly selected Girl Scouts from the region sold an average of 44.54 boxes this year. Scout leaders are concerned that the demand for Girl Scout cookies may have decreased.

a. Test at a 10% significance level whether the average number of boxes of cookies sold by all Girl Scouts in the three-county area is lower than the historical average.

b. What will your decision be in part a if the probability of a Type I error is zero? Explain.

9.30 At Farmer's Dairy, a machine is set to fill 32-ounce milk cartons. However, this machine does not put exactly 32 ounces of milk into each carton; the amount varies slightly from carton to carton but has an approximate normal distribution. It is known that when the machine is working properly, the mean net weight of these cartons is 32 ounces. The standard deviation of the milk in all such cartons is always equal to .15 ounce. The quality control inspector at this company takes a sample of 25 such cartons every week, calculates the mean net weight of these cartons, and tests the null hypothesis, $\mu = 32$ ounces, against the alternative hypothesis, $\mu \neq 32$ ounces. If the null hypothesis is rejected, the machine is stopped and adjusted. A recent sample of 25 such cartons produced a mean net weight of 31.93 ounces.

a. Calculate the p-value for this test of hypothesis. Based on this p-value, will the quality control inspector decide to stop the machine and adjust it if she chooses the maximum probability of a Type I error to be .01? What if the maximum probability of a Type I error is .05?

b. Test the hypothesis of part a using the critical-value approach and $\alpha = .01$. Does the machine need to be adjusted? What if $\alpha = .05$?

9.31 According to Moebs Services Inc., an individual checking account at major U.S. banks costs these banks between \$350 and \$450 per year (*Time*, November 21, 2011). Suppose that the average cost of individual checking accounts at major U.S. banks was \$400 for the year 2011. A bank consultant wants to determine whether the current mean cost of such checking accounts at major U.S. banks is more than \$400 a year. A recent random sample of 150 such checking accounts taken from major U.S. banks produced a mean annual cost to them of \$390. Assume that the standard deviation of annual costs to major banks of all such checking accounts is \$50.

a. Find the p-value for the test of hypothesis. Based on this p-value, would you reject the null hypothesis if the maximum probability of Type I error is to be .05? What if the maximum probability of Type I error is to be .01?

b. Test the hypothesis of part a using the critical-value approach and $\alpha = .05$. Would you reject the null hypothesis? What if $\alpha = .01$? What if $\alpha = 0$?

9.32 According to a survey by the College Board, undergraduate students at private nonprofit four-year colleges spent an average of \$1244 on books and supplies in 2014–2015 (www.collegeboard.org). A recent random sample of 200 undergraduate college students from a large private nonprofit four-year college showed that they spent an average of \$1204 on books and supplies during the last academic year. Assume that the standard deviation of annual expenditures on books and supplies by all such students at this college is \$200.

a. Find the p-value for the test of hypothesis with the alternative hypothesis that the annual mean expenditure by all such students at this college is less than \$1244. Based on this p-value, would you reject the null hypothesis if the significance level is .025?

b. Test the hypothesis of part a using the critical-value approach and $\alpha = .025$. Would you reject the null hypothesis? Explain.

9.33 A journalist claims that all adults in her city spend an average of 25 hours or more per month on general reading, such as newspapers, magazines, novels, and so forth. A recent sample of 36 adults from this city showed that they spend an average of 27 hours per month on general reading. The population of such times is known to be normally distributed with the population standard deviation of 7 hours.

  a. Using a 2.5% significance level, would you conclude that the mean time spent per month on such reading by all adults in

this city is less than 30 hours? Use both procedures—the *p*-value approach and the critical-value approach.

  b. Make the test of part a using a 1% significance level. Is your decision different from that of part a? Comment on the results of parts a and b.

9.3 | Hypothesis Tests About μ: σ Not Known

This section explains how to perform a test of hypothesis about the population mean μ when the population standard deviation σ is not known. Here, again, there are three possible cases as follows.

Case I. If the following three conditions are fulfilled:

1. The population standard deviation σ is not known
2. The sample size is small (i.e., $n < 30$)
3. The population from which the sample is selected is approximately normally distributed,

then we use the *t* distribution to perform a test of hypothesis about μ.

Case II. If the following two conditions are fulfilled:

1. The population standard deviation σ is not known
2. The sample size is large (i.e., $n \geq 30$),

then again we use the *t* distribution to perform a test of hypothesis about μ.

Case III. If the following three conditions are fulfilled:

1. The population standard deviation σ is not known
2. The sample size is small (i.e., $n < 30$)
3. The population from which the sample is selected is not normally distributed (or the shape of its distribution is unknown),

then we use a nonparametric method to perform a test of hypothesis about μ.

  The following chart summarizes the above three cases.

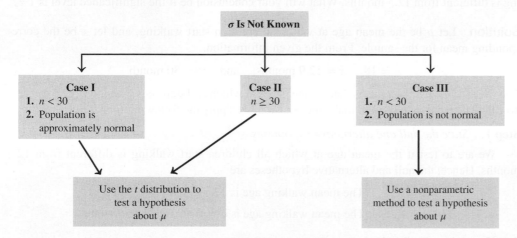

Below we discuss Cases I and II and learn how to use the *t* distribution to perform a test of hypothesis about μ when σ is not known. When the conditions mentioned for Case I or Case II are satisfied, the random variable

$$t = \frac{\bar{x} - \mu}{s_{\bar{x}}} \qquad \text{where} \qquad s_{\bar{x}} = \frac{s}{\sqrt{n}}$$

has a *t* distribution. Here, the *t* is called the **test statistic** to perform a test of hypothesis about a population mean μ.

Test Statistic   The value of the **test statistic** t for the sample mean $\bar{x}$ is computed as

$$t = \frac{\bar{x} - \mu}{s_{\bar{x}}} \quad \text{where} \quad s_{\bar{x}} = \frac{s}{\sqrt{n}}$$

The value of t calculated for $\bar{x}$ by using this formula is also called the **observed value** of t.

In Section 9.2, we discussed two procedures, the p-value approach and the critical-value approach, to test hypotheses about μ when σ is known. In this section also we will use these two procedures to test hypotheses about μ when σ is not known. The steps used in these procedures are the same as in Section 9.2. The only difference is that we will be using the t distribution in place of the normal distribution.

9.3.1   The p-Value Approach

To use the p-value approach to perform a test of hypothesis about μ using the t distribution, we will use the same four steps that we used in such a procedure in Section 9.2.1. Although the p-value can be obtained very easily by using technology, we can use Table V of Appendix B to find a **range for the p-value** when technology is not available. Note that when using the t distribution and Table V, we cannot find the exact p-value but only a range within which it falls.

Examples 9–5 and 9–6 illustrate the p-value procedure to test a hypothesis about μ using the t distribution.

EXAMPLE 9–5   **Age at Which Children Start Walking**

Finding a p-value and making a decision for a two-tailed test of hypothesis about μ: σ not known, n < 30, and population normal.

A psychologist claims that the mean age at which children start walking is 12.5 months. Carol wanted to check if this claim is true. She took a random sample of 18 children and found that the mean age at which these children started walking was 12.9 months with a standard deviation of .80 month. It is known that the ages at which all children start walking are approximately normally distributed. Find the p-value for the test that the mean age at which all children start walking is different from 12.5 months. What will your conclusion be if the significance level is 1%?

Solution   Let μ be the mean age at which all children start walking, and let $\bar{x}$ be the corresponding mean for the sample. From the given information,

$$n = 18, \quad \bar{x} = 12.9 \text{ months}, \quad \text{and} \quad s = .80 \text{ month}$$

The claim of the psychologist is that the mean age at which children start walking is 12.5 months. To calculate the p-value and to make the decision, we apply the following four steps.

Step 1.   *State the null and alternative hypotheses.*

We are to test if the mean age at which all children start walking is different from 12.5 months. Hence, the null and alternative hypotheses are

H_0: $\mu = 12.5$ (The mean walking age is 12.5 months.)

H_1: $\mu \neq 12.5$ (The mean walking age is different from 12.5 months.)

Step 2.   *Select the distribution to use.*

In this example, we do not know the population standard deviation σ, the sample size is small ($n < 30$), and the population is approximately normally distributed. Hence, it is Case I mentioned in the beginning of this section. Consequently, we will use the t distribution to find the p-value for this test.

Step 3.   *Calculate the p-value.*

The $\neq$ sign in the alternative hypothesis indicates that the test is two-tailed. To find the p-value, first we find the degrees of freedom and the t value for $\bar{x} = 12.9$ months. Then, the

p-value is equal to twice the area in the tail of the t distribution curve beyond this t value for $\bar{x} = 12.9$ months. This p-value is shown in Figure 9.11. We find this p-value as follows:

$$s_{\bar{x}} = \frac{s}{\sqrt{n}} = \frac{.80}{\sqrt{18}} = .18856181$$

$$t = \frac{\bar{x} - \mu}{s_{\bar{x}}} = \frac{12.9 - 12.5}{.18856181} = 2.121 \qquad \text{From } H_0$$

and

$$df = n - 1 = 18 - 1 = 17$$

Now we can find the range for the p-value. To do so, we go to Table V of Appendix B (the t distribution table) and find the row of $df = 17$. In this row, we find the two values of t that cover $t = 2.121$. From Table V, for $df = 17$, these two values of t are 2.110 and 2.567. The test statistic $t = 2.121$ falls between these two values. Now look in the top row of this table to find the areas in the tail of the t distribution curve that correspond to 2.110 and 2.567. These two areas are .025 and .01, respectively. In other words, the area in the upper tail of the t distribution curve for $df = 17$ and $t = 2.110$ is .025, and the area in the upper tail of the t distribution curve for $df = 17$ and $t = 2.567$ is .01. Because it is a two-tailed test, the p-value for $t = 2.121$ is between $2(.025) = .05$ and $2(.01) = .02$, which can be written as

$$.02 < p\text{-value} < .05$$

Note that by using Table V of Appendix B, we cannot find the exact p-value but only a range for it. If we have access to technology, we can find the exact p-value. If we use technology for this example, we will obtain a p-value of .049.

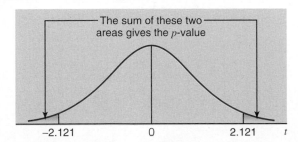

Figure 9.11 The required p-value.

Step 4. *Make a decision.*

Thus, we can state that for any α greater than or equal to .05 (the upper limit of the p-value range), we will reject the null hypothesis. For any α less than or equal to .02 (the lower limit of the p-value range), we will not reject the null hypothesis. However, if α is between .02 and .05, we cannot make a decision. Note that if we use technology, then the p-value we will obtain for this example is .049, and we can make a decision for any value of α. For our example, $\alpha = .01$, which is less than the lower limit of the p-value range of .02. As a result, we fail to reject H_0 and conclude that the mean age at which all children start walking is not significantly different from 12.5 months. As a result, we can state that the difference between the hypothesized population mean and the sample mean is so small that it may have occurred because of sampling error.

EXAMPLE 9–6   Life of Batteries

Finding a p-value and making a decision for a left-tailed test of hypothesis about μ: σ not known and $n \geq 30$.

Grand Auto Corporation produces auto batteries. The company claims that its top-of-the-line Never Die batteries are good, on average, for at least 65 months. A consumer protection agency tested 45 such batteries to check this claim. It found that the mean life of these 45 batteries is 63.4 months, and the standard deviation is 3 months. Find the p-value for the test that the mean life of all such batteries is less than 65 months. What will your conclusion be if the significance level is 2.5%?

Solution   Let μ be the mean life of all such auto batteries, and let $\bar{x}$ be the corresponding mean for the sample. From the given information,

$$n = 45, \quad \bar{x} = 63.4 \text{ months}, \quad \text{and} \quad s = 3 \text{ months}$$

The claim of the company is that the mean life of these batteries is at least 65 months. To calculate the p-value and make the decision, we apply the following four steps.

Step 1.   *State the null and alternative hypotheses.*

We are to test if the mean life of these batteries is at least 65 months. Hence, the null and alternative hypotheses are

$$H_0: \mu \geq 65 \qquad \text{(The mean life of batteries is at least 65 months.)}$$
$$H_1: \mu < 65 \qquad \text{(The mean life of batteries is less than 65 months.)}$$

Step 2.   *Select the distribution to use.*

In this example, we do not know the population standard deviation σ, and the sample size is large ($n \geq 30$). Hence, it is Case II mentioned in the beginning of this section. Consequently, we will use the t distribution to find the p-value for this test.

Step 3.   *Calculate the p-value.*

The $<$ sign in the alternative hypothesis indicates that the test is left-tailed. To find the p-value, first we find the degrees of freedom and the t value for $\bar{x} = 63.4$ months. Then, the p-value is given by the area in the tail of the t distribution curve beyond this t value for $\bar{x} = 63.4$ months. This p-value is shown in Figure 9.12. We find this p-value as follows:

$$s_{\bar{x}} = \frac{s}{\sqrt{n}} = \frac{3}{\sqrt{45}} = .44721360$$

From H_0

$$t = \frac{\bar{x} - \mu}{s_{\bar{x}}} = \frac{63.4 - 65}{.44721360} = -3.578$$

and

$$df = n - 1 = 45 - 1 = 44$$

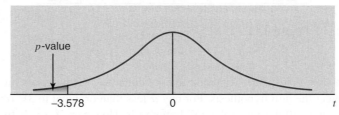

Figure 9.12   The required p-value.

Now we can find the range for the p-value. To do so, we go to Table V of Appendix B (the t distribution table) and find the row of $df = 44$. In this row, we find the two values of t that cover $t = 3.578$. Note that we use the positive value of the test statistic t, although our test statistic has a negative value. From Table V, for $df = 44$, the largest value of t is 3.286, for which the area in the tail of the t distribution is .001. This means that the area to the left of $t = -3.286$ is .001. Because -3.578 is smaller than -3.286, the area to the left of $t = -3.578$ is smaller than .001. Therefore, the p-value for $t = -3.578$ is less than .001, which can be written as

$$p\text{-value} < .001$$

Thus, here the p-value has only the upper limit of .001. In other words, the p-value for this example is less than .001. If we use technology for this example, we will obtain a p-value of .00043.

Step 4.   *Make a decision.*

Thus, we can state that for any α greater than or equal to .001 (the upper limit of the p-value range), we will reject the null hypothesis. For our example $\alpha = .025$, which is greater than the upper limit of the p-value of .001. As a result, we reject H_0 and conclude that the mean life of such batteries is less than 65 months. Therefore, we can state that the difference between the hypothesized population mean of 65 months and the sample mean of 63.4 is too large to be attributed to sampling error alone.

9.3.2 The Critical-Value Approach

In this procedure, as mentioned in Section 9.2.2, we have a predetermined value of the significance level α. The value of α gives the total area of the rejection region(s). First we find the critical value(s) of t from the t distribution table in Appendix B for the given degrees of freedom and the significance level. Then we find the value of the test statistic t for the observed value of the sample statistic $\bar{x}$. Finally we compare these two values and make a decision. Remember, if the test is one-tailed, there is only one critical value of t, and it is obtained by using the value of α, which gives the area in the left or right tail of the t distribution curve, depending on whether the test is left-tailed or right-tailed, respectively. However, if the test is two-tailed, there are two critical values of t, and they are obtained by using $\alpha/2$ area in each tail of the t distribution curve. The value of the test statistic t is obtained as mentioned earlier in this section.

Examples 9–7 and 9–8 describe the procedure to test a hypothesis about μ using the critical-value approach and the t distribution.

EXAMPLE 9–7   Age at Which Children Start Walking

Conducting a two-tailed test of hypothesis about μ using the critical-value approach: σ unknown, n < 30, and population normal.

Refer to Example 9–5. A psychologist claims that the mean age at which children start walking is 12.5 months. Carol wanted to check if this claim is true. She took a random sample of 18 children and found that the mean age at which these children started walking was 12.9 months with a standard deviation of .80 month. Using a 1% significance level, can you conclude that the mean age at which all children start walking is different from 12.5 months? Assume that the ages at which all children start walking have an approximate normal distribution.

Cohen Ostrow/Digital Vision/ Getty Images

Solution   Let μ be the mean age at which all children start walking, and let $\bar{x}$ be the corresponding mean for the sample. Then, from the given information,

$$n = 18, \quad \bar{x} = 12.9 \text{ months}, \quad s = .80 \text{ month}, \quad \text{and} \quad \alpha = .01$$

Step 1.   *State the null and alternative hypotheses.*

We are to test if the mean age at which all children start walking is different from 12.5 months. The null and alternative hypotheses are

H_0: $\mu = 12.5$ (The mean walking age is 12.5 months.)

H_1: $\mu \neq 12.5$ (The mean walking age is different from 12.5 months.)

Step 2.   *Select the distribution to use.*

In this example, the population standard deviation σ is not known, the sample size is small ($n < 30$), and the population is approximately normally distributed. Hence, it is Case I mentioned in the beginning of Section 9.3. Consequently, we will use the t distribution to perform the test in this example.

Step 3.   *Determine the rejection and nonrejection regions.*

The significance level is .01. The $\neq$ sign in the alternative hypothesis indicates that the test is two-tailed and the rejection region lies in both tails. The area of the rejection region in each tail of the t distribution curve is

$$\text{Area in each tail} = \alpha/2 = .01/2 = .005$$
$$df = n - 1 = 18 - 1 = 17$$

From the t distribution table, the critical values of t for 17 degrees of freedom and .005 area in each tail of the t distribution curve are -2.898 and 2.898. These values are shown in Figure 9.13.

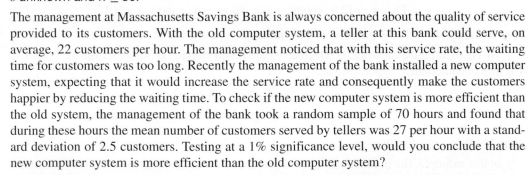

Figure 9.13　The critical values of t.

Step 4.　*Calculate the value of the test statistic.*

We calculate the value of the test statistic t for $\bar{x} = 12.9$ as follows:

$$s_{\bar{x}} = \frac{s}{\sqrt{n}} = \frac{.80}{\sqrt{18}} = .18856181$$

$$t = \frac{\bar{x} - \mu}{s_{\bar{x}}} = \frac{12.9 - 12.5}{.18856181} = 2.121$$

From H_0

Step 5.　*Make a decision.*

The value of the test statistic $t = 2.121$ falls between the two critical points, -2.898 and 2.898, which is the nonrejection region. Consequently, we fail to reject H_0. As a result, we can state that the difference between the hypothesized population mean and the sample mean is so small that it may have occurred because of sampling error. The mean age at which children start walking is not significantly different from 12.5 months.

EXAMPLE 9–8　Waiting Time for Service at a Bank

Conducting a right-tailed test of hypothesis about μ using the critical-value approach: σ unknown and $n \geq 30$.

The management at Massachusetts Savings Bank is always concerned about the quality of service provided to its customers. With the old computer system, a teller at this bank could serve, on average, 22 customers per hour. The management noticed that with this service rate, the waiting time for customers was too long. Recently the management of the bank installed a new computer system, expecting that it would increase the service rate and consequently make the customers happier by reducing the waiting time. To check if the new computer system is more efficient than the old system, the management of the bank took a random sample of 70 hours and found that during these hours the mean number of customers served by tellers was 27 per hour with a standard deviation of 2.5 customers. Testing at a 1% significance level, would you conclude that the new computer system is more efficient than the old computer system?

PhotoDisc, Inc./Getty Images

Solution　Let μ be the mean number of customers served per hour by a teller using the new system, and let $\bar{x}$ be the corresponding mean for the sample. Then, from the given information,

$$n = 70 \text{ hours}, \quad \bar{x} = 27 \text{ customers}, \quad s = 2.5 \text{ customers}, \quad \text{and} \quad \alpha = .01$$

Step 1.　*State the null and alternative hypotheses.*

We are to test whether or not the new computer system is more efficient than the old system. The new computer system will be more efficient than the old system if the mean number of customers served per hour by using the new computer system is significantly more than 22; otherwise, it will not be more efficient. The null and alternative hypotheses are

$$H_0: \mu = 22 \quad \text{(The new computer system is not more efficient.)}$$
$$H_1: \mu > 22 \quad \text{(The new computer system is more efficient.)}$$

Step 2. *Select the distribution to use.*

In this example, the population standard deviation σ is not known and the sample size is large ($n \geq 30$). Hence, it is Case II mentioned in the beginning of Section 9.3. Consequently, we will use the t distribution to perform the test for this example.

Step 3. *Determine the rejection and nonrejection regions.*

The significance level is .01. The $>$ sign in the alternative hypothesis indicates that the test is right-tailed and the rejection region lies in the right tail of the t distribution curve.

$$\text{Area in the right tail} = \alpha = .01$$
$$df = n - 1 = 70 - 1 = 69$$

From the t distribution table, the critical value of t for 69 degrees of freedom and .01 area in the right tail is 2.382. This value is shown in Figure 9.14.

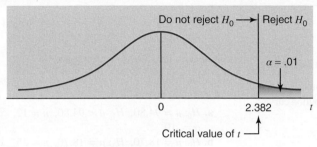

Figure 9.14 The critical value of t.

Step 4. *Calculate the value of the test statistic.*

The value of the test statistic t for $\bar{x} = 27$ is calculated as follows:

$$s_{\bar{x}} = \frac{s}{\sqrt{n}} = \frac{2.5}{\sqrt{70}} = .29880715$$

$$t = \frac{\bar{x} - \mu}{s_{\bar{x}}} = \frac{27 - 22}{.29880715} = 16.733$$

From H_0

Step 5. *Make a decision.*

The value of the test statistic $t = 16.733$ is greater than the critical value of $t = 2.382$ and falls in the rejection region. Consequently, we reject H_0. As a result, we conclude that the value of the sample mean is very large compared to the hypothesized value of the population mean, and the difference between the two may not be attributed to chance alone. The mean number of customers served per hour using the new computer system is more than 22. The new computer system is more efficient than the old computer system.

Note: What if the Sample Size Is Large and the Number of *df* Is Not in the *t* Distribution Table?

In this section when σ is not known, we used the t distribution to perform tests of hypothesis about μ in Cases I and II. Note that in Case II, the sample size is large. If we have access to technology, it does not matter how large the sample size is, we can always use the t distribution. However, if we are using the t distribution table (Table V of Appendix B), this may pose a problem. Usually such a table goes only up to a certain number of degrees of freedom. For example, Table V in Appendix B goes only up to 75 degrees of freedom. Thus, if the sample size is larger than 76 (with *df* more than 75), we cannot use Table V to find the critical value(s) of t to make a decision in this section. In such a situation when n is large and is not included in the t distribution table, there are two options:

1.   Use the t value from the last row (the row of ∞) in Table V of Appendix B.
2.   Use the normal distribution as an approximation to the t distribution.

To use the normal distribution as an approximation to the t distribution to make a test of hypothesis about μ, the procedure is exactly like the one in Section 9.2, except that now we will replace σ by s, and $\sigma_{\bar{x}}$ by $s_{\bar{x}}$.

Note that the t values obtained from the last row of the t distribution table are the same as will be obtained from the normal distribution table for the same areas in the upper tail or lower tail of the distribution. Again, note that here we can use the normal distribution as a convenience and as an approximation, but if we can, we should use the t distribution by using technology.

EXERCISES

CONCEPTS AND PROCEDURES

9.34   Briefly explain the conditions that must hold true to use the t distribution to make a test of hypothesis about the population mean.

9.35   For each of the following examples of tests of hypothesis about μ, show the rejection and nonrejection regions on the t distribution curve.

a. A two-tailed test with $\alpha = .01$ and $n = 15$
b. A left-tailed test with $\alpha = .005$ and $n = 25$
c. A right-tailed test with $\alpha = .025$ and $n = 22$

9.36   A random sample of 25 observations taken from a population that is normally distributed produced a sample mean of 58.5 and a standard deviation of 7.5. Find the ranges for the p-value and the critical and observed values of t for each of the following tests of hypotheses using $\alpha = .01$.

a. H_0: $\mu = 55$ versus H_1: $\mu > 55$
b. H_0: $\mu = 55$ versus H_1: $\mu \neq 55$

9.37   A random sample of 16 observations taken from a population that is normally distributed produced a sample mean of 42.4 and a standard deviation of 8. Find the ranges for the p-value and the critical and observed values of t for each of the following tests of hypotheses using $\alpha = .05$.

a. H_0: $\mu = 46$ versus H_1: $\mu < 46$
b. H_0: $\mu = 46$ versus H_1: $\mu \neq 46$

9.38   Consider the null hypothesis H_0: $\mu = 100$. Suppose that a random sample of 30 observations is taken from this population to perform this test. Using a significance level of .01, show the rejection and nonrejection regions and find the critical value(s) of t when the alternative hypothesis is as follows.

a. H_1: $\mu \neq 100$ **b.** H_1: $\mu > 100$ **c.** H_1: $\mu < 100$

9.39   Consider the null hypothesis H_0: $\mu = 35$ about the mean of a population. Suppose a random sample of 52 observations is taken from this population to make this test. Using $\alpha = .05$, show the rejection and nonrejection regions and find the critical value(s) of t for a

a. left-tailed test   **b.** two-tailed test   **c.** right-tailed test

9.40   Consider H_0: $\mu = 40$ versus H_1: $\mu > 40$.

a. A random sample of 64 observations taken from this population produced a sample mean of 43 and a standard deviation of 5. Using $\alpha = .025$, would you reject the null hypothesis?
b. Another random sample of 64 observations taken from the same population produced a sample mean of 41 and a standard

deviation of 7. Using $\alpha = .025$, would you reject the null hypothesis?

Comment on the results of parts a and b.

9.41   Perform the following tests of hypotheses for data coming from a normal distribution.

a. H_0: $\mu = 94.80$, H_1: $\mu < 94.80$, $n = 12$, $\bar{x} = 92.87$, $s = 5.34$, $\alpha = .10$
b. H_0: $\mu = 18.70$, H_1: $\mu \neq 18.70$, $n = 25$, $\bar{x} = 20.05$, $s = 2.99$, $\alpha = .05$
c. H_0: $\mu = 59$, H_1: $\mu > 59$, $n = 7$, $\bar{x} = 59.42$, $s = .418$, $\alpha = .01$

APPLICATIONS

9.42   The police that patrol a heavily traveled highway claim that the average driver exceeds the 65 miles per hour speed limit by more than 10 miles per hour. Eighty randomly selected cars were clocked by airplane radar. The average speed was 76.90 miles per hour, and the standard deviation of the speeds was 4.30 miles per hour. Find the range for the p-value for this test. What will your conclusion be using this p-value range and $\alpha = .02$?

9.43   The president of a university claims that the mean time spent partying by all students at this university is not more than 5 hours per week. A random sample of 40 students taken from this university showed that they spent an average of 11.0 hours partying the previous week with a standard deviation of 3.8 hours. Test at a 2.5% significance level whether the president's claim is true. Explain your conclusion in words.

9.44   The mean balance of all checking accounts at a bank on December 31, 2008, was $1050. A random sample of 60 checking accounts taken recently from this bank gave a mean balance of $910 with a standard deviation of $230. Using a 1% significance level, can you conclude that the mean balance of such accounts has decreased during this period? Explain your conclusion in words. What if $\alpha = .025$?

9.45   A paint manufacturing company claims that the mean drying time for its paints is not longer than 60 minutes. A random sample of 20 gallons of paints selected from the production line of this company showed that the mean drying time for this sample is 63.50 minutes with a standard deviation of 4 minutes. Assume that the drying times for these paints have a normal distribution.

a. Using a 1% significance level, would you conclude that the company's claim is true?

b. What is the Type I error in this exercise? Explain in words. What is the probability of making such an error?

9.46 A business school claims that students who complete a 3-month typing course can type, on average, at least 1200 words an hour. A random sample of 25 students who completed this course typed, on average, 1125 words an hour with a standard deviation of 85 words. Assume that the typing speeds for all students who complete this course have an approximate normal distribution.

    a. Suppose the probability of making a Type I error is selected to be zero. Can you conclude that the claim of the business school is true? Answer without performing the five steps of a test of hypothesis.

    b. Using a 5% significance level, can you conclude that the claim of the business school is true? Use both approaches.

9.47 According to an estimate, 2 years ago the average age of all CEOs of medium-sized companies in the United States was 57 years. Jennifer wants to check if this is still true. She took a random sample of 50 such CEOs and found their mean age to be 55 years with a standard deviation of 4 years.

    a. Suppose that the probability of making a Type I error is selected to be zero. Can you conclude that the current mean age of all CEOs of medium-sized companies in the Untied States is different from 57 years?

    b. Using a 1% significance level, can you conclude that the current mean age of all CEOs of medium-sized companies in the United States is different from 58 years? Use both approaches.

9.48 A past study claimed that adults in America spent an average of 18 hours a week on leisure activities. A researcher wanted to test this claim. She took a sample of 12 adults and asked them about the time they spend per week on leisure activities. Their responses (in hours) are as follows.

    10.4 14.0 23.2 24.6 22.9 37.7 16.4 14.5 21.5
    19.8 17.8 18.4

Assume that the times spent on leisure activities by all adults are normally distributed. Using a 10% significance level, can you conclude that the average amount of time spent on leisure activities has changed? (*Hint:* First calculate the sample mean and the sample standard deviation for these data using the formulas learned in Sections 3.1.1 and 3.2.2 of Chapter 3. Then make the test of hypothesis about μ.)

9.49 The past records of a supermarket show that its customers spend an average of $65 per visit at this store. Recently the management of the store initiated a promotional campaign according to which each customer receives points based on the total money spent at the store, and these points can be used to buy products at the store. The management expects that as a result of this campaign, the customers should be encouraged to spend more money at the store. To check whether this is true, the manager of the store took a sample of 12 customers who visited the store. The following data give the money (in dollars) spent by these customers at this supermarket during their visits.

    88 69 141 28 106 45 32 51 78 54 110 83

Assume that the money spent by all customers at this supermarket has a normal distribution. Using a 1% significance level, can you conclude that the mean amount of money spent by all customers at this supermarket after the campaign was started is more than $65? (*Hint:* First calculate the sample mean and the sample standard deviation for these data using the formulas learned in Sections 3.1.1 and 3.2.2 of Chapter 3. Then make the test of hypothesis about μ.)

9.50 According to the Kaiser Family Foundation, U.S. workers who had employer-provided health insurance paid an average premium of $4129 for family health insurance coverage during 2011 (*USA TODAY*, October 10, 2011). Suppose a recent random sample of 25 workers with employer-provided health insurance selected from a city paid an average premium of $4517 for family health insurance coverage with a standard deviation of $580. Assume that such premiums paid by all such workers in this city are normally distributed. Does the sample information support the alternative hypothesis that the average premium for such coverage paid by all such workers in this city is different from $4129? Use a 5% significance level. Use both the *p*-value approach and the critical-value approach.

*9.51 The manager of a service station claims that the mean amount spent on gas by its customers is $22.60 per visit. You want to test if the mean amount spent on gas at this station is different from $22.60 per visit. Briefly explain how you would conduct this test when σ is not known.

*9.52 A tool manufacturing company claims that its top-of-the-line machine that is used to manufacture bolts produces an average of 88 or more bolts per hour. A company that is interested in buying this machine wants to check this claim. Suppose you are asked to conduct this test. Briefly explain how you would do so when σ is not known.

9.4 | Hypothesis Tests About a Population Proportion: Large Samples

Often we want to conduct a test of hypothesis about a population proportion. For example, in a 2014 Gallup poll, 61% of adults said that upper-income people were paying too little in federal taxes. An agency may want to check if this percentage has changed since 2014. As another example, a mail-order company claims that 90% of all orders it receives are shipped within 72 hours. The company's management may want to determine from time to time whether or not this claim is true.

    This section presents the procedure to perform tests of hypotheses about the population proportion, p, for large samples. The procedures to make such tests are similar in many respects to the ones for the population mean, μ. Again, the test can be two-tailed or one-tailed. We know from Chapter 7 that when the sample size is large, the sample proportion, $\hat{p}$, is approximately normally distributed with its mean equal to p and standard deviation equal to $\sqrt{pq/n}$. Hence, we use the normal distribution to perform a test of hypothesis about the population proportion, p, for a large sample. As was mentioned in Chapters 7 and 8, in the case of a proportion, the sample size is considered to be large when np and nq are both greater than 5.

Test Statistic   The value of the *test statistic* z for the sample proportion, $\hat{p}$, is computed as

$$z = \frac{\hat{p} - p}{\sigma_{\hat{p}}} \quad \text{where} \quad \sigma_{\hat{p}} = \sqrt{\frac{pq}{n}}$$

The value of p that is used in this formula is the one from the null hypothesis. The value of q is equal to $1 - p$.

The value of z calculated for $\hat{p}$ using the above formula is also called the **observed value of** z.

In Section 9.2, we discussed two procedures, the p-value approach and the critical-value approach, to test hypotheses about μ. Here too we will use these two procedures to test hypotheses about p. The steps used in these procedures are the same as in Section 9.2. The only difference is that we will be making tests of hypotheses about p rather than about μ.

9.4.1   The p-Value Approach

To use the p-value approach to perform a test of hypothesis about p, we will use the same four steps that we used in such a procedure in Section 9.2. Although the p-value for a test of hypothesis about p can be obtained very easily by using technology, we can use Table IV of Appendix B to find this p-value when technology is not available.

Examples 9–9 and 9–10 illustrate the p-value procedure to test a hypothesis about p for a large sample.

EXAMPLE 9–9   Emotional Attachment to Alma Mater

Finding a p-value and making a decision for a two-tailed test of hypothesis about p: large sample.

According to a Gallup poll conducted February 4 to March 7, 2014, 20% of traditional college graduates (those who earned their degree before age 25) were *emotionally attached* to their alma mater (www.gallup.com). Suppose this result is true for the 2014 population of traditional college graduates. In a recent random sample of 2000 traditional college graduates, 22% said that they are emotionally attached to their alma mater. Find the p-value to test the hypothesis that the current percentage of traditional college graduates who are emotionally attached to their alma mater is different from 20%. What is your conclusion if the significance level is 5%?

Solution   Let p be the current proportion of all traditional college graduates who are emotionally attached to their alma mater, and $\hat{p}$ be the corresponding sample proportion. Then, from the given information,

$$n = 2000, \qquad \hat{p} = .22, \qquad \text{and} \qquad \alpha = .05$$

In 2014, 20% of traditional college graduates were emotionally attached to their alma mater. Hence,

$$p = .20 \qquad \text{and} \qquad q = 1 - p = 1 - .20 = .80$$

To calculate the p-value and to make a decision, we apply the following four steps.

Step 1.   *State the null and alternative hypotheses.*

The current percentage of traditional college graduates who are emotionally attached to their alma mater will not be different from 20% if $p = .20$, and the current percentage will be different from 20% if $p \neq .20$. The null and alternative hypotheses are as follows:

$$H_0\text{: } p = .20 \quad \text{(The current percentage is not different from 20\%.)}$$
$$H_1\text{: } p \neq .20 \quad \text{(The current percentage is different from 20\%.)}$$

Step 2.   *Select the distribution to use.*

To check whether the sample is large, we calculate the values of np and nq.

$$np = 2000(.20) = 400 \quad \text{and} \quad nq = 2000(.80) = 1600$$

Since np and nq are both greater than 5, we can conclude that the sample size is large. Consequently, we will use the normal distribution to find the p-value for this test.

Step 3.   *Calculate the p-value.*

The $\neq$ sign in the alternative hypothesis indicates that the test is two-tailed. The p-value is equal to twice the area in the tail of the normal distribution curve to the right of z for $\hat{p} = .22$. This p-value is shown in Figure 9.15. To find this p-value, first we find the test statistic z for $\hat{p} = .22$ as follows:

$$\sigma_{\hat{p}} = \sqrt{\frac{pq}{n}} = \sqrt{\frac{(.20)(.80)}{2000}} = .00894427$$

$$z = \frac{\hat{p} - p}{\sigma_{\hat{p}}} = \frac{.22 - .20}{.00894427} = 2.24$$

Now we find the area to the right of $z = 2.24$ from the normal distribution table. This area is $1 - .9875 = .0125$. Consequently, the p-value is

$$p\text{-value} = 2(.0125) = .0250$$

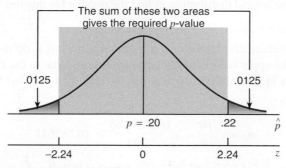

Figure 9.15   The required p-value.

Step 4.   *Make a decision.*

Thus, we can state that for any α greater than or equal to .0250 we will reject the null hypothesis, and for any α less than .0250 we will not reject the null hypothesis. In our example, $\alpha = .05$, which is greater than the p-value of .0250. As a result, we reject H_0 and conclude that the current percentage of traditional college graduates who are emotionally attached to their alma mater is different from .20. Consequently, we can state that the difference between the hypothesized population proportion of .20 and the sample proportion of .22 is too large to be attributed to sampling error alone when $\alpha = .05$.

EXAMPLE 9–10   Percentage of Defective Chips

Finding a p-value and making a decision for a right-tailed test of hypothesis about p: large sample.

When working properly, a machine that is used to make chips for calculators does not produce more than 4% defective chips. Whenever the machine produces more than 4% defective chips, it needs an adjustment. To check if the machine is working properly, the quality control department at the company often takes samples of chips and inspects them to determine if they are good or defective. One such random sample of 200 chips taken recently from the production line contained 12 defective chips. Find the p-value to test the hypothesis whether or not the machine needs an adjustment. What would your conclusion be if the significance level is 2.5%?

Solution   Let p be the proportion of defective chips in all chips produced by this machine, and let $\hat{p}$ be the corresponding sample proportion. Then, from the given information,

$$n = 200, \quad \hat{p} = 12/200 = .06, \quad \text{and} \quad \alpha = .025$$

When the machine is working properly, it does not produce more than 4% defective chips. Hence, assuming that the machine is working properly, we obtain

$$p = .04 \quad \text{and} \quad q = 1 - p = 1 - .04 = .96$$

To calculate the p-value and to make a decision, we apply the following four steps.

Step 1.   *State the null and alternative hypotheses.*

The machine will not need an adjustment if the percentage of defective chips is 4% or less, and it will need an adjustment if this percentage is greater than 4%. Hence, the null and alternative hypotheses are as follows:

$$H_0: p \leq .04 \quad \text{(The machine does not need an adjustment.)}$$
$$H_1: p > .04 \quad \text{(The machine needs an adjustment.)}$$

Step 2.   *Select the distribution to use.*

To check if the sample is large, we calculate the values of np and nq:

$$np = 200(.04) = 8 \quad \text{and} \quad nq = 200(.96) = 192$$

Since np and nq are both greater than 5, we can conclude that the sample size is large. Consequently, we will use the normal distribution to find the p-value for this test.

Step 3.   *Calculate the p-value.*

The $>$ sign in the alternative hypothesis indicates that the test is right-tailed. The p-value is given by the area in the upper tail of the normal distribution curve to the right of z for $\hat{p} = .06$. This p-value is shown in Figure 9.16. To find this p-value, first we find the test statistic z for $\hat{p} = .06$ as follows:

$$\sigma_{\hat{p}} = \sqrt{\frac{pq}{n}} = \sqrt{\frac{(.04)(.96)}{200}} = .01385641$$

$$z = \frac{\hat{p} - p}{\sigma_{\hat{p}}} = \frac{.06 - .04}{.01385641} = 1.44 \qquad \text{From } H_0$$

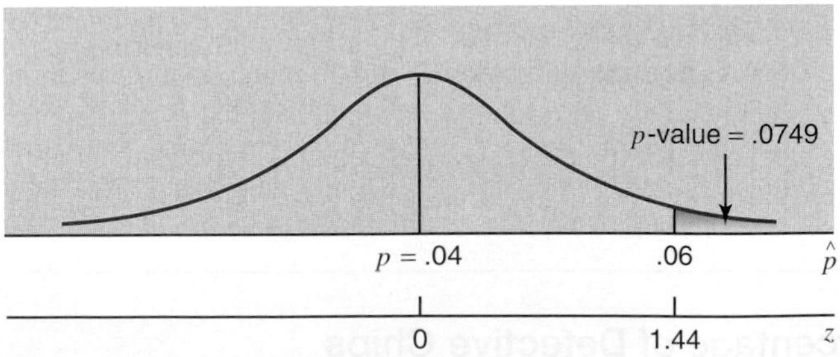

Figure 9.16   The required p-value.

Now we find the area to the right of $z = 1.44$ from the normal distribution table. This area is $1 - .9251 = .0749$. Consequently, the p-value is

$$p\text{-value} = .0749$$

Step 4.   *Make a decision.*

Thus, we can state that for any α greater than or equal to .0749 we will reject the null hypothesis, and for any α less than .0749 we will not reject the null hypothesis. For our example, $\alpha = .025$, which is less than the p-value of .0749. As a result, we fail to reject H_0 and conclude that the machine does not need an adjustment.

9.4.2   The Critical-Value Approach

In this procedure, as mentioned in Section 9.2.2, we have a predetermined value of the significance level α. The value of α gives the total area of the rejection region(s). First we find the critical value(s) of z from the normal distribution table for the given significance level. Then we find the value of the test statistic z for the observed value of the sample statistic $\hat{p}$. Finally we compare these two values and make a decision. Remember, if the test is one-tailed, there is only one critical value of z, and it is obtained by using the value of α, which gives the area in the left or right tail of the normal distribution curve, depending on whether the test is left-tailed or right-tailed, respectively. However, if the test is two-tailed, there are two critical values of z, and they are obtained by using $\alpha/2$ area in each tail of the normal distribution curve. The value of the test statistic z is obtained as mentioned earlier in this section.

Examples 9–11 and 9–12 describe the procedure to test a hypothesis about p using the critical-value approach and the normal distribution.

EXAMPLE 9–11   Emotional Attachment to Alma Mater

Making a two-tailed test of hypothesis about p using the critical-value approach:
large sample.

Refer to Example 9–9. According to a Gallup poll conducted February 4 to March 7, 2014, 20% of traditional college graduates (those who earned their degree before age 25) were *emotionally attached* to their alma mater (www.gallup.com). Suppose this result is true for the 2014 population of traditional college graduates. In a recent random sample of 2000 traditional college graduates, 22% said that they are emotionally attached to their alma mater. Using a 5% significance level, can you conclude that the current percentage of traditional college graduates who are emotionally attached to their alma mater is different from 20%?

Solution   Let p be the current proportion of all traditional college graduates who are emotionally attached to their alma mater, and $\hat{p}$ be the corresponding sample proportion. Then, from the given information,

$$n = 2000, \quad \hat{p} = .22, \quad \text{and} \quad \alpha = .05$$

In 2014, 20% of traditional college graduates were emotionally attached to their alma mater. Hence,

$$p = .20 \quad \text{and} \quad q = 1 - p = 1 - .20 = .80$$

To use the critical-value approach to perform a test of hypothesis, we apply the following five steps.

Step 1.   *State the null and alternative hypotheses.*

The current percentage of traditional college graduates who are emotionally attached to their alma mater will not be different from 20% if $p = .20$, and the current percentage will be different from 20% if $p \neq .20$. The null and alternative hypotheses are as follows:

$$H_0: p = .20 \quad \text{(The current percentage is not different from 20\%.)}$$
$$H_1: p \neq .20 \quad \text{(The current percentage is different from 20\%.)}$$

Step 2.   *Select the distribution to use.*

To check if the sample is large, we calculate the values of np and nq.

$$np = 2000(.20) = 400 \quad \text{and} \quad nq = 2000(.80) = 1600$$

Since np and nq are both greater than 5, we can conclude that the sample size is large. Consequently, we will use the normal distribution to make the test.

Step 3.   *Determine the rejection and nonrejection regions.*

The $\neq$ sign in the alternative hypothesis indicates that the test is two-tailed. The significance level is .05. Therefore, the total area of the two rejection regions is .05, and the rejection region in each tail of the sampling distribution of $\hat{p}$ is $\alpha/2 = .05/2 = .025$. The critical values of z, obtained from the standard normal distribution table, are -1.96 and 1.96, as shown in Figure 9.17.

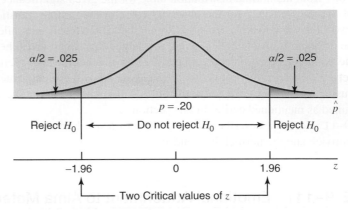

$\alpha/2 = .025$ $\alpha/2 = .025$

$p = .20$ $\hat{p}$

Reject H_0 ⟵ Do not reject H_0 ⟶ Reject H_0

-1.96 0 1.96 z

Two Critical values of z

Figure 9.17   The critical values of z.

Step 4. *Calculate the value of the test statistic.*

The value of the test statistic z for $\hat{p} = .22$ is calculated as follows.

$$\sigma_{\hat{p}} = \sqrt{\frac{pq}{n}} = \sqrt{\frac{(.20)(.80)}{2000}} = .00894427$$

From H_0

$$z = \frac{\hat{p} - p}{\sigma_{\hat{p}}} = \frac{.22 - .20}{.00894427} = 2.24$$

Step 5. *Make a decision.*

The value of the test statistic $z = 2.24$ for $\hat{p}$ falls in the rejection region. As a result, we reject H_0 and conclude that the current percentage of traditional college graduates who are emotionally attached to their alma mater is different from .20. Consequently, we can state that the difference between the hypothesized population proportion of .20 and the sample proportion of .22 is too large to be attributed to sampling error alone when $\alpha = .05$.

EXAMPLE 9–12   Mailing the Received Orders

Conducting a left-tailed test of hypothesis about p using the critical-value approach: large sample.

Direct Mailing Company sells computers and computer parts by mail. The company claims that at least 90% of all orders are mailed within 72 hours after they are received. The quality control department at the company often takes samples to check if this claim is valid. A recently taken sample of 150 orders showed that 129 of them were mailed within 72 hours. Do you think the company's claim is true? Use a 2.5% significance level.

Solution   Let p be the proportion of all orders that are mailed by the company within 72 hours, and let $\hat{p}$ be the corresponding sample proportion. Then, from the given information,

$$n = 150, \quad \hat{p} = 129/150 = .86, \quad \text{and} \quad \alpha = .025$$

The company claims that at least 90% of all orders are mailed within 72 hours. Assuming that this claim is true, the values of p and q are

$$p = .90 \quad \text{and} \quad q = 1 - p = 1 - .90 = .10$$

Step 1. *State the null and alternative hypotheses.*

The null and alternative hypotheses are

$$H_0: p \geq .90 \quad \text{(The company's claim is true.)}$$
$$H_1: p < .90 \quad \text{(The company's claim is false.)}$$

Step 2. *Select the distribution to use.*

We first check whether np and nq are both greater than 5:

$$np = 150(.90) = 135 > 5 \quad \text{and} \quad nq = 150(.10) = 15 > 5$$

Consequently, the sample size is large. Therefore, we use the normal distribution to make the hypothesis test about p.

Step 3. *Determine the rejection and nonrejection regions.*

The significance level is .025. The $<$ sign in the alternative hypothesis indicates that the test is left-tailed, and the rejection region lies in the left tail of the sampling distribution of $\hat{p}$ with its area equal to .025. As shown in Figure 9.18, the critical value of z, obtained from the normal distribution table for .0250 area in the left tail, is -1.96.

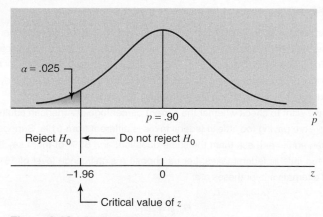

Figure 9.18 Critical value of z.

Step 4. *Calculate the value of the test statistic.*

The value of the test statistic z for $\hat{p} = .86$ is calculated as follows:

$$\sigma_{\hat{p}} = \sqrt{\frac{pq}{n}} = \sqrt{\frac{(.90)(.10)}{150}} = .02449490$$

$$z = \frac{\hat{p} - p}{\sigma_{\hat{p}}} = \frac{.86 - .90}{.02449490} = -1.63$$

(From H_0)

Step 5. *Make a decision.*

The value of the test statistic $z = -1.63$ is greater than the critical value of $z = -1.96$, and it falls in the nonrejection region. Therefore, we fail to reject H_0. We can state that the difference between the sample proportion and the hypothesized value of the population proportion is small, and this difference may have occurred owing to chance alone. Therefore, the proportion of all orders that are mailed within 72 hours is at least 90%, and the company's claim seems to be true.

Are Upper-Income People Paying Their Fair Share in Federal Taxes?

ARE UPPER-INCOME PEOPLE PAYING THEIR FAIR SHARE IN FEDERAL TAXES?

Too little **61%**

Fair share **24%**

Too much **13%**

No opinion **2%**

Data source: www.gallup.com

In a Gallup poll of Americans aged 18 and over conducted April 3–6, 2014, among other questions, adults were asked: are the upper-income people "paying their fair share in federal taxes, paying too much or paying too little?" The responses of the adults polled are shown in the above graph. Of these adults, 61% said that upper-income people were paying too little in federal taxes, 24% said that they were paying their fair share, 13% said that they were paying too much, and 2% had no opinion.

Suppose that we want to check whether the current percentage of American adults who will say that upper-income people are paying too little in federal taxes is different from 61%. Suppose we take a sample of 1600 American adults and ask them the same question, and 65% of them say that upper-income people are paying too little in federal taxes. Let us choose a significance level of 1%. The test is two-tailed. The null and alternative hypotheses are:

$$H_0: p = .61$$
$$H_1: p \neq .61$$

Here, $n = 1600$, $\hat{p} = .65$, $\alpha = .01$, and $\alpha/2 = .005$. The sample is large. (The reader should check that np and nq are both greater than 5.) Using the normal distribution for the test, the critical values of z for .0050 and .9950 areas to the left are -2.58 and 2.58, respectively. We find the observed value of z as follows:

$$\sigma_{\hat{p}} = \sqrt{\frac{pq}{n}} = \sqrt{\frac{(.61)(.39)}{1600}} = .01219375$$

$$z = \frac{\hat{p} - p}{\sigma_{\hat{p}}} = \frac{.65 - .61}{.01219375} = 3.28$$

The value of the test statistic $z = 3.28$ for $\hat{p}$ is larger than the upper critical value of $z = 2.58$, and it falls in the rejection region. Consequently, we reject H_0 and conclude that the current percentage of American adults who hold the opinion that upper-income people are paying too little in federal taxes is significantly different from 61%.

We can use the p-value approach too. From the normal distribution table, the area under the normal curve to the right of $z = 3.28$ is .0005. Therefore, the p-value is 2(.0005) = .0010. Since $\alpha = .01$ is larger than .0010, we reject the null hypothesis.

Source: http://www.gallup.com/poll/168521/taxes-rise-half-say-middle-income-pay.aspx.

EXERCISES

CONCEPTS AND PROCEDURES

9.53 Explain when a sample is large enough to use the normal distribution to make a test of hypothesis about the population proportion.

9.54 In each of the following cases, do you think the sample size is large enough to use the normal distribution to make a test of hypothesis about the population proportion? Explain why or why not.

    a. $n = 30$ and $p = .65$
    b. $n = 70$ and $p = .05$
    c. $n = 60$ and $p = .06$
    d. $n = 900$ and $p = .17$

9.55 For each of the following examples of tests of hypothesis about the population proportion, show the rejection and nonrejection regions on the graph of the sampling distribution of the sample proportion.

    a. A two-tailed test with $\alpha = .05$
    b. A left-tailed test with $\alpha = .02$
    c. A right-tailed test with $\alpha = .025$

9.56 A random sample of 200 observations produced a sample proportion equal to .60. Find the critical and observed values of z for each of the following tests of hypotheses using $\alpha = .01$.

    a. $H_0: p = .63$ versus $H_1: p < .63$
    b. $H_0: p = .63$ versus $H_1: p \neq .63$

9.57 Consider the null hypothesis $H_0: p = .85$. Suppose a random sample of 1000 observations is taken to perform this test about the population proportion. Using $\alpha = .05$, show the rejection and nonrejection regions and find the critical value(s) of z for a

    a. left-tailed test
    b. two-tailed test
    c. right-tailed test

9.58 Consider the null hypothesis $H_0: p = .20$. Suppose a random sample of 400 observations is taken to perform this test about the population proportion. Using $\alpha = .01$, show the rejection and nonrejection regions and find the critical value(s) of z for a

    a. left-tailed test
    b. two-tailed test
    c. right-tailed test

9.59 Consider $H_0: p = .45$ versus $H_1: p < .45$.

    a. A random sample of 400 observations produced a sample proportion equal to .42. Using $\alpha = .025$, would you reject the null hypothesis?
    b. Another random sample of 400 observations taken from the same population produced a sample proportion of .39. Using $\alpha = .025$, would you reject the null hypothesis?

Comment on the results of parts a and b.

9.60 Consider $H_0: p = .70$ versus $H_1: p \neq .70$.

    a. A random sample of 600 observations produced a sample proportion equal to .68. Using $\alpha = .01$, would you reject the null hypothesis?
    b. Another random sample of 600 observations taken from the same population produced a sample proportion equal to .76. Using $\alpha = .01$, would you reject the null hypothesis?

Comment on the results of parts a and b.

9.61 Make the following hypothesis tests about p.

    a. $H_0: p = .57$, $H_1: p \neq .57$, $n = 800$, $\hat{p} = .50$, $\alpha = .05$
    b. $H_0: p = .26$, $H_1: p < .26$, $n = 400$, $\hat{p} = .23$, $\alpha = .01$
    c. $H_0: p = .84$, $H_1: p > .84$, $n = 250$, $\hat{p} = .85$, $\alpha = .025$

APPLICATIONS

9.62 According to a book published in 2011, 45% of the undergraduate students in the United States show almost no gain in learning in their first 2 years of college (Richard Arum *et al., Academically Adrift*, University of Chicago Press, Chicago, 2011). A recent sample of 1500 undergraduate students showed that this percentage is 38%. Can you reject the null hypothesis at a 1% significance level in favor of the alternative that the percentage of undergraduate students in the United States who show almost no gain in learning in their first 2 years of college is currently lower than 45%? Use both the *p*-value and the critical-value approaches.

9.63 According to the U.S. Census Bureau, in 2014, 62% of Americans age 18 and older were married. A recent sample of 2000 Americans age 18 and older showed that 58% of them are married. Can you reject the null hypothesis at a 1% significance level in favor of the alternative that the percentage of current population of Americans age 18 and older who are married is lower than 62%? Use both the *p*-value and the critical-value approaches.

9.64 According to a *New York Times*/CBS News poll conducted during June 24–28, 2011, 55% of the American adults polled said that owning a home is a *very important part* of the American Dream (*The New York Times*, June 30, 2011). Suppose this result was true for the population of all American adults in 2011. In a recent poll of 1800 American adults, 61% said that owning a home is a *very important part* of the American Dream. Perform a hypothesis test to determine whether it is reasonable to conclude that the percentage of all American adults who currently hold this opinion is higher than 55%. Use a 2% significance level, and use both the *p*-value and the critical-value approaches.

9.65 Beginning in the second half of 2011, there were widespread protests in many American cities that were primarily against Wall Street corruption and the increasing gap between the rich and the poor in America. According to a *Time* Magazine/ABT SRBI poll conducted by telephone during October 9–10, 2011, 86% of adults who were familiar with those protests agreed that Wall Street and lobbyists have too much influence in Washington (*The New York Times*, October 22, 2011). Assume that 86% of all American adults in 2011 believed that Wall Street and lobbyists have too much influence in Washington. A recent random sample of 2000 American adults showed that 1780 of them believe that Wall Street and lobbyists have too much influence in Washington. Using a 5% significance level, perform a test of hypothesis to determine whether the current percentage of American adults who believe that Wall Street and lobbyists have too much influence in Washington is higher than 86%. Use both the *p*-value and the critical-value approaches.

9.66 According to a Pew Research Center nationwide telephone survey conducted between March 15 and April 24, 2011, 55% of college graduates said that college education prepared them for a job (*Time*, May 30, 2011). Suppose this result was true of all college graduates at that time. In a recent sample of 2100 college graduates, 60% said that college education prepared them for a job. Is there significant evidence at a 1% significance level to conclude that the current percentage of all college graduates who will say that college education prepared them for a job is different from 55%? Use both the *p*-value and the critical-value approaches.

9.67 A food company is planning to market a new type of frozen yogurt. However, before marketing this yogurt, the company wants to find what percentage of the people like it. The company's management has decided that it will market this yogurt only if at least 35% of the people like it. The company's research department selected a random sample of 400 persons and asked them to taste this yogurt. Of these 400 persons, 112 said they liked it.

    a. Testing at a 2.5% significance level, can you conclude that the company should market this yogurt?

    b. What will your decision be in part a if the probability of making a Type I error is zero? Explain.

    c. Make the test of part a using the *p*-value approach and $\alpha = .025$.

9.68 Brooklyn Corporation manufactures DVDs. The machine that is used to make these DVDs is known to produce not more than 5% defective DVDs. The quality control inspector selects a sample of 200 DVDs each week and inspects them for being good or defective. Using the sample proportion, the quality control inspector tests the null hypothesis $p \leq .05$ against the alternative hypothesis $p > .05$, where *p* is the proportion of DVDs that are defective. She always uses a 2.5% significance level. If the null hypothesis is rejected, the production

process is stopped to make any necessary adjustments. A recent sample of 200 DVDs contained 17 defective DVDs.

    a. Using a 2.5% significance level, would you conclude that the production process should be stopped to make necessary adjustments?

    b. Perform the test of part a using a 1% significance level. Is your decision different from the one in part a?

Comment on the results of parts a and b.

*9.69 Two years ago, 75% of the customers of a bank said that they were satisfied with the services provided by the bank. The manager of the bank wants to know if this percentage of satisfied customers has changed since then. She assigns this responsibility to you. Briefly explain how you would conduct such a test.

*9.70 A study claims that 65% of students at all colleges and universities hold off-campus (part-time or full-time) jobs. You want to check if the percentage of students at your school who hold off-campus jobs is different from 65%. Briefly explain how you would conduct such a test. Collect data from 40 students at your school on whether or not they hold off-campus jobs. Then, calculate the proportion of students in this sample who hold off-campus jobs. Using this information, test the hypothesis. Select your own significance level.

USES AND MISUSES...

FOLLOW THE RECIPE

Hypothesis testing is one of the most powerful and dangerous tools of statistics. It allows us to make statements about a population and attach a degree of uncertainty to these statements. Pick up a newspaper and flip through it; rare will be the day when the paper does not contain a story featuring a statistical result, often reported with a significance level. Given that the subjects of these reports—public health, the environment, and so on—are important to our lives, it is critical that we perform the statistical calculations and interpretations properly. The first step, one that you should look for when reading statistical results, is proper formulation/specification.

Formulation or *specification*, simply put, is the list of steps you perform when constructing a hypothesis test. In this chapter, these steps are: stating the null and alternative hypotheses; selecting the appropriate distribution; and determining the rejection and nonrejection regions. Once these steps are performed, all you need to do is to calculate the *p*-value or the test statistic to complete the hypothesis test. It is important to beware of traps in the specification.

Though it might seem obvious, stating the hypothesis properly can be difficult. For hypotheses around a population mean, the null and alternative hypotheses are mathematical statements that do not overlap and also provide no holes. Suppose that a confectioner states that the average mass of his chocolate bars is 100 grams. The null hypothesis is that the mass of the bars is 100 grams, and the alternative hypothesis is that the mass of the bars is not 100 grams. When you take a sample of chocolate bars and measure their masses, all possibilities for the sample mean will fall within one of your decision regions. The problem is a little more difficult for hypotheses based on

proportions. Make sure that you only have two categories. For example, if you are trying to determine the percentage of the population that has blonde hair, your groups are "blonde" and "not blonde." You need to decide how to categorize bald people before you conduct this experiment: Do not include bald people in the survey.

Finally, beware of numerical precision. When your sample is large and you assume that it has a normal distribution, the rejection region for a two-tailed test using the normal distribution with a significance level of 5% will be values of the sample mean that are farther than 1.96 standard deviations from the assumed mean. When you perform your calculations, the sample mean may fall on the border of your decision region. Remember that there is measurement error and sampling error that you cannot account for. In this case, it is probably best to adjust your significance level so that the sample mean falls squarely in a decision region.

THE POWER OF NEGATIVE THINKING

In the beginning of this chapter, you learned about Type I and Type II errors. If this is your first statistics class, you might think that this is your first exposure to the concepts of Type I and Type II errors, but that is not the case. As a matter of fact, if you can recall having a medical test, you must have had an interaction with a variety of concepts related to hypothesis testing, including Type I and Type II errors.

In a typical medical test, the assumption (our null hypothesis) is that you do not have the condition for which you are being tested. If the assumption is true, the doctor knows what should happen in the test. If the test results are different from the *normal* range, the doctor has data that would allow the assumption to be rejected. Whenever the null hypothesis is rejected (that is, the test results demonstrate that the person has the condition for which he or she was tested), the medical result is called a *positive* test result. If the doctor fails to reject the null hypothesis (if the test results do not demonstrate that you have the condition), the medical result is called a *negative* test result.

As with other types of hypothesis tests, medical tests are not perfect. Sometimes people are (falsely) diagnosed as having a condition or illness when they actually do not have it. In medical terminology, this Type I error is referred to as a *false positive*. Similarly, the result of a test may (wrongly) indicate that a person does not have a condition or illness when actually he or she has it. This Type II error is called a *false negative* in medical terminology.

Companies that develop medical tests perform intensive research and clinical tests to reduce the risk of making both these types of errors. Specifically, data are collected on the *sensitivity* and the *specificity* of medical tests. In the context of an illness, the sensitivity of a test is the proportion of all people with the illness who are identified by the test as actually having it. For example, suppose 100 students on a college campus have been identified by throat culture

as having strep throat. All these 100 students are tested for strep throat using another type of test. Suppose 97 of the 100 tests come back positive with the second test. Then the sensitivity of the (second) test is .97 (or 97%), and the probability of a false negative (Type II error) is one minus the sensitivity, which is .03 here.

The specificity of a test refers to how well a test identifies that a healthy person does not have a given disease. Using the strep throat reference again, suppose that 400 students have been identified by throat culture as not having strep throat. All these 400 students are given a new strep test, and 394 of them are shown to have a negative result (that is, they are identified as not having strep throat). Then the specificity of the test is 394/400 = .985 (or 98.5%), and the probability of a false positive (Type I error) is one minus the specificity, which is .015 here.

Of course, low probabilities for both types of errors are very important in medical testing. A false positive can result in an individual obtaining unnecessary, often expensive, and sometime debilitating treatment, whereas a false negative can allow a disease to progress to an advanced stage when early detection could have helped to save a person's life.

GLOSSARY

α The significance level of a test of hypothesis that denotes the probability of rejecting a null hypothesis when it actually is true. (The probability of committing a Type I error.)

Alternative hypothesis   A claim about a population parameter that will be true if the null hypothesis is false.

β The probability of not rejecting a null hypothesis when it actually is false. (The probability of committing a Type II error.)

Critical value or **critical point**   One or two values that divide the whole region under the sampling distribution of a sample statistic into rejection and nonrejection regions.

Left-tailed test   A test in which the rejection region lies in the left tail of the distribution curve.

Null hypothesis   A claim about a population parameter that is assumed to be true until proven otherwise.

Observed value of z or t   The value of z or t calculated for a sample statistic such as the sample mean or the sample proportion.

One-tailed test   A test in which there is only one rejection region, either in the left tail or in the right tail of the distribution curve.

p-value   The smallest significance level at which a null hypothesis can be rejected.

Right-tailed test   A test in which the rejection region lies in the right tail of the distribution curve.

Significance level   The value of α that gives the probability of committing a Type I error.

Test statistic   The value of z or t calculated for a sample statistic such as the sample mean or the sample proportion.

Two-tailed test   A test in which there are two rejection regions, one in each tail of the distribution curve.

Type I error   An error that occurs when a true null hypothesis is rejected.

Type II error   An error that occurs when a false null hypothesis is not rejected.

SUPPLEMENTARY EXERCISES

9.71 Consider the following null and alternative hypotheses:

$$H_0: \mu = 40 \quad \text{versus} \quad H_1: \mu \neq 40$$

A random sample of 64 observations taken from this population produced a sample mean of 38.4. The population standard deviation is known to be 6.

    a. If this test is made at a 2% significance level, would you reject the null hypothesis? Use the critical-value approach.

    b. What is the probability of making a Type I error in part a?

    c. Calculate the p-value for the test. Based on this p-value, would you reject the null hypothesis if $\alpha = .01$? What if $\alpha = .05$?

9.72 Consider the following null and alternative hypotheses:

$$H_0: p = .82 \quad \text{versus} \quad H_1: p \neq .82$$

A random sample of 600 observations taken from this population produced a sample proportion of .86.

    a. If this test is made at a 2% significance level, would you reject the null hypothesis? Use the critical-value approach.

    b. What is the probability of making a Type I error in part a?

    c. Calculate the p-value for the test. Based on this p-value, would you reject the null hypothesis if $\alpha = .025$? What if $\alpha = .005$?

9.73 Consider the following null and alternative hypotheses:

$$H_0: p = .44 \quad \text{versus} \quad H_1: p < .44$$

A random sample of 450 observations taken from this population produced a sample proportion of .39.

   a. If this test is made at a 2% significance level, would you reject the null hypothesis? Use the critical-value approach.
   b. What is the probability of making a Type I error in part a?
   c. Calculate the p-value for the test. Based on this p-value, would you reject the null hypothesis if $\alpha = .01$? What if $\alpha = .025$?

9.74 According to the American Time Use Survey, Americans watched television each weekday for an average of 151 minutes in 2011 (*Time*, July 11, 2011). Suppose that this result is true for the 2011 population of all American adults. A recent sample of 120 American adults showed that they watch television each weekday for an average of 162 minutes. Assume that the population standard deviation for times spent watching television each weekday by American adults is 30 minutes.

   a. Find the p-value for the test of hypothesis with the alternative hypothesis that the current average time spent watching television each weekday by American adults is higher than 151 minutes. What is your conclusion at $\alpha = .01$?
   b. Test the hypothesis of part a using the critical-value approach and $\alpha = .01$.

9.75 According to an article entitled "Do You Need Long-Term-Care Insurance?" published in the *Consumer Reports* magazine, the average age of admission to nursing homes is 83 years (*Consumer Reports*, November 2003). Suppose that a random sample of 500 admissions to nursing homes yielded a mean age of 83.5 years. Assume that the population standard deviation is 3.5 years.

   a. Using the critical-value approach, can you conclude that the current mean admission age to nursing homes is different from 83 years? Use $\alpha = .05$.
   b. What is the Type I error in part a? Explain. What is the probability of making this error in part a?
   c. Will your conclusion of part a change if the probability of making a Type I error is zero?
   d. Calculate the p-value for the test of part a. What is your conclusion if $\alpha = .05$?

9.76 A highway construction zone has a posted speed limit of 40 miles per hour. Workers working at the site claim that the mean speed of vehicles passing through this construction zone is at least 50 miles per hour. A random sample of 36 vehicles passing through this zone produced a mean speed of 48 miles per hour. The population standard deviation is known to be 4 miles per hour.

   a. Do you think the sample information is consistent with the workers' claim? Use $\alpha = .025$.
   b. What is the Type I error in this case? Explain. What is the probability of making this error?
   c. Will your conclusion of part a change if the probability of making a Type I error is zero?
   d. Find the p-value for the test of part a. What is your decision if $\alpha = .025$?

9.77 The customers at a bank complained about long lines and the time they had to spend waiting for service. It is known that the customers at this bank had to wait 8 minutes, on average, before being served. The management made some changes to reduce the waiting time for its customers. A sample of 60 customers taken after these changes were made produced a mean waiting time of 7.5 minutes with a standard deviation of 2.1 minutes. Using this sample mean, the bank manager displayed a huge banner inside the bank mentioning that the mean waiting time for customers has been reduced by new changes. Do you think the bank manager's claim is justifiable? Use a 2.5% significance level to answer this question. Use both approaches.

9.78 According to Moebs Services Inc., the cost of an individual checking account at U.S. community banks to these banks was between $175 and $200 in 2011 (*Time*, November 21, 2011). Suppose that the average annual cost of individual checking accounts at U.S. community banks to these banks was $190 in 2011. A recent sample of 40 individual checking accounts selected from U.S. community banks showed that they cost these banks an average of $211 per year with a standard deviation of $35.

   a. Using $\alpha = .025$, can you conclude that the current average annual cost of individual checking accounts at U.S. community banks to these banks is higher than $190? Use the critical-value approach.
   b. Find the range of the p-value for the test of part a. What is your conclusion with $\alpha = .025$?

9.79 An earlier study claimed that U.S. adults spent an average of 114 minutes per day with their family. A recently taken sample of 25 adults from a city showed that they spend an average of 109 minutes per day with their family. The sample standard deviation is 11 minutes. Assume that the times spent by adults with their families have an approximate normal distribution.

   a. Using a 1% significance level, test whether the mean time spent currently by all adults with their families in this city is different from 114 minutes a day.
   b. Suppose the probability of making a Type I error is zero. Can you make a decision for the test of part a without going through the five steps of hypothesis testing? If yes, what is your decision? Explain.

9.80 A computer company that recently introduced a new software product claims that the mean time taken to learn how to use this software is not more than 2 hours for people who are somewhat familiar with computers. A random sample of 12 such persons was selected. The following data give the times taken (in hours) by these persons to learn how to use this software.

1.75	2.25	2.40	1.90	1.50	2.75
2.15	2.25	1.80	2.20	3.25	2.60

Test at a 1% significance level whether the company's claim is true. Assume that the times taken by all persons who are somewhat familiar with computers to learn how to use this software are approximately normally distributed.

9.81 According to the U.S. Census Bureau, 69% of children under the age of 18 years in the United States lived with two parents in 2009. Suppose that in a recent sample of 2000 children, 1298 were living with two parents.

   a. Using the critical value approach and $\alpha = .05$, test whether the current percentage of all children under the age of 18 years in the United States who live with two parents is different from 69%.
   b. How do you explain the Type I error in part a? What is the probability of making this error in part a?
   c. Calculate the p-value for the test of part a. What is your conclusion if $\alpha = .05$?

9.82 In a *Time* Magazine/Aspen poll of American adults conducted by the strategic research firm, Penn Schoen Berland, these adults were asked, "In your opinion, what is more important for the U.S. to focus on in the next decade?" Eighty-three percent of the adults polled said *domestic issues* (*Time*, July 11, 2011). Assume that this percentage is true for the 2011 population of American adults. In a recent random sample of 1400 adults, 1078 held this opinion.

    a. Using the critical-value approach and $\alpha = .01$, test whether the current percentage of American adults who hold the above opinion is less than 83%.

    b. How do you explain the Type I error in part a? What is the probability of making this error in part a?

    c. Calculate the *p*-value for the test of part a. What is your conclusion if $\alpha = .01$?

9.83 A 2008 study performed by careerbuilder.com entitled *No, Really, Your Excuse is Totally Believable!* notes that 11% of workers who call in sick do so to catch up on housework. Suppose that in a survey of 675 male workers who have called in sick, 61 did so to have time to catch up on housework. At the 2% significance level, can you conclude that the proportion of all male workers who call in sick do so to catch up on housework is different from 11%?

9.84 In a poll conducted by *The New York Times* and CBS News, 44% of Americans approve of the job that the Supreme Court is doing (*The New York Times*, June 8, 2012). Assume that this percentage was true for the population of Americans at the time of this poll was conducted. A recent poll of 1300 Americans showed that 39% of them approve of the job that the Supreme Court is doing. At a 2% significance level, can you conclude that the current proportion of Americans who approve of the job that the Supreme Court is doing is different from .44?

9.85 Mong Corporation makes auto batteries. The company claims that 80% of its LL70 batteries are good for 70 months or longer. A consumer agency wanted to check if this claim is true. The agency took a random sample of 40 such batteries and found that 75% of them were good for 70 months or longer.

    a. Using a 1% significance level, can you conclude that the company's claim is false?

    b. What will your decision be in part a if the probability of making a Type I error is zero? Explain.

ADVANCED EXERCISES

9.86 Professor Hansen believes that some people have the ability to predict in advance the outcome of a spin of a roulette wheel. He takes 100 student volunteers to a casino. The roulette wheel has 38 numbers, each of which is equally likely to occur. Of these 38 numbers, 18 are red, 18 are black, and 2 are green. Each student is to place a series of five bets, choosing either a red or a black number before each spin of the wheel. Thus, a student who bets on red has an 18/38 chance of winning that bet. The same is true of betting on black.

    a. Assuming random guessing, what is the probability that a particular student will win all five of his or her bets?

    b. Suppose for each student we formulate the hypothesis test
        H_0: The student is guessing
        H_1: The student has some predictive ability
        Suppose we reject H_0 only if the student wins all five bets. What is the significance level?

    c. Suppose that 2 of the 100 students win all five of their bets. Professor Hansen says, "For these two students we can reject H_0 and conclude that we have found two students with some ability to predict." What do you make of Professor Hansen's conclusion?

9.87 The standard therapy that is used to treat a disorder cures 60% of all patients in an average of 140 visits. A health care provider considers supporting a new therapy regime for the disorder if it is effective in reducing the number of visits while retaining the cure rate of the standard therapy. A study of 200 patients with the disorder who were treated by the new therapy regime reveals that 108 of them were cured in an average of 132 visits with a standard deviation of 38 visits. What decision should be made using a .01 level of significance?

9.88 The package of Sylvania CFL 65-watt replacement bulbs that use only 16 watts claims that these bulbs have an average life of 8000 hours. Assume that the standard deviation of lives of these light bulbs is 400 hours. A skeptical consumer does not think that these light bulbs last as long as the manufacturer claims, and she decides to test 52 randomly selected light bulbs. She has set up the decision rule that if the average life of these 52 light bulbs is less than or equal to 7890

hours, then she will reject company's claim and conclude that the company has printed too high an average life on the packages, and she will write them a letter to that effect. Approximately what significance level is she using? If she decides instead on the decision rule that if the average life of these 52 light bulbs is less than or equal to 7857 hours she will reject the null hypothesis that company's claim is true, then approximately what significance level is she using? Interpret the values you get.

9.89 Before a championship football game, the referee is given a special commemorative coin to toss to decide which team will kick the ball first. Two minutes before game time, he receives an anonymous tip that the captain of one of the teams may have substituted a biased coin that has a 70% chance of showing heads each time it is tossed. The referee has time to toss the coin 10 times to test it. He decides that if it shows 8 or more heads in 10 tosses, he will reject this coin and replace it with another coin. Let *p* be the probability that this coin shows heads when it is tossed once.

    a. Formulate the relevant null and alternative hypotheses (in terms of *p*) for the referee's test.

    b. Using the referee's decision rule, find α for this test.

9.90 A statistician performs the test H_0: $\mu = 15$ versus H_1: $\mu \neq 15$ and finds the *p*-value to be .4546.

    a. The statistician performing the test does not tell you the value of the sample mean and the value of the test statistic. Despite this, you have enough information to determine the pair of *p*-values associated with the following alternative hypotheses.

        i. H_1: $\mu < 15$ **ii.** H_1: $\mu > 15$

    Note that you will need more information to determine which *p*-value goes with which alternative. Determine the pair of *p*-values. Here the value of the sample mean is the same in both cases.

    b. Suppose the statistician tells you that the value of the test statistic is negative. Match the *p*-values with the alternative hypotheses.

Note that the result for one of the two alternatives implies that the sample mean is not on the same side of $\mu = 15$ as the rejection region. Although we have not discussed this scenario in the book, it is important to recognize that there are many real-world scenarios in which this type of situation does occur. For example, suppose the EPA is to test whether or not a company is exceeding a specific pollution level. If the average discharge level obtained from the sample falls below the threshold (mentioned in the null hypothesis), then there would be no need to perform the hypothesis test.

9.91 Since 1984, all automobiles have been manufactured with a middle tail-light. You have been hired to answer the following question: Is the middle tail-light effective in reducing the number of rear-end collisions? You have available to you any information you could possibly want about all rear-end collisions involving cars built before 1984. How would you conduct an experiment to answer the question? In your answer, include things like (a) the precise meaning of the unknown parameter you are testing; (b) H_0 and H_1; (c) a detailed explanation of what sample data you would collect to draw a conclusion; and (d) any assumptions you would make, particularly about the characteristics of cars built before 1984 versus those built since 1984.

SELF-REVIEW TEST

1. A test of hypothesis is always about
  a. a population parameter
  b. a sample statistic
  c. a test statistic

2. A Type I error is committed when
  a. a null hypothesis is not rejected when it is actually false
  b. a null hypothesis is rejected when it is actually true
  c. an alternative hypothesis is rejected when it is actually true

3. A Type II error is committed when
  a. a null hypothesis is not rejected when it is actually false
  b. a null hypothesis is rejected when it is actually true
  c. an alternative hypothesis is rejected when it is actually true

4. A critical value is the value
  a. calculated from sample data
  b. determined from a table (e.g., the normal distribution table or other such tables)
  c. neither a nor b

5. The computed value of a test statistic is the value
  a. calculated for a sample statistic
  b. determined from a table (e.g., the normal distribution table or other such tables)
  c. neither a nor b

6. The observed value of a test statistic is the value
  a. calculated for a sample statistic
  b. determined from a table (e.g., the normal distribution table or other such tables)
  c. neither a nor b

7. The significance level, denoted by α, is
  a. the probability of committing a Type I error
  b. the probability of committing a Type II error
  c. neither a nor b

8. The value of β gives the
  a. probability of committing a Type I error
  b. probability of committing a Type II error
  c. power of the test

9. The value of $1 - \beta$ gives the
  a. probability of committing a Type I error
  b. probability of committing a Type II error
  c. power of the test

10. A two-tailed test is a test with
  a. two rejection regions
  b. two nonrejection regions
  c. two test statistics

11. A one-tailed test
  a. has one rejection region
  b. has one nonrejection region
  c. both a and b

12. The smallest level of significance at which a null hypothesis is rejected is called
  a. α **b.** p-value   **c.** β

13. The sign in the alternative hypothesis in a two-tailed test is always
  a. $<$ **b.** $>$ **c.** $\neq$

14. The sign in the alternative hypothesis in a left-tailed test is always
  a. $<$ **b.** $>$ **c.** $\neq$

15. The sign in the alternative hypothesis in a right-tailed test is always
  a. $<$ **b.** $>$ **c.** $\neq$

16. According to the Kaiser Family Foundation, U.S. workers who had employer-provided health insurance paid an average premium of $1170 for single (one person) health insurance coverage during 2013 (www.kff.org). Suppose that a recent random sample of 100 workers with employer-provided health insurance selected from a large city paid an average premium of $1198 for single health insurance coverage. Assume that such premiums paid by all such workers in this city have a standard deviation of $125.
  a. Using the critical-value approach and a 1% significance level, can you conclude that the current average such premium paid by all such workers in this city is different from $1170?
  b. Using the critical-value approach and a 2.5% significance level, can you conclude that the current average such premium paid by all such workers in this city is higher than $1170?
  c. What is the Type I error in parts a and b? What is the probability of making this error in each of parts a and b?
  d. Calculate the p-value for the test of part a. What is your conclusion if $\alpha = .01$?
  e. Calculate the p-value for the test of part b. What is your conclusion if $\alpha = .025$?

17. A minor league baseball executive has become concerned about the slow pace of games played in her league, fearing that it will lower attendance. She meets with the league's managers and umpires and discusses guidelines for speeding up the games. Before the meeting, the mean duration of nine-inning games was 3 hours, 5 minutes (i.e., 185 minutes). A random sample of 36 nine-inning games after the meeting showed a mean of 179 minutes with a standard deviation of 12 minutes.
  a. Testing at a 1% significance level, can you conclude that the mean duration of nine-inning games has decreased after the meeting?

b. What is the Type I error in part a? What is the probability of making this error?

c. What will your decision be in part a if the probability of making a Type I error is zero? Explain.

d. Find the range for the *p*-value for the test of part a. What is your decision based on this *p*-value?

18. An editor of a New York publishing company claims that the mean time taken to write a textbook is at least 31 months. A sample of 16 textbook authors found that the mean time taken by them to write a textbook was 25 months with a standard deviation of 7.2 months.

a. Using a 2.5% significance level, would you conclude that the editor's claim is true? Assume that the time taken to write a textbook is approximately normally distributed for all textbook authors.

b. What is the Type I error in part a? What is the probability of making this error?

c. What will your decision be in part a if the probability of making a Type I error is .001?

19. A financial advisor claims that less than 50% of adults in the United States have a will. A random sample of 1000 adults showed that 450 of them have a will.

a. At a 5% significance level, can you conclude that the percentage of people who have a will is less than 50%?

b. What is the Type I error in part a? What is the probability of making this error?

c. What would your decision be in part a if the probability of making a Type I error were zero? Explain.

d. Find the *p*-value for the test of hypothesis mentioned in part a. Using this *p*-value, will you reject the null hypothesis if $\alpha = .05$? What if $\alpha = .01$?

MINI-PROJECTS

Note: The Mini-Projects are located on the text's Web site, www.wiley.com/college/mann.

DECIDE FOR YOURSELF

Note: The Decide for Yourself feature is located on the text's Web site, www.wiley.com/college/mann.

TECHNOLOGY INSTRUCTIONS | CHAPTER 9

Note: Complete TI-84, Minitab, and Excel manuals are available for download at the textbook's Web site, www.wiley.com/college/mann.

TI-84 Color/TI-84

The TI-84 Color Technology Instructions feature of this text is written for the TI-84 Plus C color graphing calculator running the 4.0 operating system. Some screens, menus, and functions will be slightly different in older operating systems. The TI-84 and TI-84 Plus can perform all of the same functions but will not have the "Color" option referenced in some of the menus.

Testing a Hypothesis about μ, σ Known, for Example 9–2 of the Text

1. Select **STAT > TESTS > Z-Test**.

2. Use the following settings in the **Z-Test** menu (see **Screen 9.1**):

   • At the **Inpt** prompt, select **Stats**.

   Note: If you have the data in a list, select **Data** at the **Inpt** prompt.

   • At the μ_0 prompt, type 10.

   • At the σ prompt type 2.4.

   • At the $\bar{x}$ prompt, type 9.2.

   • At the n prompt, type 36.

   • At the μ prompt, select $< \mu_0$.

3. Highlight **Calculate** and press **ENTER**.

4. The output includes the test statistic and the *p*-value. (See **Screen 9.2**.)

Compare the test statistic to the critical-value of z or the *p*-value to the value of α and make a decision.

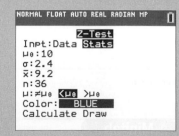

Screen 9.1

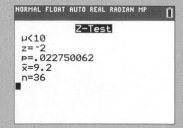

Screen 9.2

Testing a Hypothesis About μ, σ Unknown, for Example 9–5 of the Text

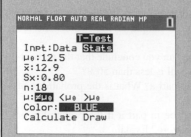

Screen 9.3

Screen 9.4

1. Select **STAT > TESTS > T-Test**.

2. Use the following settings in the **T-Test** menu (see **Screen 9.3**):

   • At the **Inpt** prompt, select **Stats**.

     Note: If you have the data in a list, select **Data** at the **Inpt** prompt.

   • At the μ_0 prompt, type 12.5.

   • At the $\bar{x}$ prompt, type 12.9.

   • At the **S**x prompt, type 0.80.

   • At the **n** prompt, type 18.

   • At the μ prompt, select $\neq \mu_0$.

3. Highlight **Calculate** and press **ENTER**.

4. The output includes the test statistic and the p-value. (See **Screen 9.4**.)

Compare the test statistic to the critical value of t or p-value to the value of α and make a decision.

Testing a Hypothesis About p for Example 9–9 of the Text

Screen 9.5

Screen 9.6

1. Select **STAT > TESTS > 1-PropZTest**.

2. Use the following settings in the **1-PropZTest** menu (see **Screen 9.5**):

   • At the p_0 prompt, type 0.20.

   • At the **x** prompt, type 440.

     Note: The value of x is the number of successes in the sample and must be a whole number or the calculator will return an error message. To find x, multiply n by $\hat{p}$ and round the result to the nearest whole number.

   • At the **n** prompt, type 2000.

   • At the **prop** prompt, select $\neq p_0$.

3. Highlight **Calculate** and press **ENTER**.

4. The output includes the test statistic and the p-value. (See **Screen 9.6**.)

Compare the test statistic to the critical value of z or the p-value to the value of α and make a decision.

Minitab

The Minitab Technology Instructions feature of this text is written for Minitab version 17. Some screens, menus, and functions will be slightly different in older versions of Minitab.

Testing a Hypothesis About μ, σ Known, for Example 9–2 of the Text

1. Select **Stat > Basic Statistics > 1-Sample Z**.

2. Use the following settings in the dialog box that appears on screen (see **Screen 9.7**):

- From the drop-down menu, select **Summarized Data**.

  Note: If you have the data in a column, select **One or more samples, each in a column**, type the column name(s) in the box, and move to step 3 below.

- In the **Sample size** box, type 36.

- In the **Sample mean** box, type 9.2.

- In the **Known standard deviation** box, type 2.4.

- Next to **Perform hypothesis test**, check the check box.

- In the **Hypothesized mean** box, type 10.

Screen 9.7

3. Select **Options**. When the new dialog box appears on screen, select **Mean < hypothesized mean** from the **Alternative hypothesis** box.

4. Click **OK** in both dialog boxes.

5. The output, including the test statistic and *p*-value, will be displayed in the Session window. (See **Screen 9.8**.)

Compare the test statistic to the critical value of *z* or the *p*-value to the value of α and make a decision.

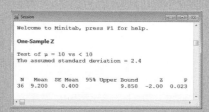

Screen 9.8

Testing a Hypothesis About μ, σ Unknown, for Example 9–5 of the Text

1. Select **Stat > Basic Statistics > 1-Sample t**.

2. Use the following settings in the dialog box that appears on screen (see **Screen 9.9**):

- From the drop-down menu, select **Summarized Data**.

  Note: If you have the data in a column, select **One or more samples, each in a column**, type the column name(s) in the box, and move to step 3 below.

- In the **Sample size** box, type 18.

- In the **Sample mean** box, type 12.9.

- In the **Standard deviation** box, type 0.80.

- Next to **Perform hypothesis test**, check the check box.

- In the **Hypothesized mean** box, type 12.5.

Screen 9.9

3. Select **Options**. When the new dialog box appears on screen, select **Mean ≠ hypothesized mean** from the **Alternative hypothesis** box.

4. Click **OK** in both dialog boxes.

5. The output, including the test statistic and *p*-value, will be displayed in the Session window. (See **Screen 9.10**.)

Compare the test statistic to the critical value of *t* or the *p*-value to the value of α and make a decision.

Screen 9.10

Testing a Hypothesis About *p* for Example 9–9 of the Text

1. Select **Stat > Basic Statistics > 1 Proportion**.

2. Use the following settings in the dialog box that appears on screen (see **Screen 9.11**):

- From the drop-down menu, select **Summarized Data**.

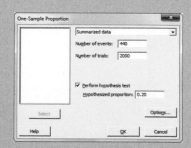

Screen 9.11

Note: If you have the data in a column, select **One or more samples, each in a column**, type the column name(s) in the box, and move to step 3 below.

- In the **Number of events** box, type 440.

  Note: The number of events is the number of successes in the sample, and this must be a whole number or Minitab will return an error message. To find the number of events, multiply n by $\hat{p}$ and round the result to the nearest whole number.

- In the **Number of trials** box, type 2000.
- Next to **Perform hypothesis test**, check the check box.
- In the **Hypothesized mean** box, type 0.20.

3. Select **Options**. Use the following settings in the dialog box that appears on screen:

- Select **Proportion ≠ hypothesized proportion** from the **Alternative hypothesis** box,
- In the Method box, select **Normal approximation** from the drop-down menu.

4. Click **OK** in both dialog boxes.

5. The output, including the test statistic and *p*-value, will be displayed in the Session window. (See **Screen 9.12**.)

Compare the test statistic to the critical value of z or the *p*-value to the value of α and make a decision.

Screen 9.12

Excel

The Excel Technology Instructions feature of this text is written for Excel 2013. Some screens, menus, and functions will be slightly different in older versions of Excel. To enable some of the advanced Excel functions, you must enable the Data Analysis Add-In for Excel. For Excel 2007 and older versions of Excel, replace the function Z.TEST with the function ZTEST, and replace NORM.S.DIST with the function NORMSDIST.

Testing a Hypothesis About μ, σ Known

1. To perform this test, Excel must have the raw data, not summary statistics, so enter these 10 data values into cells A1 through A10 (see **Screen 9.13**):

$$81, 79, 62, 98, 74, 82, 85, 72, 90, 88$$

2. Click on cell B1.

Screen 9.18

3. Suppose we wish to test H_0: $\mu = 75$ and it is known that $\sigma = 12$. Type one of the following commands into cell B1 based on your alternative hypothesis (see **Screen 9.13**):

- For H_1: $\mu > 75$, type **=Z.TEST(A1:A10,75,12)**.
- For H_1: $\mu < 75$, type **=1 − Z.TEST(A1:A10,75,12)**.
- For H_1: $\mu \neq 75$, type **=2*MIN(Z.TEST(A1:A10,75,12),1-Z.TEST(A1:A10,75,12))**.

4. The output is the *p*-value for the test.

Compare the *p*-value to the value of α and make a decision.

Testing a Hypothesis About μ, σ Unknown

1. To perform this test, Excel must have the raw data, not summary statistics, so enter these 10 data values into cells A1 through A10 (see **Screen 9.14**):

81, 79, 62, 98, 74, 82, 85, 72, 90, 88

2. Suppose we wish to test H_0: $\mu = 75$. There is no Excel function for a one-sample t-test for μ, but it is possible to trick Excel into performing these calculations for us.

3. Enter 0 into cells B1 through B10. (See **Screen 9.14**.)

4. Click **DATA** and then click **Data Analysis Tools** in the **Analysis** group.

Screen 9.14

5. Select **t-Test: Two Sample Assuming Unequal Variances** from the dialog box that appears on screen.

6. Use the following settings in the dialog box that appears on screen (see **Screen 9.14**):

   - Type **A1:A10** in the **Variable 1 Range** box.

   - Type **B1:B10** in the **Variable 2 Range** box.

   - Type **75** in the Hypothesized Mean Difference box.

   - Select **New Worksheet Ply** from the **Output Options**.

7. Click **OK**.

8. When the output appears, resize column A so that it is easier to read. (See **Screen 9.15**.)

	A	B	C
1	t-Test: Two-Sample Assuming Unequal Variances		
2			
3		Variable 1	Variable 2
4	Mean	81.1	0
5	Variance	103.4333	0
6	Observations	10	10
7	Hypothesized Mean Difference	75	
8	df	9	
9	t Stat	1.896704	
10	P(T<=t) one-tail	0.045181	
11	t Critical one-tail	1.833113	
12	P(T<=t) two-tail	0.090363	
13	t Critical two-tail	2.262157	

Screen 9.15

9. The correct formula for the p-value depends on your alternative hypothesis and the t statistic.

   - For H_1: $\mu > 75$,

     - If the t statistic is positive, use the p-value from **P(T<=t) one-tail**.

     - If the t statistic is negative, the p-value is found by subtracting **P(T<=t) one-tail** from 1.

   - For H_1: $\mu < 75$,

     - If the t statistic is positive, the p-value is found by subtracting **P(T<=t) one-tail** from 1.

     - If the t statistic is negative, use the p-value from **P(T<=t) one-tail**.

   - For H_1: $\mu \neq 75$ use the p-value from **P(T<=t) two-tail**.

Testing a Hypothesis about p for Example 9–9 of the Text

1. There is no Excel function that can be used to create a confidence interval for a population proportion, so we must enter a few calculations in sequence. To make our work easier to understand, type the following text into the corresponding spreadsheet cells (see **Screen 9.16**):

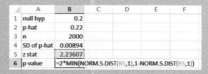

	A	B	C	D	E	F
1	null hyp	0.2				
2	p-hat	0.22				
3	n	2000				
4	SD of p-hat	0.00894				
5	z stat	2.23607				
6	p-value	=2*MIN(NORM.S.DIST(B5,1),1-NORM.S.DIST(B5,1))				

Screen 9.16

   - Type "**null hyp**" in cell A1.

   - Type "**p-hat**" in cell A2.

   - Type "**n**" in cell A3.

   - Type "**SD of p-hat**" in cell A4.

   - Type "**z stat**" in cell A5.

   - Type "**p-value**" in cell A6.

2. Click on cell B1 and type **0.20**.

3. Click on cell B2 and type **0.22**.

4. Click on cell B3 and type **2000**.

5. Click on cell B4 and type =**SQRT(B1*(1-B1)/B3)**.

6. Click on cell B5 and type =**(B2-B1)/B4**.

7. Click on cell B6. The correct formula for the p-value depends on your alternative hypothesis.

- For H_1: $p > 0.20$, type =**1−NORM.S.DIST(B5,1)**.

- For H_1: $p < 0.20$, type =**NORM.S.DIST(B5,1)**.

- For H_1: $p \neq 0.20$, type =**2*MIN(NORM.S.DIST(B5,1),1-NORM.S.DIST(B5,1))** (see **Screen 9.16**).

TECHNOLOGY ASSIGNMENTS

TA9.1 According to Freddie Mac (http://www.freddiemac.com/pmms/pmms30.htm), the average interest rate on a 30-year fixed-rate mortgage in October 2011 was 4.07%. The following data represent the interest rates on 50 randomly selected 30-year fixed-rate mortgages approved during the week of November 7–11, 2011:

4.19	4.43	4.43	4.29	4.16	4.42	3.76	4.07	3.78	3.97
4.15	4.04	3.86	3.96	4.03	4.45	4.40	4.07	4.30	3.98
3.96	3.68	4.64	3.95	3.95	3.83	4.26	4.28	4.30	3.87
3.89	4.10	4.33	3.84	3.94	3.78	4.11	4.19	4.26	3.91
3.96	4.19	4.10	3.73	4.04	4.69	3.88	4.34	3.93	4.09

a. Perform a graphical analysis to determine whether the assumption that the distribution of all 30-year fixed-rate mortgage rates is normal is a reasonable assumption.

b. Test at a 5% significance level whether the average interest rate on all 30-year fixed-rate mortgages granted during the week of November 7–11, 2011, was different from 4.07%.

TA9.2 The Wechsler adult Intelligence Scale-III (WAIS-III) is a widely used intelligence test. On this test, a score of 90-109 is considered *typical*. An administrator at a college believes that the incoming freshmen class is of higher intelligence than typical. She selects a random sample of 36 students from the incoming class and records their intelligence test scores, which are produced below.

95	123	128	114	118	102
122	98	125	120	121	132
113	117	134	100	121	130
121	99	113	111	113	115
116	103	114	107	131	128
115	129	126	118	106	115

Test if this sample data support this administrator's belief that the mean intelligence test score for all freshman is greater than 109. Use $\alpha = .05$.

TA9.3 According to an earlier study, the mean amount spent on clothes by American women is $675 per year. A researcher wanted to check if this result still holds true. A random sample of 39 women taken recently by this researcher produced the following data on the amounts they spent on clothes last year.

671	1284	328	1698	827	921	725	304	382	539
1070	854	669	328	537	849	930	1234	1195	738
341	189	867	923	721	125	298	473	876	932
973	931	460	1430	391	887	958	674	1482	

Test at a 1% significance level whether the mean expenditure on clothes for American women for last year is different from $675. Assume that the population standard deviation is $132.

TA9.4 The president of a large university claims that the mean time spent partying by all students at the university is not more than 7 hours per week. The following data give the times spent partying during the previous week by a random sample of 16 students taken from this university.

12	9	5	15	11	13	10	6
4	11	6	9	13	6	16	8

Test at a 1% significance level whether the president's claim is true. Assume that the times spent partying by all students at this university have an approximately normal distribution and the population standard deviation is known to be 2 hours.

TA9.5 In a November 2003 Lemelson-MIT survey, adult Americans were asked which invention they hated the most but could not live without. The cell phone was the most frequently named such invention, chosen by 30% of the adults (*Business Week*, March 1, 2004). In a recent random sample of 1000 adults, the same question was asked, and 363 of the respondents chose the cell phone. Test at the 5% significance level whether the current percentage of all adults who feel that the cell phone is the invention that they hate the most but cannot live without ls different from 30%.

TA9.6 Take a random sample of 150 runners from Data Set IV, which contains the Manchester Road Race Data. Use these data to perform 10 hypothesis tests on the variable that identifies whether a runner is from Connecticut, where p is the proportion of all runners who are from Connecticut. Specifically, perform a test for each of the sets of hypotheses

$$H_0: p = .70 \quad \text{versus} \quad H_1: p \neq .70$$
$$\text{through}$$
$$H_0: p = .79 \quad \text{versus} \quad H_1: p \neq .79$$

incrementing the hypothesized value by .01. Which null hypotheses are rejected at the 10% significance level? What conclusions can you make about the value of p based on the results of these tests?

TA9.7 A manufacturer of contact lenses must ensure that each contact lens is properly manufactured with no flaws, since flaws lead to poor vision or even eye damage. In a recent quality control check, 48 of 5000 lenses were found to have flaws. Using this sample information and 1% significance level, test the hypothesis $H_0: p = .01$ versus $H_1: p > .01$.

TA9.8 A mail-order company claims that at least 60% of all orders it receives are mailed within 48 hours. From time to time the quality control department at the company checks if this promise is kept. Recently, the quality control department at this company took a sample of 400 orders and found that 224 of them were mailed within 48 hours of the placement of the orders. Test at a 1% significance level whether or not the company's claim is true.

CHAPTER 10

© Westend61 GmbH/Alamy Stock Photo

Estimation and Hypothesis Testing: Two Populations

Which toothpaste do you use? Are you a loyal user of this toothpaste? Will you ever switch to another toothpaste? Many of us never switch from one brand to another brand of a product. So, how do we find which of the two toothpastes has more loyal users? In other words, how do we test that one toothpaste has a higher percentage of its users who will not switch to the other toothpaste? (See Example 10–14.)

Chapters 8 and 9 discussed the estimation and hypothesis-testing procedures for μ and p involving a single population. This chapter extends the discussion of estimation and hypothesis-testing procedures to the difference between two population means and the difference between two population proportions. For example, we may want to make a confidence interval for the difference between the mean prices of houses in California and in New York, or we may want to test the hypothesis that the mean price of houses in California is different from that in New York. As another example, we may want to make a confidence interval for the difference between the proportions of all male and female adults who abstain from drinking, or we may want to test the hypothesis that the proportion of all adult men who abstain from drinking is different from the proportion of all adult women who abstain from drinking. Constructing confidence intervals and testing hypotheses about population parameters are referred to as *making inferences*.

10.1 | Inferences About the Difference Between Two Population Means for Independent Samples: σ_1 and σ_2 Known

Let μ_1 be the mean of the first population and μ_2 be the mean of the second population. Suppose we want to make a confidence interval and test a hypothesis about the difference between these two population means, that is, $\mu_1 - \mu_2$. Let $\bar{x}_1$ be the mean of a sample taken from the first population and $\bar{x}_2$ be the mean of a sample taken from the second population. Then, $\bar{x}_1 - \bar{x}_2$ is the sample statistic that is used to make an interval estimate and to test a hypothesis about $\mu_1 - \mu_2$. This section discusses how to make confidence intervals and test hypotheses about $\mu_1 - \mu_2$ when certain conditions (to be explained later in this section) are satisfied. First we explain the concepts of independent and dependent samples.

10.1.1 Independent Versus Dependent Samples

Two samples are **independent** if they are drawn from two different populations and the elements of one sample have no relationship to the elements of the second sample. If the elements of the two samples are somehow related, then the samples are said to be **dependent**. Thus, in two independent samples, the selection of one sample has no effect on the selection of the second sample. Examples 10–1 and 10–2 illustrate independent and dependent samples, respectively.

Two samples drawn from two populations are **independent** if the selection of one sample from one population does not affect the selection of the second sample from the second population. Otherwise, the samples are **dependent**.

| EXAMPLE 10–1 | Salaries of Male and Female Executives |

Illustrating two independent samples.

Suppose we want to estimate the difference between the mean salaries of all male and all female executives. To do so, we draw two samples, one from the population of male executives and another from the population of female executives. These two samples are *independent* because they are drawn from two different populations, and the samples have no effect on each other.

| EXAMPLE 10–2 | Weights Before and After a Weight Loss Program |

Illustrating two dependent samples.

Suppose we want to estimate the difference between the mean weights of all participants before and after a weight loss program. To accomplish this, suppose we take a sample of 40 participants and measure their weights before and after the completion of this program. Note that these two samples include the same 40 participants. This is an example of two *dependent* samples.

This section and Sections 10.2, 10.3, and 10.5 discuss how to make confidence intervals and test hypotheses about the difference between two population parameters when samples are independent. Section 10.4 discusses how to make confidence intervals and test hypotheses about the difference between two population means when samples are dependent.

10.1.2 Mean, Standard Deviation, and Sampling Distribution of $\bar{x}_1 - \bar{x}_2$

Suppose we select two (independent) samples from two different populations that are referred to as population 1 and population 2. Let

μ_1 = the mean of population 1

μ_2 = the mean of population 2

σ_1 = the standard deviation of population 1

σ_2 = the standard deviation of population 2

n_1 = the size of the sample drawn from population 1

n_2 = the size of the sample drawn from population 2

$\bar{x}_1$ = the mean of the sample drawn from population 1

$\bar{x}_2$ = the mean of the sample drawn from population 2

Then, as we discussed in Chapters 8 and 9, if

1. The standard deviation σ_1 of population 1 is known
2. At least one of the following two conditions is fulfilled:
   i. The sample is large (i.e., $n_1 \geq 30$)
   ii. If the sample size is small, then the population from which the sample is drawn is approximately normally distributed

then the sampling distribution of $\bar{x}_1$ is approximately normal with its mean equal to μ_1 and the standard deviation equal to $\sigma_1/\sqrt{n_1}$, assuming that $n_1/N_1 \leq .05$.

Similarly, if

1. The standard deviation σ_2 of population 2 is known
2. At least one of the following two conditions is fulfilled:
   i. The sample is large (i.e., $n_2 \geq 30$)
   ii. If the sample size is small, then the population from which the sample is drawn is approximately normally distributed

then the sampling distribution of $\bar{x}_2$ is approximately normal with its mean equal to μ_2 and the standard deviation equal to $\sigma_2/\sqrt{n_2}$, assuming that $n_2/N_2 \leq .05$.

Using these results, we can make the following statements about the mean, the standard deviation, and the shape of the sampling distribution of $\bar{x}_1 - \bar{x}_2$.

If the following conditions are satisfied:

1. The two samples are independent
2. The standard deviations σ_1 and σ_2 of the two populations are known
3. At least one of the following two conditions is fulfilled:
   i. Both samples are large (i.e., $n_1 \geq 30$ and $n_2 \geq 30$)
   ii. If either one or both sample sizes are small, then both populations from which the samples are drawn are approximately normally distributed

then the sampling distribution of $\bar{x}_1 - \bar{x}_2$ is approximately normally distributed with its mean and standard deviation,[1] respectively,

$$\mu_{\bar{x}_1 - \bar{x}_2} = \mu_1 - \mu_2$$

and

$$\sigma_{\bar{x}_1-\bar{x}_2} = \sqrt{\frac{\sigma_1^2}{n_1} + \frac{\sigma_2^2}{n_2}}$$

In these cases, we can use the normal distribution to make a confidence interval and test a hypothesis about $\mu_1 - \mu_2$. Figure 10.1 shows the sampling distribution of $\bar{x}_1 - \bar{x}_2$ when the above conditions are fulfilled.

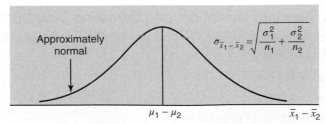

Approximately normal

$$\sigma_{\bar{x}_1-\bar{x}_2} = \sqrt{\frac{\sigma_1^2}{n_1} + \frac{\sigma_2^2}{n_2}}$$

$\mu_1 - \mu_2$

$\bar{x}_1 - \bar{x}_2$

Figure 10.1   The sampling distribution of $\bar{x}_1 - \bar{x}_2$.

[1]The formula for the standard deviation of $\bar{x}_1 - \bar{x}_2$ can also be written as

$$\sigma_{\bar{x}_1-\bar{x}_2} = \sqrt{\sigma_{\bar{x}_1}^2 + \sigma_{\bar{x}_2}^2}$$

where $\sigma_{\bar{x}_1} = \sigma_1/\sqrt{n_1}$ and $\sigma_{\bar{x}_2} = \sigma_2/\sqrt{n_2}$.

Sampling Distribution, Mean, and Standard Deviation of $\bar{x}_1 - \bar{x}_2$ When the conditions listed on the previous page are satisfied, the **sampling distribution of $\bar{x}_1 - \bar{x}_2$** is (approximately) normal with its **mean and standard deviation** as, respectively,

$$\mu_{\bar{x}_1 - \bar{x}_2} = \mu_1 - \mu_2 \quad \text{and} \quad \sigma_{\bar{x}_1 - \bar{x}_2} = \sqrt{\frac{\sigma_1^2}{n_1} + \frac{\sigma_2^2}{n_2}}$$

Note that to apply the procedures learned in this chapter, the samples selected must be simple random samples.

10.1.3   Interval Estimation of $\mu_1 - \mu_2$

By constructing a confidence interval for $\mu_1 - \mu_2$, we find the difference between the means of two populations. For example, we may want to find the difference between the mean heights of male and female adults. The difference between the two sample means, $\bar{x}_1 - \bar{x}_2$, is the point estimator of the difference between the two population means, $\mu_1 - \mu_2$. When the conditions mentioned earlier in this section hold true, we use the normal distribution to make a confidence interval for the difference between the two population means. The following formula gives the interval estimation for $\mu_1 - \mu_2$.

Confidence Interval for $\mu_1 - \mu_2$ When using the normal distribution, the $(1 - \alpha)100\%$ **confidence interval for $\mu_1 - \mu_2$** is

$$(\bar{x}_1 - \bar{x}_2) \pm z\sigma_{\bar{x}_1 - \bar{x}_2}$$

The value of z is obtained from the normal distribution table for the given confidence level. The value of $\sigma_{\bar{x}_1 - \bar{x}_2}$ is calculated as explained earlier. Here, $\bar{x}_1 - \bar{x}_2$ is the point estimator of $\mu_1 - \mu_2$.

Note that in the real world, σ_1 and σ_2 are never known. Consequently we will never use the procedures of this section, but we are discussing these procedures in this book for the information of the readers.

Example 10–3 illustrates the procedure to construct a confidence interval for $\mu_1 - \mu_2$ using the normal distribution.

EXAMPLE 10–3	**Annual Premiums for Employer-Sponsored Health Insurance**

Constructing a confidence interval for $\mu_1 - \mu_2$: σ_1 and σ_2 known, and samples are large.

According to the Kaiser Family Foundation surveys in 2014 and 2012, the average annual premiums for employer-sponsored health insurance for family coverage was $16,834 in 2014 and $15,745 in 2012 (www.kff.org). Suppose that these averages are based on random samples of 250 and 200 employees who had such employer-sponsored health insurance plans for 2014 and 2012, respectively. Further assume that the population standard deviations for 2014 and 2012 were $2160 and $1990, respectively. Let μ_1 and μ_2 be the population means for such annual premiums for the years 2014 and 2012, respectively.

  (a) What is the point estimate of $\mu_1 - \mu_2$?
  (b) Construct a 97% confidence interval for $\mu_1 - \mu_2$.

Solution   Let us refer to the employees who had employer-sponsored health insurance for family coverage in 2014 as population 1 and those for 2012 as population 2. Then the respective samples are samples 1 and 2. Let $\bar{x}_1$ and $\bar{x}_2$ be the means of the two samples, respectively. From the given information,

$$\text{For 2014:} \quad n_1 = 250, \quad \bar{x}_1 = \$16{,}834, \quad \sigma_1 = \$2160$$

$$\text{For 2012:} \quad n_2 = 200, \quad \bar{x}_2 = \$15{,}745, \quad \sigma_2 = \$1990$$

(a) The point estimate of $\mu_1 - \mu_2$ is given by the value of $\bar{x}_1 - \bar{x}_2$. Thus,

$$\text{Point estimate of } \mu_1 - \mu_2 = \$16,834 - \$15,745 = \textbf{\$1089}$$

(b) The confidence level is $1 - \alpha = .97$. From the normal distribution table, the values of z for .0150 and .9850 areas to the left are -2.17 and 2.17, respectively. Hence, we will use $z = 2.17$ in the confidence interval formula. First we calculate the standard deviation of $\bar{x}_1 - \bar{x}_2$, $\sigma_{\bar{x}_1 - \bar{x}_2}$, as follows:

$$\sigma_{\bar{x}_1 - \bar{x}_2} = \sqrt{\frac{\sigma_1^2}{n_1} + \frac{\sigma_2^2}{n_2}} = \sqrt{\frac{(2160)^2}{250} + \frac{(1990)^2}{200}} = \$196.1196064$$

Next, substituting all the values in the confidence interval formula, we obtain a 97% confidence interval for $\mu_1 - \mu_2$ as

$$(\bar{x}_1 - \bar{x}_2) \pm z\sigma_{\bar{x}_1 - \bar{x}_2} = (\$16,834 - \$15,745) \pm 2.17(196.1196064)$$

$$= 1089 \pm 425.58 = \textbf{\$663.42 to \$1514.58}$$

Thus, with 97% confidence we can state that the difference between the average annual premiums for employer-sponsored health insurance for family coverage in 2014 and 2012 is between \$663.42 and \$1514.58. The value $z\sigma_{\bar{x}_1 - \bar{x}_2} = \$425.58$ is called the margin of error for this estimate.

Note that in Example 10–3 both sample sizes were large and the population standard deviations were known. If the standard deviations of the two populations are known, at least one of the sample sizes is small, and both populations are approximately normally distributed, we use the normal distribution to make a confidence interval for $\mu_1 - \mu_2$. The procedure in this case is exactly the same as in Example 10–3.

10.1.4   Hypothesis Testing About $\mu_1 - \mu_2$

It is often necessary to compare the means of two populations. For example, we may want to know if the mean price of houses in Chicago is different from that in Los Angeles. Similarly, we may be interested in knowing if American children spend, on average, fewer hours in school than Japanese children do. In both these cases, we will perform a test of hypothesis about $\mu_1 - \mu_2$. The alternative hypothesis in a test of hypothesis may be that the means of the two populations are different, or that the mean of the first population is greater than the mean of the second population, or that the mean of the first population is less than the mean of the second population. These three situations are described next.

1. Testing an alternative hypothesis that the means of two populations are different is equivalent to $\mu_1 \neq \mu_2$, which is the same as $\mu_1 - \mu_2 \neq 0$.
2. Testing an alternative hypothesis that the mean of the first population is greater than the mean of the second population is equivalent to $\mu_1 > \mu_2$, which is the same as $\mu_1 - \mu_2 > 0$.
3. Testing an alternative hypothesis that the mean of the first population is less than the mean of the second population is equivalent to $\mu_1 < \mu_2$, which is the same as $\mu_1 - \mu_2 < 0$.

The procedure that is followed to perform a test of hypothesis about the difference between two population means is similar to the one that is used to test hypotheses about single-population parameters in Chapter 9. The procedure involves the same five steps for the critical-value approach that were used in Chapter 9 to test hypotheses about μ and p. Here, again, if the following conditions are satisfied, we will use the normal distribution to make a test of hypothesis about $\mu_1 - \mu_2$.

1. The two samples are independent.
2. The standard deviations σ_1 and σ_2 of the two populations are known.
3. At least one of the following two conditions is fulfilled:

i. Both samples are large (i.e., $n_1 \geq 30$ and $n_2 \geq 30$)

ii. If either one or both sample sizes are small, then both populations from which the samples are drawn are approximately normally distributed

---

Test Statistic z for $\bar{x}_1 - \bar{x}_2$ When using the normal distribution, the value of the **test statistic z for $\bar{x}_1 - \bar{x}_2$** is computed as

$$z = \frac{(\bar{x}_1 - \bar{x}_2) - (\mu_1 - \mu_2)}{\sigma_{\bar{x}_1 - \bar{x}_2}}$$

The value of $\mu_1 - \mu_2$ is substituted from H_0. The value of $\sigma_{\bar{x}_1 - \bar{x}_2}$ is calculated as earlier in this section.

Example 10–4 shows how to make a test of hypothesis about $\mu_1 - \mu_2$.

EXAMPLE 10–4   Annual Premiums for Employer-Sponsored Health Insurance

Making a two-tailed test of hypothesis about $\mu_1 - \mu_2$: σ_1 and σ_2 are known, and samples are large.

Refer to Example 10–3 about the average annual premiums for employer-sponsored health insurance for family coverage in 2014 and 2012. Test at a 1% significance level whether the population means for the two years are different.

Solution   From the information given in Example 10–3,

$$\text{For 2014:} \qquad n_1 = 250, \qquad \bar{x}_1 = \$16{,}834, \qquad \sigma_1 = \$2160$$

$$\text{For 2012:} \qquad n_2 = 200, \qquad \bar{x}_2 = \$15{,}745, \qquad \sigma_2 = \$1990$$

Let μ_1 and μ_2 be the population means for such annual premiums for the years 2014 and 2012, respectively. Let $\bar{x}_1$ and $\bar{x}_2$ be the corresponding sample means.

Step 1.   *State the null and alternative hypotheses.*

We are to test whether the two population means are different. The two possibilities are:

i. The mean annual premiums for the years 2014 and 2012 are not different. In other words, $\mu_1 = \mu_2$, which can be written as $\mu_1 - \mu_2 = 0$.

ii. The mean annual premiums for the years 2014 and 2012 are different. That is, $\mu_1 \neq \mu_2$, which can be written as $\mu_1 - \mu_2 \neq 0$.

Considering these two possibilities, the null and alternative hypotheses are, respectively,

$$H_0\text{: } \mu_1 - \mu_2 = 0 \qquad \text{(The two population means are not different.)}$$
$$H_0\text{: } \mu_1 - \mu_2 \neq 0 \qquad \text{(The two population means are different.)}$$

Step 2.   *Select the distribution to use.*

Here, the population standard deviations, σ_1 and σ_2, are known, and both samples are large ($n_1 \geq 30$ and $n_2 \geq 30$). Therefore, the sampling distribution of $\bar{x}_1 - \bar{x}_2$ is approximately normal, and we use the normal distribution to perform the hypothesis test.

Step 3.   *Determine the rejection and nonrejection regions.*

The significance level is given to be .01. The $\neq$ sign in the alternative hypothesis indicates that the test is two-tailed. The area in each tail of the normal distribution curve is $\alpha/2 = .01/2 = .005$. The critical values of z for .0050 and .9950 areas to the left are (approximately) -2.58 and 2.58 from Table IV of Appendix B. These values are shown in Figure 10.2.

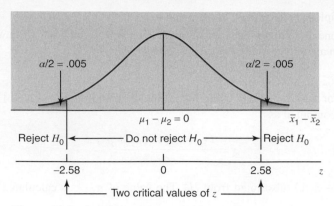

Figure 10.2   Rejection and nonrejection regions.

Step 4.   *Calculate the value of the test statistic.*

The value of the test statistic z for $\bar{x}_1 - \bar{x}_2$ is computed as follows:

$$\sigma_{\bar{x}_1 - \bar{x}_2} = \sqrt{\frac{\sigma_1^2}{n_1} + \frac{\sigma_2^2}{n_2}} = \sqrt{\frac{(2160)^2}{250} + \frac{(1990)^2}{200}} = \$196.1196064$$

$$z = \frac{(\bar{x}_1 - \bar{x}_2) - (\mu_1 - \mu_2)}{\sigma_{\bar{x}_1 - \bar{x}_2}} = \frac{(\$16{,}834 - \$15{,}745) - 0}{196.1196064} = 5.55$$

From H_0

Step 5.   *Make a decision.*

Because the value of the test statistic $z = 5.55$ falls in the rejection region, we reject the null hypothesis H_0. Therefore, we conclude that the average annual premiums for employer-sponsored health insurance for family coverage were different for 2014 and 2012.

Using the *p*-Value to Make a Decision for Example 10–4

We can use the *p*-value approach to make the above decision. To do so, we keep Steps 1 and 2. Then in Step 3 we calculate the value of the test statistic z (as done in Step 4) and find the *p*-value for this z from the normal distribution table. In Step 4, the z value for $\bar{x}_1 - \bar{x}_2$ was calculated to be 5.55. In this example, the test is two-tailed. The *p*-value is equal to twice the area under the sampling distribution of $\bar{x}_1 - \bar{x}_2$ to the right of $z = 5.55$. From the normal distribution table (Table IV in Appendix B), the area to the right of $z = 5.55$ is (approximately) zero. Therefore, the *p*-value is zero. As we know from Chapter 9, we will reject the null hypothesis for any α (significance level) that is greater than or equal to the *p*-value. Consequently, in this example, we will reject the null hypothesis for (almost) any $\alpha > 0$. Since $\alpha = .01$ in this example, which is greater than zero, we reject the null hypothesis.

EXERCISES

CONCEPTS AND PROCEDURES

10.1   Briefly explain the meaning of independent and dependent samples. Give one example of each.

10.2   Describe the sampling distribution of $\bar{x}_1 - \bar{x}_2$ for two independent samples when σ_1 and σ_2 are known and either both sample sizes are large or both populations are normally distributed. What are the mean and standard deviation of this sampling distribution?

10.3   The following information is obtained from two independent samples selected from two normally distributed populations.

$$n_1 = 20 \qquad \bar{x}_1 = 8.34 \qquad \sigma_1 = 1.89$$
$$n_2 = 15 \qquad \bar{x}_2 = 4.85 \qquad \sigma_2 = 2.72$$

a. What is the point estimate of $\mu_1 - \mu_2$?

b. Construct a 99% confidence interval for $\mu_1 - \mu_2$. Find the margin of error for this estimate.

10.4 The following information is obtained from two independent samples selected from two populations.

$$n_1 = 750 \qquad \bar{x}_1 = 1.05 \qquad \sigma_1 = 4.85$$
$$n_2 = 800 \qquad \bar{x}_2 = 1.54 \qquad \sigma_2 = 7.60$$

a. What is the point estimate of $\mu_1 - \mu_2$?

b. Construct a 95% confidence interval for $\mu_1 - \mu_2$. Find the margin of error for this estimate.

10.5 The following information is obtained from two independent samples selected from two normally distributed populations.

$$n_1 = 24 \qquad \bar{x}_1 = 5.56 \qquad \sigma_1 = 1.65$$
$$n_2 = 27 \qquad \bar{x}_2 = 4.80 \qquad \sigma_2 = 1.58$$

Test at a 5% significance level if the two population means are different.

10.6 The following information is obtained from two independent samples selected from two populations.

$$n_1 = 300 \qquad \bar{x}_1 = 22.0 \qquad \sigma_1 = 4.9$$
$$n_2 = 250 \qquad \bar{x}_2 = 27.6 \qquad \sigma_2 = 4.5$$

Test at a 5% significance level if μ_1 is less than μ_2.

APPLICATIONS

10.7 The U.S. Department of Labor collects data on unemployment insurance payments made to unemployed people in different states. Suppose that during 2011 a random sample of 1000 unemployed people in Florida received an average weekly unemployment benefit of $219.65, while a random sample of 900 unemployed people in Mississippi received an average weekly unemployment benefit of $191.47. Assume that the population standard deviations of 2011 weekly unemployment benefits paid to all unemployed workers in Florida and Mississippi were $35.15 and $28.22, respectively. (Note: A 2011 study by Daily-Finance.com (http://www.dailyfinance.com/2011/05/12/unemployment-benefits-best-worst-states/) rated Mississippi and Florida as the two worst states for unemployment benefits.)

a. Let μ_1 and μ_2 be the means of weekly unemployment benefits paid to all unemployed workers during 2011 in Florida and Mississippi, respectively. What is the point estimate of $\mu_1 - \mu_2$?

b. Construct a 96% confidence interval for $\mu_1 - \mu_2$.

c. Using a 2% significance level, can you conclude that the means of all weekly unemployment benefits paid to all unemployed workers during 2011 in Florida and Mississippi are different? Use both the p-value and the critical-value approaches to make this test.

10.8 A local college cafeteria has a self-service soft ice cream machine. The cafeteria provides bowls that can hold up to 16 ounces of ice cream. The food service manager is interested in comparing the average amount of ice cream dispensed by male students to the average amount dispensed by female students. A measurement device was placed on the ice cream machine to determine the amounts dispensed. Random samples of 90 male and 67 female students who got ice cream were selected. The sample averages were 8.32 and 6.49 ounces for the male and female students, respectively. Assume that the population standard deviations are 1.22 and 1.17 ounces, respectively.

a. Let μ_1 and μ_2 be the population means of ice cream amounts dispensed by all male and all female students at this college, respectively. What is the point estimate of $\mu_1 - \mu_2$?

b. Construct a 95% confidence interval for $\mu_1 - \mu_2$.

c. Using a 1% significance level, can you conclude that the average amount of ice cream dispensed by all male college students is larger than the average amount dispensed by all female college students? Use both approaches to make this test.

10.9 Employees of a large corporation are concerned about the declining quality of medical services provided by their group health insurance. A random sample of 100 office visits by employees of this corporation to primary care physicians during 2004 found that the doctors spent an average of 18 minutes with each patient. This year a random sample of 108 such visits showed that doctors spent an average of 15.5 minutes with each patient. Assume that the standard deviations for the two populations are 2.7 and 2.2 minutes, respectively.

a. Construct a 95% confidence interval for the difference between the two population means for these two years.

b. Using a 2.5% level of significance, can you conclude that the mean time spent by doctors with each patient is lower for this year than for 2004?

c. What would your decision be in part b if the probability of making a Type I error were zero? Explain.

10.10 The management at New Century Bank claims that the mean waiting time for all customers at its branches is less than that at the Public Bank, which is its main competitor. A business consulting firm took a sample of 200 customers from the New Century Bank and found that they waited an average of 4.7 minutes before being served. Another sample of 300 customers taken from the Public Bank showed that these customers waited an average of 5 minutes before being served. Assume that the standard deviations for the two populations are 1.2 and 1.5 minutes, respectively.

a. Make a 97% confidence interval for the difference between the two population means.

b. Test at a 2.5% significance level whether the claim of the management of the New Century Bank is true.

c. Calculate the p-value for the test of part b. Based on this p-value, would you reject the null hypothesis if $\alpha = .01$? What if $\alpha = .05$?

10.11 Maine Mountain Dairy claims that its 8-ounce low-fat yogurt cups contain, on average, fewer calories than the 8-ounce low-fat yogurt cups produced by a competitor. A consumer agency wanted to check this claim. A sample of 27 such yogurt cups produced by this company showed that they contained an average of 141 calories per cup. A sample of 25 such yogurt cups produced by its competitor showed that they contained an average of 144 calories per cup. Assume that the two populations are approximately normally distributed with population standard deviations of 5.5 and 6.4 calories, repectively.

a. Make a 98% confidence interval for the difference between the mean number of calories in the 8-ounce low-fat yogurt cups produced by the two companies.

b. Test at a 1% significance level whether Maine Mountain Dairy's claim is true.

c. Calculate the p-value for the test of part b. Based on this p-value, would you reject the null hypothesis if $\alpha = .05$? What if $\alpha = .025$?

10.2 | Inferences About the Difference Between Two Population Means for Independent Samples: σ_1 and σ_2 Unknown but Equal

This section discusses making a confidence interval and testing a hypothesis about the difference between the means of two populations, $\mu_1 - \mu_2$, assuming that the standard deviations, σ_1 and σ_2, of these populations are not known but are assumed to be equal. There are some other conditions, explained below, that must be fulfilled to use the procedures discussed in this section.

If the following conditions are satisfied,

1. The two samples are independent
2. The standard deviations σ_1 and σ_2 of the two populations are unknown, but they can be assumed to be equal, that is, $\sigma_1 = \sigma_2$
3. At least one of the following two conditions is fulfilled:
    i. Both samples are large (i.e., $n_1 \geq 30$ and $n_2 \geq 30$)
    ii. If either one or both sample sizes are small, then both populations from which the samples are drawn are approximately normally distributed

then we use the t distribution to make a confidence interval and test a hypothesis about the difference between the means of two populations, $\mu_1 - \mu_2$.

When the standard deviations of the two populations are equal, we can use σ for both σ_1 and σ_2. Because σ is unknown, we replace it by its point estimator s_p, which is called the **pooled sample standard deviation** (hence, the subscript p). The value of s_p is computed by using the information from the two samples as follows.

Pooled Standard Deviation for Two Samples   The **pooled standard deviation for two samples** is computed as

$$s_p = \sqrt{\frac{(n_1 - 1)s_1^2 + (n_2 - 1)s_2^2}{n_1 + n_2 - 2}}$$

where n_1 and n_2 are the sizes of the two samples and s_1^2 and s_2^2 are the variances of the two samples, respectively. Here s_p is an estimator of σ.

In this formula, $n_1 - 1$ are the degrees of freedom for sample 1, $n_2 - 1$ are the degrees of freedom for sample 2, and $n_1 + n_2 - 2$ are the **degrees of freedom for the two samples taken together**. Note that s_p is an estimator of the standard deviation, σ, of each of the two populations.

When s_p is used as an estimator of σ, the standard deviation of $\bar{x}_1 - \bar{x}_2$, $\sigma_{\bar{x}_1 - \bar{x}_2}$ is estimated by $s_{\bar{x}_1 - \bar{x}_2}$. The value of $s_{\bar{x}_1 - \bar{x}_2}$ is calculated by using the following formula.

Estimator of the Standard Deviation of $\bar{x}_1 - \bar{x}_2$ The **estimator of the standard deviation of $\bar{x}_1 - \bar{x}_2$** is

$$s_{\bar{x}_1 - \bar{x}_2} = s_p \sqrt{\frac{1}{n_1} + \frac{1}{n_2}}$$

Now we are ready to discuss the procedures that are used to make confidence intervals and test hypotheses about $\mu_1 - \mu_2$ for independent samples selected from two populations with unknown but equal standard deviations.

10.2.1   Interval Estimation of $\mu_1 - \mu_2$

As was mentioned earlier in this chapter, the difference between the two sample means, $\bar{x}_1 - \bar{x}_2$, is the point estimator of the difference between the two population means, $\mu_1 - \mu_2$. The following formula gives the confidence interval for $\mu_1 - \mu_2$ when the t distribution is used and the conditions mentioned earlier in this section are fulfilled.

Confidence Interval for $\mu_1 - \mu_2$ The $(1 - \alpha)100\%$ **confidence interval for $\mu_1 - \mu_2$** is

$$(\bar{x}_1 - \bar{x}_2) \pm t s_{\bar{x}_1 - \bar{x}_2}$$

where the value of t is obtained from the t distribution table for the given confidence level and $n_1 + n_2 - 2$ degrees of freedom, and $s_{\bar{x}_1 - \bar{x}_2}$ is calculated as explained earlier.

Example 10–5 describes the procedure to make a confidence interval for $\mu_1 - \mu_2$ using the t distribution.

EXAMPLE 10–5   Caffeine in Two Brands of Coffee

Constructing a confidence interval for $\mu_1 - \mu_2$: two independent samples, unknown but equal σ_1 and σ_2.

A consumer agency wanted to estimate the difference in the mean amounts of caffeine in two brands of coffee. The agency took a sample of 15 one-pound jars of Brand I coffee that showed the mean amount of caffeine in these jars to be 80 milligrams per jar with a standard deviation of 5 milligrams. Another sample of 12 one-pound jars of Brand II coffee gave a mean amount of caffeine equal to 77 milligrams per jar with a standard deviation of 6 milligrams. Construct a 95% confidence interval for the difference between the mean amounts of caffeine in one-pound jars of these two brands of coffee. Assume that the two populations are approximately normally distributed and that the standard deviations of the two populations are equal.

Solution   Let μ_1 and μ_2 be the mean amounts of caffeine per jar in all 1-pound jars of Brand I and II coffee, respectively, and let $\bar{x}_1$ and $\bar{x}_2$ be the means of the two respective samples. From the given information,

Brand I coffee: $n_1 = 15$ $\bar{x}_1 = 80$ milligrams $s_1 = 5$ milligrams

Brand II coffee: $n_2 = 12$ $\bar{x}_2 = 77$ milligrams $s_2 = 6$ milligrams

© ansonsaw/iStockphoto

The confidence level is $1 - \alpha = .95$.

Here, σ_1 and σ_2 are unknown but assumed to be equal, the samples are independent (taken from two different populations), and the sample sizes are small but the two populations are approximately normally distributed. Hence, we will use the t distribution to make the confidence interval for $\mu_1 - \mu_2$ as all conditions mentioned in the beginning of this section are satisfied.

First we calculate the standard deviation of $\bar{x}_1 - \bar{x}_2$ as follows. Note that since it is assumed that σ_1 and σ_2 are equal, we will use s_p to calculate $s_{\bar{x}_1 - \bar{x}_2}$.

$$s_p = \sqrt{\frac{(n_1 - 1)s_1^2 + (n_2 - 1)s_2^2}{n_1 + n_2 - 2}} = \sqrt{\frac{(15 - 1)(5)^2 + (12 - 1)(6)^2}{15 + 12 - 2}} = 5.46260011$$

$$s_{\bar{x}_1 - \bar{x}_2} = s_p \sqrt{\frac{1}{n_1} + \frac{1}{n_2}} = (5.46260011)\sqrt{\frac{1}{15} + \frac{1}{12}} = 2.11565593$$

Next, to find the t value from the t distribution table, we need to know the area in each tail of the t distribution curve and the degrees of freedom.

$$\text{Area in each tail} = \alpha/2 = (1 - .95)/2 = .025$$

$$\text{Degrees of freedom} = n_1 + n_2 - 2 = 15 + 12 - 2 = 25$$

The t value for $df = 25$ and .025 area in the right tail of the t distribution curve is 2.060. The 95% confidence interval for $\mu_1 - \mu_2$ is

$$(\bar{x}_1 - \bar{x}_2) \pm t s_{\bar{x}_1 - \bar{x}_2} = (80 - 77) \pm 2.060(2.11565593)$$

$$= 3 \pm 4.36 = \mathbf{-1.36 \text{ to } 7.36 \text{ milligrams}}$$

Thus, with 95% confidence we can state that based on these two sample results, the difference in the mean amounts of caffeine in 1-pound jars of these two brands of coffee lies between

−1.36 and 7.36 milligrams. Because the lower limit of the interval is negative, it is possible that the mean amount of caffeine is greater in the second brand than in the first brand of coffee.

Note that the value of $\bar{x}_1 - \bar{x}_2$, which is $80 - 77 = 3$, gives the point estimate of $\mu_1 - \mu_2$. The value of $ts_{\bar{x}_1 - \bar{x}_2}$, which is 4.36, is the margin of error.

10.2.2 Hypothesis Testing About $\mu_1 - \mu_2$

When the conditions mentioned in the beginning of Section 10.2 are satisfied, the t distribution is applied to make a hypothesis test about the difference between two population means. The test statistic in this case is t, which is calculated as follows.

Test Statistic t for $\bar{x}_1 - \bar{x}_2$ The value of the **test statistic t for $\bar{x}_1 - \bar{x}_2$** is computed as

$$t = \frac{(\bar{x}_1 - \bar{x}_2) - (\mu_1 - \mu_2)}{s_{\bar{x}_1 - \bar{x}_2}}$$

The value of $\mu_1 - \mu_2$ in this formula is substituted from the null hypothesis, and $s_{\bar{x}_1 - \bar{x}_2}$ is calculated as explained earlier in Section 10.2.1.

Examples 10–6 and 10–7 illustrate how to use the t distribution to perform a test of hypothesis about the difference between two population means for independent samples that are selected from two populations with equal standard deviations.

EXAMPLE 10–6 | Calories in Two Brands of Diet Soda

Making a two-tailed test of hypothesis about $\mu_1 - \mu_2$: two independent samples, and unknown but equal σ_1 and σ_2.

A sample of 14 cans of Brand I diet soda gave the mean number of calories of 23 per can with a standard deviation of 3 calories. Another sample of 16 cans of Brand II diet soda gave the mean number of calories of 25 per can with a standard deviation of 4 calories. At a 1% significance level, can you conclude that the mean numbers of calories per can are different for these two brands of diet soda? Assume that the calories per can of diet soda are approximately normally distributed for each of the two brands and that the standard deviations for the two populations are equal.

Solution Let μ_1 and μ_2 be the mean number of calories for all cans of Brand I and Brand II diet soda, respectively, and let $\bar{x}_1$ and $\bar{x}_2$ be the means of the respective samples. From the given information,

$$\text{Brand I diet soda:} \quad n_1 = 14 \quad \bar{x}_1 = 23 \quad s_1 = 3$$
$$\text{Brand II diet soda:} \quad n_2 = 16 \quad \bar{x}_2 = 25 \quad s_2 = 4$$

The significance level is $\alpha = .01$.

Step 1. *State the null and alternative hypotheses.*

We are to test for the difference in the mean number of calories for the two brands. The null and alternative hypotheses are, respectively,

$$H_0: \mu_1 - \mu_2 = 0 \quad \text{(The mean number of calories are not different)}$$
$$H_1: \mu_1 - \mu_2 \neq 0 \quad \text{(The mean number of calories are different)}$$

Step 2. *Select the distribution to use.*

Here, the two samples are independent, σ_1 and σ_2 are unknown but equal, and the sample sizes are small but both populations are approximately normally distributed. Hence, all conditions mentioned in the beginning of Section 10.2 are fulfilled. Consequently, we will use the t distribution.

Step 3. *Determine the rejection and nonrejection regions.*

The $\neq$ sign in the alternative hypothesis indicates that the test is two-tailed. The significance level is .01. Hence,

$$\text{Area in each tail} = \alpha/2 = .01/2 = .005$$
$$\text{Degrees of freedom} = n_1 + n_2 - 2 = 14 + 16 - 2 = 28$$

The critical values of t for $df = 28$ and .005 area in each tail of the t distribution curve are -2.763 and 2.763, as shown in Figure 10.3.

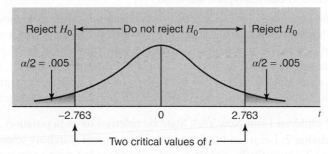

Figure 10.3   Rejection and nonrejection regions.

Step 4.   *Calculate the value of the test statistic.*

The value of the test statistic t for $\bar{x}_1 - \bar{x}_2$ is computed as follows:

$$s_p = \sqrt{\frac{(n_1 - 1)s_1^2 + (n_2 - 1)s_2^2}{n_1 + n_2 - 2}} = \sqrt{\frac{(14 - 1)(3)^2 + (16 - 1)(4)^2}{14 + 16 - 2}} = 3.57071421$$

$$s_{\bar{x}_1 - \bar{x}_2} = s_p \sqrt{\frac{1}{n_1} + \frac{1}{n_2}} = (3.57071421)\sqrt{\frac{1}{14} + \frac{1}{16}} = 1.30674760$$

$$t = \frac{(\bar{x}_1 - \bar{x}_2) - (\mu_1 - \mu_2)}{s_{\bar{x}_1 - \bar{x}_2}} = \frac{(23 - 25) - 0}{1.30674760} = -1.531 \qquad \text{From } H_0$$

Step 5.   *Make a decision.*

Because the value of the test statistic $t = -1.531$ for $\bar{x}_1 - \bar{x}_2$ falls in the nonrejection region, we fail to reject the null hypothesis. Consequently we conclude that there is no difference in the mean number of calories for the two brands of diet soda. The difference in $\bar{x}_1$ and $\bar{x}_2$ observed for the two samples may have occurred due to sampling error only.

Using the *p*-Value to Make a Decision for Example 10–6

We can use the *p*-value approach to make the above decision. To do so, we keep Steps 1 and 2 of this example. Then in Step 3, we calculate the value of the test statistic t (as done in Step 4 above) and then find the *p*-value for this t from the t distribution table (Table V of Appendix B) or by using technology. In Step 4 above, the t-value for $\bar{x}_1 - \bar{x}_2$ was calculated to be -1.531. In this example, the test is two-tailed. Therefore, the *p*-value is equal to twice the area under the t distribution curve to the left of $t = -1.531$. If we have access to technology, we can use it to find the exact *p*-value, which will be .137. If we use the t distribution table, we can only find a range for the *p*-value. From Table V of Appendix B, for $df = 28$, the two values that include 1.531 are 1.313 and 1.701. (Note that we use the positive value of t, although our t is negative.) Thus, the test statistic $t = -1.531$ falls between -1.313 and -1.701. The areas in the t distribution table that correspond to 1.313 and 1.701 are .10 and .05, respectively. Because it is a two-tailed test, the *p*-value for $t = -1.531$ is between $2(.10) = .20$ and $2(.05) = .10$, which can be written as

$$.10 < p\text{-value} < .20$$

As we know from Chapter 9, we will reject the null hypothesis for any α (significance level) that is greater than or equal to the *p*-value. Consequently, in this example, we will reject the null hypothesis for any $\alpha \geq .20$ using the above range and not reject it for $\alpha < .10$. If we use technology, we will reject the null hypothesis for $\alpha \geq .137$. Since $\alpha = .01$ in this example, which is smaller than both .10 and .137, we fail to reject the null hypothesis.

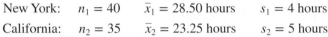

EXAMPLE 10–7 | Time Spent Watching Television by Children

Making a right-tailed test of hypothesis about $\mu_1 - \mu_2$: two independent samples, σ_1 and σ_2 unknown but equal, and both samples are large.

A sample of 40 children from New York State showed that the mean time they spend watching television is 28.50 hours per week with a standard deviation of 4 hours. Another sample of 35 children from California showed that the mean time spent by them watching television is 23.25 hours per week with a standard deviation of 5 hours. Using a 2.5% significance level, can you conclude that the mean time spent watching television by children in New York State is greater than that for children in California? Assume that the standard deviations for the two populations are equal.

© Dori OConnell/iStockphoto

Solution Let all children from New York State be referred to as population 1 and those from California as population 2. Let μ_1 and μ_2 be the mean time spent watching television by children in populations 1 and 2, respectively, and let $\bar{x}_1$ and $\bar{x}_2$ be the mean time spent watching television by children in the respective samples. From the given information,

$$\text{New York:} \quad n_1 = 40 \quad \bar{x}_1 = 28.50 \text{ hours} \quad s_1 = 4 \text{ hours}$$
$$\text{California:} \quad n_2 = 35 \quad \bar{x}_2 = 23.25 \text{ hours} \quad s_2 = 5 \text{ hours}$$

The significance level is $\alpha = .025$.

Step 1. *State the null and alternative hypotheses.*

The two possible decisions are:

1. The mean time spent watching television by children in New York State is not greater than that for children in California. This can be written as $\mu_1 = \mu_2$ or $\mu_1 - \mu_2 = 0$.
2. The mean time spent watching television by children in New York State is greater than that for children in California. This can be written as $\mu_1 > \mu_2$ or $\mu_1 - \mu_2 > 0$.

Hence, the null and alternative hypotheses are, respectively,

$$H_0: \mu_1 - \mu_2 = 0$$
$$H_1: \mu_1 - \mu_2 > 0$$

Note that the null hypothesis can also be written as $\mu_1 - \mu_2 \leq 0$.

Step 2. *Select the distribution to use.*

Here, the two samples are independent (taken from two different populations), σ_1 and σ_2 are unknown but assumed to be equal, and both samples are large. Hence, all conditions mentioned in the beginning of Section 10.2 are fulfilled. Consequently, we use the t distribution to make the test.

Step 3. *Determine the rejection and nonrejection regions.*

The $>$ sign in the alternative hypothesis indicates that the test is right-tailed. The significance level is .025.

$$\text{Area in the right tail of the } t \text{ distribution} = \alpha = .025$$
$$\text{Degrees of freedom} = n_1 + n_2 - 2 = 40 + 35 - 2 = 73$$

From the t distribution table, the critical value of t for $df = 73$ and .025 area in the right tail of the t distribution is 1.993. This value is shown in Figure 10.4.

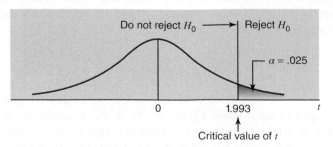

Figure 10.4 Rejection and nonrejection regions.

Step 4.   *Calculate the value of the test statistic.*

The value of the test statistic t for $\bar{x}_1 - \bar{x}_2$ is computed as follows:

$$s_p = \sqrt{\frac{(n_1 - 1)s_1^2 + (n_2 - 1)s_2^2}{n_1 + n_2 - 2}} = \sqrt{\frac{(40 - 1)(4)^2 + (35 - 1)(5)^2}{40 + 35 - 2}} = 4.49352655$$

$$s_{\bar{x}_1 - \bar{x}_2} = s_p \sqrt{\frac{1}{n_1} + \frac{1}{n_2}} = (4.49352655)\sqrt{\frac{1}{40} + \frac{1}{35}} = 1.04004930$$

$$t = \frac{(\bar{x}_1 - \bar{x}_2) - (\mu_1 - \mu_2)}{s_{\bar{x}_1 - \bar{x}_2}} = \frac{(28.50 - 23.25) - 0}{1.04004930} = 5.048 \qquad \text{From } H_0$$

Step 5.   *Make a decision.*

Because the value of the test statistic $t = 5.048$ for $\bar{x}_1 - \bar{x}_2$ falls in the rejection region (see Figure 10.4), we reject the null hypothesis H_0. Hence, we conclude that children in New York State spend more time, on average, watching TV than children in California.

Using the *p*-Value to Make a Decision for Example 10–7

To use the *p*-value approach to make the above decision, we keep Steps 1 and 2 of this example. Then in Step 3, we calculate the value of the test statistic t (as done in Step 4 above) and then find the *p*-value for this t from the t distribution table (Table V of Appendix B) or by using technology. In Step 4 above, the t-value for $\bar{x}_1 - \bar{x}_2$ was calculated to be 5.048. In this example, the test is right-tailed. Therefore, the *p*-value is equal to the area under the t distribution curve to the right of $t = 5.048$. If we have access to technology, we can use it to find the exact *p*-value, which will be .000. If we use the t distribution table, for $df = 73$, the value of the test statistic $t = 5.048$ is larger than 3.206. Therefore, the *p*-value for $t = 5.048$ is less than .001, which can be written as

$$p\text{-value} < .001$$

Since we will reject the null hypothesis for any α (significance level) greater than or equal to the *p*-value, here we reject the null hypothesis because $\alpha = .025$ is greater than both the *p*-values, .001 obtained above from the table and .000 obtained by using technology. Note that obtaining the *p*-value = .000 from technology does not mean that the *p*-value is zero. It means that when the *p*-value is rounded to three digits after the decimal, it is .000.

Note: What If the Sample Sizes Are Large and the Number of *df* Are Not in the *t* Distribution Table?

In this section, we used the t distribution to make confidence intervals and perform tests of hypothesis about $\mu_1 - \mu_2$. When both sample sizes are large, it does not matter how large the sample sizes are if we are using technology. However, if we are using the t distribution table (Table V of Appendix B), this may pose a problem if samples are too large. Table V in Appendix B goes up to only 75 degrees of freedom. Thus, if the degrees of freedom are larger than 75, we cannot use Table V to find the critical value(s) of t. As mentioned in Chapters 8 and 9, in such a situation, there are two options:

1.   Use the t value from the last row (the row of ∞) in Table V.
2.   Use the normal distribution as an approximation to the t distribution.

EXERCISES

CONCEPTS AND PROCEDURES

10.12   Explain what conditions must hold true to use the t distribution to make a confidence interval and to test a hypothesis about $\mu_1 - \mu_2$ for two independent samples selected from two populations with unknown but equal standard deviations.

10.13   The following information was obtained from two independent samples selected from two normally distributed populations with unknown but equal standard deviations.

$$n_1 = 21 \quad \bar{x}_1 = 12.82 \quad s_1 = 4.12$$
$$n_2 = 20 \quad \bar{x}_2 = 16.75 \quad s_2 = 3.46$$

a. What is the point estimate of $\mu_1 - \mu_2$?

b Construct a 95% confidence interval for $\mu_1 - \mu_2$.

10.14 The following information was obtained from two independent samples selected from two normally distributed populations with unknown but equal standard deviations.

$$n_1 = 21 \qquad \bar{x}_1 = 12.82 \qquad s_1 = 4.12$$
$$n_2 = 20 \qquad \bar{x}_2 = 16.75 \qquad s_2 = 3.46$$

Test at a 5% significance level if the two population means are different.

10.15 The following information was obtained from two independent samples selected from two populations with unknown but equal standard deviations.

$$n_1 = 50 \qquad \bar{x}_1 = 90.40 \qquad s_1 = 10.40$$
$$n_2 = 55 \qquad \bar{x}_2 = 86.30 \qquad s_2 = 9.75$$

Test at a 1% significance level if the two population means are different.

10.16 The following information was obtained from two independent samples selected from two populations with unknown but equal standard deviations.

$$n_1 = 50 \qquad \bar{x}_1 = 90.40 \qquad s_1 = 10.40$$
$$n_2 = 55 \qquad \bar{x}_2 = 86.30 \qquad s_2 = 9.75$$

Test at a 5% significance level if μ_1 is greater than μ_2.

10.17 The following information was obtained from two independent samples selected from two normally distributed populations with unknown but equal standard deviations.

Sample 1: 47.7 42.8 51.9 34.1 65.8 61.5 52.1 40.8
53.1 46.1 47.9 45.7 46.7

Sample 2: 50.0 47.4 32.7 38.6 54.0 46.3 42.5 40.8
39.0 68.2

a. Let μ_1 be the mean of population 1 and μ_2 be the mean of population 2. What is the point estimate of $\mu_1 - \mu_2$?

b. Construct a 98% confidence interval for $\mu_1 - \mu_2$.

c. Test at a 1% significance level if μ_1 is greater than μ_2.

10.18 The following information was obtained from two independent samples selected from two normally distributed populations with unknown but equal standard deviations.

Sample 1: 2.18 2.23 1.96 2.24 2.84 1.87 2.68
2.15 2.49 1.95

Sample 2: 1.82 1.86 2.00 1.89 1.73 1.34 1.43
2.05 1.54 2.50 2.08 2.13

a. Let μ_1 be the mean of population 1 and μ_2 be the mean of population 2. What is the point estimate of $\mu_1 - \mu_2$?

b. Construct a 99% confidence interval for $\mu_1 - \mu_2$.

c. Test at a 2.5% significance level if μ_1 is lower than μ_2.

APPLICATIONS

10.19 The standard recommendation for automobile oil changes is once every 3000 miles. A local mechanic is interested in determining whether people who drive more expensive cars are more likely to follow the recommendation. Independent random samples of 45 customers who drive luxury cars and 40 customers who drive compact lower-price cars were selected. The average distance driven between oil changes was 3187 miles for the luxury car owners and 3214 miles for the compact lower-price cars. The sample standard deviations were 42.40 and 50.70 miles for the luxury and compact groups, respectively. Assume that the population distributions of the distances between oil changes have the same standard deviation for the two populations.

a. Construct a 95% confidence interval for the difference in the mean distances between oil changes for all luxury cars and all compact lower-price cars.

b. Using a 1% significance level, can you conclude that the mean distance between oil changes is less for all luxury cars than that for all compact lower-price cars?

10.20 A town that recently started a single-stream recycling program provided 60-gallon recycling bins to 30 randomly selected households and 75-gallon recycling bins to 22 randomly selected households. The total volume of recycling over a 10-week period was measured for each of the households. The average total volumes were 365 and 423 gallons for the households with the 60- and 75-gallon bins, respectively. The sample standard deviations were 52.5 and 43.8 gallons, respectively. Assume that the 10-week total volumes of recycling are approximately normally distributed for both groups and that the population standard deviations are equal.

a. Construct a 98% confidence interval for the difference in the mean volumes of 10-week recycling for the households with the 60- and 75-gallon bins.

b. Using the 2% significance level, can you conclude that the average 10-week recycling volume of all households having 60-gallon containers is different from the average volume of all households that have 75-gallon containers?

10.21 A consumer organization tested two paper shredders, the Piranha and the Crocodile, designed for home use. Each of 10 randomly selected volunteers shredded 100 sheets of paper with the Piranha, and then another sample of 10 randomly selected volunteers each shredded 100 sheets with the Crocodile. The Piranha took an average of 218 seconds to shred 100 sheets with a standard deviation of 8 seconds. The Crocodile took an average of 163 seconds to shred 100 sheets with a standard deviation of 4 seconds. Assume that the shredding times for both machines are normally distributed with equal but unknown standard deviations.

a. Construct a 99% confidence interval for the difference between the two population means.

b. Using the 1% significance level, can you conclude that the mean time taken by the Piranha to shred 100 sheets is greater than that for the Crocodile?

c. What would your decision be in part b if the probability of making a Type I error were zero? Explain.

10.22 A company claims that its medicine, Brand A, provides faster relief from pain than another company's medicine, Brand B. A researcher tested both brands of medicine on two groups of randomly selected patients. The results of the test are given in the following table. The mean and standard deviation of relief times are in minutes.

Brand	Sample Size	Mean of Relief Times	Standard Deviation of Relief Times
A	30	43	12
B	28	48	7

Assume that the two populations are normally distributed with unknown but equal standard deviations.

a. Construct a 99% confidence interval for the difference between the mean relief times for the two brands of medicine.

b. Test at a 1% significance level whether the mean relief time for Brand A is less than that for Brand B.

10.23 According to the credit rating agency Equifax, credit limits on newly issued credit cards increased between January 2011 and May 2011 (*money.cnn.com/2011/08/19/pf/credit_card_issuance/index.htm*). Suppose that random samples of 400 credit cards issued in January 2011 and 500 credit cards issued in May 2011 had average credit limits of \$2635 and \$2887, respectively. Suppose that the sample standard deviations for these two samples were \$365 and \$412, respectively, and the assumption that the population standard deviations are equal for the two populations is reasonable.

a. Let μ_1 and μ_2 be the average credit limits on all credit cards issued in January 2011 and in May 2011, respectively. What is the point estimate of $\mu_1 - \mu_2$?

b. Construct a 98% confidence interval for $\mu_1 - \mu_2$.

c. Using a 1% significance level, can you conclude that the average credit limit for all new credit cards issued in January 2011 was lower than the corresponding average for all credit cards issued in May 2011? Use both the *p*-value and the critical-value approaches to make this test.

10.3 | Inferences About the Difference Between Two Population Means for Independent Samples: σ_1 and σ_2 Unknown and Unequal

Section 10.2 explained how to make inferences about the difference between two population means using the *t* distribution when the standard deviations of the two populations are unknown but equal and certain other assumptions hold true. Now, what if all other assumptions of Section 10.2 hold true, but the population standard deviations are not only unknown but also unequal? In this case, the procedures used to make confidence intervals and to test hypotheses about $\mu_1 - \mu_2$ remain similar to the ones we learned in Sections 10.2.1 and 10.2.2, except for two differences. When the population standard deviations are unknown and not equal, the degrees of freedom are no longer given by $n_1 + n_2 - 2$, and the standard deviation of $\bar{x}_1 - \bar{x}_2$ is not calculated using the pooled standard deviation s_p.

Degrees of Freedom If

1. The two samples are independent
2. The standard deviations σ_1 and σ_2 of the two populations are unknown and unequal, that is, $\sigma_1 \neq \sigma_2$
3. At least one of the following two conditions is fulfilled:

   i. Both samples are large (i.e., $n_1 \geq 30$ and $n_2 \geq 30$)

   ii. If either one or both sample sizes are small, then both populations from which the samples are drawn are approximately normally distributed

then the *t* distribution is used to make inferences about $\mu_1 - \mu_2$, and the **degrees of freedom** for the *t* distribution are given by

$$df = \frac{\left(\dfrac{s_1^2}{n_1} + \dfrac{s_2^2}{n_2}\right)^2}{\dfrac{\left(\dfrac{s_1^2}{n_1}\right)^2}{n_1 - 1} + \dfrac{\left(\dfrac{s_2^2}{n_2}\right)^2}{n_2 - 1}}$$

The number given by this formula is always rounded down for *df*.

Because the standard deviations of the two populations are not known, we use $s_{\bar{x}_1 - \bar{x}_2}$ as a point estimator of $\sigma_{\bar{x}_1 - \bar{x}_2}$. The following formula is used to calculate the standard deviation $s_{\bar{x}_1 - \bar{x}_2}$ of $\bar{x}_1 - \bar{x}_2$.

Estimate of the Standard Deviation of $\bar{X}_1 - \bar{X}_2$ The value of $s_{\bar{x}_1 - \bar{x}_2}$ is calculated as

$$s_{\bar{x}_1 - \bar{x}_2} = \sqrt{\frac{s_1^2}{n_1} + \frac{s_2^2}{n_2}}$$

10.3.1 Interval Estimation of $\mu_1 - \mu_2$

Again, the difference between the two sample means, $\bar{x}_1 - \bar{x}_2$, is the point estimator of the difference between the two population means, $\mu_1 - \mu_2$. The following formula gives the confidence interval for $\mu_1 - \mu_2$ when the t distribution is used and the conditions mentioned earlier in this section are satisfied.

Confidence Interval for $\mu_1 - \mu_2$ The $(1 - \alpha)100\%$ **confidence interval for $\mu_1 - \mu_2$** is

$$(\bar{x}_1 - \bar{x}_2) \pm t s_{\bar{x}_1 - \bar{x}_2}$$

where the value of t is obtained from the t distribution table for a given confidence level and the degrees of freedom that are given by the formula mentioned earlier, and $s_{\bar{x}_1 - \bar{x}_2}$ is also calculated as explained earlier.

Example 10–8 describes how to construct a confidence interval for $\mu_1 - \mu_2$ when the standard deviations of the two populations are unknown and unequal.

EXAMPLE 10–8	Caffeine in Two Brands of Coffee

Constructing a confidence interval for $\mu_1 - \mu_2$: two independent samples, σ_1 and σ_2 unknown and unequal.

According to Example 10–5 of Section 10.2.1, a sample of 15 one-pound jars of coffee of Brand I showed that the mean amount of caffeine in these jars is 80 milligrams per jar with a standard deviation of 5 milligrams. Another sample of 12 one-pound coffee jars of Brand II gave a mean amount of caffeine equal to 77 milligrams per jar with a standard deviation of 6 milligrams. Construct a 95% confidence interval for the difference between the mean amounts of caffeine in one-pound coffee jars of these two brands. Assume that the two populations are approximately normally distributed and that the standard deviations of the two populations are not equal.

Solution   Let μ_1 and μ_2 be the mean amounts of caffeine per jar in all 1-pound jars of Brands I and II, respectively, and let $\bar{x}_1$ and $\bar{x}_2$ be the means of the two respective samples. From the given information,

$$\text{Brand I coffee:} \quad n_1 = 15 \quad \bar{x}_1 = 80 \text{ milligrams} \quad s_1 = 5 \text{ milligrams}$$

$$\text{Brand II coffee:} \quad n_2 = 12 \quad \bar{x}_2 = 77 \text{ milligrams} \quad s_2 = 6 \text{ milligrams}$$

The confidence level is $1 - \alpha = .95$.

First, we calculate the standard deviation of $\bar{x}_1 - \bar{x}_2$ as follows:

$$s_{\bar{x}_1 - \bar{x}_2} = \sqrt{\frac{s_1^2}{n_1} + \frac{s_2^2}{n_2}} = \sqrt{\frac{(5)^2}{15} + \frac{(6)^2}{12}} = 2.16024690$$

Next, to find the t value from the t distribution table, we need to know the area in each tail of the t distribution curve and the degrees of freedom.

$$\text{Area in each tail} = \alpha/2 = (1 - .95)/2 = .025$$

$$df = \frac{\left(\dfrac{s_1^2}{n_1} + \dfrac{s_2^2}{n_2}\right)^2}{\dfrac{\left(\dfrac{s_1^2}{n_1}\right)^2}{n_1 - 1} + \dfrac{\left(\dfrac{s_2^2}{n_2}\right)^2}{n_2 - 1}} = \frac{\left(\dfrac{(5)^2}{15} + \dfrac{(6)^2}{12}\right)^2}{\dfrac{\left(\dfrac{(5)^2}{15}\right)^2}{15 - 1} + \dfrac{\left(\dfrac{(6)^2}{12}\right)^2}{12 - 1}} = 21.42 \approx 21$$

Note that the degrees of freedom are always rounded down as in this calculation. From the t distribution table, the t value for $df = 21$ and .025 area in the right tail of the t distribution curve is 2.080. The 95% confidence interval for $\mu_1 - \mu_2$ is

$$(\bar{x}_1 - \bar{x}_2) \pm t s_{\bar{x}_1 - \bar{x}_2} = (80 - 77) \pm 2.080(2.16024690)$$

$$= 3 \pm 4.49 = -1.49 \text{ to } 7.49$$

Thus, with 95% confidence we can state that based on these two sample results, the difference in the mean amounts of caffeine in 1-pound jars of these two brands of coffee is between -1.49 and 7.49 milligrams.

Comparing this confidence interval with the one obtained in Example 10–5, we observe that the two confidence intervals are very close. From this we can conclude that even if the standard deviations of the two populations are not equal and we use the procedure of Section 10.2.1 to make a confidence interval for $\mu_1 - \mu_2$, the margin of error will be small as long as the difference between the two population standard deviations is not too large.

10.3.2   Hypothesis Testing About $\mu_1 - \mu_2$

When the standard deviations of the two populations are unknown and unequal along with the other conditions of Section 10.2 holding true, we use the t distribution to make a test of hypothesis about $\mu_1 - \mu_2$. This procedure differs from the one in Section 10.2.2 only in the calculation of degrees of freedom for the t distribution and the standard deviation of $\bar{x}_1 - \bar{x}_2$. The df and the standard deviation of $\bar{x}_1 - \bar{x}_2$ in this case are given by the formulas used in Section 10.3.1.

Test Statistic t for $\overline{x}_1 - \overline{x}_2$ The value of the **test statistic t for $\overline{x}_1 - \overline{x}_2$** is computed as

$$t = \frac{(\bar{x}_1 - \bar{x}_2) - (\mu_1 - \mu_2)}{s_{\bar{x}_1 - \bar{x}_2}}$$

The value of $\mu_1 - \mu_2$ in this formula is substituted from the null hypothesis, and $s_{\bar{x}_1 - \bar{x}_2}$ is calculated as explained earlier.

Example 10–9 illustrates the procedure used to conduct a test of hypothesis about $\mu_1 - \mu_2$ when the standard deviations of the two populations are unknown and unequal.

EXAMPLE 10–9   **Calories in Two Brands of Diet Soda**

Making a two-tailed test of hypothesis about $\mu_1 - \mu_2$: two independent samples, and unknown and unequal σ_1 and σ_2.

According to Example 10–6 of Section 10.2.2, a sample of 14 cans of Brand I diet soda gave the mean number of calories per can of 23 with a standard deviation of 3 calories. Another sample of 16 cans of Brand II diet soda gave the mean number of calories of 25 per can with a standard deviation of 4 calories. Test at a 1% significance level whether the mean numbers of calories per can of diet soda are different for these two brands. Assume that the calories per can of diet soda are approximately normally distributed for each of these two brands and that the standard deviations for the two populations are not equal.

Solution   Let μ_1 and μ_2 be the mean number of calories for all cans of Brand I and Brand II diet soda, respectively, and let $\bar{x}_1$ and $\bar{x}_2$ be the means of the respective samples. From the given information,

$$\text{Brand I diet soda:} \quad n_1 = 14 \quad \bar{x}_1 = 23 \quad s_1 = 3$$
$$\text{Brand II diet soda:} \quad n_2 = 16 \quad \bar{x}_2 = 25 \quad s_2 = 4$$

The significance level is $\alpha = .01$.

Step 1.   *State the null and alternative hypotheses.*

We are to test for the difference in the mean number of calories for the two brands. The null and alternative hypotheses are, respectively,

$$H_0: \mu_1 - \mu_2 = 0 \quad \text{(The mean number of calories are not different.)}$$
$$H_1: \mu_1 - \mu_2 \neq 0 \quad \text{(The mean number of calories are different.)}$$

Step 2. *Select the distribution to use.*

Here, the two samples are independent, σ_1 and σ_2 are unknown and unequal, the sample sizes are small, but both populations are approximately normally distributed. Hence, all conditions mentioned in the beginning of Section 10.3 are fulfilled. Consequently, we use the t distribution to make the test.

Step 3. *Determine the rejection and nonrejection regions.*

The $\neq$ sign in the alternative hypothesis indicates that the test is two-tailed. The significance level is .01. Hence,

$$\text{Area in each tail} = \alpha/2 = .01/2 = .005$$

The degrees of freedom are calculated as follows:

$$df = \frac{\left(\dfrac{s_1^2}{n_1} + \dfrac{s_2^2}{n_2}\right)^2}{\dfrac{\left(\dfrac{s_1^2}{n_1}\right)^2}{n_1 - 1} + \dfrac{\left(\dfrac{s_2^2}{n_2}\right)^2}{n_2 - 1}} = \frac{\left(\dfrac{(3)^2}{14} + \dfrac{(4)^2}{16}\right)^2}{\dfrac{\left(\dfrac{(3)^2}{14}\right)^2}{14 - 1} + \dfrac{\left(\dfrac{(4)^2}{16}\right)^2}{16 - 1}} = 27.41 \approx 27$$

From the t distribution table, the critical values of t for $df = 27$ and .005 area in each tail of the t distribution curve are -2.771 and 2.771. These values are shown in Figure 10.5.

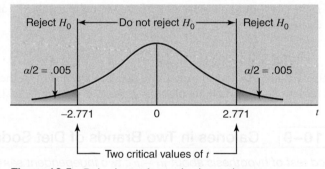

Figure 10.5 Rejection and nonrejection regions.

Step 4. *Calculate the value of the test statistic.*

The value of the test statistic t for $\bar{x}_1 - \bar{x}_2$ is computed as follows:

$$s_{\bar{x}_1 - \bar{x}_2} = \sqrt{\frac{s_1^2}{n_1} + \frac{s_2^2}{n_2}} = \sqrt{\frac{(3)^2}{14} + \frac{(4)^2}{16}} = 1.28173989$$

$$t = \frac{(\bar{x}_1 - \bar{x}_2) - (\mu_1 - \mu_2)}{s_{\bar{x}_1 - \bar{x}_2}} = \frac{(23 - 25) - 0}{1.28173989} = -1.560$$

From H_0

Step 5. *Make a decision.*

Because the value of the test statistic $t = -1.560$ for $\bar{x}_1 - \bar{x}_2$ falls in the nonrejection region, we fail to reject the null hypothesis. Hence, there is no difference in the mean numbers of calories per can for the two brands of diet soda. The difference in $\bar{x}_1$ and $\bar{x}_2$ observed for the two samples may have occurred due to sampling error only.

Using the p-Value to Make a Decision for Example 10–9

We can use the p-value approach to make the above decision. To do so, we keep Steps 1 and 2 of this example. Then in Step 3 we calculate the value of the test statistic t (as done in Step 4 above) and then find the p-value for this t from the t distribution table (Table V of Appendix B) or by using technology. In Step 4 above, the t-value for $\bar{x}_1 - \bar{x}_2$ was calculated to be -1.560. In this example, the test is two-tailed. Therefore, the p-value is equal to twice the area under the t distribution curve to the left of $t = -1.560$. If we have access to technology, we can use it to find the exact p-value, which will be .130. If we use the t distribution table, we can only find a range for the p-value. From Table V of Appendix B, for $df = 27$, the two values that include 1.560 are 1.314 and 1.703. (Note that we use the positive value of t, although our t is negative.) Thus, the test statistic $t = -1.560$ falls between -1.314 and -1.703. The areas in the t distribution table that correspond to 1.314 and 1.703 are .10 and .05, respectively. Because it is a two-tailed test, the p-value for $t = -1.560$ is between $2(.10) = .20$ and $2(.05) = .10$, which can be written as

$$.10 < p\text{-value} < .20$$

Since we will reject the null hypothesis for any α (significance level) that is greater than the p-value, we will reject the null hypothesis in this example for any $\alpha \geq .20$ using the above range and not reject for $\alpha < .10$. If we use technology, we will reject the null hypothesis for $\alpha \geq .130$. Since $\alpha = .01$ in this example, which is smaller than both .10 and .130, we fail to reject the null hypothesis.

The degrees of freedom for the procedures to make a confidence interval and to test a hypothesis about $\mu_1 - \mu_2$ learned in Sections 10.3.1 and 10.3.2 are always rounded down.   ◄ *Remember*

EXERCISES

CONCEPTS AND PROCEDURES

10.24 Assuming that the two populations are normally distributed with unequal and unknown population standard deviations, construct a 95% confidence interval for $\mu_1 - \mu_2$ for the following.

$$n_1 = 24 \quad \bar{x}_1 = 20.50 \quad s_1 = 3.90$$
$$n_2 = 16 \quad \bar{x}_2 = 22.60 \quad s_2 = 5.15$$

10.25 Assuming that the two populations have unequal and unknown population standard deviations, construct a 99% confidence interval for $\mu_1 - \mu_2$ for the following.

$$n_1 = 50 \quad \bar{x}_1 = .863 \quad s_1 = .143$$
$$n_2 = 48 \quad \bar{x}_2 = .796 \quad s_2 = .048$$

10.26 Refer to Exercise 10.24. Test at a 5% significance level if the two population means are different.

10.27 Refer to Exercise 10.25. Test at a 1% significance level if the two population means are different.

10.28 Refer to Exercise 10.24. Test at a 1% significance level if μ_1 is less than μ_2.

10.29 Refer to Exercise 10.25. Test at a 2.5% significance level if μ_1 is greater than μ_2.

APPLICATIONS

10.30 A sample of 45 customers who drive luxury cars showed that their average distance driven between oil changes was 3187 miles

with a sample standard deviation of 42.40 miles. Another sample of 40 customers who drive compact lower-price cars resulted in an average distance of 3214 miles with a standard deviation of 50.70 miles. Suppose that the standard deviations for the two populations are not equal.

a. Construct a 95% confidence interval for the difference in the mean distance between oil changes for all luxury cars and all compact lower-price cars.

b. Using a 1% significance level, can you conclude that the mean distance between oil changes is lower for all luxury cars than for all compact lower-price cars?

c. Suppose that the sample standard deviations were 28.9 and 61.4 miles, respectively. Redo parts a and b. Discuss any changes in the results.

10.31 An insurance company wants to know if the average speed at which men drive cars is higher than that of women drivers. The company took a random sample of 30 cars driven by men on a highway and found the mean speed to be 74 miles per hour with a standard deviation of 3.1 miles per hour. Another sample of 18 cars driven by women on the same highway gave a mean speed of 68 miles per hour with a standard deviation of 2.5 miles per hour. Assume that the speeds at which all men and all women drive cars on this highway are both normally distributed with unequal population standard deviations.

a. Construct a 98% confidence interval for the difference between the mean speeds of cars driven by all men and all women on this highway.

b. Test at the 1% significance level whether the mean speed of cars driven by all men drivers on this highway is higher than that of cars driven by all women drivers.

c. Suppose that the sample standard deviations were 1.9 and 3.4 miles per hour, respectively. Redo parts a and b. Discuss any changes in the results.

10.32 A high school counselor wanted to know if tenth-graders at her high school tend to have more free time than the twelfth-graders. She took random samples of 25 tenth-graders and 22 twelfth-graders. Each student was asked to record the amount of free time he or she had in a typical week. The mean for the tenth-graders was found to be 29 hours of free time per week with a standard deviation of 7.0 hours. For the twelfth-graders, the mean was 22 hours of free time per week with a standard deviation of 6.4 hours. Now assume that the two populations are normally distributed with unequal and unknown population standard deviations.

a. Make a 90% confidence interval for the difference between the corresponding population means.

b. Test at a 5% significance level whether the two population means are different.

c. Suppose that the sample standard deviations were 9.5 and 5.1 hours, respectively. Redo parts a and b. Discuss any changes in the results.

10.33 A town that recently started a single-stream recycling program provided 60-gallon recycling bins to 30 randomly selected households and 75-gallon recycling bins to 22 randomly selected households. The average total volumes of recycling over a 10-week period were 365 and 423 gallons for the two groups, respectively, with standard deviations of 52.5 and 43.8 gallons, respectively. Suppose that the standard deviations for the two populations are not equal.

a. Construct a 98% confidence interval for the difference in the mean volumes of 10-week recycling for the households with the 60- and 75-gallon bins.

b. Using the 2% significance level, can you conclude that the average 10-week recycling volume of all households having 60-gallon containers is different from the average 10-week recycling volume of all households that have 75-gallon containers?

c. Suppose that the sample standard deviations were 59.3 and 33.8 gallons, respectively. Redo parts a and b. Discuss any changes in the results.

10.34 According to a credit rating agency, credit limits on newly issued credit cards increased between January 2016 and May 2016. Suppose that random samples of 400 new credit cards issued in January 2016 and 500 new credit cards issued in May 2016 had average credit limits of $2635 and $2887, respectively. Suppose that the sample standard deviations for these two samples were $365 and $412, respectively. Now assume that the population standard deviations for the two populations are unknown and not equal.

a. Let μ_1 and μ_2 be the average credit limits on all credit cards issued in January 2016 and in May 2016, respectively. What is the point estimate $\mu_1 - \mu_2$?

b. Construct a 98% confidence interval for $\mu_1 - \mu_2$.

c. Using a 1% significance level, can you conclude that the average credit limit for all new credit cards issued in January 2016 was lower than the corresponding average for all credit cards issued in May 2016? Use both the p-value and the critical-value approaches to make this test.

10.4 | Inferences About the Mean of Paired Samples (Dependent Samples)

Sections 10.1, 10.2, and 10.3 were concerned with estimation and hypothesis testing about the difference between two population means when the two samples were drawn independently from two different populations. This section describes estimation and hypothesis testing procedures for the mean of paired samples, which are also called two dependent samples.

In a case of two dependent samples, two data values—one for each sample—are collected from the same source (or element) and, hence, these are also called **paired** or **matched samples**. For example, we may want to make inferences about the mean weight loss for members of a health club after they have gone through an exercise program for a certain period of time. To do so, suppose we select a sample of 15 members of this health club and record their weights before and after the program. In this example, both sets of data are collected from the same 15 persons, once before and once after the program. Thus, although there are two samples, they contain the same 15 persons. This is an example of paired (or dependent or matched) samples. The procedures to make confidence intervals and test hypotheses in the case of paired samples are different from the ones for independent samples discussed in earlier sections of this chapter.

As another example of paired samples, suppose an agronomist wants to measure the effect of a new brand of fertilizer on the yield of potatoes. To do so, he selects 10 pieces of land and divides each piece into two portions. Then he randomly assigns one of the two portions from each piece of land to grow potatoes without using fertilizer (or using some other brand of fertilizer). The second portion from each piece of land is used to grow potatoes with the new brand of fertilizer. Thus, he will have 10 pairs of data values. Then, using the procedure to be discussed in this

> Two samples are said to be **paired** or **matched samples** when for each data value collected from one sample there is a corresponding data value collected from the second sample, and both these data values are collected from the same source.

section, he will make inferences about the difference in the mean yields of potatoes with and without the new fertilizer.

The question arises, why does the agronomist not choose 10 pieces of land on which to grow potatoes without using the new brand of fertilizer and another 10 pieces of land to grow potatoes by using the new brand of fertilizer? If he does so, the effect of the fertilizer might be confused with the effects due to soil differences at different locations. Thus, he will not be able to isolate the effect of the new brand of fertilizer on the yield of potatoes. Consequently, the results will not be reliable. By choosing 10 pieces of land and then dividing each of them into two portions, the researcher decreases the possibility that the difference in the productivities of different pieces of land affects the results.

In paired samples, the difference between the two data values for each element of the two samples is denoted by d. This value of d is called the **paired difference**. We then treat all the values of d as one sample and make inferences applying procedures similar to the ones used for one-sample cases in Chapters 8 and 9. Note that because each source (or element) gives a pair of values (one for each of the two data sets), each sample contains the same number of values. That is, both samples are of the same size. Therefore, we denote the (common) **sample size** by n, which gives the number of paired difference values denoted by d. The **degrees of freedom** for the paired samples are $n - 1$. Let

> μ_d = the mean of the paired differences for the population
>
> σ_d = the standard deviation of the paired differences for the population, which is usually not known
>
> $\bar{d}$ = the mean of the paired differences for the sample
>
> s_d = the standard deviation of the paired differences for the sample
>
> n = the number of paired difference values

Mean and Standard Deviation of the Paired Differences for Two Samples   The values of the mean and standard deviation, $\bar{d}$ and s_d, respectively, of paired differences for two samples are calculated as[2]

$$\bar{d} = \frac{\Sigma d}{n}$$

$$s_d = \sqrt{\frac{\Sigma d^2 - \dfrac{(\Sigma d)^2}{n}}{n - 1}}$$

In paired samples, instead of using $\bar{x}_1 - \bar{x}_2$ as the sample statistic to make inferences about $\mu_1 - \mu_2$, we use the sample statistic $\bar{d}$ to make inferences about μ_d. Actually the value of $\bar{d}$ is always equal to $\bar{x}_1 - \bar{x}_2$, and the value of μ_d is always equal to $\mu_1 - \mu_2$.

---

[2]The basic formula used to calculate s_d is

$$s_d = \sqrt{\frac{\Sigma (d - \bar{d})^2}{n - 1}}$$

However, we will not use this formula to make calculations in this chapter.

Sampling Distribution, Mean, and Standard Deviation of $\overline{d}$ If σ_d is known and either the sample size is large ($n \geq 30$) or the population is normally distributed, then the **sampling distribution of $\overline{d}$** is approximately normal with its **mean and standard deviation** given as, respectively,

$$\mu_{\overline{d}} = \mu_d \quad \text{and} \quad \sigma_{\overline{d}} = \frac{\sigma_d}{\sqrt{n}}$$

Thus, if the standard deviation σ_d of the population of paired differences is known and either the sample size is large (i.e., $n \geq 30$) or the population of paired differences is approximately normally distributed (with $n < 30$), then the normal distribution can be used to make a confidence interval and to test a hypothesis about μ_d. However, usually σ_d is not known. Then, if the standard deviation σ_d of the population of paired differences is unknown and either the sample size is large (i.e., $n \geq 30$) or the population of paired differences is approximately normally distributed (with $n < 30$), then the t distribution is used to make a confidence interval and to test a hypothesis about μ_d.

Making Inferences About μ_d   If

1. The standard deviation σ_d of the population of paired differences is unknown
2. At least one of the following two conditions is fulfilled:

    i. The sample size is large (i.e., $n \geq 30$)

    ii. If the sample size is small, and the population of paired differences is approximately normally distributed

then the t distribution is used to make inferences about μ_d. The standard deviation $\sigma_{\overline{d}}$ of $\overline{d}$ is estimated by $s_{\overline{d}}$, which is calculated as

$$s_{\overline{d}} = \frac{s_d}{\sqrt{n}}$$

Sections 10.4.1 and 10.4.2 describe the procedures that are used to make a confidence interval and to test a hypothesis about μ_d under the above conditions. The inferences are made using the t distribution.

10.4.1   Interval Estimation of μ_d

The mean $\overline{d}$ of paired differences for paired samples is the point estimator of μ_d. The following formula is used to construct a confidence interval for μ_d when the t distribution is used.

Confidence Interval for μ_d   The $(1 - \alpha)100\%$ **confidence interval for μ_d** is

$$\overline{d} \pm t s_{\overline{d}}$$

where the value of t is obtained from the t distribution table for the given confidence level and $n - 1$ degrees of freedom, and $s_{\overline{d}}$ is calculated as explained earlier.

Example 10–10 illustrates the procedure to construct a confidence interval for μ_d.

EXAMPLE 10–10 | Special Diet and Systolic Blood Pressure

Constructing a confidence interval for μ_d: paired samples, σ_d unknown, n < 30, and population normal.

A researcher wanted to find the effect of a special diet on systolic blood pressure. She selected a sample of seven adults and put them on this dietary plan for 3 months. The following table gives the systolic blood pressures (in mm Hg) of these seven adults before and after the completion of this plan.

Subject	1	2	3	4	5	6	7
Before	210	180	195	220	231	199	224
After	193	186	186	223	220	183	233

Let μ_d be the mean reduction in the systolic blood pressures due to this special dietary plan for the population of all adults. Construct a 95% confidence interval for μ_d. Assume that the population of paired differences is approximately normally distributed.

Solution Because the information obtained is from paired samples, we will make the confidence interval for the paired difference mean μ_d of the population using the paired difference mean $\bar{d}$ of the sample. Let d be the difference in the systolic blood pressure of an adult before and after this special dietary plan. Then, d is obtained by subtracting the systolic blood pressure after the plan from the systolic blood pressure before the plan. The third column of Table 10.1 lists the values of d for the seven adults. The fourth column of the table records the values of d^2, which are obtained by squaring each of the d values.

Table 10.1

Before	After	Difference d	d^2
210	193	17	289
180	186	−6	36
195	186	9	81
220	223	−3	9
231	220	11	121
199	183	16	256
224	233	−9	81
		$\Sigma d = 35$	$\Sigma d^2 = 873$

The values of $\bar{d}$ and s_d are calculated as follows:

$$\bar{d} = \frac{\Sigma d}{n} = \frac{35}{7} = 5.00$$

$$s_d = \sqrt{\frac{\Sigma d^2 - \frac{(\Sigma d)^2}{n}}{n-1}} = \sqrt{\frac{873 - \frac{(35)^2}{7}}{7-1}} = 10.78579312$$

Hence, the standard deviation of $\bar{d}$ is

$$s_{\bar{d}} = \frac{s_d}{\sqrt{n}} = \frac{10.78579312}{\sqrt{7}} = 4.07664661$$

Here, σ_d is not known, the sample size is small, but the population is approximately normally distributed. Hence, we will use the t distribution to make the confidence interval. For the 95% confidence interval, the area in each tail of the t distribution curve is

$$\text{Area in each tail} = \alpha/2 = (1 - .95)/2 = .025$$

The degrees of freedom are

$$df = n - 1 = 7 - 1 = 6$$

From the t distribution table, the t value for $df = 6$ and .025 area in the right tail of the t distribution curve is 2.447. Therefore, the 95% confidence interval for μ_d is

$$\bar{d} \pm ts_{\bar{d}} = 5.00 \pm 2.447(4.07664661) = 5.00 \pm 9.98 = \textbf{-4.98 to 14.98}$$

Thus, we can state with 95% confidence that the mean difference between systolic blood pressures before and after the given dietary plan for all adult participants is between −4.98 and 14.98 mm Hg.

10.4.2   Hypothesis Testing About μ_d

A hypothesis about μ_d is tested by using the sample statistic $\bar{d}$. This section illustrates the case of the t distribution only. Earlier in this section we learned what conditions should hold true to use the t distribution to test a hypothesis about μ_d. The following formula is used to calculate the value of the test statistic t when testing a hypothesis about μ_d.

Test Statistic t for $\bar{d}$ The value of the **test statistic t for $\bar{d}$** is computed as follows:

$$t = \frac{\bar{d} - \mu_d}{s_{\bar{d}}}$$

The critical value of t is found from the t distribution table for the given significance level and $n - 1$ degrees of freedom.

Examples 10–11 and 10–12 illustrate the hypothesis-testing procedure for μ_d.

EXAMPLE 10–11   How to Be a Successful Salesperson

Conducting a left-tailed test of hypothesis about μ_d for paired samples: σ_d not known, small sample but normally distributed population.

A company wanted to know if attending a course on "how to be a successful salesperson" can increase the average sales of its employees. The company sent six of its salespersons to attend this course. The following table gives the 1-week sales of these salespersons before and after they attended this course.

Subject	1	2	3	4	5	6
Before	12	18	25	9	14	16
After	18	24	24	14	19	20

Using a 1% significance level, can you conclude that the mean weekly sales for all salespersons increase as a result of attending this course? Assume that the population of paired differences has an approximate normal distribution.

Solution   Because the data are for paired samples, we test a hypothesis about the paired differences mean μ_d of the population using the paired differences mean $\bar{d}$ of the sample.

Let

$$d = \text{(Weekly sales before the course)} - \text{(Weekly sales after the course)}$$

In Table 10.2, we calculate d for each of the six salespersons by subtracting the sales after the course from the sales before the course. The fourth column of the table lists the values of d^2.

Table 10.2

Before	After	Difference d	d^2
12	18	−6	36
18	24	−6	36
25	24	1	1
9	14	−5	25
14	19	−5	25
16	20	−4	16
		$\Sigma d = -25$	$\Sigma d^2 = 139$

The values of $\bar{d}$ and s_d are calculated as follows:

$$\bar{d} = \frac{\Sigma d}{n} = \frac{-25}{6} = -4.17$$

$$s_d = \sqrt{\frac{\Sigma d^2 - \frac{(\Sigma d)^2}{n}}{n-1}} = \sqrt{\frac{139 - \frac{(-25)^2}{6}}{6-1}} = 2.63944439$$

The standard deviation of $\bar{d}$ is

$$s_{\bar{d}} = \frac{s_d}{\sqrt{n}} = \frac{2.63944439}{\sqrt{6}} = 1.07754866$$

Step 1. *State the null and alternative hypotheses.*

We are to test if the mean weekly sales for all salespersons increase as a result of taking the course. Let μ_1 be the mean weekly sales for all salespersons before the course and μ_2 the mean weekly sales for all salespersons after the course. Then $\mu_d = \mu_1 - \mu_2$. The mean weekly sales for all salespersons will increase due to attending the course if μ_1 is less than μ_2, which can be written as $\mu_1 - \mu_2 < 0$ or $\mu_d < 0$. Consequently, the null and alternative hypotheses are, respectively,

$$H_0: \mu_d = 0 \quad (\mu_1 - \mu_2 = 0 \text{ or the mean weekly sales do not increase})$$
$$H_1: \mu_d < 0 \quad (\mu_1 - \mu_2 < 0 \text{ or the mean weekly sales do increase})$$

Note that we can also write the null hypothesis as $\mu_d \geq 0$.

Step 2. *Select the distribution to use.*

Here σ_d is unknown, the sample size is small ($n < 30$), but the population of paired differences is approximately normally distributed. Therefore, we use the t distribution to conduct the test.

Step 3. *Determine the rejection and nonrejection regions.*

The $<$ sign in the alternative hypothesis indicates that the test is left-tailed. The significance level is .01. Hence,

$$\text{Area in left tail} = \alpha = .01$$

$$\text{Degrees of freedom} = n - 1 = 6 - 1 = 5$$

The critical value of t for $df = 5$ and .01 area in the left tail of the t distribution curve is -3.365. This value is shown in Figure 10.6.

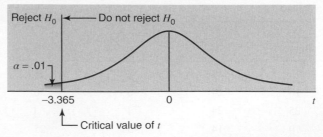

Figure 10.6 Rejection and nonrejection regions.

Step 4. *Calculate the value of the test statistic.*

The value of the test statistic t for $\bar{d}$ is computed as follows:

$$t = \frac{\bar{d} - \mu_d}{s_{\bar{d}}} = \frac{-4.17 - 0}{1.07754866} = -3.870$$

From H_0

Step 5. *Make a decision.*

Because the value of the test statistic $t = -3.870$ for $\bar{d}$ falls in the rejection region, we reject the null hypothesis. Consequently, we conclude that the mean weekly sales for all salespersons increase as a result of this course.

Using the *p*-Value to Make a Decision for Example 10–11

We can use the *p*-value approach to make the above decision. To do so, we keep Steps 1 and 2 of this example. Then in Step 3, we calculate the value of the test statistic t for $\bar{d}$ (as done in Step 4 above) and then find the *p*-value for this t from the t distribution table (Table V of Appendix B) or by using technology. If we have access to technology, we can use it to find the exact *p*-value, which will be .006. By using Table V, we can find a range for *p*-value. From Table V, for $df = 5$, the test statistic $t = -3.870$ falls between -3.365 and -4.032. The areas in the t distribution table that correspond to -3.365 and -4.032 are .01 and .005, respectively. Because it is a left-tailed test, the *p*-value is between .01 and .005, which can be written as

$$.005 < p\text{-value} < .01$$

Since we will reject the null hypothesis for any α (significance level) that is greater than or equal to the *p*-value, we will reject the null hypothesis in this example for any $\alpha \geq .006$ using the technology and $\alpha \geq .01$ using the above range. Since $\alpha = .01$ in this example, which is larger than .006 obtained from technology, we reject the null hypothesis. Also, because α is equal to .01, using the *p*-value range we reject the null hypothesis.

| **EXAMPLE 10–12** | Special Diet and Systolic Blood Pressure |

Making a two-tailed test of hypothesis about μ_d for paired samples: σ_d not known, small sample but normally distributed population.

Refer to Example 10–10. The table that gives the blood pressures (in mm Hg) of seven adults before and after the completion of a special dietary plan is reproduced here.

Subject	1	2	3	4	5	6	7
Before	210	180	195	220	231	199	224
After	193	186	186	223	220	183	233

Let μ_d be the mean of the differences between the systolic blood pressures before and after completing this special dietary plan for the population of all adults. Using a 5% significance level, can you conclude that the mean of the paired differences μ_d is different from zero? Assume that the population of paired differences is approximately normally distributed.

Solution   Table 10.3 gives d and d^2 for each of the seven adults (blood pressure values in mm Hg).

Table 10.3

Before	After	Difference d	d^2
210	193	17	289
180	186	−6	36
195	186	9	81
220	223	−3	9
231	220	11	121
199	183	16	256
224	233	−9	81
		$\Sigma d = 35$	$\Sigma d^2 = 873$

The values of $\overline{d}$ and s_d are calculated as follows:

$$\overline{d} = \frac{\Sigma d}{n} = \frac{35}{7} = 5.00$$

$$s_d = \sqrt{\frac{\Sigma d^2 - \dfrac{(\Sigma d)^2}{n}}{n-1}} = \sqrt{\frac{873 - \dfrac{(35)^2}{7}}{7-1}} = 10.78579312$$

Hence, the standard deviation of $\overline{d}$ is

$$s_{\overline{d}} = \frac{s_d}{\sqrt{n}} = \frac{10.78579312}{\sqrt{7}} = 4.07664661$$

Step 1.   *State the null and alternative hypotheses.*

H_0: $\mu_d = 0$ (The mean of the paired differences is not different from zero.)

H_1: $\mu_d \neq 0$ (The mean of the paired differences is different from zero.)

Step 2.   *Select the distribution to use.*

Here σ_d is unknown, the sample size is small, but the population of paired differences is approximately normal. Hence, we use the t distribution to make the test.

Step 3.   *Determine the rejection and nonrejection regions.*

The $\neq$ sign in the alternative hypothesis indicates that the test is two-tailed. The significance level is .05.

Area in each tail of the curve $= \alpha/2 = .05/2 = .025$

Degrees of freedom $= n - 1 = 7 - 1 = 6$

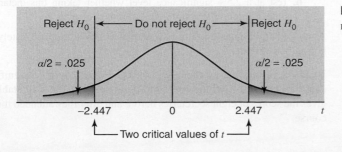

Reject H_0 ← — Do not reject H_0 — →   Reject H_0

$\alpha/2 = .025$ $\alpha/2 = .025$

−2.447   0   2.447 t

Two critical values of t

Figure 10.7   Rejection and nonrejection regions.

The two critical values of t for $df = 6$ and .025 area in each tail of the t distribution curve are -2.447 and 2.447. These values are shown in Figure 10.7.

Step 4. *Calculate the value of the test statistic.*

The value of the test statistic t for $\bar{d}$ is computed as follows:

$$t = \frac{\bar{d} - \mu_d}{s_{\bar{d}}} = \frac{5.00 - 0}{4.07664661} = 1.226 \qquad \overset{\text{From } H_0}{}$$

Step 5. *Make a decision.*

Because the value of the test statistic $t = 1.226$ for $\bar{d}$ falls in the nonrejection region, we fail to reject the null hypothesis. Hence, we conclude that the mean of the population of paired differences is not different from zero. In other words, we can state that the mean of the differences between the systolic blood pressures before and after completing this special dietary plan for the population of all adults is not different from zero.

Using the *p*-Value to Make a Decision for Example 10–12

We can use the *p*-value approach to make the above decision. To do so, we keep Steps 1 and 2 of this example. Then in Step 3, we calculate the value of the test statistic t for $\bar{d}$ (as done in Step 4 above) and then find the *p*-value for this t from the t distribution table (Table V of Appendix B) or by using technology. If we have access to technology, we can use it to find the exact *p*-value, which will be .266. By using Table V, we can find a range for the *p*-value. From Table V, for $df = 6$, the test statistic $t = 1.226$ is less than 1.440. The area in the t distribution table that corresponds to 1.440 is .10. Because it is a two-tailed test, the *p*-value is greater than $2(.10) = .20$, which can be written as

$$p\text{-value} > .20$$

Since $\alpha = .05$ in this example, which is smaller than .20 and also smaller than .266 (obtained from technology), we fail to reject the null hypothesis.

EXERCISES

CONCEPTS AND PROCEDURES

10.35 Explain when would you use the paired-samples procedure to make confidence intervals and test hypotheses.

10.36 Find the following confidence intervals for μ_d, assuming that the populations of paired differences are normally distributed.

  a. $n = 12$ $\bar{d} = 17.5$, $s_d = 6.3$,   confidence level = 99%
  b. $n = 27$, $\bar{d} = 55.9$, $s_d = 14.7$,   confidence level = 95%
  c. $n = 16$, $\bar{d} = 29.3$, $s_d = 8.3$,   confidence level = 90%

10.37 Conduct the following tests of hypotheses, assuming that the populations of paired differences are normally distributed.

  a. $H_0: \mu_d = 0$, $H_1: \mu_d \neq 0$, $n = 26$, $\bar{d} = 9.6$, $s_d = 3.9$, $\alpha = .05$
  b. $H_0: \mu_d = 0$, $H_1: \mu_d > 0$, $n = 15$, $\bar{d} = 8.8$, $s_d = 4.7$, $\alpha = .01$
  c. $H_0: \mu_d = 0$, $H_1: \mu_d < 0$, $n = 20$, $\bar{d} = -7.4$, $s_d = 2.3$, $\alpha = .10$

APPLICATIONS

10.38 Several retired bicycle racers are coaching a large group of young prospects. They randomly select seven of their riders to take part in a test of the effectiveness of a new dietary supplement that is supposed to increase strength and stamina. Each of the seven riders does a time trial on the same course. Then they all take the dietary supplement for four weeks. All other aspects of their training program remain as they were prior to the time trial. At the end of the four weeks, these riders do another time trial on the same course. The times (in minutes) recorded by each rider for these trials, before and after the four-week period, are shown in the following table.

Before	103	97	111	95	102	96	108
After	100	95	104	101	96	91	101

  a. Construct a 99% confidence interval for the mean μ_d of the population paired differences, where a paired difference is equal to the time taken before the dietary supplement minus the time taken after the dietary supplement.
  b. Test at a 2.5% significance level whether taking this dietary supplement results in faster times in the time trials.

Assume that the population of paired differences is (approximately) normally distributed.

10.39 A private agency claims that the crash course it offers significantly increases the writing speed of secretaries. The following table gives the scores of eight secretaries before and after they attended this course.

Before	81	75	89	91	65	70	90	64
After	97	72	93	110	78	69	115	72

a. Make a 90% confidence interval for the mean μ_d of the population paired differences, where a paired difference is equal to the score before attending the course minus the score after attending the course.

b. Using the 5% significance level, can you conclude that attending this course increases the writing speed of secretaries?

Assume that the population of paired differences is (approximately) normally distributed.

10.40   The manufacturer of a gasoline additive claims that the use of this additive increases gasoline mileage. A random sample of six cars was selected, and these cars were driven for 1 week without the gasoline additive and then for 1 week with the gasoline additive. The following table gives the miles per gallon for these cars without and with the gasoline additive.

Without	23.8	28.3	18.5	23.7	13.6	28.8
With	26.3	30.5	18.2	25.3	16.3	30.9

a. Construct a 99% confidence interval for the mean μ_d of the population paired differences, where a paired difference is equal to the miles per gallon without the gasoline additive minus the miles per gallon with the gasoline additive.

b. Using a 2.5% significance level, can you conclude that the use of the gasoline additive increases the gasoline mileage?

Assume that the population of paired differences is (approximately) normally distributed.

10.41   A company is considering installing new machines to assemble its products. The company is considering two types of machines, but it will buy only one type. The company selected eight assembly workers and asked them to use these two types of machines to assemble products. The following table gives the time taken (in minutes) to assemble one unit of the product on each type of machine for each of these eight workers.

Machine I	23	26	19	24	27	22	20	18
Machine II	21	24	23	25	24	28	24	23

a. Construct a 98% confidence interval for the mean μ_d of the population paired differences, where a paired difference is equal to the time taken to assemble a unit of the product on Machine I minus the time taken to assemble a unit of the product on Machine II.

b. Test at a 5% significance level whether the mean times taken to assemble a unit of the product are different for the two types of machines.

Assume that the population of paired differences is (approximately) normally distributed.

10.5 | Inferences About the Difference Between Two Population Proportions for Large and Independent Samples

Quite often we need to construct a confidence interval and test a hypothesis about the difference between two population proportions. For instance, we may want to estimate the difference between the proportions of defective items produced on two different machines. If p_1 and p_2 are the proportions of defective items produced on the first and second machine, respectively, then we are to make a confidence interval for $p_1 - p_2$. Alternatively, we may want to test the hypothesis that the proportion of defective items produced on Machine I is different from the proportion of defective items produced on Machine II. In this case, we are to test the null hypothesis $p_1 - p_2 = 0$ against the alternative hypothesis $p_1 - p_2 \neq 0$.

This section discusses how to make a confidence interval and test a hypothesis about $p_1 - p_2$ for two large and independent samples. The sample statistic that is used to make inferences about $p_1 - p_2$ is $\hat{p}_1 - \hat{p}_2$, where $\hat{p}_1$ and $\hat{p}_2$ are the proportions for two large and independent samples. As discussed in Chapter 7, we determine a sample proportion by dividing the number of elements in the sample that possess a given attribute by the sample size. Thus,

$$\hat{p}_1 = x_1/n_1 \quad \text{and} \quad \hat{p}_2 = x_2/n_2$$

where x_1 and x_2 are the number of elements that possess a given characteristic in the two samples and n_1 and n_2 are the sizes of the two samples, respectively.

10.5.1   Mean, Standard Deviation, and Sampling Distribution of $\hat{p}_1 - \hat{p}_2$

As discussed in Chapter 7, for a large sample, the sample proportion $\hat{p}$ is approximately normally distributed with mean p and standard deviation $\sqrt{pq/n}$. Hence, for two large and independent samples of sizes n_1 and n_2, respectively, their sample proportions $\hat{p}_1$ and $\hat{p}_2$ are approximately normally distributed with means p_1 and p_2 and standard deviations $\sqrt{p_1 q_1/n_1}$ and $\sqrt{p_2 q_2/n_2}$, respectively. Using these results, we can make the following statements about the shape of the sampling distribution of $\hat{p}_1 - \hat{p}_2$ and its mean and standard deviation.

Mean, Standard Deviation, and Sampling Distribution of $\hat{p}_1 - \hat{p}_2$ For two large and independent samples, the *sampling distribution* of $\hat{p}_1 - \hat{p}_2$ is approximately normal, with its **mean and standard deviation** given as

$$\mu_{\hat{p}_1-\hat{p}_2} = p_1 - p_2$$

and

$$\sigma_{\hat{p}_1-\hat{p}_2} = \sqrt{\frac{p_1 q_1}{n_1} + \frac{p_2 q_2}{n_2}}$$

respectively, where $q_1 = 1 - p_1$ and $q_2 = 1 - p_2$.

Thus, to construct a confidence interval and to test a hypothesis about $p_1 - p_2$ for large and independent samples, we use the normal distribution. As was indicated in Chapter 7, in the case of proportion, the sample is large if np and nq are both greater than 5. In the case of two samples, both sample sizes are large if $n_1 p_1$, $n_1 q_1$, $n_2 p_2$, and $n_2 q_2$ are all greater than 5.

10.5.2   Interval Estimation of $p_1 - p_2$

The difference between two sample proportions $\hat{p}_1 - \hat{p}_2$ is the point estimator for the difference between two population proportions $p_1 - p_2$. Because we do not know p_1 and p_2 when we are making a confidence interval for $p_1 - p_2$, we cannot calculate the value of $\sigma_{\hat{p}_1-\hat{p}_2}$. Therefore, we use $s_{\hat{p}_1-\hat{p}_2}$ as the point estimator of $\sigma_{\hat{p}_1-\hat{p}_2}$ in the interval estimation. We construct the confidence interval for $p_1 - p_2$ using the following formula.

Confidence Interval for $p_1 - p_2$ The $(1 - \alpha)100\%$ **confidence interval for $p_1 - p_2$** is

$$(\hat{p}_1 - \hat{p}_2) \pm z s_{\hat{p}_1-\hat{p}_2}$$

where the value of z is read from the normal distribution table for the given confidence level, and $s_{\hat{p}_1-\hat{p}_2}$ is calculated as

$$s_{\hat{p}_1-\hat{p}_2} = \sqrt{\frac{\hat{p}_1 \hat{q}_1}{n_1} + \frac{\hat{p}_2 \hat{q}_2}{n_2}}$$

Example 10–13 describes the procedure that is used to make a confidence interval for the difference between two population proportions for large samples.

EXAMPLE 10–13   Users of Two Toothpastes

Constructing a confidence interval for $p_1 - p_2$: large and independent samples.

A researcher wanted to estimate the difference between the percentages of users of two toothpastes who will never switch to another toothpaste. In a sample of 500 users of Toothpaste A taken by this researcher, 100 said that they will never switch to another toothpaste. In another sample of 400 users of Toothpaste B taken by the same researcher, 68 said that they will never switch to another toothpaste.

   (a) Let p_1 and p_2 be the proportions of all users of Toothpastes A and B, respectively, who will never switch to another toothpaste. What is the point estimate of $p_1 - p_2$?

   (b) Construct a 97% confidence interval for the difference between the proportions of all users of the two toothpastes who will never switch.

Solution   Let p_1 and p_2 be the proportions of all users of Toothpastes A and B, respectively, who will never switch to another toothpaste, and let $\hat{p}_1$ and $\hat{p}_2$ be the respective sample proportions. Let x_1 and x_2 be the number of users of Toothpastes A and B, respectively, in the two samples who said that they will never switch to another toothpaste. From the given information,

$$\text{Toothpaste A:} \quad n_1 = 500 \quad \text{and} \quad x_1 = 100$$
$$\text{Toothpaste B:} \quad n_2 = 400 \quad \text{and} \quad x_2 = 68$$

The two sample proportions are calculated as follows:

$$\hat{p}_1 = x_1/n_1 = 100/500 = .20$$
$$\hat{p}_2 = x_2/n_2 = 68/400 = .17$$

Then,

$$\hat{q}_1 = 1 - .20 = .80 \quad \text{and} \quad \hat{q}_2 = 1 - .17 = .83$$

(a) The point estimate of $p_1 - p_2$ is as follows:

$$\text{Point estimate of } p_1 - p_2 = \hat{p}_1 - \hat{p}_2 = .20 - .17 = \mathbf{.03}$$

(b) The values of $n_1\hat{p}_1$, $n_1\hat{q}_1$, $n_2\hat{p}_2$, and $n_2\hat{q}_2$ are

$$n_1\hat{p}_1 = 500(.20) = 100 \quad n_1\hat{q}_1 = 500(.80) = 400$$
$$n_2\hat{p}_2 = 400(.17) = 68 \quad n_2\hat{q}_2 = 400(.83) = 332$$

Because each of these values is greater than 5, both sample sizes are large. Consequently we use the normal distribution to make a confidence interval for $p_1 - p_2$. The standard deviation of $\hat{p}_1 - \hat{p}_2$ is

$$s_{\hat{p}_1-\hat{p}_2} = \sqrt{\frac{\hat{p}_1\hat{q}_1}{n_1} + \frac{\hat{p}_2\hat{q}_2}{n_2}} = \sqrt{\frac{(.20)(.80)}{500} + \frac{(.17)(.83)}{400}} = .02593742$$

The z value for a 97% confidence level, obtained from the normal distribution table, is 2.17. The 97% confidence interval for $p_1 - p_2$ is

$$(\hat{p}_1 - \hat{p}_2) \pm z s_{\hat{p}_1-\hat{p}_2} = (.20 - .17) \pm 2.17(.02593742)$$
$$= .03 \pm .056 = \mathbf{-.026 \text{ to } .086}$$

Thus, with 97% confidence we can state that the difference between the two population proportions is between $-.026$ and $.086$ or -2.6% and 8.6%.

Note that here $\hat{p}_1 - \hat{p}_2 = .03$ gives the point estimate of $p_1 - p_2$ and $z s_{\hat{p}_1-\hat{p}_2} = .056$ is the margin of error of the estimate.

© Andrey Armyagov/iStockphoto

10.5.3   Hypothesis Testing About $p_1 - p_2$

In this section we learn how to test a hypothesis about $p_1 - p_2$ for two large and independent samples. The procedure involves the same five steps we have used previously. Once again, we calculate the standard deviation of $\hat{p}_1 - \hat{p}_2$ as

$$\sigma_{\hat{p}_1-\hat{p}_2} = \sqrt{\frac{p_1 q_1}{n_1} + \frac{p_2 q_2}{n_2}}$$

When a test of hypothesis about $p_1 - p_2$ is performed, usually the null hypothesis is $p_1 = p_2$ and the values of p_1 and p_2 are not known. Assuming that the null hypothesis is true and $p_1 = p_2$, a common value of p_1 and p_2, denoted by $\bar{p}$, is calculated by using one of the following two formulas:

$$\bar{p} = \frac{x_1 + x_2}{n_1 + n_2} \quad \text{or} \quad \frac{n_1\hat{p}_1 + n_2\hat{p}_2}{n_1 + n_2}$$

Which of these formulas is used depends on whether the values of x_1 and x_2 or the values of $\hat{p}_1$ and $\hat{p}_2$ are known. Note that x_1 and x_2 are the number of elements in each of the two samples that possess a certain characteristic. This value of $\bar{p}$ is called the **pooled sample proportion**. Using the value of the pooled sample proportion, we compute an estimate of the standard deviation of $\hat{p}_1 - \hat{p}_2$ as follows:

$$s_{\hat{p}_1 - \hat{p}_2} = \sqrt{\bar{p}\,\bar{q}\left(\frac{1}{n_1} + \frac{1}{n_2}\right)}$$

where $\bar{q} = 1 - \bar{p}$.

Test Statistic z for $\hat{p}_1 - \hat{p}_2$ The value of the **test statistic z for $\hat{p}_1 - \hat{p}_2$** is calculated as

$$z = \frac{(\hat{p}_1 - \hat{p}_2) - (p_1 - p_2)}{s_{\hat{p}_1 - \hat{p}_2}}$$

The value of $p_1 - p_2$ is substituted from H_0, which usually is zero.

Examples 10–14 and 10–15 illustrate the procedure to test hypotheses about the difference between two population proportions for large samples.

EXAMPLE 10–14	**Users of Two Toothpastes**

Making a right-tailed test of hypothesis about $p_1 - p_2$: large and independent samples.

Reconsider Example 10–13 about the percentages of users of two toothpastes who will never switch to another toothpaste. At a 1% significance level, can you conclude that the proportion of users of Toothpaste A who will never switch to another toothpaste is higher than the proportion of users of Toothpaste B who will never switch to another toothpaste?

Solution   Let p_1 and p_2 be the proportions of all users of Toothpastes A and B, respectively, who will never switch to another toothpaste, and let $\hat{p}_1$ and $\hat{p}_2$ be the corresponding sample proportions. Let x_1 and x_2 be the number of users of Toothpastes A and B, respectively, in the two samples who said that they will never switch to another toothpaste. From the given information,

$$\text{Toothpaste A:} \quad n_1 = 500 \quad \text{and} \quad x_1 = 100$$

$$\text{Toothpaste B:} \quad n_2 = 400 \quad \text{and} \quad x_2 = 68$$

The significance level is $\alpha = .01$. The two sample proportions are calculated as follows:

$$\hat{p}_1 = x_1/n_1 = 100/500 = .20$$
$$\hat{p}_2 = x_2/n_2 = 68/400 = .17$$

Step 1.   *State the null and alternative hypotheses.*

We are to test if the proportion of users of Toothpaste A who will never switch to another toothpaste is higher than the proportion of users of Toothpaste B who will never switch to another toothpaste. In other words, we are to test whether p_1 is greater than p_2. This can be written as $p_1 - p_2 > 0$. Thus, the two hypotheses are

$$H_0: p_1 = p_2 \quad \text{or} \quad p_1 - p_2 = 0 \quad (p_1 \text{ is the same as } p_2)$$

$$H_1: p_1 > p_2 \quad \text{or} \quad p_1 - p_2 > 0 \quad (p_1 \text{ is greater than } p_2)$$

Step 2.   *Select the distribution to use.*

As shown in Example 10–13, $n_1\hat{p}_1$, $n_1\hat{q}_1$, $n_2\hat{p}_2$, and $n_2\hat{q}_2$ are all greater than 5. Consequently both samples are large, and we use the normal distribution to make the test.

Step 3.   *Determine the rejection and nonrejection regions.*

The $>$ sign in the alternative hypothesis indicates that the test is right-tailed. From the normal distribution table, for a .01 significance level, the critical value of z is 2.33 for .9900 area to the left. This is shown in Figure 10.8.

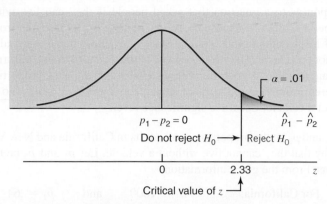

Figure 10.8   Rejection and nonrejection regions.

Step 4.   *Calculate the value of the test statistic.*

The pooled sample proportion is

$$\bar{p} = \frac{x_1 + x_2}{n_1 + n_2} = \frac{100 + 68}{500 + 400} = .187 \qquad \text{and} \qquad \bar{q} = 1 - \bar{p} = 1 - .187 = .813$$

The estimate of the standard deviation of $\hat{p}_1 - \hat{p}_2$ is

$$s_{\hat{p}_1-\hat{p}_2} = \sqrt{\bar{p}\,\bar{q}\left(\frac{1}{n_1} + \frac{1}{n_2}\right)} = \sqrt{(.187)(.813)\left(\frac{1}{500} + \frac{1}{400}\right)} = .02615606$$

The value of the test statistic z for $\hat{p}_1 - \hat{p}_2$ is

$$z = \frac{(\hat{p}_1 - \hat{p}_2) - (p_1 - p_2)}{s_{\hat{p}_1-\hat{p}_2}} = \frac{(.20 - .17) - 0}{.02615606} = 1.15$$

From H_0

Step 5.   *Make a decision.*

Because the value of the test statistic $z = 1.15$ for $\hat{p}_1 - \hat{p}_2$ falls in the nonrejection region (see Figure 10.8), we fail to reject the null hypothesis. Therefore, we conclude that the proportion of users of Toothpaste A who will never switch to another toothpaste is not greater than the proportion of users of Toothpaste B who will never switch to another toothpaste.

Using the *p*-Value to Make a Decision for Example 10–14

We can use the *p*-value approach to make the above decision. To do so, we keep Steps 1 and 2 above. Then in Step 3, we calculate the value of the test statistic z (as done in Step 4 above) and find the *p*-value for this z from the normal distribution table. In Step 4 above, the z-value for $\hat{p}_1 - \hat{p}_2$ was calculated to be 1.15. In this example, the test is right-tailed. The *p*-value is given by the area under the normal distribution curve to the right of $z = 1.15$. From the normal distribution table (Table IV of Appendix B), this area is $1 - .8749 = .1251$. Hence, the *p*-value is .1251. We reject the null hypothesis for any α (significance level) greater than or equal to the *p*-value; in this example, we will reject the null hypothesis for any $\alpha \geq .1251$ or 12.51%. Because $\alpha = .01$ here, which is less than .1251, we fail to reject the null hypothesis.

EXAMPLE 10–15 Can We Live Without a Vehicle?

Conducting a two-tailed test of hypothesis about $p_1 - p_2$: large and independent samples.

The International Organization of Motor Vehicle Manufacturers (OICA), a group that defends the interests of the vehicle manufacturers, released the findings of a study it conducted during 2015 that included responses to a question that asked people if they can live without a car (*USA TODAY*, September 17, 2015). Suppose recently a sample of 1000 adults from California and another sample of 1200 adults from New York State were selected and these adults were asked if they can live without a vehicle. Sixty-four percent of these adults from California and 61% from New York State said that they cannot live without a vehicle. Test if the proportion of adults from California is different from the proportion of adults from New York State who will say that they cannot live without a vehicle. Use a 1% significance level.

Solution   Let p_1 and p_2 be the proportion of all adults in California and New York State, respectively, who will say that they cannot live without a vehicle. Let $\hat{p}_1$ and $\hat{p}_2$ be the corresponding sample proportions. From the given information,

$$\text{For California:} \quad n_1 = 1000 \quad \text{and} \quad \hat{p}_1 = .64$$
$$\text{For New York State:} \quad n_2 = 1200 \quad \text{and} \quad \hat{p}_2 = .61$$

The significance level is $\alpha = .01$.

Step 1. *State the null and alternative hypotheses.*

The null and alternative hypotheses are, respectively,

$$H_0: p_1 - p_2 = 0 \quad \text{(The two population proportions are not different.)}$$
$$H_1: p_1 - p_2 \neq 0 \quad \text{(The two population proportions are different.)}$$

Step 2. *Select the distribution to use.*

Because the samples are large and independent, we apply the normal distribution to make the test. (The reader should check that $n_1\hat{p}_1$, $n_1\hat{q}_1$, $n_2\hat{p}_2$, and $n_2\hat{q}_2$ are all greater than 5.)

Step 3. *Determine the rejection and nonrejection regions.*

The $\neq$ sign in the alternative hypothesis indicates that the test is two-tailed. For a 1% significance level, the critical values of z are -2.58 and 2.58. Note that to find these two critical values, we look for .0050 and .9950 areas in Table IV of Appendix B. These values are shown in Figure 10.9.

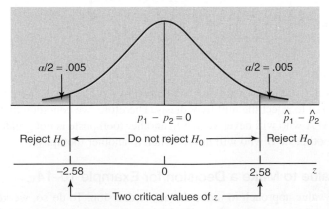

Figure 10.9 Rejection and nonrejection regions.

Step 4. *Calculate the value of the test statistic.*

The pooled sample proportion is

$$\bar{p} = \frac{n_1\hat{p}_1 + n_2\hat{p}_2}{n_1 + n_2} = \frac{1000(.64) + 1200(.61)}{1000 + 1200} = .624$$

and

$$\bar{q} = 1 - \bar{p} = 1 - .624 = .376$$

The estimate of the standard deviation of $\hat{p}_1 - \hat{p}_2$ is

$$s_{\hat{p}_1 - \hat{p}_2} = \sqrt{\bar{p}\,\bar{q}\left(\frac{1}{n_1} + \frac{1}{n_2}\right)} = \sqrt{(.624)(.376)\left(\frac{1}{1000} + \frac{1}{1200}\right)} = .02073991$$

The value of the test statistic z for $\hat{p}_1 - \hat{p}_2$ is

$$z = \frac{(\hat{p}_1 - \hat{p}_2) - (p_1 - p_2)}{s_{\hat{p}_1 - \hat{p}_2}} = \frac{(.64 - .61) - 0}{.02073991} = 1.45$$

Step 5. *Make a decision.*

Because the value of the test statistic $z = 1.45$ falls in the nonrejection region, we fail to reject the null hypothesis H_0. Therefore, we conclude that the proportions of all adults in California and New York State who will say that they cannot live without a vehicle are not different.

Using the *p*-Value to Make a Decision for Example 10–15

We can use the *p*-value approach to make the above decision. To do so, we keep Steps 1 and 2. Then in Step 3, we calculate the value of the test statistic z (as done in Step 4) and find the *p*-value for this z from the normal distribution table. In Step 4, the z-value for $\hat{p}_1 - \hat{p}_2$ was calculated to be 1.45. In this example, the test is two-tailed. The *p*-value is given by twice the area under the normal distribution curve to the right of $z = 1.45$. From the normal distribution table (Table IV of Appendix B), the area to the right of $z = 1.45$ is .0735. Hence, the *p*-value is $2(.0735) = .1470$. As we know, we will reject the null hypothesis for any α (significance level) greater than or equal to the *p*-value. Since $\alpha = .01$ in this example, which is smaller than .1470, we fail to reject the null hypothesis.

EXERCISES

CONCEPTS AND PROCEDURES

10.42 What is the shape of the sampling distribution of $\hat{p}_1 - \hat{p}_2$ for two large samples? What are the mean and standard deviation of this sampling distribution?

10.43 When are the samples considered large enough for the sampling distribution of the difference between two sample proportions to be (approximately) normal?

10.44 Construct a 95% confidence interval for $p_1 - p_2$ for the following.

$$n_1 = 300, \quad \hat{p}_1 = .55, \quad n_2 = 200, \quad \hat{p}_2 = .62$$

10.45 Construct a 99% confidence interval for $p_1 - p_2$ for the following.

$$n_1 = 100, \quad \hat{p}_1 = .81, \quad n_2 = 150, \quad \hat{p}_2 = .77$$

10.46 Consider the following information obtained from two independent samples:

$$n_1 = 300, \quad \hat{p}_1 = .55, \quad n_2 = 200, \quad \hat{p}_2 = .62$$

Test at a 1% significance level if the two population proportions are different.

10.47 Consider the following information obtained from two independent samples:

$$n_1 = 100, \quad \hat{p}_1 = .81, \quad n_2 = 150, \quad \hat{p}_2 = .77$$

Test at a 5% significance level if $p_1 - p_2$ is different from zero.

10.48 A sample of 1000 observations taken from the first population gave $x_1 = 290$. Another sample of 1200 observations taken from the second population gave $x_2 = 396$.

a. Find the point estimate of $p_1 - p_2$.
b. Make a 98% confidence interval for $p_1 - p_2$.
c. Show the rejection and nonrejection regions on the sampling distribution of $\hat{p}_1 - \hat{p}_2$ for H_0: $p_1 = p_2$ versus H_1: $p_1 < p_2$. Use a significance level of 1%.
d. Find the value of the test statistic z for the test of part c.
e. Will you reject the null hypothesis mentioned in part c at a significance level of 1%?

10.49 A sample of 500 observations taken from the first population gave $x_1 = 305$. Another sample of 600 observations taken from the second population gave $x_2 = 348$.

a. Find the point estimate of $p_1 - p_2$.
b. Make a 97% confidence interval for $p_1 - p_2$.
c. Show the rejection and nonrejection regions on the sampling distribution of $\hat{p}_1 - \hat{p}_2$ for H_0: $p_1 = p_2$ versus H_1: $p_1 > p_2$. Use a significance level of 2.5%.
d. Find the value of the test statistic z for the test of part c.
e. Will you reject the null hypothesis mentioned in part c at a significance level of 2.5%?

APPLICATIONS

10.50 The global recession has led more and more people to move in with relatives, which has resulted in a large number of multigenerational households. An October 2011 Pew Research Center poll showed that 11.5% of people living in multigenerational households were living below the poverty level, and 14.6% of people living in other types of households were living below the poverty level (www. pewsocialtrends.org/2011/10/03/fighting-poverty-in-a-bad-economy-americans-move-in-with-relatives/?src-prc-headline). Suppose that these results were based on samples of 1000 people living in multi-generational households and 2000 people living in other types of households.

a. Let p_1 be the proportion of all people in multigenerational households who live below the poverty level and p_2 be the proportion of all people in other types of households who live below the poverty level. Construct a 98% confidence interval for $p_1 - p_2$.

b. Using a 2.5% significance level, can you conclude that p_1 is less than p_2? Use the critical-value approach.

c. Repeat part b using the p-value approach.

10.51 A November 2011 Gallup poll asked American adults about their views of healthcare and the healthcare system in the United States. Although feelings about the quality of healthcare were positive, the same cannot be said about the quality of the healthcare system. According to this study, 29% of Independents and 27% of Democrats rated the healthcare system as being excellent or good (www.gallup.com/poll/150788/Americans-Maintain-Negative-View-Healthcare-Coverage.aspx). Suppose that these results were based on samples of 1200 Independents and 1300 Democrats.

a. Let p_1 and p_2 be the proportions of all Independents and all Democrats, respectively, who will rate the healthcare system as being excellent or good. Construct a 97% confidence interval for $p_1 - p_2$.

b. Using a 1% significance level, can you conclude that p_1 is different from p_2? Use both the critical-value and the p-value approaches.

10.52 According to Pew Research Center surveys, 79% of U.S. adults were using the Internet in January 2011 and 83% were using it in January 2012 (*USA TODAY*, January 26, 2012). Suppose that these percentages are based on random samples of 1800 U.S. adults in January 2011 and 1900 in January 2012.

a. Let p_1 and p_2 be the proportions of all U.S. adults who were using the Internet in January 2011 and January 2012, respectively. Construct a 98% confidence interval for $p_1 - p_2$.

b. Using a 1% significance level, can you conclude that p_1 is lower than p_2? Use both the critical-value and the p-value approaches.

10.53 Bernard Haldane Associates conducted a survey in March 2004 of 1021 workers who held white-collar jobs and had changed jobs in the previous twelve months. Of these workers, 56% of the men and 35% of the women were paid more in their new positions when they changed jobs (*USA TODAY*, April 28, 2004). Suppose that these percentages are based on random samples of 510 men and 511 women white-collar workers.

a. What is the point estimate of the difference between the two population proportions?

b. Construct a 95% confidence interval for the difference between the two population proportions.

c. Using a 2% significance level, can you conclude that the two population proportions are different? Use both the critical-value and the p-value approaches.

10.54 According to a June 2009 report (http://www.alertnet.org/thenews/newsdesk/L31011082.htm), 68% of people with "green" jobs in North America felt that they had job security, whereas 60% of people with green jobs in the United Kingdom felt that they had job security. Suppose that these results were based on samples of 305 people with green jobs from North America and 280 people with green jobs from the United Kingdom.

a. Construct a 96% confidence interval for the difference between the two population proportions.

b. Using a 2% significance level, can you conclude that the proportion of all people with green jobs in North America who feel that they have job security is higher than the corresponding proportion for the United Kingdom? Use the critical-value approach.

c. Repeat part b using the p-value approach.

10.55 A study in the July 7, 2009, issue of *USA TODAY* stated that the 401(k) participation rate among U.S. employees of Asian heritage is 76%, whereas the participation rate among U.S. employees of Hispanic heritage is 66%. Suppose that these results were based on random samples of 100 U.S. employees from each group.

a. Construct a 95% confidence interval for the difference between the two population proportions.

b. Using a 5% significance level, can you conclude that the 401(k) participation rates are different for all U.S. employees of Asian heritage and all U.S. employees of Hispanic heritage? Use the critical-value and p-value approaches.

c. Repeat parts a and b for both sample sizes of 200 instead of 100. Does your conclusion change in part b?

USES AND MISUSES...

STATISTICS AND HEALTH-RELATED STUDIES

While watching the news or browsing through a newspaper, we are likely to see or read the results of some new medical study. Often, the study result is really eye-catching, such as the following headline from *USA TODAY*: "Chocolate Lowers Heart Stroke Risk" (http://yourlife.usatoday.com/health/healthcare/studies/story/2011-08-29/Chocolate-lowers-heart-stroke-risk/50174422/1). Certainly a headline such as this one can grab your attention, especially if you are a chocolate lover. However, it is important to find out about the type of study that was conducted and determine whether it indicated an association (i.e., it showed a potential link between chocolate

consumption and stroke risk) or a causal relationship (i.e., it showed that consuming chocolate actually does lower the risk of a stroke).

Two primary types of medical studies are (1) case–control studies and (b) cohort studies. A case–control study is an observational study that essentially works *backward*. In the case of the aforementioned story, people in the study were classified into two different groups—those who had a stroke, and those who did not have a stroke. The people were interviewed and asked about their chocolate consumption habits. The study revealed that the proportion of people who consumed chocolate was higher in the nonstroke group than in the stroke group. As noted in the *USA TODAY* article, this result regarding the relationship was consistent across a number of independent case–control studies, which were combined into a single study using a process called *meta-analysis*. So, with this information, this set of case–control studies leads to a new question: Does consuming chocolate actually reduce the risk of having a stroke, or is there something else that is the cause of these results?

When a case–control study identifies a potential causal link, the next step is to perform a cohort study (i.e., a *clinical trial*). In a cohort study, individuals are selected to participate in the study, and then these individuals are randomly assigned to different groups, with each group receiving a specific *treatment*. In this case, members in one group would eat chocolate regularly, whereas the members in the other group would not. The participants would be observed over a long period of time, and the proportion of people who have a stroke in each group would be compared. If the result of this study is consistent with the result of the case–control study, more research would be conducted to determine the biological or chemical link between chocolate and stroke prevention.

One question you might ask at this point is: Why would we perform a case–control study instead of commencing with a cohort study from the start? The reason is that cohort studies, especially those involving health-related issues, are expensive and involve a great deal of time. In the chocolate-and-stroke example, the researchers will have to wait to determine whether the individuals in the study have a stroke or not. If you do not know whether there is an association between chocolate consumption and stroke risk, you would probably not want to spend the time and money to perform this study. Case–control studies, on the other hand, are relatively inexpensive and easy to perform because databases are maintained on health events, and, hence, the results of the study can be obtained by accessing databases and performing interviews. The results of a cheaper case–control study would allow us to determine whether it is worth spending the money to perform a cohort study that provides data to help make a more substantial decision.

GLOSSARY

d The difference between two matched values in two samples collected from the same source. It is called the paired difference.

$\bar{d}$ The mean of the paired differences for a sample.

Independent samples Two samples drawn from two populations such that the selection of one does not affect the selection of the other.

Paired or matched samples Two samples drawn in such a way that they include the same elements and two data values are obtained

from each element, one for each sample. Also called **dependent samples**.

μ_d The mean of the paired differences for the population.

s_d The standard deviation of the paired differences for a sample.

σ_d The standard deviation of the paired differences for the population.

SUPPLEMENTARY EXERCISES

10.56 A consulting agency was asked by a large insurance company to investigate if business majors were better salespersons than those with other majors. A sample of 20 salespersons with a business degree showed that they sold an average of 11 insurance policies per week. Another sample of 25 salespersons with a degree other than business showed that they sold an average of 9 insurance policies per week. Assume that the two populations are approximately normally distributed with population standard deviations of 1.75 and 1.25 policies per week, respectively.

    a. Construct a 99% confidence interval for the difference between the two population means.

    b. Using a 1% significance level, can you conclude that persons with a business degree are better salespersons than those who have a degree in another area?

10.57 According to an estimate, the average earnings of female workers who are not union members are $909 per week and those of female workers who are union members are $1035 per week. Suppose that these average earnings are calculated based on random samples of 1500 female workers who are not union members and 2000 female workers who are union members. Further assume that the standard deviations for the two corresponding populations are $70 and $90, respectively.

    a. Construct a 95% confidence interval for the difference between the two population means.

    b. Test at a 2.5% significance level whether the mean weekly earnings of female workers who are not union members are less than those of female workers who are union members.

10.58 In 2007, the average number of fatalities in all railroad accidents was .1994 per accident, whereas the average number of fatalities in all recreational boating accidents was .1320 per accident (*Source*: http://www.bts.gov/publications/national_transportation_statistics/#chapter_4). Suppose that random samples of railroad and recreational boating accidents for this year have average numbers of fatalities of .183 and .146 per accident, with standard deviations of .82 and .67, respectively. The railroad statistics are based on a random sample of 418 accidents, whereas the boating statistics are based on a random sample of 392 accidents. Assume that the distributions of the numbers of fatalities have the same population standard deviation for the two groups.

    a. Construct a 98% confidence interval for the difference in the average number of fatalities per accident in all railroad and all recreational boating accidents.

    b. Using a 1% significance level, can you conclude that the average number of fatalities in all railroad accidents is higher than the average number of fatalities in all recreational boating accidents?

10.59 A researcher wants to test if the mean GPAs (grade point averages) of all male and all female college students who actively participate in sports are different. She took a random sample of 28 male students and 24 female students who are actively involved in sports. She found the mean GPAs of the two groups to be 2.62 and 2.74, respectively, with the corresponding standard deviations equal to .43 and .38.

a. Test at a 5% significance level whether the mean GPAs of the two populations are different.

b. Construct a 90% confidence interval for the difference between the two population means.

Assume that the GPAs of all male and all female student athletes both are normally distributed with equal but unknown population standard deviations.

10.60 Based on a nationwide survey of adults by the Travel Industry Association, Americans expected to spend an average of $1019 on their longest vacation trips in 2005, compared to $1101 in 2004 (*USA TODAY*, May 20, 2005). Assume that the average for 2005 is based on a random sample of 1000 adults and the one for 2004 is based on a random sample of 900 adults, and that the sample standard deviations for such expenses are $320 for 2005 and $305 for 2004. Let μ_1 and μ_2 be the population means of such expenses for the years 2005 and 2004, respectively.

a. Construct a 99% confidence interval for $\mu_1 - \mu_2$.

b. Testing at a 1% level of significance, can you conclude that the mean for 2005 is lower than that for 2004?

Assume that such expenses for years 2005 and 2004 have unknown but equal population standard deviations.

10.61 According to a Bureau of Labor Statistics report released on March 25, 2015, statisticians earn an average of $84,010 a year and accountants and auditors earn an average of $73,670 a year (www.bls.gov). Suppose that these estimates are based on random samples of 2000 statisticians and 1800 accountants and auditors. Further assume that the sample standard deviations of the annual earnings of these two groups are $15,200 and $14,500, respectively, and the population standard deviations are unknown and unequal for the two groups.

a. Construct a 98% confidence interval for the difference in the mean annual earnings of the two groups, statisticians and accountants and auditors.

b. Using a 1% significance level, can you conclude that the average annual earnings of statisticians is higher than that of accountants and auditors?

10.62 The manager of a factory has devised a detailed plan for evacuating the building as quickly as possible in the event of a fire or other emergency. An industrial psychologist believes that workers actually leave the factory faster at closing time without following any system. The company holds fire drills periodically in which a bell sounds and workers leave the building according to the system. The evacuation time for each drill is recorded. For comparison, the psychologist also records the evacuation time when the bell sounds for closing time each day. A random sample of 36 fire drills showed a mean evacuation time of 5.1 minutes with a standard deviation of 1.1 minutes. A random sample of 37 days at closing time showed a mean evacuation time of 4.2 minutes with a standard deviation of 1.0 minute.

a. Construct a 99% confidence interval for the difference between the two population means.

b. Test at a 5% significance level whether the mean evacuation time is smaller at closing time than during fire drills.

Assume that the evacuation times at closing time and during fire drills have unknown and unequal population standard deviations.

10.63 In the Pew Internet and American Life Project, 809 artists and 2755 musicians were questioned on some issues involving the Internet. (Here, *artists* included people in performing and visual arts and creative writing.) In response to the question whether or not it should be illegal to burn a copy of a CD or DVD for a friend, 48% of the artists and 41% of the musicians indicated that it should be illegal (*The New York Times*, December 6, 2004). Assume that the 809 artists and 2755 musicians surveyed make random samples from their respective populations.

a. Find a 98% confidence interval for the difference between the two population proportions.

b. At a 2% significance level, can you conclude that the two population proportions are different?

10.64 According to a report in *The New York Times*, in the United States, accountants and auditors earn an average of $70,130 a year and loan officers earn $67,960 a year (Jessica Silver-Greenberg, *The New York Times*, April 22, 2012). Suppose that these estimates are based on random samples of 1650 accountants and auditors and 1820 loan officers. Further assume that the sample standard deviations of the salaries of the two groups are $14,400 and $13,600, respectively, and the population standard deviations are equal for the two groups.

a. Construct a 98% confidence interval for the difference in the mean salaries of the two groups—accountants and auditors, and loan officers.

b. Using a 1% significance level, can you conclude that the average salary of accountants and auditors is higher than that of loan officers?

10.65 In a *Newsweek* telephone poll conducted by Princeton Survey Research Associates in February 2005, adults were asked about their attitudes toward social security and possible changes in the system. Forty-two percent of 45- to 54-year-olds expected social security would be able to pay all benefits to which they were entitled under the law at that time, while only 32% of 18- to 34-year-olds held that opinion (*Newsweek*, February 14, 2005). Suppose that these percentages were based on random samples of 800 adults from each of these two age groups.

a. Construct a 95% confidence interval for the difference between the two population proportions.

b. Using the 2.5% significance level, can you conclude that the proportion of 18- to 34-year-olds who expected social security to pay all such benefits is lower than the percentage of 45- to 54-year-olds who held this opinion?

10.66 A May 2011 Harris Interactive poll asked American adult women, "How often do you think women of your age, who have no special risk factors for breast cancer, should have a mammogram to check for breast cancer?" Fifty percent of women age 40 to 49 years and 56% of women age 50 years or older said annually (www.harrisinteractive.com/NewsRoom/PressReleases/tabid/446/ctl/ReadCustomDefault/mid/1506/ArticleId/769/Default.aspx). Suppose that these results were based on samples of 1055 women age 40 to 49 years and 1240 women age 50 years or older.

a. Let p_1 and p_2 be the proportions of all women age 40 to 49 years and age 50 years or older, respectively, who will say that women of their age with no special risk factors for breast cancer should have an annual mammogram. Construct a 98% confidence interval for $p_1 - p_2$.

b. Using a 1% significance level, can you conclude that p_1 is less than p_2? Use both the critical-value and the p-value approaches.

10.67 According to a Randstad Global Work Monitor survey, 52% of men and 43% of women said that working part-time hinders their career opportunities (*USA TODAY*, October 6, 2011). Suppose that these results are based on random samples of 1350 men and 1480 women.

a. Let p_1 and p_2 be the proportions of all men and all women, respectively, who will say that working part-time hinders their

career opportunities. Construct a 95% confidence interval for $p_1 - p_2$.

b. Using a 2% significance level, can you conclude that p_1 and p_2 are different? Use both the critical-value and the *p*-value approaches.

ADVANCED EXERCISES

10.68 A new type of sleeping pill is tested against an older, standard pill. Two thousand insomniacs are randomly divided into two equal groups. The first group is given the old pill, and the second group receives the new pill. The time required to fall asleep after the pill is administered is recorded for each person. The results of the experiment are given in the following table, where $\bar{x}$ and s represent the mean and standard deviation, respectively, for the times required to fall asleep for people in each group after the pill is taken.

	Group 1 (Old Pill)	Group 2 (New Pill)
n	1000	1000
$\bar{x}$	15.4 minutes	15.0 minutes
s	3.5 minutes	3.0 minutes

Consider the test of hypothesis $H_0: \mu_1 - \mu_2 = 0$ versus $H_1: \mu_1 - \mu_2 > 0$, where μ_1 and μ_2 are the mean times required for all potential users to fall asleep using the old pill and the new pill, respectively.

a. Find the *p*-value for this test.

b. Does your answer to part a indicate that the result is statistically significant? Use $\alpha = .025$.

c. Find a 95% confidence interval for $\mu_1 - \mu_2$.

d. Does your answer to part c imply that this result is of great *practical* significance?

10.69 Manufacturers of two competing automobile models, Gofer and Diplomat, each claim to have the lowest mean fuel consumption. Let μ_1 be the mean fuel consumption in miles per gallon (mpg) for the Gofer and μ_2 the mean fuel consumption in mpg for the Diplomat. The two manufacturers have agreed to a test in which several cars of each model will be driven on a 100-mile test run. The fuel consumption, in mpg, will be calculated for each test run. The average mpg for all 100-mile test runs for each model gives the corresponding mean. Assume that for each model the gas mileages for the test runs are normally distributed with $\sigma = 2$ mpg. Note that each car is driven for one and only one 100-mile test run.

a. How many cars (i.e., sample size) for each model are required to estimate $\mu_1 - \mu_2$ with a 90% confidence level and with a margin of error of estimate of 1.5 mpg? Use the same number of cars (i.e., sample size) for each model.

b. If μ_1 is actually 33 mpg and μ_2 is actually 30 mpg, what is the probability that five cars for each model would yield $\bar{x}_1 \geq \bar{x}_2$?

10.70 Maria and Ellen both specialize in throwing the javelin. Maria throws the javelin a mean distance of 200 feet with a standard deviation of 10 feet, whereas Ellen throws the javelin a mean distance of 210 feet with a standard deviation of 12 feet. Assume that the distances each of these athletes throws the javelin are normally distributed

with these population means and standard deviations. If Maria and Ellen each throw the javelin once, what is the probability that Maria's throw is longer than Ellen's?

10.71 Does the use of cellular telephones increase the risk of brain tumors? Suppose that a manufacturer of cell phones hires you to answer this question because of concern about public liability suits. How would you conduct an experiment to address this question? Be specific. Explain how you would observe, how many observations you would take, and how you would analyze the data once you collect them. What are your null and alternative hypotheses? Would you want to use a higher or a lower significance level for the test? Explain.

10.72 Two competing airlines, Alpha and Beta, fly a route between Des Moines, Iowa, and Wichita, Kansas. Each airline claims to have a lower percentage of flights that arrive late. Let p_1 be the proportion of Alpha's flights that arrive late and p_2 the proportion of Beta's flights that arrive late.

a. You are asked to observe a random sample of arrivals for each airline to estimate $p_1 - p_2$ with a 90% confidence level and a margin of error of estimate of .05. How many arrivals for each airline would you have to observe? (Assume that you will observe the same number of arrivals, n, for each airline. To be sure of taking a large enough sample, use $p_1 = p_2 = .50$ in your calculations for n.)

b. Suppose that p_1 is actually .30 and p_2 is actually .23. What is the probability that a sample of 100 flights for each airline (200 in all) would yield $\hat{p}_1 \geq \hat{p}_2$?

10.73 The weekly weight losses of all dieters on Diet I have a normal distribution with a mean of 1.3 pounds and a standard deviation of .4 pound. The weekly weight losses of all dieters on Diet II have a normal distribution with a mean of 1.5 pounds and a standard deviation of .7 pound. A random sample of 25 dieters on Diet I and another sample of 36 dieters on Diet II are observed.

a. What is the probability that the difference between the two sample means, $\bar{x}_1 - \bar{x}_2$, will be within $-.15$ to .15, that is, $-.15 < \bar{x}_1 - \bar{x}_2 < .15$?

b. What is the probability that the average weight loss $\bar{x}_1$ for dieters on Diet I will be greater than the average weight loss $\bar{x}_2$ for dieters on Diet II?

c. If the average weight loss of the 25 dieters using Diet I is computed to be 2.0 pounds, what is the probability that the difference between the two sample means, $\bar{x}_1 - \bar{x}_2$, will be within $-.15$ to .15, that is, $-.15 < \bar{x}_1 - \bar{x}_2 < .15$?

d. Suppose you conclude that the assumption $-.15 < \mu_1 - \mu_2 < .15$ is reasonable. What does this mean to a person who chooses one of these diets?

10.74 A random sample of six cars was selected to test a gasoline additive. The six cars were driven for 1 week without the gasoline additive and then for 1 week with the additive. The data reproduced here from that exercise show miles per gallon without and with the additive.

Without	24.6	28.3	18.9	23.7	15.4	29.5
With	26.3	31.7	18.2	25.3	18.3	30.9

Suppose that instead of the study with 6 cars, a random sample of 12 cars is selected and these cars are divided randomly into two groups of 6 cars each. The cars in the first group are driven for 1 week without the additive, and the cars in the second group are driven for 1 week with the additive. Suppose that the top row of the table lists the gas mileages for the 6 cars without the additive, and the bottom row gives the gas mileages for the cars with the additive. Assume that the distributions of the gas mileages with or without the additive are (approximately) normal with equal but unknown standard deviations.

a. Would a paired sample test be appropriate in this case? Why or why not? Explain.

b. If the paired sample test is inappropriate here, carry out a suitable test of whether the mean gas mileage is lower without the additive. Use $\alpha = 0.25$.

c. Compare your conclusion in part b with the result of the hypothesis test in Exercise 10.40.

SELF-REVIEW TEST

1. To test the hypothesis that the mean blood pressure of university professors is lower than that of company executives, which of the following would you use?

    a. A left-tailed test   **b.** A two-tailed test   **c.** A right-tailed test

2. Briefly explain the meaning of independent and dependent samples. Give one example of each of these cases.

3. A company psychologist wanted to test if company executives have job-related stress scores higher than those of university professors. He took a sample of 40 executives and 50 professors and tested them for job-related stress. The sample of 40 executives gave a mean stress score of 7.6. The sample of 50 professors produced a mean stress score of 5.4. Assume that the standard deviations of the two populations are .8 and 1.3, respectively.

    a. Construct a 99% confidence interval for the difference between the mean stress scores of all executives and all professors.

    b. Test at a 2.5% significance level whether the mean stress score of all executives is higher than that of all professors.

4. A sample of 20 alcoholic fathers showed that they spend an average of 2.3 hours per week playing with their children with a standard deviation of .54 hour. A sample of 25 nonalcoholic fathers gave a mean of 4.6 hours per week with a standard deviation of .8 hour.

    a. Construct a 95% confidence interval for the difference between the mean times spent per week playing with their children by all alcoholic and all nonalcoholic fathers.

    b. Test at a 1% significance level whether the mean time spent per week playing with their children by all alcoholic fathers is less than that of nonalcoholic fathers.

Assume that the times spent per week playing with their children by all alcoholic and all nonalcoholic fathers both are normally distributed with equal but unknown standard deviations.

5. Repeat Problem 4 assuming that the times spent per week playing with their children by all alcoholic and all nonalcoholic fathers both are normally distributed with unequal and unknown standard deviations.

6. Lake City has two shops, Zeke's and Elmer's, that handle the majority of the town's auto body repairs. Seven cars that were damaged in collisions were taken to both shops for written estimates of the repair costs. These estimates (in dollars) are shown in the following table.

Zeke's	1058	544	1349	1296	676	998	1698
Elmer's	995	540	1175	1350	605	970	1520

    a. Construct a 99% confidence interval for the mean μ_d of the population of paired differences, where a paired difference is equal to Zeke's estimate minus Elmer's estimate.

    b. Test at a 5% significance level whether the mean μ_d of the population of paired differences is different from zero.

Assume that the population of paired differences is (approximately) normally distributed.

7. A sample of 500 male registered voters showed that 57% of them voted in the last presidential election. Another sample of 400 female registered voters showed that 55% of them voted in the same election.

    a. Construct a 97% confidence interval for the difference between the proportions of all male and all female registered voters who voted in the last presidential election.

    b. Test at a 1% significance level whether the proportion of all male voters who voted in the last presidential election is different from that of all female voters.

MINI-PROJECTS

Note: The Mini-Projects are located on the text's Web site, www.wiley.com/college/mann.

DECIDE FOR YOURSELF

Note: The Decide for Yourself feature is located on the text's Web site, www.wiley.com/college/mann.

Note: Complete TI-84, Minitab, and Excel manuals are available for download at the textbook's Web site, www.wiley.com/college/mann.

TI-84 Color/TI-84

The TI-84 Color Technology Instructions feature of this text is written for the TI-84 Plus C color graphing calculator running the 4.0 operating system. Some screens, menus, and functions will be slightly different in older operating systems. The TI-84 and TI-84 Plus can perform all of the same functions, but will not have the "Color" option referenced in some of the menus.

Estimating $\mu_1 - \mu_2$, σ_1 and σ_2 Known, for Example 10–3(b) of the Text

See the graphing calculator manual on the textbook Web site for specific directions on creating confidence intervals for $\mu_1 - \mu_2$ when σ_1 and σ_2 are known. The correct menu can be found by selecting **STAT > TESTS > 2-SampZInt**.

Testing a Hypothesis about $\mu_1 - \mu_2$, σ_1 and σ_2 Known, for Example 10–4 of the Text

See the graphing calculator manual on the textbook Web site for specific directions on testing a hypothesis for $\mu_1 - \mu_2$ when σ_1 and σ_2 are known. The correct menu can be found by selecting **STAT > TESTS > 2-SampZTest**.

Estimating $\mu_1 - \mu_2$, σ_1 and σ_2 Unknown But Equal, for Example 10–5 of the Text

1. Select **STAT > TESTS > 2-SampTInt**.

2. Use the following settings in the **2-SampTInt** menu:

   - Select **Stats** at the **Inpt** prompt.

     Note: If you have the data in a list, select **Data** at the **Inpt** prompt.

   - Type 80 at the $\bar{x}_1$ prompt.

   - Type 5 at the **Sx₁** prompt.

   - Type 15 at the **n₁** prompt.

   - Type 77 at the $\bar{x}_2$ prompt.

   - Type 6 at the **Sx₂** prompt.

   - Type 12 at the **n₂** prompt.

   - Type 0.95 at the **C-Level** prompt.

   - Select Yes at the **Pooled** prompt.

3. Scroll down, highlight **Calculate**, and press **ENTER**.

4. The output includes the confidence interval. (See **Screen 10.1**.)

```
NORMAL FLOAT AUTO REAL RADIAN MP      []
            2-SampTInt
(-1.357,7.3573)
df=25
x̄1=80
x̄2=77
Sx1=5
Sx2=6
SxP=5.46260011
↓n1=15
```

Screen 10.1

Testing a Hypothesis about $\mu_1 - \mu_2$, σ_1 and σ_2 Unknown But Equal, for Example 10–6 of the Text

1. Select **STAT > TESTS > 2-SampTTest**.

2. Use the following settings in the **2-SampTTest** menu:

   - Select **Stats** at the **Inpt** prompt.

Note: If you have the data in a list, select **Data** at the **Inpt** prompt.

- Type 23 at the $\bar{x}_1$ prompt.
- Type 3 at the Sx_1 prompt.
- Type 14 at the n_1 prompt.
- Type 25 at the $\bar{x}_2$ prompt.
- Type 4 at the Sx_2 prompt.
- Type 16 at the n_2 prompt.
- Select $\neq \mu_2$ at the μ_1 prompt.
- Select Yes at the **Pooled** prompt.

NORMAL FLOAT AUTO REAL RADIAN MP

2-SampTTest
$\mu_1 \neq \mu_2$
t=-1.530517448
p=.1371095581
df=28
$\bar{x}_1$=23
$\bar{x}_2$=25
Sx_1=3
↓Sx_2=4

Screen 10.2

3. Scroll down, highlight **Calculate**, and press **ENTER**.

4. The output includes the test statistic and the *p*-value. (See **Screen 10.2**.)

Compare the value of the test statistic to the critical-value of *t* or the *p*-value to α and make a decision.

Estimating $\mu_1 - \mu_2$, σ_1 and σ_2 Unknown and Unequal, for Example 10–8 of the Text

Follow the instructions given previously in the TI-84 section entitled "Estimating $\mu_1 - \mu_2$, σ_1 and σ_2 Unknown but Equal, for Example 10–5 of the Text," but select No at the **Pooled** prompt in step 2.

Testing a Hypothesis about $\mu_1 - \mu_2$, σ_1 and σ_2 Unknown and Unequal, for Example 10–9 of the Text

Follow the instructions given previously in the TI-84 section entitled "Testing a Hypothesis about $\mu_1 - \mu_2$, σ_1 and σ_2 Unknown but Equal, for Example 10–6 of the Text," but select No at the **Pooled** prompt in step 2.

Estimating μ_d, σ_d Unknown

Create a list of differences for the paired data and then follow the directions for creating a confidence interval for a single mean when the population standard deviation is unknown. Specific directions can be found in the Chapter 8 Technology Instructions.

Testing a Hypothesis about μ_d, σ_d Unknown

Create a list of differences for the paired data and then follow the directions for testing a hypothesis about a single mean when the population standard deviation is unknown. Specific directions can be found in the Chapter 9 Technology Instructions.

Estimating $p_1 - p_2$ for Example 10–13 of the Text

1. Select **STAT** > **TESTS** > **2-PropZInt**.

2. Use the following settings in the **2-PropZInt** menu:

- Type 100 at the x_1 prompt.
- Type 500 at the n_1 prompt.
- Type 68 at the x_2 prompt.
- Type 400 at the n_2 prompt.
- Type 0.97 at the **C-Level** prompt.

NORMAL FLOAT AUTO REAL RADIAN MP

2-PropZInt
(-.0263,.08629)
$\hat{p}_1$=.2
$\hat{p}_2$=.17
n_1=500
n_2=400

Screen 10.3

3. Highlight **Calculate** and press **ENTER**.

4. The output includes the confidence interval. (See **Screen 10.3**.)

Testing a Hypothesis about $p_1 - p_2$ for Example 10–14 of the Text

1. Select **STAT > TESTS > 2-PropZTest**.

2. Use the following settings in the **2-PropZTest** menu:

   - Type 100 at the x_1 prompt.

   - Type 500 at the n_1 prompt.

   - Type 68 at the x_2 prompt.

   - Type 400 at the n_2 prompt.

   - Select $>p_2$ at the p_1 prompt.

3. Highlight **Calculate** and press **ENTER**.

4. The output includes the test statistic and the p-value. (See **Screen 10.4**.)

Compare the value of the test statistic to the critical-value of z or the p-value to α and make a decision.

Screen 10.4

Minitab

The Minitab Technology Instructions feature of this text is written for Minitab version 17. Some screens, menus, and functions will be slightly different in older versions of Minitab.

Estimating $\mu_1 - \mu_2$, σ_1 and σ_2 Unknown But Equal, for Example 10–5 of the Text

1. Select **Stat > Basic Statistics > 2-Sample t**.

2. Use the following settings in the dialog box that appears on screen:

   - Select **Summarized Data** from the dropdown menu.

     Note: If you have the data in two different columns, select **Each sample is in its own column**, type the column names in the **Sample 1** and **Sample 2** boxes, and move to step 3 below.

   - Type 15 in the **Sample size** box for Sample 1.

   - Type 80 in the **Sample mean** box for Sample 1.

   - Type 5 in the **Standard deviation** box for Sample 1.

   - Type 12 in the **Sample size** box for Sample 2.

   - Type 77 in the **Sample mean** box for Sample 2.

   - Type 6 in the **Standard deviation** box for Sample 2.

3. Select **Options** and use the following settings when the dialog box appears on screen:

   - Type 95 in the **Confidence level** box.

   - Check the **Assume equal variances** checkbox.

4. Click **OK** in both dialog boxes.

5. The confidence interval will be displayed in the Session window. (See **Screen 10.5**.)

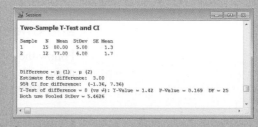

Screen 10.5

Testing a Hypothesis about $\mu_1 - \mu_2$, σ_1 and σ_2 Unknown But Equal, for Example 10–6 of the Text

1. Select **Stat > Basic Statistics > 2-Sample t**.

2. Use the following settings in the dialog box that appears on screen:
   - Select **Summarized Data** from the dropdown menu.

     Note: If you have the data in two different columns, select **Each sample is in its own column**, type the column names in the **Sample 1** and **Sample 2** boxes, and move to step 3 below.

   - Type 14 in the **Sample size** box for Sample 1.
   - Type 23 in the **Sample mean** box for Sample 1.
   - Type 3 in the **Standard deviation** box for Sample 2.
   - Type 16 in the **Sample size** box for Sample 2.
   - Type 25 in the **Sample mean** box for Sample 2.
   - Type 4 in the **Standard deviation** box for Sample 2.

3. Select **Options** and use the following settings in the dialog box that appears on screen:
   - Type 0 in the **Hypothesized difference** box.
   - Select Difference $\neq$ hypothesized difference in the **Alternative hypothesis** box.
   - Check the **Assume equal variances** box.

4. Click **OK** in both dialog boxes.

5. The output, including the test statistic and *p*-value, will be displayed in the Session window. (See **Screen 10.6**.)

Screen 10.6

Compare the value of the test statistic to the critical-value of *t* or the *p*-value to α and make a decision.

Estimating $\mu_1 - \mu_2$, σ_1 and σ_2 Unknown and Unequal, for Example 10–8 of the Text

Follow the instructions given previously in the Minitab section entitled "Estimating $\mu_1 - \mu_2$, σ_1 and σ_2 Unknown but Equal, for Example 10–5 of the Text" but do not check the **Assume equal variances** box in step 3.

Testing a Hypothesis about $\mu_1 - \mu_2$, σ_1 and σ_2 Unknown and Unequal, for Example 10–9 of the Text

Follow the instructions given previously in the Minitab section entitled "Testing a Hypothesis about $\mu_1 - \mu_2$, σ_1 and σ_2 Unknown but Equal, for Example 10–6 of the Text" but do not check the **Assume equal variances** box in step 3.

Estimating μ_d, σ_d Unknown, for Example 10–10 of the Text

1. Enter the data from Example 10–10 into C1 ("Before") and C2 ("After").

2. Select **Stat > Basic Statistics > Paired t**.

3. Use the following settings in the dialog box that appears on screen:
   - Select **Each sample is in a column** from the dropdown menu.
   - Type C1 in the **Sample 1** box.
   - Type C2 in the **Sample 2** box.

4. Select **Options** and use the following settings when the dialog box appears on screen.

- Type 95 in the **Confidence level** box

5. Click **OK** in both dialog boxes.

6. The confidence interval will be displayed in the Session window.

Testing a Hypothesis about μ_d, σ_d Unknown, for Example 10–11 of the Text

1. Enter the data from Example 10–11 into C1 ("Before") and C2 ("After").

2. Select **Stat > Basic Statistics > Paired t**.

3. Use the following settings in the dialog box that appears on screen:

- Select **Each sample is in a column** from the dropdown menu.

- Type C1 in the **Sample 1** box.

- Type C2 in the **Sample 2** box.

4. Select **Options** and use the following settings when the dialog box appears on screen.

- Type 0 in the **Hypothesized difference** box.

- Select Difference < hypothesized difference in the **Alternative hypothesis** box.

5. Click **OK** in both dialog boxes.

6. The output, including the test statistic and *p*-value, will be displayed in the Session window. (See **Screen 10.7**.)

Compare the value of the test statistic to the critical-value of *t* or the *p*-value to α and make a decision.

Screen 10.7

Estimating $p_1 - p_2$ for Example 10–13 of the Text

1. Select **Stat > Basic Statistics > 2 Proportions**.

2. Use the following settings in the dialog box that appears on screen:

- Select **Summarized Data** from the dropdown menu.

   Note: If you have the data in two different columns, select **Each sample is in its own column**, type the column names in the **Sample 1** and **Sample 2** boxes, and move to step 3 below.

- Type 100 in the **Number of events** box for Sample 1.

- Type 500 in the **Number of trials** box for Sample 1.

- Type 68 in the **Number of events** box for Sample 2.

- Type 400 in the **Number of trials** box for Sample 2.

3. Select **Options** and use the following settings when the dialog box appears on screen.

- Type 97 in the **Confidence level** box.

4. Click **OK** in both dialog boxes.

5. The confidence interval will be displayed in the Session window. (See **Screen 10.8**.)

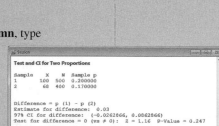

Screen 10.8

Testing a Hypothesis about $p_1 - p_2$ for Example 10–14 of the Text

1. Select **Stat > Basic Statistics > 2 Proportions**.

2. Use the following settings in the dialog box that appears on screen:

- Select **Summarized Data** from the dropdown menu.

Note: If you have the data in two different columns, select **Each sample is in its own column**, type the column names in the **Sample 1** and **Sample 2** boxes, and move to step 3 below.

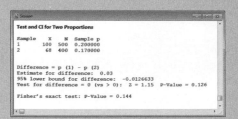

Screen 10.9

- Type 100 in the **Number of events** box for Sample 1.

- Type 500 in the **Number of trials** box for Sample 1.

- Type 68 in the **Number of events** box for Sample 2.

- Type 400 in the **Number of trials** box for Sample 2.

3. Select **Options** and use the following settings in the dialog box that appears on screen:

- Type 0 in the **Hypothesized difference** box.

- Select Difference > hypothesized difference from the **Alternative hypothesis** box.

- Select Use the pooled estimate of the proportion from the **Test method box**.

4. Click **OK** in both dialog boxes.

5. The output, including the test statistic and *p*-value, will be displayed in the Session window. (See **Screen 10.9**.)

Compare the value of the test statistic to the critical-value of *z* or the *p*-value to α and make a decision.

Excel

The Excel Technology Instructions feature of this text is written for Excel 2013. Some screens, menus, and functions will be slightly different in older versions of Excel. To enable some of the advanced Excel functions, you must enable the Data Analysis Add-In for Excel.

Testing a Hypothesis about $\mu_1 - \mu_2$, σ_1 and σ_2 Known

See the Excel manual on the textbook Web site for specific directions on creating confidence intervals for $\mu_1 - \mu_2$, when σ_1 and σ_2 are known. The correct menu can be found by clicking on **DATA** and then clicking on **Data Analysis Tools** from the **Analysis** group.

Estimating $\mu_1 - \mu_2$, σ_1 and σ_2 Unknown But Equal, for Example 10–5 of the Text

1. In a new worksheet, enter the following text into the indicated cells (see **Screen 10.10**):

- Type Sample 1 in cell A1, Sample 2 in cell A4, Sample means in cells B1 and B4, Sample standard deviations in cells B2 and B5, and Sample sizes in cells B3 and B6.

- Type xbar1 − xbar2 in cell A8, df in cell A9, t* in cell A10, Pooled s in cell A11, and std dev in cell A12.

- Type Lower bound in cell A14 and Upper bound in cell A15.

2. Enter the values of the summary statistics in the indicated cells (see **Screen 10.10**):

- Type 80 in cell C1.

- Type 5 in cell C2.

- Type 15 in cell C3.

- Type 77 in cell C4.

- Type 6 in cell C5.

- Type 12 in cell C6.

3. Enter the following formulas in the indicated cells (see **Screen 10.10**):

- Type =C1-C4 in cell B8.

- Type =C3+C6-2 in cell B9.

- Type =T.INV(0.975,B9) in cell B10.

  Note: The value 0.975 is found by $(1 - \alpha/2)$.

- Type =SQRT(((C3-1)*C2^2+(C6-1)*C5^2)/B9) in cell B11.

- Type =B11*SQRT(1/C3+1/C6) in cell B12.

- Type =B8-B10*B12 in cell B14.

- Type =B8+B10*B12 in cell B15.

4. The lower and upper bounds of the confidence interval will be in cells B14 and B15, respectively. (See **Screen 10.10**.)

	A	B	C
1	Sample 1	Sample mean	80
2		Sample std dev	5
3		Sample size	15
4	Sample 2	Sample mean	77
5		Sample std dev	6
6		Sample size	12
7			
8	xbar1 - xbar2	3	
9	df	25	
10	t*	2.059538553	
11	Pooled s	5.462600113	
12	Std dev	2.115655927	
13			
14	Lower bound	-1.357274945	
15	Upper bound	7.357274945	

Screen 10.10

Testing a Hypothesis about $\mu_1 - \mu_2$, σ_1 and σ_2 Unknown But Equal

1. To perform this test, Excel must have the raw data, not summary statistics. Suppose we wish to test $H_0: \mu_1 = \mu_2$ versus $H_1: \mu_1 > \mu_2$ at the $\alpha = 0.05$ significance level and that the population standard deviations are unknown but equal (see **Screen 10.11**):

- Enter the first set of ten data values into cells A1 through A10: 81, 79, 62, 98, 74, 82, 85, 72, 90, 88.

- Enter the second set of ten data values into cells B1 through B10: 61, 78, 85, 77, 90, 95, 64, 74, 80, 85.

2. Click **DATA** and then click **Data Analysis Tools** from the **Analysis** group.

3. Select **t-Test: Two-Sample Assuming Equal Variances** from the dialog box that appears on screen and then click OK.

4. Use the following settings in the dialog box that appears on screen (see **Screen 10.11**):

- Type A1:A10 in the **Variable 1 Range** box.

- Type B1:B10 in the **Variable 2 Range** box.

- Type 0 in the **Hypothesized Mean Difference** box.

- Type 0.05 in the **Alpha** box.

- Select New Worksheet Ply from the **Output options**.

5. Click **OK**.

6. When the output appears, resize column A so that it is easier to read.

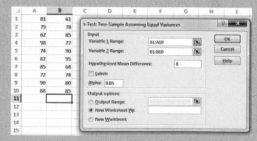

Screen 10.11

7. The correct formula for the *p*-value depends on your alternative hypothesis and the *t* statistic.

- For $H_1: \mu_1 > \mu_2$,

  - If the *t* statistic is positive, use the *p*-value from **P(T<=t) one-tail**.

  - If the *t* statistic is negative, the *p*-value is found by subtracting **P(T<=t) one-tail** from 1.

- For $H_1: \mu_1 < \mu_2$,

  - If the *t* statistic is positive, the *p*-value is found by subtracting **P(T<=t) one-tail** from 1.

  - If the *t* statistic is negative, use the *p*-value from **P(T<=t) one-tail**.

- For $H_1: \mu_1 \neq \mu_2$ use the *p*-value from **P(T<=t) two-tail**.

Estimating $\mu_1 - \mu_2$, σ_1 and σ_2 Unknown and Unequal, for Example 10–8 of the Text

1. In a new worksheet, enter the following text into the indicated cells (see **Screen 10.12**):

   - Type Sample 1 in cell A1, Sample 2 in cell A4, Sample means in cells B1 and B4, Sample standard deviations in cells B2 and B5, and Sample size in cells B4 and B6.

   - Type xbar1 − xbar2 in cell A8, df in cell A9, t* in cell A10, and std dev in cell A11.

   - Type Lower bound in cell A13 and Upper bound in cell A14.

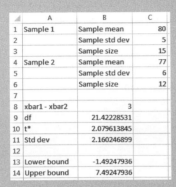

	A	B	C
1	Sample 1	Sample mean	80
2		Sample std dev	5
3		Sample size	15
4	Sample 2	Sample mean	77
5		Sample std dev	6
6		Sample size	12
7			
8	xbar1 - xbar2	3	
9	df	21.42228531	
10	t*	2.079613845	
11	Std dev	2.160246899	
12			
13	Lower bound	-1.49247936	
14	Upper bound	7.49247936	

Screen 10.12

2. Enter the values of the summary statistics in the indicated cells (see **Screen 10.12**):

   - Type 80 in cell C1.

   - Type 5 in cell C2.

   - Type 15 in cell C3.

   - Type 77 in cell C4.

   - Type 6 in cell C5.

   - Type 12 in cell C6.

3. Enter the following formulas in the indicated cells (see **Screen 10.12**):

   - Type =C1-C4 in cell B8.

   - Type =(C2^2/C3+C5^2/C6)^2/((C2^2/C3)^2/(C3-1)+(C5^2/C6)^2/(C6-1)) in cell B9.

   - Type =T.INV(0.975,B9) in cell B10.

     Note: The value 0.975 is found by $(1 - \alpha/2)$.

   - Type =SQRT(C2^2/C3+C5^2/C6) in cell B11.

   - Type =B8-B10*B11 in cell B13.

   - Type =B8+B10*B11 in cell B14.

4. The lower and upper bounds of the confidence interval will be in cells B13 and B14, respectively. (See **Screen 10.12**.)

Testing a Hypothesis about $\mu_1 - \mu_2$, σ_1 and σ_2 Unknown and Equal

1. To perform this test, Excel must have the raw data, not summary statistics. Suppose we wish to test $H_0: \mu_1 = \mu_2$ versus $H_1: \mu_1 > \mu_2$ at the $\alpha = 0.05$ significance level and that the population standard deviations are unknown but equal (see **Screen 10.13**):

   - Enter these ten data values into cells A1 through A10: 81, 79, 62, 98, 74, 82, 85, 72, 90, 88.

   - Enter these ten data values into cells B1 through B10: 61, 78, 85, 77, 90, 95, 64, 74, 80, 85.

2. Click **DATA** and then click **Data Analysis Tools** from the **Analysis** group.

3. Select **t-Test: Two-Sample Assuming Unequal Variances** from the dialog box that appears on screen and then click OK.

4. Use the following settings in the dialog box that appears on screen (see **Screen 10.13**):

   - Type A1:A10 in the **Variable 1 Range** box.

   - Type B1:B10 in the **Variable 2 Range** box.

   - Type 0 in the **Hypothesized Mean Difference** box.

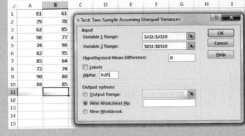

Screen 10.13

- Type 0.05 in the **Alpha** box.

- Select New Worksheet Ply from the **Output options**.

5. Click **OK**.

6. When the output appears, resize column A so that it is easier to read.

7. The correct formula for the p-value depends on your alternative hypothesis and the t statistic.

   - For $H_1: \mu_1 > \mu_2$,

      - If the t statistic is positive, use the p-value from **P(T<=t) one-tail**.

      - If the t statistic is negative, the p-value is found by subtracting **P(T<=t) one-tail** from 1.

   - For $H_1: \mu_1 < \mu_2$,

      - If the t statistic is positive, the p-value is found by subtracting **P(T<=t) one-tail** from 1.

      - If the t statistic is negative, use the p-value from **P(T<=t) one-tail**.

   - For $H_1: \mu_1 \neq \mu_2$ use the p-value from **P(T<=t) two-tail**.

Estimating μ_d, σ_d Unknown

See the Excel manual on the textbook Web site for specific directions on creating confidence intervals for μ_d when σ_d is unknown.

Testing a Hypothesis about μ_d, σ_d Unknown

See the Excel manual on the textbook Web site for specific directions on testing a hypothesis about μ_d when σ_d is unknown.

Estimating $p_1 - p_2$ for Example 10–13 of the Text

1. In a new worksheet, enter the following text into the indicated cells (see **Screen 10.14**):

   - Type Sample 1 in cell A1, Sample 2 in cell A3, Number successes in cells B1 and B3, and Number trials in cells B2 and B4.

   - Type phat1 in cell A6, phat2 in cell A7, phat1 − phat2 in cell A8, z* in cell A9, and Std dev in cell A10.

   - Type Lower bound in cell A12 and Upper bound in cell A13.

2. Enter the values of the summary statistics in the indicated cells (see **Screen 10.14**):

   - Type 100 in cell C1.

   - Type 500 in cell C2.

   - Type 68 in cell C3.

   - Type 400 in cell C4.

3. Enter the following formulas in the indicated cells (see **Screen 10.14**):

   - Type =C1/C2 in cell B6.

   - Type =C3/C4 in cell B7.

   - Type =B6-B7 in cell B8.

   - Type =NORM.INV(0.985,0,1) in cell B9.

     Note: The value 0.985 is found by $(1 - \alpha/2)$.

   - Type =SQRT(B6*(1-B6)/C2+B7*(1-B7)/C4) in cell B10.

⊿	A	B	C
1	Sample 1	Number successes	100
2		Number trials	500
3	Sample 2	Number successes	68
4		Number trials	400
5			
6	phat1	0.2	
7	phat2	0.17	
8	phat1 - phat2	0.03	
9	z*	2.170090378	
10	Std dev	0.025937425	
11			
12	Lower bound	-0.026286556	
13	Uppper bound	0.086286556	

Screen 10.14

- Type =B8-B9*B10 in cell B12.

- Type =B8+B9*B10 in cell B13.

4. The lower and upper bounds of the confidence interval will be in cells B12 and B13, respectively. (See **Screen 10.14**.)

Testing a Hypothesis about $p_1 - p_2$ for Example 10–14 of the Text

1. In a new worksheet, enter the following text into the indicated cells (see **Screen 10.15**):

   - Type Sample 1 in cell A1, Sample 2 in cell A3, Number successes in cells B1 and B3, and Number trials in cells B2 and B4.

   - Type phat1 in cell A6, phat2 in cell A7, phat1 − phat2 in cell A8, Pooled samp prop in cell A9, and Std dev in cell A10.

   - Type Test stat in cell A12 and *p*-value in cell A13.

2. Enter the values of the summary statistics in the indicated cells (see **Screen 10.15**):

   - Type 100 in cell C1.

   - Type 500 in cell C2.

   - Type 68 in cell C3.

   - Type 400 in cell C4.

	A	B	C
1	Sample 1	Number successes	100
2		Number trials	500
3	Sample 2	Number successes	68
4		Number trials	400
5			
6	phat1	0.2	
7	phat2	0.17	
8	phat1 − phat2	0.03	
9	Pooled samp prop	0.186666667	
10	Std dev	0.026138095	
11			
12	Test stat	1.147750065	
13	P-value	0.125535878	

Screen 10.15

3. Enter the following formulas in the indicated cells (see **Screen 10.15**):

   - Type =C1/C2 in cell B6.

   - Type =C3/C4 in cell B7.

   - Type =B6-B7 in cell B8.

   - Type =(C1+C3)/(C2+C4) in cell B9.

     Note: The value 0.985 is found by $(1 - \alpha/2)$.

   - Type =SQRT(B9*(1-B9)/C2+B9*(1-B9)/C4) in cell B10.

   - Type =B8/B10 in cell B12.

   - The formula for cell B13 depends on H_1:

     - For H_1: $\mu_1 > \mu_2$, type =1-NORM.S.DIST(B12,1) in cell B13.

     - For H_1: $\mu_1 < \mu_2$, type =NORM.S.DIST(B12,1) in cell B13.

     - For H_1: $\mu_1 \neq \mu_2$, type =2*(1-NORM.S.DIST(ABS(B12),1)) in cell B13.

4. The test statistic will be in cell B12 and the *p*-value will be in cell B13. (See **Screen 10.15**.)

TECHNOLOGY ASSIGNMENTS

TA10.1 Fifty randomly selected 30-year fixed-rate mortgages granted during the week of July 6 to July 10, 2009, had the following rates.

4.80	5.47	6.36	5.95	5.46	6.18	5.16	5.78	4.67	5.61
5.65	4.97	5.75	5.62	6.17	5.07	5.43	4.65	5.05	5.10
5.06	5.83	5.39	6.09	5.01	5.85	5.10	5.93	5.21	4.80
5.41	4.67	5.58	5.50	5.36	5.54	5.54	5.93	5.84	5.43
5.28	4.74	4.89	5.83	5.86	6.19	4.97	4.73	5.48	5.98

Another 45 randomly selected 20-year fixed-rate mortgages granted during the week of July 6 to July 10, 2009, had the following rates.

5.64	5.53	4.91	5.63	5.09	5.25	5.18	5.04	5.32
5.63	5.08	5.03	5.27	5.9	5.34	5.41	4.58	4.81
4.76	4.94	5.54	5.62	5.4	4.89	5.9	5.57	4.68
5.03	5.5	5.72	4.74	5.28	4.98	6.52	5.56	4.99
5.57	5.36	5.85	4.91	5.86	4.79	5.66	5.56	4.61

Assume that the population standard deviations of the interest rates on all 30-year fixed-rate mortgages and all 20-year fixed rate mortgages are the same.

a. Construct a 99% confidence interval for the difference between the mean mortgage rates on all 30-year fixed-rate mortgages and all 20-year fixed-rate mortgages for the said period.

b. Test at a 5% significance level whether the average rate on all 30-year fixed-rate mortgages granted during the week of July 6 to July 10, 2009, was higher than the average rate on all 20-year fixed-rate mortgages granted during the same time period.

TA10.2 A company recently opened two supermarkets in two different areas. The management wants to know if the mean sales per day for these two supermarkets are different. A sample of 10 days for Supermarket A produced the following data on daily sales (in thousand dollars).

47.56	57.66	51.23	58.29	43.71
49.33	52.35	50.13	47.45	53.86

A sample of 12 days for Supermarket B produced the following data on daily sales (in thousand dollars).

56.34	63.55	61.64	63.75	54.78	58.19
55.40	59.44	62.33	67.82	56.65	67.90

Assume that the daily sales of the two supermarkets are both normally distributed with equal but unknown standard deviations.

a. Construct a 99% confidence interval for the difference between the mean daily sales for these two supermarkets.

b. Test at a 1% significance level whether the mean daily sales for these two supermarkets are different.

TA10.3 Refer to Technology Assignment TA10.1. Now do that assignment without assuming that the population standard deviations are the same.

TA10.4 Refer to Technology Assignment TA10.2. Now do that assignment assuming the daily sales of the two supermarkets are both normally distributed with unequal and unknown standard deviations.

TA10.5 The following data represent the times (in seconds) taken by each contestant to eat the first Whoopie Pie and the ninth (last) Whoopie Pie in an eating contest. 13 contestants actually finished eating all nine Whoopie Pies.

Contestant	1	2	3	4	5	6	7	8	9	10	11	12	13
First pie	49	59	66	49	63	70	77	59	64	69	60	58	71
Last pie	49	74	92	93	91	73	103	59	85	94	84	87	111

a. Make a 95% confidence interval for the mean of the population paired differences, where a paired difference is equal to the time needed to eat the ninth pie minus the time needed to eat the first pie.

b. Using a 10% significance level, can you conclude that it takes at least 15 more seconds, on average, to eat the ninth pie than to eat the first pie?

Assume that the population of paired differences is (approximately) normally distributed.

TA10.6 The manufacturer of a gasoline additive claims that the use of this additive increases gasoline mileage. A random sample of six cars was selected. These cars were driven for one week without the gasoline additive and then for one week with the gasoline additive. The table gives the miles per gallon for these cars without and with gasoline additive.

Without	24.6	28.3	18.9	23.7	15.4	29.5
With	26.3	31.7	18.2	25.3	18.3	30.9

a. Construct a 99% confidence interval for the mean μ_d of the population of paired differences.

b. Test at a 1% significance level whether the use of the gasoline additive increases the gasoline mileage.

Assume that the population of paired differences is (approximately) normally distributed.

TA10.7 A company has two restaurants in two different areas of New York City. The company wants to estimate the percentages of patrons who think that the food and service at each of these restaurants are excellent. A sample of 200 patrons taken from the restaurant in Area A showed that 118 of them think that the food and service are excellent at this restaurant. Another sample of 250 patrons selected from the restaurant in Area B showed that 160 of them think that the food and service are excellent at this restaurant.

a. Construct a 97% confidence interval for the difference between the two population proportions.

b. Testing at a 2.5% significance level, can you conclude that the proportion of patrons at the restaurant in Area A who think that the food and service are excellent is lower than the corresponding proportion at the restaurant in Area B?

TA10.8 The management of a supermarket wanted to investigate whether the percentages of all men and all women who prefer to buy national brand products over the store brand products are different. A sample of 600 men shoppers at the company's supermarkets showed that 246 of them prefer to buy national brand products over the store brand products. Another sample of 700 women shoppers at the company's supermarkets showed that 266 of them prefer to buy national brand products over the store brand products.

a. Construct a 99% confidence interval for the difference between the proportions of all men and all women shoppers at these supermarkets who prefer to buy national brand products over the store brand products.

b. Testing at a 2% significance level, can you conclude that the proportions of all men and all women shoppers at these supermarkets who prefer to buy national brand products over the store brand products are different?

CHAPTER 11

mathieukor/iStockphoto

Chi-Square Tests

Are you a fan of people who work on Wall Street? Do you think that people who work on Wall Street are as honest and moral as the general public? In a Harris poll conducted in 2012, 28% of the U.S. adults polled agreed with the statement, "In general, people on Wall Street are as honest and moral as other people." Sixty-eight percent of the adults polled disagreed with this statement. (See Case Study 11–1.)

The tests of hypothesis about the mean, the difference between two means, the proportion, and the difference between two proportions were discussed in Chapters 9 and 10. The tests about proportions dealt with countable or categorical data. In the case of a proportion and the difference between two proportions in Chapters 9 and 10, the tests concerned experiments with only two categories. Recall from Chapter 5 that such experiments are called binomial experiments. This chapter describes three types of tests:

1. Tests of hypothesis for experiments with more than two categories, called goodness-of-fit tests

2. Tests of hypothesis about contingency tables, called independence and homogeneity tests

3. Tests of hypothesis about the variance and standard deviation of a single population

All of these tests are performed by using the **chi-square distribution**, which is sometimes written as χ^2 *distribution* and is read as "chi-square distribution." The symbol χ is the Greek letter *chi*, pronounced "kī." The values of a chi-square distribution are denoted by the symbol χ^2 (read as "chi-square"), just as the values of the standard normal distribution and the t distribution are denoted by z and t, respectively. Section 11.1 describes the chi-square distribution.

11.1 | The Chi-Square Distribution

Like the t distribution, the chi-square distribution has only one parameter, called the degrees of freedom (df). The shape of a specific chi-square distribution depends on the number of degrees of freedom.[1] (The degrees of freedom for a chi-square distribution are calculated by using different formulas for different tests. This will be explained when we discuss those tests.) The random variable χ^2 assumes nonnegative values only. Hence, a chi-square distribution curve starts at the origin (zero point) and lies entirely to the right of the vertical axis. Figure 11.1 shows three chi-square distribution curves. They are for 2, 7, and 12 degrees of freedom, respectively.

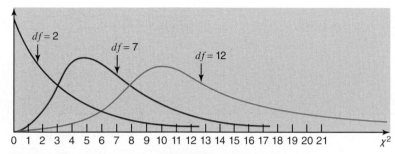

Figure 11.1   Three chi-square distribution curves.

As we can see from Figure 11.1, the shape of a chi-square distribution curve is skewed for very small degrees of freedom, and it changes drastically as the degrees of freedom increase. Eventually, for large degrees of freedom, the chi-square distribution curve looks like a normal distribution curve. The peak (or mode) of a chi-square distribution curve with 1 or 2 degrees of freedom occurs at zero and for a curve with 3 or more degrees of freedom at $df - 2$. For instance, the peak of the chi-square distribution curve with $df = 2$ in Figure 11.1 occurs at zero. The peak for the curve with $df = 7$ occurs at $7 - 2 = 5$. Finally, the peak for the curve with $df = 12$ occurs at $12 - 2 = 10$. Like all other continuous distribution curves, the total area under a chi-square distribution curve is 1.0.

If we know the degrees of freedom and the area in the right tail of a chi-square distribution curve, we can find the value of χ^2 from Table VI of Appendix B. Examples 11–1 and 11–2 show how to read that table.

> The **chi-square distribution** has only one parameter, called the degrees of freedom. The shape of a chi-square distribution curve is skewed to the right for small df and becomes symmetric for large df. The entire chi-square distribution curve lies to the right of the vertical axis. The chi-square distribution assumes nonnegative values only, and these are denoted by the symbol χ^2 (read as "chi-square").

EXAMPLE 11–1

Reading the chi-square distribution table: area in the right tail known.

Find the value of χ^2 for 7 degrees of freedom and an area of .10 in the right tail of the chi-square distribution curve.

Solution   To find the required value of χ^2, we locate 7 in the column for df and .100 in the top row in Table VI of Appendix B. The required χ^2 value is given by the entry at the intersection of

[1]The mean of a chi-square distribution is equal to its df, and the standard deviation is equal to $\sqrt{2\,df}$.

the row for 7 and the column for .100. This value is 12.017. The relevant portion of Table VI is presented as Table 11.1 here.

Table 11.1 χ^2 for $df = 7$ and .10 Area in the Right Tail

df	Area in the Right Tail Under the Chi-Square Distribution Curve				
	.995	...	.100	...	.005
1	0.000	...	2.706	...	7.879
2	0.010	...	4.605	...	10.597
.	...	...	...	...	...
.	...	...	...	...	...
.	...	...	...	...	...
7	0.989	...	12.017 ←	...	20.278
.	...	...	...	...	...
.	...	...	...	...	...
.	...	...	...	...	...
100	67.328	...	118.498	...	140.169

Required value of χ^2

As shown in Figure 11.2, for $df = 7$ and an area of .10 in the right tail of the chi-square distribution curve, the χ^2 value is **12.017**.

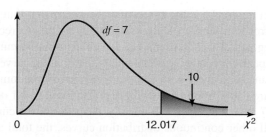

Figure 11.2   The χ^2 value.

EXAMPLE 11–2

Reading the chi-square distribution table: area in the left tail known.

Find the value of χ^2 for 12 degrees of freedom and an area of .05 in the left tail of the chi-square distribution curve.

Solution   We can read Table VI of Appendix B only when an area in the right tail of the chi-square distribution curve is known. When the given area is in the left tail, as in this example, the first step is to find the area in the right tail of the chi-square distribution curve as follows.

$$\text{Area in the right tail} = 1 - \text{Area in the left tail}$$

Therefore, for our example,

$$\text{Area in the right tail} = 1 - .05 = .95$$

Next, we locate 12 in the column for *df* and .950 in the top row in Table VI of Appendix B. The required value of χ^2, given by the entry at the intersection of the row for 12 and the column for .950, is 5.226. The relevant portion of Table VI is presented as Table 11.2 here.

Table 11.2 χ^2 for *df* = 12 and .95 Area in the Right Tail

	Area in the Right Tail Under the Chi-Square Distribution Curve				
df	**.995**	...	**.950**	...	**.005**
1	0.000	...	0.004	...	7.879
2	0.010	...	0.103	...	10.597
.	...	...	...	...	...
.	...	...	...	...	...
.	...	...	...	...	...
12	3.074	...	5.226 ←	...	28.300
.	...	...	...	...	...
.	...	...	...	...	...
.	...	...	...	...	...
100	67.328	...	77.929	...	140.169

└─ Required value of χ^2

As shown in Figure 11.3, for *df* = 12 and .05 area in the left tail, the χ^2 value is **5.226**.

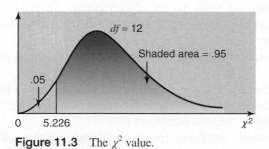

Figure 11.3 The χ^2 value.

EXERCISES

CONCEPTS AND PROCEDURES

11.1 Describe the chi-square distribution. What is the parameter (parameters) of such a distribution?

11.2 Find the value of χ^2 for 14 degrees of freedom and an area of .025 in the right tail of the chi-square distribution curve.

11.3 Determine the value of χ^2 for 16 degrees of freedom and an area of .10 in the left tail of the chi-square distribution curve.

11.4 Determine the value of χ^2 for 28 degrees of freedom and an area of .990 in the left tail of the chi-square distribution curve.

11.5 Determine the value of χ^2 for 12 degrees of freedom and

a. .025 area in the left tail of the chi-square distribution curve
b. .995 area in the right tail of the chi-square distribution curve

11.2 | A Goodness-of-Fit Test

This section explains how to make tests of hypothesis about experiments with more than two possible outcomes (or categories). Such experiments, called **multinomial experiments**, possess four characteristics. Note that a binomial experiment is a special case of a multinomial experiment.

A Multinomial Experiment An experiment with the following characteristics is called a **multinomial experiment**:

1. The experiment consists of n identical trials (repetitions).
2. Each trial results in one of k possible outcomes (or categories), where $k > 2$.
3. The trials are independent.
4. The probabilities of the various outcomes remain constant for each trial.

An experiment of many rolls of a die is an example of a multinomial experiment. It consists of many identical rolls (trials); each roll (trial) results in one of the six possible outcomes; each roll is independent of the other rolls; and the probabilities of the six outcomes remain constant for each roll.

As a second example of a multinomial experiment, suppose we select a random sample of people and ask them whether or not the quality of American cars is better than that of Japanese cars. The response of a person can be *yes*, *no*, or *does not know*. Each person included in the sample can be considered as one trial (repetition) of the experiment. There will be as many trials for this experiment as the number of persons selected. Each person can belong to any of the three categories—*yes*, *no*, or *does not know*. The response of each selected person is independent of the responses of other persons. Given that the population is large, the probabilities of a person belonging to the three categories remain the same for each trial. Consequently, this is an example of a multinomial experiment.

The frequencies obtained from the actual performance of an experiment are called the **observed frequencies**. In a **goodness-of-fit test**, we test the null hypothesis that the observed frequencies for an experiment follow a certain pattern or theoretical distribution. The test is called a goodness-of-fit test because the hypothesis tested is how *good* the observed frequencies *fit* a given pattern.

For our first example involving the experiment of many rolls of a die, we may test the null hypothesis that the given die is fair. The die will be fair if the observed frequency for each outcome is close to one-sixth of the total number of rolls.

For our second example involving opinions of people on the quality of American cars, suppose such a survey was conducted in 2015, and in that survey 41% of the people said *yes*, 48% said *no*, and 11% said *do not know*. We want to test if these percentages still hold true. Suppose we take a random sample of 1000 adults and observe that 536 of them think that the quality of American cars is better than that of Japanese cars, 362 say it is worse, and 102 have no opinion. The frequencies 536, 362, and 102 are the observed frequencies. These frequencies are obtained by actually performing the survey. Now, assuming that the 2015 percentages are still true (which will be our null hypothesis), in a sample of 1000 adults we will expect 410 to say *yes*, 480 to say *no*, and 110 to say *do not know*. These frequencies are obtained by multiplying the sample size (1000) by the 2015 proportions. These frequencies are called the **expected frequencies**. Then, we will make a decision to reject or not to reject the null hypothesis based on how large the difference between the observed frequencies and the expected frequencies is. To perform this test, we will use the chi-square distribution. Note that in this case we are testing the null hypothesis that all three percentages (or proportions) are unchanged. However, if we want to make a test for only one of the three proportions, we use the procedure learned in Section 9.4 of Chapter 9. For example, if we are testing the hypothesis that the current percentage of people who think the quality of American cars is better than that of the Japanese cars is different from 41%, then we will test the null hypothesis $H_0: p = .41$ against the alternative hypothesis $H_1: p \neq .41$. This test will be conducted using the procedure discussed in Section 9.4 of Chapter 9.

As mentioned earlier, the frequencies obtained from the performance of an experiment are called the observed frequencies. They are denoted by O. To make a goodness-of-fit test, we calculate the expected frequencies for all categories of the experiment. The expected frequency for a category, denoted by E, is given by the product of n and p, where n is the total number of trials and p is the probability for that category.

Observed and Expected Frequencies The frequencies obtained from the performance of an experiment are called the **observed frequencies** and are denoted by O. The **expected frequencies**, denoted by E, are the frequencies that we expect to obtain if the null hypothesis is true. The expected frequency for a category is obtained as

$$E = np$$

where n is the sample size and p is the probability that an element belongs to that category if the null hypothesis is true.

Degrees of Freedom for a Goodness-of-Fit Test In a goodness-of-fit test, the **degrees of freedom** are

$$df = k - 1$$

where k denotes the number of possible outcomes (or categories) for the experiment.

The procedure to make a goodness-of-fit test involves the same five steps that we used in the preceding chapters. *The chi-square goodness-of-fit test is always a right-tailed test.*

Test Statistic for a Goodness-of-Fit Test The *test statistic for a goodness-of-fit test* is χ^2, and its value is calculated as

$$\chi^2 = \sum \frac{(O - E)^2}{E}$$

where

$$O = \text{observed frequency for a category}$$

$$E = \text{expected frequency for a category} = np$$

Remember that a **chi-square goodness-of-fit test is always a right-tailed test**.

Whether or not the null hypothesis is rejected depends on how much the observed and expected frequencies differ from each other. To find how large the difference between the observed frequencies and the expected frequencies is, we do not look at just $\Sigma(O - E)$, because some of the $O - E$ values will be positive and others will be negative. The net result of the sum of these differences will always be zero. Therefore, we square each of the $O - E$ values to obtain $(O - E)^2$, and then we weight them according to the reciprocals of their expected frequencies. The sum of the resulting numbers gives the computed value of the test statistic χ^2.

To make a goodness-of-fit test, the sample size should be large enough so that the expected frequency for each category is at least 5. If there is a category with an expected frequency of less than 5, either increase the sample size or combine two or more categories to make each expected frequency at least 5.

Examples 11–3 and 11–4 describe the procedure for performing goodness-of-fit tests using the chi-square distribution.

EXAMPLE 11–3   Number of People Using an ATM Each Day

Conducting a goodness-of-fit test: equal proportions for all categories.

A bank has an ATM installed inside the bank, and it is available to its customers only from 7 AM to 6 PM Monday through Friday. The manager of the bank wanted to investigate if the number of people who use this ATM is the same for each of the 5 days (Monday through Friday) of the week. She randomly selected one week and counted the number of people who used this ATM

on each of the 5 days during that week. The information she obtained is given in the following table, where the number of users represents the number of people who used this ATM on these days.

Day	Monday	Tuesday	Wednesday	Thursday	Friday
Number of users	253	197	204	279	267

At a 1% level of significance, can we reject the null hypothesis that the number of people who use this ATM each of the 5 days of the week is the same? Assume that this week is typical of all weeks in regard to the use of this ATM.

Solution   To conduct this test of hypothesis, we proceed as follows.

Step 1.   *State the null and alternative hypotheses.*

Because there are 5 categories (days) as listed in the table, the number of ATM users will be the same for each of these 5 days if 20% of all users use the ATM each day. The null and alternative hypotheses are as follows.

H_0: The number of people using the ATM is the same for all 5 days of the week.

H_1: The number of people using the ATM is not the same for all 5 days of the week.

If the number of people using this ATM is the same for all 5 days of the week, then .20 of the users will use this ATM on any of the 5 days of the week. Let p_1, p_2, p_3, p_4, and p_5 be the proportion of people who use this ATM on Monday, Tuesday, Wednesday, Thursday, and Friday, respectively. Then, the null and alternative hypotheses can also be written as

$$H_0: p_1 = p_2 = p_3 = p_4 = p_5 = .20$$
$$H_1: \text{At least two of the five proportions are not equal to .20}$$

Step 2.   *Select the distribution to use.*

Because there are 5 categories (i.e., 5 days on which the ATM is used), this is a multinomial experiment. Consequently, we use the chi-square distribution to make this test.

Step 3.   *Determine the rejection and nonrejection regions.*

The significance level is given to be .01, and the goodness-of-fit test is always right-tailed. Therefore, the area in the right tail of the chi-square distribution curve is .01, that is,

$$\text{Area in the right tail} = \alpha = .01$$

The degrees of freedom are calculated as follows:

$$k = \text{number of categories} = 5$$
$$df = k - 1 = 5 - 1 = 4$$

From the chi-square distribution table (Table VI of Appendix B), for $df = 4$ and .01 area in the right tail of the chi-square distribution curve, the critical value of χ^2 is 13.277, as shown in Figure 11.4.

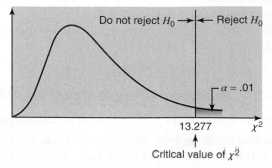

Figure 11.4   Rejection and nonrejection regions.

Step 4.   *Calculate the value of the test statistic.*

Table 11.3   Calculating the Value of the Test Statistic

Category (Day)	Observed Frequency O	p	Expected Frequency E = np	(O − E)	(O − E)²	$\frac{(O-E)^2}{E}$
Monday	253	.20	1200(.20) = 240	13	169	.704
Tuesday	197	.20	1200(.20) = 240	−43	1849	7.704
Wednesday	204	.20	1200(.20) = 240	−36	1296	5.400
Thursday	279	.20	1200(.20) = 240	39	1521	6.338
Friday	267	.20	1200(.20) = 240	27	729	3.038
	n = 1200					Sum = 23.184

All the required calculations to find the value of the test statistic χ^2 are shown in Table 11.3. The calculations made in Table 11.3 are explained as follows.

1. The first two columns of Table 11.3 list the 5 categories (days) and the observed frequencies for the 1200 persons who used the ATM during each of the 5 days of the selected week. The third column contains the probabilities for the 5 categories assuming that the null hypothesis is true.

2. The fourth column contains the expected frequencies. These frequencies are obtained by multiplying the total users ($n = 1200$) by the probabilities listed in the third column. If the null hypothesis is true (i.e., the ATM users are equally distributed over all 5 days), then we will expect 240 out of 1200 persons to use the ATM each day. Consequently, each category in the fourth column has the same expected frequency.

3. The fifth column lists the differences between the observed and expected frequencies, that is, $O − E$. These values are squared and recorded in the sixth column.

4. Finally, we divide the squared differences (that appear in the sixth column) by the corresponding expected frequencies (listed in the fourth column) and write the resulting numbers in the seventh column.

5. The sum of the seventh column gives the value of the test statistic χ^2. Thus,

$$\chi^2 = \sum \frac{(O-E)^2}{E} = 23.184$$

Step 5. *Make a decision.*

The value of the test statistic $\chi^2 = 23.184$ is larger than the critical value of $\chi^2 = 13.277$, and falls in the rejection region. Hence, we reject the null hypothesis and state that the number of persons who use this ATM is not the same for each of the 5 days of the week. In other words, we conclude that a higher number of users of this ATM use this machine on one or more of these days.

If you make this chi-square test using any of the statistical software packages, you will obtain a *p*-value of .000311 for the test. In this case you can compare the *p*-value obtained in the computer output with the level of significance and make a decision. As you know from Chapter 9, you will reject the null hypothesis if α (significance level) is greater than or equal to the *p*-value and not reject it otherwise.

EXAMPLE 11–4	Are Federal Taxes Paid by Upper-Income People Fair?

Conducting a goodness-of-fit test: testing if results of a survey fit a given distribution.

In a Gallup poll conducted April 3–6, 2014, Americans aged 18 and older were asked if upper-income people were "paying their fair share in federal taxes, paying too much or paying too little." Of the respondents, 61% said too little, 24% said fair share, 13% said too much, and 2% had no opinion (www. gallup.com). Assume that these percentages hold true for the 2014 population of Americans aged 18 and older. Recently, 1000 randomly selected Americans aged 18 and older were asked the same question. The following table lists the number of Americans in this sample who belonged to each response.

Response	Too Little	Fair Share	Too Much	No Opinion
Frequency	581	256	138	25

Test at a 2.5% level of significance whether the current distribution of opinions is different from that for 2014.

Solution We perform the following five steps for this test of hypothesis.

Step 1. *State the null and alternative hypotheses.*
The null and alternative hypotheses are

H_0: The current percentage distribution of opinions is the same as for 2014.

H_1: The current percentage distribution of opinions is different from that for 2014.

Step 2. *Select the distribution to use.*

Because this experiment has four categories as listed in the table, it is a multinomial experiment. Consequently we use the chi-square distribution to make this test.

Step 3. *Determine the rejection and nonrejection regions.*

The significance level is given to be .025, and because the goodness-of-fit test is always right-tailed, the area in the right tail of the chi-square distribution curve is .025, that is,

$$\text{Area in the right tail} = \alpha = .025$$

The degrees of freedom are calculated as follows:

$$k = \text{number of categories} = 4$$
$$df = k - 1 = 4 - 1 = 3$$

From the chi-square distribution table (Table VI of Appendix B), for $df = 3$ and .025 area in the right tail of the chi-square distribution curve, the critical value of χ^2 is 9.348, as shown in Figure 11.5.

Figure 11.5 Rejection and nonrejection regions.

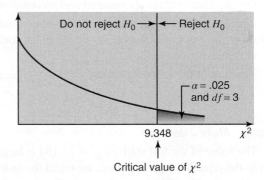

Step 4. *Calculate the value of the test statistic.*

All the required calculations to find the value of the test statistic χ^2 are shown in Table 11.4. Note that the four percentages for 2014 have been converted into probabilities and recorded in the third column of Table 11.4. The value of the test statistic χ^2 is given by the sum of the last column. Thus,

$$\chi^2 = \sum \frac{(O - E)^2}{E} = 4.188$$

Table 11.4 Calculating the Value of the Test Statistic

Category (Response)	Observed Frequency O	p	Expected Frequency $E = np$	$(O - E)$	$(O - E)^2$	$\dfrac{(O - E)^2}{E}$
Too little	581	.61	$1000(.61) = 610$	-29	841	1.379
Fair share	256	.24	$1000(.24) = 240$	16	256	1.067
Too much	138	.13	$1000(.13) = 130$	8	64	.492
No opinion	25	.02	$1000(.02) = 20$	5	25	1.250
	$n = 1000$					Sum = 4.188

Are People on Wall Street Honest and Moral?

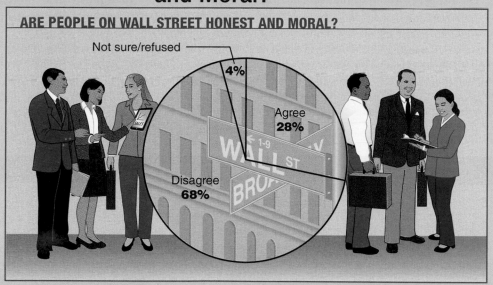

ARE PEOPLE ON WALL STREET HONEST AND MORAL?

Not sure/refused — 4%

Agree 28%

Disagree 68%

Data source: Harris Interactive telephone poll of U.S. adults conducted April 10–17, 2012.

In a Harris poll conducted by Harris Interactive between April 10 and April 17, 2012, U.S. adults aged 18 years and older were asked whether they agreed with the statement, "In general, people on Wall Street are as honest and moral as other people." (http://www.harrisinteractive.com/NewsRoom/HarrisPolls/tabid/447/ctl/Read-Custom%20Default/mid/1508/ArticleId/1018/Default.aspx.) The accompanying chart shows the percentage distribution of the responses of these adults. Twenty-eight percent of the adults polled said that they agree with this statement, 68% disagreed, and 4% were not sure or refused to answer. Assume that these percentages were true for the population of U.S. adults in 2012. Suppose that we want to test the hypothesis whether these percentages with respect to the foregoing statement are still true. Then the two hypotheses are as follows:

H_0: The current percentage distribution of opinions is the same as in 2012

H_1: The current percentage distribution of opinions is not the same as in 2012

To test this hypothesis, suppose that we currently take a sample of 2000 U.S. adults and ask them whether they agree with the foregoing statement. Suppose that 488 of them say that they agree with the statement, 1444 say that they disagree, and 68 are not sure or refuse to give an answer. Using the given information, we calculate the value of the test statistic as shown in the following table.

Category	Observed Frequency O	p	Expected Frequency $E = np$	$(O - E)$	$(O - E)^2$	$\dfrac{(O - E)^2}{E}$
Agree	488	.28	2000(.28) = 560	−72	5184	9.257
Disagree	1444	.68	2000(.68) = 1360	84	7056	5.188
Not sure/refuse	68	.04	2000(.04) = 80	−12	144	1.800
	$n = 2000$					Sum = 16.245

Suppose that we use a 1% significance level to perform this test. Then for $df = 3 - 1 = 2$ and .01 area in the right tail, the critical value of χ^2 is 9.210 from Table VI of Appendix B. Since the observed value of χ^2 is 16.245 and it is larger than the critical value of $\chi^2 = 9.210$, we reject the null hypothesis. Thus, we conclude that the current percentage distribution of opinions of U.S. adults in response to the given statement is significantly different from the distribution of those in 2012.

We can also use the *p*-value approach to make this decision. In Table VI of Appendix B, for $df = 2$, the largest value of χ^2 is 10.597, and the area to the right of $\chi^2 = 10.597$ is .005. Thus, the *p*-value for $\chi^2 = 16.245$ will be less than .005. (By using technology, we obtain the *p*-value of .0003.) Since $\alpha = .01$ in this example is greater than .005 (or .0003), we reject the null hypothesis and conclude that the current percentage distribution of opinions of U.S. adults in response to the given statement is significantly different from the distribution of those in 2012.

449

Step 5. *Make a decision.*

The observed value of the test statistic $\chi^2 = 4.188$ is smaller than the critical value of $\chi^2 = 9.348$, and falls in the nonrejection region. Hence, we fail to reject the null hypothesis, and state that the current percentage distribution of opinions seems to be the same as for 2014.

If you make this chi-square test using any of the statistical software packages, you will obtain a *p*-value of .242 for the test. In this case you can compare the *p*-value obtained in the computer output with the level of significance and make a decision. As you know from Chapter 9, you will reject the null hypothesis if α (significance level) is greater than or equal to the *p*-value and not reject it otherwise.

EXERCISES

CONCEPTS AND PROCEDURES

11.6 Describe the four characteristics of a multinomial experiment.

11.7 What is a goodness-of-fit test and when is it applied? Explain.

11.8 Explain the difference between the observed and expected frequencies for a goodness-of-fit test.

11.9 How is the expected frequency of a category calculated for a goodness-of-fit test? What are the degrees of freedom for such a test?

11.10 To make a goodness-of-fit test, what should be the minimum expected frequency for each category? What are the alternatives if this condition is not satisfied?

11.11 The following table lists the frequency distribution for 60 rolls of a die.

Outcome	1-spot	2-spot	3-spot	4-spot	5-spot	6-spot
Frequency	7	15	10	16	8	4

Test at a 5% significance level whether the null hypothesis that the given die is fair is true.

APPLICATIONS

11.12 A drug company is interested in investigating whether the color of their packaging has any impact on sales. To test this, they used five different colors (blue, green, orange, red, and yellow) for the boxes of an over-the-counter pain reliever, instead of their traditional white box. The following table shows the number of boxes of each color sold during the first month.

Box color	Blue	Green	Orange	Red	White
Number of boxes sold	340	246	295	189	284

Using a 1% significance level, test the null hypothesis that the number of boxes sold of each of these five colors is the same.

11.13 Over the last 3 years, Art's Supermarket has observed the following distribution of modes of payment in the checkout lines: cash (C) 41%, check (CK) 24%, credit or debit card (D) 26%, and other (N) 9%. In an effort to make express checkout more efficient, Art's has just begun offering a 1% discount for cash payment in the express checkout line. The following table lists the frequency distribution of the modes of payment for a sample of 600 express-line customers after the discount went into effect.

Mode of payment	C	CK	D	N
Number of customers	265	129	136	70

Test at a 1% significance level whether the distribution of modes of payment in the express checkout line changed after the discount went into effect.

11.14 Home Mail Corporation sells products by mail. The company's management wants to find out if the number of orders received at the company's office on each of the 5 days of the week is the same. The company took a sample of 400 orders received during a 4-week period. The following table lists the frequency distribution for these orders by the day of the week.

Day of the week	Mon	Tue	Wed	Thu	Fri
Number of orders received	98	68	52	86	96

Test at a 5% significance level whether the null hypothesis that the orders are evenly distributed over all days of the week is true.

11.15 Of all students enrolled at a large undergraduate university, 19% are seniors, 23% are juniors, 27% are sophomores, and 31% are freshmen. A sample of 200 students taken from this university by the student senate to conduct a survey includes 50 seniors, 46 juniors, 55 sophomores, and 49 freshmen. Using a 2.5% significance level, test the null hypothesis that this sample is a random sample. (*Hint:* This sample will be a random sample if it includes approximately 19% seniors, 23% juniors, 27% sophomores, and 31% freshmen.)

11.16 A Shopping on the Job Survey in 2016 asked employees, "During the holiday season (November and December), how much total time do you think an average employee at your enterprise spends shopping online using a work-supplied computer or smartphone?" Among those who responded, 3% said 0 hours, 24% said 1 to 2 hours, 22% said 3 to 5 hours, and 51% said 6 or more hours. Suppose that another poll conducted recently asked the same question of 215 randomly selected business executives, which produced the frequencies listed in the following table.

Response/category	0 hours	1–2 hours	3–5 hours	6 or more hours
Frequency	2	41	55	117

Test at a 2.5% significance level whether the distribution of responses for the executive survey differs from that of October 2016 survey of employees.

11.17 Clasp your hands together. Which thumb is on top? Believe it or not, the thumb that you place on top is determined by genetics. If either of your parents has the gene that *tells* you to place your left thumb on top and passes it on to you, you will place your left thumb on top. The *left-thumb* gene is called the dominant gene, which means that if either parent passes it on to you, you will have that trait. If you place your right thumb on top, you received the recessive gene from both parents. If both parents have both the left and right thumb genes (the case denoted Lr), Mendelian genetics gives the probabilities listed in the following table about the children's genes.

Child's genes	LL (left-thumbed)	Lr (left-thumbed, but also received a right-thumbed gene)	rr (right-thumbed)
Probability	.25	.50	.25

Source: http://humangenetics.suite101.com/article.cfm/dominant_human_genetic_traits.

Suppose that a random sample of 65 children whose both parents had Lr genes were tested for the genes. The following table lists the results of this experiment.

Child's genes	LL	Lr	rr
Frequency	14	31	20

Test at a 5% significance level whether the genes received by the sample of children are significantly different from what Mendelian genetics predicts.

11.3 | A Test of Independence or Homogeneity

This section is concerned with tests of independence and homogeneity, which are performed using contingency tables. Except for a few modifications, the procedure used to make such tests is almost the same as the one applied in Section 11.2 for a goodness-of-fit test.

11.3.1 A Contingency Table

Often we may have information on more than one variable for each element. Such information can be summarized and presented using a two-way classification table, which is also called a *contingency table* or *cross-tabulation*. Suppose a university has a total of 20,758 students enrolled. By classifying these students based on gender and whether these students are full-time or part-time, we can prepare Table 11.5, which provides an example of a contingency table. Table 11.5 has two rows (one for males and the second for females) and two columns (one for full-time and the second for part-time students). Hence, it is also called a 2 × 2 (read as "two by two") contingency table.

Table 11.5   Total Enrollment at a University

	Full-Time	**Part-Time**
Male	6768	2615
Female	7658	3717

Students who are male and enrolled part-time

A contingency table can be of any size. For example, it can be 2 × 3, 3 × 2, 3 × 3, or 4 × 2. Note that in these notations, the first digit refers to the number of rows in the table, and the second digit refers to the number of columns. For example, a 3 × 2 table will contain three rows and two columns. In general, an $R \times C$ table contains R rows and C columns.

Each of the four boxes that contain numbers in Table 11.5 is called a **cell**. The number of cells in a contingency table is obtained by multiplying the number of rows by the number of columns. Thus, Table 11.5 contains 2 × 2 = 4 cells. The subjects that belong to a cell of a contingency table possess two characteristics. For example, 2615 students listed in the second cell of the first row in Table 11.5 are *male* and enrolled *part-time*. The numbers written inside the cells are called the *joint frequencies*. For example, 2615 students belong to the joint category of *male* and *part-time*. Hence, it is referred to as the joint frequency of this category.

11.3.2 A Test of Independence

In a **test of independence** for a contingency table, we test the null hypothesis that the two attributes (characteristics) of the elements of a given population are not related (that is, they are independent) against the alternative hypothesis that the two characteristics are related (that is, they are dependent). For example, we may want to test if the affiliation of people with the Democratic and Republican parties is independent of their income levels. We perform such a test by using the chi-square distribution. As another example, we may want to test if there is an association between being a man or a woman and having a preference for watching sports or soap operas on television.

Degrees of Freedom for a Test of Independence A test of independence involves a test of the null hypothesis that two attributes of a population are not related. The **degrees of freedom for a test of independence** are

$$df = (R - 1)(C - 1)$$

where R and C are the number of rows and the number of columns, respectively, in the given contingency table.

The value of the test statistic χ^2 in a test of independence is obtained using the same formula as in the goodness-of-fit test described in Section 11.2.

Test Statistic for a Test of Independence The value of the **test statistic χ^2 for a test of independence** is calculated as

$$\chi^2 = \sum \frac{(O - E)^2}{E}$$

where O and E are the observed and expected frequencies, respectively, for a cell.

The null hypothesis in a test of independence is always that the two attributes are not related. The alternative hypothesis is that the two attributes are related.

The frequencies obtained from the performance of an experiment for a contingency table are called the **observed frequencies**. The procedure to calculate the **expected frequencies** for a contingency table for a test of independence is different from the one for a goodness-of-fit test. Example 11–5 describes this procedure.

EXAMPLE 11–5 | Lack of Discipline in Schools

Calculating expected frequencies for a test of independence.

Lack of discipline has become a major problem in schools in the United States. A random sample of 300 adults was selected, and these adults were asked if they favor giving more freedom to schoolteachers to punish students for lack of discipline. The two-way classification of the responses of these adults is presented in the following table.

	In Favor (F)	Against (A)	No Opinion (N)
Men (M)	93	70	12
Women (W)	87	32	6

Calculate the expected frequencies for this table, assuming that the two attributes, gender and opinions on the issue, are independent.

Solution The preceding table is reproduced as Table 11.6 here. Note that Table 11.6 includes the row and column totals.

Table 11.6 Observed Frequencies

	In Favor (F)	Against (A)	No Opinion (N)	Row Totals
Men (M)	93	70	12	175
Women (W)	87	32	6	125
Column Totals	180	102	18	300

The numbers 93, 70, 12, 87, 32, and 6 listed inside the six cells of Table 11.6 are called the **observed frequencies** of the respective cells.

As mentioned earlier, the null hypothesis in a test of independence is that the two attributes (or classifications) are independent. In a test of independence, first we assume that the null hypothesis is true and that the two attributes are independent. Assuming that the null hypothesis is true and that gender and opinions are not related in this example, we calculate the expected frequency for the cell corresponding to *Men* and *In Favor* as shown below. From Table 11.6,

$$P(\text{a person is a } Man) = P(M) = 175/300$$
$$P(\text{a person is } In\ Favor) = P(F) = 180/300$$

Because we are assuming that M and F are independent (by assuming that the null hypothesis is true), from the formula learned in Chapter 4, the joint probability of these two events is

$$P(M \text{ and } F) = P(M) \times P(F) = (175/300) \times (180/300)$$

Then, assuming that M and F are independent, the number of persons expected to be *Men* and *In Favor* in a sample of 300 is

$$E \text{ for } Men \text{ and } In\ Favor = 300 \times P(M \text{ and } F)$$
$$= 300 \times \frac{175}{300} \times \frac{180}{300} = \frac{175 \times 180}{300}$$
$$= \frac{(\text{Row total})(\text{Column total})}{\text{Sample size}}$$

Thus, the rule for obtaining the expected frequency for a cell is to divide the product of the corresponding row and column totals by the sample size.

Expected Frequencies for a Test of Independence   The expected frequency E for a cell is calculated as

$$E = \frac{(\text{Row total})(\text{Column total})}{\text{Sample size}}$$

Using this rule, we calculate the expected frequencies of the six cells of Table 11.6 as follows:

E for *Men* and *In Favor* cell = $(175)(180)/300 = \mathbf{105.00}$

E for *Men* and *Against* cell = $(175)(102)/300 = \mathbf{59.50}$

E for *Men* and *No Opinion* cell = $(175)(18)/300 = \mathbf{10.50}$

E for *Women* and *In Favor* cell = $(125)(180)/300 = \mathbf{75.00}$

E for *Women* and *Against* cell = $(125)(102)/300 = \mathbf{42.50}$

E for *Women* and *No Opinion* cell = $(125)(18)/300 = \mathbf{7.50}$

The expected frequencies are usually written in parentheses below the observed frequencies within the corresponding cells, as shown in Table 11.7.

Table 11.7   Observed and Expected Frequencies

	In Favor (F)	Against (A)	No Opinion (N)	Row Totals
Men (M)	93 (105.00)	70 (59.50)	12 (10.50)	175
Women (W)	87 (75.00)	32 (42.50)	6 (7.50)	125
Column Totals	180	102	18	300

Like a goodness-of-fit test, **a test of independence is always right-tailed**. To apply a chi-square test of independence, **the sample size should be large enough so that the expected frequency for each cell is at least 5**. If the expected frequency for a cell is not at least 5, we either increase the sample size or combine some categories. Examples 11–6 and 11–7 describe the procedure to make tests of independence using the chi-square distribution.

EXAMPLE 11–6   Lack of Discipline in Schools

Making a test of independence: 2 × 3 table.

Reconsider the two-way classification table given in Example 11–5. In that example, a random sample of 300 adults was selected, and they were asked if they favor giving more freedom to schoolteachers to punish students for lack of discipline. Based on the results of the survey, a two-way classification table was prepared and presented in Example 11–5. Does the sample provide sufficient evidence to conclude that the two attributes, gender and opinions of adults, are dependent? Use a 1% significance level.

Solution   The test involves the following five steps.

Step 1. *State the null and alternative hypotheses.*

As mentioned earlier, the null hypothesis must be that the two attributes are independent. Consequently, the alternative hypothesis is that these attributes are dependent.

$$H_0\text{: Gender and opinions of adults are independent.}$$

$$H_1\text{: Gender and opinions of adults are dependent.}$$

Step 2. *Select the distribution to use.*

We use the chi-square distribution to make a test of independence for a contingency table.

Step 3. *Determine the rejection and nonrejection regions.*

The significance level is 1%. Because a test of independence is always right-tailed, the area of the rejection region is .01 and falls in the right tail of the chi-square distribution curve. The contingency table contains two rows (*Men* and *Women*) and three columns (*In Favor*, *Against*, and *No Opinion*). Note that we do not count the row and column of totals. The degrees of freedom are

$$df = (R - 1)(C - 1) = (2 - 1)(3 - 1) = 2$$

From Table VI of Appendix B, for $df = 2$ and $\alpha = .01$, the critical value of χ^2 is 9.210. This value is shown in Figure 11.6.

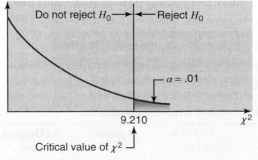

Figure 11.6 Rejection and nonrejection regions.

Step 4. *Calculate the value of the test statistic.*

Table 11.7, with the observed and expected frequencies constructed in Example 11–5, is reproduced as Table 11.8.

Table 11.8   **Observed and Expected Frequencies**

	In Favor (F)	Against (A)	No Opinion (N)	Row Totals
Men (M)	93 (105.00)	70 (59.50)	12 (10.50)	175
Women (W)	87 (75.00)	32 (42.50)	6 (7.50)	125
Column Totals	180	102	18	300

To compute the value of the test statistic χ^2, we take the difference between each pair of observed and expected frequencies listed in Table 11.8, square those differences, and then divide each of the squared differences by the respective expected frequency. The sum of the resulting numbers gives the value of the test statistic χ^2. All these calculations are made as follows:

$$\chi^2 = \sum \frac{(O - E)^2}{E}$$

$$= \frac{(93 - 105.00)^2}{105.00} + \frac{(70 - 59.50)^2}{59.50} + \frac{(12 - 10.50)^2}{10.50}$$

$$+ \frac{(87 - 75.00)^2}{75.00} + \frac{(32 - 42.50)^2}{42.50} + \frac{(6 - 7.50)^2}{7.50}$$

$$= 1.371 + 1.853 + .214 + 1.920 + 2.594 + .300 = 8.252$$

Step 5.   *Make a decision.*

The value of the test statistic $\chi^2 = 8.252$ is less than the critical value of $\chi^2 = 9.210$, and it falls in the nonrejection region. Hence, we fail to reject the null hypothesis and state that there is not enough evidence from the sample to conclude that the two characteristics, *gender* and *opinions of adults*, are dependent for this issue.

EXAMPLE 11–7	Relationship Between Gender and Owning Smart Phone

Making a test of independence: 2 × 2 table.

A researcher wanted to study the relationship between gender and owning smart phones among adults who have cell phones. She took a sample of 2000 adults and obtained the information given in the following table.

	Own Smart Phones	Do Not Own Smart Phones
Men	640	450
Women	440	470

At a 5% level of significance, can you conclude that gender and owning a smart phone are related for all adults?

Solution   We perform the following five steps to make this test of hypothesis.

Step 1.   *State the null and alternative hypotheses.*

The null and alternative hypotheses are, respectively,

H_0: Gender and owning a smart phone are not related.

H_1: Gender and owning a smart phone are related.

Step 2.   *Select the distribution to use.*

Because we are performing a test of independence, we use the chi-square distribution to make the test.

Step 3. *Determine the rejection and nonrejection regions.*

With a significance level of 5%, the area of the rejection region is .05 and falls under the right tail of the chi-square distribution curve. The contingency table contains two rows (*men* and *women*) and two columns (*own smart phones* and *do not own smart phones*). The degrees of freedom are

$$df = (R - 1)(C - 1) = (2 - 1)(2 - 1) = 1$$

From Table VI of Appendix B, the critical value of χ^2 for $df = 1$ and $\alpha = .05$ is 3.841. This value is shown in Figure 11.7.

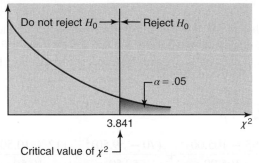

Figure 11.7 Rejection and nonrejection regions.

Step 4. *Calculate the value of the test statistic.*

The expected frequencies for the various cells are calculated as follows and are listed within parentheses in Table 11.9.

Table 11.9 **Observed and Expected Frequencies**

	Own Smart Phones (Y)	Do Not Own Smart Phones (N)	Row Totals
Men (M)	640 (588.60)	450 (501.40)	1090
Women (W)	440 (491.40)	470 (418.60)	910
Column Totals	1080	920	2000

E for *men* and *own smart phones* cell $= (1090)(1080)/2000 = 588.60$

E for *men* and *do not own smart phones* cell $= (1090)(920)/2000 = 501.40$

E for *women* and *own smart phones* cell $= (910)(1080)/2000 = 491.40$

E for *women* and *do not own smart phones* cell $= (910)(920)/2000 = 418.60$

The value of the test statistic χ^2 is calculated as follows:

$$\chi^2 = \sum \frac{(O - E)^2}{E}$$

$$= \frac{(640 - 588.60)^2}{588.60} + \frac{(450 - 501.40)^2}{501.40} + \frac{(440 - 491.40)^2}{491.40} + \frac{(470 - 418.60)^2}{418.60}$$

$$= 4.489 + 5.269 + 5.376 + 6.311 = 21.445$$

Step 5. *Make a decision.*

The value of the test statistic $\chi^2 = 21.445$ is larger than the critical value of $\chi^2 = 3.841$ and falls in the rejection region. Hence, we reject the null hypothesis and state that there is strong evidence from the sample to conclude that the two characteristics, *gender* and *owning smart phones*, are related for all adults.

11.3.3 A Test of Homogeneity

In a **test of homogeneity**, we test if two (or more) populations are homogeneous (similar) with regard to the distribution of a certain characteristic. For example, we might be interested in testing the null hypothesis that the proportions of households that belong to different income groups are the same in California and Wisconsin, or we may want to test whether or not the preferences of people in Florida, Arizona, and Vermont are similar with regard to Coke, Pepsi, and 7-Up.

Let us consider the example of testing the null hypothesis that the proportions of households in California and Wisconsin who belong to various income groups are the same. (Note that in a test of homogeneity, the null hypothesis is always that the proportions of elements with certain characteristics are the same in two or more populations. The alternative hypothesis is that these proportions are not the same.) Suppose we define three income strata: high-income group (with an income of more than $200,000), medium-income group (with an income of $70,000 to $200,000), and low-income group (with an income of less than $70,000). Furthermore, assume that we take one sample of 250 households from California and another sample of 150 households from Wisconsin, collect the information on the incomes of these households, and prepare the contingency table as in Table 11.10.

> A *test of homogeneity* involves testing the null hypothesis that the proportions of elements with certain characteristics in two or more different populations are the same against the alternative hypothesis that these proportions are not the same.

Table 11.10   Observed Frequencies

	California	Wisconsin	Row Totals
High income	70	34	104
Medium income	80	40	120
Low income	100	76	176
Column Totals	250	150	400

Note that in this example the column totals are fixed. That is, we decided in advance to take samples of 250 households from California and 150 from Wisconsin. However, the row totals (of 104, 120, and 176) are determined randomly by the outcomes of the two samples. If we compare this example to the one about lack of discipline in schools in the previous section, we will notice that neither the column nor the row totals were fixed in that example. Instead, the researcher took just one sample of 300 adults, collected the information on gender and opinions, and prepared the contingency table. Thus, in that example, the row and column totals were all determined randomly. Thus, when both the row and column totals are determined randomly, we perform a test of independence. However, when either the column totals or the row totals are fixed, we perform a test of homogeneity. In the case of income groups in California and Wisconsin, we will perform a test of homogeneity to test for the similarity of income groups in the two states.

The procedure to conduct a test of homogeneity is similar to the procedure used to make a test of independence discussed earlier. Like a test of independence, **a test of homogeneity is right-tailed**. Example 11–8 illustrates the procedure to make a homogeneity test.

| EXAMPLE 11-8 | Distribution of Households in Regard to Income Levels |

Performing a test of homogeneity.

Consider the data on income distributions for households in California and Wisconsin given in Table 11.10. Using a 2.5% significance level, test whether the distribution of households with regard to income levels is different (not homogeneous) for the two states.

Solution   We perform the following five steps to make this test of hypothesis.

Step 1. *State the null and alternative hypotheses.*

The two hypotheses are, respectively,[2]

H_0: The proportions of households that belong to different income groups are the same in both states.

H_1: The proportions of households that belong to different income groups are not the same in both states.

Step 2. *Select the distribution to use.*

We use the chi-square distribution to make a homogeneity test.

Step 3. *Determine the rejection and nonrejection regions.*

The significance level is 2.5%. Because the homogeneity test is right-tailed, the area of the rejection region is .025 and lies in the right tail of the chi-square distribution curve. The contingency table for income groups in California and Wisconsin contains three rows and two columns. Hence, the degrees of freedom are

$$df = (R - 1)(C - 1) = (3 - 1)(2 - 1) = 2$$

From Table VI of Appendix B, the value of χ^2 for $df = 2$ and .025 area in the right tail of the chi-square distribution curve is 7.378. This value is shown in Figure 11.8.

Figure 11.8 Rejection and nonrejection regions.

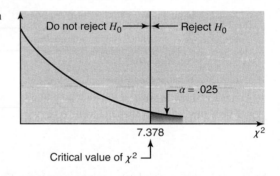

Step 4. *Calculate the value of the test statistic.*

To compute the value of the test statistic χ^2, we first need to calculate the expected frequencies. Table 11.11 lists the observed and the expected frequencies. The numbers in parentheses in this table are the expected frequencies, which are calculated using the formula

$$E = \frac{(\text{Row total})(\text{Column total})}{\text{Total of both samples}}$$

Table 11.11 Observed and Expected Frequencies

	California	Wisconsin	Row Totals
High income	70 (65)	34 (39)	104
Medium income	80 (75)	40 (45)	120
Low income	100 (110)	76 (66)	176
Column Totals	250	150	400

[2]Let p_{HC}, p_{MC}, and p_{LC} be the proportions of households in California who belong to high-, middle-, and low-income groups, respectively. Let p_{HW}, p_{MW}, and p_{LW} be the corresponding proportions for Wisconsin. Then we can also write the null hypothesis as

$$H_0: p_{HC} = p_{HW}, p_{MC} = p_{MW}, \text{ and } p_{LC} = p_{LW}$$

and the alternative hypothesis as

$$H_1: \text{At least two of the equalities mentioned in } H_0 \text{ are not true.}$$

Thus, for instance,

$$E \text{ for } High \text{ income and } California \text{ cell} = \frac{(104)(250)}{400} = 65$$

The remaining expected frequencies are calculated in the same way. Note that the expected frequencies in a test of homogeneity are calculated in the same way as in a test of independence. The value of the test statistic χ^2 is computed as follows:

$$\chi^2 = \sum \frac{(O - E)^2}{E}$$

$$= \frac{(70 - 65)^2}{65} + \frac{(34 - 39)^2}{39} + \frac{(80 - 75)^2}{75} + \frac{(40 - 45)^2}{45}$$

$$+ \frac{(100 - 110)^2}{110} + \frac{(76 - 66)^2}{66}$$

$$= .385 + .641 + .333 + .556 + .909 + 1.515 = \mathbf{4.339}$$

Step 5. *Make a decision.*

The value of the test statistic $\chi^2 = 4.339$ is less than the critical value of $\chi^2 = 7.378$ and falls in the nonrejection region. Hence, we fail to reject the null hypothesis and state that there is no evidence that the distributions of households with regard to income are different in California and Wisconsin.

EXERCISES

CONCEPTS AND PROCEDURES

11.18 Describe in your own words a test of independence and a test of homogeneity. Give one example of each.

11.19 Explain how the expected frequencies for cells of a contingency table are calculated in a test of independence or homogeneity. How do you find the degrees of freedom for such tests?

11.20 To make a test of independence or homogeneity, what should be the minimum expected frequency for each cell? What are the alternatives if this condition is not satisfied?

11.21 Consider the following contingency table, which records the results obtained for four samples of fixed sizes selected from four populations.

	Sample Selected From			
	Population 1	Population 2	Population 3	Population 4
Row 1	24	81	50	121
Row 2	46	64	91	69
Row 3	18	41	105	93

a. Write the null and alternative hypotheses for a test of homogeneity for this table.
b. Calculate the expected frequencies for all cells assuming that the null hypothesis is true.
c. For $\alpha = .025$, find the critical value of χ^2. Show the rejection and nonrejection regions on the chi-square distribution curve.
d. Find the value of the test statistic χ^2.
e. Using $\alpha = .025$, would you reject the null hypothesis?

11.22 Consider the following contingency table, which is based on a sample survey.

	Column 1	Column 2	Column 3
Row 1	143	64	105
Row 2	98	71	58
Row 3	111	92	115

a. Write the null and alternative hypotheses for a test of independence for this table.
b. Calculate the expected frequencies for all cells, assuming that the null hypothesis is true.
c. For $\alpha = .01$, find the critical value of χ^2. Show the rejection and nonrejection regions on the chi-square distribution curve.
d. Find the value of the test statistic χ^2.
e. Using $\alpha = .01$, would you reject the null hypothesis?

APPLICATIONS

11.23 Many students graduate from college deeply in debt from student loans, credit card debts, and so on. A sociologist took a random sample of 401 single persons, classified them by gender, and asked, "Would you consider marrying someone who was $25,000 or more in debt?" The results of this survey are shown in the following table.

	Yes	No	Uncertain
Women	130	59	18
men	96	82	16

Test at a 1% significance level whether gender and response are related.

11.24 The game show *Deal or No Deal* involves a series of opportunities for the contestant to either accept an amount of money from the show's *banker* or to decline it and open a specific number of briefcases in the hope of exposing and, thereby eliminating, low amounts of money from the game, which would lead the banker to increase the amount of the next offer. Suppose that 700 people aged 21 years and older were selected at random. Each of them watched an episode of the show until exactly four briefcases were left unopened. The money amounts in these four briefcases were $750, $5000, $50,000, and $400,000, respectively. The banker's offer to the contestant was $81,600 if the contestant would stop the game and accept the offer. If the contestant were to decline the offer, he or she would choose one briefcase out of these four to open, and then there would be a new offer. All 700 persons were asked whether they would accept the offer (Deal) for $81,600 or turn it down (No Deal), as well as their ages. The responses of these 700 persons are listed in the following table.

	Age Group (years)				
	21–29	**30–39**	**40–49**	**50–59**	**60 and Over**
Deal	64	86	92	84	58
No Deal	66	74	64	62	50

Test at a 5% significance level whether the decision to accept or not to accept the offer (Deal or No Deal) and age group are dependent.

11.25 During the recent economic recession, many families faced hard times financially. Some studies observed that more people stopped buying name brand products and started buying less expensive store brand products instead. Data produced by a recent sample of 700 adults on whether they usually buy store brand or name brand products are recorded in the following table.

	More Often Buy	
	Name Brand	**Store Brand**
Men	148	170
Women	155	219

Using a 1% significance level, can you reject the null hypothesis that the two attributes, gender and buying name or store brand products, are independent?

11.26 One hundred auto drivers who were stopped by police for some violation were also checked to see if they were wearing seat belts. The following table records the results of this survey.

	Wearing Seat Belt	**Not Wearing Seat Belt**
Men	35	20
Women	33	12

Test at a 2.5% significance level whether being a man or a woman and wearing or not wearing a seat belt are related.

11.27 National Electronics Company buys parts from two subsidiaries. The quality control department at this company wanted to check if the distribution of good and defective parts is the same for the supplies of parts received from both subsidiaries. The quality control inspector selected a sample of 300 parts received from Subsidiary A and a sample of 400 parts received from Subsidiary B. These parts were checked for being good or defective. The following table records the results of this investigation.

	Subsidiary A	**Subsidiary B**
Good	286	379
Defective	14	21

Using a 5% significance level, test the null hypothesis that the distributions of good and defective parts are the same for both subsidiaries.

11.28 A news poll asked, "Which of the following comes closest to your view about what government policy should be toward illegal immigrants currently in the United States?" The three options were (A) Send all illegal immigrants back to their home country, (B) Have a guest worker program that allows immigrants to remain in the United States to work but only for a limited amount of time, and (C) Allow illegal immigrants to remain in the country and eventually qualify for U.S. citizenship, but only if they meet certain requirements, such as paying back taxes, learning English, and passing a background check. The following table lists the party affiliations of the respondents and their responses. The numbers in the table are approximately the same as reported in percentages in the poll.

	A	**B**	**C**	**Unsure**
Democrat	55	43	288	8
Independent	19	25	107	6
Republican	89	52	196	7

Test at a 5% significance level whether the distributions of responses are significantly different for at least two of the political affiliations.

11.29 The following table gives the distributions of grades for three professors for a few randomly selected classes that each of them taught during the last 2 years.

		Professor		
		Miller	**Smith**	**Moore**
	A	16	36	24
	B	25	44	12
Grade	C	85	81	82
	D & F	20	12	8

Using a 2.5% significance level, test the null hypothesis that the grade distributions are homogeneous for these three professors.

11.30 A forestry official is comparing the causes of forest fires in two regions, A and B. The following table shows the causes of fire for 79 randomly selected recent fires in these two regions.

	Arson	**Accident**	**Lightning**	**Unknown**
Region A	8	9	7	12
Region B	7	18	12	6

Test at a 5% significance level whether causes of fire and regions of fires are related.

11.4 | Inferences About the Population Variance

Earlier chapters explained how to make inferences (confidence intervals and hypothesis tests) about the population mean and population proportion. However, we may often need to control the variance (or standard deviation). Consequently, there may be a need to estimate and to test a hypothesis about the population variance σ^2. Section 11.4.1 describes how to make a confidence interval for the population variance (or standard deviation). Section 11.4.2 explains how to test a hypothesis about the population variance.

As an example, suppose a machine is set up to fill packages of cookies so that the net weight of cookies per package is 32 ounces. Note that the machine will not put exactly 32 ounces of cookies into each package. Some of the packages will contain less and some will contain more than 32 ounces. However, if the variance (and, hence, the standard deviation) is too large, some of the packages will contain quite a bit less than 32 ounces of cookies, and some others will contain quite a bit more than 32 ounces. The manufacturer will not want a large variation in the amounts of cookies put into different packages. To keep this variation within some specified acceptable limit, the machine will be adjusted from time to time. Before the manager decides to adjust the machine at any time, he or she must estimate the variance or test a hypothesis or do both to find out if the variance exceeds the maximum acceptable value.

Like every sample statistic, the sample variance is a random variable, and it possesses a sampling distribution. If all the possible samples of a given size are taken from a population and their variances are calculated, the probability distribution of these variances is called the **sampling distribution of the sample variance**.

Sampling Distribution of $(n - 1)\, s^2/\sigma^2$ If the population from which the sample is selected is (approximately) normally distributed, then

$$\frac{(n - 1)\, s^2}{\sigma^2}$$

has a chi-square distribution with $n - 1$ degrees of freedom. Note that it is not s^2 but the above expression that has a chi-square distribution.

Thus, the chi-square distribution is used to construct a confidence interval and test a hypothesis about the population variance σ^2.

11.4.1   Estimation of the Population Variance

The value of the sample variance s^2 gives a point estimate of the population variance σ^2. The $(1 - \alpha)100\%$ confidence interval for σ^2 is given by the following formula.

Confidence Interval for the Population Variance σ^2   Assuming that the population from which the sample is selected is (approximately) normally distributed, we obtain the $(1 - \alpha)100\%$ **confidence interval for the population variance σ^2** as

$$\frac{(n - 1)\, s^2}{\chi^2_{\alpha/2}} \quad \text{to} \quad \frac{(n - 1)\, s^2}{\chi^2_{1-\alpha/2}}$$

where $\chi^2_{\alpha/2}$ and $\chi^2_{1-\alpha/2}$ are obtained from the chi-square distribution table for $\alpha/2$ and $1 - \alpha/2$ areas in the right tail of the chi-square distribution curve, respectively, and for $n - 1$ degrees of freedom.

The confidence interval for the population standard deviation can be obtained by simply taking the positive square roots of the two limits of the confidence interval for the population variance.

The procedure for making a confidence interval for σ^2 involves the following three steps.

1. Take a sample of size n and compute s^2 using the formula learned in Chapter 3. However, if n and s^2 are given, then perform only steps 2 and 3.
2. Calculate $\alpha/2$ and $1 - \alpha/2$. Find two values of χ^2 from the chi-square distribution table (Table VI of Appendix B): one for $\alpha/2$ area in the right tail of the chi-square distribution curve and $df = n - 1$, and the second for $1 - \alpha/2$ area in the right tail and $df = n - 1$.
3. Substitute all the values in the formula for the confidence interval for σ^2 and simplify.

Example 11–9 illustrates the estimation of the population variance and population standard deviation.

EXAMPLE 11–9 Weights of Packages of Cookies

Constructing confidence intervals for σ^2 and σ.

One type of cookie manufactured by Haddad Food Company is Cocoa Cookies. The machine that fills packages of these cookies is set up in such a way that the average net weight of these packages is 32 ounces with a variance of .015 square ounce. From time to time the quality control inspector at the company selects a sample of a few such packages, calculates the variance of the net weights of these packages, and constructs a 95% confidence interval for the population variance. If either one of the two limits of this confidence interval is not in the interval .008 to .030, the machine is stopped and adjusted. A recently taken random sample of 25 packages from the production line gave a sample variance of .029 square ounce. Based on this sample information, do you think the machine needs an adjustment? Assume that the net weights of cookies in all packages are normally distributed.

Solution The following three steps are performed to estimate the population variance and to make a decision.

Step 1. From the given information, $n = 25$ and $s^2 = .029$

Step 2. The confidence level is $1 - \alpha = .95$. Hence, $\alpha = 1 - .95 = .05$. Therefore,

$$\alpha/2 = .05/2 = .025$$
$$1 - \alpha/2 = 1 - .025 = .975$$
$$df = n - 1 = 25 - 1 = 24$$

From Table VI of Appendix B,

$$\chi^2 \text{ for 24 } df \text{ and .025 area in the right tail} = 39.364$$

$$\chi^2 \text{ for 24 } df \text{ and .975 area in the right tail} = 12.401$$

These values are shown in Figure 11.9.

Figure 11.9 The values of χ^2.

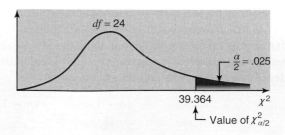

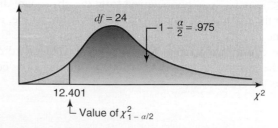

Step 3.   The 95% confidence interval for σ^2 is

$$\frac{(n-1)s^2}{\chi^2_{\alpha/2}} \quad \text{to} \quad \frac{(n-1)s^2}{\chi^2_{1-\alpha/2}}$$

$$\frac{(25-1)(.029)}{39.364} \quad \text{to} \quad \frac{(25-1)(.029)}{12.401}$$

$$\mathbf{.0177} \quad \mathbf{to} \quad \mathbf{.0561}$$

Thus, with 95% confidence, we can state that the variance for all packages of Cocoa Cookies lies between .0177 and .0561 square ounce. Note that the lower limit (.0177) of this confidence interval is between .008 and .030, but the upper limit (.0561) is larger than .030 and falls outside the interval .008 to .030. Because the upper limit is larger than .030, we can state that the machine needs to be stopped and adjusted.

We can obtain the confidence interval for the population standard deviation σ by taking the positive square roots of the two limits of the above confidence interval for the population variance. Thus, a 95% confidence interval for the population standard deviation is

$$\sqrt{.0177} \text{ to } \sqrt{.0561} \quad \text{or} \quad \mathbf{.133 \text{ to } .237}$$

Hence, the standard deviation of all packages of Cocoa Cookies is between .133 and .237 ounce at a 95% confidence level.

11.4.2   Hypothesis Tests About the Population Variance

A test of hypothesis about the population variance can be one-tailed or two-tailed. To make a test of hypothesis about σ^2, we perform the same five steps we used earlier in hypothesis-testing examples. The procedure to test a hypothesis about σ^2 discussed in this section is applied only when the population from which a sample is selected is (approximately) normally distributed.

Test Statistic for a Test of Hypothesis About σ^2   The value of the **test statistic χ^2** is calculated as

$$\chi^2 = \frac{(n-1)s^2}{\sigma^2}$$

where s^2 is the sample variance, σ^2 is the hypothesized value of the population variance, and $n-1$ represents the degrees of freedom. The population from which the sample is selected is assumed to be (approximately) normally distributed.

Examples 11–10 and 11–11 illustrate the procedure for making tests of hypothesis about σ^2.

EXAMPLE 11–10	Weights of Packages of Cookies

Performing a right-tailed test of hypothesis about σ^2.

One type of cookie manufactured by Haddad Food Company is Cocoa Cookies. The machine that fills packages of these cookies is set up in such a way that the average net weight of these packages is 32 ounces with a variance of .015 square ounce. From time to time the quality control inspector at the company selects a sample of a few such packages, calculates the variance of the net weights of these packages, and makes a test of hypothesis about the population variance. She always uses $\alpha = .01$. The acceptable value of the population variance is .015 square ounce or less. If the conclusion from the test of hypothesis is that the population variance is not within the acceptable limit, the machine is stopped and adjusted. A recently taken random sample of 25 packages from the production line gave a sample variance of .029 square ounce. Based on this sample information, do you think the machine needs an adjustment? Assume that the net weights of cookies in all packages are normally distributed.

Solution   From the given information,

$$n = 25, \quad \alpha = .01, \quad \text{and} \quad s^2 = .029$$

The population variance should not exceed .015 square ounce.

Step 1. *State the null and alternative hypotheses.*

We are to test whether or not the population variance is within the acceptable limit. The population variance is within the acceptable limit if it is less than or equal to .015; otherwise, it is not. Thus, the two hypotheses are

$$H_0: \sigma^2 \le .015 \quad \text{(The population variance is within the acceptable limit.)}$$
$$H_1: \sigma^2 > .015 \quad \text{(The population variance exceeds the acceptable limit.)}$$

Step 2. *Select the distribution to use.*

Since the population is normally distributed, we will use the chi-square distribution to test a hypothesis about σ^2.

Step 3. *Determine the rejection and nonrejection regions.*

The significance level is 1% and, because of the $>$ sign in H_1, the test is right-tailed. The rejection region lies in the right tail of the chi-square distribution curve with its area equal to .01. The degrees of freedom for a chi-square test about σ^2 are $n - 1$; that is,

$$df = n - 1 = 25 - 1 = 24$$

From Table VI of Appendix B, the critical value of χ^2 for 24 degrees of freedom and .01 area in the right tail is 42.980. This value is shown in Figure 11.10.

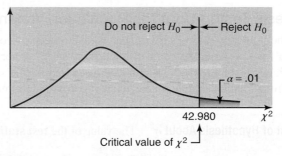

Do not reject H_0 → ← Reject H_0

$\alpha = .01$

42.980 χ^2

Critical value of χ^2

Figure 11.10   Rejection and nonrejection regions.

Step 4. *Calculate the value of the test statistic.*

The value of the test statistic χ^2 for the sample variance is calculated as follows:

$$\chi^2 = \frac{(n - 1)\,s^2}{\sigma^2} = \frac{(25 - 1)(.029)}{.015} = 46.400$$

From H_0

Step 5. *Make a decision.*

The value of the test statistic $\chi^2 = 46.400$ is greater than the critical value of $\chi^2 = 42.980$ and falls in the rejection region. Consequently, we reject H_0 and conclude that the population variance is not within the acceptable limit. The machine should be stopped and adjusted.

EXAMPLE 11–11   Variance of GPAs of Students at a University

Conducting a two-tailed test of hypothesis about σ^2.

It is known that the variance of GPAs (with a maximum GPA of 4) of all students at a large university was .24 in 2014. A professor wants to determine whether the variance of the current GPAs of students at this university is different from .24. She took a random sample of 20 students and found that the variance of their GPAs is .27. Using a 5% significance level, can you conclude that the current variance of the GPAs of students at this university is different from .24? Assume that the GPAs of all current students at this university are (approximately) normally distributed.

Solution   From the given information,

$$n = 20, \quad \alpha = .05, \quad \text{and} \quad s^2 = .27$$

The population variance was .24 in 2014.

Step 1.   *State the null and alternative hypotheses.*

The null and alternative hypotheses are, respectively,

$$H_0: \sigma^2 = .24 \quad \text{(The population variance is not different from .24.)}$$
$$H_1: \sigma^2 \neq .24 \quad \text{(The population variance is different from .24.)}$$

Corbis Digital Stock/©Corbis

Step 2.   *Select the distribution to use.*

Since the population of GPAs is normally distributed, we will use the chi-square distribution to test a hypothesis about σ^2.

Step 3.   *Determine the rejection and nonrejection regions.*

The significance level is 5%. The $\neq$ sign in H_1 indicates that the test is two-tailed. The rejection region lies in both tails of the chi-square distribution curve with its total area equal to .05. Consequently, the area in each tail of the distribution curve is .025. The values of $\alpha/2$ and $1 - \alpha/2$ are, respectively,

$$\frac{\alpha}{2} = \frac{.05}{2} = .025 \quad \text{and} \quad 1 - \frac{\alpha}{2} = 1 - .025 = .975$$

The degrees of freedom are

$$df = n - 1 = 20 - 1 = 19$$

From Table VI of Appendix B, the critical values of χ^2 for 19 degrees of freedom and for $\alpha/2$ and $1 - \alpha/2$ areas in the right tail are

$$\chi^2 \text{ for 19 } df \text{ and .025 area in the right tail} = 32.852$$
$$\chi^2 \text{ for 19 } df \text{ and .975 area in the right tail} = 8.907$$

These two values are shown in Figure 11.11.

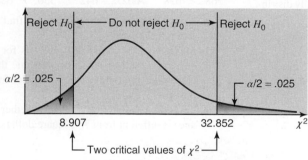

Figure 11.11   Rejection and nonrejection regions.

Step 4.   *Calculate the value of the test statistic.*

The value of the test statistic χ^2 for the sample variance is calculated as follows:

$$\chi^2 = \frac{(n-1)s^2}{\sigma^2} = \frac{(20-1)(.27)}{.24} = 21.375$$

From H_0

Step 5.   *Make a decision.*

The value of the test statistic $\chi^2 = 21.375$ is between the two critical values of χ^2, 8.907 and 32.852 and falls in the nonrejection region. Consequently, we fail to reject H_0 and conclude that the population variance of the current GPAs of students at this university does not appear to be different from .24.

Note that we can make a test of hypothesis about the population standard deviation σ using the same procedure as that for the population variance σ^2. To make a test of hypothesis about σ, the only change will be mentioning the values of σ in H_0 and H_1. The rest of the procedure remains the same as in the case of σ^2.

EXERCISES

CONCEPTS AND PROCEDURES

11.31   A sample of 25 observations selected from a normally distributed population produced a sample variance of 35. Construct a confidence interval for σ^2 for each of the following confidence levels and comment on what happens to the confidence interval of σ^2 when the confidence level decreases.

    a. $1 - \alpha = .99$ **b.** $1 - \alpha = .95$ **c.** $1 - \alpha = .90$

11.32   A sample of certain observations selected from a normally distributed population produced a sample variance of 46. Construct a 95% confidence interval for σ^2 for each of the following cases and comment on what happens to the confidence interval of σ^2 when the sample size increases.

    a. $n = 10$ **b.** $n = 15$ **c.** $n = 30$

11.33   A sample of 22 observations selected from a normally distributed population produced a sample variance of 18.

    a. Write the null and alternative hypotheses to test whether the population variance is different from 14.
    b. Using $\alpha = .05$, find the critical values of χ^2. Show the rejection and nonrejection regions on a chi-square distribution curve.
    c. Find the value of the test statistic χ^2.
    d. Using a 5% significance level, will you reject the null hypothesis stated in part a?

11.34   A sample of 18 observations selected from a normally distributed population produced a sample variance of 4.6.

    a. Write the null and alternative hypotheses to test whether the population variance is different from 2.2.
    b. Using $\alpha = .05$, find the critical value of χ^2. Show the rejection and nonrejection regions on a chi-square distribution curve.
    c. Find the value of the test statistic χ^2.
    d. Using a 5% significance level, will you reject the null hypothesis stated in part a?

APPLICATIONS

11.35   The makers of Flippin' Out Pancake Mix claim that one cup of their mix contains 11 grams of sugar. However, the mix is not uniform, so the amount of sugar varies from cup to cup. One cup of mix was taken from each of 24 randomly selected boxes. The sample variance of the sugar measurements from these 24 cups was 1.47 grams. Assume that the distribution of sugar content is approximately normal.

    a. Construct the 98% confidence intervals for the population variance and standard deviation.
    b. Test at a 1% significance level whether the variance of the sugar content per cup is greater than 1.0 gram.

11.36   An auto manufacturing company wants to estimate the variance of miles per gallon for its auto model AST727. A random sample of 24 cars of this model showed that the variance of miles per gallon for these cars is .64. Assume that the miles per gallon for all such cars are (approximately) normally distributed.

    a. Construct the 95% confidence intervals for the population variance and standard deviation.
    b. Test at a 1% significance level whether the sample result indicates that the population variance is different from .30.

11.37   The following are the prices (in dollars) of the same brand of camcorder found at eight stores in Los Angeles.

    568   628   602   642   550   688   615   604

    a. Using the formula from Chapter 3, find the sample variance, s^2, for these data.
    b. Make the 95% confidence intervals for the population variance and standard deviation. Assume that the prices of this camcorder at all stores in Los Angeles follow a normal distribution.
    c. Test at a 5% significance level whether the population variance is different from 750 square dollars.

USES AND MISUSES...

1. DO NOT FEED THE ANIMALS

You are a wildlife enthusiast studying African wildlife: gnus, zebras, and gazelles. You know that a herd of each species will visit one of three watering places in a region every day, but you do not know the distribution of choices that the animals make or whether these choices are dependent. You have observed that the animals sometimes drink together and sometimes do not. A statistician offers to help and says that he will perform a test for independence of watering place choices based on your observations of the animals' behavior over the past several months. The statistician performs some calculations and says that he has answered your question because his chi-square test of the independence of watering place choices, at a 5% significance level,

told him to reject the null hypothesis. He has also performed a goodness-of-fit test on the hypothesis that the animals are equally likely to choose any watering place, and he has rejected that hypothesis as well.

The statistician barely helped you. In the first case, you know a single piece of information: the choice of a watering place for the three groups of animals is dependent. Another way of stating the result is that your data indicate that the choice of watering places for at least one of the animals is not independent of the others. Perhaps the zebras get up early, and the gnus and gazelles follow, making the gnus and gazelles dependent on the choice of the zebras. Or perhaps the animals choose the watering place of the day independent of the other animals, but always avoid the watering place at which the lions are drinking. Regarding the goodness-of-fit test, all you know is that the hypothesis that the animals equally favor the three watering places was wrong. But you do not know what the expected distribution should be. In short, the rejection of the null hypothesis raises more questions than it answers.

2. IS THERE A GENDER BIAS IN ADMISSIONS?

Categorical data analysis methods, such as a chi-square test for independence, are used quite often in analyzing employment and admissions data in discrimination cases. One of the more famous discrimination cases involved graduate admissions at the University of California, Berkeley, in 1973. The claim was that UC Berkeley was discriminating against women in their admissions decisions, as would seem to be the case based on the data in the following table.

	Applicants	Percentage Admitted
Male	8442	44%
Female	4321	35%

The p-value for the corresponding chi-square test of independence is approximately 1.1×10^{-22}, so it would seem clear that there is statistical dependence between gender and graduate admission. However, although it is true that admission rates differed by gender, the fact that the University was found not guilty of discrimination against women might shock you.

In order to find out why the University was found not guilty, we need to introduce another variable into the study: program of study. When the various programs were considered separately, the following admission rates for the six largest programs were observed.

Department	Men Applicants	Men Admitted (%)	Women Applicants	Women Admitted (%)
A	825	62	108	82
B	560	63	25	68
C	325	37	593	34
D	417	33	375	35
E	191	28	393	24
F	272	6	341	7

As one can see, four of the six programs had nominally higher admission rates for women than for men. So, why is the overall acceptance rate higher for men than for women? Looking at Departments A and B, which had the highest acceptance rates, we notice that 1385 men (16.4% of all men) applied to those programs, while 133 women (3.1% of all women) applied to these programs. On the other hand, almost twice as many women as men applied to the programs with the lowest acceptance rates. Hence, the overall acceptance rate for each gender was affected by the programs to which they applied and the number who applied to those programs. Since a much higher percentage of men applied to programs with higher (overall) acceptance rates, the overall acceptance rate for men ended up being higher than the overall acceptance rate for women.

This result is an example of what is known as Simpson's Paradox, which occurs when the inclusion of an additional variable (characteristic) reverses the conclusion made without that variable.

Source: P. J. Bickel, E. A. Hammel, and J. W. O'Connell (1975): Sex Bias in Graduate Admissions: Data From Berkeley. *Science* 187(4175): 398–404.

GLOSSARY

Chi-square distribution   A distribution, with degrees of freedom as the only parameter, that is skewed to the right for small *df* and looks like a normal curve for large *df*.

Expected frequencies   The frequencies for different categories of a multinomial experiment or for different cells of a contingency table that are expected to occur when a given null hypothesis is true.

Goodness-of-fit test   A test of the null hypothesis that the observed frequencies for an experiment follow a certain pattern or theoretical distribution.

Multinomial experiment   An experiment with *n* trials for which (1) the trials are identical, (2) there are more than two possible

outcomes per trial, (3) the trials are independent, and (4) the probabilities of the various outcomes remain constant for each trial.

Observed frequencies   The frequencies actually obtained from the performance of an experiment.

Test of homogeneity   A test of the null hypothesis that the proportions of elements that belong to different groups in two (or more) populations are similar.

Test of independence   A test of the null hypothesis that two attributes of a population are not related.

SUPPLEMENTARY EXERCISES

11.38 One of the products produced by Branco Food Company is Total-Bran Cereal, which competes with three other brands of similar total-bran cereals. The company's research office wants to investigate if the percentage of people who consume total-bran cereal is the same for each of these four brands. Let us denote the four brands of cereal

by A, B, C, and D. A sample of 1000 persons who consume total-bran cereal was taken, and they were asked which brand they most often consume. Of the respondents, 212 said they usually consume Brand A, 284 consume Brand B, 254 consume Brand C, and 250 consume Brand D. Does the sample provide enough evidence to reject the null hypothesis that the percentage of people who consume total-bran cereal is the same for all four brands? Use $\alpha = .05$.

11.39 The percentage distribution of birth weights for all children in cases of multiple births (twins, triplets, etc.) in North Carolina during 2009 was as given in the following table:

Weight (grams)	0–500	501–1500	1501–2500	2501–8165
Percentage	1.45	11.02	49.23	38.30

Source: http://www.schs.state.nc.us/SCHS/data/births/bd.cfm.

The frequency distribution of birth weights of a sample of 587 children who shared multiple births and were born in North Carolina in 2012 is as shown in the following table:

Weight (grams)	0–500	501–1500	1501–2500	2501–8165
Frequency	2	60	305	220

Test at a 2.5% significance level whether the 2012 distribution of birth weights for all children born in North Carolina who shared multiple births is significantly different from the one for 2009.

11.40 A 2010 poll by Marist University asked people to choose their favorite classic Christmas movie from a list of five choices. The following table shows the frequencies for the various movies:

Movie	It's A Wonderful Life	A Christmas Story	Miracle on 34th Street	White Christmas	A Christmas Carol
Frequency	247	237	226	124	134

Source: www.maristpoll.marist.edu/1221-no-consensus-on-favorite-holiday-film/.

Test at a 1% significance level whether these five movies are equally preferred.

11.41 A randomly selected sample of 100 persons who suffer from allergies were asked during what season they suffer the most. The results of the survey are recorded in the following table.

Season	Fall	Winter	Spring	Summer
Persons allergic	20	3	41	36

Using a 1% significance level, test the null hypothesis that the proportions of all allergic persons are equally distributed over the four seasons.

11.42 All shoplifting cases in the town of Seven Falls are randomly assigned to either Judge Stark or Judge Rivera. A citizens group wants to know whether either of the two judges is more likely to sentence the offenders to jail time. A sample of 180 recent shoplifting cases produced the following two-way table.

	Jail	Other Sentence
Judge Stark	29	61
Judge Rivera	34	56

Test at a 5% significance level whether the type of sentence for shoplifting depends on which judge tries the case.

11.43 During a bear market, 150 investors were asked how they were adjusting their portfolios to protect themselves. Some of these investors were keeping most of their money in stocks, whereas others were shifting large amounts of money to bonds, real estate, or cash (such as money market accounts). The results of the survey are shown in the following table.

Favored choice	Stocks	Bonds	Real Estate	Cash
Number of investors	48	45	35	22

Using a 2.5% significance level, test the null hypothesis that the percentages of investors favoring the four choices are all equal.

11.44 Recent recession and bad economic conditions forced many people to hold more than one job to make ends meet. A sample of 500 persons who held more than one job produced the following two-way table.

	Single	Married	Other
Male	65	231	26
Female	38	120	20

Test at a 10% significance level whether gender and marital status are related for all people who hold more than one job.

11.45 ATVs (all-terrain vehicles) have become a source of controversy. Some people feel that their use should be tightly regulated, while others prefer fewer restrictions. Suppose a survey consisting of a random sample of 200 people aged 18 to 27 and another survey of a random sample of 210 people aged 28 to 37 was conducted, and these people were asked whether they favored more restrictions on ATVs, fewer restrictions, or no change. The results of this survey are summarized in the following table.

Age (years)	More Restrictions	Fewer Restrictions	No Change
18 to 27	40	92	68
28 to 37	55	68	87

Test at a 2.5% significance level whether the distribution of opinions in regard to ATVs are the same for both age groups.

11.46 Construct the 98% confidence intervals for the population variance and standard deviation for the following data, assuming that the respective populations are (approximately) normally distributed.

$$n = 21, s^2 = 9.2$$

Test at a 5% significance level if the population variance is different from 6.5.

11.47 Construct the 95% confidence intervals for the population variance and standard deviation for the following data, assuming that the respective populations are (approximately) normally distributed.

$$n = 10, s^2 = 7.2$$

Test at a 1% significance level if the population variance is greater than 4.2.

11.48 A sample of 17 units selected from a normally distributed population produced a variance of 1.7. Test at a 2.5% significance level if the population variance is greater than 1.1.

11.49 A sample of 18 units selected from a normally distributed population produced a variance of 14.8. Test at a 5% significance level if the population variance is different from 10.4.

11.50 The variance of the SAT scores for all students who took that test this year is 5000. The variance of the SAT scores for a random sample of 20 students from one school is equal to 3175.

    a. Test at a 2.5% significance level whether the variance of the SAT scores for students from this school is lower than 5000. Assume that the SAT scores for all students at this school are (approximately) normally distributed.

    b. Construct the 98% confidence intervals for the variance and the standard deviation of SAT scores for all students at this school.

11.51 A company manufactures ball bearings that are supplied to other companies. The machine that is used to manufacture these ball bearings produces them with a variance of diameters of .025 square millimeter or less. The quality control officer takes a sample of such ball bearings quite often and checks, using confidence intervals and tests of hypotheses, whether or not the variance of these bearings is within .025 square millimeter. If it is not, the machine is stopped and adjusted. A recently taken random sample of 23 ball bearings gave a variance of the diameters equal to .034 square millimeter.

    a. Using a 5% significance level, can you conclude that the machine needs an adjustment? Assume that the diameters of all ball bearings have a normal distribution.

    b. Construct a 95% confidence interval for the population variance.

11.52 Usually people do not like waiting in line for a long time for service. A bank manager does not want the variance of the waiting times for her customers to be greater than 4.0 square minutes. A random sample of 25 customers taken from this bank gave the variance of the waiting times equal to 8.3 square minutes.

    a. Test at a 1% significance level whether the variance of the waiting times for all customers at this bank is greater than 4.0 square minutes. Assume that the waiting times for all customers are normally distributed.

    b. Construct a 99% confidence interval for the population variance.

11.53 A sample of eight passengers boarding a domestic flight produced the following data on weights (in pounds) of their carry-on bags.

46.3	41.5	40.1	31.0	41.4	35.8	39.8	38.4

    a. Using the formula from Chapter 3, find the sample variance, s^2, for these data.

    b. Make the 98% confidence intervals for the population variance and standard deviation. Assume that the population from which this sample is selected is normally distributed.

    c. Test at a 5% significance level whether the population variance is larger than 20 square pounds.

ADVANCED EXERCISES

11.54 Suppose that you have a two-way table with the following row and column totals.

		Variable 1			
		A	**B**	**C**	**Total**
Variable 2	**X**				120
	Y				205
	Z				175
	Total	165	140	195	500

The observed values in the cells must be counts, which are nonnegative integers. Calculate the expected counts for the cells under the assumption that the two variables are independent. Based on your calculations, explain why it is impossible for the test statistic to have a value of zero.

11.55 A random sample of 100 persons was selected from each of four regions in the United States. These people were asked whether or not they support a certain farm subsidy program. The results of the survey are summarized in the following table.

	Favor	**Oppose**	**Uncertain**
Northeast	56	33	11
Midwest	73	23	4
South	67	28	5
West	59	35	6

Explain why the hypothesis test is a test of homogeneity as opposed to a test of independence. What feature of the data would change if you were to collect data in order to test for independence?

11.56 You are performing a goodness-of-fit test with four categories, all of which are supposed to be equally likely. You have a total of 100 observations. The observed frequencies are 21, 26, 31, and 22, respectively, for the four categories.

    a. Show that you would fail to reject the null hypothesis for these data for any reasonable significance level.

    b. The sum of the absolute differences (between the expected and the observed frequencies) for these data is 14 (i.e., $4 + 1 + 6 + 3 = 14$). Is it possible to have different observed frequencies keeping the sum at 14 so that you get a p-value of .10 or less?

11.57 You have collected data on a variable, and you want to determine if a normal distribution is a reasonable model for these data. The following table shows how many of the values fall within certain ranges of z values for these data.

Category	Count
z score below -2	48
z score from -2 to less than -1.5	67
z score from -1.5 to less than -1	146
z score from -1 to less than -0.5	248
z score from -0.5 to less than 0	187
z score from 0 to less than 0.5	125
z score from 0.5 to less than 1	88
z score from 1 to less than 1.5	47
z score from 1.5 to less than 2	25
z score of 2 or above	19
Total	1000

Perform a hypothesis test to determine if a normal distribution is an appropriate model for these data. Use a significance level of 5%.

SELF-REVIEW TEST

1. The random variable χ^2 assumes only
    a. positive   **b.** nonnegative   **c.** nonpositive values

2. The parameter(s) of the chi-square distribution is (are)
    a. degrees of freedom   **b.** *df* and *n*   **c.** χ^2

3. Which of the following is *not* a characteristic of a multinomial experiment?
    a. It consists of *n* identical trials.
    b. There are *k* possible outcomes for each trial and $k > 2$.
    c. The trials are random.
    d. The trials are independent.
    e. The probabilities of outcomes remain constant for each trial.

4. The observed frequencies for a goodness-of-fit test are
    a. the frequencies obtained from the performance of an experiment
    b. the frequencies given by the product of *n* and *p*
    c. the frequencies obtained by adding the results of a and b

5. The expected frequencies for a goodness-of-fit test are
    a. the frequencies obtained from the performance of an experiment
    b. the frequencies given by the product of *n* and *p*
    c. the frequencies obtained by adding the results of a and b

6. The degrees of freedom for a goodness-of-fit test are
    a. $n - 1$ **b.** $k - 1$ **c.** $n + k - 1$

7. The chi-square goodness-of-fit test is always
    a. two-tailed   **b.** left-tailed   **c.** right-tailed

8. To apply a goodness-of-fit test, the expected frequency of each category must be at least
    a. 10   **b.** 5   **c.** 8

9. The degrees of freedom for a test of independence are
    a. $(R - 1)(C - 1)$ **b.** $n - 2$ **c.** $(n - 1)(k - 1)$

10. In a Gallup poll conducted August 7–10, 2014, American adults aged 18 and older were asked, "If you were taking a new job and had your choice of a boss, would you prefer to work for a man or a woman?" Of the respondents, 33% said that they would prefer a male boss, 20% said a female boss, 46% said it would not make a difference to them, and 1% had no opinion (www.gallup.com). Suppose these results are true for the 2014 population. Recently, 1500 randomly selected American adults were asked the same question. The following table contains the frequency distribution that resulted from this survey.

Response	Male Boss	Female Boss	No Difference	No Opinion
Frequency	480	360	630	30

Test at the 1% significance level whether the current distribution of opinions is different from the 2014 distribution.

11. The following table gives the two-way classification of 1000 persons who have been married at least once. They are classified by educational level and marital status.

	Educational Level			
	Less Than High School	High School Degree	Some College	College Degree
Divorced	173	158	95	53
Never divorced	162	126	110	123

Test at a 1% significance level whether educational level and ever being divorced are dependent.

12. A researcher wanted to investigate if people who belong to different income groups are homogeneous with regard to playing lotteries. She took a sample of 600 people from the low-income group, another sample of 500 people from the middle-income group, and a third sample of 400 people from the high-income group. All these people were asked whether they play the lottery often, sometimes, or never. The results of the survey are summarized in the following table.

	Income Group		
	Low	Middle	High
Play often	174	163	90
Play sometimes	286	217	120
Never play	140	120	190

Using a 5% significance level, can you reject the null hypothesis that the percentages of people who play the lottery often, sometimes, and never are the same for each income group?

13. The owner of an ice cream parlor is concerned about consistency in the amount of ice cream his servers put in each cone. He would like the variance of all such cones to be no more than .25 square ounce. He decides to weigh each double-dip cone just before it is given to the customer. For a sample of 20 double-dip cones, the weights were found to have a variance of .48 square ounce. Assume that the weights of all such cones are (approximately) normally distributed.
    a. Construct the 99% confidence intervals for the population variance and the population standard deviation.
    b. Test at a 1% significance level whether the variance of the weights of all such cones exceeds .25 square ounce.

MINI-PROJECTS

Note: The Mini-Projects are located on the text's Web site, www.wiley.com/college/mann.

DECIDE FOR YOURSELF

Note: The Decide for Yourself feature is located on the text's Web site, www.wiley.com/college/mann.

TECHNOLOGY INSTRUCTIONS | CHAPTER 11

Note: Complete TI-84, Minitab, and Excel manuals are available for download at the textbook's Web site, www.wiley.com/college/mann.

TI-84 Color/TI-84

The TI-84 Color Technology Instructions feature of this text is written for the TI-84 Plus C color graphing calculator running the 4.0 operating system. Some screens, menus, and functions will be slightly different in older operating systems. The TI-84 and TI-84 Plus can perform all of the same functions but will not have the "Color" option referenced in some of the menus.

Performing a Chi-Square Goodness of Fit Test for Example 11–3 of the Text

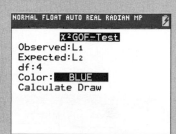

Screen 11.1

1. Enter the observed counts from Example 11–3 into list 1 and the corresponding expected counts in list 2.

2. Select **STAT > TESTS > χ^2 GOF-Test**.

3. Use the following settings in the χ^2 **GOF-Test** menu (see **Screen 11.1**):

   • Type L1 at the **Observed** prompt.

   • Type L2 at the **Expected** prompt.

   • Type 4 at the **df** prompt.

   • Select any desired color at the **Color** prompt.

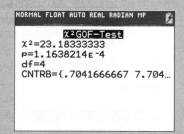

Screen 11.2

4. Highlight **Calculate**, and press **ENTER**.

5. The output includes the test statistic and the *p*-value. The values of $(O-E)^2/E$ are stored in a list named CNTRB. (See **Screen 11.2**.)

Now compare the χ^2-value with the critical value of χ^2 or the *p*-value from Screen 11.2 with α and make a decision.

Performing a Chi-Square Independence/Homogeneity Test for Example 11–6 of the Text

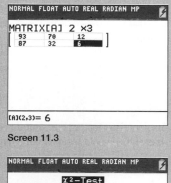

Screen 11.3

1. Select **2nd > X^{-1} > EDIT > [A]**. Press **ENTER**.

2. Use the following settings in the matrix editor (see **Screen 11.3**):

   • At the **Matrix[A]** prompt, type 2 for the number of rows and 3 for the number of columns. Use the arrow keys to navigate between these numbers.

   • Enter the observed counts from Example 11–6 into the matrix. Use **ENTER** and the arrow keys to move between cells.

3. Select **STAT > TESTS > χ^2-Test**.

4. Use the following settings in the χ^2-**Test** menu (see **Screen 11.4**):

   • Enter [A] at the **Observed** prompt.

   Note: To enter the name for the matrix [A], select **2nd > X^{-1} > NAMES > [A]**.

   • Enter [B] at the **Expected** prompt.

Screen 11.4

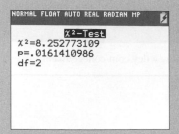

Screen 11.5

Note: To enter the name for the matrix [B], select **2nd** > X^{-1} > **NAMES** > **[B]**.

- Select any desired color at the **Color** prompt.

5. Highlight **Calculate**, and press **ENTER**.

6. The output includes the test statistic and the *p*-value. (See **Screen 11.5**.)

Now compare the χ^2-value with the critical value of χ^2 or the *p*-value from Screen 11.5 with α and make decision.

Minitab

The Minitab Technology Instructions feature of this text is written for Minitab version 17. Some screens, menus, and functions will be slightly different in older versions of Minitab.

Performing a Chi-Square Goodness of Fit Test for Example 11–3 of the Text

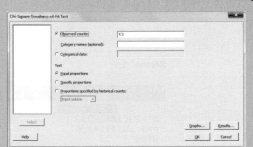

Screen 11.6

1. Enter the observed counts from Example 11–3 into C1.

2. Select **Stat > Tables > Chi-Square Goodness-of-Fit Test (One Variable)**.

3. Use the following settings in the dialog box that appears on screen (see **Screen 11.6**):

   - Select **Observed counts** and type **C1** in the box.

   - Select **Equal proportions** at the **Test** submenu.

   Note: If the alternative hypothesis does not specify equal proportions, go to the worksheet and type the proportions in C2. Return to the dialog box, select **Specific proportions** at the **Test** submenu, and type **C2** in the box.

4. Click **OK**.

5. The output, including the test statistic and *p*-value, will be displayed in the Session window. (See **Screen 11.7**.)

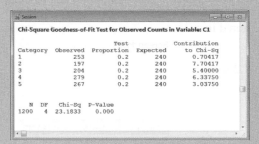

Screen 11.7

   Note: By default, Minitab will also generate two different bar graphs: one of the observed and expected counts and another of the $(O-E)^2/E$ values, which are called the contributions to the Chi-Square statistic. These graphs are not shown here.

Now compare the χ^2-value with the critical value of χ^2 or the *p*-value from Screen 11.7 with α and make a decision.

Performing a Chi-Square Independence/Homogeneity Test for Example 11–6 of the Text

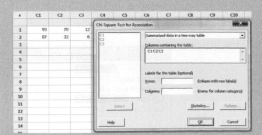

Screen 11.8

1. Enter the contingency table from Example 11–6 into the first two rows of C1 through C3. (See **Screen 11.8**.)

2. Select **Stat > Tables > Chi-Square Test for Association**.

3. Use the following settings in the dialog box that appears on screen (see **Screen 11.8**):

   - Select **Summarized data in a two-way table** from the drop-down menu.

   Note: For raw data in C1 and C2, select **Raw data (categorical variables)** from the drop-down menu, type C2 in the **Rows** box, and C1 in the **Columns** box. Then go to step 4.

   - Type C1-C3 in the **Columns containing the table** box.

4. Click **OK**.

5. The output, including the test statistic and *p*-value, will be displayed in the Session window. (See **Screen 11.9**.)

Now compare the χ^2-value with the critical value of χ^2 or the *p*-value from Screen 11.9 with α and make a decision.

Testing a Hypothesis About σ^2 for Example 11–10 of the Text

1. Select **Stat > Basic Statistics > 1 Variance**.

2. Use the following settings in the dialog box that appears on screen:

- Select **Sample variance** from the drop-down menu.

  Note: If you have the data in a column, select **One or more samples, each in a column**, type the column name in the next box, and move to step 3 below.

- Type 25 in the **Sample size** box.

- Type .029 in the **Sample variance** box.

- Check the **Perform hypothesis test** checkbox.

- Select **Hypothesized variance** from the drop-down menu.

- Type .015 in the **Value** box.

3. Select **Options** and use the following settings in the dialog box that appears on screen:

- Type 95 in the **Confidence level** box.

- Select Variance > hypothesized variance from the **Alternative hypothesis** drop-down menu.

4. Click **OK** in both dialog boxes.

5. The output, including the test statistic and *p*-value, will be displayed in the Session window. The output also includes a confidence interval, but it is a one-sided confidence interval, which is beyond the scope of this text.

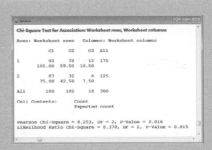

Screen 11.9

Excel

The Excel Technology Instructions feature of this text is written for Excel 2013. Some screens, menus, and functions will be slightly different in older versions of Excel. To enable some of the advanced Excel functions, you must enable the Data Analysis Add-In for Excel.

Performing a Chi-Square Goodness of Fit Test for Example 11–3 of the Text

1. In a new worksheet, enter the following text into the indicated cells (see **Screen 11.10**):

- Type Observed in cell A1, Expected in cell B1, and Contribution in cell C1.

- Type Chi-square in cell E1 and *p*-value in cell E2.

2. Enter the observed counts in cells A2-A6 and the expected counts in cells B2-B6. (See **Screen 11.10**.)

3. Enter the formula =(A2-B2)^2/B2 in cell C2. Copy the formula from cell C2 to cells C3-C6. (See **Screen 11.10**.)

4. Enter the formula =SUM(C2:C6) in cell F1. The output is the value of the test statistic. (See **Screen 11.10**.)

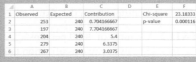

Screen 11.10

5. Enter the formula =**CHISQ.DIST.RT(F1,4)** in cell F2. The output is the *p*-value. (See **Screen 11.10**.)

Note: In this formula, 4 is the degrees of freedom. Adjust this value as needed for other tests.

Now compare the χ^2-value with the critical value of χ^2 or the *p*-value from Screen 11.10 with α and make a decision.

Performing a Chi-Square Independence/Homogeneity Test for Example 11–6 of the Text

1. In a new worksheet, enter Observed in cell A1, Expected in cell E1, and *p*-value in cell I1. (See **Screen 11.11**.)

2. Enter the observed counts in cells A2-C3 and the expected counts in cells E2-G3. (See **Screen 11.11**.)

3. Enter the formula =**CHISQ.TEST(A2:C3,E2:G3)** in cell J1. The output is the *p*-value. (See **Screen 11.11**.)

Now compare the *p*-value from Screen 11.11 with α and make a decision.

	A	B	C	D	E	F	G	H	I	J
1	Observed				Expected				p-value	0.016141
2	93	70	12		105	59.5	10.5			
3	87	32	6		75	42.5	7.5			

Screen 11.11

TECHNOLOGY ASSIGNMENTS

TA11.1 Air Quality Index (AQI) data for the city of Kitchener, Ontario, Canada, during the period January 1, 2010, to June 21, 2012, produced the following percentage distribution:

AQI	Very good	Good	Moderate	Poor
Percentage	13.10	71.81	14.76	0.33

Source: www.airqualityontario.com.

The following table gives the AQI data for a sample of 1100 readings from cities similar to Kitchener (population 150,000 to 250,000, metropolitan area population of 400,000 to 500,000):

AQI	Very good	Good	Moderate	Poor
Number of readings	127	816	150	7

Test at a 5% significance level whether the distribution of AQI for the sample data differs from the distribution for Kitchener, Ontario.

TA11.2 A sample of 4000 persons aged 18 years and older produced the following two-way classification table:

	Men	Women
Single	531	357
Married	1375	1179
Widowed	55	195
Divorced	139	169

Test at a 1% significance level whether gender and marital status are dependent for all persons aged 18 years and older.

TA11.3 Two samples, one of 3000 students from urban high schools and another of 2000 students from rural high schools, were taken. These students were asked if they have ever smoked. The following table lists the summary of the results.

	Urban	Rural
Have never smoked	1448	1228
Have smoked	1552	772

Using a 5% significance level, test the null hypothesis that the proportions of urban and rural students who have smoked and who have never smoked are homogeneous.

TA11.4 Refer to Data Set IV on the Manchester Road Race that accompanies this text (see Appendix A). Select a random sample of 500 participants. Perform a hypothesis test with the null hypothesis that the gender of a participant (listed in column 6) is independent of whether the person is from Connecticut or not from Connecticut (listed in column 9).

CHAPTER **12**

Tyler Olson/Shutterstock

Analysis of Variance

Trying something new can be risky, and there can be uncertainty about the results. Suppose a school district plans to test three different methods for teaching arithmetic. After teachers implement these different methods for a semester, administrators want to know if the mean scores of students taught with these three different methods are all the same. What data will they require and how will they test for this equality of more than two means? (See Examples 12–2 and 12–3.)

12.1 The *F* Distribution

12.2 One-Way Analysis of Variance

Chapter 10 described the procedures that are used to test hypotheses about the difference between two population means using the normal and *t* distributions. Also described in that chapter were the hypothesis-testing procedures for the difference between two population proportions using the normal distribution. Then, Chapter 11 explained the procedures that are used to test hypotheses about the equality of more than two population proportions using the chi-square distribution.

This chapter explains how to test the null hypothesis that the means of more than two populations are equal. For example, suppose that teachers at a school have devised three different methods to teach arithmetic. They want to find out if these three methods produce different mean scores. Let μ_1, μ_2, and μ_3 be the mean scores of all students who will be taught by Methods I, II, and III, respectively. To test whether or not the three teaching methods produce the same mean, we test the null hypothesis

$$H_0: \mu_1 = \mu_2 = \mu_3 \quad \text{(All three population means are equal.)}$$

against the alternative hypothesis

$$H_1: \text{Not all three population means are equal.}$$

We use the analysis of variance procedure to perform such a test of hypothesis.

Note that the analysis of variance procedure can be used to compare two population means. However, the procedures learned in Chapter 10 are more efficient for performing tests of hypothesis about the difference between two population means; the analysis of variance procedure, to be discussed in this chapter, is used to compare three or more population means.

An *analysis of variance* test is performed using the F distribution. First, the F distribution is described in Section 12.1 of this chapter. Then, Section 12.2 discusses the application of the one-way analysis of variance procedure to perform tests of hypothesis.

12.1 | The *F* Distribution

Like the chi-square distribution, the shape of a particular **F distribution**[1] curve depends on the number of degrees of freedom. However, the F distribution has *two* numbers of degrees of freedom: **degrees of freedom for the numerator** and **degrees of freedom for the denominator**. These two numbers representing two types of degrees of freedom are the *parameters of the F distribution*. Each combination of degrees of freedom for the numerator and for the denominator gives a different F distribution curve. The values of an F distribution are denoted by F, which assumes only nonnegative values. Like the normal, t, and chi-square distributions, the F distribution is a continuous distribution. The shape of an F distribution curve is skewed to the right, but the skewness decreases as both numbers of degrees of freedom increase.

> **The *F* Distribution**
>
> 1. The F *distribution* is continuous and skewed to the right.
> 2. The F distribution has two numbers of degrees of freedom: *df* for the numerator and *df* for the denominator.
> 3. The units (the values of the F-variable) of an F distribution, denoted by F, are nonnegative.

For an F distribution, degrees of freedom for the numerator and degrees of freedom for the denominator are usually written as follows:

$$df = (8, 14)$$

First number denotes the *df* for the numerator   Second number denotes the *df* for the denominator

Figure 12.1 shows three F distribution curves for three sets of degrees of freedom for the numerator and for the denominator. In the figure, the first number gives the degrees of freedom associated with the numerator, and the second number gives the degrees of freedom associated with the denominator. We can observe from this figure that as both degrees of freedom increase, the peak of the curve moves to the right; that is, the skewness decreases.

Figure 12.1 Three F distribution curves.

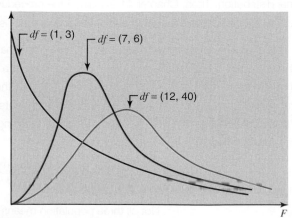

---
[1]The F distribution is named after Sir Ronald Fisher.

Table VII in Appendix B lists the values of *F* for the *F* distribution. To read Table VII, we need to know three quantities: the degrees of freedom for the numerator, the degrees of freedom for the denominator, and an area in the right tail of an *F* distribution curve. Note that the *F* distribution table (Table VII) is read only for an area in the right tail of the *F* distribution curve. Also note that Table VII has four parts. These four parts give the *F* values for areas of .01, .025, .05, and .10, respectively, in the right tail of the *F* distribution curve. We can make the *F* distribution table for other values in the right tail. Example 12–1 illustrates how to read Table VII.

EXAMPLE 12–1

Reading the F distribution table.

Find the *F* value for 8 degrees of freedom for the numerator, 14 degrees of freedom for the denominator, and .05 area in the right tail of the *F* distribution curve.

Solution To find the required value of *F*, we use the portion of Table VII of Appendix B that corresponds to .05 area in the right tail of the *F* distribution curve. The relevant portion of that table is shown here as Table 12.1. To find the required *F* value, we locate 8 in the row for degrees of freedom for the numerator (at the top of Table VII) and 14 in the column for degrees of freedom for the denominator (the first column on the left side in Table VII). The entry where the column for 8 and the row for 14 intersect gives the required *F* value. This value of *F* is **2.70**, as shown in Table 12.1 and Figure 12.2. The *F* value obtained from this table for a test of hypothesis is called the critical value of *F*.

Table 12.1   Obtaining the *F* Value From Table VII

		Degrees of Freedom for the Numerator				
		1	2	...	8	...	100
	1	161.5	199.5	...	238.9	...	253.0
Degrees of Freedom for	2	18.51	19.00	...	19.37	...	19.49
the Denominator	.	...	...	...	...	...	...
	.	...	...	...	...	...	...
	14	4.60	3.74	...	2.70	...	2.19
	.	...	...	...	...	...	...
	.	...	...	...	...	...	...
	100	3.94	3.09	...	2.03	...	1.39

The *F* value for 8 *df* for the numerator, 14 *df* for the denominator, and .05 area in the right tail

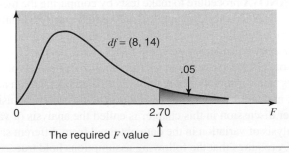

Figure 12.2 The value of *F* from Table VII for 8 *df* for the numerator, 14 *df* for the denominator, and .05 area in the right tail.

$df = (8, 14)$

.05

2.70

The required *F* value

EXERCISES

CONCEPTS AND PROCEDURES

12.1 Describe the main characteristics of an *F* distribution.

12.2 Find the critical value of *F* for the following.

   a. $df = (5, 5)$ and area in the right tail = .05
   b. $df = (5, 10)$ and area in the right tail = .05
   c. $df = (5, 40)$ and area in the right tail = .05

12.3 Find the critical value of *F* for the following.

   a. $df = (2, 8)$ and area in the right tail = .025
   b. $df = (8, 8)$ and area in the right tail = .025
   c. $df = (20, 8)$ and area in the right tail = .025

12.4 Determine the critical value of *F* for the following.

   a. $df = (4, 8)$ and area in the right tail = .01

b. $df = (4, 30)$ and area in the right tail $= .01$
c. $df = (4, 100)$ and area in the right tail $= .01$

12.5 Find the critical value of F for an F distribution with $df = (11, 5)$ and

    a. area in the right tail $= .01$
    b. area in the right tail $= .025$

12.6 Find the critical value of F for an F distribution with .01 area in the right tail and

    a. $df = (20, 20)$
    b. $df = (8, 24)$

12.2 | One-Way Analysis of Variance

As mentioned in the beginning of this chapter, the analysis of variance procedure is used to test the null hypothesis that the means of three or more populations are the same against the alternative hypothesis that not all population means are the same. The analysis of variance procedure can be used to compare two population means. However, the procedures learned in Chapter 10 are more efficient for performing tests of hypotheses about the difference between two population means; the analysis of variance procedure is used to compare three or more population means.

Reconsider the example of teachers at a school who have devised three different methods to teach arithmetic. They want to find out if these three methods produce different mean scores. Let μ_1, μ_2, and μ_3 be the mean scores of all students who are taught by Methods I, II, and III, respectively. To test if the three teaching methods produce different means, we test the null hypothesis

$$H_0\text{: } \mu_1 = \mu_2 = \mu_3 \quad \text{(All three population means are equal.)}$$

against the alternative hypothesis

$$H_1\text{: Not all three population means are equal.}$$

> **ANOVA** is a procedure that is used to test the null hypothesis that the means of three or more populations are all equal.

One method to test such a hypothesis is to test the three hypotheses $H_0\text{: } \mu_1 = \mu_2$, $H_0\text{: } \mu_1 = \mu_3$, and $H_0\text{: } \mu_2 = \mu_3$ separately using the procedure discussed in Chapter 10. Besides being time consuming, such a procedure has other disadvantages. First, if we reject even one of these three hypotheses, then we must reject the null hypothesis $H_0\text{: } \mu_1 = \mu_2 = \mu_3$. Second, combining the Type I error probabilities for the three tests (one for each test) will give a very large Type I error probability for the test $H_0\text{: } \mu_1 = \mu_2 = \mu_3$. Hence, we should prefer a procedure that can test the equality of three means in one test. The **ANOVA**, short for **analysis of variance**, provides such a procedure. It is used to compare three or more population means in a single test. Note that if the null hypothesis is rejected, it does not necessarily imply that all three of the means are different or unequal. It could imply that one mean is different from the other two means, or that all three means are different, or that two means are significantly different from each other, but neither is significantly different from the third mean.

This section discusses the **one-way ANOVA** procedure to make tests by comparing the means of several populations. By using a one-way ANOVA test, we analyze only **one factor or variable**. For instance, in the example of testing for the equality of mean arithmetic scores of students taught by each of the three different methods, we are considering only one factor, which is the effect of different teaching methods on the scores of students. Sometimes we may analyze the effects of two factors. For example, if different teachers teach arithmetic using these three methods, we can analyze the effects of teachers and teaching methods on the scores of students. This is done by using a two-way ANOVA. The procedure under discussion in this chapter is called the analysis of variance because the test is based on the analysis of variation in the data obtained from different samples. The application of one-way ANOVA requires that the following assumptions hold true.

> **Assumptions of One-Way ANOVA**   The following assumptions must hold true to use *one-way ANOVA*.
>
> 1. The populations from which the samples are drawn are approximately normally distributed.
> 2. The populations from which the samples are drawn have the same variance (or standard deviation).
> 3. The samples drawn from different populations are random and independent.

For instance, in the example about three methods of teaching arithmetic, we first assume that the scores of all students taught by each method are approximately normally distributed. Second, the means of the distributions of scores for the three teaching methods may or may not be the same, but all three distributions have the same variance, σ^2. Third, when we take samples to perform an ANOVA test, these samples are drawn independently and randomly from three different populations.

The ANOVA test is applied by calculating two estimates of the variance, σ^2, of population distributions: the **variance between samples** and the **variance within samples**. The variance between samples is also called the **mean square between samples** or **MSB**. The variance within samples is also called the **mean square within samples** or **MSW**.

The variance between samples, MSB, gives an estimate of σ^2 based on the variation among the means of samples taken from different populations. For the example of three teaching methods, MSB will be based on the values of the mean scores of three samples of students taught by three different methods. If the means of all populations under consideration are equal, the means of the respective samples will still be different, but the variation among them is expected to be small, and, consequently, the value of MSB is expected to be small. However, if the means of populations under consideration are not all equal, the variation among the means of the respective samples is expected to be large, and, consequently, the value of MSB is expected to the large.

The variance within samples, MSW, gives an estimate of σ^2 based on the variation within the data of different samples. For the example of three teaching methods, MSW will be based on the scores of individual students included in the three samples taken from three populations. The concept of MSW is similar to the concept of the pooled standard deviation, s_p, for two samples discussed in Section 10.2 of Chapter 10.

The one-way ANOVA test is always right-tailed with the rejection region in the right tail of the F distribution curve. The hypothesis-testing procedure using ANOVA involves the same five steps that were used in earlier chapters. The next subsection explains how to calculate the value of the test statistic F for an ANOVA test.

12.2.1   Calculating the Value of the Test Statistic

The value of the test statistic F for a test of hypothesis using ANOVA is given by the ratio of two variances, the variance between samples (MSB) and the variance within samples (MSW).

Test Statistic F for a One-Way Anova Test   The value of the **test statistic F** for an ANOVA test is calculated as

$$F = \frac{\text{Variance between samples}}{\text{Variance within samples}} \quad \text{or} \quad \frac{\text{MSB}}{\text{MSW}}$$

The calculation of MSB and MSW is explained in Example 12–2.

Example 12–2 describes the calculation of MSB, MSW, and the value of the test statistic F. Since the basic formulas are laborious to use, they are not presented here. We have used only the short-cut formulas to make calculations in this chapter.

EXAMPLE 12–2   Three Methods of Teaching Arithmetic

Calculating the value of the test statistic F.

Fifteen fourth-grade students were randomly assigned to three groups to experiment with three different methods of teaching arithmetic. At the end of the semester, the same test was given to all 15 students. The following table gives the scores of students in the three groups.

Method I	Method II	Method III
48	55	84
73	85	68
51	70	95
65	69	74
87	90	67

Calculate the value of the test statistic F. Assume that all the required assumptions mentioned in Section 12.2 hold true.

Solution In ANOVA terminology, the three methods used to teach arithmetic are called **treatments**. The table contains data on the scores of fourth-graders included in the three samples. Each sample of students is taught by a different method. Let

$$x = \text{the score of a student}$$
$$k = \text{the number of different samples (or treatments)}$$
$$n_i = \text{the size of sample } i$$
$$T_i = \text{the sum of the values in sample } i$$
$$n = \text{the number of values in all samples} = n_1 + n_2 + n_3 + \cdots$$
$$\Sigma x = \text{the sum of the values in all samples} = T_1 + T_2 + T_3 + \cdots$$
$$\Sigma x^2 = \text{the sum of the squares of the values in all samples}$$

To calculate MSB and MSW, we first compute the **between-samples sum of squares**, denoted by **SSB**, and the **within-samples sum of squares**, denoted by **SSW**. The sum of SSB and SSW is called the **total sum of squares** and is denoted by **SST**; that is,

$$\text{SST} = \text{SSB} + \text{SSW}$$

The values of SSB and SSW are calculated using the following formulas.

Between- and Within-Samples Sums of Squares The **between-samples sum of squares**, denoted by SSB, is calculated as

$$\text{SSB} = \left(\frac{T_1^2}{n_1} + \frac{T_2^2}{n_2} + \frac{T_3^2}{n_3} + \cdots \right) - \frac{(\Sigma x)^2}{n}$$

The **within-samples sum of squares**, denoted by SSW, is calculated as

$$\text{SSW} = \Sigma x^2 - \left(\frac{T_1^2}{n_1} + \frac{T_2^2}{n_2} + \frac{T_3^2}{n_3} + \cdots \right)$$

Table 12.2 lists the scores of 15 students who were taught arithmetic by each of the three different methods; the values of T_1, T_2, and T_3; and the values of n_1, n_2, and n_3.

In Table 12.2, T_1 is obtained by adding the five scores of the first sample. Thus, $T_1 = 48 + 73 + 51 + 65 + 87 = 324$. Similarly, the sums of the values in the second and third samples give $T_2 = 369$ and $T_3 = 388$, respectively. Because there are five observations in each sample, $n_1 = n_2 = n_3 = 5$. The values of Σx and n are, respectively,

$$\Sigma x = T_1 + T_2 + T_3 = 324 + 369 + 388 = 1081$$
$$n = n_1 + n_2 + n_3 = 5 + 5 + 5 = 15$$

Table 12.2

Method I	Method II	Method III
48	55	84
73	85	68
51	70	95
65	69	74
87	90	67
$T_1 = 324$	$T_2 = 369$	$T_3 = 388$
$n_1 = 5$	$n_2 = 5$	$n_3 = 5$

To calculate $\sum x^2$, we square all the scores included in all three samples and then add them. Thus,

$$\sum x^2 = (48)^2 + (73)^2 + (51)^2 + (65)^2 + (87)^2 + (55)^2 + (85)^2 + (70)^2$$
$$+ (69)^2 + (90)^2 + (84)^2 + (68)^2 + (95)^2 + (74)^2 + (67)^2$$
$$= 80,709$$

Substituting all the values in the formulas for SSB and SSW, we obtain the following values of SSB and SSW:

$$SSB = \left(\frac{(324)^2}{5} + \frac{(369)^2}{5} + \frac{(388)^2}{5}\right) - \frac{(1081)^2}{15} = 432.1333$$

$$SSW = 80,709 - \left(\frac{(324)^2}{5} + \frac{(369)^2}{5} + \frac{(388)^2}{5}\right) = 2372.8000$$

The value of SST is obtained by adding the values of SSB and SSW. Thus,

$$SST = 432.1333 + 2372.8000 = 2804.9333$$

The variance between samples (MSB) and the variance within samples (MSW) are calculated using the following formulas.

Calculating the Values of MSB and MSW   MSB and MSW are calculated as, respectively,

$$MSB = \frac{SSB}{k - 1} \quad \text{and} \quad MSW = \frac{SSW}{n - k}$$

where $k - 1$ and $n - k$ are, respectively, the *df* for the numerator and the *df* for the denominator for the F distribution. Remember, k is the number of different samples.

Consequently, the variance between samples is

$$MSB = \frac{SSB}{k - 1} = \frac{432.1333}{3 - 1} = 216.0667$$

The variance within samples is

$$MSW = \frac{SSW}{n - k} = \frac{2372.8000}{15 - 3} = 197.7333$$

The value of the test statistic F is given by the ratio of MSB and MSW. Therefore,

$$F = \frac{MSB}{MSW} = \frac{216.0667}{197.7333} = \mathbf{1.09}$$

For convenience, all these calculations are often recorded in a table called the *ANOVA table*. Table 12.3 gives the general form of an ANOVA table.

Table 12.3   **ANOVA Table**

Source of Variation	Degrees of Freedom	Sum of Squares	Mean Square	Value of the Test Statistic
Between	$k - 1$	SSB	MSB	
Within	$n - k$	SSW	MSW	$F = \dfrac{\text{MSB}}{\text{MSW}}$
Total	$n - 1$	SST		

Substituting the values of the various quantities into Table 12.3, we write the ANOVA table for our example as Table 12.4.

Table 12.4   **ANOVA Table for Example 12–2**

Source of Variation	Degrees of Freedom	Sum of Squares	Mean Square	Value of the Test Statistic
Between	2	432.1333	216.0667	
Within	12	2372.8000	197.7333	$F = \dfrac{216.0667}{197.7333} = 1.09$
Total	14	2804.9333		

12.2.2   One-Way ANOVA Test

Now suppose we want to test the null hypothesis that the mean scores are equal for all three groups of fourth-graders taught by three different methods of Example 12–2 against the alternative hypothesis that the mean scores of all three groups are not equal. Note that in a one-way ANOVA test, the null hypothesis is that the means for all populations are equal. The alternative hypothesis is that not all population means are equal. In other words, the alternative hypothesis states that at least one of the population means is different from the others. Example 12–3 demonstrates how we use the one-way ANOVA procedure to make such a test.

EXAMPLE 12–3	Three Methods of Teaching Arithmetic

Performing a one-way ANOVA test: all samples the same size.

Reconsider Example 12–2 about the scores of 15 fourth-grade students who were randomly assigned to three groups in order to experiment with three different methods of teaching arithmetic. At a 1% significance level, can we reject the null hypothesis that the mean arithmetic score of all fourth-grade students taught by each of these three methods is the same? Assume that all the assumptions required to apply the one-way ANOVA procedure hold true.

Solution   To make a test about the equality of the means of three populations, we follow our standard procedure with five steps.

Step 1.   *State the null and alternative hypotheses.*

Let μ_1, μ_2, and μ_3 be the mean arithmetic scores of all fourth-grade students who are taught, respectively, by Methods I, II, and III. The null and alternative hypotheses are

$H_0: \mu_1 = \mu_2 = \mu_3$ (The mean scores of the three groups are all equal.)

H_1: Not all three means are equal.

Note that the alternative hypothesis states that at least one population mean is different from the other two.

Step 2.   *Select the distribution to use.*

Because we are comparing the means for three normally distributed populations and all of the assumption required to apply ANOVA procedure are satisfied, we use the F distribution to make this test.

PhotoDisc, Inc./Getty Images

Step 3. *Determine the rejection and nonrejection regions.*

The significance level is .01. Because a one-way ANOVA test is always right-tailed, the area in the right tail of the *F* distribution curve is .01, which is the rejection region in Figure 12.3.

Next we need to know the degrees of freedom for the numerator and the denominator. In our example, the students were assigned to three different methods. As mentioned earlier, these methods are called treatments. The number of treatments is denoted by k. The total number of observations in all samples taken together is denoted by n. Then, the number of degrees of freedom for the numerator is equal to $k - 1$ and the number of degrees of freedom for the denominator is equal to $n - k$. In our example, there are 3 treatments (methods of teaching) and 15 total observations (total number of students) in all 3 samples. Thus,

$$\text{Degrees of freedom for the numerator} = k - 1 = 3 - 1 = 2$$

$$\text{Degrees of freedom for the denominator} = n - k = 15 - 3 = 12$$

From Table VII of Appendix B, we find the critical value of *F* for 2 *df* for the numerator, 12 *df* for the denominator, and .01 area in the right tail of the *F* distribution curve. This value of *F* is 6.93, as shown in Figure 12.3.

Thus, we will fail to reject H_0 if the calculated value of the test statistic *F* is less than 6.93, and we will reject H_0 if it is 6.93 or larger.

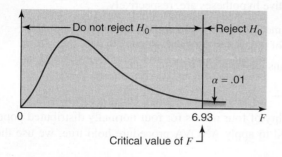

Figure 12.3 Critical value of *F* for $df = (2, 12)$ and $\alpha = .01$.

Step 4. *Calculate the value of the test statistic.*

We computed the value of the test statistic *F* for these data in Example 12–2. This value is

$$F = 1.09$$

Step 5. *Make a decision.*

Because the value of the test statistic $F = 1.09$ is less than the critical value of $F = 6.93$, it falls in the nonrejection region. Hence, we fail to reject the null hypothesis, and conclude that the means of the three populations seem to be equal. In other words, the three different methods of teaching arithmetic do not seem to affect the mean scores of students. The difference in the three mean scores in the case of our three samples may have occurred only because of sampling error.

In Example 12–3, the sample sizes were the same for all treatments. Example 12–4 describes a case in which the sample sizes are not the same for all treatments.

EXAMPLE 12–4   Bank Tellers Serving Customers

Performing a one-way ANOVA test: all samples not the same size.

From time to time, unknown to its employees, the research department at Post Bank observes various employees for their work productivity. Recently this department wanted to check whether the four tellers at a branch of this bank serve, on average, the same number of customers per hour. The research manager observed each of the four tellers for a certain number of hours. The following table gives the number of customers served by the four tellers during each of the observed hours.

Teller A	Teller B	Teller C	Teller D
19	14	11	24
21	16	14	19
26	14	21	21
24	13	13	26
18	17	16	20
	13	18	

At a 5% significance level, test the null hypothesis that the mean number of customers served per hour by each of these four tellers is the same. Assume that all the assumptions required to apply the one-way ANOVA procedure hold true.

Solution   To make a test about the equality of means of four populations, we follow our standard procedure with five steps.

Step 1.   *State the null and alternative hypotheses.*

Let μ_1, μ_2, μ_3, and μ_4 be the mean number of customers served per hour by tellers A, B, C, and D, respectively. The null and alternative hypotheses are, respectively,

$$H_0: \mu_1 = \mu_2 = \mu_3 = \mu_4 \quad \text{(The mean number of customers served per hour by each of the four tellers is the same.)}$$

H_1: Not all four population means are equal.

Step 2.   *Select the distribution to use.*

Because we are testing for the equality of four means for four normally distributed populations and all of the assumptions required to apply ANOVA procedure hold true, we use the F distribution to make the test.

Step 3.   *Determine the rejection and nonrejection regions.*

The significance level is .05, which means the area in the right tail of the F distribution curve is .05. In this example, there are 4 treatments (tellers) and 22 total observations in all four samples. Thus,

$$\text{Degrees of freedom for the numerator} = k - 1 = 4 - 1 = 3$$
$$\text{Degrees of freedom for the denominator} = n - k = 22 - 4 = 18$$

The critical value of F from Table VII for 3 *df* for the numerator, 18 *df* for the denominator, and .05 area in the right tail of the F distribution curve is 3.16. This value is shown in Figure 12.4.

Figure 12.4   Critical value of F for $df = (3, 18)$ and $\alpha = .05$.

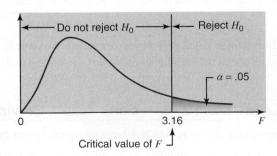

Step 4.   *Calculate the value of the test statistic.*

First we calculate SSB and SSW. Table 12.5 lists the number of customers served by each of the four tellers during the selected hours; the values of T_1, T_2, T_3, and T_4; and the values of n_1, n_2, n_3, and n_4.

Table 12.5

Teller A	Teller B	Teller C	Teller D
19	14	11	24
21	16	14	19
26	14	21	21
24	13	13	26
18	17	16	20
	13	18	
$T_1 = 108$	$T_2 = 87$	$T_3 = 93$	$T_4 = 110$
$n_1 = 5$	$n_2 = 6$	$n_3 = 6$	$n_4 = 5$

The values of $\sum x$ and n are, respectively,

$$\sum x = T_1 + T_2 + T_3 + T_4 = 108 + 87 + 93 + 110 = 398$$

$$n = n_1 + n_2 + n_3 + n_4 = 5 + 6 + 6 + 5 = 22$$

The value of $\sum x^2$ is calculated as follows:

$$\sum x^2 = (19)^2 + (21)^2 + (26)^2 + (24)^2 + (18)^2 + (14)^2 + (16)^2 + (14)^2$$
$$+ (13)^2 + (17)^2 + (13)^2 + (11)^2 + (14)^2 + (21)^2 + (13)^2$$
$$+ (16)^2 + (18)^2 + (24)^2 + (19)^2 + (21)^2 + (26)^2 + (20)^2$$
$$= 7614$$

Substituting all the values in the formulas for SSB and SSW, we obtain the following values of SSB and SSW:

$$\text{SSB} = \left(\frac{T_1^2}{n_1} + \frac{T_2^2}{n_2} + \frac{T_3^2}{n_3} + \frac{T_4^2}{n_4} \right) - \frac{(\sum x)^2}{n}$$

$$= \left(\frac{(108)^2}{5} + \frac{(87)^2}{6} + \frac{(93)^2}{6} + \frac{(110)^2}{5} \right) - \frac{(398)^2}{22} = 255.6182$$

$$\text{SSW} = \sum x^2 - \left(\frac{T_1^2}{n_1} + \frac{T_2^2}{n_2} + \frac{T_3^2}{n_3} + \frac{T_4^2}{n_4} \right)$$

$$= 7614 - \left(\frac{(108)^2}{5} + \frac{(87)^2}{6} + \frac{(93)^2}{6} + \frac{(110)^2}{5} \right) = 158.2000$$

Hence, the variance between samples MSB and the variance within samples MSW are, respectively,

$$\text{MSB} = \frac{\text{SSB}}{k - 1} = \frac{255.6182}{4 - 1} = 85.2061$$

$$\text{MSW} = \frac{\text{SSW}}{n - k} = \frac{158.2000}{22 - 4} = 8.7889$$

The value of the test statistic F is given by the ratio of MSB and MSW, which is

$$F = \frac{\text{MSB}}{\text{MSW}} = \frac{85.2061}{8.7889} = 9.69$$

Writing the values of the various quantities in the ANOVA table, we obtain Table 12.6 given on the next page.

Step 5. *Make a decision.*

Because the value of the test statistic $F = 9.69$ is greater than the critical value of $F = 3.16$, it falls in the rejection region. Consequently, we reject the null hypothesis, and conclude that the mean number of customers served per hour by each of the four tellers is not the same. In other

Table 12.6   ANOVA Table for Example 12–4

Source of Variation	Degrees of Freedom	Sum of Squares	Mean Square	Value of the Test Statistic
Between	3	255.6182	85.2061	$F = \dfrac{85.2061}{8.7889} = 9.69$
Within	18	158.2000	8.7889	
Total	21	413.8182		

words, for at least one of the tellers, the mean number of customers served per hour is different from that of the other three tellers.

Note: What if the Sample Size is Large and the Number of *df* Are Not in the *F* Distribution Table?

In this chapter, we used the *F* distribution to perform tests of hypothesis about the equality of population means for three or more populations. If we use technology to perform such tests, it does not matter how large the *df* (degrees of freedom) for the numerator and denominator are. However, when we use the *F* distribution table (Table VII in Appendix B), sometime we may not find the exact *df* for the numerator and/or for the denominator in this table, especially when either of these *df* are large. In such cases, we use the following alternative.

If the number of *df* is not given in the table, use the closest number of *df* that falls below the actual value of *df*. For example, if an ANOVA problem has 4 *df* for the numerator and 47 *df* for the denominator, we will use 4 *df* for the numerator and 40 *df* for the denominator to obtain the critical value of *F* from the table. As long as the number of *df* for the denominator is 3 or larger, the critical values of *F* become smaller as the numbers of *df* increase. Hence, whenever the observed value of *F* falls in the rejection region for a smaller number of *df*, it will fall in the rejection region for the larger number of *df* also.

EXERCISES

CONCEPTS AND PROCEDURES

12.7   Briefly explain when a one-way ANOVA procedure is used to make a test of hypothesis.

12.8   Describe the assumptions that must hold true to apply the one-way analysis of variance procedure to test hypotheses.

12.9   Consider the following data obtained for two samples selected at random from two populations that are independent and normally distributed with equal variances.

Sample I	Sample II
30	29
28	38
33	30
19	38
25	36
36	28

a. Calculate the means and standard deviations for these samples using the formulas from Chapter 3.
b. Using the procedure learned in Section 10.2 of Chapter 10, test at a 1% significance level whether the means of the populations from which these samples are drawn are equal.

c. Using the one-way ANOVA procedure, test at a 1% significance level whether the means of the populations from which these samples are drawn are equal.
d. Are the conclusions reached in parts b and c the same?

12.10   The following ANOVA table, based on information obtained for four samples selected from four independent populations that are normally distributed with equal variances, has a few missing values.

Source of Variation	Degrees of Freedom	Sum of Squares	Mean Square	Value of the Test Statistic
Between				
Within	15		9.2154	$F = \underline{\quad\quad} = 4.07$
Total	18			

a. Find the missing values and complete the ANOVA table.
b. Using $\alpha = .05$, what is your conclusion for the test with the null hypothesis that the means of the four populations are all equal against the alternative hypothesis that the means of the four populations are not all equal?

APPLICATIONS

For the following exercises assume that all the assumptions required to apply the one-way ANOVA procedure hold true.

12.11 People who have home gaming systems, such as Wii™, Playstation™, and Xbox™, are well aware of how quickly they need to replace the batteries in the remote controls. A consumer agency decided to test three major brands of alkaline batteries to determine whether they differ in their average lifetimes in these remotes. For each of the three brands, 10 sets of batteries were placed in the remotes, and people played games until the batteries died. The following ANOVA table contains some of the values from the analysis.

Source of Variation	Degrees of Freedom	Sum of Squares	Mean Square	Value of the Test Statistic
Between			25711.60	
Within		22388.25		$F = \underline{\quad} =$
Total				

Assume that the three populations are normally distributed with equal variances.

    a. Find the missing values, and complete the ANOVA table.
    b. What are the appropriate hypotheses (null and alternative) for this analysis? Using $\alpha = .05$, what is your conclusion about the equality of the three population means?

12.12 A local "pick-your-own" farmer decided to grow blueberries. The farmer purchased and planted eight plants of each of the four different varieties of highbush blueberries. The yield (in pounds) of each plant was measured in the upcoming year to determine whether the average yields were different for at least two of the four plant varieties. The yields of these plants of the four varieties are given in the following table.

Berkeley	5.13	5.36	5.20	5.15	4.96	5.14	5.54	5.22
Duke	5.31	4.89	5.09	5.57	5.36	4.71	5.13	5.30
Jersey	5.20	4.92	5.44	5.20	5.17	5.24	5.08	5.13
Sierra	5.08	5.30	5.43	4.99	4.89	5.30	5.35	5.26

    a. We are to test the null hypothesis that the mean yields for all such bushes of the four varieties are the same. Write the null and alternative hypotheses.
    b. What are the degrees of freedom for the numerator and the denominator?
    c. Calculate SSB, SSW, and SST.
    d. Show the rejection and nonrejection regions on the F distribution curve for $\alpha = .01$.
    e. Calculate the between-samples and within-samples variances.
    f. What is the critical value of F for $\alpha = .01$?
    g. What is the calculated value of the test statistic F?
    h. Write the ANOVA table for this exercise.
    i. Will you reject the null hypothesis stated in part a at a significance level of 1%?

12.13 Surfer Dude swimsuit company plans to produce a new line of quick-dry swimsuits. Three textile companies are competing for the company's quick-dry fabric contract. To check the fabrics of the three companies, Surfer Dude selected 10 random swatches of fabric from each company, soaked them with water, and then measured the amount of time (in seconds) each swatch took to dry when exposed to sun and a temperature of 80°F. The following table contains the amount of time (in seconds) each of these swatches took to dry.

Company A	756	801	760	777	772	768	808	770	805	824
Company B	791	696	761	760	741	810	770	823	815	850
Company C	773	701	733	740	780	793	794	719	766	743

Using a 5% significance level, test the null hypothesis that the mean drying times for all such fabric produced by the three companies are the same.

12.14 A student who has a 9 A.M. class on Monday, Wednesday, and Friday mornings wants to know if the mean time taken by students to find parking spaces just before 9 A.M. is the same for each of these three days of the week. He randomly selects five weeks and records the time taken to find a parking space on Monday, Wednesday, and Friday of each of these five weeks. These times (in minutes) are given in the following table. Assume that this student is representative of all students who need to find a parking space just before 9 A.M. on these three days.

Monday	Wednesday	Friday
6	9	3
12	12	2
15	5	10
14	14	7
10	13	5

At a 5% significance level, test the null hypothesis that the mean time taken to find a parking space just before 9 A.M. on Monday, Wednesday, and Friday is the same for all students.

12.15 An ophthalmologist is interested in determining whether a golfer's type of vision (far-sightedness, near-sightedness, no prescription) impacts how well he or she can judge distance. Random samples of golfers from these three groups (far-sightedness, near-sightedness, no prescription) were selected, and these golfers were blindfolded and taken to the same location on a golf course. Then each of them was asked to estimate the distance from this location to the pin at the end of the hole. The data (in yards) given in the following table represent how far off the estimates (let us call these errors) of these golfers were from the actual distance. A negative value implies that the person underestimated the distance, and a positive value implies that a person overestimated the distance.

Far-sighted	−11	−9	−8	−10	−3	−11	−8	1	−4
Near-sighted	−2	−5	−7	−8	−6	−9	2	−10	−10
No prescription	−5	1	0	4	3	−2	0	−8	

At a 1% significance level can you reject the null hypothesis that the average errors in predicting distance for all golfers of the three different vision types are the same.

USES AND MISUSES...

DO NOT BE LATE

Imagine that working at your company requires that staff travel frequently. You want to determine if the on-time performance of any one airline is sufficiently different from that of the remaining airlines to warrant a preferred status with your company. The local airport Web site publishes the scheduled and actual departure and arrival times for the four airlines that service it. You decide to perform an ANOVA test on the mean delay times for all airline carriers at the airport. The null hypothesis here is that the mean delay times for Airlines A, B, C, and D are all the same. The results of the ANOVA test tell you not to reject the null hypothesis: All airline carriers have the same mean departure and arrival delay times, so that adopting a preferred status based on the on-time performance is not warranted.

When your boss tells you to redo your analysis, you should not be surprised. The choice to study flights only at the local airport was a good one because your company should be concerned about the performance of an airline at the most convenient airport. A regional airport will have a much different on-time performance profile than a large hub airport. By mixing both arrival and departure data, however, you violated the assumption that the populations are normally distributed. For arrival data, this assumption could be valid: The influence of high-altitude winds, local weather, and the fact that the arrival time is an estimate in the first place result in a distribution of arrival times around the predicted arrival times. However, departure delays are not normally distributed. Because a flight does not leave before its departure time but can leave after, departure delays are skewed to the right. As the statistical methods become more sophisticated, so do the assumptions regarding the characteristics of the data. Careful attention to these assumptions is required.

GLOSSARY

Analysis of variance (ANOVA)   A statistical technique used to test whether the means of three or more populations are all equal.

F distribution   A continuous distribution that has two parameters: df for the numerator and df for the denominator.

Mean square between samples or **MSB**   A measure of the variation among the means of samples taken from different populations.

Mean square within samples or **MSW**   A measure of the variation within the data of all samples taken from different populations.

One-way ANOVA   The analysis of variance technique that analyzes one variable only.

SSB   The sum of squares between samples. Also called the sum of squares of the factor or treatment.

SST   The total sum of squares given by the sum of SSB and SSW.

SSW   The sum of squares within samples. Also called the sum of squares of errors.

SUPPLEMENTARY EXERCISES

For the following exercises, assume that all the assumptions required to apply the one-way ANOVA procedure hold true.

12.16   The following table lists the numbers of violent crimes reported to police on randomly selected days for this year. The data are taken from three large cities of about the same size.

City A	City B	City C
4	2	8
9	4	12
12	1	11
3	13	3
10	8	9
7	6	14
13		

Using a 5% significance level, test the null hypothesis that the mean number of violent crimes reported per day is the same for each of these three cities.

12.17   A music company collects data from customers who purchase CDs and MP3 downloads from them. Each person is asked to state his or her favorite musical genre from the following list: Classic Rock, Country, Hip-Hop/Rap, Jazz, Pop, and R&B. Random samples of customers were selected from each genre. Each customer was asked how much he or she spent (in dollars) on music purchases in the last month. The following table gives the information (in dollars) obtained from these customers.

Classic Rock	22	35	62	17	11	59	43
Country	60	36	59	27	32	56	
Hip-Hop/Rap	35	52	35	55	71	75	
Jazz	13	40	27	38	31	28	22
Pop	40	17	52	59	56	24	55
R&B	24	45	36	65	58	44	51

a. At a 10% significance level, will you reject the null hypothesis that the average monthly expenditures of all customers in each of the six genres are the same?

b. What is the Type I error in this case, and what is the probability of committing such an error? Explain.

12.18 A large company buys thousands of lightbulbs every year. The company is currently considering four brands of lightbulbs to choose from. Before the company decides which lightbulbs to buy, it wants to investigate if the mean lifetimes of the four types of lightbulbs are the same. The company's research department randomly selected a few bulbs of each type and tested them. The following table lists the number of hours (in thousands) that each of the bulbs in each brand lasted before being burned out.

Brand I	Brand II	Brand III	Brand IV
23	19	23	26
24	23	27	24
19	18	25	21
26	24	26	29
22	20	23	28
23	22	21	24
25	19	27	28

At a 2.5% significance level, test the null hypothesis that the mean lifetime of bulbs for each of these four brands is the same.

12.19 A travel magazine took a random sample of students from a midwestern university and asked whether they went to California, Texas, or Florida for 2005 spring break. The students who went somewhere else or did not go anywhere were excluded from the survey. The following table shows the amount of money spent on this trip by a sample of six students for each of these three destinations.

California	Texas	Florida
$ 850	$775	$668
904	808	810
1045	810	844
882	912	787
910	855	705
946	680	908

a. At a 1% level of significance, can you conclude that the mean amount spent by all such students who went to each of these three destinations for 2005 spring break is the same?

b. What is the Type I error in this case and what is the probability of committing such an error? Explain.

12.20 A consumer agency wanted to find out if the mean time taken by each of three brands of medicines to provide relief from a headache is the same. The first drug was administered to six randomly selected patients, the second to four randomly selected patients, and the third to five randomly selected patients. The following table gives the time (in minutes) taken by each patient to get relief from a headache after taking the medicine.

Drug I	Drug II	Drug III
25	15	44
38	21	39
42	19	54
65	25	58
47		73
52		

At a 2.5% significance level, will you reject the null hypothesis that the mean time taken to provide relief from a headache is the same for each of the three drugs?

12.21 The following table lists the prices of certain randomly selected college textbooks in statistics, psychology, economics, and business.

Statistics	Psychology	Economics	Business
102	88	75	84
81	101	99	115
93	92	80	94
78	86	105	103
112		91	87

a. Using a 5% significance level, test the null hypothesis that the mean prices of college textbooks in statistics, psychology, economics, and business are all equal.

b. If you did not reject the null hypothesis in part a, explain the Type II error that you may have made in this case. Note that you cannot calculate the probability of committing a Type II error without additional information.

ADVANCED EXERCISES

12.22 A billiards parlor in a small town is open just 4 days per week—Thursday through Sunday. Revenues vary considerably from day to day and week to week, so the owner is not sure whether some days of the week are more profitable than others. He takes random samples of 5 Thursdays, 5 Fridays, 5 Saturdays, and 5 Sundays from last year's records and lists the revenues for these 20 days. His bookkeeper finds the average revenue for each of the four samples, and then calculates $\sum x^2$. The results are shown in the following table. The value of the $\sum x^2$ came out to be 2,890,000.

Day	Mean Revenue ($)	Sample Size
Thursday	295	5
Friday	380	5
Saturday	405	5
Sunday	345	5

Assume that the revenues for each day of the week are normally distributed and that the standard deviations are equal for all four populations. At a 1% level of significance, can you reject the null hypothesis that the mean revenue is the same for each of the four days of the week?

12.23 Suppose that you are a reporter for a newspaper whose editor has asked you to compare the hourly wages of carpenters, plumbers, electricians, and masons in your city. Since many of these workers are not union members, the wages vary considerably among individuals in the same trade.

    a. What data should you gather, and how would you collect them? What statistics would you present in your article, and how would you calculate them? Assume that your newspaper is not intended for technical readers.

    b. Suppose that you must submit your findings to a technical journal that requires statistical analysis of your data. If you want to determine whether or not the mean hourly wages are the same for all four trades, briefly describe how you would analyze the data. Assume that hourly wages in each trade are normally distributed and that the four variances are equal.

12.24 Do rock music CDs and country music CDs give the consumers the same amount of music listening time? A sample of 12 randomly selected single rock music CDs and a sample of 14 randomly selected single country music CDs have the following total lengths (in minutes).

Rock Music	Country Music
43.0	45.3
44.3	40.2
63.8	42.8

32.8	33.0
54.2	33.5
51.3	37.7
64.8	36.8
36.1	34.6
33.9	33.4
51.7	36.5
36.5	43.3
59.7	31.7
	44.0
	42.7

Assume that the two populations are normally distributed with equal standard deviations.

    a. Compute the value of the test statistic t for testing the null hypothesis that the mean lengths of the rock and country music single CDs are the same against the alternative hypothesis that these mean lengths are not the same. Use the value of this t statistic to compute the (approximate) p-value.

    b. Compute the value of the (one-way ANOVA) test statistic F for performing the test of equality of the mean lengths of the rock and country music single CDs and use it to find the (approximate) p-value.

    c. How do the test statistics in parts a and b compare? How do the p-values computed in parts a and b compare? Do you think that this is a coincidence, or will this always happen?

SELF-REVIEW TEST

1. The F distribution is
    a. continuous   **b.** discrete   **c.** neither

2. The F distribution is always
    a. symmetric
    b. skewed to the right
    c. skewed to the left

3. The units of the F distribution, denoted by F, are always
    a. nonpositive   **b.** positive   **c.** nonnegative

4. The one-way ANOVA test analyzes only one
    a. variable   **b.** population   **c.** sample

5. The one-way ANOVA test is always
    a. right-tailed   **b.** left-tailed   **c.** two-tailed

6. For a one-way ANOVA with k treatments and n observations in all samples taken together, the degrees of freedom for the numerator are
    a. $k-1$ **b.** $n-k$ **c.** $n-1$

7. For a one-way ANOVA with k treatments and n observations in all samples taken together, the degrees of freedom for the denominator are
    a. $k-1$ **b.** $n-k$ **c.** $n-1$

8. The ANOVA test can be applied to compare
    a. two or more population means
    b. more than four population means only
    c. more than three population means only

9. Briefly describe the assumptions that must hold true to apply the one-way ANOVA procedure as mentioned in this chapter.

10. A small college town has four pizza parlors that make deliveries. A student doing a research paper for her business management class decides to compare how promptly the four parlors deliver. On six randomly chosen nights, she orders a large pepperoni pizza from each establishment, then records the elapsed time until the pizza is delivered to her apartment. Assume that her apartment is approximately the same distance from the four pizza parlors. The following table shows the times (in minutes) for these deliveries. Assume that all the assumptions required to apply the one-way ANOVA procedure hold true.

Tony's	Luigi's	Angelo's	Kowalski's
20.0	22.1	22.3	23.9
24.0	27.0	26.0	24.1
18.3	20.2	24.0	25.8
22.0	32.0	30.1	29.0
20.8	26.0	28.0	25.0
19.0	24.8	25.8	24.2

    a. Using a 5% significance level, test the null hypothesis that the mean delivery time is the same for each of the four pizza parlors.

    b. Is it a Type I error or a Type II error that may have been committed in part a? Explain.

MINI-PROJECTS

Note: The Mini-Projects are located on the text's Web site, www.wiley.com/college/mann.

DECIDE FOR YOURSELF

Note: The Decide for Yourself feature is located on the text's Web site, www.wiley.com/college/mann.

TECHNOLOGY INSTRUCTIONS | **CHAPTER 12**

Note: Complete TI-84, Minitab, and Excel manuals are available for download at the textbook's Web site, www.wiley.com/college/mann.

TI-84 Color/TI-84

The TI-84 Color Technology Instructions feature of this text is written for the TI-84 Plus C color graphing calculator running the 4.0 operating system. Some screens, menus, and functions will be slightly different in older operating systems. The TI-84 and TI-84 Plus can perform all of the same functions but will not have the "Color" option referenced in some of the menus.

Performing a One-Way ANOVA Test for Example 12–3 of the Text

1. Enter the data from Example 12–3 into list 1 (Method I), list 2 (Method II), and list 3 (Method III).

2. Select **STAT > TESTS > ANOVA**. This will paste the ANOVA function to the home screen.

3. Complete the command ANOVA(L1,L2,L3) by following these steps (see **Screen 12.1**):

   • Select 2nd > **STAT** > **NAMES** > **L1**.

   • Press the comma key.

   • Select 2nd > **STAT** > **NAMES** > **L2**.

   • Press the comma key.

   • Select 2nd > **STAT** > **NAMES** > **L3**.

   • Press the right parenthesis key.

4. Press **ENTER**.

5. The output includes the test statistic and the *p*-value. Scroll down to see the rest of the ANOVA table. (See **Screen 12.2**.)

Now compare the *F*-value to the critical value of *F* or the *p*-value from Screen 12.2 with α and make a decision.

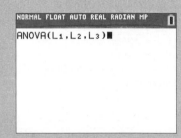
Screen 12.1

Screen 12.2

Minitab

The Minitab Technology Instructions feature of this text is written for Minitab version 17. Some screens, menus, and functions will be slightly different in older versions of Minitab.

Performing a One-Way ANOVA Test for Example 12–3 of the Text

1. Enter the data from Example 12–3 into C1 (Method I), C2 (Method II), and C3 (Method III).

2. Select **Stat > ANOVA > One-Way**.

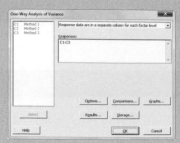

Screen 12.3

Screen 12.4

3. Use the following settings in the dialog box that appears on screen (see **Screen 12.3**):

   - Select Response data are in a separate column for each factor level from the drop-down menu.

   - Type C1-C3 in the **Responses** box.

4. Select **Options** and use the following settings in the dialog box that appears on screen:

   - Check the **Assume equal variances** checkbox.

   - Type 99 in the **Confidence level** box.

   - Select Two-sided from the **Type of confidence interval** dropdown menu.

5. Click **OK** in both dialog boxes.

6. The output, including the test statistic and p-value, will be displayed in the Session window. (See **Screen 12.4**.)

   Note: By default, Minitab will show a graph called an interval plot, which is not shown here.

Now compare the F-value to the critical value of F or the p-value from Screen 12.4 with α and make a decision.

Excel

The Excel Technology Instructions feature of this text is written for Excel 2013. Some screens, menus, and functions will be slightly different in older versions of Excel. To enable some of the advanced Excel functions, you must enable the Data Analysis Add-In for Excel.

Performing a One-Way ANOVA Test for Example 12–3 of the Text

Screen 12.5

1. In a new worksheet, enter the data from Method I in cells A1-A5, Method II in cells B1-B5, and the data from Method III in cells C1-C5. (See **Screen 12.5**.)

2. Click **DATA** and then click **Data Analysis Tools** from the **Analysis** group.

3. Select **Anova: Single Factor** from the dialog box that appears on screen and then click OK.

4. Use the following settings in the dialog box that appears on screen (see **Screen 12.5**):

   - Type A1:C5 in the **Input Range** box.

   - Select Columns in the Grouped by field.

   - Type .01 in the **Alpha** box.

   - Select New Worksheet Ply from the **Output options**.

5. Click **OK**.

6. The output includes the test statistic and the p-value.

Now compare the *F*-value to the critical value of *F* or the *p*-value from Screen 12.6 with α and make a decision.

	A	B	C	D	E	F	G
1	Anova: Single Factor						
2							
3	SUMMARY						
4	*Groups*	*Count*	*Sum*	*Average*	*Variance*		
5	Column 1	5	324	64.8	258.2		
6	Column 2	5	369	73.8	194.7		
7	Column 3	5	388	77.6	140.3		
8							
9							
10	ANOVA						
11	ce of Varia	*SS*	*df*	*MS*	*F*	*P-value*	*F crit*
12	Between (	432.1333	2	216.0667	1.092717	0.366464	3.885294
13	Within Gr	2372.8	12	197.7333			
14							
15	Total	2804.933	14				

Screen 12.6

TECHNOLOGY ASSIGNMENTS

TA12.1 A local "pick-your-own" farmer decided to grow blueberries. The farmer purchased and planted eight plants of each of the four different varieties of highbush blueberries. The yield (in pounds) of each plant was measured in the upcoming year to determine whether the average yields were different for at least two of the four plant varieties. The yields of these plants of the four varieties are given in the following table.

Berkeley	5.13	5.36	5.20	5.15	4.96	5.14	5.54	5.22
Duke	5.31	4.89	5.09	5.57	5.36	4.71	5.13	5.30
Jersey	5.20	4.92	5.44	5.20	5.17	5.24	5.08	5.13
Sierra	5.08	5.30	5.43	4.99	4.89	5.30	5.35	5.26

a. We are to test whether the mean yields for all such bushes of the four varieties are the same. Write the null and alternative hypotheses.

b. What are the degrees of freedom for the numerator and the denominator?

c. Calculate SSB, SSW, and SST.

d. Show the rejection and nonrejection regions on the *F* distribution curve for $\alpha = .01$.

e. Calculate the between-samples and within-samples variances.

f. What is the critical value of *F* for $\alpha = .01$?

g. What is the calculated value of the test statistic *F*?

h. Write the ANOVA table for this exercise.

i. Will you reject the null hypothesis stated in part a at a significance level of 1%?

TA12.2 The recommended acidity levels for sweet white wines (e.g., certain Rieslings, Port, Eiswein, Muscat) is .70% to .85% (www.grapestompers.com/articles/measure_acidity.htm). A vintner (wine-maker) takes three random samples of Riesling from casks that are 15, 20, and 25 years old, respectively, and measures the acidity of each sample. The sample results are given in the table below.

15 years	**20 years**	**25 years**
.8036	.8109	.7735
.8001	.8246	.7813
.8291	.8245	.8052
.8077	.8070	.8000
.8298	.8023	.8091
.8126	.8182	.7952
.8169	.8265	.7882
.8066	.8262	.7789
.8142	.8048	.7976
.8197	.7995	.7918
.8129	.8102	.7850
.8133	.7957	.7801
.8251	.8164	.7843

a. We are to test whether the mean acidity levels for all casks of Riesling are the same for the three different ages. Write the null and alternative hypotheses.

b. Show the rejection and nonrejection regions on the *F* distribution curve for $\alpha = .025$.

c. Calculate SSB, SSW, and SST.

d. What are the degrees of freedom for the numerator and the denominator?

e. Calculate the between-samples and within-samples variances.

f. What is the critical value of *F* for $\alpha = .025$?

g. What is the calculated value of the test statistic *F*?

h. Write the ANOVA table for this exercise.

i. Will you reject the null hypothesis stated in part a at a significance level of 2.5%?

TA12.3 A local car dealership is interested in determining how successful their salespeople are in turning a profit when selling a car. Specifically, they are interested in the average percentage of price markups earned on various car sales. The following table lists the percentages of price markups for a random sample of car sales by three salespeople at this dealership. Note that here the markups are calculated as follows. Suppose an auto dealer pays $14,000 for a car and lists the sale price as $20,000, which gives a markup of $6000. If the car is sold for $17,000, the markup percentage earned on this sale is 50% ($3000 is half of $6000).

Ira	23.2	26.9	27.3	34.1	30.7	31.6	43.8	
Jim	19.6	41.2	60.3	34.3	52.0	23.3	39.1	44.2
Kelly	52.3	50.0	53.4	37.9	26.4	41.1	25.2	41.2

a. Test at a 5% significance level whether the average markup percentage earned on all car sales is the same for Ira, Jim, and Kelly.

b. What is the Type I error in this case, and what is the probability of committing such an error? Explain.

CHAPTER **13**

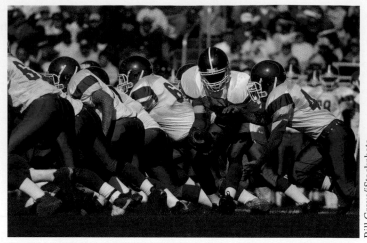

Bill Grove/iStockphoto

Simple Linear Regression

Are the heights and weights of persons related? Does a person's weight depend on his/her height? If yes, what is the change in the weight of a person, on average, for every 1 inch increase in height? What is this rate of change for National Football League players? (See Case Study 13–1.)

This chapter considers the relationship between two variables in two ways: (1) by using regression analysis and (2) by computing the correlation coefficient. By using a regression model, we can evaluate the magnitude of change in one variable due to a certain change in another variable. For example, an economist can estimate the amount of change in food expenditure due to a certain change in the income of a household by using a regression model. A sociologist may want to estimate the increase in the crime rate due to a particular increase in the unemployment rate. Besides answering these questions, a regression model also helps predict the value of one variable for a given value of another variable. For example, by using the regression line, we can predict the (approximate) food expenditure of a household with a given income.

The correlation coefficient, on the other hand, simply tells us how strongly two variables are related. It does not provide any information about the size of the change in one variable as a result of a certain change in the other variable. For example, the correlation coefficient tells us how strongly income and food expenditure or crime rate and unemployment rate are related.

13.1 | Simple Linear Regression

Only simple linear regression will be discussed in this chapter.[1] In the next two subsections the meaning of the words *simple* and *linear* as used in *simple linear regression* is explained.

13.1.1 Simple Regression

Let us return to the example of an economist investigating the relationship between food expenditure and income. What factors or variables does a household consider when deciding how much money it should spend on food every week or every month? Certainly, income of the household is one factor. However, many other variables also affect food expenditure. For instance, the assets owned by the household, the size of the household, the preferences and tastes of household members, and any special dietary needs of household members are some of the variables that influence a household's decision about food expenditure. These variables are called **independent** or **explanatory variables** because they all vary independently, and they explain the variation in food expenditures among different households. In other words, these variables explain why different households spend different amounts of money on food. Food expenditure is called the **dependent variable** because it depends on the independent variables. Studying the effect of two or more independent variables on a dependent variable using regression analysis is called **multiple regression**. However, if we choose only one (usually the most important) independent variable and study the effect of that single variable on a dependent variable, it is called a **simple regression**. Thus, a simple regression includes only two variables: one independent and one dependent. Note that whether it is a simple or a multiple regression analysis, it always includes one and only one dependent variable. It is the number of independent variables that changes in simple and multiple regressions.

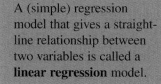

A regression model is a mathematical equation that describes the relationship between two or more variables. A **simple regression** model includes only two variables: one independent and one dependent. The dependent variable is the one being explained, and the independent variable is the one that explains the variation in the dependent variable.

13.1.2 Linear Regression

The relationship between two variables in a regression analysis is expressed by a mathematical equation called a **regression equation** or **model**. A regression equation, when plotted, may assume one of many possible shapes, including a straight line. A regression equation that gives a straight-line relationship between two variables is called a **linear regression model**; otherwise, the model is called a **nonlinear regression model**. In this chapter, only linear regression models are studied.

A (simple) regression model that gives a straight-line relationship between two variables is called a **linear regression** model.

The two diagrams in Figure 13.1 show a linear and a nonlinear relationship between the dependent variable food expenditure and the independent variable income. A linear relationship between income and food expenditure, shown in Figure 13.1a, indicates that as income increases, the food expenditure always increases at a constant rate. A nonlinear relationship between income and food expenditure, as depicted in Figure 13.1b, shows that as income increases, the food expenditure increases, although, after a point, the rate of increase in food expenditure is lower for every subsequent increase in income.

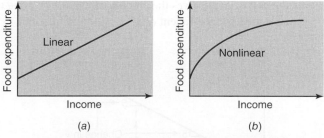

(a)   (b)

Figure 13.1   Relationship between food expenditure and income. (*a*) Linear relationship. (*b*) Nonlinear relationship.

---

[1]The term *regression* was first used by Sir Francis Galton (1822–1911), who studied the relationship between the heights of children and the heights of their parents.

The **equation of a linear relationship** between two variables x and y is written as

$$y = a + bx$$

Each set of values of a and b gives a different straight line. For instance, when $a = 50$ and $b = 5$, this equation becomes

$$y = 50 + 5x$$

To plot a straight line, we need to know two points that lie on that line. We can find two points on a line by assigning any two values to x and then calculating the corresponding values of y. For the equation $y = 50 + 5x$:

1. When $x = 0$, then $y = 50 + 5(0) = 50$.
2. When $x = 10$, then $y = 50 + 5(10) = 100$.

These two points are plotted in Figure 13.2. By joining these two points, we obtain the line representing the equation $y = 50 + 5x$.

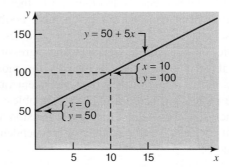

Figure 13.2 Plotting a linear equation.

Note that in Figure 13.2 the line intersects the y (vertical) axis at 50. Consequently, 50 is called the **y-intercept**. The y-intercept is given by the constant term in the equation. It is the value of y when x is zero.

In the equation $y = 50 + 5x$, 5 is called the **coefficient of x** or the **slope** of the line. It gives the amount of change in y due to a change of one unit in x. For example:

$$\text{If } x = 10, \text{ then } y = 50 + 5(10) = 100.$$
$$\text{If } x = 11, \text{ then } y = 50 + 5(11) = 105.$$

Hence, as x increases by 1 unit (from 10 to 11), y increases by 5 units (from 100 to 105). This is true for any value of x. Such changes in x and y are shown in Figure 13.3.

In general, when an equation is written in the form

$$y = a + bx$$

a gives the y-intercept and b represents the slope of the line. In other words, a represents the point where the line intersects the y-axis, and b gives the amount of change in y due to a change of one unit in x. Note that b is also called the coefficient of x.

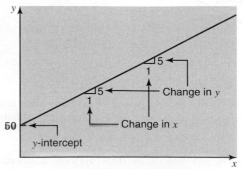

Figure 13.3 y-intercept and slope of a line.

13.1.3 Simple Linear Regression Model

In a regression model, the independent variable is usually denoted by x, and the dependent variable is usually denoted by y. The x variable, with its coefficient, is written on the right side of the = sign, whereas the y variable is written on the left side of the = sign. The y-intercept and the slope, which we earlier denoted by a and b, respectively, can be represented by any of the many commonly used symbols. Let us denote the y-intercept (which is also called the *constant term*) by A, and the slope (or the coefficient of the x variable) by **B**. Then, our simple linear regression model is written as

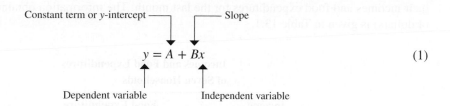

$$y = A + Bx \qquad (1)$$

In model (1), A gives the value of y for $x = 0$, and B gives the change in y due to a change of one unit in x.

Model (1) is called a **deterministic model**. It gives an **exact relationship** between x and y. This model simply states that y is determined exactly by x, and for a given value of x there is one and only one (unique) value of y.

However, in many cases the relationship between variables is not exact. For instance, if y is food expenditure and x is income, then model (1) would state that food expenditure is determined by income only and that all households with the same income spend the same amount on food. As mentioned earlier, however, food expenditure is determined by many variables, only one of which is included in model (1). In reality, different households with the same income spend different amounts of money on food because of the differences in the sizes of the household, the assets they own, and their preferences and tastes. Hence, to take these variables into consideration and to make our model complete, we add another term to the right side of model (1). This term is called the **random error term**. It is denoted by ε (Greek letter *epsilon*). The complete regression model is written as

$$y = A + Bx + \varepsilon \qquad (2)$$

Random error term

The regression model (2) is called a **probabilistic model** or a **statistical relationship**. The random error term ε is included in the model to represent the following two phenomena:

1. *Missing or omitted variables.* As mentioned earlier, food expenditure is affected by many variables other than income. The random error term ε is included to capture the effect of all those missing or omitted variables that have not been included in the model.

2. *Random variation.* Human behavior is unpredictable. For example, a household may have many parties during one month and spend more than usual on food during that month. The same household may spend less than usual during another month because it spent quite a bit of money to buy furniture. The variation in food expenditure for such reasons may be called random variation.

In model (2), A and B are the **population parameters**. The regression line obtained for model (2) by using the population data is called the **population regression line**. The values of A and B in the population regression line are called the **true values of the y-intercept and slope**, respectively.

However, population data are difficult to obtain. As a result, we almost always use sample data to estimate model (2). The values of the y-intercept and slope calculated from sample data

> In the **regression model** $y = A + Bx + \varepsilon$, A is called the y-intercept or constant term, B is the slope, and ε is the random error term. The dependent and independent variables are y and x, respectively.

on x and y are called the **estimated values of A and B** and are **denoted by a and b**, respectively. Using a and b, we write the estimated regression model as

$$\hat{y} = a + bx \qquad (3)$$

where $\hat{y}$ (read as *y hat*) is the **estimated** or **predicted value of y** for a given value of x. Equation (3) is called the **estimated regression model**; it gives the **regression of y on x**.

> In the model $\hat{y} = a + bx$, a and b, which are calculated using sample data, are called the **estimates of A and B**, respectively.

13.1.4   Scatter Diagram

Suppose we take a sample of seven households from a small city and collect information on their incomes and food expenditures for the last month. The information obtained (in hundreds of dollars) is given in Table 13.1.

Table 13.1   Incomes and Food Expenditures of Seven Households

Income	Food Expenditure
55	14
83	24
38	13
61	16
33	9
49	15
67	17

> A plot of paired observations is called a **scatter diagram**.

In Table 13.1, we have a pair of observations for each of the seven households. Each pair consists of one observation on income and a second on food expenditure. For example, the first household's income for the last month was $5500 and its food expenditure was $1400. By plotting all seven pairs of values, we obtain a **scatter diagram** or **scatterplot**. Figure 13.4 gives the scatter diagram for the data of Table 13.1. Each dot in this diagram represents one household. A scatter diagram is helpful in detecting a relationship between two variables. For example, by looking at the scatter diagram of Figure 13.4, we can observe that there exists a strong linear relationship between food expenditure and income. If a straight line is drawn through the points, the points will be scattered closely around the line.

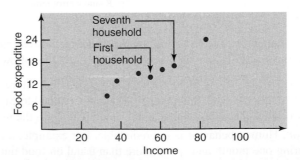

Figure 13.4   Scatter diagram.

As shown in Figure 13.5, a large number of straight lines can be drawn through the scatter diagram of Figure 13.4. Each of these lines will give different values for a and b of model (3).

In regression analysis, we try to find a line that best fits the points in the scatter diagram. Such a line provides a best possible description of the relationship between the dependent and independent variables. The **least squares method**, discussed in the next section, gives such a line. The line obtained by using the least squares method is called the **least squares regression line**.

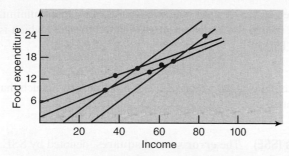

Figure 13.5   Scatter diagram and straight lines.

13.1.5   Least Squares Regression Line

The value of y obtained for a member from the survey is called the **observed or actual value of y**. As mentioned earlier in this section, the value of y, denoted by $\hat{y}$, obtained for a given x by using the regression line is called the **predicted value of y**. The random error ε denotes the difference between the actual value of y and the predicted value of y for population data. For example, for a given household, ε is the difference between what this household actually spent on food during the last month and what is predicted using the population regression line. The ε is also called the *residual* because it measures the surplus (positive or negative) of actual food expenditure over what is predicted by using the regression model. If we estimate model (2) by using sample data, the difference between the actual y and the predicted y based on this estimation cannot be denoted by ε. **The random error for the sample regression model is denoted by e.** Thus, e is an estimator of ε. If we estimate model (2) using sample data, then the value of e is given by

$$e = \text{Actual food expenditure} - \text{Predicted food expenditure} = y - \hat{y}$$

In Figure 13.6, e is the vertical distance between the actual position of a household and the point on the regression line. Note that in such a diagram, we always measure the dependent variable on the vertical axis and the independent variable on the horizontal axis.

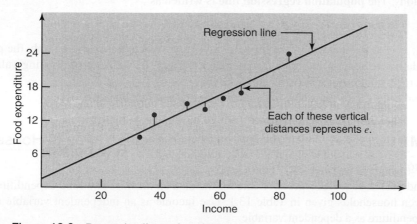

Figure 13.6   Regression line and random errors.

The value of an error is positive if the point that gives the actual food expenditure is above the regression line and negative if it is below the regression line. *The sum of these errors is always zero.* In other words, the sum of the actual food expenditures for seven households included in the sample will be the same as the sum of the food expenditures predicted by the regression model. Thus,

$$\Sigma e = \Sigma (y - \hat{y}) = 0$$

Hence, to find the line that best fits the scatter of points, we cannot minimize the sum of errors. Instead, we minimize the **error sum of squares**, denoted by **SSE**, which is obtained by adding the squares of errors. Thus,

$$\text{SSE} = \Sigma e^2 = \Sigma (y - \hat{y})^2$$

The least squares method gives the values of a and b for model (3) such that the sum of squared errors (SSE) is minimum.

Error Sum of Squares (SSE)   The **error sum of squares**, denoted by SSE, is

$$\text{SSE} = \Sigma e^2 = \Sigma (y - \hat{y})^2$$

The values of a and b that give the minimum SSE are called the **least squares estimates** of A and B, and the regression line obtained with these estimates is called the **least squares regression line**.

The Least Squares Line   For the least squares regression line $\hat{y} = a + bx$,

$$b = \frac{\text{SS}_{xy}}{\text{SS}_{xx}} \quad \text{and} \quad a = \bar{y} - b\bar{x}$$

where

$$\text{SS}_{xy} = \Sigma xy - \frac{(\Sigma x)(\Sigma y)}{n} \quad \text{and} \quad \text{SS}_{xx} = \Sigma x^2 - \frac{(\Sigma x)^2}{n}$$

and SS stands for "sum of squares." The least squares regression line $\hat{y} = a + bx$ is also called the regression of y on x.

The least squares values of a and b are computed using the formulas given above.[2] These formulas are for estimating a sample regression line. Suppose we have access to a population data set. We can find the population regression line by using the same formulas with a little adaptation. If we have access to population data, we replace a by A, b by B, and n by N in these formulas, and use the values of Σx, Σy, Σxy, and Σx^2 calculated for population data to make the required computations. The **population regression line** is written as

$$\mu_{y|x} = A + Bx$$

where $\mu_{y|x}$ is read as "the mean value of y for a given x." When plotted on a graph, the points on this population regression line give the average values of y for the corresponding values of x. These average values of y are denoted by $\mu_{y|x}$.

Example 13–1 illustrates how to estimate a regression line for sample data.

EXAMPLE 13–1   **Incomes and Food Expenditures of Households**

Estimating the least squares regression line.

Find the least squares regression line for the data on incomes and food expenditures of the seven households given in Table 13.1. Use income as an independent variable and food expenditure as a dependent variable.

Solution   We are to find the values of a and b for the regression model $\hat{y} = a + bx$. Table 13.2 shows the calculations required for the computation of a and b. We denote the independent variable (income) by x and the dependent variable (food expenditure) by y, both in hundreds of dollars.

© Troels Graugaard/iStockphoto

---

[2]The values of SS_{xy} and SS_{xx} can also be obtained by using the following basic formulas:

$$\text{SS}_{xy} = \Sigma (x - \bar{x})(y - \bar{y}) \quad \text{and} \quad \text{SS}_{xx} = \Sigma (x - \bar{x})^2$$

However, these formulas take longer to make calculations.

Table 13.2

Income	Food Expenditure		
x	y	xy	x^2
55	14	770	3025
83	24	1992	6889
38	13	494	1444
61	16	976	3721
33	9	297	1089
49	15	735	2401
67	17	1139	4489
$\Sigma x = 386$	$\Sigma y = 108$	$\Sigma xy = 6403$	$\Sigma x^2 = 23{,}058$

The following steps are performed to compute a and b.

Step 1. *Compute Σx, Σy, $\bar{x}$, and $\bar{y}$.*

$$\Sigma x = 386, \qquad \Sigma y = 108$$
$$\bar{x} = \Sigma x/n = 386/7 = 55.1429$$
$$\bar{y} = \Sigma y/n = 108/7 = 15.4286$$

Step 2. *Compute Σxy and Σx^2.*

To calculate Σxy, we multiply the corresponding values of x and y. Then, we sum all these products. The products of x and y are recorded in the third column of Table 13.2. To compute Σx^2, we square each of the x values and then add them. The squared values of x are listed in the fourth column of Table 13.2. From these calculations,

$$\Sigma xy = 6403 \quad \text{and} \quad \Sigma x^2 = 23{,}058$$

Step 3. *Compute SS_{xy} and SS_{xx}.*

$$SS_{xy} = \Sigma xy - \frac{(\Sigma x)(\Sigma y)}{n} = 6403 - \frac{(386)(108)}{7} = 447.5714$$

$$SS_{xx} = \Sigma x^2 - \frac{(\Sigma x)^2}{n} = 23{,}058 - \frac{(386)^2}{7} = 1772.8571$$

Step 4. *Compute a and b:*

$$b = \frac{SS_{xy}}{SS_{xx}} = \frac{447.5714}{1772.8571} = .2525$$

$$a = \bar{y} - b\bar{x} = 15.4286 - (.2525)(55.1429) = 1.5050$$

Thus, our estimated regression model $\hat{y} = a + bx$ is

$$\hat{y} = 1.5050 + .2525x$$

This regression line is called the least squares regression line. It gives the *regression of food expenditure on income.*

Note that we have rounded all calculations to four decimal places. We can round the values of a and b in the regression equation to two decimal places, but we do not do this here because we will use this regression equation for prediction and estimation purposes later.

Using this estimated regression model, we can find the predicted value of y for any specific value of x. For instance, suppose we randomly select a household whose monthly income is $6100, so that $x = 61$ (recall that x denotes income in hundreds of dollars). The predicted value of food expenditure for this household is

$$\hat{y} = 1.5050 + (.2525)(61) = \$16.9075 \text{ hundred} = \$1690.75$$

In other words, based on our regression line, we predict that a household with a monthly income of \$6100 is expected to spend \$1690.75 per month on food. This value of $\hat{y}$ can also be interpreted as a point estimator of the mean value of y for $x = 61$. Thus, we can state that, on average, all households with a monthly income of \$6100 spend about \$1690.75 per month on food.

In our data on seven households, there is one household whose income is \$6100. The actual food expenditure for that household is \$1600 (see Table 13.1). The difference between the actual and predicted values gives the error of prediction. Thus, the error of prediction for this household, which is shown in Figure 13.7, is

$$e = y - \hat{y} = 16 - 16.9075 = -\$.9075 \text{ hundred} = -\$90.75$$

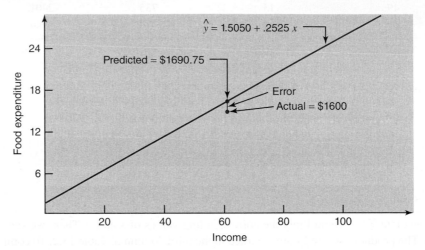

Figure 13.7 Error of prediction.

Therefore, the error of prediction is $-\$90.75$. The negative error indicates that the predicted value of y is greater than the actual value of y. Thus, if we use the regression model, this household's food expenditure is overestimated by \$90.75.

13.1.6 Interpretation of *a* and *b*

How do we interpret $a = 1.5050$ and $b = .2525$ obtained in Example 13–1 for the regression of food expenditure on income? A brief explanation of the y-intercept and the slope of a regression line was given in Section 13.1.2. Below we explain the meaning of a and b in more detail.

Interpretation of *a*

Consider a household with zero income. Using the estimated regression line obtained in Example 13–1, we get the predicted value of y for $x = 0$ as

$$\hat{y} = 1.5050 + .2525(0) = \$1.5050 \text{ hundred} = \$150.50$$

Thus, we can state that a household with no income is expected to spend \$150.50 per month on food. Alternatively, we can also state that the point estimate of the average monthly food expenditure for all households with zero income is \$150.50. Note that here we have used $\hat{y}$ as a point estimate of $\mu_{y|x}$. Thus, $a = 150.50$ gives the predicted or mean value of y for $x = 0$ based on the regression model estimated for the sample data.

However, we should be very careful when making this interpretation of a. In our sample of seven households, the incomes vary from a minimum of \$3300 to a maximum of \$8300. (Note that in Table 13.1, the minimum value of x is 33 and the maximum value is 83.) Hence, our regression line is valid only for the values of x between 33 and 83. If we predict y for a value of x outside this range, the prediction usually will not hold true. Thus, since $x = 0$ is outside the range of household incomes that we have in the sample data, the prediction that a household with zero income spends \$150.50 per month on food does not carry much credibility. The same is true if we try to predict y for an income greater than \$8300, which is the maximum value of x in Table 13.1.

Interpretation of b

The value of b in a regression model gives the change in the predicted value of y (dependent variable) due to a change of one unit in x (independent variable). For example, by using the regression equation obtained in Example 13–1, we see:

$$\text{When } x = 50, \quad \hat{y} = 1.5050 + .2525(50) = 14.1300$$
$$\text{When } x = 51, \quad \hat{y} = 1.5050 + .2525(51) = 14.3825$$

Hence, when x increased by one unit, from 50 to 51, $\hat{y}$ increased by $14.3825 - 14.1300 = .2525$, which is the value of b. Because our unit of measurement is hundreds of dollars, we can state that, on average, a \$100 increase in income will result in a \$25.25 increase in food expenditure. We can also state that, on average, a \$1 increase in income of a household will increase the food expenditure by \$.2525. Note the phrase "on average" in these statements. The regression line is seen as a measure of the mean value of y for a given value of x. If one household's income is increased by \$100, that household's food expenditure may or may not increase by \$25.25. However, if the incomes of all households are increased by \$100 each, the average increase in their food expenditures will be very close to \$25.25.

Note that when b is positive, an increase in x will lead to an increase in y, and a decrease in x will lead to a decrease in y. In other words, when b is positive, the movements in x and y are in the same direction. Such a relationship between x and y is called a **positive linear relationship**. The regression line in this case slopes upward from left to right. On the other hand, if the value of b is negative, an increase in x will lead to a decrease in y, and a decrease in x will cause an increase in y. The changes in x and y in this case are in opposite directions. Such a relationship between x and y is called a **negative linear relationship**. The regression line in this case slopes downward from left to right. The two diagrams in Figure 13.8 show these two cases.

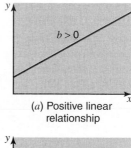

(a) Positive linear relationship

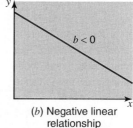

(b) Negative linear relationship

Figure 13.8 Positive and negative linear relationships between x and y.

For a regression model, b is computed as $b = SS_{xy}/SS_{xx}$. The value of SS_{xx} is always positive, and that of SS_{xy} can be positive or negative. Hence, the sign of b depends on the sign of SS_{xy}. If SS_{xy} is positive (as in our example on the incomes and food expenditures of seven households), then b will be positive, and if SS_{xy} is negative, then b will be negative.

◀ *Remember*

Case Study 13–1 illustrates the difference between the population regression line and a sample regression line.

Regression of Weights on Heights for NFL Players

Data Set VIII that accompanies this text lists many characteristics of National Football League (NFL) players who were on the rosters of all NFL teams during the 2014 season. These data comprise the population of NFL players for that season. We postulate the following simple linear regression model for these data:

$$y = A + Bx + \varepsilon$$

where y is the weight (in pounds) and x is the height (in inches) of an NFL player.

Using the population data that contain 1965 players, we obtain the following regression line:

$$\mu_{y|x} = -640.9 + 11.94x$$

This equation gives the population regression line because it is obtained by using the population data. (Note that in the population regression line we write $\mu_{y|x}$ instead of $\hat{y}$.) Thus, the true values of A and B are, respectively,

$$A = -640.9 \quad \text{and} \quad B = 11.94$$

The value of B indicates that for every 1-inch increase in the height of an NFL player, weight increases on average by 11.94 pounds. However, $A = -640.9$ does not make any sense. It states that the weight of a player with zero height is −640.9 pounds. (Recall from Section 13.1.6 that we must be very careful if and when we apply the regression equation to predict y for values of x outside the range of data used to find the regression line.) Figure 13.9 gives the scatter diagram and the regression line for the heights and weights of all NFL players.

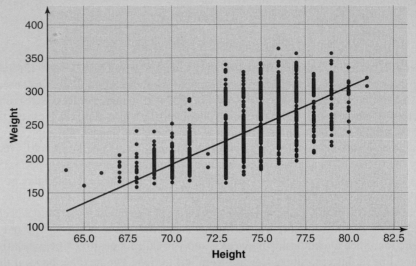

Figure 13.9 Scatter diagram and regression line for the data on heights and weights of all NFL players.

Next, we selected a random sample of 50 players and estimated the regression model for this sample. The estimated regression line for this sample is

$$\hat{y} = -610 + 11.50x$$

The values of a and b are

$$a = -610 \quad \text{and} \quad b = 11.50$$

These values of a and b give the estimates of A and B based on sample data. The scatter diagram and the regression line for the sample observations on heights and weights are given in Figure 13.10. Note that this figure does not show exactly 50 dots because some points/dots may be exactly the same or very close to each other.

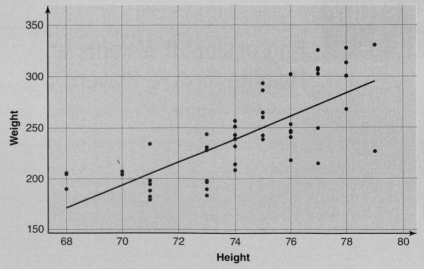

Figure 13.10 Scatter diagram and regression line for the data on heights and weights of 50 NFL players.

As we can observe from Figures 13.9 and 13.10, the scatter diagrams for population and sample data both show a (positive) linear relationship between the heights and weights of NFL players, although not a very strong positive relationship.

Source: www.sportscity.com/NFL-salaries and www.nfl.com/teams.

13.1.7   Assumptions of the Regression Model

Like any other theory, the linear regression analysis is also based on certain assumptions. Consider the population regression model

$$y = A + Bx + \varepsilon \qquad (4)$$

Four assumptions are made about this model. These assumptions are explained next with reference to the example on incomes and food expenditures of households. Note that these assumptions are made about the population regression model and not about the sample regression model.

Assumption 1:   The random error term ε has a mean equal to zero for each x. In other words, among all households with the same income, some spend more than the predicted food expenditure (and, hence, have positive errors) and others spend less than the predicted food expenditure (and, consequently, have negative errors). This assumption simply states that the sum of the positive errors is equal to the sum of the negative errors, so that the mean of errors for all households with the same income is zero. Thus, when the mean value of ε is zero, the mean value of y for a given x is equal to $A + Bx$, and it is written as

$$\mu_{y|x} = A + Bx$$

As mentioned earlier in this chapter, $\mu_{y|x}$ is read as "the mean value of y for a given value of x." When we find the values of A and B for model (4) using the population data, the points on the regression line give the average values of y, denoted by $\mu_{y|x}$, for the corresponding values of x.

Assumption 2:   The errors associated with different observations are independent. According to this assumption, the errors for any two households in our example are independent. In other words, all households decide independently how much to spend on food.

Assumption 3:   For any given x, the distribution of errors is normal. The corollary of this assumption is that the food expenditures for all households with the same income are normally distributed.

Assumption 4:   The distribution of population errors for each x has the same (constant) standard deviation, which is denoted by σ_ε. This assumption indicates that the spread of points around the regression line is similar for all x values.

Figure 13.11 illustrates the meanings of the first, third, and fourth assumptions for households with incomes of $4000 and $7500 per month. The same assumptions hold true for any other income level. In the population of all households, there will be many households with a monthly income of $4000. Using the population regression line, if we calculate the errors for all these households and prepare the distribution of these errors, it will look like the distribution given in Figure 13.11a. Its standard deviation will be σ_ε. Similarly, Figure 13.11b gives the distribution of errors for all those households in the population whose monthly income is $7500. Its standard deviation is also σ_ε. Both of these distributions are identical. Note that the mean of both of these distributions is $E(\varepsilon) = 0$.

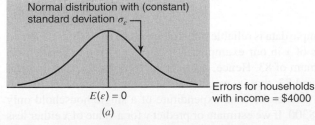

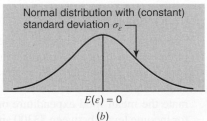

Figure 13.11   (a) Errors for households with an income of $4000 per month. (b) Errors for households with an income of $7500 per month.

Figure 13.12 shows how the distributions given in Figure 13.11 look when they are plotted on the same diagram with the population regression line. The points on the vertical line through $x = 40$ give the food expenditures for various households in the population, each of which has the same monthly income of \$4000. The same is true about the vertical line through $x = 75$ or any other vertical line for some other value of x.

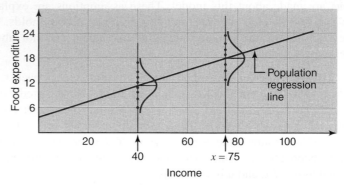

Figure 13.12   Distribution of errors around the population regression line.

13.1.8   Cautions in Using Regression

When carefully applied, regression is a very helpful technique for making predictions and estimations about one variable for a certain value of another variable. However, we need to be cautious when using the regression analysis, for it can give us misleading results and predictions. The following are the two most important points to remember when using regression.

(a) A Note on the Use of Simple Linear Regression

We should apply linear regression with caution. When we use simple linear regression, we assume that the relationship between two variables is described by a straight line. In the real world, the relationship between variables may not be linear. Hence, before we use a simple linear regression, it is better to construct a scatter diagram and look at the plot of the data points. We should estimate a linear regression model only if the scatter diagram indicates such a relationship. The scatter diagrams of Figure 13.13 give two examples for which the relationship between x and y is not linear. Consequently, using linear regression in such cases would be wrong.

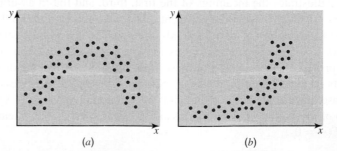

Figure 13.13   Nonlinear relationship between x and y.

(b) Extrapolation

The regression line estimated for the sample data is reliable only for the range of x values observed in the sample. For example, the values of x in our example on incomes and food expenditures vary from a minimum of 33 to a maximum of 83. Hence, our estimated regression line is applicable only for values of x between 33 and 83; that is, we should use this regression line to estimate the mean food expenditure or to predict the food expenditure of a single household only for income levels between \$3300 and \$8300. If we estimate or predict y for a value of x either less than 33 or greater than 83, it is called *extrapolation*. This does not mean that we should never use the regression line for extrapolation. Instead, we should interpret such predictions cautiously and not attach much importance to them.

Similarly, if the data used for the regression estimation are time-series data, the predicted values of y for periods outside the time interval used for the estimation of the regression line should be interpreted very cautiously. When using the estimated regression line for extrapolation, we are assuming that the same linear relationship between the two variables holds true for values of x outside the given range. It is possible that the relationship between the two variables may not be linear outside that range. Nonetheless, even if it is linear, adding a few more observations at either end will probably give a new estimation of the regression line.

EXERCISES

CONCEPTS AND PROCEDURES

13.1 Explain the meaning of the words *simple* and *linear* as used in *simple linear regression*.

13.2 Explain the meaning of independent and dependent variables for a regression model.

13.3 Explain the difference between exact and nonexact relationships between two variables. Give one example of each.

13.4 Explain the difference between linear and nonlinear relationships between two variables.

13.5 Explain the difference between a simple and a multiple regression model.

13.6 Briefly explain the difference between a deterministic and a probabilistic regression model.

13.7 Why is the random error term included in a regression model?

13.8 Explain the least squares method and least squares regression line. Why are they called by these names?

13.9 Explain the meaning and concept of SSE. You may use a graph for illustration purposes.

13.10 Explain the difference between y and $\hat{y}$.

13.11 Two variables x and y have a negative linear relationship. Explain what happens to the value of y when x increases. Give one example of a negative relationship between two variables.

13.12 Two variables x and y have a positive linear relationship. Explain what happens to the value of y when x increases. Give one example of a positive relationship between two variables.

13.13 Explain the following.

a. Population regression line
b. Sample regression line
c. True values of A and B
d. Estimated values of A and B that are denoted by a and b, respectively

13.14 Briefly explain the assumptions of the population regression model.

13.15 Plot the following straight lines. Give the values of the y-intercept and slope for each of these lines and interpret them. Indicate whether each of the lines gives a positive or a negative relationship between x and y.

a. $y = 100 + 6x$ b. $y = -40 + 9x$

13.16 A population data set produced the following information.

$$N = 300, \quad \Sigma x = 9880, \quad \Sigma y = 1324, \quad \Sigma xy = 85,080,$$
$$\Sigma x^2 = 500,350$$

Find the population regression line.

13.17 The following information is obtained from a sample data set.

$$n = 10, \quad \Sigma x = 100, \quad \Sigma y = 280, \quad \Sigma xy = 3120, \quad \Sigma x^2 = 1140$$

Find the estimated regression line.

APPLICATIONS

13.18 Bob's Pest Removal Service specializes in removing wild creatures (skunks, bats, reptiles, etc.) from private homes. He charges $50 to go to a house plus $30 per hour for his services. Let y be the total amount (in dollars) paid by a household using Bob's services and x the number of hours Bob spends capturing and removing the animal(s). The equation for the relationship between x and y is

$$y = 50 + 30x$$

a. Bob spent 4 hours removing a coyote from under Alice's house. How much will he be paid?
b. Suppose nine persons called Bob for assistance during a week. Strangely enough, each of these jobs required exactly 4 hours. Will each of these clients pay Bob the same amount, or do you expect each one to pay a different amount? Explain.
c. Is the relationship between x and y exact or nonexact?

13.19 A researcher took a sample of 25 electronics companies and found the following relationship between x and y, where x is the amount of money (in millions of dollars) spent on advertising by a company in 2009 and y represents the total gross sales (in millions of dollars) of that company for 2009.

$$\hat{y} = 3.6 + 11.75x$$

a. An electronics company spent $3 million on advertising in 2009. What are its expected gross sales for 2009?
b. Suppose four electronics companies spent $3 million each on advertising in 2009. Do you expect these four companies to have the same actual gross sales for 2009? Explain.
c. Is the relationship between x and y exact or nonexact?

13.20 An auto manufacturing company wanted to investigate how the price of one of its car models depreciates with age. The research department at the company took a sample of eight cars of this model and collected the following information on the ages (in years) and prices (in hundreds of dollars) of these cars.

Age	8	3	6	9	2	5	6	2
Price	38	220	95	33	267	134	112	245

a. Construct a scatter diagram for these data. Does the scatter diagram exhibit a linear relationship between ages and prices of cars?
b. Find the regression line with price as a dependent variable and age as an independent variable.
c. Give a brief interpretation of the values of a and b calculated in part b.
d. Plot the regression line on the scatter diagram of part a and show the errors by drawing vertical lines between scatter points and the regression line.
e. Predict the price of a 7-year-old car of this model.
f. Estimate the price of an 18-year-old car of this model. Comment on this finding.

13.21 The following table contains information on the amount of time that each of 12 students spends each day (on average) on social networks (Facebook, Twitter, etc.) and the Internet for social or entertainment purposes and his or her grade point average (GPA).

Time (hours per day)	4.4	6.2	4.2	1.6	4.7	5.4	1.3
GPA	3.22	2.21	3.13	3.69	2.7	2.2	3.69

Time (hours per day)	2.1	6.1	3.3	4.4	3.5
GPA	3.25	2.66	2.89	2.71	3.36

a. Construct a scatter diagram for these data. Does the scatter diagram exhibit a linear relationship between grade point average and time spent on social networks and the Internet?
b. Find the predictive regression line of GPA on time.
c. Give a brief interpretation of the values of a and b calculated in part b.
d. Plot the predictive regression line on the scatter diagram of part a, and show the errors by drawing vertical lines between scatter points and the predictive regression line.
e. Calculate the predicted GPA for a college student who spends 3.8 hours per day on social networks and the Internet for social or entertainment purposes.
f. Calculate the predicted GPA for a college student who spends 16 hours per day on social networks and the Internet for social or entertainment purposes. Comment on this finding.

13.22 While browsing through the magazine rack at a bookstore, a statistician decides to examine the relationship between the price of a magazine and the percentage of the magazine space that contains advertisements. The data are given in the following table.

Percentage containing ads	37	43	55	49	70	28	65	32
Price ($)	5.50	7.00	4.95	5.75	3.95	9.00	5.50	6.50

a. Construct a scatter diagram for these data. Does the scatter diagram exhibit a linear relationship between the percentage of a magazine's space containing ads and the price of the magazine?
b. Find the estimated regression equation of price on the percentage containing ads.
c. Give a brief interpretation of the values of a and b calculated in part b.
d. Plot the estimated regression line on the scatter diagram of part a, and show the errors by drawing vertical lines between scatter points and the predictive regression line.
e. Predict the price of a magazine with 50% of its space containing ads.
f. Estimate the price of a magazine with 99% of its space containing ads. Comment on this finding.

13.23 The following table gives the total payroll (in millions of dollars) on the opening day of the 2011 season and the percentage of games won during the 2011 season by each of the American League baseball teams.

Team	Total Payroll (millions of dollars)	Percentage of Games Won
Baltimore Orioles	85.30	42.6
Boston Red Sox	161.40	55.6
Chicago White Sox	129.30	48.8
Cleveland Indians	49.20	49.4
Detroit Tigers	105.70	58.6
Kansas City Royals	36.10	43.8
Los Angeles Angels	139.00	53.1
Minnesota Twins	112.70	38.9
New York Yankees	201.70	59.9
Oakland Athletics	66.60	45.7
Seattle Mariners	86.40	41.4
Tampa Bay Rays	41.90	56.2
Texas Rangers	92.30	59.3
Toronto Blue Jays	62.50	50.0

Source: http://baseball.about.com/od/newsrumors/a/2011-Baseball-Team-Payrolls.htm.

a. Find the least squares regression line with total payroll as the independent variable and percentage of games won as the dependent variable.
b. Is the equation of the regression line obtained in part a the population regression line? Why or why not? Do the values of the y-intercept and the slope of the regression line give A and B or a and b?
c. Give a brief interpretation of the values of the y-intercept and the slope obtained in part a.
d. Predict the percentage of games won by a team with a total payroll of $100 million.

13.24 The following table gives the total payroll (in millions of dollars) on the opening day of the 2011 season and the percentage of games won during the 2011 season by each of the National League baseball teams.

Team	Total Payroll (millions of dollars)	Percentage of Games Won
Arizona Diamondbacks	53.60	58.0
Atlanta Braves	87.00	54.9
Chicago Cubs	125.50	43.8
Cincinnati Reds	76.20	48.8
Colorado Rockies	88.00	45.1
Houston Astros	70.70	34.6
Los Angeles Dodgers	103.80	50.9
Miami Marlins	56.90	44.4
Milwaukee Brewers	85.50	59.3
New York Mets	120.10	47.5
Philadelphia Phillies	173.00	63.0
Pittsburgh Pirates	46.00	44.4

San Diego Padres	45.90	43.8
San Francisco Giants	118.20	53.1
St. Louis Cardinals	105.40	55.6
Washington Nationals	63.70	49.7

Source: http://baseball.about.com/od/newsrumors/a/2011-Baseball-Team-Payrolls.htm.

a. Find the least squares regression line with total payroll as the independent variable and percentage of games won as the dependent variable.

b. Is the equation of the regression line obtained in part a the population regression line? Why or why not? Do the values of the y-intercept and the slope of the regression line give A and B or a and b?

c. Give a brief interpretation of the values of the y-intercept and the slope obtained in part a.

d. Predict the percentage of games won by a team with a total payroll of \$100 million.

13.2 | Standard Deviation of Errors and Coefficient of Determination

In this section we discuss two concepts related to regression analysis. First we discuss the concept of the standard deviation of random errors and its calculation. Then we learn about the concept of the coefficient of determination and its calculation.

13.2.1 Standard Deviation of Errors

When we consider incomes and food expenditures, all households with the same income are expected to spend different amounts on food. Consequently, the random error ε will assume different values for these households. The standard deviation σ_ε measures the spread of these errors around the population regression line. The **standard deviation of errors** tells us how widely the errors and, hence, the values of y are spread for a given x. In Figure 13.12, which is reproduced as Figure 13.14, the points on the vertical line through $x = 40$ give the monthly food expenditures for all households with a monthly income of \$4000. The distance of each dot from the point on the regression line gives the value of the corresponding error. The standard deviation of errors σ_ε measures the spread of such points around the population regression line. The same is true for $x = 75$ or any other value of x.

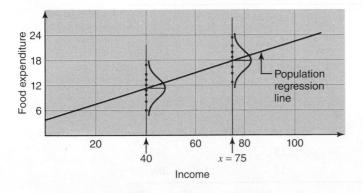

Figure 13.14 Spread of errors for $x = 40$ and $x = 75$.

Note that σ_ε denotes the standard deviation of errors for the population. However, usually σ_ε is unknown. In such cases, it is estimated by s_e, which is the standard deviation of errors for the sample data. The following is the basic formula to calculate s_e:

$$s_e = \sqrt{\frac{SSE}{n-2}} \qquad \text{where} \qquad SSE = \Sigma(y - \hat{y})^2$$

In this formula, $n - 2$ represents the **degrees of freedom** for the regression model. The reason $df = n - 2$ is that we lose one degree of freedom to calculate $\bar{x}$ and one for $\bar{y}$.

Degrees of Freedom for a Simple Linear Regression Model The **degrees of freedom** for a simple linear regression model are

$$df = n - 2$$

For computational purposes, it is more convenient to use the following formula to calculate the standard deviation of errors s_e.

Standard Deviation of Errors The **standard deviation of errors** is calculated as[3]

$$s_e = \sqrt{\frac{SS_{yy} - b\,SS_{xy}}{n - 2}}$$

where

$$SS_{yy} = \Sigma y^2 - \frac{(\Sigma y)^2}{n}$$

The calculation of SS_{xy} was discussed earlier in this chapter.[4]

Like the value of SS_{xx}, the value of SS_{yy} is always positive.

Example 13–2 illustrates the calculation of the standard deviation of errors for the data of Table 13.1.

EXAMPLE 13–2 | **Incomes and Food Expenditures of Households**

Calculating the standard deviation of errors.

Compute the standard deviation of errors s_e for the data on monthly incomes and food expenditures of the seven households given in Table 13.1.

Solution To compute s_e, we need to know the values of SS_{yy}, SS_{xy}, and b. In Example 13–1, we computed SS_{xy} and b. These values are

$$SS_{xy} = 447.5714 \quad \text{and} \quad b = .2525$$

To compute SS_{yy}, we calculate Σy^2 as shown in Table 13.3.

Table 13.3

Income x	Food Expenditure y	y^2
55	14	196
83	24	576
38	13	169
61	16	256
33	9	81
49	15	225
67	17	289
$\Sigma x = 386$	$\Sigma y = 108$	$\Sigma y^2 = 1792$

---

[3]If we have access to population data, the value of σ_ε is calculated using the formula

$$\sigma_\varepsilon = \sqrt{\frac{SS_{yy} - B\,SS_{xy}}{N}}$$

[4]The basic formula to calculate SS_{yy} is $\Sigma(y - \bar{y})^2$.

The value of SS_{yy} is

$$SS_{yy} = \Sigma y^2 - \frac{(\Sigma y)^2}{n} = 1792 - \frac{(108)^2}{7} = 125.7143$$

Hence, the standard deviation of errors is

$$s_e = \sqrt{\frac{SS_{yy} - b\,SS_{xy}}{n-2}} = \sqrt{\frac{125.7143 - .2525(447.5714)}{7-2}} = \mathbf{1.5939}$$

13.2.2 Coefficient of Determination

We may ask the question: How good is the regression model? In other words: How well does the independent variable explain the dependent variable in the regression model? The *coefficient of determination* is one concept that answers this question.

For a moment, assume that we possess information only on the food expenditures of households and not on their incomes. Hence, in this case, we cannot use the regression line to predict the food expenditure for any household. As we did in earlier chapters, in the absence of a regression model, we use $\bar{y}$ to estimate or predict every household's food expenditure. Consequently, the error of prediction for each household is now given by $y - \bar{y}$, which is the difference between the actual food expenditure of a household and the mean food expenditure. If we calculate such errors for all households in the sample and then square and add them, the resulting sum is called the **total sum of squares** and is denoted by **SST**. Actually SST is the same as SS_{yy} and is defined as

$$SST = SS_{yy} = \Sigma(y - \bar{y})^2$$

However, for computational purposes, SST is calculated using the following formula.

Total Sum of Squares (SST)   The **total sum of squares**, denoted by SST, is calculated as

$$SST = \Sigma y^2 - \frac{(\Sigma y)^2}{n}$$

Note that this is the same formula that we used to calculate SS_{yy}.

The value of SS_{yy}, which is 125.7143, was calculated in Example 13–2. Consequently, the value of SST is

$$SST = 125.7143$$

From Example 13–1, $\bar{y} = 15.4286$. Figure 13.15 shows the error for each of the seven households in our sample using the scatter diagram of Figure 13.4 and using $\bar{y}$.

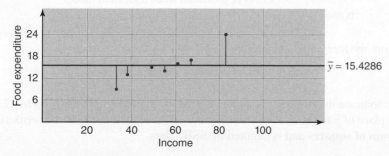

Figure 13.15   Total errors.

Now suppose we use the simple linear regression model to predict the food expenditure of each of the seven households in our sample. In this case, we predict each household's food expenditure by using the regression line we estimated earlier in Example 13–1, which is

$$\hat{y} = 1.5050 + .2525x$$

The predicted food expenditures, denoted by $\hat{y}$, for the seven households are listed in Table 13.4. Also given are the errors and error squares.

Table 13.4

x	y	$\hat{y} = 1.5050 + .2525x$	$e = y - \hat{y}$	$e^2 = (y - \hat{y})^2$
55	14	15.3925	−1.3925	1.9391
83	24	22.4625	1.5375	2.3639
38	13	11.1000	1.9000	3.6100
61	16	16.9075	−.9075	.8236
33	9	9.8375	−.8375	.7014
49	15	13.8775	1.1225	1.2600
67	17	18.4225	−1.4225	2.0235

$$\Sigma e^2 = \Sigma(y - \hat{y})^2 = 12.7215$$

We calculate the values of $\hat{y}$ (given in the third column of Table 13.4) by substituting the values of x in the estimated regression model. For example, the value of x for the first household is 55. Substituting this value of x in the regression equation, we obtain

$$\hat{y} = 1.5050 + .2525(55) = 15.3925$$

Similarly we find the other values of $\hat{y}$. The error sum of squares SSE is given by the sum of the fifth column in Table 13.4. Thus,

$$SSE = \Sigma(y - \hat{y})^2 = 12.7215$$

The errors of prediction for the regression model for the seven households are shown in Figure 13.16.

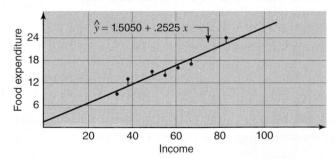

Figure 13.16 Errors of prediction when regression model is used.

Thus, from the foregoing calculations,

$$SST = 125.7143 \quad \text{and} \quad SSE = 12.7215$$

These values indicate that the sum of squared errors decreased from 125.7143 to 12.7215 when we used $\hat{y}$ in place of $\bar{y}$ to predict food expenditures. This reduction in squared errors is called the **regression sum of squares** and is denoted by **SSR**. Thus,

$$SSR = SST - SSE = 125.7143 - 12.7215 = 112.9928$$

The value of SSR can also be computed by using the formula

$$SSR = \Sigma(\hat{y} - \bar{y})^2$$

Regression Sum of Squares (SSR)   The **regression sum of squares**, denoted by SSR, is

$$SSR = SST - SSE$$

Thus, SSR is the portion of SST that is explained by the use of the regression model, and SSE is the portion of SST that is not explained by the use of the regression model. The sum of SSR and SSE is always equal to SST. Thus,

$$SST = SSR + SSE$$

The ratio of SSR to SST gives the **coefficient of determination**. The coefficient of determination calculated for population data is denoted by ρ^2 (ρ is the Greek letter *rho*), and the one calculated for sample data is denoted by r^2. The coefficient of determination gives the proportion of SST that is explained by the use of the regression model. The value of the coefficient of determination always lies in the range zero to one. The coefficient of determination can be calculated by using the formula

$$r^2 = \frac{SSR}{SST} \quad \text{or} \quad \frac{SST - SSE}{SST}$$

However, for computational purposes, the formula given on the next page is more efficient to use to calculate the coefficient of determination.

Coefficient of Determination   The **coefficient of determination**, denoted by r^2, represents the proportion of SST that is explained by the use of the regression model. The computational formula for r^2 is[5]

$$r^2 = \frac{b\,SS_{xy}}{SS_{yy}}$$

and

$$0 \le r^2 \le 1$$

Example 13–3 illustrates the calculation of the coefficient of determination for a sample data set.

| **EXAMPLE 13–3** | **Incomes and Food Expenditures of Households** |

Calculating the coefficient of determination.

For the data of Table 13.1 on monthly incomes and food expenditures of seven households, calculate the coefficient of determination.

Solution   From earlier calculations made in Examples 13–1 and 13–2,

$$b = .2525, \quad SS_{xy} = 447.5714, \quad \text{and} \quad SS_{yy} = 125.7143$$

---

[5]If we have access to population data, the value of ρ^2 is calculated using the formula

$$\rho^2 = \frac{B\,SS_{xy}}{SS_{yy}}$$

The values of SS_{xy} and SS_{yy} used here are calculated for the population data set.

Hence,

$$r^2 = \frac{b \, SS_{xy}}{SS_{yy}} = \frac{(.2525)(447.5714)}{125.7143} = .8990 = .90$$

Thus, we can state that SST is reduced by approximately 90% (from 125.7143 to 12.7215) when we use $\hat{y}$, instead of $\bar{y}$, to predict the food expenditures of households. Note that r^2 is usually rounded to two decimal places.

The total sum of squares SST is a measure of the total variation in food expenditures, the regression sum of squares SSR is the portion of total variation explained by the regression model (or by income), and the error sum of squares SSE is the portion of total variation not explained by the regression model. Hence, for Example 13–3 we can state that 90% of the total variation in food expenditures of households occurs because of the variation in their incomes, and the remaining 10% is due to randomness and other variables.

Usually, the higher the value of r^2, the better is the regression model. This is so because if r^2 is larger, a greater portion of the total errors is explained by the included independent variable, and a smaller portion of errors is attributed to other variables and randomness.

EXERCISES

CONCEPTS AND PROCEDURES

13.25 What are the degrees of freedom for a simple linear regression model?

13.26 Explain the meaning of coefficient of determination.

13.27 Explain the meaning of SST and SSR. You may use graphs for illustration purposes.

13.28 A population data set produced the following information.

$N = 250$, $\Sigma x = 9680$, $\Sigma y = 1456$, $\Sigma xy = 82,050$,

$\Sigma x^2 = 485,870$,   and $\Sigma y^2 = 140,895$

Find the values of σ_ε and ρ^2.

13.29 A population data set produced the following information.

$N = 460$, $\Sigma x = 3920$, $\Sigma y = 2650$, $\Sigma xy = 26,570$,

$\Sigma x^2 = 48,530$,   and $\Sigma y^2 = 39,347$

Find the values of σ_ε and ρ^2.

13.30 The following information is obtained from a sample data set.

$n = 15$, $\Sigma x = 82$, $\Sigma y = 602$, $\Sigma xy = 2534$,

$\Sigma x^2 = 412$,   and $\Sigma y^2 = 60,123$

Find the values of s_e and r^2.

APPLICATIONS

13.31 The following table provides information on the speed at takeoff (in meters per second) and distance traveled (in meters) by a random sample of 10 world-class long jumpers.

Speed	8.5	8.8	9.3	8.9	8.2	8.6	8.7	9.0	8.7	9.1
Distance	7.72	7.91	8.33	7.93	7.39	7.65	7.95	8.28	7.86	8.14

With distance traveled as the dependent variable and speed at takeoff as the independent variable, find the following:

a. SS_{xx}, SS_{yy}, and SS_{xy}

b. Standard deviation of errors

c. SST, SSE, and SSR

d. Coefficient of determination

13.32 The following table gives the ages (in years) and prices (in hundreds of dollars) of eight cars of a specific model.

Age	8	3	6	9	2	5	6	2
Price	38	220	95	33	267	134	112	245

a. Calculate the standard deviation of errors.

b. Compute the coefficient of determination and give a brief interpretation of it.

13.33 The following table gives information on the amount of sugar (in grams) and the calorie count in one serving of a sample of 13 varieties of Kellogg's cereal.

Sugar (grams)	4	15	12	11	8	6	7
Calories	120	200	140	110	120	80	190

Sugar (grams)	2	7	14	20	3	13
Calories	100	120	190	190	110	120

Source: kelloggs.com.

a. Determine the standard deviation of errors.

b. Find the coefficient of determination and give a brief interpretation of it.

13.34 The following table lists the percentages of space for eight magazines that contain advertisements and the prices of these magazines.

Percentage containing ads	37	43	55	49
Price ($)	5.50	7.00	4.95	5.75

Percentage containing ads	70	28	65	32
Price ($)	3.95	9.00	5.50	6.50

a. Find the standard deviation of errors.

b. Compute the coefficient of determination. What percentage of the variation in price is explained by the least squares regression of price on the percentage of magazine space containing ads? What percentage of this variation is not explained?

13.35 Refer to data given in Exercise 13.23 on the total 2011 payroll and the percentage of games won during the 2011 season by each of the American League baseball teams.

a. Find the standard deviation of errors, σ_ε. (Note that this data set belongs to a population.)

b. Compute the coefficient of determination, ρ^2.

13.3 | Inferences About *B*

This section is concerned with estimation and tests of hypotheses about the population regression slope *B*. We can also make a confidence interval and test hypothesis about the *y*-intercept *A* of the population regression line. However, making inferences about *A* is beyond the scope of this text.

13.3.1   Sampling Distribution of *b*

One of the main purposes for determining a regression line is to find the true value of the slope *B* of the population regression line. However, in almost all cases, the regression line is estimated using sample data. Then, based on the sample regression line, inferences are made about the population regression line. The slope *b* of a sample regression line is a point estimator of the slope *B* of the population regression line. The different sample regression lines estimated for different samples taken from the same population will give different values of *b*. If only one sample is taken and the regression line for that sample is estimated, the value of *b* will depend on which elements are included in the sample. Thus, *b* is a random variable, and it possesses a probability distribution that is more commonly called its sampling distribution. The shape of the sampling distribution of *b*, its mean, and standard deviation are given next.

> **The Sampling Distribution of *b* and its Mean and Standard Deviation**   Because of the assumption of normally distributed random errors, the sampling distribution of *b* is normal. The mean and standard deviation of *b*, denoted by μ_b and σ_b, respectively, are
>
> $$\mu_b = B \quad \text{and} \quad \sigma_b = \frac{\sigma_\varepsilon}{\sqrt{SS_{xx}}}$$

However, usually the standard deviation of population errors σ_ε is not known. Hence, the sample standard deviation of errors s_e is used to estimate σ_ε. In such a case, when σ_ε is unknown, the standard deviation of *b* is estimated by s_b, which is calculated as

$$s_b = \frac{s_e}{\sqrt{SS_{xx}}}$$

If σ_ε is known, the normal distribution can be used to make inferences about *B*. However, if σ_ε is not known, the normal distribution is replaced by the *t* distribution to make inferences about *B*.

13.3.2   Estimation of *B*

The value of *b* obtained from the sample regression line is a point estimate of the slope *B* of the population regression line. As mentioned in Section 13.3.1, if σ_ε is not known, the *t* distribution is used to make a confidence interval for *B*.

> **Confidence Interval for *B***   The $(1 - \alpha)100\%$ **confidence interval for *B*** is given by
>
> $$b \pm t s_b$$
>
> where
>
> $$s_b = \frac{s_e}{\sqrt{SS_{xx}}}$$
>
> and the value of *t* is obtained from the *t* distribution table, Table V of Appendix B, for $\alpha/2$ area in the right tail of the *t* distribution and $n - 2$ degrees of freedom.

Example 13–4 describes the procedure for making a confidence interval for *B*.

EXAMPLE 13–4 **Incomes and Food Expenditures of Households**

Constructing a confidence interval for B.

Construct a 95% confidence interval for B for the data on incomes and food expenditures of seven households given in Table 13.1.

Solution From the given information and earlier calculations in Examples 13–1 and 13–2,

$$n = 7, \quad b = .2525, \quad SS_{xx} = 1772.8571, \quad \text{and} \quad s_e = 1.5939$$

The confidence level is 95%. We have

$$s_b = \frac{s_e}{\sqrt{SS_{xx}}} = \frac{1.5939}{\sqrt{1772.8571}} = .0379$$

$$df = n - 2 = 7 - 2 = 5$$

$$\alpha/2 = (1 - .95)/2 = .025$$

From the t distribution table, Table V of Appendix B, the value of t for 5 df and .025 area in the right tail of the t distribution curve is 2.571. The 95% confidence interval for B is

$$b \pm ts_b = .2525 \pm 2.571(.0379) = .2525 \pm .0974 = \textbf{.155 to .350}$$

Thus, we are 95% confident that the slope B of the population regression line is between .155 and .350.

13.3.3 Hypothesis Testing About *B*

Testing a hypothesis about B when the null hypothesis is $B = 0$ (that is, the slope of the regression line is zero) is equivalent to testing that x does not determine y and that the regression line is of no use in predicting y for a given x. However, we should remember that we are testing for a linear relationship between x and y. It is possible that x may determine y nonlinearly. Hence, a nonlinear relationship may exist between x and y.

To test the hypothesis that x does not determine y linearly, we will test the null hypothesis that the slope of the regression line is zero; that is, $B = 0$. The alternative hypothesis can be: (1) x determines y, that is, $B \neq 0$; (2) x determines y positively, that is, $B > 0$; or (3) x determines y negatively, that is, $B < 0$.

The procedure used to make a hypothesis test about B is similar to the one used in earlier chapters. It involves the same five steps. Of course, we can use the p-value approach too.

Test Statistic for *b* The value of the **test statistic *t* for *b*** is calculated as

$$t = \frac{b - B}{s_b}$$

The value of B is substituted from the null hypothesis.

Example 13–5 illustrates the procedure for testing a hypothesis about B.

EXAMPLE 13–5 **Incomes and Food Expenditures of Households**

Conducting a test of hypothesis about B.

Test at the 1% significance level whether the slope of the regression line for the example on incomes and food expenditures of seven households is positive.

Solution From the given information and earlier calculations in Examples 13–1 and 13–4,

$$n = 7, \quad b = .2525, \quad \text{and} \quad s_b = .0379$$

Step 1.   *State the null and alternative hypotheses.*

We are to test whether or not the slope *B* of the population regression line is positive. Hence, the two hypotheses are

$$H_0: B = 0 \quad \text{(The slope is zero)}$$
$$H_1: B > 0 \quad \text{(The slope is positive)}$$

Note that we can also write the null hypothesis as $H_0: B \leq 0$, which states that the slope is either zero or negative.

Step 2.   *Select the distribution to use.*

Here, σ_e is not known. All assumptions for the population regression model are assumed to hold true. Hence, we will use the *t* distribution to make the test about *B*.

Step 3.   *Determine the rejection and nonrejection regions.*

The significance level is .01. The > sign in the alternative hypothesis indicates that the test is right-tailed. Therefore,

$$\text{Area in the right tail of the } t \text{ distribution} = \alpha = .01$$
$$df = n - 2 = 7 - 2 = 5$$

From the *t* distribution table, Table V of Appendix B, the critical value of *t* for 5 *df* and .01 area in the right tail of the *t* distribution is 3.365, as shown in Figure 13.17.

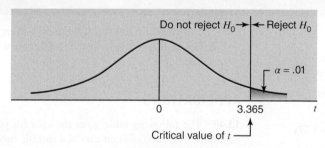

Figure 13.17   Rejection and nonrejection regions.

Step 4.   *Calculate the value of the test statistic.*

The value of the test statistic *t* for *b* is calculated as follows:

$$t = \frac{b - B}{s_b} = \frac{.2525 - 0}{.0379} = 6.662$$

where the $\overset{\text{From } H_0}{.2525 - 0}$

Step 5.   *Make a decision.*

The value of the test statistic $t = 6.662$ is greater than the critical value of $t = 3.365$, and it falls in the rejection region. Hence, we reject the null hypothesis and conclude that *x* (income) determines *y* (food expenditure) positively. That is, food expenditure increases with an increase in income and it decreases with a decrease in income.

Using the *p*-Value to Make a Decision

We can find the range for the *p*-value (as we did in Chapters 9 and 10) from the *t* distribution table, Table V of Appendix B, and make a decision by comparing that *p*-value with the significance level. For this example, $df = 5$, and the observed value of *t* is 6.662. From Table V (the *t* distribution table) in the row of $df = 5$, the largest value of *t* is 5.893 for which the area in the right tail of the *t* distribution is .001. Since our observed value of $t = 6.662$ is larger than 5.893, the *p*-value for $t = 6.662$ is less than .001, that is,

$$p\text{-value} < .001$$

Note that if we use technology to find this p-value, we will obtain a p-value of .000. Thus, we can state that for any α equal to or higher than .001 (the upper limit of the p-value range), we will reject the null hypothesis. For our example, $\alpha = .01$, which is larger than the p-value of .001. As a result, we reject the null hypothesis.

Note that the null hypothesis does not always have to be $B = 0$. We may test the null hypothesis that B is equal to a certain value.

A Note on Regression and Causality

The regression line does not prove causality between two variables; that is, it does not predict that a change in y is *caused* by a change in x. The information about causality is based on theory or common sense. A regression line describes only whether or not a significant quantitative relationship between x and y exists. Significant relationship means that we reject the null hypothesis H_0: $B = 0$ at a given significance level. The estimated regression line gives the change in y due to a change of one unit in x. Note that it does not indicate that the reason y has changed is that x has changed. In our example on incomes and food expenditures, it is economic theory and common sense, not the regression line, that tell us that food expenditure depends on income. The regression analysis simply helps determine whether or not this dependence is significant.

EXERCISES

CONCEPTS AND PROCEDURES

13.36 Describe the mean, standard deviation, and shape of the sampling distribution of the slope b of the simple linear regression model.

13.37 The following information is obtained for a sample of 36 observations taken from a population.

$$SS_{xx} = 274.600, \quad s_e = .953, \quad \text{and} \quad \hat{y} = 280.56 - 3.77x$$

a. Make a 95% confidence interval for B.
b. Using a significance level of .01, test whether B is negative.
c. Testing at a 5% significance level, can you conclude that B is different from zero?
d. Test if B is different from -5.20. Use $\alpha = .01$.

13.38 The following information is obtained for a sample of 16 observations taken from a population.

$$SS_{xx} = 340.700, \quad s_e = 1.951, \quad \text{and} \quad \hat{y} = 12.45 + 6.32x$$

a. Make a 99% confidence interval for B.
b. Using a significance level of .025, can you conclude that B is positive?
c. Using a significance level of .01, can you conclude that B is different from zero?
d. Using a significance level of .02, test whether B is different from 4.50. (*Hint:* The null hypothesis here will be H_0: $B = 4.50$, and the alternative hypothesis will be H_1: $B \neq 4.50$. Notice that the value of $B = 4.50$ will be used to calculate the value of the test statistic t.)

13.39 The following information is obtained for a sample of 100 observations taken from a population.

$$SS_{xx} = 524.884 \quad s_e = 1.464, \quad \text{and} \quad \hat{y} = 5.48 + 2.50x$$

a. Make a 98% confidence interval for B.
b. Test at a 2.5% significance level whether B is positive.

c. Can you conclude that B is different from zero? Use $\alpha = .01$.
d. Using a significance level of .01, test whether B is greater than 1.75.

APPLICATIONS

13.40 The following table gives the ages (in years) and prices (in hundreds of dollars) of eight cars of a specific model.

Age	8	3	6	9	2	5	6	2
Price	38	220	95	33	267	134	112	245

a. Construct a 95% confidence interval for B.
b. Test at the 5% significance level whether B is negative.

13.41 The following table gives information on the amount of sugar (in grams) and the calorie count in one serving of a sample of 13 varieties of Kellogg's cereal.

Sugar (grams)	4	15	12	11	8	6	7
Calories	120	200	140	110	120	80	180
Sugar (grams)	2	7	14	20	3	13	
Calories	80	120	190	190	100	120	

Source: kelloggs.com.

a. Make a 95% confidence interval for B.
b. It is well known that each additional gram of carbohydrate adds 4 calories. Sugar is one type of carbohydrate. Using regression equation for the data in the table, test at a 1% significance level whether B is different from 4.

13.42 The following table contains information on the amount of time spent each day (on average) on social networks and the Internet for social or entertainment purposes and the grade point average for a random sample of 12 college students.

Time (hours per day)	4.4	6.2	4.2	1.6	4.7	5.4
GPA	3.22	2.21	3.13	3.69	2.7	2.2
Time (hours per day)	1.3	2.1	6.1	3.3	4.4	3.5
GPA	3.69	3.25	2.66	2.89	2.71	3.36

a. Construct a 98% confidence interval for B.
b. Test at a 1% significance level whether B is negative.

13.43 The following table lists the percentages of space for eight magazines that contain advertisements and the prices of these magazines.

Percentage containing ads	37	43	55	49
Price ($)	5.50	7.00	4.95	5.75
Percentage containing ads	70	28	65	32
Price ($)	3.95	9.00	5.50	6.50

a. Construct a 98% confidence interval for B.
b. Testing at a 5% significance level, can you conclude that B is different from zero?

13.44 The following data give the experience (in years) and monthly salaries (in hundreds of dollars) of nine randomly selected secretaries.

Experience	14	3	5	6	4	9	18	5	16
Monthly salary	62	29	37	43	35	60	67	32	60

a. Find the least squares regression line with experience as an independent variable and monthly salary as a dependent variable.
b. Construct a 98% confidence interval for B.
c. Test at a 2.5% significance level whether B is greater than zero.

13.4 | Linear Correlation

This section describes the meaning and calculation of the linear correlation coefficient and the procedure to conduct a test of hypothesis about it.

13.4.1 Linear Correlation Coefficient

Another measure of the relationship between two variables is the correlation coefficient. This section describes the simple linear correlation, for short **linear correlation**, which measures the strength of the linear association between two variables. In other words, the linear correlation coefficient measures how closely the points in a scatter diagram are spread around the regression line. The correlation coefficient calculated for the population data is denoted by ρ (Greek letter *rho*) and the one calculated for sample data is denoted by r. (Note that the square of the correlation coefficient is equal to the coefficient of determination.)

Value of the Correlation Coefficient   The **value of the correlation coefficient** always lies in the range -1 to 1; that is,

$$-1 \le \rho \le 1 \quad \text{and} \quad -1 \le r \le 1$$

Although we can explain the linear correlation using the population correlation coefficient ρ, we will do so using the sample correlation coefficient r.

If $r = 1$, it is said to be a **perfect positive linear correlation**. In such a case, all points in the scatter diagram lie on a straight line that slopes upward from left to right, as shown in Figure 13.18a. If $r = -1$, the correlation is said to be a **perfect negative linear correlation**. In this case, all points in the scatter diagram fall on a straight line that slopes downward from left to right, as shown in Figure 13.18b. If the points are scattered all over the diagram, as shown in Figure 13.18c, then there is **no linear correlation** between the two variables, and consequently r is close to 0. Note that here r is *not* equal to zero but is very *close* to zero.

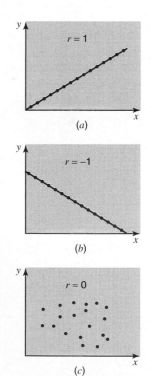

Figure 13.18   Linear correlation between two variables. (a) Perfect positive linear correlation, $r = 1$. (b) Perfect negative linear correlation, $r = -1$. (c) No linear correlation, $r \approx 0$.

We do not usually encounter an example with perfect positive or perfect negative correlation (unless the relationship between variables is exact). What we observe in real-world problems is either a positive linear correlation with $0 < r < 1$ (that is, the correlation coefficient is greater than zero but less than 1) or a negative linear correlation with $-1 < r < 0$ (that is, the correlation coefficient is greater than -1 but less than zero).

If the correlation between two variables is positive and close to 1, we say that the variables have a **strong positive linear correlation**. If the correlation between two variables is positive but close to zero, then the variables have a **weak positive linear correlation**. In contrast, if the correlation between two variables is negative and close to -1, then the variables are said to have a **strong negative linear correlation**. If the correlation between two variables is negative but close to zero, there exists a **weak negative linear correlation** between the variables. Graphically, a strong correlation indicates that the points in the scatter diagram are very close to the regression line, and a weak correlation indicates that the points in the scatter diagram are widely spread around the regression line. These four cases are shown in Figure 13.19a–d.

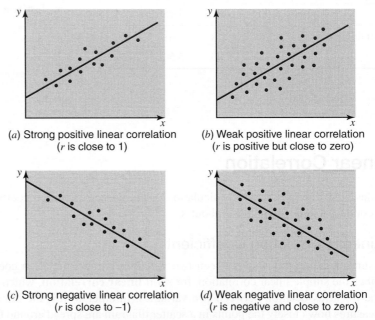

(a) Strong positive linear correlation
(r is close to 1)

(b) Weak positive linear correlation
(r is positive but close to zero)

(c) Strong negative linear correlation
(r is close to -1)

(d) Weak negative linear correlation
(r is negative and close to zero)

Figure 13.19   Linear correlation between two variables.

The linear correlation coefficient is calculated by using the following formula. (This correlation coefficient is also called the *Pearson product moment correlation coefficient*.)

Linear Correlation Coefficient   The **simple linear correlation coefficient**, denoted by r, measures the strength of the linear relationship between two variables for a sample and is calculated as[6]

$$r = \frac{\text{SS}_{xy}}{\sqrt{\text{SS}_{xx}\,\text{SS}_{yy}}}$$

Because both SS_{xx} and SS_{yy} are always positive, the sign of the correlation coefficient r depends on the sign of SS_{xy}. If SS_{xy} is positive, then r will be positive, and if SS_{xy} is negative, then r will be negative. Another important observation to remember is that *r and b, calculated for the same sample, will always have the same sign. That is, both r and b are either positive or negative.* This is so because both r and b provide information about the relationship between x and y. Likewise, the corresponding population parameters ρ and B will always have the same sign.

[6]If we have access to population data, the value of ρ is calculated using the formula

$$\rho = \frac{\text{SS}_{xy}}{\sqrt{\text{SS}_{xx}\,\text{SS}_{yy}}}$$

Here the values of SS_{xy}, SS_{xx}, and SS_{yy} are calculated using the population data.

Example 13–6 illustrates the calculation of the linear correlation coefficient r.

EXAMPLE 13–6   Incomes and Food Expenditures of Households

Calculating the linear correlation coefficient.

Calculate the correlation coefficient for the example on incomes and food expenditures of seven households.

Solution   From earlier calculations made in Examples 13–1 and 13–2,

$$SS_{xy} = 447.5714, \quad SS_{xx} = 1772.8571, \quad \text{and} \quad SS_{yy} = 125.7143$$

Substituting these values in the formula for r, we obtain

$$r = \frac{SS_{xy}}{\sqrt{SS_{xx}\, SS_{yy}}} = \frac{447.5714}{\sqrt{(1772.8571)(125.7143)}} = .9481 = \mathbf{.95}$$

Thus, the linear correlation coefficient is .95. The correlation coefficient is usually rounded to two decimal places.

The linear correlation coefficient simply tells us how strongly the two variables are (linearly) related. The correlation coefficient of .95 for incomes and food expenditures of seven households indicates that income and food expenditure are very strongly and positively correlated. This correlation coefficient does not, however, provide us with any more information.

The square of the correlation coefficient gives the coefficient of determination, which was explained in Section 13.2.2. Thus, $(.95)^2$ is .90, which is the value of r^2 calculated in Example 13–3.

Sometimes the calculated value of r may indicate that the two variables are very strongly linearly correlated, but in reality they may not be. For example, if we calculate the correlation coefficient between the price of a haircut and the size of families in the United States using data for the last 30 years, we will find a strong negative linear correlation. Over time, the price of a haircut has increased and the size of families has decreased. This finding does not mean that family size and the price of a haircut are related. As a result, before we calculate the correlation coefficient, we must seek help from a theory or from common sense to postulate whether or not the two variables have a causal relationship.

Another point to note is that in a simple regression model, one of the two variables is categorized as an independent (also known as an explanatory or predictor) variable and the other is classified as a dependent (also known as a response) variable. However, no such distinction is made between the two variables when the correlation coefficient is calculated.

13.4.2   Hypothesis Testing About the Linear Correlation Coefficient

This section describes how to perform a test of hypothesis about the population correlation coefficient ρ using the sample correlation coefficient r. We use the t distribution to make this test. However, to use the t distribution, both variables should be normally distributed.

Usually (although not always), the null hypothesis is that the linear correlation coefficient between the two variables is zero, that is, $\rho = 0$. The alternative hypothesis can be one of the following: (1) the linear correlation coefficient between the two variables is less than zero, that is, $\rho < 0$; (2) the linear correlation coefficient between the two variables is greater than zero, that is, $\rho > 0$; or (3) the linear correlation coefficient between the two variables is not equal to zero, that is, $\rho \neq 0$.

Test Statistic for r   If both variables are normally distributed and the null hypothesis is H_0: $\rho = 0$, then the value of the test statistic t is calculated as

$$t = r\sqrt{\frac{n - 2}{1 - r^2}}$$

Here $n - 2$ gives the degrees of freedom.

Example 13–7 describes the procedure to perform a test of hypothesis about the linear correlation coefficient.

EXAMPLE 13-7 Incomes and Food Expenditures of Households

Performing a test of hypothesis about the correlation coefficient.

Using a 1% level of significance and the data from Example 13–1, test whether the linear correlation coefficient between incomes and food expenditures is positive. Assume that the populations of both variables are normally distributed.

Solution From Examples 13–1 and 13–6,

$$n = 7 \quad \text{and} \quad r = .9481$$

Below we use the five steps to perform this test of hypothesis.

Step 1. *State the null and alternative hypotheses.*

We are to test whether the linear correlation coefficient between incomes and food expenditures is positive. Hence, the null and alternative hypotheses are, respectively,

$$H_0: \rho = 0 \quad \text{(The linear correlation coefficient is zero.)}$$
$$H_1: \rho > 0 \quad \text{(The linear correlation coefficient is positive.)}$$

Step 2. *Select the distribution to use.*

The population distributions for both variables are normally distributed. Hence, we use the t distribution to perform this test about the linear correlation coefficient.

Step 3. *Determine the rejection and nonrejection regions.*

The significance level is 1%. From the alternative hypothesis we know that the test is right-tailed. Hence,

$$\text{Area in the right tail of the } t \text{ distribution} = .01$$
$$df = n - 2 = 7 - 2 = 5$$

From the t distribution table, Table V of Appendix B, the critical value of t is 3.365. The rejection and nonrejection regions for this test are shown in Figure 13.20.

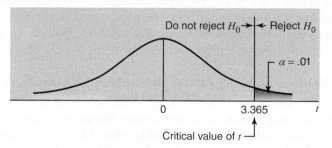

Figure 13.20 Rejection and nonrejection regions.

Step 4. *Calculate the value of the test statistic.*

The value of the test statistic t for r is calculated as follows:

$$t = r\sqrt{\frac{n-2}{1-r^2}} = .9481\sqrt{\frac{7-2}{1-(.9481)^2}} = 6.667$$

Step 5. *Make a decision.*

The value of the test statistic $t = 6.667$ is greater than the critical value of $t = 3.365$, and it falls in the rejection region. Hence, we reject the null hypothesis and conclude that there is a positive linear relationship between incomes and food expenditures.

Using the *p*-Value to Make a Decision

We can find the range for the *p*-value from the *t* distribution table (Table V of Appendix B) and make a decision by comparing that *p*-value with the significance level. For this example, $df = 5$, and the observed value of *t* is 6.667. From Table V (the *t* distribution table) of Appendix B in the row of $df = 5$ the largest value of *t* is 5.893 for which the area in the right tail of the *t* distribution is .001. Since our observed value of $t = 6.667$ is larger than 5.893, the *p*-value for $t = 6.667$ is less than .001, that is,

$$p\text{-value} < .001$$

Thus, we can state that for any α equal to or greater than .001 (the upper limit of the *p*-value range), we will reject the null hypothesis. For our example, $\alpha = .01$, which is greater than the *p*-value of .001. As a result, we reject the null hypothesis.

EXERCISES

CONCEPTS AND PROCEDURES

13.45 What does a linear correlation coefficient tell about the relationship between two variables? Within what range can a correlation coefficient assume a value?

13.46 What is the difference between ρ and r? Explain.

13.47 Explain each of the following concepts. You may use graphs to illustrate each concept.

    a. Perfect positive linear correlation
    b. Perfect negative linear correlation
    c. Strong positive linear correlation
    d. Strong negative linear correlation
    e. Weak positive linear correlation
    f. Weak negative linear correlation
    g. No linear correlation

13.48 Can the values of B and ρ calculated for the same population data have different signs? Explain.

13.49 For a sample data set, the linear correlation coefficient r has a positive value. Which of the following is true about the slope b of the regression line estimated for the same sample data?

    a. The value of b will be positive.
    b. The value of b will be negative.
    c. The value of b can be positive or negative.

13.50 For a sample data set, the slope b of the regression line has a negative value. Which of the following is true about the linear correlation coefficient r calculated for the same sample data?

    a. The value of r will be positive.
    b. The value of r will be negative.
    c. The value of r can be positive or negative.

13.51 For a sample data set on two variables, the value of the linear correlation coefficient is (close to) zero. Does this mean that these variables are not related? Explain.

13.52 Will you expect a positive, zero, or negative linear correlation between the two variables for each of the following examples?

    a. Grade of a student and hours spent studying
    b. Incomes and entertainment expenditures of households
    c. Ages of women and makeup expenses per month

    d. Price of a computer and consumption of Coca-Cola
    e. Price and consumption of wine

13.53 Will you expect a positive, zero, or negative linear correlation between the two variables for each of the following examples?

    a. SAT scores and GPAs of students
    b. Stress level and blood pressure of individuals
    c. Amount of fertilizer used and yield of corn per acre
    d. Ages and prices of houses
    e. Heights of husbands and incomes of their wives

13.54 A population data set produced the following information.

$$N = 460, \quad \Sigma x = 3920, \quad \Sigma y = 2650, \quad \Sigma xy = 26{,}570,$$
$$\Sigma x^2 = 48{,}530, \quad \text{and} \quad \Sigma y^2 = 39{,}347$$

Find the linear correlation coefficient ρ.

13.55 A population data set produced the following information.

$$N = 250, \quad \Sigma x = 9880, \quad \Sigma y = 1456, \quad \Sigma xy = 90{,}980,$$
$$\Sigma x^2 = 485{,}870, \quad \text{and} \quad \Sigma y^2 = 140{,}875$$

Find the linear correlation coefficient ρ.

13.56 A sample data set produced the following information.

$$n = 12, \quad \Sigma x = 66, \quad \Sigma y = 588, \quad \Sigma xy = 2244,$$
$$\Sigma x^2 = 396, \quad \text{and} \quad \Sigma y^2 = 58{,}734$$

    a. Calculate the linear correlation coefficient r.
    b. Using a 1% significance level, can you conclude that ρ is negative?

APPLICATIONS

13.57 The data on ages (in years) and prices (in hundreds of dollars) for eight cars of a specific model are reproduced from that exercise.

Age	8	3	6	9	2	5	6	2
Price	38	220	95	33	267	134	112	245

    a. Do you expect the ages and prices of cars to be positively or negatively related? Explain.

b. Calculate the linear correlation coefficient.

c. Test at a 2.5% significance level whether ρ is negative.

13.58 The following table lists the midterm and final exam scores for seven students in a statistics class.

Midterm score	79	95	81	66	87	94	59
Final exam score	85	97	78	76	94	84	67

a. Do you expect the midterm and final exam scores to be positively or negatively related?

b. Plot a scatter diagram. By looking at the scatter diagram, do you expect the correlation coefficient between these two variables to be close to zero, 1, or −1?

c. Find the correlation coefficient. Is the value of r consistent with what you expected in parts a and b?

d. Using a 1% significance level, test whether the linear correlation coefficient is positive.

13.59 The following table gives information on the amount of sugar (in grams) and the calorie count in one serving of a sample of 13 varieties of Kellogg's cereal.

Sugar (grams)	4	15	12	11	8	6	7	2	7	14	20	3	13
Calories	120	200	140	110	120	80	180	80	120	190	190	100	120

Source: kelloggs.com.

a. Find the correlation coefficient.

b. Test at a 1% significance level whether the linear correlation coefficient between the two variables listed in the table is positive.

13.60 The following data give the ages (in years) of husbands and wives for six couples.

Husband's age	43	57	28	19	35	39
Wife's age	37	51	32	20	33	38

a. Do you expect the ages of husbands and wives to be positively or negatively related?

b. Plot a scatter diagram. By looking at the scatter diagram, do you expect the correlation coefficient between these two variables to be close to zero, 1, or −1?

c. Find the correlation coefficient. Is the value of r consistent with what you expected in parts a and b?

d. Using a 5% significance level, test whether the correlation coefficient is different from zero.

13.61 Refer to data given in Exercise 13.24 on the total 2011 payroll and the percentage of games won during the 2011 season by each of the National League baseball teams. Compute the linear correlation coefficient, ρ. Does it make sense to make a confidence interval and to test a hypothesis about ρ here? Explain.

13.62 The following table gives information on the calorie count and grams of fat for 11 types of bagels produced by Panera Bread.

Bagel	Calories	Fat (grams)
Asiago Cheese	330	6.0
Blueberry	330	1.5
Chocolate Chip	370	6.0
Cinnamon Crunch	430	8.0
Cinnamon Swirl & Raisin	320	2.5
Everything	300	2.5
French Toast	350	5.0
Jalapeno & Cheddar	310	3.0
Plain	290	1.5
Sesame	310	3.0
Sweet Onion & Poppyseed	390	7.0

a. Find the correlation coefficient.

b. Test at a 1% significance level whether ρ is different from zero.

13.5 | Regression Analysis: A Complete Example

This section works out an example that includes all the topics we have discussed so far in this chapter.

EXAMPLE 13–8 Driving Experience and Monthly Auto Insurance

A complete example of regression analysis.

A random sample of eight drivers selected from a small town insured with a company and having similar minimum required auto insurance policies was selected. The table on the next page lists their driving experiences (in years) and monthly auto insurance premiums (in dollars):

Driving Experience (years)	Monthly Auto Insurance Premium ($)
5	64
2	87
12	50
9	71
15	44
6	56
25	42
16	60

(a) Does the insurance premium depend on the driving experience, or does the driving experience depend on the insurance premium? Do you expect a positive or a negative relationship between these two variables?

(b) Compute SS_{xx}, SS_{yy}, and SS_{xy}.

(c) Find the least squares regression line by choosing appropriate dependent and independent variables based on your answer in part a.

(d) Interpret the meaning of the values of a and b calculated in part c.

(e) Plot the scatter diagram and the regression line.

(f) Calculate r and r^2, and explain what they mean.

(g) Predict the monthly auto insurance premium for a driver with 10 years of driving experience.

(h) Compute the standard deviation of errors.

(i) Construct a 90% confidence interval for B.

(j) Test at a 5% significance level whether B is negative.

(k) Using $\alpha = .05$, test whether ρ is different from zero.

Solution

(a) Based on theory and intuition, we expect the insurance premium to depend on driving experience. Consequently, the insurance premium is the dependent variable (variable y) and driving experience is the independent variable (variable x) in the regression model. A new driver is considered a high risk by the insurance companies, and he or she has to pay a higher premium for auto insurance. On average, the insurance premium is expected to decrease with an increase in the years of driving experience. Therefore, we expect a negative relationship between these two variables. In other words, both the population correlation coefficient ρ and the population regression slope B are expected to be negative.

(b) Table 13.5 shows the calculation of Σx, Σy, Σxy, Σx^2, and Σy^2.

Table 13.5

Experience x	Premium y	xy	x^2	y^2
5	64	320	25	4096
2	87	174	4	7569
12	50	600	144	2500
9	71	639	81	5041
15	44	660	225	1936
6	56	336	36	3136
25	42	1050	625	1764
16	60	960	256	3600
$\Sigma x = 90$	$\Sigma y = 474$	$\Sigma xy = 4739$	$\Sigma x^2 = 1396$	$\Sigma y^2 = 29,642$

The values of $\bar{x}$ and $\bar{y}$ are

$$\bar{x} = \Sigma x/n = 90/8 = 11.25$$
$$\bar{y} = \Sigma y/n = 474/8 = 59.25$$

The values of SS_{xy}, SS_{xx}, and SS_{yy} are computed as follows:

$$SS_{xy} = \Sigma xy - \frac{(\Sigma x)(\Sigma y)}{n} = 4739 - \frac{(90)(474)}{8} = -593.5000$$

$$SS_{xx} = \Sigma x^2 - \frac{(\Sigma x)^2}{n} = 1396 - \frac{(90)^2}{8} = 383.5000$$

$$SS_{yy} = \Sigma y^2 - \frac{(\Sigma y)^2}{n} = 29{,}642 - \frac{(474)^2}{8} = 1557.5000$$

(c) To find the regression line, we calculate a and b as follows:

$$b = \frac{SS_{xy}}{SS_{xx}} = \frac{-593.5000}{383.5000} = -1.5476$$

$$a = \bar{y} - b\bar{x} = 59.25 - (-1.5476)(11.25) = 76.6605$$

Thus, our estimated regression line $\hat{y} = a + bx$ is

$$\hat{y} = 76.6605 - 1.5476x$$

(d) The value of $a = 76.6605$ gives the value of $\hat{y}$ for $x = 0$; that is, it gives the monthly auto insurance premium for a driver with no driving experience. However, as mentioned earlier in this chapter, we should not attach much importance to this statement because the sample contains drivers with only 2 or more years of experience. The value of b gives the change in $\hat{y}$ due to a change of one unit in x. Thus, $b = -1.5476$ indicates that, on average, for every extra year of driving experience, the monthly auto insurance premium decreases by $1.55. Note that when b is negative, y decreases as x increases.

(e) Figure 13.21 shows the scatter diagram and the regression line for the data on eight auto drivers. Note that the regression line slopes downward from left to right. This result is consistent with the negative relationship we anticipated between driving experience and insurance premium.

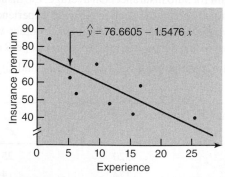

Figure 13.21   Scatter diagram and the regression line.

(f) The values of r and r^2 are computed as follows:

$$r = \frac{SS_{xy}}{\sqrt{SS_{xx}\, SS_{yy}}} = \frac{-593.5000}{\sqrt{(383.5000)(1557.5000)}} = -.7679 = \mathbf{-.77}$$

$$r^2 = \frac{b\, SS_{xy}}{SS_{yy}} = \frac{(-1.5476)(-593.5000)}{1557.5000} = .5897 = \mathbf{.59}$$

The value of $r = -.77$ indicates that the driving experience and the monthly auto insurance premium are negatively related. The (linear) relationship is strong but not very strong. The value of $r^2 = .59$ states that 59% of the total variation in insurance premiums is explained by years of driving experience, and 41% is not. The low value of r^2 indicates that there may be many other important variables that contribute to the determination of auto insurance premiums. For example, the premium is expected to depend on the driving record of a driver and the type and age of the car.

(g) Using the estimated regression line, we find the predicted value of y for $x = 10$ as:

$$\hat{y} = 76.6605 - 1.5476x = 76.6605 - 1.5476(10) = \mathbf{\$61.18}$$

Thus, we expect the monthly auto insurance premium of a driver with 10 years of driving experience to be $61.18.

(h) The standard deviation of errors is

$$s_e = \sqrt{\frac{SS_{yy} - b\,SS_{xy}}{n-2}} = \sqrt{\frac{1557.5000 - (-1.5476)(-593.5000)}{8-2}} = \mathbf{10.3199}$$

(i) To construct a 90% confidence interval for B, first we calculate the standard deviation of b:

$$s_b = \frac{s_e}{\sqrt{SS_{xx}}} = \frac{10.3199}{\sqrt{383.5000}} = .5270$$

For a 90% confidence level, the area in each tail of the t distribution is

$$\alpha/2 = (1 - .90)/2 = .05$$

The degrees of freedom are

$$df = n - 2 = 8 - 2 = 6$$

From the t distribution table, the t value for .05 area in the right tail of the t distribution and 6 df is 1.943. The 90% confidence interval for B is

$$b \pm ts_b = -1.5476 \pm 1.943(.5270)$$
$$= -1.5476 \pm 1.0240 = \mathbf{-2.57 \text{ to } -.52}$$

Thus, we can state with 90% confidence that B lies in the interval -2.57 to $-.52$. That is, on average, the monthly auto insurance premium of a driver decreases by an amount between $.52 and $2.57 for every extra year of driving experience.

(j) We perform the following five steps to test the hypothesis about B.

Step 1. *State the null and alternative hypotheses.*

The null and alternative hypotheses are, respectively,

$$H_0: B = 0 \quad (B \text{ is not negative.})$$
$$H_1: B < 0 \quad (B \text{ is negative.})$$

Note that the null hypothesis can also be written as $H_0: B \geq 0$.

Step 2. *Select the distribution to use.*

Because σ_ε is not known, we use the t distribution to make the hypothesis test.

Step 3. *Determine the rejection and nonrejection regions.*

The significance level is .05. The $<$ sign in the alternative hypothesis indicates that it is a left-tailed test.

$$\text{Area in the left tail of the } t \text{ distribution} = \alpha = .05$$
$$df = n - 2 = 8 - 2 = 6$$

From the t distribution table, the critical value of t for .05 area in the left tail of the t distribution and 6 df is -1.943, as shown in Figure 13.22.

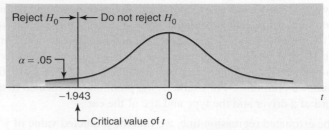

Figure 13.22   Rejection and nonrejection regions.

Step 4.   *Calculate the value of the test statistic.*

The value of the test statistic t for b is calculated as follows:

$$t = \frac{b - B}{s_b} = \frac{-1.5476 - 0}{.5270} = -2.937$$

From H_0

Step 5.   *Make a decision.*

The value of the test statistic $t = -2.937$ falls in the rejection region. Hence, we reject the null hypothesis and conclude that B is negative. That is, the monthly auto insurance premium decreases with an increase in years of driving experience.

Using the *p*-Value to Make a Decision

We can find the range for the p-value from the t distribution table (Table V of Appendix B) and make a decision by comparing that p-value with the significance level. For this example, $df = 6$ and the observed value of t is -2.937. From Table V (the t distribution table) in the row of $df = 6$, 2.937 is between 2.447 and 3.143. The corresponding areas in the right tail of the t distribution are .025 and .01, respectively. Our test is left-tailed, however, and the observed value of t is negative. Thus, $t = -2.937$ lies between -2.447 and -3.143. The corresponding areas in the left tail of the t distribution are .025 and .01. Therefore the range of the p-value is

$$.01 < p\text{-value} < .025$$

Thus, we can state that for any α equal to or greater than .025 (the upper limit of the p-value range), we will reject the null hypothesis. For our example, $\alpha = .05$, which is greater than the upper limit of the p-value of .025. As a result, we reject the null hypothesis.

Note that if we use technology to find this p-value, we will obtain a p-value of .013. Then we can reject the null hypothesis for any $\alpha \geq .013$.

(k) We perform the following five steps to test the hypothesis about the linear correlation coefficient ρ.

Step 1.   *State the null and alternative hypotheses.*

The null and alternative hypotheses are, respectively,

$H_0: \rho = 0$ (The linear correlation coefficient is zero.)

$H_1: \rho \neq 0$ (The linear correlation coefficient is different from zero.)

Step 2.   *Select the distribution to use.*

Assuming that variables x and y are normally distributed, we will use the t distribution to perform this test about the linear correlation coefficient.

Step 3. *Determine the rejection and nonrejection regions.*

The significance level is 5%. From the alternative hypothesis we know that the test is two-tailed. Hence,

$$\text{Area in each tail of the } t \text{ distribution} = .05/2 = .025$$
$$df = n - 2 = 8 - 2 = 6$$

From the t distribution table, Table V of Appendix B, the critical values of t are -2.447 and 2.447. The rejection and nonrejection regions for this test are shown in Figure 13.23.

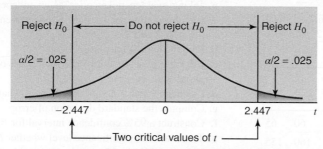

Figure 13.23 Rejection and nonrejection regions.

Step 4. *Calculate the value of the test statistic.*

The value of the test statistic t for r is calculated as follows:

$$t = r\sqrt{\frac{n-2}{1-r^2}} = (-.7679)\sqrt{\frac{8-2}{1-(-.7679)^2}} = -2.936$$

Step 5. *Make a decision.*

The value of the test statistic $t = -2.936$ falls in the rejection region. Hence, we reject the null hypothesis and conclude that the linear correlation coefficient between driving experience and auto insurance premium is different from zero.

Using the *p*-Value to Make a Decision

We can find the range for the p-value from the t distribution table and make a decision by comparing that p-value with the significance level. For this example, $df = 6$ and the observed value of t is -2.936. From Table V (the t distribution table) in the row of $df = 6$, $t = 2.936$ is between 2.447 and 3.143. The corresponding areas in the right tail of the t distribution curve are $.025$ and $.01$, respectively. Since the test is two tailed, the range of the p-value is

$$2(.01) < p\text{-value} < 2(.025) \qquad \textbf{or} \qquad .02 < p\text{-value} < .05$$

Thus, we can state that for any α equal to or greater than $.05$ (the upper limit of the p-value range), we will reject the null hypothesis. For our example, $\alpha = .05$, which is equal to the upper limit of the p-value. As a result, we reject the null hypothesis.

EXERCISES

APPLICATIONS

13.63 The owner of a small factory that produces working gloves is concerned about the high cost of air conditioning in the summer but is afraid that keeping the temperature in the factory too high will lower productivity. During the summer, he experiments with temperature settings from 68°F to 81°F and measures each day's productivity. The following table gives the temperature and the number of pairs of gloves (in hundreds) produced on each of the 8 randomly selected days.

Temperature (°F)	72	71	78	75	81	77	68	76
Pairs of gloves	37	37	32	36	33	35	39	34

a. Do the pairs of gloves produced depend on temperature, or does temperature depend on pairs of gloves produced? Do you expect a positive or a negative relationship between these two variables?

b. Taking temperature as an independent variable and pairs of gloves produced as a dependent variable, compute SS_{xx}, SS_{yy}, and SS_{xy}.

c. Find the least squares regression line.

d. Interpret the meaning of the values of a and b calculated in part c.

e. Plot the scatter diagram and the regression line.

f. Calculate r and r^2, and explain what they mean.

g. Compute the standard deviation of errors.

h. Predict the number of pairs of gloves produced when $x = 74$.

i. Construct a 99% confidence interval for B.

j. Test at a 5% significance level whether B is negative.

k. Using $\alpha = .01$ can you conclude that ρ is negative?

13.64 The following table gives information on the limited tread warranties (in thousands of miles) and the prices of 12 randomly selected tires at a national tire retailer as of July 2012.

Warranty (thousands of miles)	60	70	75	50	80	55
Price per tire ($)	95	135	94	90	121	70
Warranty (thousands of miles)	65	65	70	65	60	65
Price per tire ($)	140	80	92	125	160	155

a. Taking warranty length as an independent variable and price per tire as a dependent variable, compute SS_{xx}, SS_{yy}, and SS_{xy}.

b. Find the regression of price per tire on warranty length.

c. Briefly explain the meaning of the values of a and b calculated in part b.

d. Calculate r and r^2 and explain what they mean.

e. Plot the scatter diagram and the regression line.

f. Predict the price of a tire with a warranty length of 73,000 miles.

g. Compute the standard deviation of errors.

h. Construct a 95% confidence interval for B.

i. Test at a 5% significance level if B is positive.

j. Using $\alpha = .025$, can you conclude that the linear correlation coefficient is positive?

13.65 The following data give information on the average ticket prices (in U.S. dollars) and the average percentage of capacity filled for seven hockey teams during the 2011–2012 National Hockey League regular season. (Note: Capacity levels exceeding 100.0% imply standing-room-only attendees.)

Team	Anaheim	Vancouver	Dallas	Edmonton	New Jersey	Toronto	Philadelphia
Average ticket price ($)	36.94	68.38	29.95	70.13	45.86	123.27	66.89
Percentage capacity filled	86.4	102.5	76.8	100.0	87.4	103.7	107.4

Source: http://espn.go.com/blog/dallas/stars/post/_/id/13315/stars-have-cheapest-ticket-in-nhl and http://espn.go.com/nhl/attendance.

a. Taking average ticket price as an independent variable and percentage of capacity filled as a dependent variable, compute SS_{xx}, SS_{yy}, and SS_{xy}.

b. Find the least squares regression line.

c. Briefly explain the meaning of the values of a and b calculated in part b.

d Calculate r and r^2 and briefly explain what they mean.

e. Compute the standard deviation of errors.

f. Construct a 95% confidence interval for B.

g. Test at a 2.5% significance level whether B is positive.

h. Using a 2.5% significance level, test whether ρ is positive.

13.66 The following data give information on the lowest cost ticket price (in dollars) and the average attendance (rounded to the nearest thousand) for the last year for six football teams.

Ticket price	38.50	26.50	34.00	45.50	59.50	36.00
Attendance	56	65	71	69	55	42

a. Taking ticket price as an independent variable and attendance as a dependent variable, compute SS_{xx}, SS_{yy}, and SS_{xy}.

b. Find the least squares regression line.

c. Briefly explain the meaning of the values of a and b calculated in part b.

d. Calculate r and r^2 and briefly explain what they mean.

e. Compute the standard deviation of errors.

f. Construct a 90% confidence interval for B.

g. Test at a 2.5% significance level whether B is negative.

h. Using a 2.5% significance level, test whether ρ is negative.

13.6 | Using the Regression Model

Let us return to the example on incomes and food expenditures to discuss two major uses of a regression model:

1. Estimating the mean value of y for a given value of x. For instance, we can use our food expenditure regression model to estimate the mean food expenditure of all households with a specific income (say, $5500 per month).

2. Predicting a particular value of y for a given value of x. For instance, we can determine the expected food expenditure of a randomly selected household with a particular monthly income (say, $5500) using our food expenditure regression model.

13.6.1 Using the Regression Model for Estimating the Mean Value of y

Our population regression model is

$$y = A + Bx + \varepsilon$$

As mentioned earlier in this chapter, the mean value of y for a given x is denoted by $\mu_{y|x}$, read as "the mean value of y for a given value of x." Because of the assumption that the mean value of ε is zero, the mean value of y is given by

$$\mu_{y|x} = A + Bx$$

Our objective is to estimate this mean value. The value of $\hat{y}$, obtained from the sample regression line by substituting the value of x, gives the **point estimate of $\mu_{y|x}$** for that x.

For our example on incomes and food expenditures, the estimated sample regression line (from Example 13–1) is

$$\hat{y} = 1.5050 + .2525x$$

Suppose we want to estimate the mean food expenditure for all households with a monthly income of $5500. We will denote this population mean by $\mu_{y|x=55}$ or $\mu_{y|55}$. Note that we have written $x = 55$ and not $x = 5500$ in $\mu_{y|55}$ because the units of measurement for the data used to estimate the above regression line in Example 13–1 were hundreds of dollars. Using the regression line, we find that the point estimate of $\mu_{y|55}$ is

$$\hat{y} = 1.5050 + .2525(55) = \$15.3925 \text{ hundred}$$

Thus, based on the sample regression line, the point estimate for the mean food expenditure $\mu_{y|55}$ for all households with a monthly income of $5500 is $1539.25 per month.

However, suppose we take a second sample of seven households from the same population and estimate the regression line for this sample. The point estimate of $\mu_{y|55}$ obtained from the regression line for the second sample is expected to be different. All possible samples of the same size taken from the same population will give different regression lines as shown in Figure 13.24, and, consequently, a different point estimate of $\mu_{y|x}$. Therefore, a confidence interval constructed for $\mu_{y|x}$ based on one sample will give a more reliable estimate of $\mu_{y|x}$ than will a point estimate.

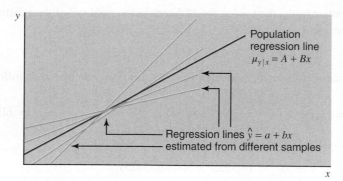

Figure 13.24   Population and sample regression lines.

To construct a confidence interval for $\mu_{y|x}$, we must know the mean, the standard deviation, and the shape of the sampling distribution of its point estimator $\hat{y}$.

The point estimator $\hat{y}$ of $\mu_{y|x}$ is normally distributed with a mean of $A + Bx$ and a standard deviation of

$$\sigma_{\hat{y}_m} = \sigma_\varepsilon \sqrt{\frac{1}{n} + \frac{(x_0 - \bar{x})^2}{SS_{xx}}}$$

where $\sigma_{\hat{y}_m}$ is the standard deviation of $\hat{y}$ when it is used to estimate $\mu_{y|x}$, x_0 is the value of x for which we are estimating $\mu_{y|x}$, and σ_ε is the population standard deviation of ε.

However, usually σ_ε is not known. Rather, it is estimated by the standard deviation of sample errors s_e. In this case, we replace σ_ε by s_e and $\sigma_{\hat{y}_m}$ by $s_{\hat{y}_m}$ in the foregoing expression. To make a confidence interval for $\mu_{y|x}$, we use the t distribution because σ_ε is not known.

Confidence Interval for $\mu_{y|x}$ The $(1 - \alpha)100\%$ **confidence interval for** $\mu_{y|x}$ for $x = x_0$ is

$$\hat{y} \pm t s_{\hat{y}_m}$$

where the value of t is obtained from the t distribution table for $\alpha/2$ area in the right tail of the t distribution curve and $df = n - 2$. The value of $s_{\hat{y}_m}$ is calculated as follows:

$$s_{\hat{y}_m} = s_e \sqrt{\frac{1}{n} + \frac{(x_0 - \bar{x})^2}{SS_{xx}}}$$

Example 13–9 illustrates how to make a confidence interval for the mean value of y, $\mu_{y|x}$.

EXAMPLE 13–9   Incomes and Food Expenditures of Households

Constructing a confidence interval for the mean value of y for a given x.

Refer to Example 13–1 on incomes and food expenditures. Find a 99% confidence interval for the mean food expenditure for all households with a monthly income of $5500.

Solution   Using the regression line estimated in Example 13–1, we find the point estimate of the mean food expenditure for $x = 55$ as

$$\hat{y} = 1.5050 + .2525(55) = \$15.3925 \text{ hundred}$$

The confidence level is 99%. Hence, the area in each tail of the t distribution is

$$\alpha/2 = (1 - .99)/2 = .005$$

The degrees of freedom are

$$df = n - 2 = 7 - 2 = 5$$

From the t distribution table, the t value for .005 area in the right tail of the t distribution and 5 df is 4.032. From calculations in Examples 13–1 and 13–2, we know that

$$s_e = 1.5939, \quad \bar{x} = 55.1429, \quad \text{and} \quad SS_{xx} = 1772.8571$$

The standard deviation of $\hat{y}$ as an estimate of $\mu_{y|x}$ for $x = 55$ is calculated as follows:

$$s_{\hat{y}_m} = s_e \sqrt{\frac{1}{n} + \frac{(x_0 - \bar{x})^2}{SS_{xx}}} = (1.5939) \sqrt{\frac{1}{7} + \frac{(55 - 55.1429)^2}{1772.8571}} = .6025$$

Hence, the 99% confidence interval for $\mu_{y|55}$ is

$$\hat{y} \pm t s_{\hat{y}_m} = 15.3925 \pm 4.032(.6025)$$
$$= 15.3925 \pm 2.4293 = \mathbf{12.9632 \text{ to } 17.8218}$$

Thus, with 99% confidence we can state that the mean food expenditure for all households with a monthly income of $5500 is between $1296.32 and $1782.18.

13.6.2   Using the Regression Model for Predicting a Particular Value of y

The second major use of a regression model is to predict a particular value of y for a given value of x—say, x_0. For example, we may want to predict the food expenditure of a randomly selected household with a monthly income of $5500. In this case, we are not interested in the mean food expenditure of all households with a monthly income of $5500 but in the food expenditure of one particular household with a monthly income of $5500. This predicted value of y is denoted by $\mathbf{y_p}$. Again, to predict a single value of y for $x = x_0$ from the estimated sample regression line, we use the value of $\hat{y}$ **as a point estimate of** $\mathbf{y_p}$. Using the estimated regression line, we find that $\hat{y}$ for $x = 55$ is

$$\hat{y} = 1.5050 + .2525(55) = \$15.3925 \text{ hundred}$$

Thus, based on our regression line, the point estimate for the food expenditure of a given household with a monthly income of $5500 is $1539.25 per month. Note that $\hat{y} = 1539.25$ is the point estimate for the mean food expenditure for all households with $x = 55$ as well as for the predicted value of food expenditure of one household with $x = 55$.

Different regression lines estimated by using different samples of seven households each taken from the same population will give different values of the point estimator for the predicted value of y for $x = 55$. Hence, a confidence interval constructed for y_p based on one sample will give a more reliable estimate of y_p than will a point estimate. The confidence interval constructed for y_p is more commonly called a **prediction interval**.

The procedure for constructing a prediction interval for y_p is similar to that for constructing a confidence interval for $\mu_{y|x}$ except that the standard deviation of $\hat{y}$ is larger when we predict a single value of y than when we estimate $\mu_{y|x}$.

The point estimator $\hat{y}$ of y_p is normally distributed with a mean of $A + Bx$ and a standard deviation of

$$\sigma_{\hat{y}_p} = \sigma_\varepsilon \sqrt{1 + \frac{1}{n} + \frac{(x_0 - \bar{x})^2}{SS_{xx}}}$$

where $\sigma_{\hat{y}_p}$ is the standard deviation of the predicted value of y, x_0 is the value of x for which we are predicting y, and σ_ε is the population standard deviation of ε.

However, usually σ_ε is not known. In this case, we replace σ_ε by s_e and $\sigma_{\hat{y}_p}$ by $s_{\hat{y}_p}$ in the foregoing expression. To make a prediction interval for y_p, we use the t distribution when σ_ε is not known.

Prediction Interval for y_p   The $(1 - \alpha)100\%$ **prediction interval** for the predicted value of y, denoted by y_p, for $x = x_0$ is

$$\hat{y} \pm t s_{\hat{y}_p}$$

where the value of t is obtained from the t distribution table for $\alpha/2$ area in the right tail of the t distribution curve and $df = n - 2$. The value of $s_{\hat{y}_p}$ is calculated as follows:

$$s_{\hat{y}_p} = s_e \sqrt{1 + \frac{1}{n} + \frac{(x_0 - \bar{x})^2}{SS_{xx}}}$$

Example 13–10 illustrates the procedure to make a prediction interval for a particular value of y.

EXAMPLE 13–10   Incomes and Food Expenditures of Households

Making a prediction interval for a particular value of y for a given x.

Refer to Example 13–1 on incomes and food expenditures. Find a 99% prediction interval for the predicted food expenditure for a randomly selected household with a monthly income of $5500.

Solution   Using the regression line estimated in Example 13–1, we find the point estimate of the predicted food expenditure for $x = 55$:

$$\hat{y} = 1.5050 + .2525(55) = \$15.3925 \text{ hundred}$$

The area in each tail of the t distribution for a 99% confidence level is

$$\alpha/2 = (1 - .99)/2 = .005$$

The degrees of freedom are

$$df = n - 2 = 7 - 2 = 5$$

From the t distribution table, the t value for .005 area in the right tail of the t distribution curve and 5 df is 4.032. From calculations in Examples 13–1 and 13–2,

$$s_e = 1.5939, \quad \bar{x} = 55.1429, \quad \text{and} \quad SS_{xx} = 1772.8571$$

The standard deviation of $\hat{y}$ as an estimator of y_p for $x = 55$ is calculated as follows:

$$s_{\hat{y}_p} = s_e \sqrt{1 + \frac{1}{n} + \frac{(x_0 - \bar{x})^2}{SS_{xx}}}$$

$$= (1.5939) \sqrt{1 + \frac{1}{7} + \frac{(55 - 55.1429)^2}{1772.8571}} = 1.7040$$

Hence, the 99% prediction interval for y_p for $x = 55$ is

$$\hat{y} \pm t s_{\hat{y}_p} = 15.3925 \pm 4.032(1.7040)$$

$$= 15.3925 \pm 6.8705 = \mathbf{8.5220 \text{ to } 22.2630}$$

Thus, with 99% confidence we can state that the predicted food expenditure of a household with a monthly income of $5500 is between $852.20 and $2226.30.

As we can observe in Example 13–10, this interval is much wider than the one for the mean value of y for $x = 55$ calculated in Example 13–9, which was $1296.32 to $1782.18. This is always true. The prediction interval for predicting a single value of y is always larger than the confidence interval for estimating the mean value of y for a certain value of x.

EXERCISES

CONCEPTS AND PROCEDURES

13.67 Briefly explain the difference between estimating the mean value of y and predicting a particular value of y using a regression model.

13.68 Construct a 95% confidence interval for the mean value of y and a 95% prediction interval for the predicted value of y for the following.

   a. $\hat{y} = 13.40 + 2.58x$ for $x = 8$ given $s_e = 1.29$, $\bar{x} = 11.30$, $SS_{xx} = 210.45$, and $n = 12$
   b. $\hat{y} = -8.6 + 3.72x$ for $x = 24$ given $s_e = 1.89$, $\bar{x} = 19.70$, $SS_{xx} = 315.40$, and $n = 10$

13.69 Construct a 99% confidence interval for the mean value of y and a 99% prediction interval for the predicted value of y for the following.

   a. $\hat{y} = 3.25 + .80x$ for $x = 15$ given $s_e = .954$, $\bar{x} = 18.52$, $SS_{xx} = 144.65$, and $n = 10$
   b. $\hat{y} = -27 + 7.67x$ for $x = 12$ given $s_e = 2.46$, $\bar{x} = 13.43$, $SS_{xx} = 369.77$, and $n = 10$

APPLICATIONS

13.70 The following data give the experience (in years) and monthly salaries (in hundreds of dollars) of nine randomly selected secretaries.

Experience	14	3	5	6	4	9	18	5	16
Monthly salary	62	29	37	43	35	60	67	32	60

Construct a 90% confidence interval for the mean monthly salary of all secretaries with 10 years of experience. Construct a 90% prediction interval for the monthly salary of a randomly selected secretary with 10 years of experience.

13.71 The following table gives information on the incomes (in thousands of dollars) and charitable contributions (in hundreds of dollars) for the last year for a random sample of 10 households.

Income	Charitable Contributions
76	15
57	4
140	42
97	33
75	5
107	32
65	10
77	18
102	28
53	4

Construct a 95% confidence interval for the mean charitable contributions made by all households with an income of $84,000. Make a 95% prediction interval for the charitable contributions made by a randomly selected household with an income of $84,000.

13.72 The owner of a small factory that produces work gloves is concerned about the high cost of air conditioning in the summer, but he is afraid that keeping the temperature in the factory too high will lower productivity. During the summer, he experiments with temperature settings from 68°F to 81°F and measures each day's productivity. The following table gives the temperature and the number of pairs of gloves (in hundreds) produced on each of the 8 randomly selected days.

Temperature (°F)	72	71	78	75	81	77	68	76
Pairs of gloves	37	37	32	36	33	35	39	34

Construct a 99% confidence interval for $\mu_{y|x}$ for $x = 77$ and a 99% prediction interval for y_p for $x = 77$. Here pairs of gloves is the dependent variable.

13.73 The following table gives information on the limited tread warranties (in thousands of miles) and the prices of 12 randomly selected tires at a national tire retailer as of July 2012.

Warranty (thousands of miles)	60	70	75	50	80	55	65
Price per tire ($)	95	135	94	90	121	70	140
Warranty (thousands of miles)	65	70	65	60	65		
Price per tire ($)	80	92	125	160	155		

Construct a 95% confidence interval for the mean price of all tires that have a 65,000-mile limited tread warranty. Construct a 95% prediction interval for the price of a randomly selected tire that has a 65,000-mile limited tread warranty.

USES AND MISUSES...

1. PROCESSING ERRORS

Stuck on the far right side of the linear regression model is the Greek letter epsilon, ε. Despite its diminutive size, proper respect for the error term is critical to good linear regression modeling and analysis.

One interpretation of the error term is that it is a process. Imagine you are a chemist and you have to weigh a number of chemicals for an experiment. The balance that you use in your laboratory is very accurate—so accurate, in fact, that the shuffling of your feet, your exhaling near it, or the rumbling of trucks on the road outside can cause the reading to fluctuate. Because the value of the measurement that you take will be affected by a number of factors out of your control, you must make several measurements for each chemical, note each measurement, and then take the means and standard deviations of your samples. The distribution of measurements around a mean is the result of a random error process dependent on a number of factors out of your control; each time you use the balance, the measurement you take is the sum of the actual mass of the chemical and a "random" error. In this example, the measurements will most likely be normally distributed around the mean.

Linear regression analysis makes the same assumption about the two variables you are comparing: The value of the dependent variable is a linear function of the independent variable, plus a little bit of error that you cannot control. Unfortunately, when working with economic or survey data, you rarely can duplicate an experiment to identify the error model. As a statistician, however, you can use the errors to help you refine your model of the relationship among the variables and to guide your collection of new data. For example, if the errors are skewed to the right for moderate values of the independent variable and skewed to the left for small and large values of the independent variable, you can modify your model to account for this difference. Or you can think about other relationships among the variables that might explain this particular distribution of errors. A detailed analysis of the error in your model can be just as instructive as analysis of the slope and y-intercept of the identified model.

2. OUTLIERS AND CORRELATION

In Chapter 3 we learned that outliers can affect the values of some of the summary measures such as the mean, standard deviation, and range. Note that although outliers do affect many other summary measures, these three are affected substantially. Here we will see that just looking at a number that represents the correlation coefficient does not provide the entire story. A very famous data set for demonstrating this concept was created by F. J. Anscombe (Anscombe, F. J., Graphs in Statistical Analysis, *American Statistician*, 27, pp. 17–21). He created four pairs of data sets on x and y variables, each of which has a correlation of .816. To the novice, it may seem that the scatterplots for these four data sets should look virtually the same, but that may not be true. Look at the four scatterplots shown in Figure 13.25.

No two of these scatterplots are even remotely close to being the same or even similar. The data used in the upper left plot are linearly associated, as are the data in the lower left plot. However the plot of $y3$ versus $x3$ contains an outlier. Without this outlier, the correlation between $x3$ and $y3$ would be 1. On the other hand, there is

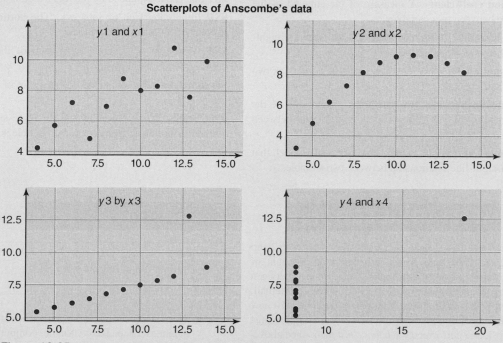

Figure 13.25 Four scatterplots with the same correlation coefficient.

much more variability in the relationship between $x1$ and $y1$. As far as $x4$ and $y4$ are concerned, the strong correlation is defined by the single point in the upper right corner of the scatterplot. Without this point, there would be no variability among the $x4$ values, and the correlation would be undefined. Lastly, the scatterplot of $y2$ versus $x2$ reveals that there is an extremely well-defined relationship between these variables, but it is not linear. Being satisfied that the correlation

coefficient is close to 1.0 between variables $x2$ and $y2$ implies that there is a strong linear association between the variables when actually we are fitting a line to a set of data that should be represented by another type of mathematical function.

As we have mentioned before, the process of making a graph may seem trivial, but the importance of graphs in our analysis can never be overstated.

GLOSSARY

Coefficient of determination A measure that gives the proportion (or percentage) of the total variation in a dependent variable that is explained by a given independent variable.

Degrees of freedom for a simple linear regression model Sample size minus 2; that is, $n - 2$.

Dependent variable The variable to be predicted or explained.

Deterministic model A model in which the independent variable determines the dependent variable exactly. Such a model gives an exact relationship between two variables.

Estimated or **predicted value of y** The value of the dependent variable, denoted by $\hat{y}$, that is calculated for a given value of x using the estimated regression model.

Independent or **explanatory variable** The variable included in a model to explain the variation in the dependent variable.

Least squares estimates of A and B The values of a and b that are calculated by using the sample data.

Least squares method The method used to fit a regression line through a scatter diagram such that the error sum of squares is minimum.

Least squares regression line A regression line obtained by using the least squares method.

Linear correlation coefficient A measure of the strength of the linear relationship between two variables.

Linear regression model A regression model that gives a straight-line relationship between two variables.

Multiple regression model A regression model that contains two or more independent variables.

Negative relationship between two variables The value of the slope in the regression line and the correlation coefficient between two variables are both negative.

Nonlinear (simple) regression model A regression model that does not give a straight-line relationship between two variables.

Population parameters for a simple regression model The values of A and B for the regression model $y = A + Bx + \varepsilon$ that are obtained by using population data.

Positive relationship between two variables The value of the slope in the regression line and the correlation coefficient between two variables are both positive.

Prediction interval The confidence interval for a particular value of y for a given value of x.

Probabilistic or **statistical model** A model in which the independent variable does not determine the dependent variable exactly.

Random error term (ε) The difference between the actual and predicted values of y.

Scatter diagram or **scatterplot** A plot of the paired observations of x and y.

Simple linear regression A regression model with one dependent and one independent variable that assumes a straight-line relationship.

Slope The coefficient of x in a regression model that gives the change in y for a change of one unit in x.

SSE (error sum of squares) The sum of the squared differences between the actual and predicted values of y. It is the portion of the SST that is not explained by the regression model.

SSR (regression sum of squares) The portion of the SST that is explained by the regression model.

SST (total sum of squares) The sum of the squared differences between actual y values and $\bar{y}$.

Standard deviation of errors A measure of spread for the random errors.

y-intercept The point at which the regression line intersects the vertical axis on which the dependent variable is marked. It is the value of y when x is zero.

SUPPLEMENTARY EXERCISES

13.74 The recommended air pressure in a basketball is between 7 and 9 pounds per square inch (psi). When dropped from a height of 6 feet, a properly inflated basketball should bounce upward between 52 and 56 inches (http://www.bestsoccerbuys.com/balls-basketball. html). The basketball coach at a local high school purchased 10 new basketballs for the upcoming season, inflated the balls to pressures between 7 and 9 psi, and performed the *bounce test* mentioned above. The data obtained are given in the following table.

Pressure (psi)	7.8	8.1	8.3	7.4	8.9	7.2	8.6	7.5	8.1	8.5
Bounce height (inches)	54.1	54.3	55.2	53.3	55.4	52.2	55.7	54.6	54.8	55.3

a. With the pressure as an independent variable and bounce height as a dependent variable, compute SS_{xx}, SS_{yy}, and SS_{xy}.

b. Find the least squares regression line.

c. Interpret the meaning of the values of a and b calculated in part b.

d. Calculate r and r^2 and explain what they mean.

e. Compute the standard deviation of errors.

f. Predict the bounce height of a basketball for $x = 8.0$.

g. Construct a 98% confidence interval for B.

h. Test at a 5% significance level whether B is different from zero.

i. Using $\alpha = .05$, can you conclude that ρ is different from zero?

13.75 The management of a supermarket wants to find if there is a relationship between the number of times a specific product is promoted on the intercom system in the store and the number of units of that product sold. To experiment, the management selected a product and promoted it on the intercom system for 7 days. The following table gives the number of times this product was promoted each day and the number of units sold.

Number of Promotions per Day	Number of Units Sold per Day (hundreds)
15	11
22	22
42	30
30	26
18	17
12	15
38	23

a. With the number of promotions as an independent variable and the number of units sold as a dependent variable, what do you expect the sign of B in the regression line $y = A + Bx + \varepsilon$ will be?

b. Find the least squares regression line $\hat{y} = a + bx$. Is the sign of b the same as you hypothesized for B in part a?

c. Give a brief interpretation of the values of a and b calculated in part b.

d. Compute r and r^2 and explain what they mean.

e. Predict the number of units of this product sold on a day with 35 promotions.

f. Compute the standard deviation of errors.

g. Construct a 98% confidence interval for B.

h. Testing at a 1% significance level, can you conclude that B is positive?

i. Using $\alpha = .02$, can you conclude that the correlation coefficient is different from zero?

13.76 The following table provides information on the living area (in square feet) and price (in thousands of dollars) of 10 randomly selected houses listed for sale in a city.

Living area	3008	2032	2272	1840	2579	2583	1650	3932	2978	2176
Price	275	220	255	189	260	284	172	370	295	260

a. Find the least squares regression line $\hat{y} = a + bx$. Take living area as an independent variable and price as a dependent variable.

b. Give a brief interpretation of the values of a and b.

c. Compute r and r^2 and explain what they mean.

d. Predict the price of a house with 2700 square feet of living area.

e. Compute the standard deviation of errors.

f. Construct a 99% confidence interval for B.

g. Testing at a 1% significance level, can you conclude that B is different from zero?

h. Using $\alpha = .01$, can you conclude that the correlation coefficient is different from zero?

13.77 A local ice cream parlor wants to determine whether the temperature has an effect on its business. The following table contains data on the temperature at 7 PM on 10 rain-free weekend days during the summer and the number of customers served by this ice cream parlor.

Temperature (°F)	68	63	74	72	79	78	71	71	69	66
Customers served	317	355	463	419	507	482	433	388	362	340

a. With temperature as an independent variable and number of customers as a dependent variable, compute SS_{xx}, SS_{yy}, and SS_{xy}.

b. Construct a scatter diagram for these data. Does the scatter diagram exhibit a positive linear relationship between temperature and the number of customers served?

c. Find the regression equation $\hat{y} = a + bx$.

d. Give a brief interpretation of the values of a and b calculated in part c.

e. Compute the correlation coefficient r.

f. Predict the number of customers served on a rain-free weekend summer day when the temperature is 73°F. Returning to part b, how reliable do you think this prediction will be? Explain.

13.78 The recommended air pressure in a basketball is between 7 and 9 pounds per square inch (psi). When dropped from a height of 6 feet, a properly inflated basketball should bounce upward between 52 and 56 inches (http://www.bestsoccerbuys.com/balls-basketball.html). The basketball coach at a local high school purchased 10 new basketballs for the upcoming season, inflated the balls to pressures between 7 and 9 psi, and performed the *bounce test* mentioned above. The data obtained are given in the following table.

Pressure (psi)	7.8	8.1	8.3	7.4	8.9	7.2	8.6	7.5	8.1	8.5
Bounce height (inches)	54.1	54.3	55.2	53.3	55.4	52.2	55.7	54.6	54.8	55.3

Construct a 99% confidence interval for the mean bounce height of all basketballs that are inflated to 8.5 psi. Construct a 99% prediction interval for the bounce height of a randomly selected basketball that is inflated to 8.5 psi. Here bounce height is the dependent variable.

13.79 The following table gives information on GPAs and starting salaries (rounded to the nearest thousand dollars) of seven recent college graduates.

GPA	2.90	3.81	3.20	2.42	3.94	2.05	2.25
Starting salary	48	53	50	37	65	32	37

Construct a 98% confidence interval for the mean starting salary of recent college graduates with a GPA of 3.15. Construct a 98%

prediction interval for the starting salary of a randomly selected recent college graduate with a GPA of 3.15.

13.80 Refer to the data given in Exercise 13.76 on the living area (in square feet) and price (in thousands of dollars) of 10 randomly

selected houses listed for sale in a city. Construct a 98% confidence interval for the mean price of all houses with living areas of 2400 square feet. Construct a 98% prediction interval for the price of a randomly selected house with a living area of 2400 square feet.

ADVANCED EXERCISES

13.81 Suppose that you work part-time at a bowling alley that is open daily from noon to midnight. Although business is usually slow from noon to 6 PM, the owner has noticed that it is better on hotter days during the summer, perhaps because the premises are comfortably air-conditioned. The owner shows you some data that she gathered last summer. This data set includes the maximum temperature and the number of lines bowled between noon and 6 PM for each of 20 days. (The maximum temperatures ranged from 77°F to 95°F during this period.) The owner would like to know if she can estimate tomorrow's business from noon to 6 PM by looking at tomorrow's weather forecast. She asks you to analyze the data. Let x be the maximum temperature for a day and y the number of lines bowled between noon and 6 PM on that day. The computer output based on the data for 20 days provided the following results:

$$\hat{y} = -432 + 7.7x, \quad s_e = 29.17, \quad SS_{xx} = 607, \quad \text{and} \quad \bar{x} = 87.5$$

Assume that the weather forecasts are reasonably accurate.

   a. Does the maximum temperature seem to be a useful predictor of bowling activity between noon and 6 PM? Use an appropriate statistical procedure based on the information given. Use $\alpha = .05$.
   b. The owner wants to know how many lines of bowling she can expect, on average, for days with a maximum temperature of 90°. Answer using a 95% confidence level.
   c. The owner has seen tomorrow's weather forecast, which predicts a high of 90°F. About how many lines of bowling can she expect? Answer using a 95% confidence level.
   d. Give a brief commonsense explanation to the owner for the difference in the interval estimates of parts b and c.
   e. The owner asks you how many lines of bowling she could expect if the high temperature were 100°F. Give a point estimate, together with an appropriate warning to the owner.

13.82 For the past 25 years Burton Hodge has been keeping track of how many times he mows his lawn and the average size of the ears of corn in his garden. Hearing about the Pearson correlation coefficient from a statistician buddy of his, Burton decides to substantiate his

suspicion that the more often he mows his lawn, the bigger are the ears of corn. He does so by computing the correlation coefficient. Lo and behold, Burton finds a .93 coefficient of correlation! Elated, he calls his friend the statistician to thank him and announce that next year he will have prize-winning ears of corn because he plans to mow his lawn every day. Do you think Burton's logic is correct? If not, how would you explain to Burton the mistake he is making in his presumption (without eroding his new opinion of statistics)? Suggest what Burton could do next year to make the ears of corn large, and relate this to the Pearson correlation coefficient.

13.83 Consider the formulas for calculating a prediction interval for a new (specific) value of y. For each of the changes mentioned in parts a through c below, state the effect on the width of the confidence interval (increase, decrease, or no change) and why it happens. Note that besides the change mentioned in each part, everything else such as the values of a, b, $\bar{x}$, s_e, and SS_{xx} remains unchanged.
   a. The confidence level is increased.
   b. The sample size is increased.
   c. The value of x_0 is moved farther away from the value of $\bar{x}$.
   d. What will the value of the margin of error be if x_0 equals $\bar{x}$?

13.84 It seems reasonable that the more hours per week a full-time college student works at a job, the less time he or she will have to study and, consequently, the lower his or her GPA would be.
   a. Assuming a linear relationship, suggest specifically what the equation relating x and y would be, where x is the average number of hours a student works per week and y represents the student's GPA. Try several values of x and see if your equation gives reasonable values of y.
   b. Using the following observations taken from 10 randomly selected students, compute the regression equation and compare it to yours of part a.

Average number of hours worked	20	28	10	35	5	14	0	40	8	23
GPA	2.8	3.2	3.1	2.4	3.4	3.3	2.8	2.1	3.8	1.8

SELF-REVIEW TEST

1. A simple regression is a regression model that contains
   a. only one independent variable
   b. only one dependent variable
   c. more than one independent variable
   d. both a and b

2. The relationship between independent and dependent variables represented by the (simple) linear regression is that of
   a. a straight line   **b.** a curve   **c.** both a and b

3. A deterministic regression model is a model that
   a. contains the random error term
   b. does not contain the random error term
   c. gives a nonlinear relationship

4. A probabilistic regression model is a model that
   a. contains the random error term
   b. does not contain the random error term
   c. shows an exact relationship

5. The least squares regression line minimizes the sum of
   a. errors   **b.** squared errors   **c.** predictions

6. The degrees of freedom for a simple regression model are
   a. $n-1$ **b.** $n-2$ **c.** $n-5$

7. Is the following statement true or false?

   The coefficient of determination gives the proportion of total squared errors (SST) that is explained by the use of the regression model.

8. Is the following statement true or false?

   The linear correlation coefficient measures the strength of the linear association between two variables.

9. The value of the coefficient of determination is always in the range
   a. 0 to 1   **b.** -1 to 1   **c.** -1 to 0

10. The value of the correlation coefficient is always in the range
    a. 0 to 1   **b.** -1 to 1   **c.** -1 to 0

11. Explain why the random error term ε is added to the regression model.

12. Explain the difference between A and a and between B and b for a regression model.

13. Briefly explain the assumptions of a regression model.

14. Briefly explain the difference between the population regression line and a sample regression line.

15. The following table gives the temperatures (in degrees Fahrenheit) at 6 PM and the attendance (rounded to hundreds) at a minor league baseball team's night games on 7 randomly selected evenings in May.

Temperature	61	70	50	65	48	75	53
Attendance	10	16	12	15	8	20	18

a. Do you think temperature depends on attendance or attendance depends on temperature?

b. With temperature as an independent variable and attendance as a dependent variable, what is your hypothesis about the sign of B in the regression model?

c. Construct a scatter diagram for these data. Does the scatter diagram exhibit a linear relationship between the two variables?

d. Find the least squares regression line. Is the sign of b the same as the one you hypothesized for B in part b?

e. Give a brief interpretation of the values of the y-intercept and slope calculated in part d.

f. Compute r and r^2, and explain what they mean.

g. Predict the attendance at a night game in May for a temperature of 60°F.

h. Compute the standard deviation of errors.

i. Construct a 99% confidence interval for B.

j. Testing at a 1% significance level, can you conclude that B is positive?

k. Construct a 95% confidence interval for the mean attendance at a night game in May when the temperature is 60°F.

l. Make a 95% prediction interval for the attendance at a night game in May when the temperature is 60°F.

m. Using a 1% significance level, can you conclude that the linear correlation coefficient is positive?

MINI-PROJECTS

Note: The Mini-Projects are located on the text's Web site, www.wiley.com/college/mann.

DECIDE FOR YOURSELF

Note: The Decide for Yourself feature is located on the text's Web site, www.wiley.com/college/mann.

TECHNOLOGY INSTRUCTIONS   CHAPTER 13

Note: Complete TI-84, Minitab, and Excel manuals are available for download at the textbook's Web site, www.wiley.com/college/mann.

TI-84 Color/TI-84

The TI-84 Color Technology Instructions feature of this text is written for the TI-84 Plus C color graphing calculator running the 4.0 operating system. Some screens, menus, and functions will be slightly different in older operating systems. The TI-84 and TI-84 Plus can perform all of the same functions but will not have the "Color" option referenced in some of the menus.

Finding the Regression Equation, r^2, and r for Example 13–1 of the Text

1. Enter the data from Example 13–1 into L1 (Income) and L2 (Food Expenditure).

2. Select **STAT > CALC > LinReg(a+bx)**.

Screen 13.1

Screen 13.2

3. Use the following settings in the **LinReg(a+bx)** menu (see **Screen 13.1**):

- Enter L1 at the **Xlist** prompt.

- Enter L2 at the **Ylist** prompt.

- Leave the **FreqList** prompt blank.

    Note: If you have another list with the frequency for each ordered pair, enter the name of that list at this prompt.

- Enter Y_1 at the **Store RegEQ** prompt.

    Note: To get Y_1, select **VARS > Y-VARS > Function > Y_1**.

4. Highlight **Calculate** and press **ENTER**.

5. The output includes the least squares regression equation, the coefficient of determination, and the correlation coefficient. In addition, the regression equation will be stored as function Y_1. (See **Screen 13.2**.)

    Note: If you did not see the values of r and r^2, select $2^{nd} > 0 >$ **DiagnosticOn** and press **ENTER** once the function **DiagnosticOn** is pasted onto the home screen. You should see **Done** once the calculator has executed this command. This needs to be done only once.

Constructing a Confidence Interval for Slope for Example 13–4 of the Text

1. Enter the data from Table 13.1 into L1 (Income) and L2 (Food Expenditure).

2. Select **STAT > TESTS > LinRegTInt**.

3. Use the following settings in the **LinRegTInt** menu (see **Screen 13.3**):

- Enter L1 at the **Xlist** prompt.

- Enter L2 at the **Ylist** prompt.

- Type 1 at the **Freq** prompt.

    Note: If you have another list with the frequency for each ordered pair, enter the name of that list at this prompt.

- Type 0.95 at the **C-Level** prompt.

- Enter Y_1 at the **RegEQ** prompt.

    Note: To get Y_1, select **VARS > Y-VARS > Function > Y_1**.

4. Highlight **Calculate** and press **ENTER**.

5. The output includes the confidence interval, the value of b, the degrees of freedom, the regression equation coefficients, the standard deviation of errors, the coefficient of determination, and the correlation coefficient. In addition, the regression equation will be stored as function Y_1. (See **Screen 13.4**.)

Screen 13.3

Screen 13.4

Testing a Hypothesis About Slope or Correlation for Example 13–5 of the Text

1. Enter the data from Table 13.1 into L1 (Income) and L2 (Food Expenditure).

2. Select **STAT > TESTS > LinRegTTest**.

3. Use the following settings in the **LinRegTTest** menu (see **Screen 13.5**):

- Enter L1 at the **Xlist** prompt.

- Enter L2 at the **Ylist** prompt.

Screen 13.5

- Type 1 at the **Freq** prompt.

    Note: If you have another list with the frequency for each ordered pair, enter the name of that list at this prompt.

- Select **>0** at the β **&** ρ prompt.

- Enter Y$_1$ at the **RegEQ** prompt.

    Note: To get Y$_1$, select **VARS > Y-VARS > Function > Y$_1$**.

4. Highlight **Calculate** and press **ENTER**.

5. The output includes the test statistic, the *p*-value, the degrees of freedom, the regression equation coefficients, the standard deviation of errors, the coefficient of determination, and the correlation coefficient (you must scroll down to see *r* and r^2). In addition, the regression equation will be stored as function Y$_1$. (See **Screen 13.6**.)

Now compare the *t*-value to the critical value of *t* or the *p*-value from Screen 13.6 with α and make a decision.

Screen 13.6

Minitab

The Minitab Technology Instructions feature of this text is written for Minitab version 17. Some screens, menus, and functions will be slightly different in older versions of Minitab.

Finding the Regression Equation and r^2 for Example 13–1 of the Text

1. Enter the data from Example 13–1 into C1 (Income) and C2 (Food Expenditure).

2. Select **Stat > Regression > Regression > Fit Regression Model**.

3. Use the following settings in the dialog box that appears on screen:

- Type C2 in the **Responses** box.

- Type C1 in the **Continuous predictors** box.

- Click on the **Options** button and enter 95 in the **Confidence level for all intervals** box. Click **OK**.

- Click on the **Results** button and select Expanded tables from the **Display of results** drop-down menu. Click **OK**.

4. Click **OK**.

5. The output will be displayed in the Session window. The output includes the value of r^2 and the regression equation. (See **Screen 13.7**.)

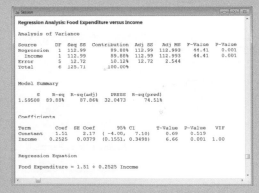

Screen 13.7

Finding *r* for Example 13–6 of the Text

1. Enter the data from Table 13.1 into C1 (Income) and C2 (Food Expenditure).

2. Select **Stat > Basic Statistics > Correlation**. (See **Screen 13.8**.)

3. Use the following settings in the dialog box that appears on screen:

- Type C1 C2 in the **Variables** box. Be sure to separate C1 and C2 with a space.

- Uncheck the **Display *p*-values** check box.

4. Click **OK**.

5. The output will be displayed in the Session window.

Screen 13.8

Constructing a Confidence Interval for Slope for Example 13–4 of the Text

Follow the directions shown above for finding the regression equation. The output includes the confidence interval in the **Coefficients** table in the **Income** row. (See **Screen 13.7**.)

Testing a Hypothesis About Slope or Correlation for Example 13–5 of the Text

Follow the directions shown above for finding the regression equation. The output also includes the test statistic t and p-value in the **Coefficients** table in the **Income** row. The p-value is for a two-sided test, so if a one-sided test is required, be sure to adjust the p-value accordingly. (See **Screen 13.7**.)

Now compare the t-value to the critical value of t or the p-value from Screen 13.7 with α and make a decision.

Constructing a Prediction Interval for Slope for Example 13–10 of the Text

1. Enter the data from Table 13.1 into C1 (Income) and C2 (Food Expenditure).

2. To create a prediction interval, you must first find the regression equation (fit a regression model) as described above. If you do not fit the regression model, you will not get the correct output.

3. Select **Stat > Regression > Regression > Predict**.

4. Use the following settings in the dialog box that appears on screen (see **Screen 13.9**):

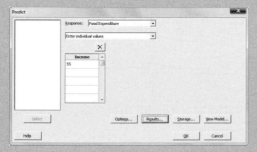

   - Select **Food Expenditure (or C2)** from the **Response** drop-down menu.

   - Select **Enter individual values** from second drop-down menu.

   - Type 55 in the list below **Income (or C1)**.

   - Click on the **Options** button and enter 99 in the **Confidence level** box. Click **OK**.

5. Click **OK**.

6. The output, including the prediction interval, will be displayed in the Session window.

Screen 13.9

Excel

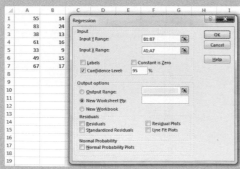

Screen 13.10

The Excel Technology Instructions feature of this text is written for Excel 2013. Some screens, menus, and functions will be slightly different in older versions of Excel. To enable some of the advanced Excel functions, you must enable the Data Analysis Add-In for Excel.

Finding the Regression Equation, r^2, and r for Example 13–1 of the Text

1. In a new worksheet, enter the data from Example 13–1 into cells A1-A7 (Income) and cells B1-B7 (Food Expenditure). (See **Screen 13.10**.)

2. Click **DATA** and then click **Data Analysis Tools** from the **Analysis** group.

3. Select **Regression** from the dialog box that appears on screen and then click **OK**.

4. Use the following settings in the dialog box that appears on screen (see **Screen 13.10**):

- Type B1:B7 in the **Input Y Range** box.

- Type A1:A7 in the **Input X Range** box.

- Check the **Confidence Level** check box and type 95 in the **Confidence Level** box.

- Select New Worksheet Ply from the **Output options**.

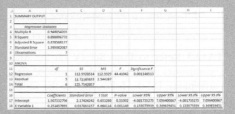

Screen 13.11

5. Click **OK**.

6. When the output appears, resize the columns so that they are easier to read. The output includes the coefficients of the least squares regression equation, the coefficient of determination (R square), and the correlation coefficient (Multiple R). (See **Screen 13.11**.)

Constructing a Confidence Interval for Slope for Example 13–4 of the Text

Follow the directions shown above for finding the regression equation. The output also includes the lower and upper bounds of the confidence interval in the last table of the output in the **X Variable 1** row. (See **Screen 13.11**.)

Testing a Hypothesis About Slope or Correlation for Example 13–5 of the Text

Follow the directions shown above for finding the regression equation. The output also includes the test statistic t and p-value in the last table of the output in the **X Variable 1** row. The p-value is for a two-sided test, so if a one-sided test is required, be sure to adjust the p-value accordingly. (See **Screen 13.11**.)

Now compare the t-value to the critical value of t or the p-value from Screen 13.11 with α and make a decision.

TECHNOLOGY ASSIGNMENTS

TA13.1 In a rainy coastal town in the Pacific Northwest, the local TV weatherman is often criticized for making inaccurate forecasts for daily precipitation. On each of 30 randomly selected days last winter, his precipitation forecast (x) for the next day was recorded along with the actual precipitation (y) for that day. These data are shown in the following table (in inches of rain).

x	y	x	y	x	y
1.0	.6	0	0	.4	.2
0	.1	0	.1	.2	.5
.2	0	.1	.2	.1	.1
0	0	.2	.2	0	.2
.5	.3	.1	0	.1	0
1.0	1.4	2.0	2.1	.2	.1
.5	.3	.4	.2	1.4	1.2
.1	.1	.2	.1	.5	1.0
0	.1	0	0	0	.5
2.0	.3	.3	.2	0	0

a. Construct a scatter diagram for these data.

b. Find the correlation coefficient between the two variables.

c. Find the regression line with actual precipitation as a dependent variable and predicted precipitation as an independent variable.

d. Make a 95% confidence interval for B.

e. Test at a 1% significance level whether B is positive.

f. Using a 1% significance level, can you conclude that the linear correlation coefficient is positive?

TA13.2 The following data give information on the ages (in years) and the number of breakdowns during the last month for a sample of seven machines at a large company.

Age (in years)	12	7	2	8	13	9	4
Number of breakdowns	10	5	1	4	12	7	2

a. Construct a scatter diagram for these data.

b. Find the least squares regression line with age as an independent variable and the number of breakdowns as a dependent variable.

c. Compute the correlation coefficient.

d. Construct a 99% confidence interval for *B*.

e. Test at a 2.5% significance level whether *B* is positive.

TA13.3 Refer to Data Set IV on the Manchester Road Race that accompanies this text (see Appendix A). Take a random sample of 45 participants. Do the following for the variables on ages and times (in minutes) for this sample.

a. Construct a scatter diagram for these data, using age as the independent variable. Discuss whether it is appropriate to fit a linear regression model to these data.

b. Find the correlation coefficient for these two variables.

c. Find the equation of the regression line with age as the independent variable and time as the dependent variable.

d. Make a 98% confidence interval for *B*. Explain what this interval means with regard to a person's time for each additional year of age.

e. Test at a 1% significance level whether *B* is positive.

f. Test at a 1% significance level whether *B* is greater than .10.

APPENDIX A

Explanation of Data Sets

This textbook is accompanied by 13 large data sets that can be used for statistical analysis using technology. These data sets are

Data Set I	City Data
Data Set II	Data on States
Data Set III	NFL Data
Data Set IV	Manchester Road Race Data
Data Set V	Data on Movies
Data Set VI	Standard & Poor's 500 Index Data
Data Set VII	McDonald's Data
Data Set VIII	Data on NFL Kickers
Data Set IX	Subway Menu Items Data
Data Set X	Major League Baseball Data
Data Set XI	Data on Airline Flights
Data Set XII	Data on Coffee Maker Ratings
Data Set XIII	Simulated Data
Data Set XIV	Billboard Data

These data sets are available in Minitab, Excel, and a few other formats on the Web site for this text, www.wiley.com/college/mann. Once you are on this Web site, click on the companion sites next to the cover of the book. Choose one of the two sites. Search for Data Sets on the drop-down menus and then click on Data Sets. These data sets can be downloaded from this Web site. If you need more information on these data sets, you may either contact the John Wiley area representative or send an email to the author (see Preface). The Web site contains the following files:

1. CITYDATA (This file contains Data Set I)
2. STATEDATA (This file contains Data Set II)
3. NFL (This file contains Data Set III)
4. ROADRACE (This file contains Data Set IV)
5. MOVIEDATA (This file contains Data Set V)
6. S&PDATA (This file contains Data Set VI)
7. MCDONALDSDATA (This file contains Data Set VII)
8. NFLKICKERS (This file contains Data Set VIII)
9. SUBWAYMENU (This file contains Data Set IX)
10. BASEBALL (This file contains Data Set X)
11. AIRLINEFLIGHTS (This file contains Data Set XI)
12. COFFEEMAKERRATINGS (This file contains Data Set XII)
13. SIMULATED (This file contains Data Set XIII)
14. BILLBOARD DATA (This file contains Data Set XIV)

The following are the explanations of these data sets.

Data Set I: City Data[1]

This data set contains prices (in dollars) of selected products and services for selected urban areas across the United States. This data set is reproduced from the C2ER Cost of Living Index (COLI) survey and shows the average prices for the first quarter of 2015. It is reproduced with the permission of the Center for Regional Economic Competitiveness. This data set has 21 columns that contain the following variables:

C1   Name of the urban area

C2   Name of the state

C3   Price of T-bone steak per pound

C4   Price of 1 pound of ground beef (minimum 80% lean)

C5   Price per pound of a whole fryer chicken

C6   Price of one half-gallon carton of whole milk

C7   Price of one dozen large eggs, grade A/AA

C8   Price of a loaf of white bread

C9   Price of fresh orange juice, 64 ounces, Tropicana or Florida Natural brand

C10   Price of an 11.5-ounce can or brick of coffee

C11   Price of 2-liter Coca-Cola, excluding any deposit

C12   Monthly rent of an unfurnished two-bedroom apartment (excluding all utilities except water), 1.5 or 2 baths, 950 square feet

C13   Price of 1 gallon of regular unleaded gas, natural brand, including all taxes; cash price at self-service pump, if available

C14   Cost of a visit to the doctor's office for a routine examination for a problem with low to moderate severity

C15   Cost of a teeth-cleaning visit to the dentist's office (established patients only)

C16   Price of an 11- to 12-inch thin-crust regular cheese pizza (no extra cheese) at Pizza Hut and/or Pizza Inn

C17   Price of a man's barbershop haircut, no styling

C18   Price of a woman's haircut with shampoo and blow-dry at beauty salons that make appointments and allow customer to select stylist

C19   Price of a movie ticket; first run (new release), indoor 6 to 10 PM, Saturday evening rates, no discounts

C20   Price of Heineken's beer; six-pack of 12-ounce containers, excluding any deposit

C21   Price of wine, Livingston Cellars or Gallo brand Chablis or Chenin Blanc, 1.5-liter bottle

Data Set II: Data on States

This data set contains information on different variables for all 50 states of the United States. These data have been obtained from various Web sites such as the Web sites of the census bureau, insurance institute for highway safety, Bureau of Labor Statistics, and National Center for Education Statistics. This data set has 14 columns that contain the following variables:

C1   Name of the state

C2   Total population, 2013

C3   Area of the state in square miles

C4   Median household income (in dollars), 2013

[1]We are thankful to Jennie Allison (assistant project manager) and Erol Yildirim (vice president of product development) at the Center for Regional Economic Competitiveness for providing this data set to us.

C5   Total miles traveled by vehicles in millions of miles, 2013

C6   Traffic fatalities, 2013

C7   Average salary (in dollars) of teachers teaching in the K–12 public school system, 2012–2013

C8   Total revenue of the state in thousands of dollars, 2013

C9   Total state expenditures on education in thousands of dollars, 2013

C10 Total state expenditure on public welfare in thousands of dollars, 2013

C11 Total state expenditure on hospitals in thousands of dollars, 2013

C12 Total state expenditure on health in thousands of dollars, 2013

C13 Total state expenditure on highways in thousands of dollars, 2013

C14 Total state expenditure on police protection in thousands of dollars, 2013

Data Set III: NFL Data

This data set contains information on the NFL (National Football League) players who played for NFL teams during the 2014 season. These data were obtained from the Web site www.pro-football-reference.com. Data that were missing on this Web site were supplemented from other Internet sources. If a player played for more than one team during the 2014 season, the last team for whom he played is recorded, but the total number of games for all teams is recorded. This data set has 10 columns that contain the following variables:

C1   Name of the player

C2   Team for whom the player played

C3   Age of the player in years as of December 31, 2014

C4   Number of games played during the 2014 season

C5   Weight in pounds

C6   Height in inches

C7   College/university that the player attended

C8   Birth date

C9   Number of years played in the NFL prior to the 2014 season

C10 Whether the player was nominated to the Pro Bowl during the 2014 season (Yes/No)

Data Set IV: Manchester Road Race Data

This data set contains information on 11,682 runners who participated in and completed the 78th annual Manchester Road Race held on November 27, 2014 (Thanksgiving Day) in Manchester, Connecticut. Twelve of the 11,682 runners have missing information on one or more variables. This road race is held every year on Thanksgiving Day. The length of the race is 4.748 miles. This data set was obtained from the Web site www.coolrunning.com. This data set has nine columns that contain the following variables:

C1   Overall place in which the runner finished

C2   Division in which the runner competed, classified by gender and age

C3   Net time (in minutes) to complete the race

C4   Average pace of the runner in minutes per mile

C5   Age of the runner in years

C6   Gender (M/F)

C7   City of residence

C8   State of residence

C9   Whether the runner is a Connecticut resident (Yes/No)

Data Set V: Data on Movies

This data set contains information on the top 200 films of 2014 in terms of total gross domestic box office earnings until the movie's closing date in first-run theaters. For the eight movies that did not have a first-run theater close date at the time of this writing, the total gross domestic box office earnings was recorded as of October 13, 2015. Earnings do not include additional sources of revenue, such as home entertainment sales and rentals, television rights, and product placement fees. This data set was obtained from the Web site www.boxofficemojo.com. This data set has eight columns that contain the following variables:

C1   Movie title

C2   Studio that produced the movie

C3   Gross earnings in dollars from box office sales in the United States and Canada from the release date until the closing date of the movie

C4   Number of theater locations at which the movie was shown during the length of its release

C5   Gross earnings in dollars from box office sales in the United States and Canada during the opening weekend of the movie

C6   Number of theater locations at which the movie was shown during the opening weekend of the movie

C7   Genre of the movie as reported by IMDB, the Internet Movie Data Base

C8   Length of the film in minutes

Data Set VI: Standard & Poor's 500 Index Data

This data set contains information on the 500 stocks in the Standard & Poor's 500 index at the close of the financial markets on Friday, June 26, 2015. This data set was obtained from the Web site www.barchart.com. This data set has eight columns that contain the following variables:

C1   Stock symbol

C2   Company name

C3   Closing stock price for the day in dollars

C4   Difference between the closing stock price for the current day and the previous day in dollars

C5   Difference between the closing stock price for the current day and the previous day as a percentage of the previous day closing price

C6   Highest trade price for the day in dollars

C7   Lowest trade price for the day in dollars

C8   Total number of contracts traded for the day

Data Set VII: McDonald's Data

This data set contains information on 228 menu items available at McDonald's restaurants and is obtained from the Web site www.nutrition-charts.com. This data set has 14 columns that contain the following variables:

C1   Menu item name

C2   Calories

C3   Protein in grams

C4   Total fat in grams

C5   Total carbohydrates in grams

C6	Sodium in milligrams
C7	Calories from fat
C8	Saturated fat in grams
C9	Trans fat in grams
C10	Cholesterol in milligrams
C11	Fiber in grams
C12	Sugars in grams
C13	Weight Watchers points
C14	Menu category (Beverages/Breakfast/Burgers and Sandwiches/Chicken and Fish/Desserts and Shakes/Salads/Snacks and Sides)

Data Set VIII: Data on NFL Kickers

This data set contains information on the performance of 31 placekickers who had at least one attempt per game during the 2014 NFL (National Football League) regular season. This data set was obtained from the Web site espn.go.com/nfl/statistics/. It has nine columns that contain the following variables:

C1	Name of the player
C2	Name of the team
C3	Field goals made during the season
C4	Field goals attempted during the season
C5	Field goal completion percentage during the season
C6	Longest field goal made during the season in yards
C7	Extra points made during the season
C8	Extra points attempted during the season
C9	Extra point completion percentage during the season

Data Set IX: Subway Menu Items Data

This data set contains information on the nutritional characteristics of Subway food items. Breads, condiments, sides, and beverages are not included. This data set, obtained from the Web site www.subway.com, has 19 columns that contain the following variables:

C1	Menu item name
C2	Item type (Sandwich/Salad/Soup/Dessert)
C3	Whether the item is a breakfast item (Yes/No)
C4	Serving size in grams
C5	Total calories
C6	Calories from fat
C7	Total fat in grams
C8	Saturated fat in grams
C9	Trans fat in grams
C10	Cholesterol in milligrams
C11	Sodium in milligrams
C12	Carbohydrates in grams
C13	Dietary fiber in grams

C14 Sugars in grams

C15 Protein in grams

C16 Vitamin A percent daily value

C17 Vitamin C percent daily value

C18 Calcium percent daily value

C19 Iron percent daily value

Data Set X: Major League Baseball Data

This data set contains information on the offensive performance of all 443 Major League Baseball players who had at least 100 visits to the plate during the 2014 regular season. This data set, obtained from the Web site mlb.mlb.com, has 20 columns that contain the following variables:

C1 Name of the player

C2 Name of the team

C3 The fielding position played by the player

C4 The number of games in which a player appeared

C5 The number of official at bats by a batter, with official at bat defined as the number of visits to the plate minus sacrifices, walks, and hits by pitches

C6 The number of times a base runner safely reached home plate

C7 The number of times a batter hit the ball and reached base safely without the aid of an error or fielder's choice

C8 The number of times a batter hit the ball and reached second base without the aid of an error or fielder's choice

C9 The number of times a batter hit the ball and reached third base without the aid of an error or fielder's choice

C10 The number of times a batter hit the ball and reached home plate, scoring a run either by hitting the ball out of play in fair territory or without the aid of an error or fielder's choice

C11 The number of runs scored safely due to a batter hitting a ball or drawing a base on balls

C12 The number of walks by a batter

C13 The number of strikeouts by a batter

C14 The number of times a player stole a base

C15 The number of times a player was thrown out attempting to steal a base

C16 The average number of hits by a batter defined by hits divided by at bats

C17 The on-base percentage, defined as the number of times each batter reached base by hit, walk, or hit by a pitch, divided by plate appearances including at bats, walks, hit by a pitch, and sacrifice flies

C18 The slugging percentage, defined as total bases divided by at bats

C19 The number of times a batter was hit by a pitch

C20 The number of times a runner tagged up and scored after a batter's fly out

Data Set XI: Data on Airline Flights

This data set contains information on airline flights departing Newark (New Jersey) Liberty International Airport during May 2015. No arriving flights are included in these data, and flights with missing data were removed, leaving 8946 flights in the data. This data set, obtained from the Web site www.transtats.bts.gov, has 13 columns that contain the following variables:

C1   Day of the week

C2   Date of the flight

C3   Airline carrier of the flight

C4   Destination airport IATA code

C5   Destination airport city

C6   Local departure time

C7   Departure delay in minutes (a negative number indicates early departure)

C8   Whether the departure delay was 15 minutes or more (Yes/No)

C9   Local arrival time

C10   Arrival delay in minutes (a negative number indicates early arrival)

C11   Whether the arrival delay was 15 minutes or more (Yes/No)

C12   Total airtime in minutes from the time the wheels left the ground to the time they touched ground again

C13   Distance between airports in miles

Data Set XII: Data on Coffee Maker Ratings

This data set contains information on the *Consumer Reports* ratings of 54 drip coffee makers with carafe and an 8- to 14-cup capacity. The ratings were current as of Monday, July 13, 2015. This data set, obtained from the Web site www.consumerreports.org, has seven columns that contain the following variables:

C1   Coffeemaker model name

C2   Manufacturer's suggested retail price in dollars

C3   Overall score as computed by *Consumer Reports* on a 100-point scale, with 100 points representing the best possible score

C4   Whether this model was recommended by *Consumer Reports* (Yes/No)

C5   Brew performance as rated by *Consumer Reports* (Excellent/Very Good/Good/Fair/Poor)

C6   Convenience as rated by *Consumer Reports* (Excellent/Very Good/Good/Fair/Poor)

C7   Carafe handling as rated by *Consumer Reports* (Excellent/Very Good/Good/Fair/Poor)

Data Set XIII: Simulated Data

This data set contains four columns of simulated data from four different probability distributions. There are 1000 observations in each of the four columns, and these columns contain the following information:

C1   Simulated data from probability distribution 1

C2   Simulated data from probability distribution 2

C3   Simulated data from probability distribution 3

C4   Simulated data from probability distribution 4

Data Set XIV: Billboard Data

This data set contains 100 observations in two columns, and was collected from the Billboard Hot 100 Popular Music charts for the week of July 9, 2011. The two columns contain the following information:

C1 Number of weeks (for each song) in the Hot 100 chart

C2 Ranking group for the week of July 9, 2011 (1-50 or 51-100)

Statistical Tables

Note: The following tables are on the Web site of the text along with Chapters 14 and 15.

Table I   **Table of Binomial Probabilities**

							p					
n	x	.05	.10	.20	.30	.40	.50	.60	.70	.80	.90	.95
1	0	.9500	.9000	.8000	.7000	.6000	.5000	.4000	.3000	.2000	.1000	.0500
	1	.0500	.1000	.2000	.3000	.4000	.5000	.6000	.7000	.8000	.9000	.9500
2	0	.9025	.8100	.6400	.4900	.3600	.2500	.1600	.0900	.0400	.0100	.0025
	1	.0950	.1800	.3200	.4200	.4800	.5000	.4800	.4200	.3200	.1800	.0950
	2	.0025	.0100	.0400	.0900	.1600	.2500	.3600	.4900	.6400	.8100	.9025
3	0	.8574	.7290	.5120	.3430	.2160	.1250	.0640	.0270	.0080	.0010	.0001
	1	.1354	.2430	.3840	.4410	.4320	.3750	.2880	.1890	.0960	.0270	.0071
	2	.0071	.0270	.0960	.1890	.2880	.3750	.4320	.4410	.3840	.2430	.1354
	3	.0001	.0010	.0080	.0270	.0640	.1250	.2160	.3430	.5120	.7290	.8574
4	0	.8145	.6561	.4096	.2401	.1296	.0625	.0256	.0081	.0016	.0001	.0000
	1	.1715	.2916	.4096	.4116	.3456	.2500	.1536	.0756	.0256	.0036	.0005
	2	.0135	.0486	.1536	.2646	.3456	.3750	.3456	.2646	.1536	.0486	.0135
	3	.0005	.0036	.0256	.0756	.1536	.2500	.3456	.4116	.4096	.2916	.1715
	4	.0000	.0001	.0016	.0081	.0256	.0625	.1296	.2401	.4096	.6561	.8145
5	0	.7738	.5905	.3277	.1681	.0778	.0312	.0102	.0024	.0003	.0000	.0000
	1	.2036	.3280	.4096	.3602	.2592	.1562	.0768	.0284	.0064	.0005	.0000
	2	.0214	.0729	.2048	.3087	.3456	.3125	.2304	.1323	.0512	.0081	.0011
	3	.0011	.0081	.0512	.1323	.2304	.3125	.3456	.3087	.2048	.0729	.0214
	4	.0000	.0004	.0064	.0283	.0768	.1562	.2592	.3601	.4096	.3281	.2036
	5	.0000	.0000	.0003	.0024	.0102	.0312	.0778	.1681	.3277	.5905	.7738
6	0	.7351	.5314	.2621	.1176	.0467	.0156	.0041	.0007	.0001	.0000	.0000
	1	.2321	.3543	.3932	.3025	.1866	.0937	.0369	.0102	.0015	.0001	.0000
	2	.0305	.0984	.2458	.3241	.3110	.2344	.1382	.0595	.0154	.0012	.0001
	3	.0021	.0146	.0819	.1852	.2765	.3125	.2765	.1852	.0819	.0146	.0021
	4	.0001	.0012	.0154	.0595	.1382	.2344	.3110	.3241	.2458	.0984	.0305
	5	.0000	.0001	.0015	.0102	.0369	.0937	.1866	.3025	.3932	.3543	.2321
	6	.0000	.0000	.0001	.0007	.0041	.0156	.0467	.1176	.2621	.5314	.7351
7	0	.6983	.4783	.2097	.0824	.0280	.0078	.0016	.0002	.0000	.0000	.0000
	1	.2573	.3720	.3670	.2471	.1306	.0547	.0172	.0036	.0004	.0000	.0000
	2	.0406	.1240	.2753	.3177	.2613	.1641	.0774	.0250	.0043	.0002	.0000
	3	.0036	.0230	.1147	.2269	.2903	.2734	.1935	.0972	.0287	.0026	.0002
	4	.0002	.0026	.0287	.0972	.1935	.2734	.2903	.2269	.1147	.0230	.0036
	5	.0000	.0002	.0043	.0250	.0774	.1641	.2613	.3177	.2753	.1240	.0406
	6	.0000	.0000	.0004	.0036	.0172	.0547	.1306	.2471	.3670	.3720	.2573
	7	.0000	.0000	.0000	.0002	.0016	.0078	.0280	.0824	.2097	.4783	.6983
8	0	.6634	.4305	.1678	.0576	.0168	.0039	.0007	.0001	.0000	.0000	.0000
	1	.2793	.3826	.3355	.1977	.0896	.0312	.0079	.0012	.0001	.0000	.0000

Table I Table of Binomial Probabilities   B3

Table I   **Table of Binomial Probabilities**   (continued)

n	x	.05	.10	.20	.30	.40	.50	.60	.70	.80	.90	.95
	2	.0515	.1488	.2936	.2965	.2090	.1094	.0413	.0100	.0011	.0000	.0000
	3	.0054	.0331	.1468	.2541	.2787	.2187	.1239	.0467	.0092	.0004	.0000
	4	.0004	.0046	.0459	.1361	.2322	.2734	.2322	.1361	.0459	.0046	.0004
	5	.0000	.0004	.0092	.0467	.1239	.2187	.2787	.2541	.1468	.0331	.0054
	6	.0000	.0000	.0011	.0100	.0413	.1094	.2090	.2965	.2936	.1488	.0515
	7	.0000	.0000	.0001	.0012	.0079	.0312	.0896	.1977	.3355	.3826	.2793
	8	.0000	.0000	.0000	.0001	.0007	.0039	.0168	.0576	.1678	.4305	.6634
9	0	.6302	.3874	.1342	.0404	.0101	.0020	.0003	.0000	.0000	.0000	.0000
	1	.2985	.3874	.3020	.1556	.0605	.0176	.0035	.0004	.0000	.0000	.0000
	2	.0629	.1722	.3020	.2668	.1612	.0703	.0212	.0039	.0003	.0000	.0000
	3	.0077	.0446	.1762	.2668	.2508	.1641	.0743	.0210	.0028	.0001	.0000
	4	.0006	.0074	.0661	.1715	.2508	.2461	.1672	.0735	.0165	.0008	.0000
	5	.0000	.0008	.0165	.0735	.1672	.2461	.2508	.1715	.0661	.0074	.0006
	6	.0000	.0001	.0028	.0210	.0743	.1641	.2508	.2668	.1762	.0446	.0077
	7	.0000	.0000	.0003	.0039	.0212	.0703	.1612	.2668	.3020	.1722	.0629
	8	.0000	.0000	.0000	.0004	.0035	.0176	.0605	.1556	.3020	.3874	.2985
	9	.0000	.0000	.0000	.0000	.0003	.0020	.0101	.0404	.1342	.3874	.6302
10	0	.5987	.3487	.1074	.0282	.0060	.0010	.0001	.0000	.0000	.0000	.0000
	1	.3151	.3874	.2684	.1211	.0403	.0098	.0016	.0001	.0000	.0000	.0000
	2	.0746	.1937	.3020	.2335	.1209	.0439	.0106	.0014	.0001	.0000	.0000
	3	.0105	.0574	.2013	.2668	.2150	.1172	.0425	.0090	.0008	.0000	.0000
	4	.0010	.0112	.0881	.2001	.2508	.2051	.1115	.0368	.0055	.0001	.0000
	5	.0001	.0015	.0264	.1029	.2007	.2461	.2007	.1029	.0264	.0015	.0001
	6	.0000	.0001	.0055	.0368	.1115	.2051	.2508	.2001	.0881	.0112	.0010
	7	.0000	.0000	.0008	.0090	.0425	.1172	.2150	.2668	.2013	.0574	.0105
	8	.0000	.0000	.0001	.0014	.0106	.0439	.1209	.2335	.3020	.1937	.0746
	9	.0000	.0000	.0000	.0001	.0016	.0098	.0403	.1211	.2684	.3874	.3151
	10	.0000	.0000	.0000	.0000	.0001	.0010	.0060	.0282	.1074	.3487	.5987
11	0	.5688	.3138	.0859	.0198	.0036	.0005	.0000	.0000	.0000	.0000	.0000
	1	.3293	.3835	.2362	.0932	.0266	.0054	.0007	.0000	.0000	.0000	.0000
	2	.0867	.2131	.2953	.1998	.0887	.0269	.0052	.0005	.0000	.0000	.0000
	3	.0137	.0710	.2215	.2568	.1774	.0806	.0234	.0037	.0002	.0000	.0000
	4	.0014	.0158	.1107	.2201	.2365	.1611	.0701	.0173	.0017	.0000	.0000
	5	.0001	.0025	.0388	.1321	.2207	.2256	.1471	.0566	.0097	.0003	.0000
	6	.0000	.0003	.0097	.0566	.1471	.2256	.2207	.1321	.0388	.0025	.0001
	7	.0000	.0000	.0017	.0173	.0701	.1611	.2365	.2201	.1107	.0158	.0014
	8	.0000	.0000	.0002	.0037	.0234	.0806	.1774	.2568	.2215	.0710	.0137
	9	.0000	.0000	.0000	.0005	.0052	.0269	.0887	.1998	.2953	.2131	.0867

Table I   **Table of Binomial Probabilities**   (continued)

n	x	.05	.10	.20	.30	.40	.50	.60	.70	.80	.90	.95
							p					
	10	.0000	.0000	.0000	.0000	.0007	.0054	.0266	.0932	.2362	.3835	.3293
	11	.0000	.0000	.0000	.0000	.0000	.0005	.0036	.0198	.0859	.3138	.5688
12	0	.5404	.2824	.0687	.0138	.0022	.0002	.0000	.0000	.0000	.0000	.0000
	1	.3413	.3766	.2062	.0712	.0174	.0029	.0003	.0000	.0000	.0000	.0000
	2	.0988	.2301	.2835	.1678	.0639	.0161	.0025	.0002	.0000	.0000	.0000
	3	.0173	.0852	.2362	.2397	.1419	.0537	.0125	.0015	.0001	.0000	.0000
	4	.0021	.0213	.1329	.2311	.2128	.1208	.0420	.0078	.0005	.0000	.0000
	5	.0002	.0038	.0532	.1585	.2270	.1934	.1009	.0291	.0033	.0000	.0000
	6	.0000	.0005	.0155	.0792	.1766	.2256	.1766	.0792	.0155	.0005	.0000
	7	.0000	.0000	.0033	.0291	.1009	.1934	.2270	.1585	.0532	.0038	.0002
	8	.0000	.0000	.0005	.0078	.0420	.1208	.2128	.2311	.1329	.0213	.0021
	9	.0000	.0000	.0001	.0015	.0125	.0537	.1419	.2397	.2362	.0852	.0173
	10	.0000	.0000	.0000	.0002	.0025	.0161	.0639	.1678	.2835	.2301	.0988
	11	.0000	.0000	.0000	.0000	.0003	.0029	.0174	.0712	.2062	.3766	.3413
	12	.0000	.0000	.0000	.0000	.0000	.0002	.0022	.0138	.0687	.2824	.5404
13	0	.5133	.2542	.0550	.0097	.0013	.0001	.0000	.0000	.0000	.0000	.0000
	1	.3512	.3672	.1787	.0540	.0113	.0016	.0001	.0000	.0000	.0000	.0000
	2	.1109	.2448	.2680	.1388	.0453	.0095	.0012	.0001	.0000	.0000	.0000
	3	.0214	.0997	.2457	.2181	.1107	.0349	.0065	.0006	.0000	.0000	.0000
	4	.0028	.0277	.1535	.2337	.1845	.0873	.0243	.0034	.0001	.0000	.0000
	5	.0003	.0055	.0691	.1803	.2214	.1571	.0656	.0142	.0011	.0000	.0000
	6	.0000	.0008	.0230	.1030	.1968	.2095	.1312	.0442	.0058	.0001	.0000
	7	.0000	.0001	.0058	.0442	.1312	.2095	.1968	.1030	.0230	.0008	.0000
	8	.0000	.0000	.0011	.0142	.0656	.1571	.2214	.1803	.0691	.0055	.0003
	9	.0000	.0000	.0001	.0034	.0243	.0873	.1845	.2337	.1535	.0277	.0028
	10	.0000	.0000	.0000	.0006	.0065	.0349	.1107	.2181	.2457	.0997	.0214
	11	.0000	.0000	.0000	.0001	.0012	.0095	.0453	.1388	.2680	.2448	.1109
	12	.0000	.0000	.0000	.0000	.0001	.0016	.0113	.0540	.1787	.3672	.3512
	13	.0000	.0000	.0000	.0000	.0000	.0001	.0013	.0097	.0550	.2542	.5133
14	0	.4877	.2288	.0440	.0068	.0008	.0001	.0000	.0000	.0000	.0000	.0000
	1	.3593	.3559	.1539	.0407	.0073	.0009	.0001	.0000	.0000	.0000	.0000
	2	.1229	.2570	.2501	.1134	.0317	.0056	.0005	.0000	.0000	.0000	.0000
	3	.0259	.1142	.2501	.1943	.0845	.0222	.0033	.0002	.0000	.0000	.0000
	4	.0037	.0349	.1720	.2290	.1549	.0611	.0136	.0014	.0000	.0000	.0000
	5	.0004	.0078	.0860	.1963	.2066	.1222	.0408	.0066	.0003	.0000	.0000
	6	.0000	.0013	.0322	.1262	.2066	.1833	.0918	.0232	.0020	.0000	.0000
	7	.0000	.0002	.0092	.0618	.1574	.2095	.1574	.0618	.0092	.0002	.0000
	8	.0000	.0000	.0020	.0232	.0918	.1833	.2066	.1262	.0322	.0013	.0000
	9	.0000	.0000	.0003	.0066	.0408	.1222	.2066	.1963	.0860	.0078	.0004
	10	.0000	.0000	.0000	.0014	.0136	.0611	.1549	.2290	.1720	.0349	.0037

Table I Table of Binomial Probabilities   B5

Table I   **Table of Binomial Probabilities**   (continued)

n	x	.05	.10	.20	.30	.40	.50	.60	.70	.80	.90	.95
	11	.0000	.0000	.0000	.0002	.0033	.0222	.0845	.1943	.2501	.1142	.0259
	12	.0000	.0000	.0000	.0000	.0005	.0056	.0317	.1134	.2501	.2570	.1229
	13	.0000	.0000	.0000	.0000	.0001	.0009	.0073	.0407	.1539	.3559	.3593
	14	.0000	.0000	.0000	.0000	.0000	.0001	.0008	.0068	.0440	.2288	.4877
15	0	.4633	.2059	.0352	.0047	.0005	.0000	.0000	.0000	.0000	.0000	.0000
	1	.3658	.3432	.1319	.0305	.0047	.0005	.0000	.0000	.0000	.0000	.0000
	2	.1348	.2669	.2309	.0916	.0219	.0032	.0003	.0000	.0000	.0000	.0000
	3	.0307	.1285	.2501	.1700	.0634	.0139	.0016	.0001	.0000	.0000	.0000
	4	.0049	.0428	.1876	.2186	.1268	.0417	.0074	.0006	.0000	.0000	.0000
	5	.0006	.0105	.1032	.2061	.1859	.0916	.0245	.0030	.0001	.0000	.0000
	6	.0000	.0019	.0430	.1472	.2066	.1527	.0612	.0116	.0007	.0000	.0000
	7	.0000	.0003	.0138	.0811	.1771	.1964	.1181	.0348	.0035	.0000	.0000
	8	.0000	.0000	.0035	.0348	.1181	.1964	.1771	.0811	.0138	.0003	.0000
	9	.0000	.0000	.0007	.0116	.0612	.1527	.2066	.1472	.0430	.0019	.0000
	10	.0000	.0000	.0001	.0030	.0245	.0916	.1859	.2061	.1032	.0105	.0006
	11	.0000	.0000	.0000	.0006	.0074	.0417	.1268	.2186	.1876	.0428	.0049
	12	.0000	.0000	.0000	.0001	.0016	.0139	.0634	.1700	.2501	.1285	.0307
	13	.0000	.0000	.0000	.0000	.0003	.0032	.0219	.0916	.2309	.2669	.1348
	14	.0000	.0000	.0000	.0000	.0000	.0005	.0047	.0305	.1319	.3432	.3658
	15	.0000	.0000	.0000	.0000	.0000	.0000	.0005	.0047	.0352	.2059	.4633
16	0	.4401	.1853	.0281	.0033	.0003	.0000	.0000	.0000	.0000	.0000	.0000
	1	.3706	.3294	.1126	.0228	.0030	.0002	.0000	.0000	.0000	.0000	.0000
	2	.1463	.2745	.2111	.0732	.0150	.0018	.0001	.0000	.0000	.0000	.0000
	3	.0359	.1423	.2463	.1465	.0468	.0085	.0008	.0000	.0000	.0000	.0000
	4	.0061	.0514	.2001	.2040	.1014	.0278	.0040	.0002	.0000	.0000	.0000
	5	.0008	.0137	.1201	.2099	.1623	.0667	.0142	.0013	.0000	.0000	.0000
	6	.0001	.0028	.0550	.1649	.1983	.1222	.0392	.0056	.0002	.0000	.0000
	7	.0000	.0004	.0197	.1010	.1889	.1746	.0840	.0185	.0012	.0000	.0000
	8	.0000	.0001	.0055	.0487	.1417	.1964	.1417	.0487	.0055	.0001	.0000
	9	.0000	.0000	.0012	.0185	.0840	.1746	.1889	.1010	.0197	.0004	.0000
	10	.0000	.0000	.0002	.0056	.0392	.1222	.1983	.1649	.0550	.0028	.0001
	11	.0000	.0000	.0000	.0013	.0142	.0666	.1623	.2099	.1201	.0137	.0008
	12	.0000	.0000	.0000	.0002	.0040	.0278	.1014	.2040	.2001	.0514	.0061
	13	.0000	.0000	.0000	.0000	.0008	.0085	.0468	.1465	.2463	.1423	.0359
	14	.0000	.0000	.0000	.0000	.0001	.0018	.0150	.0732	.2111	.2745	.1463
	15	.0000	.0000	.0000	.0000	.0000	.0002	.0030	.0228	.1126	.3294	.3706
	16	.0000	.0000	.0000	.0000	.0000	.0000	.0003	.0033	.0281	.1853	.4401
17	0	.4181	.1668	.0225	.0023	.0002	.0000	.0000	.0000	.0000	.0000	.0000
	1	.3741	.3150	.0957	.0169	.0019	.0001	.0000	.0000	.0000	.0000	.0000
	2	.1575	.2800	.1914	.0581	.0102	.0010	.0001	.0000	.0000	.0000	.0000

Table I   **Table of Binomial Probabilities**   (continued)

n	x	.05	.10	.20	.30	.40	.50	.60	.70	.80	.90	.95
	3	.0415	.1556	.2393	.1245	.0341	.0052	.0004	.0000	.0000	.0000	.0000
	4	.0076	.0605	.2093	.1868	.0796	.0182	.0021	.0001	.0000	.0000	.0000
	5	.0010	.0175	.1361	.2081	.1379	.0472	.0081	.0006	.0000	.0000	.0000
	6	.0001	.0039	.0680	.1784	.1839	.0944	.0242	.0026	.0001	.0000	.0000
	7	.0000	.0007	.0267	.1201	.1927	.1484	.0571	.0095	.0004	.0000	.0000
	8	.0000	.0001	.0084	.0644	.1606	.1855	.1070	.0276	.0021	.0000	.0000
	9	.0000	.0000	.0021	.0276	.1070	.1855	.1606	.0644	.0084	.0001	.0000
	10	.0000	.0000	.0004	.0095	.0571	.1484	.1927	.1201	.0267	.0007	.0000
	11	.0000	.0000	.0001	.0026	.0242	.0944	.1839	.1784	.0680	.0039	.0001
	12	.0000	.0000	.0000	.0006	.0081	.0472	.1379	.2081	.1361	.0175	.0010
	13	.0000	.0000	.0000	.0001	.0021	.0182	.0796	.1868	.2093	.0605	.0076
	14	.0000	.0000	.0000	.0000	.0004	.0052	.0341	.1245	.2393	.1556	.0415
	15	.0000	.0000	.0000	.0000	.0001	.0010	.0102	.0581	.1914	.2800	.1575
	16	.0000	.0000	.0000	.0000	.0000	.0001	.0019	.0169	.0957	.3150	.3741
	17	.0000	.0000	.0000	.0000	.0000	.0000	.0002	.0023	.0225	.1668	.4181
18	0	.3972	.1501	.0180	.0016	.0001	.0000	.0000	.0000	.0000	.0000	.0000
	1	.3763	.3002	.0811	.0126	.0012	.0001	.0000	.0000	.0000	.0000	.0000
	2	.1683	.2835	.1723	.0458	.0069	.0006	.0000	.0000	.0000	.0000	.0000
	3	.0473	.1680	.2297	.1046	.0246	.0031	.0002	.0000	.0000	.0000	.0000
	4	.0093	.0700	.2153	.1681	.0614	.0117	.0011	.0000	.0000	.0000	.0000
	5	.0014	.0218	.1507	.2017	.1146	.0327	.0045	.0002	.0000	.0000	.0000
	6	.0002	.0052	.0816	.1873	.1655	.0708	.0145	.0012	.0000	.0000	.0000
	7	.0000	.0010	.0350	.1376	.1892	.1214	.0374	.0046	.0001	.0000	.0000
	8	.0000	.0002	.0120	.0811	.1734	.1669	.0771	.0149	.0008	.0000	.0000
	9	.0000	.0000	.0033	.0386	.1284	.1855	.1284	.0386	.0033	.0000	.0000
	10	.0000	.0000	.0008	.0149	.0771	.1669	.1734	.0811	.0120	.0002	.0000
	11	.0000	.0000	.0001	.0046	.0374	.1214	.1892	.1376	.0350	.0010	.0000
	12	.0000	.0000	.0000	.0012	.0145	.0708	.1655	.1873	.0816	.0052	.0002
	13	.0000	.0000	.0000	.0002	.0045	.0327	.1146	.2017	.1507	.0218	.0014
	14	.0000	.0000	.0000	.0000	.0011	.0117	.0614	.1681	.2153	.0700	.0093
	15	.0000	.0000	.0000	.0000	.0002	.0031	.0246	.1046	.2297	.1680	.0473
	16	.0000	.0000	.0000	.0000	.0000	.0006	.0069	.0458	.1723	.2835	.1683
	17	.0000	.0000	.0000	.0000	.0000	.0001	.0012	.0126	.0811	.3002	.3763
	18	.0000	.0000	.0000	.0000	.0000	.0000	.0001	.0016	.0180	.1501	.3972
19	0	.3774	.1351	.0144	.0011	.0001	.0000	.0000	.0000	.0000	.0000	.0000
	1	.3774	.2852	.0685	.0093	.0008	.0000	.0000	.0000	.0000	.0000	.0000
	2	.1787	.2852	.1540	.0358	.0046	.0003	.0000	.0000	.0000	.0000	.0000
	3	.0533	.1796	.2182	.0869	.0175	.0018	.0001	.0000	.0000	.0000	.0000
	4	.0112	.0798	.2182	.1491	.0467	.0074	.0005	.0000	.0000	.0000	.0000

Table I Table of Binomial Probabilities **B7**

Table I **Table of Binomial Probabilities** (continued)

n	x	.05	.10	.20	.30	.40	.50	.60	.70	.80	.90	.95
							p					
	5	.0018	.0266	.1636	.1916	.0933	.0222	.0024	.0001	.0000	.0000	.0000
	6	.0002	.0069	.0955	.1916	.1451	.0518	.0085	.0005	.0000	.0000	.0000
	7	.0000	.0014	.0443	.1525	.1797	.0961	.0237	.0022	.0000	.0000	.0000
	8	.0000	.0002	.0166	.0981	.1797	.1442	.0532	.0077	.0003	.0000	.0000
	9	.0000	.0000	.0051	.0514	.1464	.1762	.0976	.0220	.0013	.0000	.0000
	10	.0000	.0000	.0013	.0220	.0976	.1762	.1464	.0514	.0051	.0000	.0000
	11	.0000	.0000	.0003	.0077	.0532	.1442	.1797	.0981	.0166	.0002	.0000
	12	.0000	.0000	.0000	.0022	.0237	.0961	.1797	.1525	.0443	.0014	.0000
	13	.0000	.0000	.0000	.0005	.0085	.0518	.1451	.1916	.0955	.0069	.0002
	14	.0000	.0000	.0000	.0001	.0024	.0222	.0933	.1916	.1636	.0266	.0018
	15	.0000	.0000	.0000	.0000	.0005	.0074	.0467	.1491	.2182	.0798	.0112
	16	.0000	.0000	.0000	.0000	.0001	.0018	.0175	.0869	.2182	.1796	.0533
	17	.0000	.0000	.0000	.0000	.0000	.0003	.0046	.0358	.1540	.2852	.1787
	18	.0000	.0000	.0000	.0000	.0000	.0000	.0008	.0093	.0685	.2852	.3774
	19	.0000	.0000	.0000	.0000	.0000	.0000	.0001	.0011	.0144	.1351	.3774
20	0	.3585	.1216	.0115	.0008	.0000	.0000	.0000	.0000	.0000	.0000	.0000
	1	.3774	.2702	.0576	.0068	.0005	.0000	.0000	.0000	.0000	.0000	.0000
	2	.1887	.2852	.1369	.0278	.0031	.0002	.0000	.0000	.0000	.0000	.0000
	3	.0596	.1901	.2054	.0716	.0123	.0011	.0000	.0000	.0000	.0000	.0000
	4	.0133	.0898	.2182	.1304	.0350	.0046	.0003	.0000	.0000	.0000	.0000
	5	.0022	.0319	.1746	.1789	.0746	.0148	.0013	.0000	.0000	.0000	.0000
	6	.0003	.0089	.1091	.1916	.1244	.0370	.0049	.0002	.0000	.0000	.0000
	7	.0000	.0020	.0545	.1643	.1659	.0739	.0146	.0010	.0000	.0000	.0000
	8	.0000	.0004	.0222	.1144	.1797	.1201	.0355	.0039	.0001	.0000	.0000
	9	.0000	.0001	.0074	.0654	.1597	.1602	.0710	.0120	.0005	.0000	.0000
	10	.0000	.0000	.0020	.0308	.1171	.1762	.1171	.0308	.0020	.0000	.0000
	11	.0000	.0000	.0005	.0120	.0710	.1602	.1597	.0654	.0074	.0001	.0000
	12	.0000	.0000	.0001	.0039	.0355	.1201	.1797	.1144	.0222	.0004	.0000
	13	.0000	.0000	.0000	.0010	.0146	.0739	.1659	.1643	.0545	.0020	.0000
	14	.0000	.0000	.0000	.0002	.0049	.0370	.1244	.1916	.1091	.0089	.0003
	15	.0000	.0000	.0000	.0000	.0013	.0148	.0746	.1789	.1746	.0319	.0022
	16	.0000	.0000	.0000	.0000	.0003	.0046	.0350	.1304	.2182	.0898	.0133
	17	.0000	.0000	.0000	.0000	.0000	.0011	.0123	.0716	.2054	.1901	.0596
	18	.0000	.0000	.0000	.0000	.0000	.0002	.0031	.0278	.1369	.2852	.1887
	19	.0000	.0000	.0000	.0000	.0000	.0000	.0005	.0068	.0576	.2702	.3774
	20	.0000	.0000	.0000	.0000	.0000	.0000	.0000	.0008	.0115	.1216	.3585
21	0	.3406	.1094	.0092	.0006	.0000	.0000	.0000	.0000	.0000	.0000	.0000
	1	.3764	.2553	.0484	.0050	.0003	.0000	.0000	.0000	.0000	.0000	.0000
	2	.1981	.2837	.1211	.0215	.0020	.0001	.0000	.0000	.0000	.0000	.0000

Table I   **Table of Binomial Probabilities**   (continued)

n	x	.05	.10	.20	.30	.40	.50	.60	.70	.80	.90	.95
	3	.0660	.1996	.1917	.0585	.0086	.0006	.0000	.0000	.0000	.0000	.0000
	4	.0156	.0998	.2156	.1128	.0259	.0029	.0001	.0000	.0000	.0000	.0000
	5	.0028	.0377	.1833	.1643	.0588	.0097	.0007	.0000	.0000	.0000	.0000
	6	.0004	.0112	.1222	.1878	.1045	.0259	.0027	.0001	.0000	.0000	.0000
	7	.0000	.0027	.0655	.1725	.1493	.0554	.0087	.0005	.0000	.0000	.0000
	8	.0000	.0005	.0286	.1294	.1742	.0970	.0229	.0019	.0000	.0000	.0000
	9	.0000	.0001	.0103	.0801	.1677	.1402	.0497	.0063	.0002	.0000	.0000
	10	.0000	.0000	.0031	.0412	.1342	.1682	.0895	.0176	.0008	.0000	.0000
	11	.0000	.0000	.0008	.0176	.0895	.1682	.1342	.0412	.0031	.0000	.0000
	12	.0000	.0000	.0002	.0063	.0497	.1402	.1677	.0801	.0103	.0001	.0000
	13	.0000	.0000	.0000	.0019	.0229	.0970	.1742	.1294	.0286	.0005	.0000
	14	.0000	.0000	.0000	.0005	.0087	.0554	.1493	.1725	.0655	.0027	.0000
	15	.0000	.0000	.0000	.0001	.0027	.0259	.1045	.1878	.1222	.0112	.0004
	16	.0000	.0000	.0000	.0000	.0007	.0097	.0588	.1643	.1833	.0377	.0028
	17	.0000	.0000	.0000	.0000	.0001	.0029	.0259	.1128	.2156	.0998	.0156
	18	.0000	.0000	.0000	.0000	.0000	.0006	.0086	.0585	.1917	.1996	.0660
	19	.0000	.0000	.0000	.0000	.0000	.0001	.0020	.0215	.1211	.2837	.1981
	20	.0000	.0000	.0000	.0000	.0000	.0000	.0003	.0050	.0484	.2553	.3764
	21	.0000	.0000	.0000	.0000	.0000	.0000	.0000	.0006	.0092	.1094	.3406
22	0	.3235	.0985	.0074	.0004	.0000	.0000	.0000	.0000	.0000	.0000	.0000
	1	.3746	.2407	.0406	.0037	.0002	.0000	.0000	.0000	.0000	.0000	.0000
	2	.2070	.2808	.1065	.0166	.0014	.0001	.0000	.0000	.0000	.0000	.0000
	3	.0726	.2080	.1775	.0474	.0060	.0004	.0000	.0000	.0000	.0000	.0000
	4	.0182	.1098	.2108	.0965	.0190	.0017	.0001	.0000	.0000	.0000	.0000
	5	.0034	.0439	.1898	.1489	.0456	.0063	.0004	.0000	.0000	.0000	.0000
	6	.0005	.0138	.1344	.1808	.0862	.0178	.0015	.0000	.0000	.0000	.0000
	7	.0001	.0035	.0768	.1771	.1314	.0407	.0051	.0002	.0000	.0000	.0000
	8	.0000	.0007	.0360	.1423	.1642	.0762	.0144	.0009	.0000	.0000	.0000
	9	.0000	.0001	.0140	.0949	.1703	.1186	.0336	.0032	.0001	.0000	.0000
	10	.0000	.0000	.0046	.0529	.1476	.1542	.0656	.0097	.0003	.0000	.0000
	11	.0000	.0000	.0012	.0247	.1073	.1682	.1073	.0247	.0012	.0000	.0000
	12	.0000	.0000	.0003	.0097	.0656	.1542	.1476	.0529	.0046	.0000	.0000
	13	.0000	.0000	.0001	.0032	.0336	.1186	.1703	.0949	.0140	.0001	.0000
	14	.0000	.0000	.0000	.0009	.0144	.0762	.1642	.1423	.0360	.0007	.0000
	15	.0000	.0000	.0000	.0002	.0051	.0407	.1314	.1771	.0768	.0035	.0001
	16	.0000	.0000	.0000	.0000	.0015	.0178	.0862	.1808	.1344	.0138	.0005
	17	.0000	.0000	.0000	.0000	.0004	.0063	.0456	.1489	.1898	.0439	.0034
	18	.0000	.0000	.0000	.0000	.0001	.0017	.0190	.0965	.2108	.1098	.0182
	19	.0000	.0000	.0000	.0000	.0000	.0004	.0060	.0474	.1775	.2080	.0726

Table I Table of Binomial Probabilities   B9

Table I   **Table of Binomial Probabilities**   (continued)

n	x	.05	.10	.20	.30	.40	.50	.60	.70	.80	.90	.95
							p					
	20	.0000	.0000	.0000	.0000	.0000	.0001	.0014	.0166	.1065	.2808	.2070
	21	.0000	.0000	.0000	.0000	.0000	.0000	.0002	.0037	.0406	.2407	.3746
	22	.0000	.0000	.0000	.0000	.0000	.0000	.0000	.0004	.0074	.0985	.3235
23	0	.3074	.0886	.0059	.0003	.0000	.0000	.0000	.0000	.0000	.0000	.0000
	1	.3721	.2265	.0339	.0027	.0001	.0000	.0000	.0000	.0000	.0000	.0000
	2	.2154	.2768	.0933	.0127	.0009	.0000	.0000	.0000	.0000	.0000	.0000
	3	.0794	.2153	.1633	.0382	.0041	.0002	.0000	.0000	.0000	.0000	.0000
	4	.0209	.1196	.2042	.0818	.0138	.0011	.0000	.0000	.0000	.0000	.0000
	5	.0042	.0505	.1940	.1332	.0350	.0040	.0002	.0000	.0000	.0000	.0000
	6	.0007	.0168	.1455	.1712	.0700	.0120	.0008	.0000	.0000	.0000	.0000
	7	.0001	.0045	.0883	.1782	.1133	.0292	.0029	.0001	.0000	.0000	.0000
	8	.0000	.0010	.0442	.1527	.1511	.0584	.0088	.0004	.0000	.0000	.0000
	9	.0000	.0002	.0184	.1091	.1679	.0974	.0221	.0016	.0000	.0000	.0000
	10	.0000	.0000	.0064	.0655	.1567	.1364	.0464	.0052	.0001	.0000	.0000
	11	.0000	.0000	.0019	.0332	.1234	.1612	.0823	.0142	.0005	.0000	.0000
	12	.0000	.0000	.0005	.0142	.0823	.1612	.1234	.0332	.0019	.0000	.0000
	13	.0000	.0000	.0001	.0052	.0464	.1364	.1567	.0655	.0064	.0000	.0000
	14	.0000	.0000	.0000	.0016	.0221	.0974	.1679	.1091	.0184	.0002	.0000
	15	.0000	.0000	.0000	.0004	.0088	.0584	.1511	.1527	.0442	.0010	.0000
	16	.0000	.0000	.0000	.0001	.0029	.0292	.1133	.1782	.0883	.0045	.0001
	17	.0000	.0000	.0000	.0000	.0008	.0120	.0700	.1712	.1455	.0168	.0007
	18	.0000	.0000	.0000	.0000	.0002	.0040	.0350	.1332	.1940	.0505	.0042
	19	.0000	.0000	.0000	.0000	.0000	.0011	.0138	.0818	.2042	.1196	.0209
	20	.0000	.0000	.0000	.0000	.0000	.0002	.0041	.0382	.1633	.2153	.0794
	21	.0000	.0000	.0000	.0000	.0000	.0000	.0009	.0127	.0933	.2768	.2154
	22	.0000	.0000	.0000	.0000	.0000	.0000	.0001	.0027	.0339	.2265	.3721
	23	.0000	.0000	.0000	.0000	.0000	.0000	.0000	.0003	.0059	.0886	.3074
24	0	.2920	.0798	.0047	.0002	.0000	.0000	.0000	.0000	.0000	.0000	.0000
	1	.3688	.2127	.0283	.0020	.0001	.0000	.0000	.0000	.0000	.0000	.0000
	2	.2232	.2718	.0815	.0097	.0006	.0000	.0000	.0000	.0000	.0000	.0000
	3	.0862	.2215	.1493	.0305	.0028	.0001	.0000	.0000	.0000	.0000	.0000
	4	.0238	.1292	.1960	.0687	.0099	.0006	.0000	.0000	.0000	.0000	.0000
	5	.0050	.0574	.1960	.1177	.0265	.0025	.0001	.0000	.0000	.0000	.0000
	6	.0008	.0202	.1552	.1598	.0560	.0080	.0004	.0000	.0000	.0000	.0000
	7	.0001	.0058	.0998	.1761	.0960	.0206	.0017	.0000	.0000	.0000	.0000
	8	.0000	.0014	.0530	.1604	.1360	.0438	.0053	.0002	.0000	.0000	.0000
	9	.0000	.0003	.0236	.1222	.1612	.0779	.0141	.0008	.0000	.0000	.0000
	10	.0000	.0000	.0088	.0785	.1612	.1169	.0318	.0026	.0000	.0000	.0000
	11	.0000	.0000	.0028	.0428	.1367	.1488	.0608	.0079	.0002	.0000	.0000

Table I   Table of Binomial Probabilities   (continued)

n	x	.05	.10	.20	.30	.40	.50	.60	.70	.80	.90	.95
	12	.0000	.0000	.0008	.0199	.0988	.1612	.0988	.0199	.0008	.0000	.0000
	13	.0000	.0000	.0002	.0079	.0608	.1488	.1367	.0428	.0028	.0000	.0000
	14	.0000	.0000	.0000	.0026	.0318	.1169	.1612	.0785	.0088	.0000	.0000
	15	.0000	.0000	.0000	.0008	.0141	.0779	.1612	.1222	.0236	.0003	.0000
	16	.0000	.0000	.0000	.0002	.0053	.0438	.1360	.1604	.0530	.0014	.0000
	17	.0000	.0000	.0000	.0000	.0017	.0206	.0960	.1761	.0998	.0058	.0001
	18	.0000	.0000	.0000	.0000	.0004	.0080	.0560	.1598	.1552	.0202	.0008
	19	.0000	.0000	.0000	.0000	.0001	.0025	.0265	.1177	.1960	.0574	.0050
	20	.0000	.0000	.0000	.0000	.0000	.0006	.0099	.0687	.1960	.1292	.0238
	21	.0000	.0000	.0000	.0000	.0000	.0001	.0028	.0305	.1493	.2215	.0862
	22	.0000	.0000	.0000	.0000	.0000	.0000	.0006	.0097	.0815	.2718	.2232
	23	.0000	.0000	.0000	.0000	.0000	.0000	.0001	.0020	.0283	.2127	.3688
	24	.0000	.0000	.0000	.0000	.0000	.0000	.0000	.0002	.0047	.0798	.2920
25	0	.2774	.0718	.0038	.0001	.0000	.0000	.0000	.0000	.0000	.0000	.0000
	1	.3650	.1994	.0236	.0014	.0000	.0000	.0000	.0000	.0000	.0000	.0000
	2	.2305	.2659	.0708	.0074	.0004	.0000	.0000	.0000	.0000	.0000	.0000
	3	.0930	.2265	.1358	.0243	.0019	.0001	.0000	.0000	.0000	.0000	.0000
	4	.0269	.1384	.1867	.0572	.0071	.0004	.0000	.0000	.0000	.0000	.0000
	5	.0060	.0646	.1960	.1030	.0199	.0016	.0000	.0000	.0000	.0000	.0000
	6	.0010	.0239	.1633	.1472	.0442	.0053	.0002	.0000	.0000	.0000	.0000
	7	.0001	.0072	.1108	.1712	.0800	.0143	.0009	.0000	.0000	.0000	.0000
	8	.0000	.0018	.0623	.1651	.1200	.0322	.0031	.0001	.0000	.0000	.0000
	9	.0000	.0004	.0294	.1336	.1511	.0609	.0088	.0004	.0000	.0000	.0000
	10	.0000	.0001	.0118	.0916	.1612	.0974	.0212	.0013	.0000	.0000	.0000
	11	.0000	.0000	.0040	.0536	.1465	.1328	.0434	.0042	.0001	.0000	.0000
	12	.0000	.0000	.0012	.0268	.1140	.1550	.0760	.0115	.0003	.0000	.0000
	13	.0000	.0000	.0003	.0115	.0760	.1550	.1140	.0268	.0012	.0000	.0000
	14	.0000	.0000	.0001	.0042	.0434	.1328	.1465	.0536	.0040	.0000	.0000
	15	.0000	.0000	.0000	.0013	.0212	.0974	.1612	.0916	.0118	.0001	.0000
	16	.0000	.0000	.0000	.0004	.0088	.0609	.1511	.1336	.0294	.0004	.0000
	17	.0000	.0000	.0000	.0001	.0031	.0322	.1200	.1651	.0623	.0018	.0000
	18	.0000	.0000	.0000	.0000	.0009	.0143	.0800	.1712	.1108	.0072	.0001
	19	.0000	.0000	.0000	.0000	.0002	.0053	.0442	.1472	.1633	.0239	.0010
	20	.0000	.0000	.0000	.0000	.0000	.0016	.0199	.1030	.1960	.0646	.0060
	21	.0000	.0000	.0000	.0000	.0000	.0004	.0071	.0572	.1867	.1384	.0269
	22	.0000	.0000	.0000	.0000	.0000	.0001	.0019	.0243	.1358	.2265	.0930
	23	.0000	.0000	.0000	.0000	.0000	.0000	.0004	.0074	.0708	.2659	.2305
	24	.0000	.0000	.0000	.0000	.0000	.0000	.0000	.0014	.0236	.1994	.3650
	25	.0000	.0000	.0000	.0000	.0000	.0000	.0000	.0001	.0038	.0718	.2774

Table II Values of $e^{-\lambda}$ B11

Table II   Values of $e^{-\lambda}$

λ	$e^{-\lambda}$	λ	$e^{-\lambda}$
0.0	1.00000000	3.9	.02024191
0.1	.90483742	4.0	.01831564
0.2	.81873075	4.1	.01657268
0.3	.74081822	4.2	.01499558
0.4	.67032005	4.3	.01356856
0.5	.60653066	4.4	.01227734
0.6	.54881164	4.5	.01110900
0.7	.49658530	4.6	.01005184
0.8	.44932896	4.7	.00909528
0.9	.40656966	4.8	.00822975
1.0	.36787944	4.9	.00744658
1.1	.33287108	5.0	.00673795
1.2	.30119421	5.1	.00609675
1.3	.27253179	5.2	.00551656
1.4	.24659696	5.3	.00499159
1.5	.22313016	5.4	.00451658
1.6	.20189652	5.5	.00408677
1.7	.18268352	5.6	.00369786
1.8	.16529889	5.7	.00334597
1.9	.14956862	5.8	.00302755
2.0	.13533528	5.9	.00273944
2.1	.12245643	6.0	.00247875
2.2	.11080316	6.1	.00224287
2.3	.10025884	6.2	.00202943
2.4	.09071795	6.3	.00183630
2.5	.08208500	6.4	.00166156
2.6	.07427358	6.5	.00150344
2.7	.06720551	6.6	.00136037
2.8	.06081006	6.7	.00123091
2.9	.05502322	6.8	.00111378
3.0	.04978707	6.9	.00100779
3.1	.04504920	7.0	.00091188
3.2	.04076220	7.1	.00082510
3.3	.03688317	7.2	.00074659
3.4	.03337327	7.3	.00067554
3.5	.03019738	7.4	.00061125
3.6	.02732372	7.5	.00055308
3.7	.02472353	7.6	.00050045
3.8	.02237077	7.7	.00045283

Table II   Values of $e^{-\lambda}$ (continued)

λ	$e^{-\lambda}$	λ	$e^{-\lambda}$
7.8	.00040973	9.5	.00007485
7.9	.00037074	9.6	.00006773
8.0	.00033546	9.7	.00006128
8.1	.00030354	9.8	.00005545
8.2	.00027465	9.9	.00005017
8.3	.00024852	10.0	.00004540
8.4	.00022487	11.0	.00001670
8.5	.00020347	12.0	.00000614
8.6	.00018411	13.0	.00000226
8.7	.00016659	14.0	.00000083
8.8	.00015073	15.0	.00000031
8.9	.00013639	16.0	.00000011
9.0	.00012341	17.0	.00000004
9.1	.00011167	18.0	.000000015
9.2	.00010104	19.0	.000000006
9.3	.00009142	20.0	.000000002
9.4	.00008272		

Table III Table of Poisson Probabilities   B13

Table III   Table of Poisson Probabilities

					λ					
x	0.1	0.2	0.3	0.4	0.5	0.6	0.7	0.8	0.9	1.0
0	.9048	.8187	.7408	.6703	.6065	.5488	.4966	.4493	.4066	.3679
1	.0905	.1637	.2222	.2681	.3033	.3293	.3476	.3595	.3659	.3679
2	.0045	.0164	.0333	.0536	.0758	.0988	.1217	.1438	.1647	.1839
3	.0002	.0011	.0033	.0072	.0126	.0198	.0284	.0383	.0494	.0613
4	.0000	.0001	.0003	.0007	.0016	.0030	.0050	.0077	.0111	.0153
5	.0000	.0000	.0000	.0001	.0002	.0004	.0007	.0012	.0020	.0031
6	.0000	.0000	.0000	.0000	.0000	.0000	.0001	.0002	.0003	.0005
7	.0000	.0000	.0000	.0000	.0000	.0000	.0000	.0000	.0000	.0001

					λ					
x	1.1	1.2	1.3	1.4	1.5	1.6	1.7	1.8	1.9	2.0
0	.3329	.3012	.2725	.2466	.2231	.2019	.1827	.1653	.1496	.1353
1	.3662	.3614	.3543	.3452	.3347	.3230	.3106	.2975	.2842	.2707
2	.2014	.2169	.2303	.2417	.2510	.2584	.2640	.2678	.2700	.2707
3	.0738	.0867	.0998	.1128	.1255	.1378	.1496	.1607	.1710	.1804
4	.0203	.0260	.0324	.0395	.0471	.0551	.0636	.0723	.0812	.0902
5	.0045	.0062	.0084	.0111	.0141	.0176	.0216	.0260	.0309	.0361
6	.0008	.0012	.0018	.0026	.0035	.0047	.0061	.0078	.0098	.0120
7	.0001	.0002	.0003	.0005	.0008	.0011	.0015	.0020	.0027	.0034
8	.0000	.0000	.0001	.0001	.0001	.0002	.0003	.0005	.0006	.0009
9	.0000	.0000	.0000	.0000	.0000	.0000	.0001	.0001	.0001	.0002

					λ					
x	2.1	2.2	2.3	2.4	2.5	2.6	2.7	2.8	2.9	3.0
0	.1225	.1108	.1003	.0907	.0821	.0743	.0672	.0608	.0550	.0498
1	.2572	.2438	.2306	.2177	.2052	.1931	.1815	.1703	.1596	.1494
2	.2700	.2681	.2652	.2613	.2565	.2510	.2450	.2384	.2314	.2240
3	.1890	.1966	.2033	.2090	.2138	.2176	.2205	.2225	.2237	.2240
4	.0992	.1082	.1169	.1254	.1336	.1414	.1488	.1557	.1622	.1680
5	.0417	.0476	.0538	.0602	.0668	.0735	.0804	.0872	.0940	.1008
6	.0146	.0174	.0206	.0241	.0278	.0319	.0362	.0407	.0455	.0504
7	.0044	.0055	.0068	.0083	.0099	.0118	.0139	.0163	.0188	.0216
8	.0011	.0015	.0019	.0025	.0031	.0038	.0047	.0057	.0068	.0081
9	.0003	.0004	.0005	.0007	.0009	.0011	.0014	.0018	.0022	.0027
10	.0001	.0001	.0001	.0002	.0002	.0003	.0004	.0005	.0006	.0008
11	.0000	.0000	.0000	.0000	.0000	.0001	.0001	.0001	.0002	.0002
12	.0000	.0000	.0000	.0000	.0000	.0000	.0000	.0000	.0000	.0001

Table III   Table of Poisson Probabilities   (continued)

					λ					
x	3.1	3.2	3.3	3.4	3.5	3.6	3.7	3.8	3.9	4.0
0	.0450	.0408	.0369	.0334	.0302	.0273	.0247	.0224	.0202	.0183
1	.1397	.1304	.1217	.1135	.1057	.0984	.0915	.0850	.0789	.0733
2	.2165	.2087	.2008	.1929	.1850	.1771	.1692	.1615	.1539	.1465
3	.2237	.2226	.2209	.2186	.2158	.2125	.2087	.2046	.2001	.1954
4	.1733	.1781	.1823	.1858	.1888	.1912	.1931	.1944	.1951	.1954
5	.1075	.1140	.1203	.1264	.1322	.1377	.1429	.1477	.1522	.1563
6	.0555	.0608	.0662	.0716	.0771	.0826	.0881	.0936	.0989	.1042
7	.0246	.0278	.0312	.0348	.0385	.0425	.0466	.0508	.0551	.0595
8	.0095	.0111	.0129	.0148	.0169	.0191	.0215	.0241	.0269	.0298
9	.0033	.0040	.0047	.0056	.0066	.0076	.0089	.0102	.0116	.0132
10	.0010	.0013	.0016	.0019	.0023	.0028	.0033	.0039	.0045	.0053
11	.0003	.0004	.0005	.0006	.0007	.0009	.0011	.0013	.0016	.0019
12	.0001	.0001	.0001	.0002	.0002	.0003	.0003	.0004	.0005	.0006
13	.0000	.0000	.0000	.0000	.0001	.0001	.0001	.0001	.0002	.0002
14	.0000	.0000	.0000	.0000	.0000	.0000	.0000	.0000	.0000	.0001

					λ					
x	4.1	4.2	4.3	4.4	4.5	4.6	4.7	4.8	4.9	5.0
0	.0166	.0150	.0136	.0123	.0111	.0101	.0091	.0082	.0074	.0067
1	.0679	.0630	.0583	.0540	.0500	.0462	.0427	.0395	.0365	.0337
2	.1393	.1323	.1254	.1188	.1125	.1063	.1005	.0948	.0894	.0842
3	.1904	.1852	.1798	.1743	.1687	.1631	.1574	.1517	.1460	.1404
4	.1951	.1944	.1933	.1917	.1898	.1875	.1849	.1820	.1789	.1755
5	.1600	.1633	.1662	.1687	.1708	.1725	.1738	.1747	.1753	.1755
6	.1093	.1143	.1191	.1237	.1281	.1323	.1362	.1398	.1432	.1462
7	.0640	.0686	.0732	.0778	.0824	.0869	.0914	.0959	.1002	.1044
8	.0328	.0360	.0393	.0428	.0463	.0500	.0537	.0575	.0614	.0653
9	.0150	.0168	.0188	.0209	.0232	.0255	.0281	.0307	.0334	.0363
10	.0061	.0071	.0081	.0092	.0104	.0118	.0132	.0147	.0164	.0181
11	.0023	.0027	.0032	.0037	.0043	.0049	.0056	.0064	.0073	.0082
12	.0008	.0009	.0011	.0014	.0016	.0019	.0022	.0026	.0030	.0034
13	.0002	.0003	.0004	.0005	.0006	.0007	.0008	.0009	.0011	.0013
14	.0001	.0001	.0001	.0001	.0002	.0002	.0003	.0003	.0004	.0005
15	.0000	.0000	.0000	.0000	.0001	.0001	.0001	.0001	.0001	.0002

					λ					
x	5.1	5.2	5.3	5.4	5.5	5.6	5.7	5.8	5.9	6.0
0	.0061	.0055	.0050	.0045	.0041	.0037	.0033	.0030	.0027	.0025
1	.0311	.0287	.0265	.0244	.0225	.0207	.0191	.0176	.0162	.0149

Table III Table of Poisson Probabilities   B15

Table III   Table of Poisson Probabilities   (continued)

					λ					
x	5.1	5.2	5.3	5.4	5.5	5.6	5.7	5.8	5.9	6.0
2	.0793	.0746	.0701	.0659	.0618	.0580	.0544	.0509	.0477	.0446
3	.1348	.1293	.1239	.1185	.1133	.1082	.1033	.0985	.0938	.0892
4	.1719	.1681	.1641	.1600	.1558	.1515	.1472	.1428	.1383	.1339
5	.1753	.1748	.1740	.1728	.1714	.1697	.1678	.1656	.1632	.1606
6	.1490	.1515	.1537	.1555	.1571	.1584	.1594	.1601	.1605	.1606
7	.1086	.1125	.1163	.1200	.1234	.1267	.1298	.1326	.1353	.1377
8	.0692	.0731	.0771	.0810	.0849	.0887	.0925	.0962	.0998	.1033
9	.0392	.0423	.0454	.0486	.0519	.0552	.0586	.0620	.0654	.0688
10	.0200	.0220	.0241	.0262	.0285	.0309	.0334	.0359	.0386	.0413
11	.0093	.0104	.0116	.0129	.0143	.0157	.0173	.0190	.0207	.0225
12	.0039	.0045	.0051	.0058	.0065	.0073	.0082	.0092	.0102	.0113
13	.0015	.0018	.0021	.0024	.0028	.0032	.0036	.0041	.0046	.0052
14	.0006	.0007	.0008	.0009	.0011	.0013	.0015	.0017	.0019	.0022
15	.0002	.0002	.0003	.0003	.0004	.0005	.0006	.0007	.0008	.0009
16	.0001	.0001	.0001	.0001	.0001	.0002	.0002	.0002	.0003	.0003
17	.0000	.0000	.0000	.0000	.0000	.0001	.0001	.0001	.0001	.0001

					λ					
x	6.1	6.2	6.3	6.4	6.5	6.6	6.7	6.8	6.9	7.0
0	.0022	.0020	.0018	.0017	.0015	.0014	.0012	.0011	.0010	.0009
1	.0137	.0126	.0116	.0106	.0098	.0090	.0082	.0076	.0070	.0064
2	.0417	.0390	.0364	.0340	.0318	.0296	.0276	.0258	.0240	.0223
3	.0848	.0806	.0765	.0726	.0688	.0652	.0617	.0584	.0552	.0521
4	.1294	.1249	.1205	.1162	.1118	.1076	.1034	.0992	.0952	.0912
5	.1579	.1549	.1519	.1487	.1454	.1420	.1385	.1349	.1314	.1277
6	.1605	.1601	.1595	.1586	.1575	.1562	.1546	.1529	.1511	.1490
7	.1399	.1418	.1435	.1450	.1462	.1472	.1480	.1486	.1489	.1490
8	.1066	.1099	.1130	.1160	.1188	.1215	.1240	.1263	.1284	.1304
9	.0723	.0757	.0791	.0825	.0858	.0891	.0923	.0954	.0985	.1014
10	.0441	.0469	.0498	.0528	.0558	.0588	.0618	.0649	.0679	.0710
11	.0244	.0265	.0285	.0307	.0330	.0353	.0377	.0401	.0426	.0452
12	.0124	.0137	.0150	.0164	.0179	.0194	.0210	.0227	.0245	.0263
13	.0058	.0065	.0073	.0081	.0089	.0099	.0108	.0119	.0130	.0142
14	.0025	.0029	.0033	.0037	.0041	.0046	.0052	.0058	.0064	.0071
15	.0010	.0012	.0014	.0016	.0018	.0020	.0023	.0026	.0029	.0033
16	.0004	.0005	.0005	.0006	.0007	.0008	.0010	.0011	.0013	.0014
17	.0001	.0002	.0002	.0002	.0003	.0003	.0004	.0004	.0005	.0006
18	.0000	.0001	.0001	.0001	.0001	.0001	.0001	.0002	.0002	.0002
19	.0000	.0000	.0000	.0000	.0000	.0000	.0001	.0001	.0001	.0001

Table III   **Table of Poisson Probabilities**   (continued)

					λ					
x	7.1	7.2	7.3	7.4	7.5	7.6	7.7	7.8	7.9	8.0
0	.0008	.0007	.0007	.0006	.0006	.0005	.0005	.0004	.0004	.0003
1	.0059	.0054	.0049	.0045	.0041	.0038	.0035	.0032	.0029	.0027
2	.0208	.0194	.0180	.0167	.0156	.0145	.0134	.0125	.0116	.0107
3	.0492	.0464	.0438	.0413	.0389	.0366	.0345	.0324	.0305	.0286
4	.0874	.0836	.0799	.0764	.0729	.0696	.0663	.0632	.0602	.0573
5	.1241	.1204	.1167	.1130	.1094	.1057	.1021	.0986	.0951	.0916
6	.1468	.1445	.1420	.1394	.1367	.1339	.1311	.1282	.1252	.1221
7	.1489	.1486	.1481	.1474	.1465	.1454	.1442	.1428	.1413	.1396
8	.1321	.1337	.1351	.1363	.1373	.1381	.1388	.1392	.1395	.1396
9	.1042	.1070	.1096	.1121	.1144	.1167	.1187	.1207	.1224	.1241
10	.0740	.0770	.0800	.0829	.0858	.0887	.0914	.0941	.0967	.0993
11	.0478	.0504	.0531	.0558	.0585	.0613	.0640	.0667	.0695	.0722
12	.0283	.0303	.0323	.0344	.0366	.0388	.0411	.0434	.0457	.0481
13	.0154	.0168	.0181	.0196	.0211	.0227	.0243	.0260	.0278	.0296
14	.0078	.0086	.0095	.0104	.0113	.0123	.0134	.0145	.0157	.0169
15	.0037	.0041	.0046	.0051	.0057	.0062	.0069	.0075	.0083	.0090
16	.0016	.0019	.0021	.0024	.0026	.0030	.0033	.0037	.0041	.0045
17	.0007	.0008	.0009	.0010	.0012	.0013	.0015	.0017	.0019	.0021
18	.0003	.0003	.0004	.0004	.0005	.0006	.0006	.0007	.0008	.0009
19	.0001	.0001	.0001	.0002	.0002	.0002	.0003	.0003	.0003	.0004
20	.0000	.0000	.0001	.0001	.0001	.0001	.0001	.0001	.0001	.0002
21	.0000	.0000	.0000	.0000	.0000	.0000	.0000	.0000	.0001	.0001

					λ					
x	8.1	8.2	8.3	8.4	8.5	8.6	8.7	8.8	8.9	9.0
0	.0003	.0003	.0002	.0002	.0002	.0002	.0002	.0002	.0001	.0001
1	.0025	.0023	.0021	.0019	.0017	.0016	.0014	.0013	.0012	.0011
2	.0100	.0092	.0086	.0079	.0074	.0068	.0063	.0058	.0054	.0050
3	.0269	.0252	.0237	.0222	.0208	.0195	.0183	.0171	.0160	.0150
4	.0544	.0517	.0491	.0466	.0443	.0420	.0398	.0377	.0357	.0337
5	.0882	.0849	.0816	.0784	.0752	.0722	.0692	.0663	.0635	.0607
6	.1191	.1160	.1128	.1097	.1066	.1034	.1003	.0972	.0941	.0911
7	.1378	.1358	.1338	.1317	.1294	.1271	.1247	.1222	.1197	.1171
8	.1395	.1392	.1388	.1382	.1375	.1366	.1356	.1344	.1332	.1318
9	.1255	.1269	.1280	.1290	.1299	.1306	.1311	.1315	.1317	.1318
10	.1017	.1040	.1063	.1084	.1104	.1123	.1140	.1157	.1172	.1186
11	.0749	.0775	.0802	.0828	.0853	.0878	.0902	.0925	.0948	.0970
12	.0505	.0530	.0555	.0579	.0604	.0629	.0654	.0679	.0703	.0728

Table III Table of Poisson Probabilities B17

Table III Table of Poisson Probabilities (continued)

x	8.1	8.2	8.3	8.4	8.5	8.6	8.7	8.8	8.9	9.0
13	.0315	.0334	.0354	.0374	.0395	.0416	.0438	.0459	.0481	.0504
14	.0182	.0196	.0210	.0225	.0240	.0256	.0272	.0289	.0306	.0324
15	.0098	.0107	.0116	.0126	.0136	.0147	.0158	.0169	.0182	.0194
16	.0050	.0055	.0060	.0066	.0072	.0079	.0086	.0093	.0101	.0109
17	.0024	.0026	.0029	.0033	.0036	.0040	.0044	.0048	.0053	.0058
18	.0011	.0012	.0014	.0015	.0017	.0019	.0021	.0024	.0026	.0029
19	.0005	.0005	.0006	.0007	.0008	.0009	.0010	.0011	.0012	.0014
20	.0002	.0002	.0002	.0003	.0003	.0004	.0004	.0005	.0005	.0006
21	.0001	.0001	.0001	.0001	.0001	.0002	.0002	.0002	.0002	.0003
22	.0000	.0000	.0000	.0000	.0001	.0001	.0001	.0001	.0001	.0001

x	9.1	9.2	9.3	9.4	9.5	9.6	9.7	9.8	9.9	10
0	.0001	.0001	.0001	.0001	.0001	.0001	.0001	.0001	.0001	.0000
1	.0010	.0009	.0009	.0008	.0007	.0007	.0006	.0005	.0005	.0005
2	.0046	.0043	.0040	.0037	.0034	.0031	.0029	.0027	.0025	.0023
3	.0140	.0131	.0123	.0115	.0107	.0100	.0093	.0087	.0081	.0076
4	.0319	.0302	.0285	.0269	.0254	.0240	.0226	.0213	.0201	.0189
5	.0581	.0555	.0530	.0506	.0483	.0460	.0439	.0418	.0398	.0378
6	.0881	.0851	.0822	.0793	.0764	.0736	.0709	.0682	.0656	.0631
7	.1145	.1118	.1091	.1064	.1037	.1010	.0982	.0955	.0928	.0901
8	.1302	.1286	.1269	.1251	.1232	.1212	.1191	.1170	.1148	.1126
9	.1317	.1315	.1311	.1306	.1300	.1293	.1284	.1274	.1263	.1251
10	.1198	.1209	.1219	.1228	.1235	.1241	.1245	.1249	.1250	.1251
11	.0991	.1012	.1031	.1049	.1067	.1083	.1098	.1112	.1125	.1137
12	.0752	.0776	.0799	.0822	.0844	.0866	.0888	.0908	.0928	.0948
13	.0526	.0549	.0572	.0594	.0617	.0640	.0662	.0685	.0707	.0729
14	.0342	.0361	.0380	.0399	.0419	.0439	.0459	.0479	.0500	.0521
15	.0208	.0221	.0235	.0250	.0265	.0281	.0297	.0313	.0330	.0347
16	.0118	.0127	.0137	.0147	.0157	.0168	.0180	.0192	.0204	.0217
17	.0063	.0069	.0075	.0081	.0088	.0095	.0103	.0111	.0119	.0128
18	.0032	.0035	.0039	.0042	.0046	.0051	.0055	.0060	.0065	.0071
19	.0015	.0017	.0019	.0021	.0023	.0026	.0028	.0031	.0034	.0037
20	.0007	.0008	.0009	.0010	.0011	.0012	.0014	.0015	.0017	.0019
21	.0003	.0003	.0004	.0004	.0005	.0006	.0006	.0007	.0008	.0009
22	.0001	.0001	.0002	.0002	.0002	.0002	.0003	.0003	.0004	.0004
23	.0000	.0001	.0001	.0001	.0001	.0001	.0001	.0001	.0002	.0002
24	.0000	.0000	.0000	.0000	.0000	.0000	.0000	.0001	.0001	.0001

Table III   Table of Poisson Probabilities   (continued)

x	11	12	13	14	λ 15	16	17	18	19	20
0	.0000	.0000	.0000	.0000	.0000	.0000	.0000	.0000	.0000	.0000
1	.0002	.0001	.0000	.0000	.0000	.0000	.0000	.0000	.0000	.0000
2	.0010	.0004	.0002	.0001	.0000	.0000	.0000	.0000	.0000	.0000
3	.0037	.0018	.0008	.0004	.0002	.0001	.0000	.0000	.0000	.0000
4	.0102	.0053	.0027	.0013	.0006	.0003	.0001	.0001	.0000	.0000
5	.0224	.0127	.0070	.0037	.0019	.0010	.0005	.0002	.0001	.0001
6	.0411	.0255	.0152	.0087	.0048	.0026	.0014	.0007	.0004	.0002
7	.0646	.0437	.0281	.0174	.0104	.0060	.0034	.0019	.0010	.0005
8	.0888	.0655	.0457	.0304	.0194	.0120	.0072	.0042	.0024	.0013
9	.1085	.0874	.0661	.0473	.0324	.0213	.0135	.0083	.0050	.0029
10	.1194	.1048	.0859	.0663	.0486	.0341	.0230	.0150	.0095	.0058
11	.1194	.1144	.1015	.0844	.0663	.0496	.0355	.0245	.0164	.0106
12	.1094	.1144	.1099	.0984	.0829	.0661	.0504	.0368	.0259	.0176
13	.0926	.1056	.1099	.1060	.0956	.0814	.0658	.0509	.0378	.0271
14	.0728	.0905	.1021	.1060	.1024	.0930	.0800	.0655	.0514	.0387
15	.0534	.0724	.0885	.0989	.1024	.0992	.0906	.0786	.0650	.0516
16	.0367	.0543	.0719	.0866	.0960	.0992	.0963	.0884	.0772	.0646
17	.0237	.0383	.0550	.0713	.0847	.0934	.0963	.0936	.0863	.0760
18	.0145	.0255	.0397	.0554	.0706	.0830	.0909	.0936	.0911	.0844
19	.0084	.0161	.0272	.0409	.0557	.0699	.0814	.0887	.0911	.0888
20	.0046	.0097	.0177	.0286	.0418	.0559	.0692	.0798	.0866	.0888
21	.0024	.0055	.0109	.0191	.0299	.0426	.0560	.0684	.0783	.0846
22	.0012	.0030	.0065	.0121	.0204	.0310	.0433	.0560	.0676	.0769
23	.0006	.0016	.0037	.0074	.0133	.0216	.0320	.0438	.0559	.0669
24	.0003	.0008	.0020	.0043	.0083	.0144	.0226	.0328	.0442	.0557
25	.0001	.0004	.0010	.0024	.0050	.0092	.0154	.0237	.0336	.0446
26	.0000	.0002	.0005	.0013	.0029	.0057	.0101	.0164	.0246	.0343
27	.0000	.0001	.0002	.0007	.0016	.0034	.0063	.0109	.0173	.0254
28	.0000	.0000	.0001	.0003	.0009	.0019	.0038	.0070	.0117	.0181
29	.0000	.0000	.0001	.0002	.0004	.0011	.0023	.0044	.0077	.0125
30	.0000	.0000	.0000	.0001	.0002	.0006	.0013	.0026	.0049	.0083
31	.0000	.0000	.0000	.0000	.0001	.0003	.0007	.0015	.0030	.0054
32	.0000	.0000	.0000	.0000	.0001	.0001	.0004	.0009	.0018	.0034
33	.0000	.0000	.0000	.0000	.0000	.0001	.0002	.0005	.0010	.0020
34	.0000	.0000	.0000	.0000	.0000	.0000	.0001	.0002	.0006	.0012
35	.0000	.0000	.0000	.0000	.0000	.0000	.0000	.0001	.0003	.0007
36	.0000	.0000	.0000	.0000	.0000	.0000	.0000	.0001	.0002	.0004
37	.0000	.0000	.0000	.0000	.0000	.0000	.0000	.0000	.0001	.0002
38	.0000	.0000	.0000	.0000	.0000	.0000	.0000	.0000	.0000	.0001
39	.0000	.0000	.0000	.0000	.0000	.0000	.0000	.0000	.0000	.0001

Table IV Standard Normal Distribution Table   B19

Table IV   Standard Normal Distribution Table

The entries in the table on this page give the cumulative area under the standard normal curve to the left of z with the values of z equal to 0 or negative.

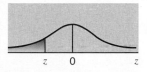

z	.00	.01	.02	.03	.04	.05	.06	.07	.08	.09
−3.4	.0003	.0003	.0003	.0003	.0003	.0003	.0003	.0003	.0003	.0002
−3.3	.0005	.0005	.0005	.0004	.0004	.0004	.0004	.0004	.0004	.0003
−3.2	.0007	.0007	.0006	.0006	.0006	.0006	.0006	.0005	.0005	.0005
−3.1	.0010	.0009	.0009	.0009	.0008	.0008	.0008	.0008	.0007	.0007
−3.0	.0013	.0013	.0013	.0012	.0012	.0011	.0011	.0011	.0010	.0010
−2.9	.0019	.0018	.0018	.0017	.0016	.0016	.0015	.0015	.0014	.0014
−2.8	.0026	.0025	.0024	.0023	.0023	.0022	.0021	.0021	.0020	.0019
−2.7	.0035	.0034	.0033	.0032	.0031	.0030	.0029	.0028	.0027	.0026
−2.6	.0047	.0045	.0044	.0043	.0041	.0040	.0039	.0038	.0037	.0036
−2.5	.0062	.0060	.0059	.0057	.0055	.0054	.0052	.0051	.0049	.0048
−2.4	.0082	.0080	.0078	.0075	.0073	.0071	.0069	.0068	.0066	.0064
−2.3	.0107	.0104	.0102	.0099	.0096	.0094	.0091	.0089	.0087	.0084
−2.2	.0139	.0136	.0132	.0129	.0125	.0122	.0119	.0116	.0113	.0110
−2.1	.0179	.0174	.0170	.0166	.0162	.0158	.0154	.0150	.0146	.0143
−2.0	.0228	.0222	.0217	.0212	.0207	.0202	.0197	.0192	.0188	.0183
−1.9	.0287	.0281	.0274	.0268	.0262	.0256	.0250	.0244	.0239	.0233
−1.8	.0359	.0351	.0344	.0336	.0329	.0322	.0314	.0307	.0301	.0294
−1.7	.0446	.0436	.0427	.0418	.0409	.0401	.0392	.0384	.0375	.0367
−1.6	.0548	.0537	.0526	.0516	.0505	.0495	.0485	.0475	.0465	.0455
−1.5	.0668	.0655	.0643	.0630	.0618	.0606	.0594	.0582	.0571	.0559
−1.4	.0808	.0793	.0778	.0764	.0749	.0735	.0721	.0708	.0694	.0681
−1.3	.0968	.0951	.0934	.0918	.0901	.0885	.0869	.0853	.0838	.0823
−1.2	.1151	.1131	.1112	.1093	.1075	.1056	.1038	.1020	.1003	.0985
−1.1	.1357	.1335	.1314	.1292	.1271	.1251	.1230	.1210	.1190	.1170
−1.0	.1587	.1562	.1539	.1515	.1492	.1469	.1446	.1423	.1401	.1379
−0.9	.1841	.1814	.1788	.1762	.1736	.1711	.1685	.1660	.1635	.1611
−0.8	.2119	.2090	.2061	.2033	.2005	.1977	.1949	.1922	.1894	.1867
−0.7	.2420	.2389	.2358	.2327	.2296	.2266	.2236	.2206	.2177	.2148
−0.6	.2743	.2709	.2676	.2643	.2611	.2578	.2546	.2514	.2483	.2451
−0.5	.3085	.3050	.3015	.2981	.2946	.2912	.2877	.2843	.2810	.2776
−0.4	.3446	.3409	.3372	.3336	.3300	.3264	.3228	.3192	.3156	.3121
−0.3	.3821	.3783	.3745	.3707	.3669	.3632	.3594	.3557	.3520	.3483
−0.2	.4207	.4168	.4129	.4090	.4052	.4013	.3974	.3936	.3897	.3859
−0.1	.4602	.4562	.4522	.4483	.4443	.4404	.4364	.4325	.4286	.4247
−0.0	.5000	.4960	.4920	.4880	.4840	.4801	.4761	.4721	.4681	.4641

Table IV   Standard Normal Distribution Table   (continued)

The entries in the table on this page give the cumulative area under the standard normal curve to the left of z with the values of z equal to 0 or positive.

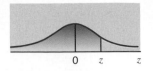

z	.00	.01	.02	.03	.04	.05	.06	.07	.08	.09
0.0	.5000	.5040	.5080	.5120	.5160	.5199	.5239	.5279	.5319	.5359
0.1	.5398	.5438	.5478	.5517	.5557	.5596	.5636	.5675	.5714	.5753
0.2	.5793	.5832	.5871	.5910	.5948	.5987	.6026	.6064	.6103	.6141
0.3	.6179	.6217	.6255	.6293	.6331	.6368	.6406	.6443	.6480	.6517
0.4	.6554	.6591	.6628	.6664	.6700	.6736	.6772	.6808	.6844	.6879
0.5	.6915	.6950	.6985	.7019	.7054	.7088	.7123	.7157	.7190	.7224
0.6	.7257	.7291	.7324	.7357	.7389	.7422	.7454	.7486	.7517	.7549
0.7	.7580	.7611	.7642	.7673	.7704	.7734	.7764	.7794	.7823	.7852
0.8	.7881	.7910	.7939	.7967	.7995	.8023	.8051	.8078	.8106	.8133
0.9	.8159	.8186	.8212	.8238	.8264	.8289	.8315	.8340	.8365	.8389
1.0	.8413	.8438	.8461	.8485	.8508	.8531	.8554	.8577	.8599	.8621
1.1	.8643	.8665	.8686	.8708	.8729	.8749	.8770	.8790	.8810	.8830
1.2	.8849	.8869	.8888	.8907	.8925	.8944	.8962	.8980	.8997	.9015
1.3	.9032	.9049	.9066	.9082	.9099	.9115	.9131	.9147	.9162	.9177
1.4	.9192	.9207	.9222	.9236	.9251	.9265	.9279	.9292	.9306	.9319
1.5	.9332	.9345	.9357	.9370	.9382	.9394	.9406	.9418	.9429	.9441
1.6	.9452	.9463	.9474	.9484	.9495	.9505	.9515	.9525	.9535	.9545
1.7	.9554	.9564	.9573	.9582	.9591	.9599	.9608	.9616	.9625	.9633
1.8	.9641	.9649	.9656	.9664	.9671	.9678	.9686	.9693	.9699	.9706
1.9	.9713	.9719	.9726	.9732	.9738	.9744	.9750	.9756	.9761	.9767
2.0	.9772	.9778	.9783	.9788	.9793	.9798	.9803	.9808	.9812	.9817
2.1	.9821	.9826	.9830	.9834	.9838	.9842	.9846	.9850	.9854	.9857
2.2	.9861	.9864	.9868	.9871	.9875	.9878	.9881	.9884	.9887	.9890
2.3	.9893	.9896	.9898	.9901	.9904	.9906	.9909	.9911	.9913	.9916
2.4	.9918	.9920	.9922	.9925	.9927	.9929	.9931	.9932	.9934	.9936
2.5	.9938	.9940	.9941	.9943	.9945	.9946	.9948	.9949	.9951	.9952
2.6	.9953	.9955	.9956	.9957	.9959	.9960	.9961	.9962	.9963	.9964
2.7	.9965	.9966	.9967	.9968	.9969	.9970	.9971	.9972	.9973	.9974
2.8	.9974	.9975	.9976	.9977	.9977	.9978	.9979	.9979	.9980	.9981
2.9	.9981	.9982	.9982	.9983	.9984	.9984	.9985	.9985	.9986	.9986
3.0	.9987	.9987	.9987	.9988	.9988	.9989	.9989	.9989	.9990	.9990
3.1	.9990	.9991	.9991	.9991	.9992	.9992	.9992	.9992	.9993	.9993
3.2	.9993	.9993	.9994	.9994	.9994	.9994	.9994	.9995	.9995	.9995
3.3	.9995	.9995	.9995	.9996	.9996	.9996	.9996	.9996	.9996	.9997
3.4	.9997	.9997	.9997	.9997	.9997	.9997	.9997	.9997	.9997	.9998

Table V The *t* Distribution Table   B21

Table V   The *t* Distribution Table

The entries in this table give the critical values of *t* for the specified number of degrees of freedom and areas in the right tail.

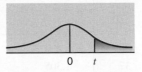

0   *t*

	Area in the Right Tail Under the *t* Distribution Curve					
df	.10	.05	.025	.01	.005	.001
1	3.078	6.314	12.706	31.821	63.657	318.309
2	1.886	2.920	4.303	6.965	9.925	22.327
3	1.638	2.353	3.182	4.541	5.841	10.215
4	1.533	2.132	2.776	3.747	4.604	7.173
5	1.476	2.015	2.571	3.365	4.032	5.893
6	1.440	1.943	2.447	3.143	3.707	5.208
7	1.415	1.895	2.365	2.998	3.499	4.785
8	1.397	1.860	2.306	2.896	3.355	4.501
9	1.383	1.833	2.262	2.821	3.250	4.297
10	1.372	1.812	2.228	2.764	3.169	4.144
11	1.363	1.796	2.201	2.718	3.106	4.025
12	1.356	1.782	2.179	2.681	3.055	3.930
13	1.350	1.771	2.160	2.650	3.012	3.852
14	1.345	1.761	2.145	2.624	2.977	3.787
15	1.341	1.753	2.131	2.602	2.947	3.733
16	1.337	1.746	2.120	2.583	2.921	3.686
17	1.333	1.740	2.110	2.567	2.898	3.646
18	1.330	1.734	2.101	2.552	2.878	3.610
19	1.328	1.729	2.093	2.539	2.861	3.579
20	1.325	1.725	2.086	2.528	2.845	3.552
21	1.323	1.721	2.080	2.518	2.831	3.527
22	1.321	1.717	2.074	2.508	2.819	3.505
23	1.319	1.714	2.069	2.500	2.807	3.485
24	1.318	1.711	2.064	2.492	2.797	3.467
25	1.316	1.708	2.060	2.485	2.787	3.450
26	1.315	1.706	2.056	2.479	2.779	3.435
27	1.314	1.703	2.052	2.473	2.771	3.421
28	1.313	1.701	2.048	2.467	2.763	3.408
29	1.311	1.699	2.045	2.462	2.756	3.396
30	1.310	1.697	2.042	2.457	2.750	3.385
31	1.309	1.696	2.040	2.453	2.744	3.375
32	1.309	1.694	2.037	2.449	2.738	3.365
33	1.308	1.692	2.035	2.445	2.733	3.356
34	1.307	1.691	2.032	2.441	2.728	3.348
35	1.306	1.690	2.030	2.438	2.724	3.340

Table V   The t Distribution Table   (continued)

df	Area in the Right Tail Under the t Distribution Curve					
	.10	.05	.025	.01	.005	.001
36	1.306	1.688	2.028	2.434	2.719	3.333
37	1.305	1.687	2.026	2.431	2.715	3.326
38	1.304	1.686	2.024	2.429	2.712	3.319
39	1.304	1.685	2.023	2.426	2.708	3.313
40	1.303	1.684	2.021	2.423	2.704	3.307
41	1.303	1.683	2.020	2.421	2.701	3.301
42	1.302	1.682	2.018	2.418	2.698	3.296
43	1.302	1.681	2.017	2.416	2.695	3.291
44	1.301	1.680	2.015	2.414	2.692	3.286
45	1.301	1.679	2.014	2.412	2.690	3.281
46	1.300	1.679	2.013	2.410	2.687	3.277
47	1.300	1.678	2.012	2.408	2.685	3.273
48	1.299	1.677	2.011	2.407	2.682	3.269
49	1.299	1.677	2.010	2.405	2.680	3.265
50	1.299	1.676	2.009	2.403	2.678	3.261
51	1.298	1.675	2.008	2.402	2.676	3.258
52	1.298	1.675	2.007	2.400	2.674	3.255
53	1.298	1.674	2.006	2.399	2.672	3.251
54	1.297	1.674	2.005	2.397	2.670	3.248
55	1.297	1.673	2.004	2.396	2.668	3.245
56	1.297	1.673	2.003	2.395	2.667	3.242
57	1.297	1.672	2.002	2.394	2.665	3.239
58	1.296	1.672	2.002	2.392	2.663	3.237
59	1.296	1.671	2.001	2.391	2.662	3.234
60	1.296	1.671	2.000	2.390	2.660	3.232
61	1.296	1.670	2.000	2.389	2.659	3.229
62	1.295	1.670	1.999	2.388	2.657	3.227
63	1.295	1.669	1.998	2.387	2.656	3.225
64	1.295	1.669	1.998	2.386	2.655	3.223
65	1.295	1.669	1.997	2.385	2.654	3.220
66	1.295	1.668	1.997	2.384	2.652	3.218
67	1.294	1.668	1.996	2.383	2.651	3.216
68	1.294	1.668	1.995	2.382	2.650	3.214
69	1.294	1.667	1.995	2.382	2.649	3.213
70	1.294	1.667	1.994	2.381	2.648	3.211
71	1.294	1.667	1.994	2.380	2.647	3.209
72	1.293	1.666	1.993	2.379	2.646	3.207
73	1.293	1.666	1.993	2.379	2.645	3.206
74	1.293	1.666	1.993	2.378	2.644	3.204
75	1.293	1.665	1.992	2.377	2.643	3.202
∞	1.282	1.645	1.960	2.326	2.576	3.090

Table VI Chi-Square Distribution Table   B23

Table VI   Chi-Square Distribution Table

The entries in this table give the critical values of χ^2 for the specified number of degrees of freedom and areas in the right tail.

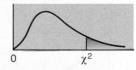

	Area in the Right Tail Under the Chi-Square Distribution Curve									
df	.995	.990	.975	.950	.900	.100	.050	.025	.010	.005
1	0.000	0.000	0.001	0.004	0.016	2.706	3.841	5.024	6.635	7.879
2	0.010	0.020	0.051	0.103	0.211	4.605	5.991	7.378	9.210	10.597
3	0.072	0.115	0.216	0.352	0.584	6.251	7.815	9.348	11.345	12.838
4	0.207	0.297	0.484	0.711	1.064	7.779	9.488	11.143	13.277	14.860
5	0.412	0.554	0.831	1.145	1.610	9.236	11.070	12.833	15.086	16.750
6	0.676	0.872	1.237	1.635	2.204	10.645	12.592	14.449	16.812	18.548
7	0.989	1.239	1.690	2.167	2.833	12.017	14.067	16.013	18.475	20.278
8	1.344	1.646	2.180	2.733	3.490	13.362	15.507	17.535	20.090	21.955
9	1.735	2.088	2.700	3.325	4.168	14.684	16.919	19.023	21.666	23.589
10	2.156	2.558	3.247	3.940	4.865	15.987	18.307	20.483	23.209	25.188
11	2.603	3.053	3.816	4.575	5.578	17.275	19.675	21.920	24.725	26.757
12	3.074	3.571	4.404	5.226	6.304	18.549	21.026	23.337	26.217	28.300
13	3.565	4.107	5.009	5.892	7.042	19.812	22.362	24.736	27.688	29.819
14	4.075	4.660	5.629	6.571	7.790	21.064	23.685	26.119	29.141	31.319
15	4.601	5.229	6.262	7.261	8.547	22.307	24.996	27.488	30.578	32.801
16	5.142	5.812	6.908	7.962	9.312	23.542	26.296	28.845	32.000	34.267
17	5.697	6.408	7.564	8.672	10.085	24.769	27.587	30.191	33.409	35.718
18	6.265	7.015	8.231	9.390	10.865	25.989	28.869	31.526	34.805	37.156
19	6.844	7.633	8.907	10.117	11.651	27.204	30.144	32.852	36.191	38.582
20	7.434	8.260	9.591	10.851	12.443	28.412	31.410	34.170	37.566	39.997
21	8.034	8.897	10.283	11.591	13.240	29.615	32.671	35.479	38.932	41.401
22	8.643	9.542	10.982	12.338	14.041	30.813	33.924	36.781	40.289	42.796
23	9.260	10.196	11.689	13.091	14.848	32.007	35.172	38.076	41.638	44.181
24	9.886	10.856	12.401	13.848	15.659	33.196	36.415	39.364	42.980	45.559
25	10.520	11.524	13.120	14.611	16.473	34.382	37.652	40.646	44.314	46.928
26	11.160	12.198	13.844	15.379	17.292	35.563	38.885	41.923	45.642	48.290
27	11.808	12.879	14.573	16.151	18.114	36.741	40.113	43.195	46.963	49.645
28	12.461	13.565	15.308	16.928	18.939	37.916	41.337	44.461	48.278	50.993
29	13.121	14.256	16.047	17.708	19.768	39.087	42.557	45.722	49.588	52.336
30	13.787	14.953	16.791	18.493	20.599	40.256	43.773	46.979	50.892	53.672
40	20.707	22.164	24.433	26.509	29.051	51.805	55.758	59.342	63.691	66.766
50	27.991	29.707	32.357	34.764	37.689	63.167	67.505	71.420	76.154	79.490
60	35.534	37.485	40.482	43.188	46.459	74.397	79.082	83.298	88.379	91.952
70	43.275	45.442	48.758	51.739	55.329	85.527	90.531	95.023	100.425	104.215
80	51.172	53.540	57.153	60.391	64.278	96.578	101.879	106.629	112.329	116.321
90	59.196	61.754	65.647	69.126	73.291	107.565	113.145	118.136	124.116	128.299
100	67.328	70.065	74.222	77.929	82.358	118.498	124.342	129.561	135.807	140.169

Table VII The *F* Distribution Table

The entries in the table on this page give the critical values of *F* for .01 area in the right tail under the *F* distribution curve and specified degrees of freedom for the numerator and denominator.

Degrees of Freedom for the Numerator

	1	2	3	4	5	6	7	8	9	10	11	12	15	20	25	30	40	50	100
1	4052	5000	5403	5625	5764	5859	5928	5981	6022	6056	6083	6106	6157	6209	6240	6261	6287	6303	6334
2	98.50	99.00	99.17	99.25	99.30	99.33	99.36	99.37	99.39	99.40	99.41	99.42	99.43	99.45	99.46	99.47	99.47	99.48	99.49
3	34.12	30.82	29.46	28.71	28.24	27.91	27.67	27.49	27.35	27.23	27.13	27.05	26.87	26.69	26.58	26.50	26.41	26.35	26.24
4	21.20	18.00	16.69	15.98	15.52	15.21	14.98	14.80	14.66	14.55	14.45	14.37	14.20	14.02	13.91	13.84	13.75	13.69	13.58
5	16.26	13.27	12.06	11.39	10.97	10.67	10.46	10.29	10.16	10.05	9.96	9.89	9.72	9.55	9.45	9.38	9.29	9.24	9.13
6	13.75	10.92	9.78	9.15	8.75	8.47	8.26	8.10	7.98	7.87	7.79	7.72	7.56	7.40	7.30	7.23	7.14	7.09	6.99
7	12.25	9.55	8.45	7.85	7.46	7.19	6.99	6.84	6.72	6.62	6.54	6.47	6.31	6.16	6.06	5.99	5.91	5.86	5.75
8	11.26	8.65	7.59	7.01	6.63	6.37	6.18	6.03	5.91	5.81	5.73	5.67	5.52	5.36	5.26	5.20	5.12	5.07	4.96
9	10.56	8.02	6.99	6.42	6.06	5.80	5.61	5.47	5.35	5.26	5.18	5.11	4.96	4.81	4.71	4.65	4.57	4.52	4.41
10	10.04	7.56	6.55	5.99	5.64	5.39	5.20	5.06	4.94	4.85	4.77	4.71	4.56	4.41	4.31	4.25	4.17	4.12	4.01
11	9.65	7.21	6.22	5.67	5.32	5.07	4.89	4.74	4.63	4.54	4.46	4.40	4.25	4.10	4.01	3.94	3.86	3.81	3.71
12	9.33	6.93	5.95	5.41	5.06	4.82	4.64	4.50	4.39	4.30	4.22	4.16	4.01	3.86	3.76	3.70	3.62	3.57	3.47
13	9.07	6.70	5.74	5.21	4.86	4.62	4.44	4.30	4.19	4.10	4.02	3.96	3.82	3.66	3.57	3.51	3.43	3.38	3.27
14	8.86	6.51	5.56	5.04	4.69	4.46	4.28	4.14	4.03	3.94	3.86	3.80	3.66	3.51	3.41	3.35	3.27	3.22	3.11
15	8.68	6.36	5.42	4.89	4.56	4.32	4.14	4.00	3.89	3.80	3.73	3.67	3.52	3.37	3.28	3.21	3.13	3.08	2.98
16	8.53	6.23	5.29	4.77	4.44	4.20	4.03	3.89	3.78	3.69	3.62	3.55	3.41	3.26	3.16	3.10	3.02	2.97	2.86
17	8.40	6.11	5.18	4.67	4.34	4.10	3.93	3.79	3.68	3.59	3.52	3.46	3.31	3.16	3.07	3.00	2.92	2.87	2.76
18	8.29	6.01	5.09	4.58	4.25	4.01	3.84	3.71	3.60	3.51	3.43	3.37	3.23	3.08	2.98	2.92	2.84	2.78	2.68
19	8.18	5.93	5.01	4.50	4.17	3.94	3.77	3.63	3.52	3.43	3.36	3.30	3.15	3.00	2.91	2.84	2.76	2.71	2.60
20	8.10	5.85	4.94	4.43	4.10	3.87	3.70	3.56	3.46	3.37	3.29	3.23	3.09	2.94	2.84	2.78	2.69	2.64	2.54
21	8.02	5.78	4.87	4.37	4.04	3.81	3.64	3.51	3.40	3.31	3.24	3.17	3.03	2.88	2.79	2.72	2.64	2.58	2.48
22	7.95	5.72	4.82	4.31	3.99	3.76	3.59	3.45	3.35	3.26	3.18	3.12	2.98	2.83	2.73	2.67	2.58	2.53	2.42
23	7.88	5.66	4.76	4.26	3.94	3.71	3.54	3.41	3.30	3.21	3.14	3.07	2.93	2.78	2.69	2.62	2.54	2.48	2.37
24	7.82	5.61	4.72	4.22	3.90	3.67	3.50	3.36	3.26	3.17	3.09	3.03	2.89	2.74	2.64	2.58	2.49	2.44	2.33
25	7.77	5.57	4.68	4.18	3.85	3.63	3.46	3.32	3.22	3.13	3.06	2.99	2.85	2.70	2.60	2.54	2.45	2.40	2.29
30	7.56	5.39	4.51	4.02	3.70	3.47	3.30	3.17	3.07	2.98	2.91	2.84	2.70	2.55	2.45	2.39	2.30	2.25	2.13
40	7.31	5.18	4.31	3.83	3.51	3.29	3.12	2.99	2.89	2.80	2.73	2.66	2.52	2.37	2.27	2.20	2.11	2.06	1.94
50	7.17	5.06	4.20	3.72	3.41	3.19	3.02	2.89	2.78	2.70	2.63	2.56	2.42	2.27	2.17	2.10	2.01	1.95	1.82
100	6.90	4.82	3.98	3.51	3.21	2.99	2.82	2.69	2.59	2.50	2.43	2.37	2.22	2.07	1.97	1.89	1.80	1.74	1.60

Degrees of Freedom for the Denominator

Table VII   The F Distribution Table   (continued)

The entries in the table on this page give the critical values of F for .025 area in the right tail under the F distribution curve and specified degrees of freedom for the numerator and denominator.

	Degrees of Freedom for the Numerator																		
	1	2	3	4	5	6	7	8	9	10	11	12	15	20	25	30	40	50	100
1	647.8	799.5	864.2	899.6	921.8	937.1	948.2	956.7	963.3	968.6	973.0	976.7	984.9	993.1	998.1	1001	1006	1008	1013
2	38.51	39.00	39.17	39.25	39.30	39.33	39.36	39.37	39.39	39.40	39.41	39.41	39.43	39.45	39.46	39.46	39.47	39.48	39.49
3	17.44	16.04	15.44	15.10	14.88	14.73	14.62	14.54	14.47	14.42	14.37	14.34	14.25	14.17	14.12	14.08	14.04	14.01	13.96
4	12.22	10.65	9.98	9.61	9.36	9.20	9.07	8.98	8.90	8.84	8.79	8.75	8.66	8.56	8.50	8.46	8.41	8.38	8.32
5	10.01	8.43	7.76	7.39	7.15	6.98	6.85	6.76	6.68	6.62	6.57	6.52	6.43	6.33	6.27	6.23	6.18	6.14	6.08
6	8.81	7.26	6.60	6.23	5.99	5.82	5.70	5.60	5.52	5.46	5.41	5.37	5.27	5.17	5.11	5.07	5.01	4.98	4.92
7	8.07	6.54	5.89	5.52	5.29	5.12	4.99	4.90	4.82	4.76	4.71	4.67	4.57	4.47	4.40	4.36	4.31	4.28	4.21
8	7.57	6.06	5.42	5.05	4.82	4.65	4.53	4.43	4.36	4.30	4.24	4.20	4.10	4.00	3.94	3.89	3.84	3.81	3.74
9	7.21	5.72	5.08	4.72	4.48	4.32	4.20	4.10	4.03	3.96	3.91	3.87	3.77	3.67	3.60	3.56	3.51	3.47	3.40
10	6.94	5.46	4.83	4.47	4.24	4.07	3.95	3.85	3.78	3.72	3.66	3.62	3.52	3.42	3.35	3.31	3.26	3.22	3.15
11	6.72	5.26	4.63	4.28	4.04	3.88	3.76	3.66	3.59	3.53	3.47	3.43	3.33	3.23	3.16	3.12	3.06	3.03	2.96
12	6.55	5.10	4.47	4.12	3.89	3.73	3.61	3.51	3.44	3.37	3.32	3.28	3.18	3.07	3.01	2.96	2.91	2.87	2.80
13	6.41	4.97	4.35	4.00	3.77	3.60	3.48	3.39	3.31	3.25	3.20	3.15	3.05	2.95	2.88	2.84	2.78	2.74	2.67
14	6.30	4.86	4.24	3.89	3.66	3.50	3.38	3.29	3.21	3.15	3.09	3.05	2.95	2.84	2.78	2.73	2.67	2.64	2.56
15	6.20	4.77	4.15	3.80	3.58	3.41	3.29	3.20	3.12	3.06	3.01	2.96	2.86	2.76	2.69	2.64	2.59	2.55	2.47
16	6.12	4.69	4.08	3.73	3.50	3.34	3.22	3.12	3.05	2.99	2.93	2.89	2.79	2.68	2.61	2.57	2.51	2.47	2.40
17	6.04	4.62	4.01	3.66	3.44	3.28	3.16	3.06	2.98	2.92	2.87	2.82	2.72	2.62	2.55	2.50	2.44	2.41	2.33
18	5.98	4.56	3.95	3.61	3.38	3.22	3.10	3.01	2.93	2.87	2.81	2.77	2.67	2.56	2.49	2.44	2.38	2.35	2.27
19	5.92	4.51	3.90	3.56	3.33	3.17	3.05	2.96	2.88	2.82	2.76	2.72	2.62	2.51	2.44	2.39	2.33	2.30	2.22
20	5.87	4.46	3.86	3.51	3.29	3.13	3.01	2.91	2.84	2.77	2.72	2.68	2.57	2.46	2.40	2.35	2.29	2.25	2.17
21	5.83	4.42	3.82	3.48	3.25	3.09	2.97	2.87	2.80	2.73	2.68	2.64	2.53	2.42	2.36	2.31	2.25	2.21	2.13
22	5.79	4.38	3.78	3.44	3.22	3.05	2.93	2.84	2.76	2.70	2.65	2.60	2.50	2.39	2.32	2.27	2.21	2.17	2.09
23	5.75	4.35	3.75	3.41	3.18	3.02	2.90	2.81	2.73	2.67	2.62	2.57	2.47	2.36	2.29	2.24	2.18	2.14	2.06
24	5.72	4.32	3.72	3.38	3.15	2.99	2.87	2.78	2.70	2.64	2.59	2.54	2.44	2.33	2.26	2.21	2.15	2.11	2.02
25	5.69	4.29	3.69	3.35	3.13	2.97	2.85	2.75	2.68	2.61	2.56	2.51	2.41	2.30	2.23	2.18	2.12	2.08	2.00
30	5.57	4.18	3.59	3.25	3.03	2.87	2.75	2.65	2.57	2.51	2.46	2.41	2.31	2.20	2.12	2.07	2.01	1.97	1.88
40	5.42	4.05	3.46	3.13	2.90	2.74	2.62	2.53	2.45	2.39	2.33	2.29	2.18	2.07	1.99	1.94	1.88	1.83	1.74
50	5.34	3.97	3.39	3.05	2.83	2.67	2.55	2.46	2.38	2.32	2.26	2.22	2.11	1.99	1.92	1.87	1.80	1.75	1.66
100	5.18	3.83	3.25	2.92	2.70	2.54	2.42	2.32	2.24	2.18	2.12	2.08	1.97	1.85	1.77	1.71	1.64	1.59	1.48

Degrees of Freedom for the Denominator

Table VII   The F Distribution Table   (continued)

The entries in the table on this page give the critical values of F for .05 area in the right tail under the F distribution curve and specified degrees of freedom for the numerator and denominator.

Degrees of Freedom for the Numerator

	1	2	3	4	5	6	7	8	9	10	11	12	15	20	25	30	40	50	100
1	161.5	199.5	215.7	224.6	230.2	234.0	236.8	238.9	240.5	241.9	243.0	243.9	246.0	248.0	249.3	250.1	251.1	251.8	253.0
2	18.51	19.00	19.16	19.25	19.30	19.33	19.35	19.37	19.38	19.40	19.40	19.41	19.43	19.45	19.46	19.46	19.47	19.48	19.49
3	10.13	9.55	9.28	9.12	9.01	8.94	8.89	8.85	8.81	8.79	8.76	8.74	8.70	8.66	8.63	8.62	8.59	8.58	8.55
4	7.71	6.94	6.59	6.39	6.26	6.16	6.09	6.04	6.00	5.96	5.94	5.91	5.86	5.80	5.77	5.75	5.72	5.70	5.66
5	6.61	5.79	5.41	5.19	5.05	4.95	4.88	4.82	4.77	4.74	4.70	4.68	4.62	4.56	4.52	4.50	4.46	4.44	4.41
6	5.99	5.14	4.76	4.53	4.39	4.28	4.21	4.15	4.10	4.06	4.03	4.00	3.94	3.87	3.83	3.81	3.77	3.75	3.71
7	5.59	4.74	4.35	4.12	3.97	3.87	3.79	3.73	3.68	3.64	3.60	3.57	3.51	3.44	3.40	3.38	3.34	3.32	3.27
8	5.32	4.46	4.07	3.84	3.69	3.58	3.50	3.44	3.39	3.35	3.31	3.28	3.22	3.15	3.11	3.08	3.04	3.02	2.97
9	5.12	4.26	3.86	3.63	3.48	3.37	3.29	3.23	3.18	3.14	3.10	3.07	3.01	2.94	2.89	2.86	2.83	2.80	2.76
10	4.96	4.10	3.71	3.48	3.33	3.22	3.14	3.07	3.02	2.98	2.94	2.91	2.85	2.77	2.73	2.70	2.66	2.64	2.59
11	4.84	3.98	3.59	3.36	3.20	3.09	3.01	2.95	2.90	2.85	2.82	2.79	2.72	2.65	2.60	2.57	2.53	2.51	2.46
12	4.75	3.89	3.49	3.26	3.11	3.00	2.91	2.85	2.80	2.75	2.72	2.69	2.62	2.54	2.50	2.47	2.43	2.40	2.35
13	4.67	3.81	3.41	3.18	3.03	2.92	2.83	2.77	2.71	2.67	2.63	2.60	2.53	2.46	2.41	2.38	2.34	2.31	2.26
14	4.60	3.74	3.34	3.11	2.96	2.85	2.76	2.70	2.65	2.60	2.57	2.53	2.46	2.39	2.34	2.31	2.27	2.24	2.19
15	4.54	3.68	3.29	3.06	2.90	2.79	2.71	2.64	2.59	2.54	2.51	2.48	2.40	2.33	2.28	2.25	2.20	2.18	2.12
16	4.49	3.63	3.24	3.01	2.85	2.74	2.66	2.59	2.54	2.49	2.46	2.42	2.35	2.28	2.23	2.19	2.15	2.12	2.07
17	4.45	3.59	3.20	2.96	2.81	2.70	2.61	2.55	2.49	2.45	2.41	2.38	2.31	2.23	2.18	2.15	2.10	2.08	2.02
18	4.41	3.55	3.16	2.93	2.77	2.66	2.58	2.51	2.46	2.41	2.37	2.34	2.27	2.19	2.14	2.11	2.06	2.04	1.98
19	4.38	3.52	3.13	2.90	2.74	2.63	2.54	2.48	2.42	2.38	2.34	2.31	2.23	2.16	2.11	2.07	2.03	2.00	1.94
20	4.35	3.49	3.10	2.87	2.71	2.60	2.51	2.45	2.39	2.35	2.31	2.28	2.20	2.12	2.07	2.04	1.99	1.97	1.91
21	4.32	3.47	3.07	2.84	2.68	2.57	2.49	2.42	2.37	2.32	2.28	2.25	2.18	2.10	2.05	2.01	1.96	1.94	1.88
22	4.30	3.44	3.05	2.82	2.66	2.55	2.46	2.40	2.34	2.30	2.26	2.23	2.15	2.07	2.02	1.97	1.94	1.91	1.85
23	4.28	3.42	3.03	2.80	2.64	2.53	2.44	2.37	2.32	2.27	2.24	2.20	2.13	2.05	2.00	1.96	1.91	1.88	1.82
24	4.26	3.40	3.01	2.78	2.62	2.51	2.42	2.36	2.30	2.25	2.22	2.18	2.11	2.03	1.97	1.94	1.89	1.86	1.80
25	4.24	3.39	2.99	2.76	2.60	2.49	2.40	2.34	2.28	2.24	2.20	2.16	2.09	2.01	1.96	1.92	1.87	1.84	1.78
30	4.17	3.32	2.92	2.69	2.53	2.42	2.33	2.27	2.21	2.16	2.13	2.09	2.01	1.93	1.88	1.84	1.79	1.76	1.70
40	4.08	3.23	2.84	2.61	2.45	2.34	2.25	2.18	2.12	2.08	2.04	2.00	1.92	1.84	1.78	1.74	1.69	1.66	1.59
50	4.03	3.18	2.79	2.56	2.40	2.29	2.20	2.13	2.07	2.03	1.99	1.95	1.87	1.78	1.73	1.69	1.63	1.60	1.52
100	3.94	3.09	2.70	2.46	2.31	2.19	2.10	2.03	1.97	1.93	1.89	1.85	1.77	1.68	1.62	1.57	1.52	1.48	1.39

Degrees of Freedom for the Denominator

Table VII   The *F* Distribution Table   (continued)

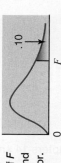

The entries in the table on this page give the critical values of *F* for .10 area in the right tail under the *F* distribution curve and specified degrees of freedom for the numerator and denominator.

							Degrees of Freedom for the Numerator												
	1	2	3	4	5	6	7	8	9	10	11	12	15	20	25	30	40	50	100
1	39.86	49.50	53.59	55.83	57.24	58.20	58.91	59.44	59.86	60.19	60.47	60.71	61.22	61.74	62.05	62.26	62.53	62.69	63.01
2	8.53	9.00	9.16	9.24	9.29	9.33	9.35	9.37	9.38	9.39	9.40	9.41	9.42	9.44	9.45	9.46	9.47	9.47	9.48
3	5.54	5.46	5.39	5.34	5.31	5.28	5.27	5.25	5.24	5.23	5.22	5.22	5.20	5.18	5.17	5.17	5.16	5.15	5.14
4	4.54	4.32	4.19	4.11	4.05	4.01	3.98	3.95	3.94	3.92	3.91	3.90	3.87	3.84	3.83	3.82	3.80	3.80	3.78
5	4.06	3.78	3.62	3.52	3.45	3.40	3.37	3.34	3.32	3.30	3.28	3.27	3.24	3.21	3.19	3.17	3.16	3.15	3.13
6	3.78	3.46	3.29	3.18	3.11	3.05	3.01	2.98	2.96	2.94	2.92	2.90	2.87	2.84	2.81	2.80	2.78	2.77	2.75
7	3.59	3.26	3.07	2.96	2.88	2.83	2.78	2.75	2.72	2.70	2.68	2.67	2.63	2.59	2.57	2.56	2.54	2.52	2.50
8	3.46	3.11	2.92	2.81	2.73	2.67	2.62	2.59	2.56	2.54	2.52	2.50	2.46	2.42	2.40	2.38	2.36	2.35	2.32
9	3.36	3.01	2.81	2.69	2.61	2.55	2.51	2.47	2.44	2.42	2.40	2.38	2.34	2.30	2.27	2.25	2.23	2.22	2.19
10	3.29	2.92	2.73	2.61	2.52	2.46	2.41	2.38	2.35	2.32	2.30	2.28	2.24	2.20	2.17	2.16	2.13	2.12	2.09
11	3.23	2.86	2.66	2.54	2.45	2.39	2.34	2.30	2.27	2.25	2.23	2.21	2.17	2.12	2.10	2.08	2.05	2.04	2.01
12	3.18	2.81	2.61	2.48	2.39	2.33	2.28	2.24	2.21	2.19	2.17	2.15	2.10	2.06	2.03	2.01	1.99	1.97	1.94
13	3.14	2.76	2.56	2.43	2.35	2.28	2.23	2.20	2.16	2.14	2.12	2.10	2.05	2.01	1.98	1.96	1.93	1.92	1.88
14	3.10	2.73	2.52	2.39	2.31	2.24	2.19	2.15	2.12	2.10	2.07	2.05	2.01	1.96	1.93	1.91	1.89	1.87	1.83
15	3.07	2.70	2.49	2.36	2.27	2.21	2.16	2.12	2.09	2.06	2.04	2.02	1.97	1.92	1.89	1.87	1.85	1.83	1.79
16	3.05	2.67	2.46	2.33	2.24	2.18	2.13	2.09	2.06	2.03	2.01	1.99	1.94	1.89	1.86	1.84	1.81	1.79	1.76
17	3.03	2.64	2.44	2.31	2.22	2.15	2.10	2.06	2.03	2.00	1.98	1.96	1.91	1.86	1.83	1.81	1.78	1.76	1.73
18	3.01	2.62	2.42	2.29	2.20	2.13	2.08	2.04	2.00	1.98	1.95	1.93	1.89	1.84	1.80	1.78	1.75	1.74	1.70
19	2.99	2.61	2.40	2.27	2.18	2.11	2.06	2.02	1.98	1.96	1.93	1.91	1.86	1.81	1.78	1.76	1.73	1.71	1.67
20	2.97	2.59	2.38	2.25	2.16	2.09	2.04	2.00	1.96	1.94	1.91	1.89	1.84	1.79	1.76	1.74	1.71	1.69	1.65
21	2.96	2.57	2.36	2.23	2.14	2.08	2.02	1.98	1.95	1.92	1.90	1.87	1.83	1.78	1.74	1.72	1.69	1.67	1.63
22	2.95	2.56	2.35	2.22	2.13	2.06	2.01	1.97	1.93	1.90	1.88	1.86	1.81	1.76	1.73	1.70	1.67	1.65	1.61
23	2.94	2.55	2.34	2.21	2.11	2.05	1.99	1.95	1.92	1.89	1.87	1.84	1.80	1.74	1.71	1.69	1.66	1.64	1.59
24	2.93	2.54	2.33	2.19	2.10	2.04	1.98	1.94	1.91	1.88	1.85	1.83	1.78	1.73	1.70	1.67	1.64	1.62	1.58
25	2.92	2.53	2.32	2.18	2.09	2.02	1.97	1.93	1.89	1.87	1.84	1.82	1.77	1.72	1.68	1.66	1.63	1.61	1.56
30	2.88	2.49	2.28	2.14	2.05	1.98	1.93	1.88	1.85	1.82	1.79	1.77	1.72	1.67	1.63	1.61	1.57	1.55	1.51
40	2.84	2.44	2.23	2.09	2.00	1.93	1.87	1.83	1.79	1.76	1.74	1.71	1.66	1.61	1.57	1.54	1.51	1.48	1.43
50	2.81	2.41	2.20	2.06	1.97	1.90	1.84	1.80	1.76	1.73	1.70	1.68	1.63	1.57	1.53	1.50	1.46	1.44	1.39
100	2.76	2.36	2.14	2.00	1.91	1.83	1.78	1.73	1.69	1.66	1.64	1.61	1.56	1.49	1.45	1.42	1.38	1.35	1.29

Degrees of Freedom for the Denominator

APPENDIX
B

Statistical Tables on the Web Site

Note: The following tables are on the Web site of the text along with Chapters 14 and 15.

Answers to Selected Exercises and Self-Review Tests

(Note: Due to differences in rounding, the answers obtained by readers may differ slightly from the ones given here.)

Chapter 1

1.3 inferential statistics

1.10 **a.** Qualitative **b.** Quantitative; continuous
c. Quantitative; discrete **d.** Qualitative

1.22 **a.** random sample **b.** simple random sample
c. no systematic error

1.26 voluntary response error and nonresponse error

1.32 **a.** Since the patients were randomly selected from the population of all people suffering from compulsive behavior and were randomly assigned to treatment and control groups, the two groups should be comparable and representative of the entire population. The patients did not know whether or not they were getting the treatment, so any improvement in their condition should be due to the medicine and not merely to the power of suggestion. Thus, the conclusion is justified.

b. designed experiment

c. not double-blind since the doctors knew who received the medicine

1.33 observational study

1.35 The conclusion is unjustified. The families volunteered; they were not randomly selected from the population of all families on welfare, thus they may not be representative of the entire population.

1.39 **a.** $\Sigma x = 856$ **b.** $(\Sigma x)^2 = 732{,}736$
c. $\Sigma x^2 = 157{,}574$

1.45 **a.** $\Sigma m = 59$ **b.** $\Sigma f^2 = 2662$
c. $\Sigma mf = 1508$ **d.** $\Sigma m^2 f = 24{,}884$
e. $\Sigma m^2 = 867$

1.47 **a.** This is an observational study since each participant decided how much meat to consume. Thus, the treatment is not controlled by the experimenters.

b. Because this is an observational study, no cause-and-effect relationship between meat consumption and cholesterol level may be inferred. The effect of meat consumption on cholesterol level may be confounded with other variables such as other dietary habits, amount

of exercise, and other features of the participants' lifestyles.

1.48 **a.** designed experiment **b.** not double blind

1.51 **a.** We would expect \$61,200 to be a biased estimate of the current mean annual income for all 5432 alumni because only 1240 of the 5432 alumni answered the income question. These 1240 are unlikely to be representative of the entire group of 5432.

b. The following types of bias are likely to be present:

Nonresponse error: Alumni with low incomes may be ashamed to respond. Thus, the 1240 who actually returned their questionnaires and answered the income question would tend to have higher than average incomes.

Response error: Some of those who answered the income question may give a value that is higher than their actual income in order to appear more successful.

c. We would expect the estimate of \$61,200 to be above the current mean annual income of all 5432 alumni, for the given reasons in part b.

Self-Review Test

1. b

2. c

3. **a.** Sample without replacement
b. Sample with replacement

8. **a.** convenience and nonrandom sample
b. not likely representative of entire population
c. convenience sample

9. **a.** no
b. voluntary response error, selection error, and response error

10. **a.** designed experiment **b.** yes **c.** no

11. observational study

12. randomized experiment

13. **a.** $\Sigma x = 54$ **b.** $\Sigma y = 113.6$ **c.** $\Sigma x = 472$
d. $\Sigma xy = 742.7$ **e.** $\Sigma x^2 y = 5686.7$ **f.** $(\Sigma y)^2 = 2172.46$

Chapter 2

2.3 **a. & b.**

Category	Frequency	Relative Frequency	Percentage
P	23	23/40 = 0.575	57.5
Q	13	13/40 = 0.325	32.5
R	4	4/40 = 0.100	10.0

c. 57.5%   **d.** 42.5%

2.13 **d.** Boundaries of the fourth class are $4200.5 and $5600.5; width = $1400.

2.14 **d.** 60.7%

2.38 **a. & b.**

Category	Frequency	Relative Frequency	Percentage
PI	9	9/36 = 0.25	25
S	8	8/36 = 0.222	22.2
V	13	13/36 = 0.361	36.1
PO	3	3/36 = 0.083	8.3
B	1	1/36 = 0.028	2.8
C	2	2/36 = 0.056	5.6

c. 50%

2.39 **d.** 50%

2.40 **a.** 369 customers were served.

b. Each class has a width of 4.

d. 57.2%

e. The number of customers who purchased 10 gallons or less cannot be determined exactly because 10 is not a boundary value.

2.48 No. The older group may drive more miles per week than the younger group.

Self-Review Test

2. **a.** 5   **b.** 7   **c.** 17   **d.** 6.5
     e. 13.5   **f.** 90   **g.** .30

4. **a. & b.**

Net Worth vs. $200,000	Frequency	Relative Frequency	Percentage
M	12	12/36 = 0.333	33.3
L	18	18/36 = 0.500	50.0
N	6	6/36 = 0.167	16.7

c. 18/36 = 50%

5. **a. & b.**

Monthly Expense on Gas (in dollars)	Frequency	Relative Frequency	Percentage
50 to 149	9	9/48 = 0.188	18.8
150 to 249	13	13/48 = 0.271	27.1
250 to 349	11	11/48 = 0.229	22.9
350 to 449	9	9/48 = 0.188	18.8
450 to 549	6	6/48 = 0.125	12.5

d. $(9 + 6)/48 = 31.25\%$

6. **b.** 40 dollars
     d. $(6 + 8 + 7)/30 = 70\%$

7.

```
0 | 4 6 7 8
1 | 0 2 2 3 4 4 5 6 6 6 7 8 9
2 | 0 1 2 2 5 9
3 | 2
```

8. 30   33   37   42   44   46   47   49   51   53   53   56   60
     67   67   71   79

Chapter 3

3.13 **a.** mean = 1803; median = 1270

b. outlier = 5490; when the outlier is dropped: mean = 1467.8; median = 1166; mean changes by a larger amount

c. median

3.14 **a.** mean = 2.92 power outages; median = 2.5 power outages; mode = 2 power outages

b. For the 25% trimmed mean, we must remove 0.25(12) = 3 values from each end of the ranked data set and average the remaining 14 values to get 2.67.

3.17 **a.** mean = 429.80 thousands of dollars, median = 103.5 thousands of dollars

b. no mode

c. 124 thousand dollars

d. Median and trimmed mean are good measures to use where there is an outlier.

3.27 weighted mean = 77.5

3.29 no

3.34 **a.** 9 sum of the deviations from the mean is zero

b. range = 12; $s^2 = 14.2857$; $s = 3.78$

3.36 **a.** range = 38; $s^2 = 151.7778$; $s = 12.3198$

b. $CV = 27.38\%$

3.38 **a.** range = 30; $s^2 = 107.4286$; $s = 10.36$

b. $CV = 103.6\%$

3.40   range = 7 stings; $s^2 = 4.5769$; $s = 2.14$ stings

3.49 $\bar{x} = 5.08$ inches, $s^2 = 6.8506$, $s = 2.62$ inches (sample statistics)

3.53 $\bar{x} = 36.80$ minutes; $s^2 = 597.7143$; $s = 24.45$ minutes

3.55   The empirical rule is applied to a bell–shaped distribution. According to this rule, approximately

- 68% of the observations lie within one standard deviation of the mean.
- 95% of the observations lie within two standard deviations of the mean.
- 99.7% of the observations lie within three standard deviations of the mean

3.57 $k = 2$: at least 75% fall in (148, 312)
$k = 2.5$: at least 84% fall in (127.5, 332.5)
$k = 3$: at least 89% fall in (107, 353)

3.60   **a.**   at least 75%   **b.**   at least 84%   **c.**   at least 89%

3.62   **a.**   99.7%   **b.**   68%   **c.**   95%

3.63   **a.**   **i.**   99.7%   **ii.**   68%   **b.**   66 to 78 mph

3.64   **a.**   **i.**   at least 84% of all workers have commuting times between 14 and 54 minutes.
ii.   at least 75% of all workers have commuting times between 18 and 50 minutes.

   b.   interval is 10 to 58.

3.66   To find the three quartiles:

1. Rank the given data set in increasing order.
2. Find the median using the procedure in Section 3.1.2. The median is the second quartile, Q_2.
3. The first quartile, Q_1, is the value of the middle term among the (ranked) observations that are less than Q_2.
4. The third quartile, Q_3, is the value of the middle term among the (ranked) observations that are greater that Q_2.

3.68   Given a data set of n values, to find the kth percentile (P_k):

1. Rank the given data in increasing order.
2. Calculate $kn/100$. Then, P_k is the term that is approximately ($kn/100$) in the ranking. If $kn/100$ falls between two consecutive integers a and b, it may be necessary to average the ath and b^{th} values in the ranking to obtain P_k.

3.71   **a.** $Q_1 = 69$; $Q_2 = 72.5$; $Q_3 = 76$; $IQR = 7$

   b. $P_{35} = 70$ **c.**   28.57%

3.72   **a.** $Q_1 = 25$; $Q_2 = 28.5$; $Q_3 = 33$; $IQR = 8$

   b. $P_{65} = 31$ **c.**   33.33%

3.73   **a.** $Q_1 = 533$; $Q_2 = 626.5$; $Q_3 = 728$; $IQR = 195$

   b. $P_{30} = 572$ **c.**   22.73%

3.91   The data set is skewed slightly to the right; 135 is an outlier.

3.93   **a.**   new mean = 76.4 inches; new median = 78 inches; new range = 13 inches

   b.   new mean = 75.2 inches

3.95   **a.**   trimmed mean = 9.5   **b.**   14.3%

3.98   **a.**   mean = 30   **b.**   mean = 50

3.100   **a.**   at least 55.56%

   b.   1 to 11 inches   **c.**   2.66 to 9.34 inches

3.104   **a.**   For men: mean = 174.91 lbs = 76,189.05 grams = 12.49 stone, median = 179 lbs = 77,970.61 grams = 12.79 stone, and st. dev. = 19.12 lbs = 8328.48 grams = 1.37 stone. For women: mean = 124.95 lbs = 54,426.97 grams = 8.93 stone, median = 123 lbs = 53,577.57 grams = 8.79 stone, st. dev. = 17.48 lbs = 7614.11 grams = 1.25 stone.

   b.   See answer to a, as answers are identical.

   c.   yes   **d. & e.**   Smaller unit has more variability.

3.105   108 to 111

3.106   **a.** $\bar{x} = 20.80$ thousand miles, Median = 15 thousand miles, and Mode = 15 thousand miles

   b.   Range = 59 thousand miles, $s^2 = 249.03$, $s = 15.78$ thousand miles

   c. $Q_1 = 10$ thousand miles and $Q_3 = 26$ thousand miles

   d.   IQR = 16 thousand miles

   Since the interquartile range is based on the middle 50% of the observations it is not affected by outliers. The standard deviation, however, is strongly affected by outliers. Thus, the interquartile range is preferable in applications in which a measure of variation is required that is unaffected by extreme values.

Self-Review Test

1.   b   **2.**   a and d   **3.**   c

4.   c   **5.**   b   **6.**   b

7.   a   **8.**   a   **9.**   b

10.   a   **11.**   b   **12.**   c

13.   a   **14.**   a

15.   **a.** $\bar{x} = 21$ times, Median = 13.5 times
The modes are 5, 8, and 14.

   b.   15.6875 times

   c.   Range = 90 times
$s^2 = 534.63$, $s = 23.12$ times

   d.   Coefficient of variation = 110.11%

   e.   sample statistics

16. $1,657.914

17.   mean of all scores = 75.43 points
If we drop the outlier (22), then
mean = 84.33 points.

18.   Range of all scores = 73 points. If we drop the outlier (22), then range = 22 points.

19.   The value of the standard deviation is zero when all the values in a data set are the same.

20.   **a.**   The frequency column gives the number of weeks for which the number of computers sold was in the corresponding class.

   b.   For the given data: $n = 25$, $\Sigma mf = 486.50$, and $\Sigma m^2 f = 10{,}524.25$
$\bar{x} = 19.46$ computers
$s^2 = 44.0400$
$s = 6.64$ computers

21. a. i. about 75% of the members spent between 59.4 and 124.2 minutes at the health club.

   ii. about 84% of the members spent between 51.3 and 132.3 minutes at the health club.

   b. 43.2 minutes to 140.4 minutes.

22. a. i. about 68% of the cars are 5.1 to 9.5 years old.

   ii. about 99.7% of the cars are 0.7 to 13.9 years.

   b. 2.9 to 11.7 years.

23. a. Median = 57, Q_1 = 52, and Q_3 = 66, IQR = 14,

   The value 54 lies between Q_1 and the Median, so it is in the second 25% group from the bottom of the ranked data set. This means that 25% of the data is less than 54 and that at least 50% of the data is larger than 54.

   b. 10.8; the 60th percentile can be approximated by the 11th term in the ranked data. So, P_{60} = 61. This means that 60% of the values in the data set are less than 61.

   c. Twelve values in the given data set are less than 64. Hence, the percentile rank of 64 = (12/18) × 100 = 66.7% or about 67%.

24. The data is skewed to the right.

25. n_1 = 15, n_2 = 20, $\bar{x}_1$ = $1035, $\bar{x}_2$ = $1090, $\bar{x}$ = $1066.43

26. 3.17

27. a. For Data Set I: $\bar{x}$ = 19.75

   For Data Set II: $\bar{x}$ = 16.75

   b. For Data Set I: Σx = 79, Σx^2 = 1945, and n = 4; s = 11.32

   c. For Data Set II: Σx = 67, Σx^2 = 1507, and n = 4; s = 11.32

Chapter 4

4.5 four possible outcomes; S = {LL, LI, IL, II}

4.7 four possible outcomes; S = {DD, DG, GD, GG}

4.8 S = {HHH, HHT, HTH, HTT, THH, THT, TTH, TTT}

4.16 not equally likely outcomes; use relative frequency approach

4.18 subjective probability

4.19 a. .48   **b.** .52   **4.24 a.** .25   **b.** .75

4.26 .9094;   .0906

4.27 a. .086   **b.** .506

4.34 a. i. .6250   **ii.** .6000   **iii.** .4375   **iv.** .6400

   b. Events "male" and "female" are mutually exclusive. Events "have shopped" and "male" are not mutually exclusive.

   c. Events "female" and "have shopped" are dependent.

4.36 a. i. .3900   **ii.** .5450   **iii.** .3448   **iv.** .3846

   b. Events "male" and "in favor" are not mutually exclusive. Events "in favor" and "against" are mutually exclusive.

   c. Events "female" and "no opinion" are dependent.

4.37 Events "female" and "pediatrician" are dependent but not mutually exclusive.

4.38 dependent

4.40 $P(A)$ = .6667; $P(\bar{A})$ = .3333

4.59 .3529

4.61 a. .0064   **b.** .8464   **4.64** .5120

4.66 0.400

4.74 a. .800   **b.** .775   **c.** 1.0

4.79 .80   **4.82** .344

4.91 384   **4.92** 240

4.93 $_{25}C_4$ = 12,650; $_{25}P_4$ = 303,600

4.96 a. i. .3640   **ii.** .4880   **iii.** .4297
   iv. .6541   **v.** .2920   **vi.** .7080

   b. Events "female" and "prefers watching sports" are dependent but not mutually exclusive.

4.98 a. i. .700   **ii.** .6296   **iii.** .275   **iv.** .725

   b. Events "athlete" and "should be paid" are dependent but not mutually exclusive.

4.100 a. .5118   **b.** .4882

4.105 a. 1/195,249,054 = .0000000051

   b. 1/5,138,133 = .00000019

4.106 a. .5000   **b.** .3333

   c. the sixth toss is independent of the first five tosses. Equivalent to part a.

4.107 a. .50   **b.** .50

4.110 a. .8333   **b.** .1667

4.111 a. .0001

   b. i. .0024   **ii.** .0012   **iii.** .0006   **iv.** 0004

Self-Review Test

1. a   **2.** b   **3.** c   **4.** a

5. a   **6.** b   **7.** c   **8.** b

9. b   **10.** c   **11.** b

12. 120

13. a. 0.3333   **b.** 0.6667

14. a. P(out-of-state) = 125/200 = 0.6250 and P(out of state | female) = 70/110 = 0.6364. Since these two probabilities are not equal, the two events are dependent. Events "female" and "out of state" are not mutually exclusive because they can occur together.

   b. i. 0.4500   **ii.** 0.6364

15. 0.825   **16.** 0.3894   **17.** 0.4225   **18.** 0.600

19. a. 0.279   **b.** 0.829

20. a. i. 0.3577   **ii.** 0.4047
   iii. 0.2352   **iv.** 0.5593

   b. $P(W)$ = 257/506 = 0.5079 and $P(W|Y)$ = 104/181 = 0.5746. Since these two probabilities are not equal, the events "woman" and "yes" are dependent. The events "woman" and "yes" are not mutually exclusive because they can occur together.

Chapter 5

5.3   discrete random variable

5.10 **b.** **i.** .51   **ii.** .235   **iii.** .285   **iv.** .305

5.11

x	0	1	2
$P(x)$	.7039	.2702	.0259

5.12

x	0	1	2
$P(x)$	.5271	.3978	.0751

5.13

x	0	1	2
$P(x)$	.9274	.0712	.0014

5.14

x	0	1	2
$P(x)$	.4789	.4422	.0789

5.18 $\mu = \Sigma x P(x) = 1.91$ root canals

$\sigma = \sqrt{\Sigma x^2 P(x) - \mu^2} = \sqrt{5.29 - (1.91)^2} = 1.281$ root canals

Dr. Sharp performs an average of 1.91 root canals on Monday.

5.20 $\mu = 1.00$ head; $\sigma = .707$ head

5.23   The contractor is expected to make an average of $3.9 million profit with a standard deviation of $3.015 million.

5.34 **a.** 0, 1, 2, 3, 4, 5, 6, 7, 8, 9, 10, 11, 12, 13, 14, 15, 16, 17

b. .1540

5.36 **a.** .7095   **b.** .7332   **c.** .5000

5.38 **a.** .3771   **b.** .1762

5.39 **a.** $\mu = 5.600$ customers; $\sigma = 1.296$ customers

b. .0467

5.44 **a.** 0.2466

b.

x	$P(x)$
0	0.2466
1	0.3452
2	0.2417
3	0.1128
4	0.0395
5	0.0111
6	0.0026
7	0.0005
8	0.0001

5.51 .1607

5.54 **a.** .3033   **b.** **i.** .0900   **ii.** .0018
         iii. .9098

5.55 **a.** .0031   **b.** **i.** .0039   **ii.** .4911

5.56 **a.** **i.** .0629   **ii.** .0722
         b **i.** .9719   **ii.** .6400   **iii.** .5718

5.58 **a.** .2466   **c.** $\mu = 1.4$ $\sigma^2 = 1.4$
         $\sigma = 1.183$

5.59 **a.** .0511   **b.** **i.** .0001   **ii.** .0070   **iii.** .0460

5.60 $\mu = \Sigma x P(x) = 0.440$ error
         $\sigma = \sqrt{\Sigma x^2 P(x) - \mu^2} = \sqrt{0.92 - (0.44)^2} = 0.852$ error

5.63 **a.** .0912   **b.** **i.** .5502   **ii.** .0817
         iii. .2933

5.66 **a.** 0.1078   **b.** 0.5147   **c.** 0.8628

5.67 **a.** 0.4091   **b.** 0.5455   **c.** 0.0455

5.68 **a.** .8643   **b.** .1357

5.71   The value of $\Sigma x P(x) = -2.22$ indicates that your expected "gain" is –$2.22, so you should not accept this offer. This game is not fair to you since you are expected to lose an average of $2.22 per play.

5.72 **a.** .0625   **b.** .125   **c.** .3125

5.73 **a.** 0.1841

5.76 **a.** 35   **b.** 10   **c.** .2857

Self-Review Test

2.   The probability distribution table.

3.   a

4.   b

6.   b   **7.** a   **8.** b   **9.** a   **10.** c

12. a

14. $\mu = \Sigma x P(x) = 2.04$ homes

$\sigma = \sqrt{\Sigma x^2 P(x) - \mu^2} = \sqrt{6.26 - (2.04)^2} = 1.449$ homes

The four real estate agents sell an average of 2.04 homes per week.

15. **a.** **i.** 0.2128
             ii. 0.8418
             iii. 0.0153

         b. $\mu = np = 12(0.60) = 7.2$ adults
             $\sigma = \sqrt{npq} = \sqrt{12(0.60)(0.40)} = 1.697$ adults

16. **a.** 0.4525   **b.** 0.0646   **c.** 0.0666

17. **a.** **i.** 0.0521
             ii. 0.2203
             iii. 0.2013

         b.

x	$P(x)$	x	$P(x)$	x	$P(x)$
0	0.0000	9	0.1251	17	0.0128
1	0.0005	10	0.1251	18	0.0071
2	0.0023	11	0.1137	19	0.0037
3	0.0076	12	0.0948	20	0.0019
4	0.0189	13	0.0729	21	0.0009
5	0.0378	14	0.0521	22	0.0004
6	0.0631	15	0.0347	23	0.0002
7	0.0901	16	0.0217	24	0.0001
8	0.1126				

Chapter 6

6.25 **a.** .0764 **b.** .0062

6.27 **a.** 89.44% **b.** .62%

6.31 **a.** .0359 **b.** .1515

6.32 **a.** .11% **b.** 49.06% **c.** .69% **d.** 47.78%

6.33 **a.** .0838 **b.** .7026 **6.34** 0%

6.38 **a.** $z = 1.65$ approximately **b.** $z = -1.96$

c. $z = -2.33$ approximately **d.** $z = 2.58$ approximately

6.39 **a.** 208.50 **b.** 241.25 **c.** 178.50
d. 145.75 **e.** 158.25 **f.** 251.25

6.43 $94 approximately **6.49** .0901

6.51 **a.** .0568 **b.** .9671 **c.** .8903

6.52 **a.** .0332 **b.** 0 **c.** .9664

6.55 **a.** 8304 hours **b.** 8132 hours approximately

6.56 $121,660

6.57 Jenn must leave by approximately 7:50 AM, 40 minutes before she is due to arrive at work.

6.58 **a.** 0.7549 **b.** 0.2451 **6.63** .0637

6.67 **a.** 106.32 **b.** .0808

Self-Review Test

1. a **2.** a **3.** d **4.** b

5. a **6.** c **7.** b **8.** b

9. **a.** 0.1878 **b.** 0.9304 **c.** 0.0985 **d.** 0.7704

10. **a.** $z = -1.28$ approximately **b.** $z = 0.61$
c. $z = 1.65$ approximately **d.** $z = -1.07$ approximately

11. **a.** 0.5608 **b.** 0.0015 **c.** 0.0170 **d.** 0.1165

12. **a.** 48,658 miles **b.** 40,162 miles

13. **a.** **i.** 0.0318 **ii.** 0.9453 **iii.** .9099
iv. 0.0268 **v.** 0.4632

b. 0.7054 **c.** 0.3986

Chapter 7

7.5

x	$P(x)$	$xP(x)$	x^2	$x^2P(x)$
70	0.20	14.00	4900	980.00
78	0.20	15.60	6084	1216.80
80	0.40	32.00	6400	2560.00
95	0.20	19.00	9025	1805.00
		$\Sigma xP(x) = 80.60$		$\Sigma x^2P(x) = 6561.80$

$\mu = 80.60$ $\sigma = 8.09$

7.14 $\mu_{\bar{x}} = \$520$; $\sigma_{\bar{x}} = \$14.40$ **7.16** $n = 256$

7.17 **a.** $\Sigma \bar{x}P(\bar{x}) = 80.60$ **b.** $\sigma_{\bar{x}} = \sqrt{\Sigma \bar{x}^2 P(\bar{x}) - \mu_{\bar{x}}^2} = 3.302$
c. $\sigma / \sqrt{n} = 4.67$ is not equal to $\sigma_{\bar{x}} = 3.30$ in this case because $n/N = 3/5 = 0.60 > 0.05$.

d. $\sigma_{\bar{x}} = \dfrac{\sigma}{\sqrt{n}} \sqrt{\dfrac{N-n}{N-1}} = 3.302$

7.24 $\mu_{\bar{x}} = 20.20$ hours; $\sigma_{\bar{x}} = .613$ hours; the normal distribution

7.26 $n = 25$: $\mu_{\bar{x}} = 28.2$ years; $\sigma_{\bar{x}} = 1.2$ years; skewed to the right
$n = 100$: $\mu_{\bar{x}} = 28.2$ years; $\sigma_{\bar{x}} = .6$ years; approximately normal distribution

7.27 $\mu_{\bar{x}} = 200$ pieces; $\sigma_{\bar{x}} = 15.821$ pieces; approximately normal distribution; no, sample size ≥ 30

7.34 **a.** .2743 **b.** .1314 **c.** .5098

7.36 **a.** .8203 **b.** .9750

7.40 **a.** .1032 **b.** .3172 **c.** .0016 **d.** .9049

7.56 **a.** $p = .600$ **b.** 5 **d.** $-.10, .15, .15, -.10, -.10$

7.57 $\mu_{\hat{p}} = .561$; $\sigma_{\hat{p}} = .027$; approximately normal distribution

7.63 **a.** .0721 **b.** .1798

7.64 **a.** .1251 **b.** .1147 **7.65** .1515

7.66 **a.** **i.** 0.9382 **ii.** 0.8283
b. 0.7994 **c.** 0.0618

7.67 $\mu_{\bar{x}} = 750$ hours; $\sigma_{\bar{x}} = 11$ hours; the normal distribution

7.68 **a.** .0838 **b.** .0991 **c.** .8968 **d.** .0301

7.70 $\mu_{\hat{p}} = .88$; $\sigma_{\hat{p}} = .036$; approximately normal distribution

7.74 .6778 **7.75** 10 approximately

Self-Review Test

1. b **2.** b **3.** a **4.** a **5.** b **6.** b
7. c **8.** a **9.** a **10.** a **11.** a **12.** a

14. **a.** $\mu_{\bar{x}} = \mu = \$4823$ and $\sigma_{\bar{x}} = \sigma / \sqrt{n} = \$156.53$
b. $\mu_{\bar{x}} = \mu = \$4823$ and $\sigma_{\bar{x}} = \sigma / \sqrt{n} = \$70$
c. $\mu_{\bar{x}} = \mu = \$4823$ and $\sigma_{\bar{x}} = \sigma / \sqrt{n} = \$24.75$

In all cases the sampling distribution of $\bar{x}$ is approximately normal because the population has an approximate normal distribution, and the approximation improves (in that the variance gets smaller) as n gets larger.

15. **a.** $4472.14

We can draw no conclusion about the shape of the sampling distribution of $\bar{x}$.

b. $2,000

The sampling distribution of $\bar{x}$ is approximately normal.

c. $707.11

The sampling distribution of $\bar{x}$ is approximately normal.

16. **a.** 0.1109 **b.** 0.7698 **c.** 0.1515 **d.** 0.2006
e. 0.2090 **f.** 0.7852 **g.** 0.0559 **h.** 0.7636

17. **a.** **i.** 0.1203 **ii.** 0.1335 **iii.** 0.7486
b. 0.9736 **c.** 0.0013

18. The sampling distribution of $\hat{p}$ is approximately normal for a, b, and c.

19. **a.** **i.** 0.0869 **ii.** 0.8924 **iii.** 0.0207
iv. 0.1452 **v.** 0.7517 **vi.** 0.7517
b. 0.9108 **c.** 0.0414 **d.** 0.0869

Chapter 8

8.20 $283,762.77 to $290,937.23

8.22 The machine needs an adjustment.

8.27 $n = 61$ **8.28** **a.** 63.73 to 76.27 hours

8.43 40.04 to 42.36 bushels

8.47 18.64 to 25.36 minutes

8.52 **a.** 6.18 years

     b. 5.85 to 6.51 years; margin of error: $\pm$.33 year

8.69 **a.** 20.3% to 55.7%

8.72 0.117 to 0.683 The corresponding interval for the population percentage is 11.7% to 68.3%.

8.74 $n = 173$

8.78 The machine does not need an adjustment.

8.81 12.82 to 17.56 hours

8.82 21.76 to 26.24 minutes

8.84 144.33 to 158.47 calories

8.85 **a.** 44%   **b.** 40.3% to 47.7%

8.86 6.1% to 56.4%   **8.87** 359   **8.89** 221

8.93 **a.** $n = 20$ days   **b.** 90%   **c.** 75 cars

8.96 1. While σ is constant, the sample standard deviation, s, is not. The sample standard deviation will change with each sample.
     2. If σ is unknown, the confidence interval is calculated with a t value. This value will change as the sample size changes because the degrees of freedom will change.

Self-Review Test

2. b   **3.** a   **4.** a   **5.** c   **6.** b

7. **a.** $\bar{x} = \$159{,}000$

     b. $147,390 to $170,610 $E = z\sigma_{\bar{x}} = 2.58(4500) = \$11{,}610$

8. $571,283.30 to $649,566.70

9. **a.** $\hat{p} = 0.37$ **b.** 0.343 to 0.397

10. 83

11. 273

12. 229

13. The width of the confidence interval can be reduced by:
     1. Lowering the confidence level
     2. Increasing the sample size
     The second alternative is better because lowering the confidence level results in a less reliable estimate for μ.

Chapter 9

9.7 **a.** H_0: $\mu = 20$ hours; H_1: $\mu \neq 20$ hours; a two-tailed test

     b. H_0: $\mu = 8$ hours; H_1: $\mu > 8$ hours; a right-tailed test

     c. H_0: $\mu = 5$ years; H_1: $\mu \neq 5$ years; a two-tailed test

     d. H_0: $\mu = \$1000$; H_1: $\mu < \$1000$; a left-tailed test

     e. H_0: $\mu = 15$ minutes; H_1: $\mu < 15$ minutes; a left-tailed test

9.25 **a.** H_0: $\mu = 45$ months; H_1: $\mu < 45$ months; p value $= .0901$; do not reject H_0

     b. z is -1.96; do not reject

9.26 **a.** H_0: $\mu = \$1038$; H_1: $> \$1038$; p-value $= .0030$; if $\alpha = .025$, reject H_0

     b. Critical value: $z = 1.96$; observed value: $z = 2.75$: reject H_0.

9.29 **a.** H_0: $\mu = 45.38$ boxes; H_1: $\mu < 45.38$ boxes; z is -1.28. $-.65$; Do not reject H_0

     b. do not reject H_0.

9.44 H_0: $\mu = \$1050$; H_1: $\mu < \$1050$; critical value of t is -2.391; $t = -4.715$; reject H_0; if $\alpha = .025$, the critical value for t is -2.001; reject H_0

9.47 **a.** 58 years

     b. H_0: $\mu = 57$ years, H_1: $\mu \neq 57$ years; $t = -3.536$; reject H_0: p-value $< .01$; for $\alpha = .01$, the critical values of t -2.680 and 2.680; $t = -3.536$; reject H_0

9.49 H_0: $\mu = \$65$; H_1: $\mu > \$65$; critical value: $t = 2.718$; test statistic: $t = .889$; do not reject H_0

9.64 H_0: $p = .55$; H_1: $p > .55$; critical value: $z = 2.05$; test statistic: $z = 5.12$; reject H_0; p-value $= 0$; for $\alpha = .02$, reject H_0

9.67 **a.** H_0: $p \geq .35$; H_1: $p < .35$; critical value: $z = -1.96$; test statistic: $z = -2.94$; reject H_0

     b. do not reject H_0

     c. $\alpha = .025$; p-value $= .0016$; reject H_0

9.68 **a.** critical value: $z = 1.96$; test statistic: $z = 2.27$; reject H_0; adjust machine

     b. critical value: $z = 2.33$; test statistic: $z = 2.27$; do not reject H_0; do not adjust the machine

9.72 **a.** critical values: $z = -2.33$ and 2.33; test statistic: $z = 2.55$; reject H_0 **b.** P(Type I error) $= .02$

     c. p-value $= .0108$; reject H_0 if $\alpha = .025$; do not reject H_0 if $\alpha = .005$

9.74 **a.** H_0: $\mu = 151$ minutes; H_1: $\mu > 151$ minutes; test statistic: $z = 4.02$; p-value $= .000$; if $\alpha = .05$, reject H_0

     b. critical value: $z = 2.33$; test statistic: $z = 4.02$; reject H_0

9.76 **a.** H_0: $\mu \geq 50$; H_1: $\mu < 50$; critical value of $z = -1.96$; test statistic: $z = -3.00$; reject H_0

     b. P(Type I error) $= .025$ **c.** do not reject H_0

     d. p-value $= .0013$; for $\alpha = .025$, reject H_0

9.80 **a.** H_0: $\mu \leq 2$ hours; H_1: $\mu > 2$ hours; critical value: $t = 2.718$; test statistic: $t = 1.682$; do not reject H_0

9.81 **a.** H_0: $p = .69$; H_1: $p \neq .69$; critical values: $z = -1.96$ and 1.96; text statistic: $z = -3.96$; reject H_0

     b. P(Type I error) $= .05$

     c. $\alpha = .05$; p-value $= 0$; reject H_0.

9.85 **a.** H_0: $p = .80$; H_1: $p < .80$; critical value: $z = -2.33$; test statistic: $z = -.79$; do not reject H_0

     b. do not reject H_0

9.86   **a.** .0238   **b.** $\alpha = .0238$

9.88 H_0: $\mu = 8000$ hours; H_1: $\mu < 8000$ hours; reject H_0 if $\bar{x} < 7890$: $\alpha = .0239$; reject H_0 if $\bar{x} < 7857$; $\alpha = .0049$

9.90   **a.** One of the p-values is $0.4546/2 = 0.2273$. The other p-value is the complement, $1 - 0.2273 = 0.7727$.

      b. Since the value of the test statistic is negative, the p-value of 0.2273 is for the left-tailed test, H_1: $\mu < 15$. The p-value for the right-tailed test, H_1: $\mu > 15$, is 0.7727.

Self-Review Test

1.	a	**2.**	b	**3.**	a
4.	b	**5.**	a	**6.**	a
7.	a	**8.**	b	**9.**	c
10.	a	**11.**	c	**12.**	b
13.	c	**14.**	a	**15.**	b

16.   **a.** Do not reject H_0. Conclude that the average premium is $1170.

      b. Reject H_0. Conclude that the average premium is greater than $1170.

      c. A Type I error would occur if the we reject H_0 when it is true, that is, to conclude that the average premium is not $1170 when it is. A Type I error in part b is to conclude that the average premium is more than $1170 when it isn't. The maximum probability of making such an error is $\alpha = 0.01$ in part a and $\alpha = 0.025$ in part b.

      d. For $\alpha = 0.01$, do not reject H_0 since $0.01 < 0.025$.

      e. For $\alpha = 0.025$, reject H_0 since $0.0125 < 0.25$.

17.   **a.** Reject H_0 since $-3.000 < -2.448$. Conclude that the mean duration of nine-inning games has decreased after the meeting.

      b. A Type I error would be to conclude that the mean durations of games have decreased after the meeting when they are actually equal to the duration of games before the meeting. The maximum probability of making such an error is $\alpha = 0.01$.

      c. Do not reject H_0.

      d. Reject H_0.

18.   **a.** Reject H_0. Conclude that the mean time it takes to write a textbook is less than 31 months.

      b. A Type I error would be to conclude that the editor's claim is false when it is actually true. The maximum probability of making such an error is $\alpha = 0.025$.

      c. Do not reject H_0.

19.   **a.** Reject H_0 since $-3.16 < -1.65$. Conclude that the percentage of people who have a will is less than 50%.

      b. A Type I error would be to conclude that the percentage of adults with wills was less than 50% when it is actually 50%. The maximum probability of making such an error is $\alpha = 0.05$.

      c. Do not reject H_0.

      d. Reject H_0.

Chapter 10

10.8   **a.** 1.83 ounce

      b. 1.45 to 2.21 ounces

      c. H_0: $\mu_1 - \mu_2 = 0$; H_1: $\mu_1 - \mu_2 > 0$; p-value ≈ 0; for $\alpha = .01$, the critical value of z is 2.33; $z = 9.52$; reject H_0

10.11   **a.** -6.87 to 0.87 calories

      b. Do not reject H_0 since $-1.81 > -2.33$. Maine Mountain Dairy's claim is false.

      c. For $\alpha = 0.05$, reject H_0 since $0.0351 < 0.05$. For $\alpha = 0.025$, do not reject H_0 since $0.0351 > 0.025$.

10.19   **a.** -46.80 to -7.20 miles;

      b. H_0: $\mu_1 - \mu_2 = 0$; H_1: $\mu_1 - \mu_2 < 0$; critical value: $t = -2.326$; test statistic: $t = -2.67$; reject H_0

10.22   **a.** -11.94 to 1.94 minutes

      b. H_0: $\mu_1 - \mu_2 = 0$, H_1: $\mu_1 - \mu_2 < 0$; critical value of t is -2.395; $t = -1.92$; do not reject H_0

10.30   **a.** -47.01 to -6.99 miles;

      b. H_0: $\mu_1 - \mu_2 = 0$; H_1: $\mu_1 - \mu_2 < 0$; critical value: $t = -2.326$; test statistic: $t = -2.64$; reject H_0

      c. -48.30 to -5.70; critical value: $t = -2.397$; test statistic: $t = -2.54$; reject H_0

10.31   **a.** 4.02 to 7.98 mph

      b. H_0: $\mu_1 - \mu_2 = 0$, H_1: $\mu_1 - \mu_2 > 0$; critical value of t is 2.418; $t = 7.34$; reject H_0

      c. 3.82 to 8.18 mph; critical value of t is 2.500; $t = 6.87$; reject H_0

10.38   **a.** -2.98 to 9.84 minutes

      b. H_0: $\mu_d = 0$; H_1: $\mu_d > 0$; critical value: $t = 2.447$; test statistic: $t = 1.983$; do not reject H_0

10.41   **a.** -5.43 to 2.17 minutes

      b. H_0: $\mu_d = 0$; H_1: $\mu_d \neq 0$ critical values: $t = -2.365$ and 2.365; test statistic: $t = -1.287$; do not reject H_0

10.51   **a.** $-.019$ to .059

      b. H_0: $p_1 - p_2 = 0$; H_1: $p_1 - p_2 \neq 0$; critical values: $z = -2.58$ and 2.58; test statistic: $z = 1.11$; do not reject H_0; p-value $= .2670$; for $\alpha = .01$, do not reject H_0

10.55   **a.** $-.025$ to .225

      b. H_0: $p_1 - p_2 = 0$; H_1: $p_1 - p_2 < 0$; critical values: $z = 1.96$ and -1.96; test statistic: $z = 1.56$; do not reject H_0; p value $= .1188$; for $\alpha = .025$, do not reject H_0

      c. .012 to .188; critical values: $z = -1.96$ and 1.96; test statistic: $z = 2.20$; reject H_0; p value $= .0278$; for $\alpha = .05$, reject H_0

10.57   **a.** $-\$131.30$ to $-\$120.70$

      b. H_0: $\mu_1 - \mu_2 = 0$; H_1: $\mu_1 - \mu_2 < 0$; critical value: $z = -1.96$; test statistic: $z = -46.58$; reject H_0

10.58   **a.** $-.086$ to .160 fatalities

      b. H_0: $\mu_1 - \mu_2 = 0$; H_1: $\mu_1 - \mu_2 > 0$; critical value: $t = 2.326$; test statistic: $t = .70$; do not reject H_0

10.60   **a.** $-\$119.04$ to $-\$44.96$

      b. H_0: $\mu_1 - \mu_2 = 0$; H_1: $\mu_1 - \mu_2 < 0$; critical value: $t = -2.326$; test statistic: $t = -5.702$; reject H_0

10.61 **a.** \$9216.94 to \$11,463.06

b. Reject H_0 since $21.45 > 2.33$.
Conclude that the average salary of statisticians is higher than that of accountants and auditors.

10.64 **a.** \$1061.95 to \$3278.05

b. H_0: $\mu_1 - \mu_2 = 0$; H_1: $\mu_1 - \mu_2 > 0$; critical value: $t = 2.326$; test statistic: $t = 4.56$; reject H_0

10.65 **a.** .053 to .147

b. H_0: $p_1 - p_2 = 0$; H_1: $p_1 - p_2 > 0$; critical value: $z = 1.96$; test statistic: $z = 4.14$; reject H_0

10.67 **a.** .053 to .127

b. H_0: $p_1 - p_2 = 0$; H_1: $p_1 - p_2 \neq 0$; critical values: $z = -2.33$ and $z = 2.33$; test statistic: $z = 4.79$; reject H_0; p-value: 0; for $\alpha = .02$, reject H_0

10.70 0.2611

Self-Review Test

1. a

3. **a.** 1.62 to 2.78

b. Reject H_0 since $9.86 > 1.96$.
Conclude that the mean stress score of all executives is higher than that of all professors.

4. **a.** -2.72 to -1.88 hours

b. Reject H_0 since $-10.997 < -2.416$.
Conclude that the mean time spent per week playing with their children by all alcoholic fathers is less than that of non-alcoholic fathers.

5. **a.** -2.70 to -1.90 hours

b. Reject H_0 since $-11.474 < -2.421$.
Conclude that the mean time spent per week playing with their children by all alcoholic fathers is less than that of non-alcoholic fathers.

6. **a.** $-\$53.60$ to \$186.18

b. Do not reject H_0 since $2.050 < 2.447$.
Conclude that the mean estimate for repair costs at Zeke's is not different than the mean estimate at Elmer's.

7. **a.** -0.052 to 0.092

b. Do not reject H_0 since $0.60 < 2.58$.
Conclude that the proportion of all male voters who voted in the last presidential election is not different from that of all female voters.

Chapter 11

11.12 critical value: $\chi^2 = 13.277$; test statistic: $\chi^2 = 47.469$; reject H_0

11.14 critical value: $\chi^2 = 9.488$; test statistic: $\chi^2 = 19.300$; reject H_0

11.15 Do not reject H_0 since $6.534 < 9.348$.
Conclude that the sample is random.

11.16 critical value: $\chi^2 = 9.348$; test statistic: $\chi^2 = 6.994$; do not reject H_0

11.26 critical value: $\chi^2 = 5.024$; test statistic: $\chi^2 = 1.069$; do not reject H_0

11.28 critical value: $\chi^2 = 12.592$; test statistic: $\chi^2 = 30.663$; reject H_0

11.30 critical value: $\chi^2 = 7.815$; test statistic: $\chi^2 = 5.806$; do not reject H_0

11.35 **a.** .8120 to 3.3160; .9011 to 1.8210

b. H_0: $\sigma^2 \leq 1.0$; H_1: $\sigma^2 > 1.0$; critical value: $\chi^2 = 41.638$; test statistic: $\chi^2 = 33.810$; do not reject H_0

11.37 **a.** 1840.696429

b. 28.37 to 87.32

c. Reject H_0 since $17.180 > 16.013$.
Conclude that the population variance is different from 750 square dollars.

11.38 critical value: $\chi^2 = 7.815$; test statistic: $\chi^2 = 10.464$; reject H_0

11.40 critical value: $\chi^2 = 13.277$; test statistic: $\chi^2 = 73.25$; reject H_0

11.41 critical value: $\chi^2 = 11.345$; test statistic: $\chi^2 = 35.440$; reject H_0

11.44 critical value: $\chi^2 = 9.210$; test statistic: $\chi^2 = 1.623$; do not reject H_0

11.48 H_0: $\sigma^2 = 1.1$; H_1: $\sigma^2 > 1.1$; critical value: $\chi^2 = 28.845$; test statistic: $\chi^2 = 24.727$; do not reject H_0

11.49 H_0: $\sigma^2 = 10.4$; H_1: $\sigma^2 \neq 10.4$; critical values: $\chi^2 = 7.564$ and 30.191; test statistic: $\chi^2 = 24.192$; do not reject H_0

11.50 **a.** H_0: $\sigma^2 = 5000$; H_1: $\sigma^2 < 5000$; critical value: $\chi^2 = 8.907$; test statistic: $\chi^2 = 12.065$; do not reject H_0

b. 1666.8509 to 7903.1835; 40.827 to 88.900

11.51 **a.** Do not reject H_0 since $29.920 < 33.924$.
Conclude that the population variance is not greater than 0.025 square millimeter, and the machine does not need an adjustment.

b. 0.0203 to 0.0681

11.56 **a.** test statistic: $\chi^2 = 2.480$ **b.** no; p-value $> .10$

11.57 critical value: $\chi^2 = 16.919$; test statistic: $\chi^2 = 215.568$; reject H_0

Self-Review Test

1. b   **2.** a   **3.** c

4. a   **5.** b   **6.** b

7. c   **8.** b   **9.** a

10. Reject H_0 since $32.672 > 11.345$.
Conclude that the current distribution of opinions is different from the 2014 distribution.

11. Reject H_0 since $31.187 > 11.345$.
Conclude that educational level and ever being divorced are dependent.

12. Reject H_0 since $82.450 > 9.488$.
Conclude that the percentages of people who play the lottery often, sometimes, and never are not the same for each income group.

13. **a.** 0.486 to 1.154

b. Reject H_0 since $36.480 > 36.191$.
Conclude that the population variance exceeds 0.25 square ounce.

Chapter 12

12.11 **a.** numerator: $df = 2$; denominator: $df = 27$; $SSB = 51,423.2$; $MSW = 829.1944$; $F = 31.01$

b. H_0: $\mu_1 = \mu_2 = \mu_3$; H_1: all three population means are not equal; critical value: $F = 3.35$; reject H_0

12.17 **a.** critical value: $F = 2.05$; test statistic: $F = 2.12$; reject H_0

b. .10

12.18 Reject H_0 since $5.44 > 3.72$.
Conclude that the mean life of bulbs for each of these four brands is not the same.

12.22 Do not reject H_0 since $0.57 < 5.29$.
Conclude that the mean revenue is the same for all four days of the week.

Self-Review Test

1. a   **2.** b   **3.** c   **4.** a   **5.** a

6. a   **7.** b   **8.** a

10. **a.** Reject H_0 since $4.46 > 3.10$.
Conclude that the mean prices for all four pizza parlors are not the same.

b. By rejecting H_0, we may have committed a Type I error.

Chapter 13

13.19 **a.** 38.85   **b.** differ   **c.** nonexact relationship

13.20 **b.** $\hat{y} = 317.6959 - 34.0870x$ **e.** \$3408.70

f. -295.8701

13.21 **b.** $\hat{y} = 4.0327 - .2687x$ **e.** 3.01   **f.** $-.27$

13.24 **a.** $\mu_{y|x} = 41.5821 + .0927x$

b. population regression line because data set includes all 16 National League teams; values of A and B

d. 50.852%

13.31 **a.** $SS_{xx} = .8960$; $SS_{yy} = .7444$; $SS_{xy} = .7782$

b. $s_e = .0926$

c. $SST = .7444$; $SSE = .0686$; $SSR = .6758$

d. $r^2 = .91$

13.33 **a.** $s_e = 31.2410$ **b.** $r^2 = .45$

13.34 **a.** $s_e = .9930$ **b.** $r^2 = .63$

13.35 **a.** $\sigma_\varepsilon = 6.2590$ **b.** $\rho^2 = .15$

13.40 **a.** -39.4166 to -28.7574

b. H_0: $B = 0$; H_1: $B < 0$; critical value of t is -1.943; -15.651; reject H_0

13.44 **a.** $\hat{y} = 25.5536 + 2.4377x$ **b.** 1.331 to 3.5443

c. H_0: $B = 0$; H_1: $B > 0$; critical value: $t = 2.365$; test statistic: $t = 6.6042$; reject H_0

13.61 **a.** $r = .88$

b. H_0: $\rho = 0$; H_1: $\rho \neq 0$; critical values: $t = -3.250$ and 3.250; test statistic: $t = 5.558$; reject H_0

13.63 **a.** $SS_{xx} = 750$; $SS_{yy} = 9986.9167$; $SS_{xy} = 565$

b. $\hat{y} = 64.119 + .7533x$ **d.** $r = .21$; $r^2 = .04$

f. \$119.11   **g.** $s_e = 30.9213$

h. -1.7623 to 3.2689

i. H_0: $B = 0$; H_1: $B > 0$; critical value: $t = 1.812$; test statistic: $t = .6672$; do not reject H_0

j. H_0: $\rho = 0$; H_1: $\rho > 0$; critical value: $t = 2.228$; test statistic: $t = .679$; do not reject H_0

13.70 \$1518.85 to \$2212.88; \$715.60 to \$3016.13

13.72 93.1957 to 132.9709; 41.3776 to 184.7890

13.74 **a.** We expect B to be positive.

b. $\hat{y} = 7.8304 + 0.5039x$
The sign of $b = 0.5039$ is positive, which is consistent with what we expected.

c. The value of $a = 7.8304$ represents the number of units (in hundreds) sold if there are no promotions. The value of $b = 0.5039$ means that the sales are expected to increase by about 50 units per day for each additional promotion.

d. The value of $r = 0.8861$ indicates that the two variables have a strong positive correlation. The value of $r^2 = 0.7853$ means that approximately 78% of the total squared errors (SST) is explained by the regression model.

e. We expect sales of about 2547 units in a day with 35 promotions.

f. $s_e = 3.3527$

g. 0.1073 to 0.9004

h. Reject H_0 since $4.2759 > 3.365$.
Conclude that B is positive.

i. Reject H_0 since $4.2759 > 3.365$.
Conclude that ρ is different from zero.

13.76 **a.** $SS_{xx} = 224.9$; $SS_{yy} = 37,258.4$; $SS_{xy} = 2616.4$

b. yes   **c.** $\hat{y} = -420.5490 + 11.6336x$

e. $r = .90$ **f.** 429

13.77 The 99% confidence interval for $\mu_{y|8.5}$ is 54.5013 to 56.0353
The 99% prediction interval for y_p for $x = 8.5$ is 53.2831 to 57.2534

13.79 233.0455 to 266.2175; 195.2831 to 303.9799

13.80 **a.** yes   **b.** 246.4670 to 275.5330 lines

c. 200.0567 to 321.9433 lines   **e.** 338 lines

Self-Review Test

1. d   **2.** a   **3.** b   **4.** a

5. b   **6.** b   **7.** True   **8.** True

9. a   **10.** b

INDEX

- Test statistic for a goodness-of-fit test and a test of independence or homogeneity:

$$\chi^2 = \Sigma \frac{(O - E)^2}{E}$$

- Confidence interval for the population variance σ^2:

$$\frac{(n - 1)s^2}{\chi^2_{\alpha/2}} \quad \text{to} \quad \frac{(n - 1)s^2}{\chi^2_{1-\alpha/2}}$$

- Test statistic for a test of hypothesis about σ^2:

$$\chi^2 = \frac{(n - 1)s^2}{\sigma^2}$$

Chapter 12 • Analysis of Variance

Let:

k = the number of different samples (or treatments)

n_i = the size of sample i

T_i = the sum of the values in sample i

n = the number of values in all samples
$\quad = n_1 + n_2 + n_3 + \cdots$

Σx = the sum of the values in all samples
$\quad = T_1 + T_2 + T_3 + \cdots$

Σx^2 = the sum of the squares of values in all samples

- For the F distribution:

  Degrees of freedom for the numerator $= k - 1$

  Degrees of freedom for the denominator $= n - k$

- Between-samples sum of squares:

$$\text{SSB} = \left(\frac{T_1^2}{n_1} + \frac{T_2^2}{n_2} + \frac{T_3^2}{n_3} + \cdots \right) - \frac{(\Sigma x)^2}{n}$$

- Within-samples sum of squares:

$$\text{SSW} = \Sigma x^2 - \left(\frac{T_1^2}{n_1} + \frac{T_2^2}{n_2} + \frac{T_3^2}{n_3} + \cdots \right)$$

- Total sum of squares:

$$\text{SST} = \text{SSB} + \text{SSW} = \Sigma x^2 - \frac{(\Sigma x)^2}{n}$$

- Variance between samples: $\text{MSB} = \text{SSB}/(k - 1)$
- Variance within samples: $\text{MSW} = \text{SSW}/(n - k)$
- Test statistic for a one-way ANOVA test:

$$F = \text{MSB}/\text{MSW}$$

Chapter 13 • Simple Linear Regression

- Simple linear regression model: $y = A + Bx + \epsilon$
- Estimated simple linear regression model: $\hat{y} = a + bx$

- Sum of squares of xy, xx, and yy:

$$\text{SS}_{xy} = \Sigma xy - \frac{(\Sigma x)(\Sigma y)}{n}$$

$$\text{SS}_{xx} = \Sigma x^2 - \frac{(\Sigma x)^2}{n} \quad \text{and} \quad \text{SS}_{yy} = \Sigma y^2 - \frac{(\Sigma y)^2}{n}$$

- Least squares estimates of A and B:

$$b = \text{SS}_{xy} / \text{SS}_{xx} \quad \text{and} \quad a = \bar{y} - b\bar{x}$$

- Standard deviation of the sample errors:

$$s_e = \sqrt{\frac{\text{SS}_{yy} - b\,\text{SS}_{xy}}{n - 2}}$$

- Error sum of squares: $\text{SSE} = \Sigma e^2 = \Sigma (y - \hat{y})^2$
- Total sum of squares: $\text{SST} = \Sigma y^2 - \dfrac{(\Sigma y)^2}{n}$
- Regression sum of squares: $\text{SSR} = \text{SST} - \text{SSE}$
- Coefficient of determination: $r^2 = b\,\text{SS}_{xy}/\text{SS}_{yy}$
- Confidence interval for B:

$$b \pm ts_b \quad \text{where} \quad s_b = s_e/\sqrt{\text{SS}_{xx}}$$

- Test statistic for a test of hypothesis about B: $\quad t = \dfrac{b - B}{s_b}$

- Linear correlation coefficient: $r = \dfrac{\text{SS}_{xy}}{\sqrt{\text{SS}_{xx}\,\text{SS}_{yy}}}$

- Test statistic for a test of hypothesis about ρ:

$$t = r\sqrt{\frac{n - 2}{1 - r^2}}$$

- Confidence interval for $\mu_{y|x}$:

$$\hat{y} \pm ts_{\hat{y}_m} \quad \text{where} \quad s_{\hat{y}_m} = s_e\sqrt{\frac{1}{n} + \frac{(x_0 - \bar{x})^2}{\text{SS}_{xx}}}$$

- Prediction interval for y_p:

$$\hat{y} \pm ts_{\hat{y}_p} \quad \text{where} \quad s_{\hat{y}_p} = s_e\sqrt{1 + \frac{1}{n} + \frac{(x_0 - \bar{x})^2}{\text{SS}_{xx}}}$$

Chapter 14 • Multiple Regression

Formulas for Chapter 14 along with the chapter are on the Web site for the text.

Chapter 15 • Nonparametric Methods

Formulas for Chapter 15 along with the chapter are on the Web site for the text.

Table IV Standard Normal Distribution Table

The entries in the table on this page give the cumulative area under the standard normal curve to the left of z with the values of z equal to 0 or negative.

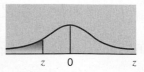

z	.00	.01	.02	.03	.04	.05	.06	.07	.08	.09
−3.4	.0003	.0003	.0003	.0003	.0003	.0003	.0003	.0003	.0003	.0002
−3.3	.0005	.0005	.0005	.0004	.0004	.0004	.0004	.0004	.0004	.0003
−3.2	.0007	.0007	.0006	.0006	.0006	.0006	.0006	.0005	.0005	.0005
−3.1	.0010	.0009	.0009	.0009	.0008	.0008	.0008	.0008	.0007	.0007
−3.0	.0013	.0013	.0013	.0012	.0012	.0011	.0011	.0011	.0010	.0010
−2.9	.0019	.0018	.0018	.0017	.0016	.0016	.0015	.0015	.0014	.0014
−2.8	.0026	.0025	.0024	.0023	.0023	.0022	.0021	.0021	.0020	.0019
−2.7	.0035	.0034	.0033	.0032	.0031	.0030	.0029	.0028	.0027	.0026
−2.6	.0047	.0045	.0044	.0043	.0041	.0040	.0039	.0038	.0037	.0036
−2.5	.0062	.0060	.0059	.0057	.0055	.0054	.0052	.0051	.0049	.0048
−2.4	.0082	.0080	.0078	.0075	.0073	.0071	.0069	.0068	.0066	.0064
−2.3	.0107	.0104	.0102	.0099	.0096	.0094	.0091	.0089	.0087	.0084
−2.2	.0139	.0136	.0132	.0129	.0125	.0122	.0119	.0116	.0113	.0110
−2.1	.0179	.0174	.0170	.0166	.0162	.0158	.0154	.0150	.0146	.0143
−2.0	.0228	.0222	.0217	.0212	.0207	.0202	.0197	.0192	.0188	.0183
−1.9	.0287	.0281	.0274	.0268	.0262	.0256	.0250	.0244	.0239	.0233
−1.8	.0359	.0351	.0344	.0336	.0329	.0322	.0314	.0307	.0301	.0294
−1.7	.0446	.0436	.0427	.0418	.0409	.0401	.0392	.0384	.0375	.0367
−1.6	.0548	.0537	.0526	.0516	.0505	.0495	.0485	.0475	.0465	.0455
−1.5	.0668	.0655	.0643	.0630	.0618	.0606	.0594	.0582	.0571	.0559
−1.4	.0808	.0793	.0778	.0764	.0749	.0735	.0721	.0708	.0694	.0681
−1.3	.0968	.0951	.0934	.0918	.0901	.0885	.0869	.0853	.0838	.0823
−1.2	.1151	.1131	.1112	.1093	.1075	.1056	.1038	.1020	.1003	.0985
−1.1	.1357	.1335	.1314	.1292	.1271	.1251	.1230	.1210	.1190	.1170
−1.0	.1587	.1562	.1539	.1515	.1492	.1469	.1446	.1423	.1401	.1379
−0.9	.1841	.1814	.1788	.1762	.1736	.1711	.1685	.1660	.1635	.1611
−0.8	.2119	.2090	.2061	.2033	.2005	.1977	.1949	.1922	.1894	.1867
−0.7	.2420	.2389	.2358	.2327	.2296	.2266	.2236	.2206	.2177	.2148
−0.6	.2743	.2709	.2676	.2643	.2611	.2578	.2546	.2514	.2483	.2451
−0.5	.3085	.3050	.3015	.2981	.2946	.2912	.2877	.2843	.2810	.2776
−0.4	.3446	.3409	.3372	.3336	.3300	.3264	.3228	.3192	.3156	.3121
−0.3	.3821	.3783	.3745	.3707	.3669	.3632	.3594	.3557	.3520	.3483
−0.2	.4207	.4168	.4129	.4090	.4052	.4013	.3974	.3936	.3897	.3859
−0.1	.4602	.4562	.4522	.4483	.4443	.4404	.4364	.4325	.4286	.4247
−0.0	.5000	.4960	.4920	.4880	.4840	.4801	.4761	.4721	.4681	.4641

Table IV   Standard Normal Distribution Table   (continued)

The entries in the table on this page give the cumulative area under the standard normal curve to the left of z with the values of z equal to 0 or positive.

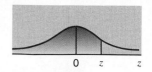

z	.00	.01	.02	.03	.04	.05	.06	.07	.08	.09
0.0	.5000	.5040	.5080	.5120	.5160	.5199	.5239	.5279	.5319	.5359
0.1	.5398	.5438	.5478	.5517	.5557	.5596	.5636	.5675	.5714	.5753
0.2	.5793	.5832	.5871	.5910	.5948	.5987	.6026	.6064	.6103	.6141
0.3	.6179	.6217	.6255	.6293	.6331	.6368	.6406	.6443	.6480	.6517
0.4	.6554	.6591	.6628	.6664	.6700	.6736	.6772	.6808	.6844	.6879
0.5	.6915	.6950	.6985	.7019	.7054	.7088	.7123	.7157	.7190	.7224
0.6	.7257	.7291	.7324	.7357	.7389	.7422	.7454	.7486	.7517	.7549
0.7	.7580	.7611	.7642	.7673	.7704	.7734	.7764	.7794	.7823	.7852
0.8	.7881	.7910	.7939	.7967	.7995	.8023	.8051	.8078	.8106	.8133
0.9	.8159	.8186	.8212	.8238	.8264	.8289	.8315	.8340	.8365	.8389
1.0	.8413	.8438	.8461	.8485	.8508	.8531	.8554	.8577	.8599	.8621
1.1	.8643	.8665	.8686	.8708	.8729	.8749	.8770	.8790	.8810	.8830
1.2	.8849	.8869	.8888	.8907	.8925	.8944	.8962	.8980	.8997	.9015
1.3	.9032	.9049	.9066	.9082	.9099	.9115	.9131	.9147	.9162	.9177
1.4	.9192	.9207	.9222	.9236	.9251	.9265	.9279	.9292	.9306	.9319
1.5	.9332	.9345	.9357	.9370	.9382	.9394	.9406	.9418	.9429	.9441
1.6	.9452	.9463	.9474	.9484	.9495	.9505	.9515	.9525	.9535	.9545
1.7	.9554	.9564	.9573	.9582	.9591	.9599	.9608	.9616	.9625	.9633
1.8	.9641	.9649	.9656	.9664	.9671	.9678	.9686	.9693	.9699	.9706
1.9	.9713	.9719	.9726	.9732	.9738	.9744	.9750	.9756	.9761	.9767
2.0	.9772	.9778	.9783	.9788	.9793	.9798	.9803	.9808	.9812	.9817
2.1	.9821	.9826	.9830	.9834	.9838	.9842	.9846	.9850	.9854	.9857
2.2	.9861	.9864	.9868	.9871	.9875	.9878	.9881	.9884	.9887	.9890
2.3	.9893	.9896	.9898	.9901	.9904	.9906	.9909	.9911	.9913	.9916
2.4	.9918	.9920	.9922	.9925	.9927	.9929	.9931	.9932	.9934	.9936
2.5	.9938	.9940	.9941	.9943	.9945	.9946	.9948	.9949	.9951	.9952
2.6	.9953	.9955	.9956	.9957	.9959	.9960	.9961	.9962	.9963	.9964
2.7	.9965	.9966	.9967	.9968	.9969	.9970	.9971	.9972	.9973	.9974
2.8	.9974	.9975	.9976	.9977	.9977	.9978	.9979	.9979	.9980	.9981
2.9	.9981	.9982	.9982	.9983	.9984	.9984	.9985	.9985	.9986	.9986
3.0	.9987	.9987	.9987	.9988	.9988	.9989	.9989	.9989	.9990	.9990
3.1	.9990	.9991	.9991	.9991	.9992	.9992	.9992	.9992	.9993	.9993
3.2	.9993	.9993	.9994	.9994	.9994	.9994	.9994	.9995	.9995	.9995
3.3	.9995	.9995	.9995	.9996	.9996	.9996	.9996	.9996	.9996	.9997
3.4	.9997	.9997	.9997	.9997	.9997	.9997	.9997	.9997	.9997	.9998

This is Table IV of Appendix B.

Table V The *t* Distribution Table

The entries in this table give the critical values of *t* for the specified number of degrees of freedom and areas in the right tail.

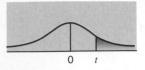

df	\.10	.05	.025	.01	.005	.001
			Area in the Right Tail Under the *t* Distribution Curve			
1	3.078	6.314	12.706	31.821	63.657	318.309
2	1.886	2.920	4.303	6.965	9.925	22.327
3	1.638	2.353	3.182	4.541	5.841	10.215
4	1.533	2.132	2.776	3.747	4.604	7.173
5	1.476	2.015	2.571	3.365	4.032	5.893
6	1.440	1.943	2.447	3.143	3.707	5.208
7	1.415	1.895	2.365	2.998	3.499	4.785
8	1.397	1.860	2.306	2.896	3.355	4.501
9	1.383	1.833	2.262	2.821	3.250	4.297
10	1.372	1.812	2.228	2.764	3.169	4.144
11	1.363	1.796	2.201	2.718	3.106	4.025
12	1.356	1.782	2.179	2.681	3.055	3.930
13	1.350	1.771	2.160	2.650	3.012	3.852
14	1.345	1.761	2.145	2.624	2.977	3.787
15	1.341	1.753	2.131	2.602	2.947	3.733
16	1.337	1.746	2.120	2.583	2.921	3.686
17	1.333	1.740	2.110	2.567	2.898	3.646
18	1.330	1.734	2.101	2.552	2.878	3.610
19	1.328	1.729	2.093	2.539	2.861	3.579
20	1.325	1.725	2.086	2.528	2.845	3.552
21	1.323	1.721	2.080	2.518	2.831	3.527
22	1.321	1.717	2.074	2.508	2.819	3.505
23	1.319	1.714	2.069	2.500	2.807	3.485
24	1.318	1.711	2.064	2.492	2.797	3.467
25	1.316	1.708	2.060	2.485	2.787	3.450
26	1.315	1.706	2.056	2.479	2.779	3.435
27	1.314	1.703	2.052	2.473	2.771	3.421
28	1.313	1.701	2.048	2.467	2.763	3.408
29	1.311	1.699	2.045	2.462	2.756	3.396
30	1.310	1.697	2.042	2.457	2.750	3.385
31	1.309	1.696	2.040	2.453	2.744	3.375
32	1.309	1.694	2.037	2.449	2.738	3.365
33	1.308	1.692	2.035	2.445	2.733	3.356
34	1.307	1.691	2.032	2.441	2.728	3.348
35	1.306	1.690	2.030	2.438	2.724	3.340